SATELLITE TECHNOLOGY

SATELLITE TECHNOLOGY
PRINCIPLES AND APPLICATIONS

Third Edition

Anil K. Maini
Varsha Agrawal

Both of Laser Science and Technology Centre,
Defence Research and Development Organization,
Ministry of Defence, India

This edition first published 2014
© 2014 John Wiley & Sons Ltd

Registered office
John Wiley & Sons Ltd, The Atrium, Southern Gate, Chichester, West Sussex, PO19 8SQ, United Kingdom

For details of our global editorial offices, for customer services and for information about how to apply for permission to reuse the copyright material in this book please see our website at www.wiley.com.

The right of the author to be identified as the author of this work has been asserted in accordance with the Copyright, Designs and Patents Act 1988.

All rights reserved. No part of this publication may be reproduced, stored in a retrieval system, or transmitted, in any form or by any means, electronic, mechanical, photocopying, recording or otherwise, except as permitted by the UK Copyright, Designs and Patents Act 1988, without the prior permission of the publisher.

Wiley also publishes its books in a variety of electronic formats. Some content that appears in print may not be available in electronic books.

Designations used by companies to distinguish their products are often claimed as trademarks. All brand names and product names used in this book are trade names, service marks, trademarks or registered trademarks of their respective owners. The publisher is not associated with any product or vendor mentioned in this book.

Limit of Liability/Disclaimer of Warranty: While the publisher and author have used their best efforts in preparing this book, they make no representations or warranties with respect to the accuracy or completeness of the contents of this book and specifically disclaim any implied warranties of merchantability or fitness for a particular purpose. It is sold on the understanding that the publisher is not engaged in rendering professional services and neither the publisher nor the author shall be liable for damages arising herefrom. If professional advice or other expert assistance is required, the services of a competent professional should be sought.

Library of Congress Cataloging-in-Publication Data

Maini, Anil Kumar.
  Satellite technology : principles and applications / Anil K. Maini, Varsha Agrawal. – Third edition.
     1 online resource.
  Includes bibliographical references and index.
  Description based on print version record and CIP data provided by publisher; resource not viewed.
  ISBN 978-1-118-63637-4 (ePub) – ISBN 978-1-118-63641-1 (Adobe PDF) – ISBN 978-1-118-63647-3 (cloth) 1. Artificial satellites. 2. Scientific satellites. I. Agrawal, Varsha. II. Title.
  TL796
  629.46–dc23
                                                   2014001139

A catalogue record for this book is available from the British Library.

ISBN: 9781118636473

Set in 10/12 TimesLTStd by Thomson Digital, Noida, India

1 2014

In loving memory of my father-in-law
Shri Ramji Bazaz

Anil K. Maini

In loving memory of my mother, **Mrs Kussum Agrawal**
Who has always been with me in Spirit and in my Heart

Varsha Agrawal

Contents

Preface xxi

PART I SATELLITE TECHNOLOGY

1 Introduction to Satellites and their Applications 3

    1.1 Ever-expanding Application Spectrum 3
    1.2 What is a Satellite? 4
    1.3 History of the Evolution of Satellites 7
        1.3.1 Era of Hot Air Balloons and Sounding Rockets 7
        1.3.2 Launch of Early Artificial Satellites 8
        1.3.3 Satellites for Communications, Meteorology and Scientific Exploration -- Early Developments 10
        1.3.4 Non-geosynchronous Communication Satellites: Telstar and Relay Programmes 11
        1.3.5 Emergence of Geosynchronous Communication Satellites 12
        1.3.6 International Communication Satellite Systems 15
        1.3.7 Domestic Communication Satellite Systems 16
        1.3.8 Satellites for other Applications also made Rapid Progress 19
        1.3.9 Small or Miniature Satellites 22
    1.4 Evolution of Launch Vehicles 27
    1.5 Future Trends 33
        1.5.1 Communication Satellites 33
        1.5.2 Weather Forecasting Satellites 33
        1.5.3 Earth Observation Satellites 33
        1.5.4 Navigational Satellites 34
        1.5.5 Military Satellites 35
    Further Reading 35
    Glossary 35

2 Satellite Orbits and Trajectories 37

    2.1 Definition of an Orbit and a Trajectory 37

	2.2	Orbiting Satellites – Basic Principles	37
		2.2.1 Newton's Law of Gravitation	39
		2.2.2 Newton's Second Law of Motion	40
		2.2.3 Kepler's Laws	41
	2.3	Orbital Parameters	44
	2.4	Injection Velocity and Resulting Satellite Trajectories	61
	2.5	Types of Satellite Orbits	67
		2.5.1 Orientation of the Orbital Plane	67
		2.5.2 Eccentricity of the Orbit	68
		2.5.3 Distance from Earth	70
		2.5.4 Sun-synchronous Orbit	73
	Further Readings	76
	Glossary	76
3	**Satellite Launch and In-orbit Operations**	**79**
	3.1	Acquiring the Desired Orbit	79
		3.1.1 Parameters Defining the Satellite Orbit	80
		3.1.2 Modifying the Orbital Parameters	83
	3.2	Launch Sequence	95
		3.2.1 Types of Launch Sequence	95
	3.3	Launch Vehicles	100
		3.3.1 Introduction	100
		3.3.2 Classification	100
		3.3.3 Anatomy of a Launch Vehicle	104
		3.3.4 Principal Parameters	106
		3.3.5 Major Launch Vehicles	108
	3.4	Space Centres	127
		3.4.1 Location Considerations	127
		3.4.2 Constituent Parts of a Space Centre	128
		3.4.3 Major Space Centres	129
	3.5	Orbital Perturbations	144
	3.6	Satellite Stabilization	146
		3.6.1 Spin Stabilization	146
		3.6.2 Three-axis or Body Stabilization	147
		3.6.3 Comparison between Spin-stabilized and Three-axis Stabilized Satellites	149
		3.6.4 Station Keeping	149
	3.7	Orbital Effects on Satellite's Performance	149
		3.7.1 Doppler Shift	149
		3.7.2 Variation in the Orbital Distance	150
		3.7.3 Solar Eclipse	150
		3.7.4 Sun Transit Outrage	150
	3.8	Eclipses	150
	3.9	Look Angles of a Satellite	154
		3.9.1 Azimuth Angle	154

	3.9.2 Elevation Angle	155
	3.9.3 Computing the Slant Range	156
	3.9.4 Computing the Line-of-Sight Distance between Two Satellites	158
3.10	Earth Coverage and Ground Tracks	166
	3.10.1 Satellite Altitude and the Earth Coverage Area	166
	3.10.2 Satellite Ground Tracks	167
	3.10.3 Orbit Inclination and Latitude Coverage	170
Further Readings		172
Glossary		172

4 Satellite Hardware — 174

4.1	Satellite Subsystems	174
4.2	Mechanical Structure	175
	4.2.1 Design Considerations	176
	4.2.2 Typical Structure	176
4.3	Propulsion Subsystem	177
	4.3.1 Basic Principle	178
	4.3.2 Types of Propulsion System	178
4.4	Thermal Control Subsystem	185
	4.4.1 Sources of Thermal Inequilibrium	186
	4.4.2 Mechanism of Heat Transfer	186
	4.4.3 Types of Thermal Control	187
4.5	Power Supply Subsystem	189
	4.5.1 Types of Power System	189
	4.5.2 Solar Energy Driven Power Systems	190
	4.5.3 Batteries	195
4.6	Attitude and Orbit Control	199
	4.6.1 Attitude Control	200
	4.6.2 Orbit Control	200
4.7	Tracking, Telemetry and Command Subsystem	201
4.8	Payload	203
4.9	Antenna Subsystem	205
	4.9.1 Antenna Parameters	207
	4.9.2 Types of Antennas	210
4.10	Space Qualification and Equipment Reliability	224
	4.10.1 Space Qualification	224
	4.10.2 Reliability	225
Further Readings		226
Glossary		227

5 Communication Techniques — 229

5.1	Types of Information Signals	229
	5.1.1 Voice Signals	230
	5.1.2 Data Signals	230
	5.1.3 Video Signals	230

	5.2	Amplitude Modulation	231
		5.2.1 Frequency Spectrum of the AM Signal	232
		5.2.2 Power in the AM Signal	233
		5.2.3 Noise in the AM Signal	233
		5.2.4 Different Forms of Amplitude Modulation	235
	5.3	Frequency Modulation	241
		5.3.1 Frequency Spectrum of the FM Signal	243
		5.3.2 Narrow Band and Wide Band FM	245
		5.3.3 Noise in the FM Signal	246
		5.3.4 Generation of FM Signals	250
		5.3.5 Detection of FM Signals	252
	5.4	Pulse Communication Systems	259
		5.4.1 Analogue Pulse Communication Systems	259
		5.4.2 Digital Pulse Communication Systems	261
	5.5	Sampling Theorem	265
	5.6	Shannon–Hartley Theorem	266
	5.7	Digital Modulation Techniques	267
		5.7.1 Amplitude Shift Keying (ASK)	268
		5.7.2 Frequency Shift Keying (FSK)	268
		5.7.3 Phase Shift Keying (PSK)	269
		5.7.4 Differential Phase Shift Keying (DPSK)	270
		5.7.5 Quadrature Phase Shift Keying (QPSK)	271
		5.7.6 Offset QPSK	273
		5.7.7 8PSK and 16PSK	274
		5.7.8 Quadrature Amplitude Modulation (QAM)	274
		5.7.9 Amplitude Phase Shift Keying (APSK)	276
	5.8	Multiplexing Techniques	277
		5.8.1 Frequency Division Multiplexing	277
		5.8.2 Time Division Multiplexing	279
		5.8.3 Code Division Multiplexing	281
	Further Readings		282
	Glossary		283
6	**Multiple Access Techniques**		**286**
	6.1	Introduction to Multiple Access Techniques	286
		6.1.1 Transponder Assignment Modes	287
	6.2	Frequency Division Multiple Access (FDMA)	288
		6.2.1 Demand Assigned FDMA	290
		6.2.2 Pre-assigned FDMA	290
		6.2.3 Calculation of C/N Ratio	290
	6.3	Single Channel Per Carrier (SCPC) Systems	293
		6.3.1 SCPC/FM/FDMA System	293
		6.3.2 SCPC/PSK/FDMA System	294
	6.4	Multiple Channels Per Carrier (MCPC) Systems	295
		6.4.1 MCPC/FDM/FM/FDMA System	295
		6.4.2 MCPC/PCM-TDM/PSK/FDMA System	296

6.5	Time Division Multiple Access (TDMA)	297
6.6	TDMA Frame Structure	297
	6.6.1	Reference Burst	298
	6.6.2	Traffic Burst	298
	6.6.3	Guard Time	299
6.7	TDMA Burst Structure	299
	6.7.1	Carrier and Clock Recovery Sequence	299
	6.7.2	Unique Word	299
	6.7.3	Signalling Channel	300
	6.7.4	Traffic Information	301
6.8	Computing Unique Word Detection Probability	301
6.9	TDMA Frame Efficiency	302
6.10	Control and Coordination of Traffic	303
6.11	Frame Acquisition and Synchronization	305
	6.11.1	Extraction of Traffic Bursts from Receive Frames	305
	6.11.2	Transmission of Traffic Bursts	305
	6.11.3	Frame Synchronization	305
6.12	FDMA vs. TDMA	307
	6.12.1	Advantages of TDMA over FDMA	308
	6.12.2	Disadvantages of TDMA over FDMA	308
6.13	Code Division Multiple Access (CDMA)	308
	6.13.1	DS-CDMA Transmission and Reception	309
	6.13.2	Frequency Hopping CDMA (FH-CDMA) System	311
	6.13.3	Time Hopping CDMA (TH-CDMA) System	313
	6.13.4	Comparison of DS-CDMA, FH-CDMA and TH-CDMA Systems	314
6.14	Space Domain Multiple Access (SDMA)	316
	6.14.1	Frequency Re-use in SDMA	316
	6.14.2	SDMA/FDMA System	317
	6.14.3	SDMA/TDMA System	318
	6.14.4	SDMA/CDMA System	319
Further Readings		319
Glossary		320

7 Satellite Link Design Fundamentals — 322

7.1	Transmission Equation	322
7.2	Satellite Link Parameters	324
	7.2.1	Choice of Operating Frequency	324
	7.2.2	Propagation Considerations	324
	7.2.3	Noise Considerations	325
	7.2.4	Interference-related Problems	325
7.3	Frequency Considerations	326
	7.3.1	Frequency Allocation and Coordination	326
7.4	Propagation Considerations	330
	7.4.1	Free-space Loss	330
	7.4.2	Gaseous Absorption	331

		7.4.3	Attenuation due to Rain	333
		7.4.4	Cloud Attenuation	334
		7.4.5	Signal Fading due to Refraction	334
		7.4.6	Ionosphere-related Effects	335
		7.4.7	Fading due to Multipath Signals	338
	7.5	Techniques to Counter Propagation Effects		341
		7.5.1	Attenuation Compensation Techniques	341
		7.5.2	Depolarization Compensation Techniques	342
	7.6	Noise Considerations		342
		7.6.1	Thermal Noise	342
		7.6.2	Noise Figure	343
		7.6.3	Noise Temperature	344
		7.6.4	Noise Figure and Noise Temperature of Cascaded Stages	345
		7.6.5	Antenna Noise Temperature	346
		7.6.6	Overall System Noise Temperature	350
	7.7	Interference-related Problems		353
		7.7.1	Intermodulation Distortion	354
		7.7.2	Interference between the Satellite and Terrestrial Links	357
		7.7.3	Interference due to Adjacent Satellites	357
		7.7.4	Cross-polarization Interference	361
		7.7.5	Adjacent Channel Interference	361
	7.8	Antenna Gain-to-Noise Temperature (G/T) Ratio		365
	7.9	Link Design		367
		7.9.1	Link Design Procedure	368
		7.9.2	Link Budget	368
	7.10	Multiple Spot Beam Technology		371
	Further Readings			374
	Glossary			375
8	**Earth Station**			**378**
	8.1	Earth Station		378
	8.2	Types of Earth Station		380
		8.2.1	Fixed Satellite Service (FSS) Earth Station	381
		8.2.2	Broadcast Satellite Service (BSS) Earth Stations	382
		8.2.3	Mobile Satellite Service (MSS) Earth Stations	383
		8.2.4	Single Function Stations	384
		8.2.5	Gateway Stations	385
		8.2.6	Teleports	386
	8.3	Earth Station Architecture		386
	8.4	Earth Station Design Considerations		387
		8.4.1	Key Performance Parameters	388
		8.4.2	Earth Station Design Optimization	390
		8.4.3	Environmental and Site Considerations	391
	8.5	Earth Station Testing		392
		8.5.1	Unit and Subsystem Level Testing	392
		8.5.2	System Level Testing	392

8.6	Earth Station Hardware		398
	8.6.1	RF Equipment	398
	8.6.2	IF and Baseband Equipment	408
	8.6.3	Terrestrial Interface	409
8.7	Satellite Tracking		412
	8.7.1	Satellite Tracking System – Block Diagram	412
	8.7.2	Tracking Techniques	412
8.8	Some Representative Earth Stations		419
	8.8.1	Goonhilly Satellite Earth Station	419
	8.8.2	Madley Communications Centre	421
	8.8.3	Madrid Deep Space Communications Complex	421
	8.8.4	Canberra Deep Space Communications Complex	422
	8.8.5	Goldstone Deep Space Communications Complex	423
	8.8.6	Honeysuckle Creek Tracking Station	424
	8.8.7	Kaena Point Satellite Tracking Station	426
	8.8.8	Bukit Timah Satellite Earth Station	426
	8.8.9	INTELSAT Teleport Earth Stations	426
	8.8.10	SUPARCO Satellite Ground Station	428
	8.8.11	Makarios Satellite Earth Station	428
	8.8.12	Raisting Earth Station	428
	8.8.13	Indian Deep Space Network	429
	Glossary		430

9 Networking Concepts — 433

9.1	Introduction		433
9.2	Network Characteristics		433
	9.2.1	Availability	434
	9.2.2	Reliability	434
	9.2.3	Security	435
	9.2.4	Throughput	436
	9.2.5	Scalability	437
	9.2.6	Topology	437
	9.2.7	Cost	437
9.3	Applications and Services		437
	9.3.1	Satellite and Network Services	438
	9.3.2	Satellite Services	438
	9.3.3	Network Services	438
	9.3.4	Internet Services	439
9.4	Network Topologies		442
	9.4.1	Bus Topology	442
	9.4.2	Star Topology	443
	9.4.3	Ring Topology	444
	9.4.4	Mesh Topology	444
	9.4.5	Tree Topology	445
	9.4.6	Hybrid Topology	446

9.5	Network Technologies	447
	9.5.1 Circuit Switched Networks	447
	9.5.2 Packet Switched Networks	448
	9.5.3 Circuit Switched versus Packet Switched Networks	449
9.6	Networking Protocols	450
	9.6.1 Common Networking Protocols	450
	9.6.2 The Open Systems Interconnect (OSI) Reference Model	453
	9.6.3 Internet Protocol (IP)	456
	9.6.4 Transmission Control Protocol (TCP)	457
	9.6.5 Hyper Text Transfer Protocol (HTTP)	457
	9.6.6 File Transfer Protocol (FTP)	457
	9.6.7 Simple Mail Transfer Protocol (SMTP)	458
	9.6.8 User Datagram Protocol (UDP)	458
	9.6.9 Asynchronous Transfer Mode (ATM)	459
9.7	Satellite Constellations	459
	9.7.1 Constellation Geometry	459
	9.7.2 Major Satellite Constellations	460
9.8	Internetworking with Terrestrial Networks	465
	9.8.1 Repeaters, Bridges, Switches and Routers	465
	9.8.2 Protocol Translation, Stacking and Tunnelling	466
	9.8.3 Quality of Service	466
Further Readings		467
Glossary		467

PART II SATELLITE APPLICATIONS

10 Communication Satellites   473

10.1	Introduction to Communication Satellites	473
10.2	Communication-related Applications of Satellites	474
	10.2.1 Geostationary Satellite Communication Systems	475
	10.2.2 Non-geostationary Satellite Communication Systems	475
10.3	Frequency Bands	475
10.4	Payloads	475
	10.4.1 Types of Transponders	477
	10.4.2 Transponder Performance Parameters	478
10.5	Satellite versus Terrestrial Networks	479
	10.5.1 Advantages of Satellites Over Terrestrial Networks	479
	10.5.2 Disadvantages of Satellites with Respect to Terrestrial Networks	480
10.6	Satellite Telephony	481
	10.6.1 Point-to-Point Trunk Telephone Networks	482
	10.6.2 Mobile Satellite Telephony	482
10.7	Satellite Television	484
	10.7.1 A Typical Satellite TV Network	484
	10.7.2 Satellite–Cable Television	485

	10.7.3	Satellite–Local Broadcast TV Network	486
	10.7.4	Direct-to-Home Satellite Television	487
	10.7.5	Digital Video Broadcasting (DVB)	490
	10.7.6	DVB-S and DVB-S2 Standards	491
	10.7.7	DVB-RCS and DVB-RCS2 Standards	493
	10.7.8	DVB-T and DVB-T2 Standards	493
	10.7.9	DVB-H and DVB-SH Standards	494
10.8	Satellite Radio		496
10.9	Satellite Data Communication Services		496
	10.9.1	Satellite Data Broadcasting	496
	10.9.2	VSATs (Very Small Aperture Terminals)	497
10.10	Important Missions		502
	10.10.1	International Satellite Systems	502
	10.10.2	Regional Satellite Systems	512
	10.10.3	National Satellite Systems	513
10.11	Future Trends		514
	10.11.1	Development of Satellite Constellations in LEO Orbits	516
	10.11.2	Development of Personal Communication Services (PCS)	516
	10.11.3	Use of Higher Frequency Bands	517
	10.11.4	Development of Light Quantum Communication Techniques	517
	10.11.5	Development of Broadband Services to Mobile Users	517
	10.11.6	Development of Hybrid Satellite/Terrestrial Networks	517
	10.11.7	Advanced Concepts	518
Further Readings			519
Glossary			521

11 Remote Sensing Satellites — 524

11.1	Remote Sensing – An Overview		524
	11.1.1	Aerial Remote Sensing	525
	11.1.2	Satellite Remote Sensing	525
11.2	Classification of Satellite Remote Sensing Systems		526
	11.2.1	Optical Remote Sensing Systems	526
	11.2.2	Thermal Infrared Remote Sensing Systems	528
	11.2.3	Microwave Remote Sensing Systems	529
11.3	Remote Sensing Satellite Orbits		531
11.4	Remote Sensing Satellite Payloads		531
	11.4.1	Classification of Sensors	531
	11.4.2	Sensor Parameters	534
11.5	Passive Sensors		535
	11.5.1	Passive Scanning Sensors	536
	11.5.2	Passive Non-scanning Sensors	539
11.6	Active Sensors		540
	11.6.1	Active Non-scanning Sensors	540
	11.6.2	Active Scanning Sensors	540

11.7	Types of Images		542
	11.7.1 Primary Images		542
	11.7.2 Secondary Images		542
11.8	Image Classification		545
11.9	Image Interpretation		546
	11.9.1 Interpreting Optical and Thermal Remote Sensing Images		546
	11.9.2 Interpreting Microwave Remote Sensing Images		547
	11.9.3 GIS in Remote Sensing		547
11.10	Applications of Remote Sensing Satellites		548
	11.10.1 Land Cover Classification		548
	11.10.2 Land Cover Change Detection		549
	11.10.3 Water Quality Monitoring and Management		550
	11.10.4 Flood Monitoring		551
	11.10.5 Urban Monitoring and Development		552
	11.10.6 Measurement of Sea Surface Temperature		552
	11.10.7 Deforestation		553
	11.10.8 Global Monitoring		553
	11.10.9 Predicting Disasters		555
	11.10.10 Other Applications		558
11.11	Major Remote Sensing Missions		558
	11.11.1 Landsat Satellite System		558
	11.11.2 SPOT Satellite System		561
	11.11.3 Radarsat Satellite System		564
	11.11.4 Indian Remote Sensing Satellite System		565
11.12	Future Trends		573
	Further Readings		574
	Glossary		575

12 Weather Satellites — 577

12.1	Weather Forecasting – An Overview		577
12.2	Weather Forecasting Satellite Fundamentals		580
12.3	Images from Weather Forecasting Satellites		580
	12.3.1 Visible Images		580
	12.3.2 IR Images		582
	12.3.3 Water Vapour Images		583
	12.3.4 Microwave Images		584
	12.3.5 Images Formed by Active Probing		585
12.4	Weather Forecasting Satellite Orbits		586
12.5	Weather Forecasting Satellite Payloads		587
	12.5.1 Radiometer		588
	12.5.2 Active Payloads		589
12.6	Image Processing and Analysis		592
	12.6.1 Image Enhancement Techniques		592
12.7	Weather Forecasting Satellite Applications		593
	12.7.1 Measurement of Cloud Parameters		594

	12.7.2	Rainfall	594
	12.7.3	Wind Speed and Direction	595
	12.7.4	Ground-level Temperature Measurements	596
	12.7.5	Air Pollution and Haze	596
	12.7.6	Fog	596
	12.7.7	Oceanography	596
	12.7.8	Severe Storm Support	597
	12.7.9	Fisheries	598
	12.7.10	Snow and Ice Studies	598
12.8	Major Weather Forecasting Satellite Missions		599
	12.8.1	GOES Satellite System	599
	12.8.2	Meteosat Satellite System	605
	12.8.3	Advanced TIROS-N (ATN) NOAA Satellites	608
12.9	Future of Weather Forecasting Satellite Systems		612
Further Readings			612
Glossary			613

13 Navigation Satellites — 614

13.1	Development of Satellite Navigation Systems		614
	13.1.1	Doppler Effect based Satellite Navigation Systems	615
	13.1.2	Trilateration-based Satellite Navigation Systems	615
13.2	Global Positioning System (GPS)		621
	13.2.1	Space Segment	621
	13.2.2	Control Segment	622
	13.2.3	User Segment	623
13.3	Working Principle of the GPS		625
	13.3.1	Principle of Operation	625
	13.3.2	GPS Signal Structure	627
	13.3.3	Pseudorange Measurements	628
	13.3.4	Determination of the Receiver Location	629
13.4	GPS Positioning Services and Positioning Modes		631
	13.4.1	GPS Positioning Services	631
	13.4.2	GPS Positioning Modes	632
13.5	GPS Error Sources		634
13.6	GLONASS Satellite System		637
	13.6.1	GLONASS Segments	638
	13.6.2	GLONASS Signal Structure	639
13.7	GPS-GLONASS Integration		641
13.8	EGNOS Satellite Navigation System		642
13.9	Galileo Satellite Navigation Systems		645
	13.9.1	Three-Phase Development Programme	645
	13.9.2	Services	646
13.10	Indian Regional Navigational Satellite System (IRNSS)		647
13.11	Compass Satellite Navigation System		648
13.12	Hybrid Navigation Systems		648

		13.13 Applications of Satellite Navigation Systems	650
		13.13.1 Military Applications	650
		13.13.2 Civilian Applications	651
		13.14 Future of Satellite Navigation Systems	654
		Further Readings	655
		Glossary	656

14 Scientific Satellites — 658

14.1 Satellite-based versus Ground-based Scientific Techniques — 658
14.2 Payloads on Board Scientific Satellites — 659
    14.2.1 Payloads for Studying Earth's Geodesy — 659
    14.2.2 Payloads for Earth Environment Studies — 660
    14.2.3 Payloads for Astronomical Studies — 661
14.3 Applications of Scientific Satellites — Study of Earth — 665
    14.3.1 Space Geodesy — 665
    14.3.2 Tectonics and Internal Geodynamics — 669
    14.3.3 Terrestrial Magnetic Fields — 670
14.4 Observation of the Earth's Environment — 670
    14.4.1 Study of the Earth's Ionosphere and Magnetosphere — 671
    14.4.2 Study of the Earth's Upper Atmosphere (Aeronomy) — 677
    14.4.3 Study of the Interaction between Earth and its Environment — 679
14.5 Astronomical Observations — 680
    14.5.1 Observation of the Sun — 681
14.6 Missions for Studying Planets of the Solar System — 686
    14.6.1 Mercury — 691
    14.6.2 Venus — 692
    14.6.3 Mars — 694
    14.6.4 Outer Planets — 697
    14.6.5 Moon — 703
    14.6.6 Asteroids — 705
    14.6.7 Comets — 706
14.7 Missions Beyond the Solar System — 707
14.8 Other Fields of Investigation — 710
    14.8.1 Microgravity Experiments — 710
    14.8.2 Life Sciences — 711
    14.8.3 Material Sciences — 712
    14.8.4 Cosmic Ray and Fundamental Physics Research — 713
14.9 Future Trends — 714
Further Readings — 715
Glossary — 715

15 Military Satellites — 717

15.1 Military Satellites – An Overview — 717
    15.1.1 Applications of Military Satellites — 718
15.2 Military Communication Satellites — 718

15.3	Development of Military Communication Satellite Systems		719
	15.3.1	American Systems	720
	15.3.2	Russian Systems	724
	15.3.3	Satellites Launched by other Countries	725
15.4	Frequency Spectrum Utilized by Military Communication Satellite Systems		726
15.5	Dual-use Military Communication Satellite Systems		727
15.6	Reconnaisance Satellites		728
	15.6.1	Image Intelligence or IMINT Satellites	728
15.7	SIGINT Satellites		732
	15.7.1	Development of SIGINT Satellites	733
15.8	Early Warning Satellites		735
	15.8.1	Major Early Warning Satellite Programmes	736
15.9	Nuclear Explosion Satellites		738
15.10	Military Weather Forecasting Satellites		738
15.11	Military Navigation Satellites		739
15.12	Space Weapons		739
	15.12.1	Classification of Space Weapons	740
15.13	Strategic Defence Initiative		745
	15.13.1	Ground Based Programmes	746
	15.13.2	Directed Energy Weapon Programmes	749
	15.13.3	Space Programmes	751
	15.13.4	Sensor Programmes	752
15.14	Directed Energy Laser Weapons		752
	15.14.1	Advantages	753
	15.14.2	Limitations	753
	15.14.3	Directed Energy Laser Weapon Components	754
	15.14.4	Important Design Parametres	755
	15.14.5	Important Laser Sources	756
	15.14.6	Beam Control Technology	763
15.15	Advanced Concepts		764
	15.15.1	New Surveillance Concepts Using Satellites	765
	15.15.2	Long Reach Non-lethal Laser Dazzler	765
	15.15.3	Long Reach Laser Target Designator	766
	Further Readings		767
	Glossary		767

16 Emerging Trends — 769

16.1	Introduction		769
16.2	Space Tethers		769
	16.2.1	Space Tethers – Different Types	770
	16.2.2	Applications	774
	16.2.3	Space Tether Missions	775
	16.2.4	Space Elevator	779
16.3	Aerostat Systems		781
	16.3.1	Components of an Aerostat System	782

	16.3.2 Types of Aerostat Systems	782
	16.3.3 Applications	783
16.4	Millimetre Wave Satellite Communication	784
	16.4.1 Millimetre Wave Band	784
	16.4.2 Advantages	785
	16.4.3 Propagation Considerations	787
	16.4.4 Applications	788
	16.4.5 Millimetre Wave Satellite Missions	789
16.5	Space Stations	793
	16.5.1 Importance of Space Stations	794
	16.5.2 Space Stations of the Past	794
	16.5.3 Currently Operational Systems	797
	16.5.4 Planned Space Stations	799
	16.5.5 Emerging Space Station Concepts	801
Further Readings		803
Glossary		804

Index   **807**

Preface

The word 'satellite' is in common use today. It sounds very familiar to all of us irrespective of our educational and professional backgrounds. It is no longer the prerogative of a few select nations and is not a topic of research and discussion that is confined to the premises of big academic institutes and research organizations. Today, satellite technology is not only one of the main subjects taught at undergraduate, graduate and postgraduate level, it is the bread and butter for a large percentage of electronics, communication and IT professionals working for academic institutes, science and technology organizations and industry. Most of the books on satellite technology and its applications cover only communications-related applications of satellites, with either occasional or no reference to other important applications, including remote sensing, weather forecasting, scientific, navigational and military applications and also other topics related to space science and technology. In addition, space encyclopedias mainly cover the satellite missions and their applications with not much information on the technological aspects.

Satellite Technology: Principles and Applications is a concise and yet comprehensive reference book on the subject of satellite technology and its applications, covering in one volume both communication and non-communication applications. The third edition is a thoroughly updated and enlarged version of the second edition. All existing chapters of the second edition have been updated and two new chapters, Networking Concepts and Satellite Technology – Emerging Trends, have been added. A number of new topics have been included in other chapters as well to make the book more comprehensive and up-to-date, covering all the developmental technologies and trends in the field of satellites.

The intended audience for this book includes undergraduate and graduate level students and electronics, telecommunication and IT professionals looking for a compact and comprehensive reference book on satellite technology and its applications. The book is logically divided into two parts, namely satellite technology fundamentals covered in Chapters 1 to 9, followed by satellite applications in Chapters 10 to 16. The first introductory chapter begins with a brief account of the historical evolution of satellite technology, different types of satellite missions and areas of application of satellite technology. A new addition to the first chapter includes a detailed discussion on small and miniature satellites covering mini, micro, nano, pico and femto satellites. The next two chapters focus on orbital dynamics and related topics. The study of orbits and trajectories of satellites and satellite launch vehicles is the most fundamental topic of satellite technology and also perhaps the most important one. It is important because it gives an insight into the operational aspects of this wonderful piece of technology. An understanding of orbital dynamics will put us on a sound footing to address issues like types of orbit and their suitability for a given application, orbit stabilization, orbit correction and station-keeping, launch requirements and typical launch trajectories for various orbits, Earth coverage and so on. These two chapters are well supported by the required mathematics and design illustrations. Comprehensive coverage of major launch vehicles and international space complexes in terms of their features and facilities is a new addition to the third chapter and another highlight of the book.

After addressing the fundamental issues related to the operational principle of satellites, the dynamics of satellite orbits, launch procedures and various in-orbit operations, the focus in Chapter 4 is on satellite hardware, irrespective of its intended application. Different subsystems of a typical satellite and issues like the major functions performed by each one of these subsystems along with a brief discussion of their operational considerations are covered in this chapter.

After an introduction to the evolution of satellites, satellite orbital dynamics and hardware in the first four chapters, the focus shifts to topics that relate mainly to communication satellites in the three chapters thereafter. The topics covered in the first of these three chapters, Chapter 5, mainly include communication fundamentals with particular reference to satellite communication followed by multiple access techniques in the next chapter. Chapter 7 focuses on satellite link design related aspects. Emerging digital modulation techniques and a brief description of multiple spot beam technology have been added to Chapters 5 and 7, respectively.

Chapter 8 is on Earth station design and discusses the different types of Earth stations used for varied applications, Earth station architecture and design considerations, key performance parameters of an Earth station, Earth station testing, and some representative Earth stations. Communication satellites account for more than 80% of the total number of satellites in operation. This is one of the most widely exploited applications of satellites. Major earth station facilities have been briefly covered in the third edition of the book.

A chapter on Networking Concepts (Chapter 9) is a new addition to the third edition. The chapter comprehensively covers various networking concepts such as characteristics, applications and services of networks, network topologies, technologies and protocols. Internetworking issues are also briefly covered.

Satellite applications are in the second part of the book in Chapters 10 to 15. Based on the intended applications, the satellites are broadly classified as communication satellites, navigation satellites, weather forecasting satellites, Earth observation satellites, scientific satellites and military satellites. We intend to focus on this ever-expanding vast arena of satellite applications. The emphasis is on the underlying principles, the application potential, their contemporary status and future trends.

The first chapter on satellite applications covers all the communication-related applications of satellites, which mainly include satellite telephony, satellite radio, satellite television and data broadcasting services. Major international communication satellite missions have also been described at length. The future trends in the field of communication satellites are also highlighted at the end of the chapter. A brief discussion on digital video broadcasting and its different variants has been included in the third edition of the book. Also all data related to the launch of communication satellites has been updated.

Remote sensing is a technology used for obtaining information about the characteristics of an object through an analysis of the data acquired from it at a distance. Satellites play an important role in remote sensing. In Chapter 11, various topics related to remote sensing satellites are covered, including their principle of operation, payloads on board these satellites and their use to acquire images, processing and analysis of these images using various digital imaging techniques, and finally interpreting these images for studying various features of Earth for varied applications. We also introduce some of the major remote sensing satellite systems used for the purpose and the recent trends in the field towards the end of the chapter. A new section on the Indian Remote Sensing System has been included. The data on all remote sensing satellites have also been updated.

The use of satellites for weather forecasting and prediction of related phenomena has become indispensable. There is a permanent demand from the media with the requirement of short term weather forecasts for the general public, reliable prediction of the movements of tropical cyclones allow re-routing of shipping and a preventive action in zones through which hurricanes pass. Meteorological information is also of considerable importance for the conduct of military operations such as reconnaissance missions. In Chapter 12, we take a closer look at various aspects related to evolution, operation and use of weather satellites. Some of the major weather satellite missions are covered towards the end of the chapter. Like previous chapters on satellite applications, this chapter also contains a large number of illustrative photographs. All data on meteorological satellites have been updated.

Navigation is the art of determining the position of a platform or an object at any specified time. Satellite-based navigation systems represent a breakthrough in this field, which has revolutionized the very concept and application potential of navigation. These systems have grown from a relatively humble beginning as a support technology to that of a critical player used in the vast array of economic, scientific, civilian and military applications. Chapter 13 gives a brief outline of the development of satellite-based navigation systems and a descriptive view of the fundamentals underlying the operation of the GPS and the GLONASS navigation systems, their functioning and applications. The GALILEO navigation system and other developmental trends are also covered in the chapter. This new edition of the book also covers satellite based navigation systems hitherto not covered in previous editions. These include EGNOS, COMPASS and the Indian Satellite Navigation System. Hybrid navigation systems are also introduced in the chapter.

The use of satellites for scientific research has removed the constraints like attenuation and blocking of radiation by the Earth's atmosphere, gravitational effects on measurements and difficulty in making *in situ* studies imposed by the Earth-based observations. Moreover, space based scientific research is global by nature and helps to give an understanding of the various phenomena at a global level. Chapter 14 focuses on the scientific applications of satellites, covering in detail the contributions made by these satellites to Earth sciences, solar physics, astronomy and astrophysics.

Military systems of today rely heavily on the use of satellites both during war as well as in peacetime. Many of the military satellites perform roles similar to their civilian counterparts, mainly including telecommunication services, weather forecasting, navigation and Earth observation applications. Though some satellite missions are exclusively military in nature, many contemporary satellite systems are dual-use satellites that are used both for military as well as civilian applications. In Chapter 15 of the book we deliberate on various facets of military satellites related to their development and application potential. We begin the chapter with an overview of military satellites, followed by a description of various types of military satellites depending upon their intended application and a detailed discussion on space weapons.

A chapter on the emerging trends in satellite technology and applications (Chapter 16) is another new addition in the third edition of the book. The chapter mainly covers some unconventional but futuristic space related topics such as space tethers, space elevators, aerostats etc. The chapter also includes a brief account of millimetre wave communication satellites and emerging space station concepts. A new chapter was considered necessary as these topics could not have been included in any of the previous chapters.

As an extra resource, the companion website for our book www.wiley.com/go/maini contains a complete compendium of the features and facilities of satellites and satellite launch vehicles from past, present and planned futuristic satellite missions for various applications. Colour versions of some of the figures within the book are also available.

The motivation to write the proposed book and the selection of topics covered lay in the absence of any book which in one volume would cover all the important aspects of satellite technology and its applications. There are space encyclopedias that provide detailed information/technical data on the satellites launched by various countries for various applications, but contain virtually no information on the principles of satellite technology. There are a host of books on satellite communications, which discuss satellite technology with a focus on communications-related applications. We have made an honest attempt to offer to our intended audience, mainly electronics, telecommunication and IT professionals, a concise yet comprehensive up-to-date reference book covering in one volume both the technology as well as the application-related aspects of satellites.

Anil K. Maini
Varsha Agrawal
Laser Science and Technology Centre,
Defence Research and Development Organization,
Ministry of Defence,
India

Part I
Satellite Technology

1

Introduction to Satellites and their Applications

The word 'Satellite' is a household name today. It sounds so familiar to everyone irrespective of educational and professional background. It is no longer the prerogative of a few select nations and not a topic of research and discussion that is confined to the premises of big academic institutes and research organizations. It is a subject of interest and discussion not only to electronics and communication engineers, scientists and technocrats; it fascinates hobbyists, electronics enthusiasts and to a large extent, everyone.

In the present chapter, the different stages of evolution of satellites and satellite launch vehicles will be briefly discussed, beginning with the days of hot air balloons and sounding rockets of the late 1940s/early 1950s to the contemporary status.

1.1 Ever-expanding Application Spectrum

What has made this dramatic transformation possible is the manifold increase in the application areas where the satellites have been put to use. The horizon of satellite applications has extended far beyond providing intercontinental communication services and satellite television. Some of the most significant and talked about applications of satellites are in the fields of remote sensing and Earth observation. Atmospheric monitoring and space exploration are the other major frontiers where satellite usage has been exploited a great deal. Then there are the host of defence related applications, which include secure communications, navigation, spying and so on.

The areas of application are multiplying and so is the quantum of applications in each of those areas. For instance, in the field of communication related applications, it is not only the long distance telephony and video and facsimile services that are important; satellites are playing an increasing role in newer communication services such as data communication, mobile communication, and so on. Today, in addition to enabling someone to talk to another person thousands of miles away from the comfort of home or bringing cultural, sporting or political events from all over the globe live on television, satellites have made it possible for all to talk to anyone anywhere in the world, with both people being able to talk while being mobile. Video conferencing, where different people at different locations, no matter how far the distance is between these locations, can hold meetings in real time to exchange ideas or take important decisions, is a reality today in big establishments. The Internet and the revolutionary services it has brought are known to all of us. Satellites are the backbone of all these happenings.

A satellite is often referred to as an 'orbiting radio star' for reasons that can be easily appreciated. These so-called orbiting radio stars assist ships and aircraft to navigate safely in all weather conditions. It is interesting to learn that even some categories of medium to long range ballistic and cruise missiles need the assistance of a satellite to hit their intended targets precisely. The satellite-based global positioning system (GPS) is used as an aid to navigate safely and securely in unknown territories.

Earth observation and remote sensing satellites give information about the weather, ocean conditions, volcanic eruptions, earthquakes, pollution and health of agricultural crops and forests. Another class of satellites keeps watch on military activity around the world and helps to some extent in enforcing or policing arms control agreements.

Although mankind is yet to travel beyond the moon, satellites have crossed the solar system to investigate all planets. These satellites for astrophysical applications have giant telescopes on board and have sent data that has led to many new discoveries, throwing new light on the universe. It is for this reason that almost all developed nations including the United States, the United Kingdom, France, Japan, Germany, Russia and major developing countries like India have a fully-fledged and heavily funded space programme, managed by organizations with massive scientific and technical manpower and infrastructure.

1.2 What is a Satellite?

A satellite in general is any natural or artificial body moving around a celestial body such as a planet or a star. In the present context, reference is made only to artificial satellites orbiting the planet Earth. These satellites are put into the desired orbit and have payloads depending upon the intended application.

The idea of a geostationary satellite originated from a paper published by Arthur C. Clarke, a science fiction writer, in *Wireless World* magazine in the year 1945. In that proposal, he emphasized the importance of this orbit whose radius from the centre of Earth was such that the orbital period equalled the time taken by Earth to complete one rotation around its axis. He also highlighted the importance of an artificial satellite in this orbit having the required instrumentation to provide intercontinental communication services because such a satellite would appear to be stationary with respect to an observer on the surface of Earth. Though the idea of a satellite originated from the desire to put an object in space that would appear to be stationary with respect to Earth's surface, thus making possible a host of communication services, there are many other varieties of satellites where they need not be stationary with respect to an observer on Earth to perform the intended function.

A satellite while in orbit performs its designated role throughout its lifetime. A communication satellite (Figure 1.1) is a kind of repeater station that receives signals from the ground, processes them and then retransmits them back to Earth. An Earth observation satellite (Figure 1.2) is equipped with a camera to take photographs of regions of interest during its

Figure 1.1 Communication satellite

periodic motion. A weather forecasting satellite (Figure 1.3) takes photographs of clouds and monitors other atmospheric parameters, thus assisting the weatherman in making timely and accurate forecasts.

A satellite could effectively do the job of a spy in the case of some purpose-built military satellites (Figure 1.4) or of an explorer when suitably equipped and launched for astrophysical applications (Figure 1.5).

Figure 1.2 Earth observation satellite

Figure 1.3 Weather forecasting satellite (Courtesy: NOAA and NASA)

Figure 1.4 Military satellite (Courtesy: Lockheed Martin Corporation)

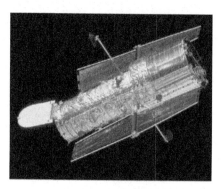

Figure 1.5 Scientific satellite (Courtesy: NASA and STScI)

1.3 History of the Evolution of Satellites

It all began with an article by Arthur C. Clarke published in the October 1945 issue of *Wireless World*, which theoretically proposed the feasibility of establishing a communication satellite in a geostationary orbit. In that article, he discussed how a geostationary orbit satellite would look static to an observer on Earth within the satellite's coverage, thus providing an uninterrupted communication service across the globe. This marked the beginning of the satellite era. The scientists and technologists started to look seriously at such a possibility and the revolution it was likely to bring along with it.

1.3.1 Era of Hot Air Balloons and Sounding Rockets

The execution of the mission began with the advent of hot air balloons and sounding rockets used for the purpose of the aerial observation of planet Earth from the upper reaches of Earth's atmosphere. The 1945–1955 period was dominated by launches of experimental sounding rockets to penetrate increasing heights of the upper reaches of the atmosphere. These rockets carried a variety of instruments to carry out their respective mission objectives.

A-4 (V-2) rockets used extensively during the Second World War for delivering explosive warheads attracted the attention of the users of these rockets for the purpose of scientific investigation of the upper atmosphere by means of a high altitude rocket. With this started the exercise of modifying these rockets so that they could carry scientific instruments. The first of these A-4 rockets to carry scientific instruments to the upper atmosphere was launched in May 1946 (Figure 1.6). The rocket carried an instrument to record cosmic ray flux from an altitude of 112 km. The launch was followed by several more during the same year.

The Soviets, in the meantime, made some major modifications to A-4 rockets to achieve higher performance levels as sounding rockets. The last V-2A rocket (the Soviet version of the modified A-4 rocket), made its appearance in 1949. It carried a payload of 860 kg and attained a height of 212 km.

Figure 1.6 First A-4 rocket to be launched (Courtesy: NASA)

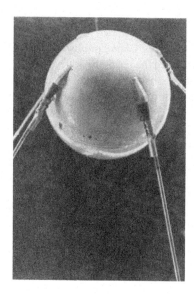

Figure 1.7 Sputnik-1 (Courtesy: NASA)

1.3.2 Launch of Early Artificial Satellites

The United States and Russia were the first two countries to draw plans for artificial satellites in 1955. Both countries announced their proposals to construct and launch artificial satellites. It all happened very quickly. Within a span of just two years, Russians accomplished the feat and the United States followed quickly thereafter.

Sputnik-1 (Figure 1.7) was the first artificial satellite that brought the space age to life. Launched on 4 October 1957 by Soviet R7 ICBM from Baikonur Cosmodrome, it orbited Earth once every 96 minutes in an elliptical orbit of 227 km × 941 km inclined at 65.1° and was designed to provide information on density and temperature of the upper atmosphere. After 92 successful days in orbit, it burned as it fell from orbit into the atmosphere on 4 January 1958.

Sputnik-2 and Sputnik-3 followed Sputnik-1. Sputnik-2 was launched on 3 November 1957 in an elliptical orbit of 212 km × 1660 km inclined at 65.33°. The satellite carried an animal, a female dog named Laika, in flight. Laika was the first living creature to orbit Earth. The mission provided information on the biological effect of the orbital flight. Sputnik-3, launched on 15 May 1958, was a geophysical satellite that provided information on Earth's ionosphere, magnetic field, cosmic rays and meteoroids. The orbital parameters of Sputnik-3 were 217 km (perigee), 1864 km (apogee) and 65.18° (orbital inclination).

The launches of Sputnik-1 and Sputnik-2 had both surprised and embarrassed the Americans as they had no successful satellite launch to their credit till then. They were more than eager to catch up. Explorer-1 (Figure 1.8) was the first satellite to be successfully launched by the United States. It was launched on 31 January 1958 by Jupiter-C rocket from Cape Canaveral. The satellite orbital parameters were 360 km (perigee), 2534 km (apogee) and 33.24° (orbital inclination). Explorer's design was pencil-shaped, which allowed it to spin like a bullet as it orbited the Earth. The spinning motion provided stability to the satellite while in orbit. Incidentally, spin stabilization is one of the established techniques of satellite stabilization.

History of the Evolution of Satellites

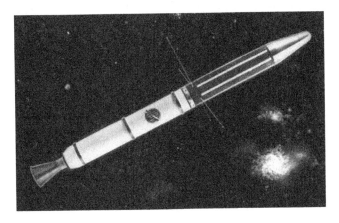

Figure 1.8 Explorer-1 (Courtesy: NASA/JPL-Caltech)

During its mission, it discovered that Earth is girdled by a radiation belt trapped by the magnetic field.

After the successful launch of Explorer-1, there followed in quick succession the launches of Vanguard-1 on 5 February 1958, Explorer-2 on 5 March 1958 and Vanguard-1 (TV-4) on 17 March 1958 (Figure 1.9). The Vanguard-1 and Explorer-2 launches were unsuccessful. The Vanguard-1 (TV-4) launch was successful. It was the first satellite to employ solar cells to charge the batteries. The orbital parameters were 404 km (perigee), 2465 km (apogee) and 34.25° (orbital inclination). The mission carried out geodetic studies and revealed that Earth was pear-shaped.

Figure 1.9 Vanguard-1 (TV-4) (Courtesy: NASA)

1.3.3 Satellites for Communications, Meteorology and Scientific Exploration -- Early Developments

Soviet experiences with the series of Sputnik launches and American experiences with the launches of the Vanguard and Explorer series of satellites had taken satellite and satellite launch technology to sufficient maturity. The two superpowers by then were busy extending the use of satellites to other possible areas such as communications, weather forecasting, navigation and so on. The 1960–1965 period saw the launches of experimental satellites for the above-mentioned applications. 1960 was a very busy year for the purpose. It saw the successful launches of the first weather satellite in the form of TIROS-1 (television and infrared observation satellite) (Figure 1.10) on 1 April 1960, the first experimental navigation satellite Transit-1B on 13 April 1960, the first experimental infrared surveillance satellite MIDAS-2 on 24 May 1960, the first experimental passive communications satellite Echo-1 (Figure 1.11) on 14 August 1960 and the active repeater communications satellite Courier-1B (Figure 1.12) on 4 October 1960. In addition, that year also saw successful launches of Sputnik-5 and Sputnik-6 satellites in August and December respectively.

While the TIROS-1 satellite with two vidicon cameras on board provided the first pictures of Earth, the Transit series of satellites was designed to provide navigational aids to the US Navy with positional accuracy approaching 160 m. The Echo series of satellites, which were aluminized Mylar balloons acting as passive reflectors to be more precise, established how two distantly located stations on Earth could communicate with each other through a space-borne passive reflector. It was followed by Courier-1B, which established the active repeater concept. The MIDAS (missile defense alarm system) series of early warning satellites established beyond any doubt the importance of surveillance from space-borne platforms to locate and identify the strategic weapon development programme of an adversary. Sputnik-5 and Sputnik-6 satellites further studied the biological effect of orbital flights. Each spacecraft had carried two dog passengers.

Figure 1.10 TIROS-1 (Courtesy: NASA)

History of the Evolution of Satellites

Figure 1.11   Echo-1 (Courtesy: NASA)

1.3.4 Non-geosynchronous Communication Satellites: Telstar and Relay Programmes

Having established the concept of passive and active repeater stations to relay communication signals, the next important phase in satellite history was the use of non-geostationary satellites for intercontinental communication services. The process was initiated by the American Telephone and Telegraph (AT&T) seeking permission from the Federal Communications Commission (FCC) to launch an experimental communications satellite. This gave birth to the Telstar series of satellites. The Relay series of satellites that followed the Telstar series also belonged to the same class.

Figure 1.12   Courier-1B (Courtesy: US Army)

Figure 1.13 Telstar-1 (Courtesy: NASA)

In the Telstar series, Telstar-1 (Figure 1.13), the first true communications satellite and also the first commercially funded satellite, was launched on 10 July 1962, followed a year later by Telstar-2 on 7 May 1963. Telstar-2 had a higher orbit to reduce exposure to the damaging effect of the radiation belt. The Telstar-1 with its orbit at 952 km (perigee) and 5632 km (apogee) and an inclination of 44.79° began the revolution in global TV communication from a non-geosynchronous orbit. It linked the United States and Europe.

Telstar-1 was followed by Relay-1 (NASA prototype of an operational communication satellite) launched on 13 December 1962. Relay-2, the next satellite in the series, was launched on 21 January 1964. The orbital parameters of Relay-1 were 1322 km (perigee), 7439 km (apogee) and 47.49° (inclination). The mission objectives were to test the transmissions of television, telephone, facsimile and digital data.

It is worthwhile mentioning here that both the Telstar and Relay series of satellites were experimental vehicles designed to discover the limits of satellite performance and were just a prelude to much bigger events to follow. For instance, through Telstar missions, scientists came to discover how damaging the radiation could be to solar cells. Though the problem has been largely overcome through intense research, it still continues to be the limiting factor on satellite life.

1.3.5 Emergence of Geosynchronous Communication Satellites

The next major milestone in the history of satellite technology was Arthur C. Clarke's idea becoming a reality. The golden era of geosynchronous satellites began with the advent of the SYNCOM (an acronym for synchronous communication satellite) series of satellites developed by the Hughes Aircraft Company. This compact spin-stabilized satellite was first shown at the Paris Air Show in 1961. SYNCOM-1 was launched in February 1963 but the mission failed shortly after. SYNCOM-2 (Figure 1.14), launched on 26 July 1963, became the first operational geosynchronous communication satellite. It was followed by SYNCOM-3,

Figure 1.14 SYNCOM-2 (Courtesy: NASA)

which was placed directly over the equator near the international date line on 19 August 1964. It was used to broadcast live the opening ceremonies of the Tokyo Olympics. That was the first time the world began to see the words 'live via satellite' on their television screens.

Another significant development during this time was the formation of INTELSAT (International Telecommunications Satellite Organization) in August 1964 with COMSAT (Communication Satellite Corporation) as its operational arm. INTELSAT achieved a major milestone with the launch of the Intelsat-1 satellite, better known as 'Early Bird' (Figure 1.15),

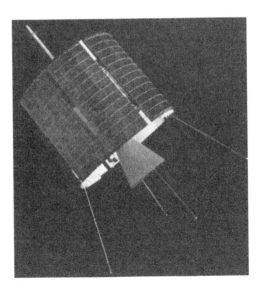

Figure 1.15 Intelsat-1 (Reproduced by permission of © Intelsat)

Figure 1.16 Molniya series satellite

on 5 April 1965 from Cape Canaveral. Early Bird was the first geostationary communications satellite in commercial service. It went into regular service in June 1965 and provided 240 telephone circuits for connectivity between Europe and North America. Though designed for an expected life span of only 18 months, it remained in service for more than three years.

While the Americans established their capability in launching communications satellites through launches of SYNCOM series of satellites and Early Bird satellite during the 1960–1965 era, the Soviets did so through their Molniya series of satellites beginning April 1965. The Molniya series of satellites (Figure 1.16) were unique in providing uninterrupted 24 hours a day communications services without being in the conventional geostationary orbit. These satellites pursued highly inclined and elliptical orbits, known as the Molniya orbit (Figure 1.17), with apogee and perigee distances of about 40 000 km and 500 km and orbit inclination of 65°. Two or three such satellites aptly spaced apart in the orbit provided uninterrupted service. Satellites in such an orbit with a 12 hour orbital period remained over the countries of the

Figure 1.17 Molniya orbit

former Soviet bloc in the northern hemisphere for more than 8 hours. The Molniya-1 series was followed later by the Molniya-2 (in 1971) and the Molniya-3 series (in 1974).

1.3.6 International Communication Satellite Systems

The Intelsat-1 satellite was followed by the Intelsat-2 series of satellites. Four Intelsat-2 satellites were launched in a span of one year from 1966 to 1967. The next major milestone *vis-à-vis* communication satellites was achieved with the Intelsat-3 series of satellites (Figure 1.18) becoming fully operational. The first satellite in the Intelsat-3 series was launched in 1968. These satellites were positioned over three main oceanic regions, namely the Atlantic, the Pacific and the Indian Oceans, and by 1969 they were providing global coverage for the first time. The other new concept tried successfully with these satellites was the use of a de-spun antenna structure, which allowed the use of a highly directional antenna on a spin-stabilized satellite. The satellites in the Intelsat-1 and Intelsat-2 series had used omnidirectional antennas.

Figure 1.18 Intelsat-3 (Reproduced by permission of © Intelsat)

The communication satellites' capabilities continued to increase with almost every new venture. With the Intelsat-4 satellites (Figure 1.19), the first of which was launched in 1971, the satellite capacity got a big boost. Intelsat-4A series introduced the concept of frequency re-use. The frequency re-use feature was taken to another dimension in the Intelsat-5 series with the use of polarization discrimination. While frequency re-use, i.e. use of the same frequency band, was possible when two footprints were spatially apart, dual polarization allowed the re-use of the same frequency band within the same footprint. The Intelsat-5 satellites (Figure 1.20), the first of which was launched in 1980, used both C band and Ku band transponders and were three-axis stabilized. The satellite transponder capacity has continued to increase through the Intelsat-6, Intelsat-7 and Intelsat-8 series of satellites launched during the 1980s and 1990s. Intelsat-9 and Intelsat-10 series were launched in the first decade of the new millennium. These series of satellites were followed by a number of more launches. As of October 2013, Intelsat operated 28 satellites and supports more than 30 DTH platforms world-wide.

Figure 1.19 Intelsat-4 (Reproduced by permission of © Intelsat)

The Russians have also continued their march towards development and launching of communication satellites after their success with the Molniya series. The Raduga series (International designation: Statsionar-1), the Ekran series (international designation: Statsionar-T), shown in Figure 1.21, Ekspress-AM series, Molniya series and the Gorizont series (international designation: Horizon) are the latest in communication satellites from the Russians. All three employ the geostationary orbit.

1.3.7 Domestic Communication Satellite Systems

Beginning in 1965, the Molniya series of satellites established the usefulness of a domestic communications satellite system when it provided communications connectivity to a large number of republics spread over the enormous land-mass of the former Soviet Union. Such

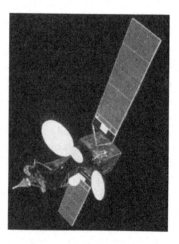

Figure 1.20 Intelsat-5 (Reproduced by permission of © Intelsat)

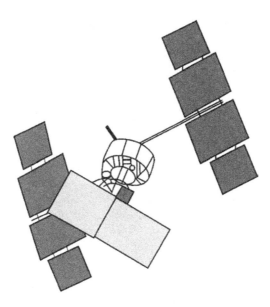

Figure 1.21 Ekran series

a system was particularly attractive to countries having a vast territory. Canada was the first non-Soviet country to have a dedicated domestic satellite system with the launch of the Anik-A series of satellites (Figure 1.22), beginning in 1972. The capabilities of these satellites were subsequently augmented with the successive series of Anik satellites, named Anik-B (beginning 1978), Anik-C (beginning 1982), Anik-D (also beginning 1982), Anik-E (beginning 1991), Anik-F (beginning 2000) and Anik-G (beginning 2013). Anik-G1, first satellite in the series, is a multi-mission satellite designed to provide direct-to-home (DTH) television service in Canada and also broadband voice, data and video services in South America. The satellite was launched on April 16, 2013 by Proton/Breeze-M rocket.

The United States began its campaign for development of domestic satellite communication systems with the launch of Westar satellite in 1974, Satcom satellite in 1975 and Comstar satellite in 1976. Satcom was also incidentally the first three-axis body-stabilized geostationary satellite. These were followed by many more ventures. Europe began with the European communications satellite (ECS series) and followed it with the Eutelsat-II series (Figure 1.23) and Eutelsat-W series of satellites. In addition to the Eutelsat satellites, other series of satellites, namely the Hot Bird, Eurobird and Atlantic Bird series, were launched to expand the horizon of the services offered and the coverage area of the satellites of the EUTELSAT organization. In 2012, EUTELSAT renamed all the satellite series under the brand name of Eutelsat.

Indonesia was the first developing nation to recognize the potential of a domestic communication satellite system and had the first of the Palapa satellites placed in orbit in 1977 to link her scattered island nation. The Palapa series of satellites have so far seen four generations named Palapa-A (beginning 1977), Palapa-B (beginning 1984), Palapa-C (beginning 1991) and Palapa-D (beginning 2009). Palapa-D was launched on August 31, 2009 aboard Chinese Long March 3B rocket.

India, China, Saudi Arabia, Brazil, Mexico and Japan followed suit with their respective domestic communication satellite systems. India began with the INSAT-1 series of satellites

Figure 1.22 Anik-A (Courtesy: Telesat Canada)

Figure 1.23 Eutelsat-II (Reproduced by permission of © Eutelsat)

in 1981 and has already entered the fourth generation of satellites with the INSAT-4 series. INSAT-4CR (Figure 1.24) was launched in September 2007. The latest in the INSAT-4 series is the INSAT-4G satellite, launched on 21 May 2011 aboard the Ariane-5 rocket. However, the latest in the INSAT-3 series is the INSAT-3D satellite launched on 26 July 2013 aboard the Ariane-5 rocket. Arabsat, which links the countries of the Arab League, has also entered the fifth generation with the Arabsat-5 series of satellites. Three satellites Arabsat-5A, -5B and -5C have been launched in this series.

Figure 1.24 INSAT-4A (Courtesy: ISRO)

1.3.8 Satellites for other Applications also made Rapid Progress

The intention to use satellites for applications other than communications was very evident, even in the early stages of development of satellites. A large number of satellites were launched mainly by the former Soviet Union and the United States for meteorological studies, navigation, surveillance and Earth observation during the 1960s.

Making a modest beginning with the TIROS series, meteorological satellites have come a long way both in terms of the number of satellites launched for the purpose and also advances in the technology of sensors used on these satellites. Both low Earth as well as geostationary orbits have been utilized in the case of satellites launched for weather forecasting applications. Major non-geostationary weather satellite systems that have evolved over the years include the TIROS (television and infrared observation satellite) series and the Nimbus series beginning around 1960, the ESSA (Environmental Science Service Administration) series (Figure 1.25) beginning in 1966, the NOAA (National Oceanic and Atmospheric Administration) series beginning in 1970, the DMSP (Defense Meteorological Satellite Program) series initiated in 1965 (all from the United States), the Meteor series beginning in 1969 from Russia and the

Figure 1.25 ESSA satellites (Courtesy: NASA)

Figure 1.26 GOES satellite (Courtesy: NOAA and NASA)

Feng Yun series (FY-1 and FY-3) beginning 1988 from China. Major meteorological satellites in the geostationary category include the GMS (geostationary meteorological satellite) series from Japan since 1977, the GOES (geostationary operational environmental satellite) series from the United States (Figure 1.26) since 1975, the METEOSAT (meteorological satellite) series from Europe since 1977 (Figure 1.27), the INSAT (Indian satellite) series from India since 1982 (Figure 1.28) and the Feng Yun series (FY-2) from China since 1997.

Sensors used on these satellites have also seen many technological advances, both in types and numbers of sensors used as well as in their performance levels. While early TIROS series satellites used only television cameras, a modern weather forecasting satellite has a variety of sensors with each one having a dedicated function to perform. These satellites provide very high resolution images of cloud cover and Earth in both visible and infrared parts of the spectrum and thus help generate data on cloud formation, tropical storms, hurricanes, likelihood of forest fires, temperature profiles, snow cover and so on.

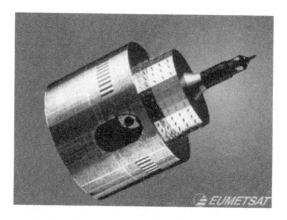

Figure 1.27 METEOSAT series (Reproduced by permission of © EUMETSAT)

Figure 1.28 INSAT series (Courtesy: ISRO)

Remote sensing satellites have also come a long way since the early 1970s with the launch of the first of the series of Landsat satellites that gave detailed attention to various aspects of observing the planet Earth from a space based platform. In fact, the initial ideas of having satellites for this purpose came from the black and white television images of Earth beneath the cloud cover as sent by the TIROS weather satellite back in 1960, followed by stunning observations revealed by Astronaut Gordon Cooper during his flight in a Mercury capsule in 1963 when he claimed to have seen roads, buildings and even smoke coming out of chimneys from an altitude of more than 160 km. His claims were subsequently verified during successive exploratory space missions.

Over the years, with significant advances in various technologies, the application spectrum of Earth observation or remote sensing satellites has expanded very rapidly from just terrain mapping called cartography to forecasting agricultural crop yield, forestry, oceanography, pollution monitoring, ice reconnaissance and so on. The Landsat series from the United States, the SPOT (satellite pour l'observation de la terre) series from France and the IRS (Indian remote sensing satellite) series from India are some of the major Earth observation satellites. The Landsat programme, beginning with Landsat-I in 1972, has progressed to Landsat-8 through Landsat-2, -3, -4, -5, -6 and -7 (Figure 1.29). Landsat-8, the most recent in the Landsat series,

Figure 1.29 Landsat-7 (Courtesy: NASA)

was launched on 11 February 2013. The SPOT series has also come a long way, beginning with SPOT-1 in 1986 to SPOT-6 launched in 2012 through SPOT-2, -3, -4 and -5 (Figure 1.30). IRS series launches began in 1988 with the launch of IRS-1A and the most recent satellites launched in the series are IRS-P6 called Resourcesat 1 (Figure1.31) launched in 2003 and IRS-P5 called Cartosat 1 launched in 2005. Cartosat 2, Cartosat 2A, Cartosat 2B and Resouresat 2 launched in 2007, 2008, 2010 and 2011 respectively are other remote sensing satellites of India. Sensors on board modern Earth observation satellites include high resolution TV cameras, multispectral scanners (MSS), very high resolution radiometers (VHRR), thematic mappers (TM), and synthetic aperture radar (SAR). RISAT-1 launched on April 26, 2012 is the recent satellite whose all weather radar images will facilitate agriculture and disaster management.

Figure 1.30 SPOT-5 (Reproduced by permission of © CNES/ill.D.DUCROS, 2002)

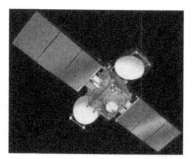

Figure 1.31 Resourcesat (Courtesy: ISRO)

1.3.9 Small or Miniature Satellites

One of the methods of classifying satellites is on the basis of the in-orbit mass of the satellite. Based on this criterion, satellites are generally classified as large, medium, mini, micro, nano,

Table 1.1 Classification of satellites based on wet mass

Satellite class	Wet mass (kg)
Large	>1000
Medium	500–1000
Mini	100–500
Micro	10–100
Nano	1–10
Pico	0.1–1
Femto	<0.1

pico or femto. Mini, micro, nano, pico and femto satellites are collectively categorized as *small or miniature satellites*. Table 1.1 shows the classification of satellites based on wet mass, i.e. the mass of the satellite including fuel. The commercial space sector today is identified by geostationary communications satellites. One of the major driving forces responsible for the relevance of small satellites over the years has been the need to enable missions that larger satellites could not have accomplished. These included constellations for low data rate communications, in-orbit inspection of larger satellites and university-related research. Other facilitating factors have been the requirement for smaller and cheaper launch vehicles as against larger rockets capable of producing much greater thrust and consequently greater financial cost for larger and heavier satellites. Also, smaller satellites can be launched in multiple numbers and as piggybacks using excess capacity on larger launch vehicles. They have a lower cost of manufacture, ease of mass production and faster building times, making them the ideal test bed for new technologies. Small satellites are not the exclusive prerogative of departments of defence and other major R&D organizations. They are also a big attraction for commercial industry and universities. According to an estimate, close to 1000 satellites will be launched between 2000 and 2020 in the category of small satellites, including mini, micro, nano, pico and femto satellites.

1.3.9.1 Medium Satellites

Medium satellites have wet mass in the range 500–1000 kg. Medium satellites, although smaller and simpler than large satellites, use the same technologies as those used in large satellites. A large number of satellites designed for remote sensing and weather forecasting applications fall into the category of medium satellites. These are discussed in Chapters 10 and 11.

1.3.9.2 Mini Satellites

Mini satellites have wet mass in the range 100–500 kg. A large number of satellites intended for military surveillance and intelligence, scientific studies and some satellites designed for weather forecasting and earth observation applications belong to this category. Some examples of mini satellites include Jason-1, Jason-2 and SARAL satellites for remote sensing applications, the SMART-1 (Small Missions for Advanced Research in Technology) satellite for scientific studies, ELISA-1 to ELISA-4 (Electronic Intelligence by Satellite) and SPIRALE (Système Préparatoire Infra Rouge pour l'Alerte) for military intelligence, and the

METEOSAT series for global weather forecasting. Jason-1 and Jason-2 satellites, respectively launched in 2001 and 2008, are a joint project between NASA (USA) and CNES (France) and are intended to monitor global ocean circulation and global climate forecasts, and to measure ocean surface topography. SARAL (Satellite with CityArgos and Altika) is an altimetry technology mission by ISRO (India) and CNES launched in February 2013 from Indian PSLV-C20. The mission is complementary to Jason-2 satellite. ELISA-1 to ELISA-4 are a suit of French military satellites. The ELISA programme is a demonstration system that paves the way for a planned radar monitoring system. SPIRALE launched in February 2009 and is intended to detect the flights of ballistic missiles in the boost phase by using infrared satellite imagery. SMART-1 from the European Space Agency (Figure 1.32) was launched in September 2003 and is intended to orbit the moon. This satellite carried instruments for lunar imaging, identification of chemical elements on the lunar surface, identification of mineral spectra of olivine and pyroxene. The satellite was deliberately made to crash into the moon's surface in September 2006.

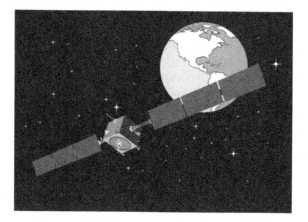

Figure 1.32 SMART-1

1.3.9.3 Micro Satellites

Micro satellites fall into the 10–100 kg wet mass category. Many of the well-known early satellites launched during the 1950s to 1970s belong to the micro satellite category primarily because of the limited launcher capacity available during this period. Some common examples are Sputnik (1957), Vanguard-1 (1958), Telstar-1 (1962), Syncom-1 (1963), Early Bird (1965) and Apollo-P and F1 (1971). With advances in technology enabling sophisticated payloads to be developed into smaller volumes, there has been a renewed interest in the scientific commercial market for micro satellites. The trend began with the launch of UoSAT-1 in 1981, which was the first satellite to carry a microprocessor. Modern micro satellites carry an onboard computer enabling them to carry out in-orbit programmable operations outside the range of the ground station. Some examples of micro satellites launched after UoSAT-1 during the 1980s include the Cosmos series, which forms part of military tactical communications constellations, the Iskra series of amateur radio relay satellites, the Fuji-1 Japanese radio relay satellite and Rohini-3, which is intended for remote sensing experiments. This trend has continued through the 1990s and 2000s. Some well-known examples include Astrid-1 (1995) and Astrid-2 (1998), intended for scientific studies, FalconSat-1 (2000) for technology demonstration to carry out analysis of ion current collection in plasma wake, UNISAT-2 (2002) for scientific research, HAMSAT (2005) for amateur radio communications, ANDE (Atmospheric Neutral Density Experiment) (2009) to measure density and composition of the low earth orbit atmosphere while being tracked from ground, WNISAT (2012) for remote sensing applications and NEOSSat (Near Earth Object Surveillance Satellite) (2013) (Figure 1.33) for discovering and tracking asteroids that might pose a threat to Earth and for tracking satellites and orbital debris.

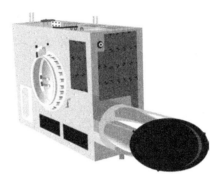

Figure 1.33 NEOSSat

1.3.9.4 Nano Satellites

Nano satellites have wet mass between 1 and 10 kg. They are launched both as individual satellites and as groups of satellites designed to work together to perform the intended tasks. Some designs require a larger satellite, generally called the 'mother satellite', for communication with ground controllers or for launching and docking with nano satellites. The term 'satellite swarm' is commonly used to designate a group of these satellites. With advances in electronics technology, particularly in size miniaturization for a given functional capability, nano satellites are rapidly occupying the space for applications that previously required the use of micro and mini satellites. One such example is that of a constellation of 35 nano satellites each weighing 8 kg replacing a constellation of five RapidEye satellites each weighing 156 kg for Earth imaging applications at the same mission cost and significantly increased revisit time of 3.5 hours as against 24 hours. Some other examples of nano satellite missions include UniBRITE and BRITE, which form part of the BRITE constellation of six satellites, designed to make precise measurements of the brightness variations of a large number of bright stars, AAUSAT-2 and AAUSAT-3, developed by Aalborg University in Denmark, and STRAND (Surrey Training Research and Nano satellite Demonstrator), developed by Surrey University and the ELaNa (Educational Launch of Nano satellites) (Figure 1.34) mission of NASA in partnership with different universities to launch small satellites for research purposes.

A term commonly used with nano and pico satellites is 'CubeSat'. CubeSat is a type of small satellite of 10 cm cube dimensions and a mass that is not greater than 1.33 kg. The standard

Figure 1.34 ELaNa (Courtesy: NASA)

10 × 10 × 10 cm basic CubeSat is also called a 1U CubeSat, where 1U stands for one unit. CubeSats are scalable in increments of 1U along one axis only. Consequently, 2U, 3U and 4U CubeSats will have dimensions of 20 × 10 × 10 cm, 30 × 10 × 10 cm and 40 × 10 × 10 cm, respectively. Since all CubeSats are 10 × 10 cm irrespective of length, they can all be launched and deployed from a common deployment system.

1.3.9.5 Pico Satellites

Pico satellites have wet mass between 0.1 and 1 kg. They are usually launched as a group of satellites ('satellite swarm') to work in formation for the intended mission objectives. As for nano satellites, some designs of pico satellites also require a larger mother satellite for communication with ground controllers or for launching and docking with pico satellites. The standard CubeSat design described in the previous section is an example of the largest pico satellite or the smallest nano satellite.

Pico satellites offer an excellent way for graduate and PhD students to get hands-on experience with satellite system design. One such example is the pico satellite UWE-1 (University of Würzburg's experimental satellite) with a mass of less than 1 kg. The satellite was developed and built by students of the University of Würzburg and Fachhochschule Weingarten, Germany. It was launched in 2005 to test adaptations of internet protocols to the space environment, characterized by significant signal propagation delays due to the large distances and much higher noise levels compared to terrestrial links. Pico satellites and launch opportunities for sub-kg class pico satellites are now commercially available.

1.3.9.6 Femto Satellites

Femto satellites have wet mass between 10 and 100 g. Again, like nano and pico satellites, some designs require a larger mother satellite for communication with ground controllers. With advances in micro and nano technologies, it is today feasible to build satellite subsystems and even the entire satellite itself on a chip. Such satellites are known as chipsats.

Three prototype chip satellites were launched to the International Space Station (ISS) on Space Shuttle Endeavor on its final mission in May 2011. The tiny chip satellites, nicknamed 'Sprite', were mounted on the Materials International Space Station Experiment (MISSE-8) pallet, which in turn was attached to the ISS to test how well they functioned in that harsh environment. These fingernail sized satellites were designed to collect the solar wind's chemistry, radiation and particle-impact data. In another example, KickSat mission plans to launch 250 of these tiny Sprite satellites into low Earth orbit (Figure 1.35). The launch is planned on a SpaceX Falcon-9 launch vehicle in early 2014. The mission will provide an opportunity for individuals to have personal satellites at a cost of about US$300 per satellite. Sprite is the size of a postage stamp with its hardware comprising of solar cells, a radio transceiver and a microcontroller with memory and sensors on a smaller scale. The first version is capable of transmitting its name and a few bits of data. Future versions, however, could include other types of sensors for enhanced functionality.

Figure 1.35 KickSat mission

1.4 Evolution of Launch Vehicles

Satellite launch vehicles have also seen various stages of evolution in order to meet launch demands of different categories of satellites. Both smaller launch vehicles capable of launching satellites in low Earth orbits and giant sized launch vehicles that can deploy multiple satellites in geostationary transfer orbit have seen improvements in their design over the last five decades of their history. The need to develop launch vehicles by countries like the United States and Russia was in the earlier stages targeted to acquire technological superiority in space technology. This led them to use the missile technology developed during the Second World War era to build launch vehicles. This was followed by their desire to have the capability to launch bigger satellites to higher orbits. The next phase was to innovate and improve the technology to an extent that these vehicles became economically viable, which meant that the attainment of mission objectives justified the costs involved in building the launch vehicle. The technological maturity in launch vehicle design backed by an ever-increasing success rate led to these vehicles being used for offering similar services to other nations who did not possess them.

The situation in different countries involved in developing launch vehicles is different. On the one hand, there are nations keen to become self-reliant and attain a certain level of autonomy in this area; there are others whose commercial activities complement a significant part of their national activity.

Beginning with a one-stage R-7 rocket (named Semyorka) that launched Sputnik-1 into its orbit in 1957, Russia has developed a large number of launch vehicles for various applications. Some of the prominent ones include the Vostok series, the Molniya series (Figure 1.36), the Soyuz series, the Proton series (Figure 1.37), the Zenit series and Energia series (Figure 1.38). Energia is capable of placing a payload of 65 to 200 tonnes in a low Earth orbit.

Figure 1.36 Molniya series (Reproduced by permission of © Mark Wade)

Figure 1.37 Proton series (Courtesy: NASA)

Evolution of Launch Vehicles

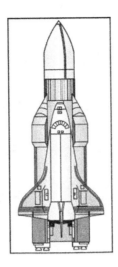

Figure 1.38 Energia series (Reproduced by permission of © Mark Wade)

Important launch vehicles developed by the United States include the Delta series (Figure 1.39), the Atlas series, the Titan series (Figure 1.40), the Pegasus series and the re-usable famous Space Shuttle (Figure 1.41). Buran (Figure 1.42) from Russia is another re-usable vehicle similar in design and even dimensions to the American Space Shuttle. The main difference between the two lies in the fact that Buran does not have its own propulsion system

Figure 1.39 Delta series

Figure 1.40 Titan series (Courtesy: NASA/JPL-Caltech)

Figure 1.41 Space Shuttle (Courtesy: NASA)

Evolution of Launch Vehicles

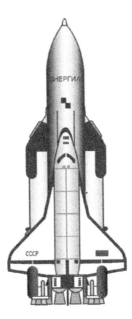

Figure 1.42 Buran series

and is launched into orbit by Energia launch vehicle. The Ariane launch vehicle from the European Space Agency (ESA) has entered the fifth generation with the Ariane-5 heavy launch vehicle. Ariane-5 ECB (Enhanced Capability-B) (Figure 1.43) has the capacity of launching 12 tonnes to the geostationary transfer orbit.

Figure 1.43 Ariane-5ECA (Reproduced by permission of © ESA-D. DUCROS)

Long March (Figure 1.44) from China, the PSLV (polar satellite launch vehicle) and the GSLV (geostationary satellite launch vehicle) (Figure 1.45) from India and the H-2 series from Japan are some of the other operational launch vehicles. Prominent launch vehicles, both expendable and re-usable, are discussed in detail in section 3.3 of Chapter 3.

Figure 1.44 Long March

Figure 1.45 GSLV (Courtesy: ISRO)

1.5 Future Trends

The technological advances in the field of satellites will be directed with an aim of reducing the cost and size of the satellites as well as improving the quality of the services provided. One of the main technological trends is to develop satellites with a longer mission life. Smaller satellites are being developed as they can be launched using smaller launchers, thereby cutting the overall mission expenditure.

1.5.1 Communication Satellites

In the case of communication satellites, key technologies include development of large-scale multi-beam antennas to allow intensive reuse of frequencies, USAT terminals to replace VSAT terminals, ground user terminals, development of signal processing algorithms to perform intelligent functions on-board the satellite including signal regeneration, overcoming the signal fading problem due to rain and allowing use of smaller antennas. Flexible cross-link communication between satellites will be developed to allow better distribution of traffic between the satellites. The trend will be to use millimetre or EHF bands of the spectrum to cope with the increased demand for broadband services. This will require the development of technologies to cope with rain-fade problems in these bands. Newer LEO and MEO satellite constellations will be launched mainly for enhancing land-mobile services.

1.5.2 Weather Forecasting Satellites

Future weather forecasting satellites will carry advanced payloads including multispectral imagers, sounders and scatterometers with better resolution. Hyper-spectral measurements from newly developed interferometers will be possible in the near future. These instruments will have more than a thousand channels over a wide spectral range. Also, the satellite data download rates are expected to exceed several terabytes per day.

The GOES-R satellite planned to be launched in the year 2015 will carry several sophisticated instruments including the Advanced Baseline Imager (ABI), Space Environment In-Situ Suite (SEISS), Solar Imaging Suite (SIS), Geostationary Lightning Mapper (GLM) and Magnetometer. SEISS further comprises two Magnetospheric Particle Sensors (MPS-HI and MPS-LO), an Energetic Heavy Ion Sensor (EHIS) and a Solar and Galactic Proton Sensor (SGPS). The SIS payload has a Solar Ultraviolet Imager (SUVI), a Solar X-Ray Sensor (XRS) and an Extreme Ultraviolet Sensor (EUVS).

1.5.3 Earth Observation Satellites

For Earth observation satellites, technological advancements will lead to better resolution, increase in observation area and reduction in access time, i.e time taken between the request of an image by the user and its delivery. Plans for future missions and instruments include entirely new types of measurement technology, such as hyper-spectral sensors, cloud radars, lidars and polarimetric sensors that will provide new insights into key parameters of atmospheric temperature and moisture, soil moisture and ocean salinity. Several new gravity field missions aimed at more precise determination of the marine geoid will also be launched in the future. These missions will also focus on disaster management and studies of key Earth System processes — the water cycle, carbon cycle, cryosphere, the role of clouds and aerosols in global climate change and sea level rise.

1.5.4 Navigational Satellites

Satellite based navigation systems are being further modernized so as to provide more accurate and reliable services. The modernization process includes launch of new more powerful satellites, use of new codes, enhancement of ground system, and so on. In fact satellite based systems will be integrated with other navigation systems so as to increase their application potential.

The GPS system is being modernized so as to provide more accurate, reliable and integrated services to the users. The first efforts in modernization began with the discontinuation of the selective availability feature, so as to improve the accuracy of the civilian receivers. In continuation of this step, Block IIRM satellites now carry a new civilian code on the L2 frequency. This helps in further improving accuracy by compensating for atmospheric delays and will ensure greater navigation security. Moreover, these satellites carry a new military code (M-code) on both the L1 and L2 frequencies. This provides increased resistance to jamming.

Block-IIF satellites have a third carrier signal, L5, at 1176.45 MHz. The first Block-II F satellite was launched in May 2010. As of August 2013, there are four operational Block-II F satellites in GPS constellation. They have larger design life, fast processors with more memory, and a new civil signal. The GPS-III phase of satellites is at the planning stage. These satellites will employ spot beams. The use of spot beams results in increased signal power, enabling the system to be more reliable and accurate, with system accuracy approaching a metre. As of August 2013, Block III satellites are in production and deployment phase. As many as 32 Block III satellites with a design life of 15 years have been planned. First Block-III satellite (Block IIIA-1) is scheduled for launch in 2014. As far as the GLONASS system is concerned, efforts are being made to make the complete system operational in order to exploit its true application potential. Second and third generation GLONASS satellites have improved lifetimes over first generation GLONASS satellites (GLONASS-M has a design lifetime of seven years and GLONASS-K of 10 to 12 years). GLONASS-K satellites will offer an additional L-band navigational signal.

Another satellite navigation system that is being developed is the European Galileo system. Galileo navigation system development has been structured into the following three phases. The first phase has been the experimental phase to experiment with and verify the critical technologies needed for the Galileo system to operate in the medium earth orbit (MEO) environment. Two experimental satellites namely GIOVE-A (Galileo In-Orbit Validation Element-A) and GIOVE-B respectively launched on December 28, 2005 and April 27, 2008. The second phase is intended to be the In-Orbit Validation (IOV) phase. The main objective of the IOV phase is validation of system design using a scaled down constellation of only four satellites, which along with a limited number of ground stations is the minimum number needed to provide positioning and timing data. The four satellites were launched in pairs. The first pair of two satellites was launched on October 21, 2011 and the second pair was launched on October 12, 2012. The third phase will achieve fully operational Galileo system comprising 30 satellites (27 operational and three active spares), positioned in three circular Medium Earth Orbit (MEO) planes at 23 222 km altitude above the Earth, and with each orbital plane inclined at 56 degrees to the equatorial plane. The system will be operational in the near future. Indian IRNSS (Indian Regional Navigational Satellite System) and Chinese Compass are the other satellite navigation systems under development.

Navigation satellite services will improve as the services provided by the three major navigation satellite systems (GPS, GLONASS, and GALILEO) will be integrated and the user will be able to obtain position information with the same receiver from any of the satellites from the three systems.

1.5.5 Military Satellites

The sphere of application of military satellites will expand further to provide a variety of services ranging from communication services to gathering intelligence imagery data, from weather forecasting to early warning applications, from providing navigation information to providing timing data. They have become an integral component of the military planning of various developed countries, especially of the USA and Russia. Developing countries are designing their military satellites so as to protect their territory. The concept of space based lasers is evolving wherein the satellites carrying onboard high power lasers will act as a nuclear deterrent. These satellites will destroy the nuclear missile in its boost phase within the country that is launching it.

Further Reading

Labrador, V. and Galace, P. (2005) *Heavens Fill with Commerce: A Brief History of the Communications Satellite Industry*, Satnews Publishers, California.

Internet Sites

1. http://electronics.howstuffworks.com/satellite.htm/printable
2. http://www.aero.org/publications/gilmore/gilmore-1.html
3. http://www.thetech.org/exhibits_events/online/satellite/home.html
4. www.intelsat.com
5. www.isro.org
6. www.nasa.gov

Glossary

Ariane: European Space Agency's launch vehicle
Astronaut: A space traveller, i.e. a person who flies in space either as a crew member or a passenger
Astrophysics: Study of the physical and chemical nature of celestial bodies and their environments
Buran: A re-usable launch vehicle, Russian counterpart of a space shuttle
Early Bird: Other name for Intelsat-1. First geostationary communications satellite in commercial service
Explorer-1: First successful satellite from the United States
Femto satellite: Satellite with wet mass < 100 g
Footprint: The area of coverage of a satellite
Geostationary orbit: An equatorial circular orbit in which the satellite moves from west to east with a velocity such that it remains stationary with respect to a point on the Earth. Also known as the Clarke

orbit after the name of the science fiction writer who first proposed this orbit

GPS: An abbreviation for the global positioning system. It is a satellite-based navigation system that allows you to know your position coordinates with the help of a receiver anywhere in the world under any weather condition

GSLV: Abbreviation for geostationary satellite launch vehicle. Launch vehicle from India

INTELSAT: Acronym for International Telecommunications Satellite Consortium operating satellites internationally for both domestic and international telecommunication services

Landsat: First remote sensing satellite series in the world from USA

Large satellite: Satellite with wet mass of > 1000 kg

Medium satellite: Satellite with wet mass in the range 500–1000 kg

Micro satellite: Satellite with wet mass in the range 10–100 kg

Mini satellite: Satellite with wet mass in the range 100–500 kg

Molniya orbit: A highly inclined and elliptical orbit used by Russian satellites with apogee and perigee distances of about 40 000 and 500 km and an orbit inclination of 65°. Two or three such satellites aptly spaced apart in the orbit provide an uninterrupted communication service

Multispectral scanner (MSS): A multispectral scanning device that uses an oscillating mirror to continuously scan Earth passing beneath the spacecraft

NASA: National Aeronautics and Space Administration

Nano satellite: Satellite with wet mass in the range 1–10 kg

Palapa: First domestic communication satellite from a developing country, Indonesia

Payload: Useful cargo-like satellite being a payload of a launch vehicle

Pico satellite: Satellite with wet mass in the range 0.1–1 kg

Satellite: A natural or artificial body moving around a celestial body

Sounding rocket: A research rocket used to obtain data from the upper atmosphere

Space shuttle: A re-usable launch vehicle from the United States

Spin-stabilized satellite: A satellite whose attitude stabilization is achieved by the spinning motion of the satellite. It employs the gyroscopic or spinning top principle

Sputnik-1: First artificial satellite launched by any country. Launched on 4 October 1957 by erstwhile Soviet Union

Thematic mapper: A type of scanning sensor used on Earth observation satellites

Three-axis stabilized satellite: A satellite whose attitude is stabilized by an active control system that applies small forces to the body of the spacecraft to correct any undesired changes in its orientation

TIROS: First series of weather forecast satellites, launched by United States

Transponder: A piece of radio equipment that receives a signal from the Earth station at the uplink frequency, amplifies it and then retransmits the same signal at the downlink frequency

Westar: First domestic communication satellite from the United States

2

Satellite Orbits and Trajectories

The study of orbits and trajectories of satellites and satellite launch vehicles is the most fundamental topic of the subject of satellite technology and perhaps also the most important one. It is important because it gives an insight into the operational aspects of this wonderful piece of technology. An understanding of the orbital dynamics would give a sound footing to address issues like types of orbit and their suitability for a given application, orbit stabilization, orbit correction and station keeping, launch requirements and typical launch trajectories for various orbits, Earth coverage and so on. This chapter and the one after this focus on all these issues and illustrate various concepts with the help of necessary mathematics and a large number of solved problems.

2.1 Definition of an Orbit and a Trajectory

While a trajectory is a path traced by a moving body, an orbit is a trajectory that is periodically repeated. While the path followed by the motion of an artificial satellite around Earth is an orbit, the path followed by a launch vehicle is called the launch trajectory. The motion of different planets of the solar system around the sun and the motion of artificial satellites around Earth (Figure 2.1) are examples of orbital motion.

The term 'trajectory', on the other hand, is associated with a path that is not periodically revisited. The path followed by a rocket on its way to the right position for a satellite launch (Figure 2.2) or the path followed by orbiting satellites when they move from an intermediate orbit to their final destined orbit (Figure 2.3) are examples of trajectories.

2.2 Orbiting Satellites – Basic Principles

The motion of natural and artificial satellites around Earth is governed by two forces. One of them is the centripetal force directed towards the centre of the Earth due to the gravitational force of attraction of Earth and the other is the centrifugal force that acts outwards from the

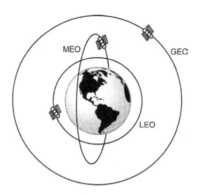

Figure 2.1 Example of orbital motion – satellites revolving around Earth

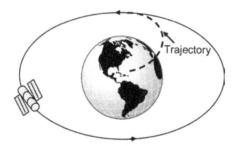

Figure 2.2 Example of trajectory – path followed by a rocket on its way during satellite launch

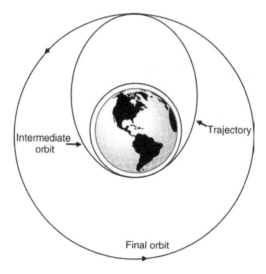

Figure 2.3 Example of trajectory – motion of a satellite from the intermediate orbit to the final orbit

Orbiting Satellites – Basic Principles

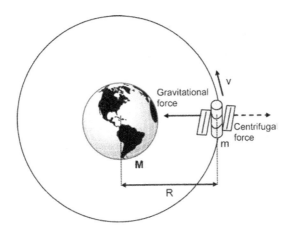

Figure 2.4 Gravitational force and the centrifugal force acting on bodies orbiting Earth

centre of the Earth (Figure 2.4). It may be mentioned here that the centrifugal force is the force exerted during circular motion, by the moving object upon the other object around which it is moving. In the case of a satellite orbiting Earth, the satellite exerts a centrifugal force. However, the force that is causing the circular motion is the centripetal force. In the absence of this centripetal force, the satellite would have continued to move in a straight line at a constant speed after injection. The centripetal force directed at right angles to the satellite's velocity towards the centre of the Earth transforms the straight line motion to the circular or elliptical one, depending upon the satellite velocity. Centripetal force further leads to a corresponding acceleration called centripetal acceleration as it causes a change in the direction of the satellite's velocity vector. The centrifugal force is simply the reaction force exerted by the satellite in a direction opposite to that of the centripetal force. This is in accordance with Newton's third law of motion, which states that for every action there is an equal and opposite reaction. This implies that there is a centrifugal acceleration acting outwards from the centre of the Earth due to the centripetal acceleration acting towards the centre of the Earth. The only radial force acting on the satellite orbiting Earth is the centripetal force. The centrifugal force is not acting on the satellite; it is only a reaction force exerted by the satellite.

The two forces can be explained from Newton's law of gravitation and Newton's second law of motion as outlined in the following paragraphs.

2.2.1 Newton's Law of Gravitation

According to Newton's law of gravitation, every particle irrespective of its mass attracts every other particle with a gravitational force whose magnitude is directly proportional to the product of the masses of the two particles and inversely proportional to the square of the distance between them and written as

$$F = \frac{Gm_1 m_2}{r^2} \tag{2.1}$$

where

m_1, m_2 = masses of the two particles
r = distance between the two particles
G = gravitational constant = 6.67×10^{-11} m^3/kg s^2

The force with which the particle with mass m_1 attracts the particle with mass m_2 equals the force with which the particle with mass m_2 attracts the particle with mass m_1. The forces are equal in magnitude but opposite in direction (Figure 2.5). The acceleration, which is force per unit mass, experienced by the two particles, however, would depend upon their masses. A larger mass experiences lesser acceleration. Newton also explained that although the law strictly applied to particles, it is applicable to real objects as long as their sizes are small compared to the distance between them. He also explained that a uniform spherical shell of matter would behave as if the entire mass of it were concentrated at its centre.

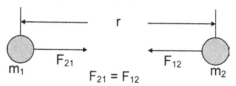

Figure 2.5 Newton's law of gravitation

2.2.2 Newton's Second Law of Motion

According to Newton's second law of motion, the force equals the product of mass and acceleration. In the case of a satellite orbiting Earth, if the orbiting velocity is v, then the acceleration, called centripetal acceleration, experienced by the satellite at a distance r from the centre of the Earth would be v^2/r. If the mass of satellite is m, it would experience a reaction force of mv^2/r. This is the centrifugal force directed outwards from the centre of the Earth and for a satellite is equal in magnitude to the gravitational force.

If the satellite orbited Earth with a uniform velocity v, which would be the case when the satellite orbit is a circular one, then equating the two forces mentioned above would lead to an expression for the orbital velocity v as follows:

$$\frac{Gm_1 m_2}{r^2} = \frac{m_2 v^2}{r} \qquad (2.2)$$

$$v = \sqrt{\left(\frac{Gm_1}{r}\right)} = \sqrt{\left(\frac{\mu}{r}\right)} \qquad (2.3)$$

where

m_1 = mass of Earth
m_2 = mass of the satellite
$\mu = Gm_1 = 3.986\,013 \times 10^5$ km^3/s^2 = $3.986\,013 \times 10^{14}$ N m^2/kg

The orbital period in such a case can be computed from

$$T = \frac{2\pi r^{3/2}}{\sqrt{\mu}} \qquad (2.4)$$

Orbiting Satellites – Basic Principles

In the case of an elliptical orbit, the forces governing the motion of the satellite are the same. The velocity at any point on an elliptical orbit at a distance d from the centre of the Earth is given by the formula

$$v = \sqrt{\left[\mu\left(\frac{2}{d} - \frac{1}{a}\right)\right]} \qquad (2.5)$$

where
a = semi-major axis of the elliptical orbit

The orbital period in the case of an elliptical orbit is given by

$$T = \frac{2\pi a^{3/2}}{\sqrt{\mu}} \qquad (2.6)$$

The movement of a satellite in an orbit is governed by three Kepler's laws, explained below.

2.2.3 Kepler's Laws

Johannes Kepler, based on his lifetime study, gave a set of three empirical expressions that explained planetary motion. These laws were later vindicated when Newton gave the law of gravitation. Though given for planetary motion, these laws are equally valid for the motion of natural and artificial satellites around Earth or for any body revolving around another body. Here, these laws will be discussed with reference to the motion of artificial satellites around Earth.

2.2.3.1 Kepler's First Law

The orbit of a satellite around Earth is elliptical with the centre of the Earth lying at one of the foci of the ellipse (Figure 2.6). The elliptical orbit is characterized by its semi-major axis a and eccentricity e. Eccentricity is the ratio of the distance between the centre of the ellipse and either of its foci ($= ae$) to the semi-major axis of the ellipse a. A circular orbit is a special

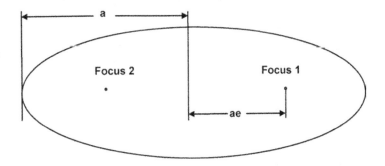

Figure 2.6 Kepler's first law

case of an elliptical orbit where the foci merge together to give a single central point and the eccentricity becomes zero. Other important parameters of an elliptical satellite orbit include its apogee (farthest point of the orbit from the Earth's centre) and perigee (nearest point of the orbit from the Earth's centre) distances. These are described in subsequent paragraphs.

For any elliptical motion, the law of conservation of energy is valid at all points on the orbit. The law of conservation of energy states that energy can neither be created nor destroyed; it can only be transformed from one form to another. In the context of satellites, it means that the sum of the kinetic and the potential energy of a satellite always remain constant. The value of this constant is equal to $-Gm_1m_2/(2a)$, where

m_1 = mass of Earth
m_2 = mass of the satellite
 a = semi-major axis of the orbit

The kinetic and potential energies of a satellite at any point at a distance r from the centre of the Earth are given by

$$\text{Kinetic energy} = \frac{1}{2}(m_2 v^2) \tag{2.7}$$

$$\text{Potential energy} = -\frac{Gm_1 m_2}{r}. \tag{2.8}$$

Therefore,

$$\frac{1}{2}(m_2 v^2) - \frac{Gm_1 m_2}{r} = -\frac{Gm_1 m_2}{2a} \tag{2.9}$$

$$v^2 = Gm_1\left(\frac{2}{r} - \frac{1}{a}\right) \tag{2.10}$$

$$v = \sqrt{\left[\mu\left(\frac{2}{r} - \frac{1}{a}\right)\right]} \tag{2.11}$$

2.2.3.2 Kepler's Second Law

The line joining the satellite and the centre of the Earth sweeps out equal areas in the plane of the orbit in equal time intervals (Figure 2.7); i.e. the rate (dA/dt) at which it sweeps area A is constant. The rate of change of the swept-out area is given by

$$\frac{dA}{dt} = \frac{\text{angular momentum of the satellite}}{2m} \tag{2.12}$$

where m is the mass of the satellite. Hence, Kepler's second law is also equivalent to the law of conservation of momentum, which implies that the angular momentum of the orbiting satellite given by the product of the radius vector and the component of linear momentum perpendicular to the radius vector is constant at all points on the orbit.

Orbiting Satellites – Basic Principles

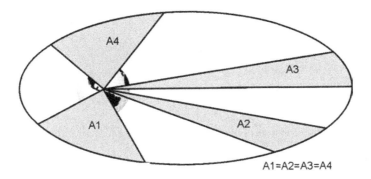

Figure 2.7 Kepler's second law

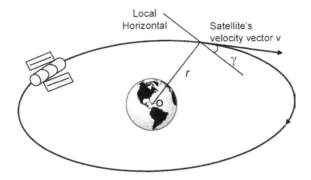

Figure 2.8 Satellite's position at any given time

The angular momentum of the satellite of mass m is given by $mr^2\omega$, where ω is the angular velocity of the satellite. This further implies that the product $mr^2\omega = (m\omega r)(r) = mv'r$ remains constant. Here v' is the component of the satellite's velocity v in the direction perpendicular to the radius vector and is expressed as $v\cos\gamma$, where γ is the angle between the direction of motion of the satellite and the local horizontal, which is in the plane perpendicular to the radius vector r (Figure 2.8). This leads to the conclusion that the product $rv\cos\gamma$ is constant. The product reduces to rv in the case of circular orbits and also at apogee and perigee points in the case of elliptical orbits due to angle γ becoming zero. It is interesting to note here that the velocity component v' is inversely proportional to the distance r. Qualitatively, this implies that the satellite is at its lowest speed at the apogee point and the highest speed at the perigee point. In other words, for any satellite in an elliptical orbit, the dot product of its velocity vector and the radius vector at all points is constant. Hence,

$$v_p r_p = v_a r_a = vr\cos\gamma \qquad (2.13)$$

where

v_p = velocity at the perigee point
r_p = perigee distance
v_a = velocity at the apogee point
r_a = apogee distance
v = satellite velocity at any point in the orbit
r = distance of the point
γ = angle between the direction of motion of the satellite and the local horizontal

2.2.3.3 Kepler's Third Law

According to the Kepler's third law, also known as the law of periods, the square of the time period of any satellite is proportional to the cube of the semi-major axis of its elliptical orbit. The expression for the time period can be derived as follows. A circular orbit with radius r is assumed. Remember that a circular orbit is only a special case of an elliptical orbit with both the semi-major axis and semi-minor axis equal to the radius. Equating the gravitational force with the centrifugal force gives

$$\frac{Gm_1 m_2}{r^2} = \frac{m_2 v^2}{r} \qquad (2.14)$$

Replacing v by ωr in the above equation gives

$$\frac{Gm_1 m_2}{r^2} = \frac{m_2 \omega^2 r^2}{r} = m_2 \omega^2 r \qquad (2.15)$$

which gives $\omega^2 = Gm_1/r^3$. Substituting $\omega = 2\pi/T$ gives

$$T^2 = \left(\frac{4\pi^2}{Gm_1}\right) r^3 \qquad (2.16)$$

This can also be written as

$$T = \left(\frac{2\pi}{\sqrt{\mu}}\right) r^{3/2} \qquad (2.17)$$

The above equation holds good for elliptical orbits provided r is replaced by the semi-major axis a. This gives the expression for the time period of an elliptical orbit as

$$T = \left(\frac{2\pi}{\sqrt{\mu}}\right) a^{3/2} \qquad (2.18)$$

2.3 Orbital Parameters

The satellite orbit, which in general is elliptical, is characterized by a number of parameters. These not only include the geometrical parameters of the orbit but also parameters that define its orientation with respect to Earth. The orbital elements and parameters will be discussed in

Orbital Parameters

the following paragraphs:

1. Ascending and descending nodes
2. Equinoxes
3. Solstices
4. Apogee
5. Perigee
6. Eccentricity
7. Semi-major axis
8. Right ascension of the ascending node
9. Inclination
10. Argument of the perigee
11. True anomaly of the satellite
12. Angles defining the direction of the satellite

1. *Ascending and descending nodes.* The satellite orbit cuts the equatorial plane at two points: the first, called the descending node (N1), where the satellite passes from the northern hemisphere to the southern hemisphere, and the second, called the ascending node (N2), where the satellite passes from the southern hemisphere to the northern hemisphere (Figure 2.9).

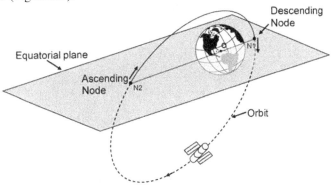

Figure 2.9 Ascending and descending nodes

2. *Equinoxes.* The inclination of the equatorial plane of Earth with respect to the direction of the sun, defined by the angle formed by the line joining the centre of the Earth and the sun with the Earth's equatorial plane follows a sinusoidal variation and completes one cycle of sinusoidal variation over a period of 365 days (Figure 2.10). The sinusoidal variation of the angle of inclination is defined by

$$\text{Inclination angle (in degrees)} = 23.4 \sin\left(\frac{2\pi t}{T}\right) \qquad (2.19)$$

where T is 365 days. This expression indicates that the inclination angle is zero for $t = T/2$ and T. This is observed to occur on 20-21 March, called the spring equinox, and 22-23 September, called the autumn equinox. The two equinoxes are understandably spaced six months apart. During the equinoxes, it can be seen that the equatorial plane of Earth will be

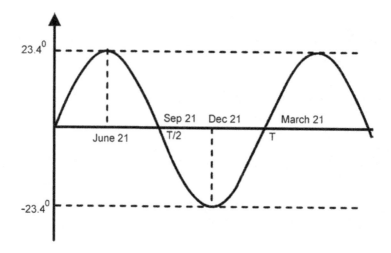

Figure 2.10 Yearly variation of angular inclination of Earth with the sun

aligned with the direction of the sun. Also, the line of intersection of the Earth's equatorial plane and the Earth's orbital plane that passes through the centre of the Earth is known as the line of equinoxes. The direction of this line with respect to the direction of the sun on 20-21 March determines a point at infinity called the vernal equinox (Y) (Figure 2.11).

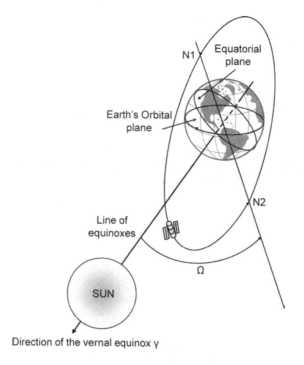

Figure 2.11 Vernal equinox

Orbital Parameters

3. *Solstices.* Solstices are the times when the inclination angle is at its maximum, i.e. 23.4°. These also occur twice during a year on 20-21 June, called the summer solstice, and 21-22 December, called the winter solstice.
4. *Apogee.* Apogee is the point on the satellite orbit that is at the farthest distance from the centre of the Earth (Figure 2.12). The apogee distance can be computed from the known values of the orbit eccentricity e and the semi-major axis a from

$$\text{Apogee distance} = a(1+e) \qquad (2.20)$$

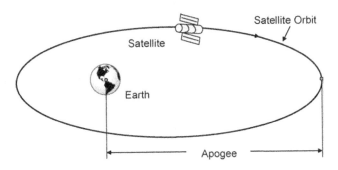

Figure 2.12 Apogee

The apogee distance can also be computed from the known values of the perigee distance and velocity at the perigee V_p from

$$V_p = \sqrt{\left(\frac{2\mu}{\text{Perigee distance}} - \frac{2\mu}{\text{Perigee distance} + \text{Apogee distance}}\right)} \qquad (2.21)$$

where

$$V_p = V\left(\frac{d \cos \gamma}{\text{Perigee distance}}\right)$$

with V being the velocity of the satellite at a distance d from the centre of the Earth.

5. *Perigee.* Perigee is the point on the orbit that is nearest to the centre of the Earth (Figure 2.13). The perigee distance can be computed from the known values of orbit

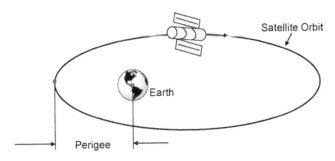

Figure 2.13 Perigee

eccentricity e and the semi-major axis a from

$$\text{Perigee distance} = a(1-e) \tag{2.22}$$

6. *Eccentricity.* The orbit eccentricity e is the ratio of the distance between the centre of the ellipse and the centre of the Earth to the semi-major axis of the ellipse. It can be computed from any of the following expressions:

$$e = \frac{\text{apogee} - \text{perigee}}{\text{apogee} + \text{perigee}} \tag{2.23}$$

$$e = \frac{\text{apogee} - \text{perigee}}{2a} \tag{2.24}$$

Thus $e = \sqrt{(a^2 - b^2)}/a$, where a and b are semi-major and semi-minor axes respectively.

7. *Semi-major axis.* This is a geometrical parameter of an elliptical orbit. It can, however, be computed from known values of apogee and perigee distances as

$$a = \frac{\text{apogee} + \text{perigee}}{2} \tag{2.25}$$

8. *Right ascension of the ascending node.* The right ascension of the ascending node tells about the orientation of the line of nodes, which is the line joining the ascending and descending nodes, with respect to the direction of the vernal equinox. It is expressed as an angle Ω measured from the vernal equinox towards the line of nodes in the direction of rotation of Earth (Figure 2.14). The angle could be anywhere from 0° to 360°.

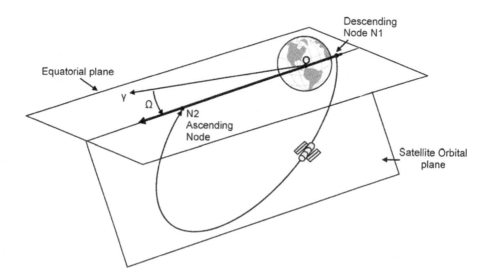

Figure 2.14 Right ascension of the ascending node

Acquisition of the correct angle of right ascension of the ascending node (Ω) is important to ensure that the satellite orbits in the given plane. This can be achieved by choosing an appropriate injection time depending upon the longitude. Angle Ω can be computed as the difference between two angles. One is the angle α between the direction of the vernal equinox and the longitude of the injection point and the other is the angle β between the line of nodes and the longitude of the injection point, as shown in Figure 2.15. Angle β can be computed from

$$\sin \beta = \frac{\cos i \, \sin l}{\cos l \, \sin i} \quad (2.26)$$

where $\angle i$ is the orbit inclination and l is the latitude at the injection point.

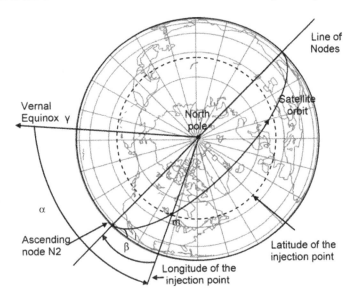

Figure 2.15 Computation of the right ascension of the ascending node

9. *Inclination.* Inclination is the angle that the orbital plane of the satellite makes with the Earths's equatorial plane. It is measured as follows. The line of nodes divides both the Earth's equatorial plane as well as the satellite's orbital plane into two halves. Inclination is measured as the angle between that half of the satellite's orbital plane containing the trajectory of the satellite from the descending node to the ascending node to that half of the Earth's equatorial plane containing the trajectory of a point on the equator from n1 to n2, where n1 and n2 are respectively the points vertically below the descending and ascending nodes (Figure 2.16). The inclination angle can be determined from the latitude l at the injection point and the angle A_z between the projection of the satellite's velocity vector on the local horizontal and north. It is given by

$$\cos i = \sin A_z \cos l \quad (2.27)$$

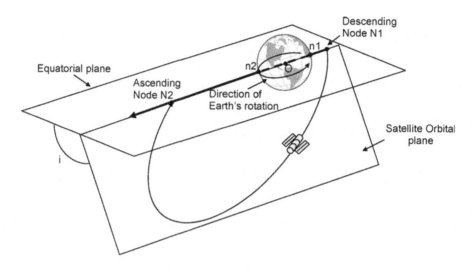

Figure 2.16 Angle of inclination

10. *Argument of the perigee.* This parameter defines the location of the major axis of the satellite orbit. It is measured as the angle ω between the line joining the perigee and the centre of the Earth and the line of nodes from the ascending node to the descending node in the same direction as that of the satellite orbit (Figure 2.17).

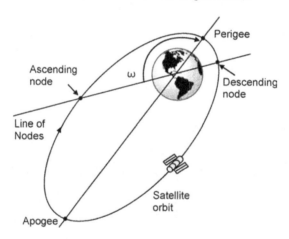

Figure 2.17 Argument of perigee

11. *True anomaly of the satellite.* This parameter is used to indicate the position of the satellite in its orbit. This is done by defining an angle θ, called the true anomaly of the satellite, formed by the line joining the perigee and the centre of the Earth with the line joining the satellite and the centre of the Earth (Figure 2.18).
12. *Angles defining the direction of the satellite.* The direction of the satellite is defined by two angles, the first by angle γ between the direction of the satellite's velocity

Orbital Parameters

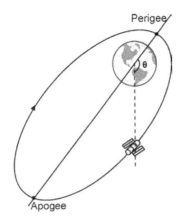

Figure 2.18 True anomaly of a satellite

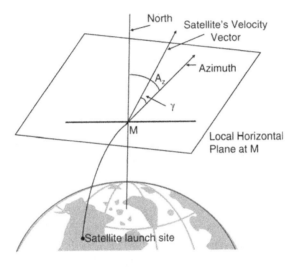

Figure 2.19 Angles defining the direction of the satellite

vector and its projection in the local horizontal and the second by angle A_z between the north and the projection of the satellite's velocity vector on the local horizontal (Figure 2.19).

Problem 2.1
A satellite is orbiting Earth in a uniform circular orbit at a height of 630 km from the surface of Earth. Assuming the radius of Earth and its mass to be 6370 km and 5.98×10^{24} kg respectively, determine the velocity of the satellite (Take the gravitational constant $G = 6.67 \times 10^{-11}$ N m^2/kg^2).

Solution:
Orbit radius $R = 6370 + 630 = 7000$ km $= 7\,000\,000$ m
Also, constant $\mu = GM = 6.67 \times 10^{-11} \times 5.98 \times 10^{24}$
$\qquad\qquad = 39.8 \times 10^{13}$ N m^2/kg
$\qquad\qquad = 39.8 \times 10^{13}$ m^3/s^2

The velocity of the satellite can be computed from

$$V = \sqrt{\left(\frac{\mu}{R}\right)} = \sqrt{\left(\frac{39.8 \times 10^{13}}{7\,000\,000}\right)} = 7.54 \text{ km/s}$$

Problem 2.2
The apogee and perigee distances of a satellite orbiting in an elliptical orbit are respectively 45 000 km and 7000 km. Determine the following:

1. Semi-major axis of the elliptical orbit
2. Orbit eccentricity
3. Distance between the centre of the Earth and the centre of the elliptical orbit

Solution:

1. Semi-major axis of the elliptical orbit $a = \dfrac{\text{apogee} + \text{perigee}}{2}$

$$= \frac{45\,000 + 7000}{2} = 26\,000 \text{ km}$$

2. Eccentricity $e = \dfrac{\text{apogee} - \text{perigee}}{2a} = \dfrac{45\,000 - 7000}{2 \times 26\,000} = \dfrac{38\,000}{52\,000} = 0.73$

3. Distance between the centre of the Earth and the centre of the ellipse $= ae$

$$= 26\,000 \times 0.73$$
$$= 18\,980 \text{ km}$$

Problem 2.3
A satellite is moving in an elliptical orbit with the major axis equal to 42 000 km. If the perigee distance is 8000 km, find the apogee and the orbit eccentricity.

Solution:
Major axis $= 42\,000$ km
Also, major axis $=$ apogee $+$ perigee $= 42\,000$ km
Therefore apogee $= 42\,000 - 8000 = 34\,000$ km

Also, eccentricity $e = \dfrac{\text{apogee} - \text{perigee}}{\text{major axis}}$

$$= \frac{34\,000 - 8000}{42\,000}$$

$$= \frac{26\,000}{42\,000} = 0.62$$

Orbital Parameters

Problem 2.4
Refer to the satellite orbit of Figure 2.20. Determine the apogee and perigee distances if the orbit eccentricity is 0.6.

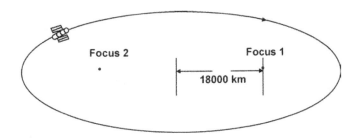

Figure 2.20 Figure for Problem 2.4

Solution: If e is the orbit eccentricity and a the semi-major axis of the elliptical orbit, then the distance between the centre of the Earth and the centre of the ellipse is equal to ae.
Therefore $ae = 18\,000$ km
This gives $a = 18\,000/e = 18\,000/0.6 = 30\,000$ km
Apogee distance $= a(1 + e) = 30\,000 \times (1 + 0.6) = 48\,000$ km
Perigee distance $= a(1 - e) = 30\,000 \times (1 - 0.6) = 12\,000$ km

Problem 2.5
The difference between the furthest and the closest points in a satellite's elliptical orbit from the surface of the Earth is 30 000 km and the sum of the distances is 50 000 km. If the mean radius of the Earth is considered to be 6400 km, determine orbit eccentricity.

Solution: Apogee − Perigee = 30 000 km as the radius of the Earth will cancel in this case

$$\text{Apogee} + \text{Perigee} = 50\,000 + 2 \times 6400 = 62\,800 \text{ km}$$

Orbit eccentricity = (Apogee − Perigee)/(Apogee + Perigee) = 30 000/62 800 = 0.478

Problem 2.6
Refer to Figure 2.21. Satellite A is orbiting Earth in a near-Earth circular orbit of radius 7000 km. Satellite B is orbiting Earth in an elliptical orbit with apogee and perigee distances of 47 000 km and 7000 km respectively. Determine the velocities of the two satellites at point X. (Take $\mu = 39.8 \times 10^{13}$ m^3/s^2.)

Solution: The velocity of a satellite moving in a circular orbit is constant throughout the orbit and is given by

$$V = \sqrt{\left(\frac{\mu}{R}\right)}$$

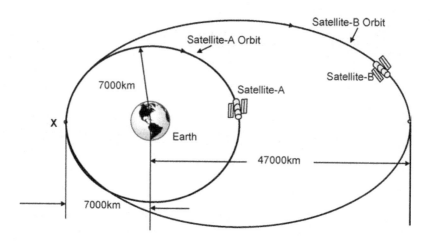

Figure 2.21 Figure for Problem 2.6

Therefore the velocity of satellite A at point X = $\sqrt{\left(\dfrac{39.8 \times 10^{13}}{7\,000\,000}\right)}$

$= 7.54$ km/s

The velocity of the satellite at any point in an elliptical orbit is given by

$$V = \sqrt{\left[\mu\left(\dfrac{2}{R} - \dfrac{1}{a}\right)\right]}$$

where a is the semi-major axis and R is the distance of the point in question from the centre of the Earth. Here $R = 7000$ km and $a = (47\,000 + 7000)/2 = 27\,000$ km. Therefore, velocity of satellite B at point X is given by

$$V = \sqrt{\left[(39.8 \times 10^{13}) \times \left(\dfrac{2}{7\,000\,000} - \dfrac{1}{27\,000\,000}\right)\right]} = 9.946 \text{ km/s}$$

Problem 2.7
Refer to Figure 2.22. Satellite A is orbiting Earth in an equatorial circular orbit of radius 42 000 km. Satellite B is orbiting Earth in an elliptical orbit with apogee and perigee distances of 42 000 km and 7000 km respectively. Determine the velocities of the two satellites at point X. (Take $\mu = 39.8 \times 10^{13}$ m^3/s^2.)

Solution: The velocity of a satellite moving in a circular orbit is constant throughout the orbit and is given by

$$V = \sqrt{\left(\dfrac{\mu}{R}\right)}$$

Orbital Parameters

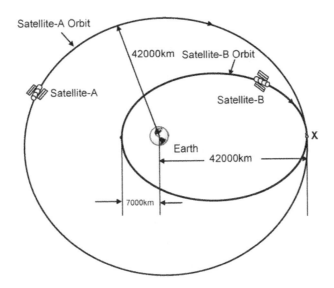

Figure 2.22 Figure for Problem 2.7

Therefore, the velocity of satellite A at point X = $\sqrt{\left(\dfrac{39.8 \times 10^{13}}{42\,000\,000}\right)}$

= 3.078 km/s

The velocity of the satellite at any point in an elliptical orbit is given by

$$v = \sqrt{\left[\mu\left(\dfrac{2}{R} - \dfrac{1}{a}\right)\right]}$$

where a is the semi-major axis and R is the distance of the point in question from the centre of the Earth. Here $R = 42\,000$ km and $a = (42\,000 + 7000)/2 = 24\,500$ km. Therefore, velocity of satellite B at point X is given by

$$v = \sqrt{\left[(39.8 \times 10^{13}) \times \left(\dfrac{2}{42\,000\,000} - \dfrac{1}{24\,500\,000}\right)\right]} = 1.645 \text{ km/s}$$

Problem 2.8
Refer to Figure 2.23. Satellite A is orbiting Earth in a circular orbit of radius 25 000 km. Satellite B is orbiting Earth in an elliptical orbit with apogee and perigee distances of 43 000 km and 7000 km respectively. Determine the velocities of the two satellites at the indicated points X and Y. (Take $\mu = 39.8 \times 10^{13}$ m³/s².)

Solution: The velocity of a satellite moving in a circular orbit is constant throughout the orbit and is given by

$$V = \sqrt{\left(\dfrac{\mu}{R}\right)}$$

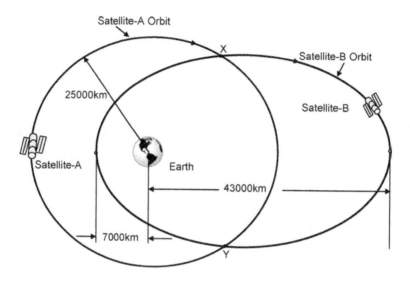

Figure 2.23 Figure for Problem 2.8

Therefore the velocity of satellite A at points X and Y = $\sqrt{\left(\dfrac{39.8 \times 10^{13}}{25\,000\,000}\right)}$ = 3.989 km/s

The velocity of the satellite at any point in an elliptical orbit is given by

$$v = \sqrt{\left[\mu\left(\dfrac{2}{R} - \dfrac{1}{a}\right)\right]}$$

where a is the semi-major axis and R is the distance of the point in question from the centre of the Earth. For satellite B, at points X and Y, R = 25 000 km and a = (43 000 + 7000)/2 = 25 000 km. Therefore, velocity of satellite B at points X and Y

$$v = \sqrt{\left[(39.8 \times 10^{13}) \times \left(\dfrac{2}{25\,000\,000} - \dfrac{1}{25\,000\,000}\right)\right]} = 3.989 \text{ m/s}$$

Please note: In Problems 2.6, 2.7 and 2.8, three different cases of satellites moving in circular and elliptical orbits have been considered. Note that the velocity of the satellite at a point in the elliptical orbit is more than the velocity of the satellite in a circular orbit if the distance of the satellite from the centre of the Earth happens to be less than the radius of the circular orbit. It will be less than the velocity in the circular orbit if the distance of the satellite in the elliptical orbit happens to be more than the radius of the circular orbit. The two velocities are the same at a point where the distance of the satellite from the centre of the Earth in the elliptical orbit equals the radius of the circular orbit provided that the semi-major axis of the ellipse also equals the radius of the circular orbit.

Orbital Parameters

Problem 2.9
Calculate the orbital period of a satellite moving in an elliptical orbit having a major axis of 50 000 km. (Take $\mu = 39.8 \times 10^{13}$ Nm²/kg).

Solution: Semi-major axis, a = 50 000/2 = 25 000 km
Orbital time period, $T = 2\pi \times \sqrt{(a^3/\mu)} = 2 \times 3.14 \times \sqrt{(25\,000\,000)^3/(39.8 \times 10^{13})}$
$$= 39\,344 \text{ seconds} = 10 \text{ hours } 55 \text{ minutes } 44 \text{ seconds}$$

Problem 2.10
Refer to Figure 2.24. The semi-major axes of the two satellites shown in the figure are 18 000 km (satellite 1) and 24 000 km (satellite 2). Determine the relationship between their orbital periods.

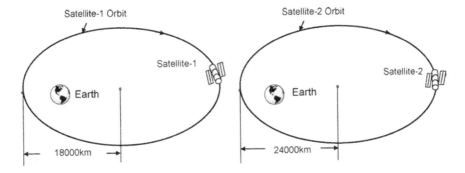

Figure 2.24 Figure for Problem 2.10

Solution: The orbital time period $T = 2\pi \times \sqrt{(a^3/\mu)}$ where a is the semi-major axis. If a_1 and a_2 are the semi-major axes in the two cases and T_1 and T_2 are their time periods, then

$$T_1 = 2\pi \times \sqrt{\left(\frac{a_1^3}{\mu}\right)}$$

and

$$T_2 = 2\pi \times \sqrt{\left(\frac{a_2^3}{\mu}\right)}$$

This gives

$$\frac{T_2}{T_1} = \left(\frac{a_2}{a_1}\right)^{3/2} = \left(\frac{24\,000}{18\,000}\right)^{3/2} = 1.54$$

The orbital period of satellite 2 is 1.54 times that of satellite 1.

Problem 2.11
The elliptical eccentric orbit of a satellite has its semi-major and semi-minor axes as 25 000 km and 18 330 km respectively. Determine the apogee and perigee distances.

Solution: If r_a and r_p are the apogee and perigee distances respectively, a and b are the semi-major and semi-minor axis respectively and e is the eccentricity, then

$$a = \frac{r_a + r_p}{2}$$

and

$$b = \sqrt{(r_a r_p)}$$

Now $(r_a + r_p)/2 = 25\,000$, which gives $r_a + r_p = 50\,000$
and $\sqrt{(r_a r_p)} = 18\,330$, which gives $r_a r_p = 335\,988\,900$
Substituting the value of r_p from these equations gives

$$r_a(50\,000 - r_a) = 335\,988\,900$$

which gives

$$r_a^2 - 50\,000 r_a + 335\,988\,900 = 0$$

Solving this quadratic equation gives two solutions:

$$r_a = 42\,000 \text{ km}, \ 8000 \text{ km}$$

If the semi-major axis is 25 000 km, r_a cannot be 8000 km. Therefore

$$\text{Apogee distance } r_a = 42\,000 \text{ km}$$

and

$$\text{Perigee distance } r_p = 50\,000 - 42\,000 = 8000 \text{ km}$$

Alternative solution:

$$e = \frac{\sqrt{(a^2 - b^2)}}{a} = e = \frac{\sqrt{[(25\,000)^2 - (18\,330)^2]}}{25\,000} = 0.68$$

This gives
Apogee distance $= a(1 + e) = 25\,000 \times 1.68 = 42\,000$ km
and Perigee distance $= a(1 - e) = 25\,000 \times 0.32 = 8000$ km
Therefore, from both the methods,
Apogee distance $= 42\,000$ km and Perigee distance $= 8000$ km

Problem 2.12
Refer to Figure 2.25. The satellite is moving in an elliptical orbit with its semi-major and semi-minor axes as a and b respectively and an eccentricity of 0.6. The satellite takes

Orbital Parameters

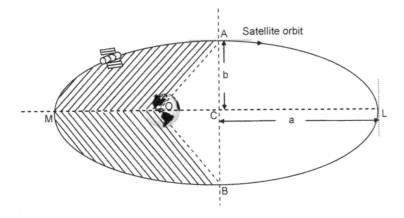

Figure 2.25 Figure for Problem 2.12

3 hours to move from point B to point A. How much time will it take to move from point A to point B?

Solution: This problem can be solved by applying Kepler's law of areas, which says that the line joining the satellite and centre of the Earth spans equal areas of the ellipse in equal times. Now, the area spanned when the satellite moves from point B to point A (area of the shaded region) is given by

$$\frac{1}{2} \times \text{(area of ellipse)} - \text{(area of } \triangle \text{ AOB)} = \frac{\pi ab}{2} - bOC$$

$$= \frac{\pi ab}{2} - bae \quad \text{(where } OC = ae\text{)}$$

$$= 1.57ab - 0.6ab$$

$$= 0.97ab$$

The area spanned when the satellite moves from point A to point B (area of the unshaded region) is given by

$$\frac{1}{2} \times \text{(area of ellipse)} + \text{(area of } \triangle \text{ AOB)} = 1.57ab + 0.6ab$$

$$= 2.17\, ab$$

The ratio of the two areas is given by $2.17/0.97 = 2.237$. The time taken by the satellite to move from point A to point B should therefore be 2.237 times the time taken by the satellite to move from point B to point A. Therefore

$$\text{Time taken} = 3 \times 2.237 = 6.711 \text{ hours}$$

Problem 2.13

The apogee and perigee distances of a certain elliptical satellite orbit are 42 000 km and 8000 km respectively. If the velocity at the perigee point is 9.142 km/s, what would be the velocity at the apogee point?

Solution: The product of the distance of a satellite from the centre of the Earth and the component of velocity vector in the direction of local horizontal at that point is constant at all possible points on the orbit; i.e. $dv \cos \gamma$ = constant, where γ is the angle between the direction of motion of the satellite and the local horizontal. This has been explained earlier. The angle γ is zero at the apogee and perigee points. The expression thus reduces to dv = constant for the apogee and perigee points. Therefore

$$\text{Velocity at the apogee} = \frac{8000 \times 9.142}{42\,000} = 1.741 \text{ km/s}$$

Problem 2.14

In the satellite orbit of Problem 2.13, at a certain point in the orbit at a distance of 16 000 km from the centre of the Earth, the direction of the satellite makes an angle of 56.245° with the local horizontal at that point (Figure 2.26). Determine the satellite velocity at that point.

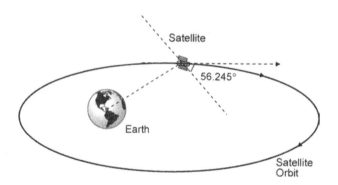

Figure 2.26 Figure for Problem 2.14

Solution: The product $dv \cos \gamma = 8000 \times 9.142$, which gives

$$V = \frac{8000 \times 9.142}{16\,000 \times \cos 56.245°}$$

$$= \frac{73\,136}{16\,000 \times 0.555} = \frac{73\,136}{8880} = 8.236 \text{ km/s}$$

Therefore

$$V = 8.236 \text{ km/s}$$

2.4 Injection Velocity and Resulting Satellite Trajectories

The horizontal velocity with which a satellite is injected into space by the launch vehicle with the intention of imparting a specific trajectory to the satellite has a direct bearing on the satellite trajectory. The phenomenon is best explained in terms of the three cosmic velocities. The general expression for the velocity of a satellite at the perigee point (V_P), assuming an elliptical orbit, is given by

$$V_P = \sqrt{\left[\left(\frac{2\mu}{r}\right) - \left(\frac{2\mu}{R+r}\right)\right]} \qquad (2.28)$$

where

R = apogee distance
r = perigee distance
$\mu = GM$ = constant

The first cosmic velocity V_1 is the one at which apogee and perigee distances are equal, i.e. $R = r$, and the orbit is circular. The above expression then reduces to

$$V_1 = \sqrt{\left(\frac{\mu}{r}\right)} \qquad (2.29)$$

Thus, irrespective of the distance r of the satellite from the centre of the Earth, if the injection velocity is equal to the first cosmic velocity, also sometimes called the first orbital velocity, the satellite follows a circular orbit (Figure 2.27) and moves with a uniform velocity equal to $\sqrt{(\mu/r)}$. A simple calculation shows that for a satellite at 35 786 km above the surface of the Earth, the first cosmic velocity turns out to be 3.075 km/s and the orbital period is 23 hours 56 minutes, which is equal to the time period of one sidereal day – the time taken by Earth to complete one full rotation around its axis with reference to distant stars. This confirms why a geostationary satellite needs to be at a height of 35 786 km above the surface of the Earth. Different types of orbits are discussed at length in the following pages.

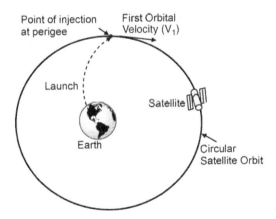

Figure 2.27 Satellite's path where the injection velocity is equal to the first orbital velocity

If the injection velocity happens to be less than the first cosmic velocity, the satellite follows a ballistic trajectory and falls back to Earth. In fact, in this case, the orbit is elliptical and the injection point is at the apogee and not the perigee. If the perigee lies in the atmosphere or exists only virtually below the surface of the Earth, the satellite accomplishes a ballistic flight and falls back to Earth (Figure 2.28).

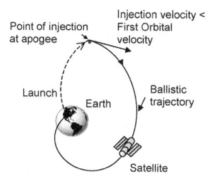

Figure 2.28 Satellite's path where the injection velocity is less than the first orbital velocity

For injection velocity greater than the first cosmic velocity and less than the second cosmic velocity, i.e. $V > \sqrt{(\mu/r)}$ and $V < \sqrt{(2\mu/r)}$, the orbit is elliptical and eccentric. The orbit eccentricity is between 0 and 1. The injection point in this case is the perigee and the apogee distance attained in the resultant elliptical orbit depends upon the injection velocity. The higher the injection velocity, the greater is the apogee distance. The apogee distance can also be computed from the known value of injection velocity, which is also the velocity at the perigee point as the perigee coincides with the injection point, and the velocity v at any other point in the orbit distant d from the centre of the Earth using

$$V_p = \sqrt{\left[(2\mu/r) - \left(\frac{2\mu}{R+r}\right)\right]} = \frac{vd \cos \gamma}{r} \qquad (2.30)$$

When the injection velocity equals $\sqrt{(2\mu/r)}$, the apogee distance R becomes infinite and the orbit takes the shape of a parabola (Figure 2.29) and the orbit eccentricity is 1. This is the second cosmic velocity v_2. At this velocity, the satellite escapes Earth's gravitational pull. For an injection velocity greater than the second cosmic velocity, the trajectory is hyperbolic within the solar system and the orbit eccentricity is greater than 1.

If the injection velocity is increased further, a stage is reached where the satellite succeeds in escaping from the solar system. This is known as the third cosmic velocity and is related to the motion of planet Earth around the sun. The third cosmic velocity (V_3) is mathematically expressed as

$$V_3 = \sqrt{\left[\frac{2\mu}{r} - V_t^2(3 - 2\sqrt{2})\right]} \qquad (2.31)$$

where V_t is the speed of Earth's revolution around the sun.

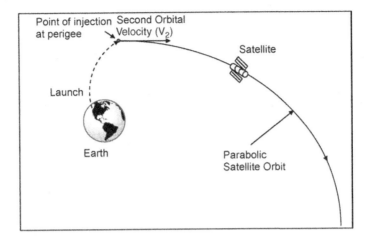

Figure 2.29 Satellite's path where the injection velocity is equal to the second orbital velocity

Beyond the third cosmic velocity, there is a region of hyperbolic flights outside the solar system. Coming back to elliptical orbits, the greater the injection velocity from the first cosmic velocity, the greater is the apogee distance. This is evident from the generalized expression for the velocity of the satellite in elliptical orbits according to which

$$V_p = \text{velocity at the perigee point}$$

$$= \sqrt{\left[(2\mu)\left(\frac{1}{r} - \frac{1}{R+r}\right)\right]} \qquad (2.32)$$

For a given perigee distance r, a higher velocity at the perigee point, which is also the injection velocity, necessitates that the apogee distance R is greater. Figure 2.30 shows a family of curves that can be used to find out the attained apogee height for a given value of injection velocity at the perigee point. The dashed line shows the relationship between the injection velocity and altitude for a circular orbit. For example, for a perigee height of 1000 km above the surface of the Earth, the injection velocity for a circular orbit as seen from the curve is about 7.3 km/s. If the injection velocity is increased to 8 km/s for this perigee height, the orbit attains an apogee height of about 4200 km. If it is increased further to 9 km/s, the apogee height goes up to 16 000 km. It can also be seen that the second cosmic velocity for this perigee height is about 10.3 km/s, as given by the vertical line to which the 1000 km perigee height curve approaches asymptotically.

In fact, for a given perigee distance r, it can be proved that the injection velocities and corresponding apogee distances are related by

$$\left(\frac{v_2}{v_1}\right)^2 = \frac{1+r/R_1}{1+r/R_2} \qquad (2.33)$$

Refer to Problem 2.15 for the derivation of this expression.

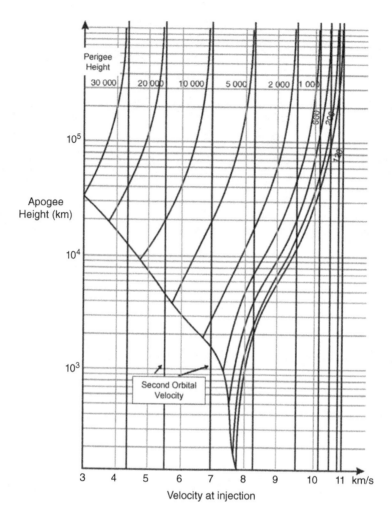

Figure 2.30 Family of curves showing the apogee height for various values of the injection velocity

Problem 2.15

A satellite launched with an injection velocity of v_1 from a point above the surface of the Earth at a distance r from the centre of the Earth attains an elliptical orbit with an apogee distance of R_1. The same satellite when launched with an injection velocity of v_2 from the same perigee distance attains an elliptical orbit with an apogee distance of R_2. Derive the relationship between v_1 and v_2 in terms of r, R_1 and R_2.

Solution: Refer to Figure 2.31. Expressions for the velocity at perigee points can be written as

$$v_1 = \sqrt{\left[(2\mu)\left(\frac{1}{r} - \frac{1}{R_1 + r}\right)\right]}$$

$$v_2 = \sqrt{\left[(2\mu)\left(\frac{1}{r} - \frac{1}{R_2 + r}\right)\right]}$$

Injection Velocity and Resulting Satellite Trajectories

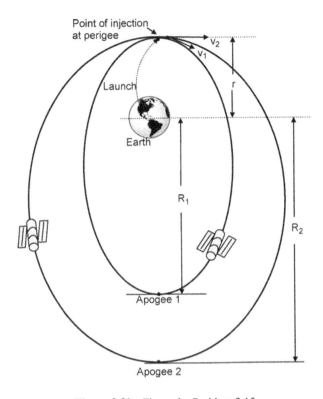

Figure 2.31 Figure for Problem 2.15

Squaring the two expressions and then taking the ratio of the two yields

$$\left(\frac{v_2}{v_1}\right)^2 = \frac{(1/r) - 1/(R_2 + r)}{(1/r) - 1/(R_1 + r)}$$

$$= \frac{R_2/r(R_2 + r)}{R_1/r(R_1 + r)}$$

$$= \left(\frac{R_2}{R_1}\right)\left(\frac{R_1 + r}{R_2 + r}\right)$$

$$= \frac{1 + r/R_1}{1 + r/R_2}$$

Therefore,

$$\left(\frac{v_2}{v_1}\right)^2 = \frac{1 + r/R_1}{1 + r/R_2}$$

Problem 2.16

A satellite launched with an injection velocity of (v_1) from a point above the surface of the Earth at a distance of (r) of 8000 km from the centre of the Earth attains an elliptical orbit with an apogee distance of 12 000 km. The same satellite when launched with an injection velocity of (v_2) that is 20 % higher than (v_1) from the same perigee distance attains an elliptical orbit with an apogee distance of (R_2). Determine the new apogee distance.

Solution:
We know that: $(v_2/v_1)^2 = [(1 + r/R_1)/(1 + r/R_2)]$
Substituting $v_2 = 1.2v_1$, we get $1.44 = [(1 + 8000/12\,000)/(1 + 8000/R_2)]$
or $1.44 \times 3 \times (1 + 8000/R_2) = 5$ or $(1 + 8000/R_2) = 1.1574$
$R_2 = 8000/0.1574 = 50\,826$ km

Problem 2.17
A rocket injects a satellite with a horizontal velocity of 8 km/s from a height of 1620 km from the surface of the Earth. What will be the velocity of the satellite at a point distant 10 000 km from the centre of the Earth, if the direction of the satellite makes an angle of 30° with the local horizontal at that point? Assume radius of the Earth to be 6380 km.

Solution:
We know that $V_2 = \sqrt{[\{2\mu/r\} - \{2\mu/(R+r)\}]} = (v.d. \cos \gamma)/r$
Here $V_2 = 8$ km/s, $r = 1620 + 6380 = 8000$ km, $d = 10\,000$ km and $\gamma = 30°$
Therefore, $[v \times 10\,000 \times \cos 30°/8000] = 8$
or $v = (8 \times 8000)/(10\,000 \times 0.866) = 7.39$ km/s

Problem 2.18
A rocket injects a satellite with a certain horizontal velocity from a height of 620 km from the surface of the Earth. The velocity of the satellite at a point distant 9000 km from the centre of the Earth is observed to be 8 km/s. If the direction of the satellite makes an angle of 30° with the local horizontal at that point, determine the apogee distance of the satellite orbit. (Assume that the radius of the Earth is 6380 km and $\mu = 39.8 \times 10^{13}$ m³/s².)

Solution:
It is known that $V_p = \sqrt{\left[\left(\dfrac{2\mu}{r}\right) - \left(\dfrac{2\mu}{R+r}\right)\right]} = \dfrac{vd \cos \gamma}{r}$

Here,
$r = 620 + 6380 = 7000$ km
$d = 9000$ km
$v = 8$ km/s
$\gamma = 30°$
$R = $ apogee distance $= ?$
Therefore,

$$\sqrt{\left[\left(\dfrac{2\mu}{r}\right) - \left(\dfrac{2\mu}{R+r}\right)\right]} = 8 \times 9000 \times \dfrac{0.866}{7000} = 8.90 \text{ km/s} = 8.90 \times 10^3 \text{ m/s}$$

or

$$\dfrac{2\mu}{r} - \dfrac{2\mu}{R+r} = (8.90 \times 10^3)^2 = 79.34 \times 10^6$$

or

$$\frac{2\mu}{R+r} = \frac{2 \times 39.8 \times 10^{13}}{7\,000\,000} - (79.34 \times 10^6) = 34.36 \times 10^6$$

This gives

$$R + r = \frac{2 \times 39.8 \times 10^{13}}{34.36 \times 10^6} = 2.317 \times 10^7 \text{ m}$$

or $R = (2.317 \times 10^7 - 7\,000\,000)$ m = 16 170 km

2.5 Types of Satellite Orbits

As described at length in the earlier pages of this chapter, satellites travel around Earth along predetermined repetitive paths called orbits. The orbit is characterized by its elements or parameters, which have been covered at length in earlier sections. The orbital elements of a particular satellite depend upon its intended application. The satellite orbits can be classified on the basis of:

1. Orientation of the orbital plane
2. Eccentricity
3. Distance from Earth

2.5.1 Orientation of the Orbital Plane

The orbital plane of the satellite can have various orientations with respect to the equatorial plane of Earth. The angle between the two planes is called the angle of inclination of the satellite. On this basis, the orbits can be classified as equatorial orbits, polar orbits and inclined orbits.

In the case of an equatorial orbit, the angle of inclination is zero, i.e. the orbital plane of the satellite coincides with the Earth's equatorial plane (Figure 2.32). A satellite in the equatorial orbit has a latitude of 0°. For an angle of inclination equal to 90°, the satellite is said to be in the polar orbit (Figure 2.33). For an angle of inclination between 0° and 180°, the orbit is said to be an inclined orbit.

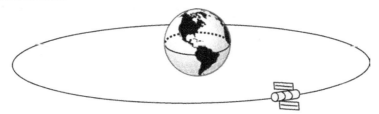

Figure 2.32 Equatorial orbit

For inclinations between 0° and 90°, the satellite travels in the same direction as the direction of rotation of the Earth. The orbit in this case is referred to as a direct or prograde

Figure 2.33 Polar orbit

Figure 2.34 Prograde orbit

orbit (Figure 2.34). For inclinations between 90° and 180°, the satellite orbits in a direction opposite to the direction of rotation of the Earth and the orbit in this case is called a retrograde orbit (Figure 2.35).

2.5.2 Eccentricity of the Orbit

On the basis of eccentricity, the orbits are classified as elliptical (Figure 2.36 (a)) and circular (Figure 2.36 (b)) orbits. Needless to say, when the orbit eccentricity lies between 0 and 1, the orbit is elliptical with the centre of the Earth lying at one of the foci of the ellipse. When the eccentricity is zero, the orbit becomes circular. It may be mentioned here that all circular orbits are eccentric to some extent. As an example, the eccentricity of orbit of geostationary

Types of Satellite Orbits

Figure 2.35 Retrograde orbit

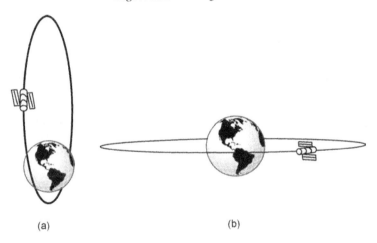

Figure 2.36 (a) Elliptical orbit and (b) circular orbit

satellite INSAT-3B, an Indian satellite in the INSAT series providing communication and meteorological services, is 0.000 252 6. Eccentricity figures for orbits of GOES-9 and Meteosat-7 geostationary satellites, both offering weather forecasting services, are 0.000 423 3 and 0.000 252 6 respectively.

2.5.2.1 Molniya Orbit

Highly eccentric, inclined and elliptical orbits are used to cover higher latitudes, which are otherwise not covered by geostationary orbits. A practical example of this type of orbit is the Molniya orbit (Figure 2.37). It is widely used by Russia and other countries of the former Soviet Union to provide communication services. Typical eccentricity and orbit inclination

Figure 2.37 Molniya orbit

figures for the Molniya orbit are 0.75 and 65° respectively. The apogee and perigee points are about 40 000 km and 400 km respectively from the surface of the Earth.

The Molniya orbit serves the purpose of a geosynchronous orbit for high latitude regions. It is a 12 hour orbit and a satellite in this orbit spends about 8 hours above a particular high latitude station before diving down to a low level perigee at an equally high southern latitude. Usually, three satellites at different phases of the same Molniya orbit are capable of providing an uninterrupted service.

2.5.3 Distance from Earth

Again depending upon the intended mission, satellites may be placed in orbits at varying distances from the surface of the Earth. Depending upon the distance, these are classified as low Earth orbits (LEOs), medium Earth orbits (MEOs) and geostationary Earth orbits (GEOs), as shown in Figure 2.38.

Satellites in the low Earth orbit (LEO) circle Earth at a height of around 160 to 500 km above the surface of the Earth. These satellites, being closer to the surface of the Earth, have much shorter orbital periods and smaller signal propagation delays. A lower propagation delay makes them highly suitable for communication applications. Due to lower propagation paths, the power required for signal transmission is also less, with the result that the satellites are of small physical size and are inexpensive to build. However, due to a shorter orbital period, of the order of an hour and a half or so, these satellites remain over a particular ground station for a short time. Hence, several of these satellites are needed for 24 hour coverage. One important application of LEO satellites for communication is the project Iridium, which is a global communication system conceived by Motorola (Figure 2.39). A total of 66 satellites are arranged in a distributed architecture, with each satellite carrying 1/66 of the total system capacity. The system is intended to provide a variety of telecommunication

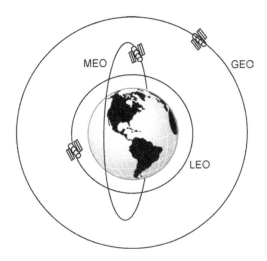

Figure 2.38 LEO, MEO and GEO

Figure 2.39 Iridium constellation of satellites

services at the global level. The project is named 'Iridium' as earlier the constellation was proposed to have 77 satellites and the atomic number of iridium is 77. Other applications where LEO satellites can be put to use are surveillance, weather forecasting, remote sensing and scientific studies.

Medium Earth orbit (MEO) satellites orbit at a distance of approximately 10 000 to 20 000 km above the surface of the Earth. They have an orbital period of 6 to 12 hours. These satellites stay in sight over a particular region of Earth for a longer time. The transmission distance and propagation delays are greater than those for LEO satellites. These orbits are generally polar in nature and are mainly used for communication and navigation applications.

A geosynchronous Earth orbit is a prograde orbit whose orbital period is equal to Earth's rotational period. If such an orbit were in the plane of the equator and circular, it would remain stationary with respect to a given point on the Earth. These orbits are referred to as the geostationary Earth orbits (GEOs). For the satellite to have such an orbital velocity, it needs to be at a height of about 36 000 km, 35 786 km to be precise, above the surface of the Earth.

To be more precise and technical, in order to remain above the same point on the Earth's surface, a satellite must fulfil the following conditions:

1. It must have a constant latitude, which is possible only at 0° latitude.
2. The orbit inclination should be zero.
3. It should have a constant longitude and thus have a uniform angular velocity, which is possible when the orbit is circular.
4. The orbital period should be equal to 23 hours 56 minutes, which implies that the satellite must orbit at a height of 35 786 km above the surface of the Earth.
5. The satellite motion must be from west to east.

In the case where these conditions are fulfilled, then as the satellite moves from a position O_1 to O_2 in its orbit, a point vertically below on the equator moves with the same angular velocity and moves from E_1 to E_2, as shown in Figure 2.40. Satellites in geostationary orbits play an essential role in relaying communication and TV broadcast signals around the globe. They also perform meteorological and military surveillance functions very effectively.

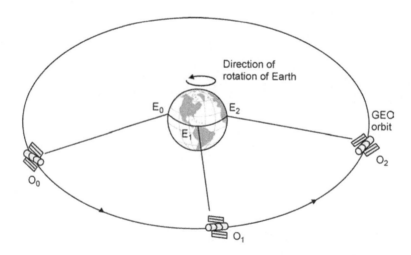

Figure 2.40 GEO satellites appear stationary with respect to a point on Earth

2.5.4 Sun-synchronous Orbit

Another type of satellite orbit, which could have been categorized as an LEO on the basis of distance from the surface of the Earth, needs a special mention and treatment because of its particular importance to satellites intended for remote sensing and military reconnaissance applications. A sun-synchronous orbit, also known as a helio-synchronous orbit, is one that lies in a plane that maintains a fixed angle with respect to the Earth–sun direction. In other words, the orbital plane has a fixed orientation with respect to the Earth–sun direction and the angle between the orbital plane and the Earth–sun line remains constant throughout the year, as shown in Figure 2.41. Satellites in sun-synchronous orbits are particularly suited to applications like passive remote sensing, meteorological, military reconnaissance and atmospheric studies.

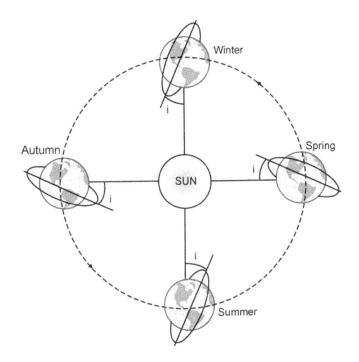

Figure 2.41 Sun-synchronous orbit

As a result of this property, sun-synchronous orbits ensure that:

1. The satellite passes over a given location on Earth every time at the same local solar time, thereby guaranteeing almost the same illumination conditions, varying only with seasons.
2. The satellite ensures coverage of the whole surface of the Earth, being quasi-polar in nature.

Every time a sun-synchronous satellite completes one revolution around Earth, it traverses a thin strip on the surface of the Earth. During the next revolution it traverses another strip

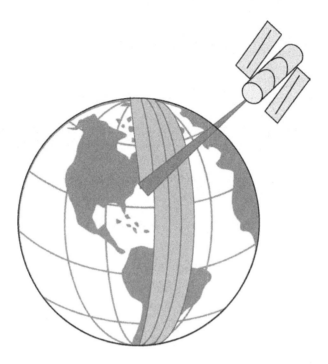

Figure 2.42 Earth coverage of sun-synchronous satellites

shifted westwards and the process of shift continues with successive revolutions, as shown in Figure 2.42. Depending upon the orbital parameters and speed of rotation of Earth, after making a certain number of revolutions around Earth, it comes back close to the first strip that it had traversed. It may not exactly overlap the first strip, as the mean distance between the two strips, called the tracking interval, may not be an integral multiple of the equatorial perimeter. However, the number of revolutions required before the satellite repeats the same strip sequence can certainly be calculated. This is called one complete orbital cycle, which is basically the time that elapses before the satellite revisits a given location in the same direction. To be more precise, orbital cycle means the whole number of orbital revolutions that a satellite must describe in order to be once again flying in the same direction over the same point on the Earth's surface. Landsat-1, -2, and -3 satellites, for instance, have an orbital cycle of 18 days. IRS-1A and IRS-P2 satellites of India have orbital cycles of 22 days and 24 days respectively.

The concept of an orbital cycle will be further illustrated with the help of an example. Take the case of Landsat-1, -2, and -3 satellites each having an orbital cycle of 18 days. During the course of a given revolution, say revolution n, the satellite crosses the equatorial plane at the descending node above a certain point on the surface of the Earth. During the next adjacent pass or revolution, the track shifts to the west by a distance known as the tracking interval. The tracking interval depends upon the speed of the Earth's rotation and the nodal precession of the orbit concerned for one revolution. Fourteen revolutions later, which take about 24 hours and a few minutes, the track is found to be located slightly to the west of the first revolution, designated revolution n. Due to the cumulative effect of these small drifts, after every 14 revolutions it is found that the satellite flies over the same point on Earth in the same direction once every 251 revolutions. The figure of 251 comes from $18 \times 14 - 1$, thus explaining an 18 day orbital cycle for Landsat-1, -2, and -3 satellites.

Another parameter of relevance to satellites in a sun-synchronous orbit is the crossing time for descending and ascending nodes. Satellites of the Landsat series, the SPOT series and the IRS series move along their descending trajectory above the sunward face and their ascending trajectory at night. Descending node-crossing times for Landsat-1, -2, and -3 satellites is 9 h 30 min and for Landsat-4 and -5 it is 9 h 37 min. In the case of IRS series and SPOT series satellites, it is 10 h 30 min.

Problem 2.19
Verify that a geostationary satellite needs to be at a height of about 35 780 km above the surface of the Earth. Assume the radius of the Earth to be 6380 km and $\mu = 39.8 \times 10^{13}$ Nm²/kg

Solution:
The orbital period of a satellite in circular orbit is given by $T = 2\pi r^{3/2}/\sqrt{\mu}$
Orbital period of a geostationary satellite is equal to 23 hours, 56 minutes, which is equal to 86 160 seconds.
Therefore, $2 \times 3.14 \times (r)^{3/2}/\sqrt{(39.8 \times 10^{13})} = 86\,160$
$6.28 \times (r)^{3/2} = 86\,160 \times \sqrt{(39.8 \times 10^{13})} = 1\,718\,886 \times 10^6$
or $(r)^{3/2} = 273\,708 \times 10^6$
or $r = (273\,708 \times 10^6)^{2/3} = 4215.5 \times 10^4$ m = 42 155 km
Hence, height of satellite orbit above the surface of the Earth = 42 155 − 6380 = 35 775 km

Problem 2.20
A typical Molniya orbit has perigee and apogee heights above the surface of Earth as 400 km and 40 000 km respectively. Verify that the orbit has a 12 hour time period assuming the radius of the Earth to be 6380 km and $\mu = 39.8 \times 10^{13}$ N m²/kg.

Solution: The orbital period of a satellite in elliptical orbit is given by

$$T = \frac{2\pi a^{3/2}}{\sqrt{\mu}}$$

where a = semi-major axis of the elliptical orbit. Now $a = (40\,000 + 400 + 6380 + 6380)/2 = 26\,580$ km. Therefore

$$T = 6.28 \times \frac{(26\,580 \times 10^3)^{3/2}}{\sqrt{(39.8 \times 10^{13})}}$$

$$= 43\,137 \text{ seconds} = 11.98 \text{ hours} \approx 12 \text{ m hours}$$

Further Readings

Capderou, M. and Lyle, S. (translator) (2005) *Satellites: Orbits and Missions*, Springer-Verlag, France.
Gatland, K. (1990) *Illustrated Encyclopedia of Space Technology*, Crown, New York.
Inglis, A.F. (1997) *Satellite Technology: An Introduction*, Butterworth Heinemann, Massachusetts.
Logsdon, T. (1998) *Orbital Mechanics: Theory and Applications*, John Wiley & Sons, Inc., New York.
Luther, A.C. and Inglis, A.F. (1997) *Satellite Technology: Introduction*, Focal Press, Boston, Massachusetts.
Montenbruck, O. and Gill, E. (2000) *Satellite Orbits: Models, Methods, Applications* (2000) Spinger-Verlag, Berlin, Heidelberg, New York.
Pattan, B. (1993) *Satellite Systems: Principles and Technologies*, Van Nostrand Reinhold, New York.
Richharia, M. (1999) *Satellite Communication Systems*, Macmillan Press Ltd.
Soop, E.M. (1994) *Handbook of Geostationary Orbits*, Kluwer Academic Publishers, Dordrecht, The Netherlands.
Verger, F., Sourbes-Verger, I., Ghirardi, R., Pasco, X., Lyle, S., and Reilly, P. (2003) *The Cambridge Encyclopedia of Space*, Cambridge University Press.

Internet Sites

1. http://electronics.howstuffworks.com/satellite.htm/printable
2. http://spaceinfo.jaxa.jp/note/eisei/e/eis9910_strctrplan_a_e.html
3. http://www.windows.ucar.edu/spaceweather/types_orbits.html
4. http://en.wikipedia.org/wiki/Astrodynamics
5. http://en.wikipedia.org/wiki/Orbit
6. http://www.thetech.org/exhibits_events/online/satellite/home.html

Glossary

Apogee: Point on the satellite orbit farthest from the centre of the Earth. The apogee distance is the distance of the apogee point from the centre of the Earth
Argument of perigee: This parameter defines the location of the major axis of the satellite orbit. It is measured as the angle between the line joining the perigee and the centre of the Earth and the line of nodes from the ascending to the descending node in the same direction as that of the satellite orbit
Ascending node: The point where the satellite orbit cuts the Earth's equatorial plane, when it passes from the southern hemisphere to the northern hemisphere
Centrifugal force: The force acting outwards from the centre of the Earth on any body orbiting it
Centripetal force: A force that is directed towards the centre of the Earth due to the gravitational force of attraction of Earth
Descending node: The point where the satellite orbit cuts the Earth's equatorial plane, when it passes from the northern hemisphere to the southern hemisphere
Eccentricity: Referring to an elliptical orbit, it is the ratio of the distance between the centre of the Earth and the centre of the ellipse to the semi-major axis of the ellipse. It is zero for a circular orbit and between 0 and 1 for an elliptical orbit

Equatorial orbit: An orbit in which the satellite's orbital plane coincides with the Earth's equatorial plane

Equinox: An equinox is said to occur when the angle of inclination of the Earth's equatorial plane with respect to the direction of the sun is zero. Such a situation occurs twice a year, one on 21 March called the spring equinox and the other on 21 September called the autumn equinox

First cosmic velocity: This is the injection velocity at which the apogee and perigee distances are equal, with the result that the satellite orbit is circular

Geostationary Earth orbit (GEO): A satellite orbit with an orbit height at 35 786 km above the surface of the Earth. This height makes the satellite's orbital velocity equal to the speed of rotation of Earth, thus making the satellite look stationary from a given point on the surface of the Earth

Inclination: Inclination is the angle that the orbital plane of the satellite makes with the Earth's equatorial plane

Inclined orbit: An orbit having an angle of inclination between 0° and 180°

Injection velocity: This is the horizontal velocity with which a satellite is injected into space by the launch vehicle with the intention of imparting a specific trajectory to the satellite

Kepler's first law: The orbit of an artificial satellite around Earth is elliptical with the centre of the Earth lying at one of its foci

Kepler's second law: The line joining the satellite and the centre of the Earth sweeps out equal areas in the plane of the orbit in equal times

Kepler's third law: The square of the time period of any satellite is proportional to the cube of the semi-major axis of its elliptical orbit

Low Earth orbit (LEO): A satellite orbit with an orbital height of around 150 km to 500 km above the surface of Earth. These orbits have lower orbital periods, shorter propagation delays and lower propagation losses

Medium Earth orbit (MEO): A satellite orbit with an orbital height around 10 000 km to 20 000 km above the surface of the Earth

Molniya orbit: A highly inclined and eccentric orbit used by Russia and other countries of the erstwhile Soviet Union for providing communication services

Orbit: A trajectory that is periodically repeated

Perigee: A point on a satellite orbit closest to the centre of the Earth. The perigee distance is the distance of the perigee point from the centre of the Earth

Polar orbit: An orbit having an angle of inclination equal to 90°

Prograde orbit: Also called a direct orbit, an orbit where the satellite travels in the same direction as the direction of rotation of Earth. This orbit has an angle of inclination between 0° and 90°

Project Iridium: Project Iridium is a global communication system conceived by Motorola that makes use of satellites in low Earth orbits. A total of 66 satellites are arranged in a distributed architecture with each satellite carrying 1/66 of the total system capacity

Retrograde orbit: An orbit where the satellite travels in a direction opposite to the direction of rotation of Earth. This orbit has an angle of inclination between 90° and 180°

Right ascension of the ascending node: The right ascension of the ascending node indicates the orientation of the line of nodes, which is the line joining the ascending and descending nodes, with respect to the direction of the vernal equinox. It is expressed as an angle (Ω) measured from the vernal equinox towards the line of nodes in the direction of rotation of Earth. The angle could be anywhere from 0° to 360°

Second cosmic velocity: This is the injection velocity at which the apogee distance becomes infinite and the orbit takes the shape of a parabola. It equals $\sqrt{2}$ times the first cosmic velocity

Solstices: Solstices are said to occur when the angle of inclination of the Earth's equatorial plane with respect to the direction of the sun is at its maximum, i.e. 23.4°. These are like equinoxes and also occur twice during the year, one on 21 June called the summer solstice and the other on 21 December called the winter solstice

Sun-synchronous orbit: A sun-synchronous orbit, also known as a helio-synchronous orbit, is one that lies in a plane that maintains a fixed angle with respect to the Earth–sun direction

Third cosmic velocity: This is the injection velocity at which the satellite succeeds in escaping from the solar system. It is related to the motion of Earth around the sun. For injection velocities beyond the third cosmic velocity, there is a region of hyperbolic flights outside the solar system

Trajectory: A path traced by a moving body

True anomaly of a satellite: This parameter is used to indicate the position of the satellite in its orbit. This is done by defining an angle (θ), called the true anomaly of the satellite, formed by the line joining the perigee and the centre of the Earth with the line joining the satellite and the centre of the Earth

3

Satellite Launch and In-orbit Operations

Fundamental issues such as laws governing motion of artificial satellites around Earth, different orbital parameters, types of orbits and their suitability for a given application, and so on, related to orbital dynamics were addressed in the previous chapter. The next obvious step is to understand the launch requirements to acquire the desired orbit. This should then lead to various in-orbit operations such as orbit stabilization, orbit correction and station keeping that are necessary for keeping the satellite in the desired orbit. Launch vehicles and launch sites play an important role in launching satellites and subsequently controlling their operation in orbit. The present chapter will address all these issues, including a detailed description of major launch vehicle systems and space centres, and also other related issues like Earth coverage, eclipses and pointing towards a given satellite from Earth. Again the chapter is amply illustrated with mathematics wherever necessary and a large number of solved problems.

3.1 Acquiring the Desired Orbit

In order to ensure that the satellite acquires the desired orbit, that is the orbit with desired values of orbital elements/parameters such as orbital plane, apogee and perigee distances, and so on, it is important that correct conditions are established at the satellite injection point. For instance, in order to ensure that the satellite orbits within a given plane, the satellite must be injected at a certain specific time, depending upon the longitude of the injection point, at which the line of nodes makes the required angle with the direction of the vernal equinox. Put in simple words, for a given orbital plane, the line of nodes will have a specific angle with the vernal equinox. To acquire this angle, the satellite must be injected at the desired time depending upon the longitude of the injection point.

Satellite Technology: Principles and Applications, Third Edition. Anil K. Maini and Varsha Agrawal.
© 2014 John Wiley & Sons, Ltd. Published 2014 by John Wiley & Sons, Ltd. Companion Website: www.wiley.com/go/maini3

3.1.1 Parameters Defining the Satellite Orbit

The satellite orbit is completely defined or specified by the following parameters:

1. Right ascension of the ascending node
2. Inclination angle
3. Position of the major axis of the orbit
4. Shape of the elliptical orbit
5. Position of the satellite in its orbit

1. *Right ascension of the ascending node.* The angle Ω defining the right ascension of the ascending node is basically the difference between two angles, θ_1 and θ_2, where θ_1 is the angle made by the longitude of the injection point at the time of launch with the direction of vernal equinox and θ_2 is the angle made by the longitude of the injection point at the time of launch with the line of nodes, as shown in Figure 3.1. Angle θ_2 can be computed from

$$\sin \theta_2 = \frac{\cos i \, \sin L}{\cos L \, \sin i} \qquad (3.1)$$

where i = angle of inclination and L = longitude of the injection point. Thus, for a known angle of inclination, the time of launch and hence the longitude of the injection point can be so chosen as to get the desired angle (Ω).

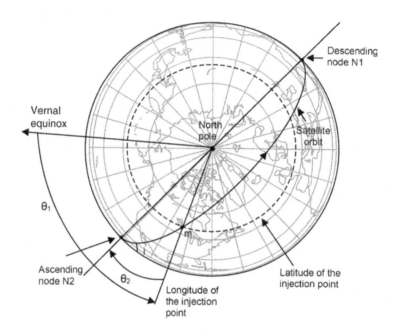

Figure 3.1 Right ascension of the ascending node

2. *Inclination angle.* The angle of inclination (i) of the orbital plane can be determined from the known values of the angle of azimuth A_z and the latitude of the injection point l using the expression

$$\cos i = \sin A_z \cos l. \qquad (3.2)$$

The azimuth angle at a given point in a satellite orbit was defined in the previous chapter as the angle made by the projection of the satellite velocity vector at that point in the local horizontal with the north. It is obvious from the above expression that for the angle of inclination to be zero, the right-hand side of the expression must equal unity. This is possible only for $A_z = 90°$ and $l = 0°$. Other inferences that can be drawn from the above expression are:

(a) For $A_z = 90°, i = l$.
(b) For $A_z < 90°, \sin A_z < 1$ and therefore $i > l$.

Thus it can be concluded that the satellite will tend to orbit in a plane which will be inclined to the equatorial plane at an angle equal to or greater than the latitude of the injection point. This will be evident if some of the prominent satellite launch sites and the corresponding latitudes of those locations are looked at.

The launch site at Kourou in French Guiana is at a latitude of 5.2°N. When the typical launch sequences are discussed in the latter part of this chapter, it will be seen that the orbital plane acquired by the satellite launched from Kourou immediately after injection is 7°, which is later corrected to get an inclination of zero or near zero through manoeuvres. The latitude of another prominent launch site at Baikonur in Russia is 45.9°. A study of the launch sequence for satellites launched from this base shows that the satellite acquires an initial orbital inclination of 51°. Both these examples confirm the axiom or the concept outlined in an earlier section. A similar result will be found if other prominent launch sites at Cape Canaveral, Sriharikota, Vandenberg and Xichang are examined.

3. *Position of the major axis of the orbit.* The position of the major axis of the satellite orbit is defined by argument of the perigee ω. Refer to Figure 3.2. If the injection point happens

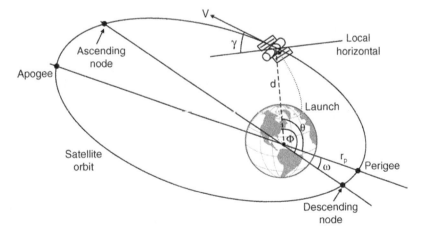

Figure 3.2 Argument of the perigee

to be different from the perigee point, then ω is the difference of two angles, ϕ and θ, as shown in Figure 3.2. Angle ϕ can be computed from

$$\sin\phi = \frac{\sin l}{\sin i} \qquad (3.3)$$

and θ can be computed from

$$\cos\theta = \frac{dV^2 \cos^2\gamma - \mu}{e\mu} \qquad (3.4)$$

The terms have their usual meaning and are indicated in Figure 3.2.

In the case where the injection point is the same as the perigee point (Figure 3.3), $\gamma = 0°$ and $\theta = 0°$. This gives

$$\sin\omega = \sin\phi = \frac{\sin l}{\sin i} \qquad (3.5)$$

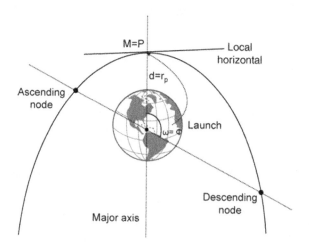

Figure 3.3 Argument of the perigee when the injection point is the same as the perigee point

4. *Shape of the elliptical orbit.* The shape of the orbit is defined by the orbit eccentricity e, the semi-major axis a, the apogee distance r_a and the perigee distance r_p. The elliptical orbit can be completely defined by either a and e or by r_a and r_p. The orbit is usually defined by the apogee and perigee distances. The perigee distance can be computed from known values of the distance d from the focus (which in this case is the centre of the Earth) at any point in the orbit and the angle γ that the velocity vector V makes with the local horizontal. The relevant expression to compute r_p is

$$r_p = \frac{V^2 d^2 \cos^2\gamma}{\mu(1+e)} \qquad (3.6)$$

The apogee distance r_a can be computed with the help of the following expression:

$$r_a = \frac{V^2 d^2 \cos^2\gamma}{\mu(1-e)} \qquad (3.7)$$

5. *Position of the satellite in its orbit.* The position of the satellite in its orbit can be defined by a time parameter t, which is the time that elapsed after a time instant t_0 when the satellite last passed through a reference point. The reference point is usually the perigee. The time t that has elapsed after the satellite last passed through the perigee point can be computed from

$$t = \left(\frac{T}{2\pi}\right)(u - e \sin u) \tag{3.8}$$

where T is the orbital period and the angle u is the eccentric anomaly of the current location of the satellite (Figure 3.4).

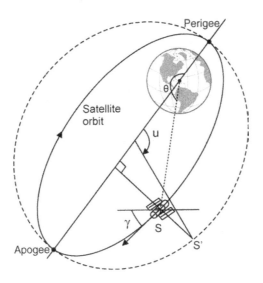

Figure 3.4 Position of the satellite in its orbit

In this section, the parameters that define an orbit as well as the factors upon which each one of these parameters is dependent, have been discussed, with the help of relevant mathematical expressions. However, when it comes to acquiring a desired orbit, this is perhaps not enough. The discussion would be complete only if it was known how each one of these parameters could be modified if needed, with the help of commands transmitted to the satellite from the ground control station. In the section that follows, this process will be discussed.

3.1.2 Modifying the Orbital Parameters

1. *Right ascension of the ascending node.* This parameter, as mentioned before, defines the orientation of the orbital plane of the satellite. Due to the Earth's equatorial bulge, there is nodal precession, that is a change in the orientation of the orbital plane with time as Earth revolves around the sun. It will be seen that this perturbation to the satellite orbital plane depends upon the orbit inclination angle i and also the apogee and perigee distances. The

rotational perturbation experienced by the satellite orbital plane (in degrees) during one orbit period is given by

$$\Delta\Omega = -0.58 \left(\frac{D}{r_a + r_p}\right)^2 \left(\frac{1}{1-e^2}\right)^2 \cos i \qquad (3.9)$$

where D = diameter of the Earth. It is clear from the above expression that the perturbation is zero only in the case of the inclination angle being 90°. Also, more the deviation of the inclination angle from 90° and the shorter the apogee and perigee distances, the larger is the perturbation. Any experimental manoeuvre that changes the orientation is a very expensive affair. It is always preferable to depend upon the natural perturbation for this purpose.

When the orbit inclination angle $i < 90°$, $\Delta\Omega$ is negative; that is the satellite orbital plane rotates in a direction opposite to the direction of the rotation of Earth (Figure 3.5). When the orbit inclination angle $i > 90°$, $\Delta\Omega$ is positive; that is the satellite orbital plane rotates in the same direction as the direction of the rotation of Earth (Figure 3.6).

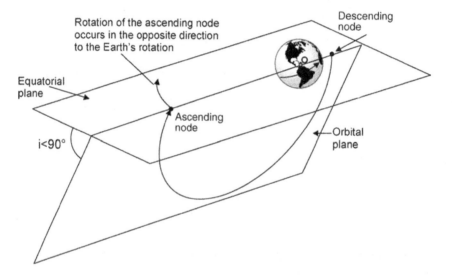

Figure 3.5 Rotation of the orbital plane when the inclination angle is less than 90°

2. *Inclination angle.* Though there is a natural perturbation experienced by the satellite orbital plane inclination w.r.t. the Earth's equatorial plane due to the sun–moon attraction phenomenon, this change is small enough to be considered negligible. A small change Δi to the inclination angle i can be affected externally by applying a thrust Δv to the velocity vector V at an angle of $90° + \Delta i/2$ to the direction of the satellite, as shown in Figure 3.7. The thrust Δv is given by

$$\Delta v = 2V \sin\left(\frac{\Delta i}{2}\right) \qquad (3.10)$$

The thrust is applied at either of the two nodes.

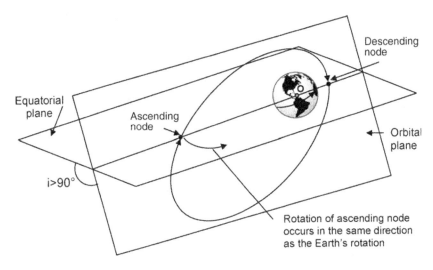

Figure 3.6 Rotation of the orbital plane when the inclination angle is greater than 90°

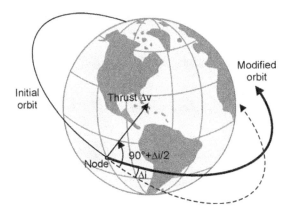

Figure 3.7 Changing the inclination angle

3. *Position of the major axis of the orbit.* The position of the major axis, as mentioned in the previous section, is defined by the parameter called the argument of the perigee ω. This parameter, like the right ascension of the ascending node, also undergoes natural perturbations due to the equatorial bulge of Earth. The phenomenon is known as apsidal precession. The further the angle of inclination from 63° 26′, the larger is the rotation of the perigee. Also, closer the satellite to the centre of the Earth, the larger is the rotation. Moreover, the rotation of the perigee occurs in the direction opposite to the satellite motion if the inclination angle is greater than 63° 26′ and in the same direction as the satellite motion if the inclination angle is less than 63° 26′. This is illustrated in Figure 3.8 (a) and (b).

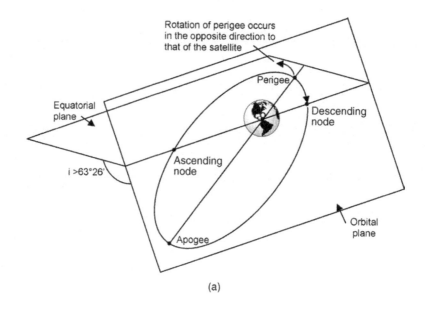

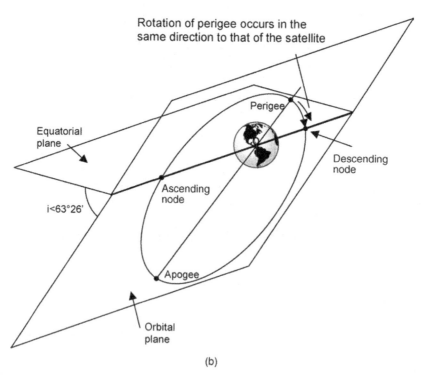

Figure 3.8 (a) Rotation of the perigee when the inclination angle is greater than 63° 26′. (b) Rotation of the perigee when the inclination angle is less than 63° 26′

The rotation experienced by ω in degrees due to natural perturbation in one orbit is given by

$$\Delta\omega = 0.29 \left[\frac{4 - 5 \sin^2 i}{(1 - e^2)^2}\right] \left[\left(\frac{D}{r_a + r_p}\right)^2\right] \quad (3.11)$$

In the case of a circular orbit, $r_a = r_p$ and $e = 0$. In order to affect an intentional change ($\Delta\omega$), a thrust Δv is applied at a point in the orbit where the line joining this point and the centre of the Earth makes an angle ($\Delta\omega/2$) with the major axis. The thrust is applied in a direction towards the centre of the Earth, as shown in Figure 3.9. The thrust Δv can be computed from

$$\Delta v = 2\sqrt{\left(\frac{\mu}{r_p}\right) \left[\frac{e}{\sqrt{(1+e)}}\right]} \sin\left(\frac{\omega}{2}\right) \quad (3.12)$$

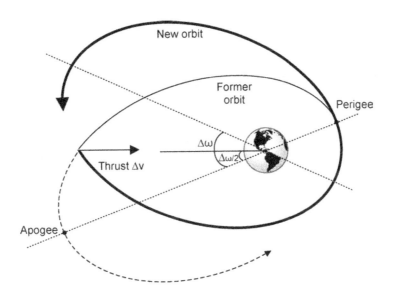

Figure 3.9 Changing the argument of the perigee

4. *Shape of the elliptical orbit.* The apogee and perigee distances fully specify the shape of the orbit as the other parameters such as eccentricity e and the semi-major axis a can be computed from this pair of distances. The apogee distance becomes affected; in fact, it reduces due to atmospheric drag and to a lesser extent by the solar radiation pressure. In fact, every elliptical orbit tends to become circular with time, with a radius equal to the perigee distance. It is interesting to note that the probable lifetime of a satellite in a circular orbit at a height of 200 km is only a few days and that of a satellite at a height of 800 km is about 100 years. For a geostationary satellite, it would be more than a million years.

Returning to modifying the apogee distance, it can be intentionally increased or decreased by applying a thrust Δv at the perigee point in the direction of motion of the satellite or

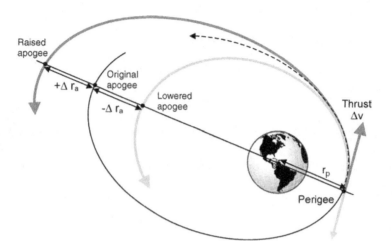

Figure 3.10 Changing the apogee distance

opposite to the direction of motion of the satellite respectively (Figure 3.10). Thus Δr_a can be computed from

$$\Delta v = \Delta r_a \frac{\mu}{V_p(r_a + r_p)^2} \qquad (3.13)$$

where V_p is the velocity at the perigee point. The perigee distance can similarly be increased or decreased by applying a thrust Δv at the apogee point in the direction of motion of the satellite or opposite to the direction of motion of the satellite respectively (Figure 3.11).

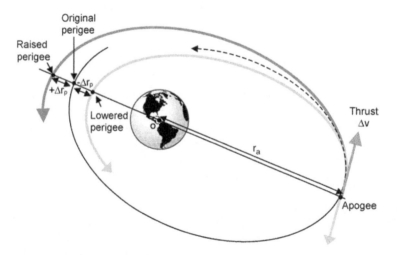

Figure 3.11 Changing the perigee distance

Thus Δr_p can be computed from

$$\Delta v = \Delta r_p \frac{\mu}{V_a(r_a + r_p)^2} \qquad (3.14)$$

where V_a is the velocity at the apogee point.

So far modifying the apogee and perigee distances has been discussed. A very economical manoeuvre that can be used to change the radius of a circular orbit is the Homann transfer, which uses an elliptical trajectory tangential to both the current and modified circular orbits. The process requires application of two thrusts, Δv and $\Delta v'$. The thrust is applied in the direction of motion of the satellite if the orbit radius is to be increased (Figure 3.12) and opposite to the direction of motion of the satellite if the orbit radius is to be decreased (Figure 3.13). The two thrusts can be computed from

$$\Delta v = \sqrt{\left[\frac{2\mu R'}{R(R + R')}\right]} - \sqrt{\left(\frac{\mu}{R}\right)} \qquad (3.15)$$

$$\Delta v' = \sqrt{\left(\frac{\mu}{R'}\right)} - \sqrt{\left[\frac{2\mu R}{R'(R + R')}\right]} \qquad (3.16)$$

where R = radius of the initial orbit and R' = radius of the final orbit.

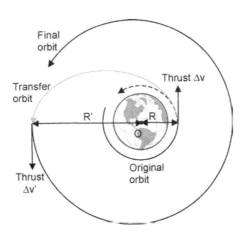

Figure 3.12 Increasing the radius of the circular orbit

5. *Position of the satellite in its orbit.* The position of the satellite was defined earlier in terms of its passage time at the perigee point. Modifying the apogee and perigee distances also changes the time of a satellite's passage at the perigee point due to the change in velocity of the satellite. A similar manoeuvre can be used to modify the longitudinal location of a geostationary satellite.

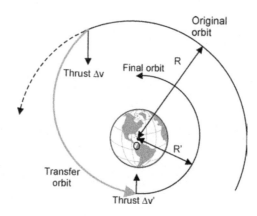

Figure 3.13 Decreasing the radius of the circular orbit

Problem 3.1

A geostationary satellite was launched from Kourou launch site in French Guiana. The satellite was injected into a highly eccentric elliptical transfer orbit from a height of 200 km. The projection of the injection velocity vector in the local horizontal plane made an angle of 85° with the north. Determine the inclination angle attained by this transfer orbit given that latitude and longitude of the Kourou site are 5.2°N and 52.7°W respectively.

Solution: The inclination angle can be computed from

$$\cos i = \sin A_z \cos l$$

where A_z = azimuth angle of the injection point = 85° and l = latitude of the launch site = 5.2°. This gives

$$\cos i = \sin 85° \cos 5.2° = 0.9962 \times 0.9959 = 0.9921$$
$$i = 7.2°.$$

Problem 3.2

A geostationary satellite was launched into an eccentric elliptical transfer orbit having an orbital plane inclination of 7° and its apogee at the geostationary height. This orbit was then circularized by an appropriate thrust manoeuvre without affecting any change to the inclination. If the velocity of the satellite in the circularized orbit is 3 km/s, determine the velocity thrust required to make the inclination 0°. Also show the relevant diagram indicating the direction of thrust to be applied.

Solution: The required thrust can be computed from

$$\Delta v = 2V \sin\left(\frac{\Delta i}{2}\right)$$

Acquiring the Desired Orbit

where $V = 3000$ m/s and $\Delta i = 7°$. Therefore

$$\Delta v = 2V \sin\left(\frac{\Delta i}{2}\right) = 2 \times 3000 \times \sin 3.5° = 6000 \times 0.061 = 366 \text{ m/s}$$

The thrust is applied when the satellite is at either of the nodes and is applied at an angle of $(90° + \Delta i/2) = 93.5°$ to the direction of motion of the satellite, as shown in Figure 3.14.

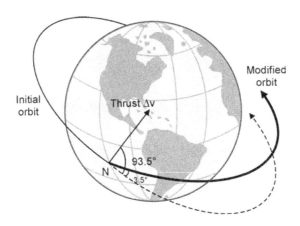

Figure 3.14 Figure for Problem 3.2

Problem 3.3

A satellite is launched into an eccentric ($e = 0.69$) elliptical orbit. The line joining the perigee and centre of the Earth makes an angle of 60° with the line of nodes. It is intended to rotate the perigee of the orbit so as to make this angle equal to 50°. If the perigee distance is 7000 km, determine the velocity thrust required to bring about the change. Also illustrate the application of thrust vis-a-vis the point of application and its direction. (Assume that $\mu = 39.8 \times 10^{13}$ N m²/kg.)

Solution: The velocity thrust required to rotate the perigee point by the desired amount can be computed from

$$\Delta v = 2\sqrt{\left(\frac{\mu}{r_p}\right)\left[\frac{e}{\sqrt{(1+e)}}\right]} \sin\left(\frac{\omega}{2}\right)$$

Now $r_p = 7000$ km $= 7\,000\,000$ m, $e = 0.69$ and $\omega = 60°$. Substituting these values gives

$$\Delta v = 2\sqrt{\left(\frac{39.8 \times 10^{13}}{7\,000\,000}\right)\left(\frac{0.69}{\sqrt{1.69}}\right)} \sin 30°$$

$$= 15080.74 \times 0.53 \times 0.5 = 3996.4 \text{ m/s} = 3.996 \text{ km/s}$$

The point of application of this thrust and its direction are shown in Figure 3.15.

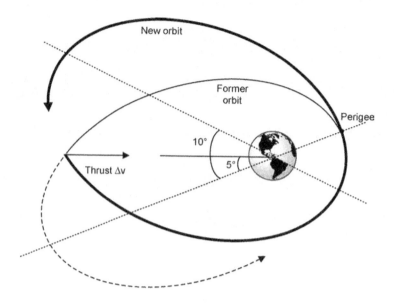

Figure 3.15 Figure for Problem 3.3

Problem 3.4

A certain satellite is moving in an elliptical orbit with apogee and perigee distances of 15 000 km and 7000 km respectively. It is intended to circularize the orbit with a radius equal to the perigee distance. Determine the magnitude and direction of the velocity thrust required to achieve the objective. Also indicate the point in the orbit where the thrust would be applied. Take the value of the constant μ as 39.8×10^{13} Nm²/kg.

Solution: Stated in other words, the objective is to reduce the apogee distance to 7000 km from 15 000 km. The required velocity thrust Δv can be computed from

$$\Delta v = \Delta r_a \frac{\mu}{V_p(r_a + r_p)^2}$$

where Δr_a = required change in the apogee distance = 8000 km. Also, the velocity at the perigee point can be computed from

$$V_p = \sqrt{\left(\frac{2\mu}{\text{perigee distance}} - \frac{2\mu}{\text{perigee distance} + \text{apogee distance}}\right)}$$

$$= \sqrt{\left(\frac{2 \times 39.8 \times 10^{13}}{7 \times 10^6} - \frac{2 \times 39.8 \times 10^{13}}{7 \times 10^6 + 15 \times 10^6}\right)}$$

$$= 10^3 \sqrt{\left(\frac{796}{7} - \frac{796}{22}\right)}$$

$$= 8.805 \text{ km/s}$$

Acquiring the Desired Orbit

Therefore,

$$\Delta v = 8\,000\,000 \times \frac{39.8 \times 10^{13}}{8805 \times (15\,000\,000 + 7\,000\,000)^2} = 747.1 \text{ m/s}.$$

The thrust will be applied at the perigee point and in the direction opposite to the motion of the satellite, as illustrated in Figure 3.16.

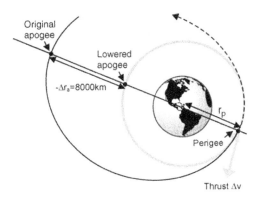

Figure 3.16 Figure for Problem 3.4

Problem 3.5

A certain satellite is moving in an elliptical orbit with apogee and perigee distances of 15 000 km and 7000 km respectively. It is intended to increase the perigee distance to 8000 km. Determine the magnitude and direction of the velocity thrust required to achieve the objective. Also indicate the point in orbit where the thrust would be applied. Take the value of constant (μ) as 39.8×10^{13} Nm^2/kg and radius of the Earth as 6378 km.

Solution:

The objective is to increase the perigee distance to 8000 km.

The required velocity thrust Δv can be computed from $\Delta v = \dfrac{\Delta r_p \mu}{[v_a(r_a + r_p)^2]}$ where $\Delta r_p =$

Required change in perigee distance = 1000 km
Also velocity at the perigee point can be computed from:

$V_p = \sqrt{[\{2\mu/(\text{Perigee distance})\} - \{2\mu/(\text{Perigee distance} + \text{Apogee distance})\}]}$
$= \sqrt{[\{2\mu/(7\,000\,000)\} - \{2\mu/(7\,000\,000 + 15\,000\,000)\}]}$
$= \sqrt{(77\,532\,469.6)} = 8.805$ km/s.

The velocity at the apogee $V_a = (V_p \times$ Distance of perigee from center of the Earth$)/$(Distance of apogee from center of the Earth).

$V_a = 8.805 \times (7\,000\,000)/(15\,000\,000)$ km/s = 4.109 km/s.
Therefore, $\delta v = 1\,000\,000 \times 39.8 \times 10^{13}/[4109 \times (15\,000\,000 + 7\,000\,000)^2] = 200.1$ m/s
The thrust will be applied at the apogee point and in the direction of motion of the satellite.

Problem 3.6

A certain satellite is moving in a circular orbit with an altitude of 500 km. The orbit altitude is intended to be increased to 800 km. Suggest a suitable manoeuvre to do this and calculate the magnitudes of the thrusts required. Show the point/s of application of the thrust/s. Take the value of the constant μ as 39.8×10^{13} N m²/kg.

Solution: The objective is to increase the orbit altitude to 800 km. This can be achieved by making use of the Homann transfer method by first applying a thrust Δv to transform the current circular orbit to an elliptical one, with its perigee at the current orbit radius and apogee at the desired orbit radius. Since the altitude of desired orbit is greater than the current one, the thrust is applied in the direction of motion of the satellite. A second thrust $\Delta v'$ is then applied at the apogee of the elliptical orbit in the direction of the motion of the satellite in order to circularize the orbit with the desired radius. The two thrusts can be computed from

$$\Delta v = \sqrt{\left[\frac{2\mu R'}{R(R+R')}\right]} - \sqrt{\left(\frac{\mu}{R}\right)}$$

$$\Delta v' = \sqrt{\left(\frac{\mu}{R'}\right)} - \sqrt{\left[\frac{2\mu R}{R'(R+R')}\right]}$$

where R = radius of the initial orbit and R' = radius of the final orbit.
Now $R = 6878$ km and $R' = 7178$ km. The two thrusts are computed as

$$\Delta v = \sqrt{\left[\frac{2\mu R'}{R(R+R')}\right]} - \sqrt{\left(\frac{\mu}{R}\right)}$$

$$= \sqrt{\left[\frac{2 \times 39.8 \times 10^{13} \times 7\,178\,000}{6\,878\,000 \times (6\,878\,000 + 7\,178\,000)}\right]} - \sqrt{\left(\frac{39.8 \times 10^{13}}{6\,878\,000}\right)}$$

$$= 7687.70 - 7606.95 = 80.75 \text{ m/s}$$

$$\Delta v' = \sqrt{\left(\frac{\mu}{R'}\right)} - \sqrt{\left[\frac{2\mu R}{R'(R+R')}\right]}$$

$$= \sqrt{\left(\frac{39.8 \times 10^{13}}{7\,178\,000}\right)} - \sqrt{\left[\frac{2 \times 39.8 \times 10^{13} \times 6\,878\,000}{7\,178\,000 \times (6\,878\,000 + 7\,178\,000)}\right]}$$

$$= 7446.29 - 7366.40 = 79.89 \text{ m/s}$$

Thus, the two thrusts to be applied are $\Delta v = 80.75$ m/s and $\Delta v' = 79.89$ m/s.

Figure 3.17 shows the point and direction of application of the two thrusts.

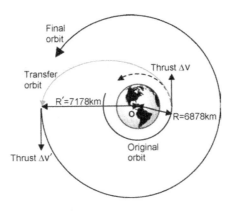

Figure 3.17 Figure for Problem 3.6

3.2 Launch Sequence

In the last chapter, the various types of satellite orbits were discussed, the prominent among them being the low Earth circular orbits, elliptical eccentric orbits and geostationary orbits. The suitability of each type of orbit for a given application was also discussed. In the present chapter, we have continued with a basic discussion on orbits and trajectories. The various factors important for acquiring the desired orbit, as well as various natural causes that cause perturbations to certain orbital parameters and external manoeuvres used to bring about intended changes in them have been discussed. In the present section, the discussion will centre on typical launch sequences employed worldwide for putting satellites in the geostationary orbit, as putting a satellite in a lower circular or elliptical orbit is only a step towards achieving the geostationary orbit.

3.2.1 Types of Launch Sequence

There are two broad categories of launch sequence, one that is employed by expendable launch vehicles such as Ariane of the European Space Agency and Atlas Centaur and Thor Delta of the United States and the other that is employed by a reusable launch vehicle such as the Space Shuttle of the United States and Buran of Russia.

Irrespective of whether a satellite is launched by a reusable launch vehicle or an expendable vehicle, the satellite heading for a geostationary orbit is first placed in a transfer orbit. The transfer orbit is elliptical in shape with its perigee at an altitude between 200 km and 300 km and its apogee at the geostationary altitude.

In some cases, the launch vehicle injects the satellite directly into a transfer orbit of this type. Following this, an apogee manoeuvre circularizes the orbit at the geostationary altitude. The last step is then to correct the orbit for its inclination. This type of launch sequence is illustrated

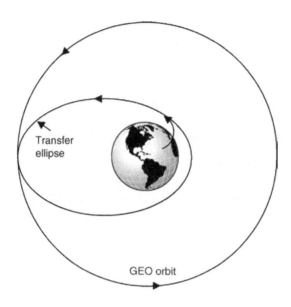

Figure 3.18 Possible geostationary satellite launch sequence

in Figure 3.18. In the second case, the satellite is first injected into a low Earth circular orbit. In the second step, the low Earth circular orbit is transformed into an elliptical transfer orbit with a perigee manoeuvre. Circularization of the transfer orbit and then correction of the orbit inclination follow this. This type of sequence is illustrated in Figure 3.19.

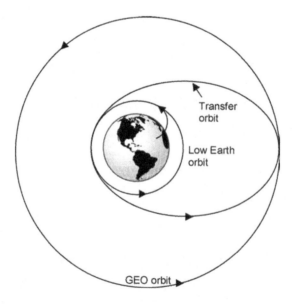

Figure 3.19 Another possible geostationary satellite launch sequence

Launch Sequence

In the following paragraphs, we will discuss some typical launch sequences observed in the case of various launch vehicles used to deploy geostationary satellites from some of the prominent launch sites all over the world. The cases presented here include the launch of geostationary satellites from Kourou in French Guiana and Cape Canaveral in the United States, both situated towards the eastern coast of America, and also Baikonur in Russia. A typical Space Shuttle launch will also be discussed. The two types of launch sequence briefly described above will be amply evident in the real-life illustrations that follow. A detailed account of these launch sites and many more is presented in Section 3.4 of the chapter.

3.2.1.1 Launch from Kourou

A typical Ariane launch of a geostationary satellite from Kourou in French Guiana will be illustrated. Different steps involved in the entire process are:

1. The launch vehicle takes the satellite to a point that is intended to be the perigee of the transfer orbit, at a height of about 200 km above the surface of the Earth. The satellite along with its apogee boost motor is injected before the launch vehicle crosses the equatorial plane, as shown in Figure 3.20. The injection velocity is such that the injected satellite attains an eccentric elliptical orbit with an apogee altitude at about 36 000 km. The orbit is inclined at about 7°, which is expected, as the latitude of the launch site is 5.2°.

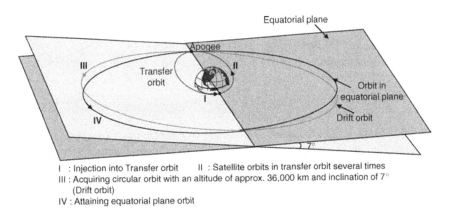

I : Injection into Transfer orbit   II : Satellite orbits in transfer orbit several times
III : Acquiring circular orbit with an altitude of approx. 36,000 km and inclination of 7°
   (Drift orbit)
IV : Attaining equatorial plane orbit

Figure 3.20   Typical launch of a geostationary satellite from Kourou

2. In the second step, after the satellite has completed several revolutions in the transfer orbit, the apogee boost motor is fired during the passage of the satellite at the apogee point. The resulting thrust gradually circularizes the orbit. The orbit now is a circular orbit with an altitude of 36 000 km.
3. Further thrust is applied at the apogee point to bring the inclination to 0°, thus making the orbit a true circular and equatorial orbit.
4. The last step is to attain the correct longitude and attitude. This is also achieved by applying thrust either tangential or normal to the orbit.

3.2.1.2 Launch from Cape Canaveral

Different steps involved in the process of the launch of a geostationary satellite from Cape Canaveral are:

1. The launch vehicle takes the satellite to a point that is intended to be the perigee of the transfer orbit, at a height of about 300 km above the surface of the Earth, and injects the satellite first into a circular orbit called the parking orbit. The orbit is inclined at an angle of 28.5° with the equatorial plane, as shown in Figure 3.21. Reasons for this inclination angle have been explained earlier.

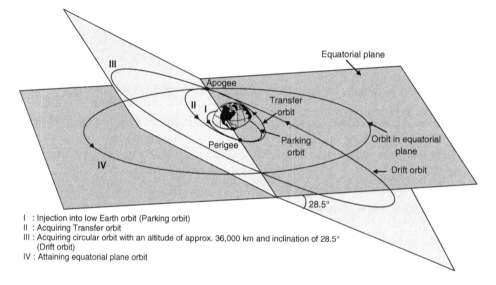

I : Injection into low Earth orbit (Parking orbit)
II : Acquiring Transfer orbit
III : Acquiring circular orbit with an altitude of approx. 36,000 km and inclination of 28.5° (Drift orbit)
IV : Attaining equatorial plane orbit

Figure 3.21 Typical launch of a geostationary satellite from Cape Canaveral

2. In the second step, a perigee manoeuvre associated with the firing of a perigee boost motor transforms the circular parking orbit to an eccentric elliptical transfer orbit with perigee and apogee altitudes of 300 km and 36 000 km respectively.
3. In the third step, an apogee manoeuvre similar to the one used in the case of the Kourou launch circularizes the transfer orbit. Till now, the orbit inclination is 28.5°. In another apogee manoeuvre, the orbit inclination is brought to 0°. The thrust required in this manoeuvre is obviously much larger than that required in the case of inclination correction in the Kourou launch.
4. In the fourth and last step, several small manoeuvres are used to put the satellite in the desired longitudinal position.

3.2.1.3 Launch from Baikonur

The launch procedure for a geostationary satellite from Baikonur is similar to the one described in case of the launch from Cape Canaveral. Different steps involved in the process of the launch of a geostationary satellite from Baikonur are:

Launch Sequence

1. The launch vehicle injects the satellite in a circular orbit with an altitude of 200 km and an inclination of 51°.
2. In the second step, during the first passage of the satellite through the intended perigee, a manoeuvre puts the satellite in the transfer orbit with an apogee altitude of approximately 36 000 km. The orbit inclination now is 47°.
3. In the third step, the transfer orbit is circularized and the inclination corrected at the descending node.
4. In the fourth and last step, the satellite drifts to its final longitudinal position. Different steps are depicted in Figure 3.22.

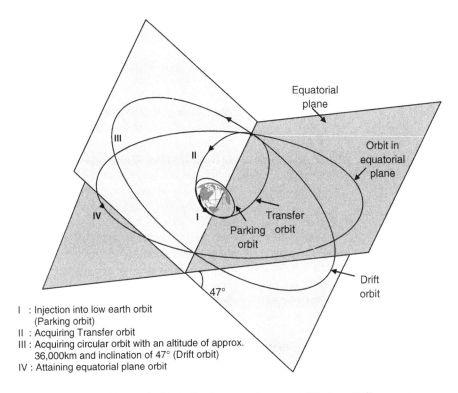

I : Injection into low earth orbit (Parking orbit)
II : Acquiring Transfer orbit
III : Acquiring circular orbit with an altitude of approx. 36,000km and inclination of 47° (Drift orbit)
IV : Attaining equatorial plane orbit

Figure 3.22 Typical launch of a geostationary satellite from Baikonour

3.2.1.4 Space Shuttle Launch

The Space Shuttle launch is also similar to the launch procedure described in the case of a launch from Cape Canaveral and Baikonur. The Space Shuttle injects the satellite along with its perigee motor in a circular parking orbit at an altitude of 100 km. After several revolutions, the perigee motor is fired by signals from an onboard timing mechanism and the thrust generated thereby puts the satellite in a transfer orbit with an apogee at the geostationary height. The perigee motor is then jettisoned. The orbit is then circularized with an apogee manoeuvre.

3.3 Launch Vehicles

Launch vehicles are rocket powered vehicles that are used to carry payloads from the Earth's surface into outer space either into an orbit around Earth or to some other destination in outer space. The payload can be an unmanned spacecraft or a space probe or an artificial satellite. The overall launch system includes the launch vehicle, the launch pad and other infrastructure. Launch vehicles have been in use since the early 1950s. Some of the prominent families of launch vehicles that have evolved and have been in use over the last six decades include the Soyuz and Proton launchers of Russia, the Ariane series of Europe, and the space shuttle, Atlas, Delta, and Titan families of launch vehicles of the United States, Long March of China and the polar satellite launch vehicle (PSLV) and geostationary satellite launch vehicle (GSLV) of India. Launch vehicles are described in detail in the following paragraphs.

3.3.1 Introduction

The pioneers of space exploration realized the importance of developing launch vehicles if one were to successfully gain an access to outer space. Most launch vehicles have their origin in the ballistic missiles developed for military use during the 1950s and early 1960s. Those missiles in turn were based on the ideas first developed by the Soviet rocket scientist and pioneer of astronautic theory Konstantin Eduardovich Tsiolkovsky in Russia (1857–1935), the American physicist and inventor Robert Hutching Goddard (1882–1945), who is credited with development of the world's first liquid fuelled rocket, and German physicist Hermann Julius Oberth (1894–1989). These three are considered to be the founding fathers of rocketry and astronautics.

The launch vehicle needs to accelerate the payload to a velocity in excess of 28 000 km per hour in order to launch it into an Earth orbit and in excess of approximately 40 000 km per hour if the intended destination of the payload is Mars or the moon. In addition, the acceleration must be provided quickly, typically in 8–10 minutes, to minimize the time spent by the launch vehicle in stressful atmospheric conditions. It also minimizes the time duration for which the vehicle's rocket engines and other systems are forced to operate near their performance limits. Another important requirement for launch vehicle designers is to maximize the payload carrying capacity, which is achieved by reducing the weight of the launch vehicle. The weight of the launch vehicle depends on the structural weight and the weight of the fuel and oxidizer. Since rapid acceleration requires one or more rocket engines burning large quantities of propellant at a high rate, thereby contributing to most of the launch vehicle weight, the vehicle's structural weight is kept as low as possible. This makes design of reliable launch vehicles a very challenging task. Launch vehicles and launch centres are a prerequisite if a particular country wants to carry out an independent space program. Some of the prominent countries/agencies running their own space programmes include Russia, the United States, Japan, China, the European Space Agency (ESA), Israel, India, and Iran. Many more are aspiring to have this capability.

3.3.2 Classification

Launch vehicles are broadly classified as expendable and reusable launch vehicles. They are further classified on the basis of launch capacity and are also frequently identified by the nation or space agency responsible for the launch, and the company or consortium that manufactures the vehicle.

Launch Vehicles

Expendable launch vehicles (ELV) are designed to be used only once and their components are not recovered after launch. They mostly comprise multi-stage rockets and the job of each stage is to provide the desired orbital manoeuvre. As the job of the stage is completed, the different stages of the rocket are expended. The process goes on until the satellite is placed into the desired trajectory. The number of rocket stages can be as many as five. It may be mentioned here that some launch vehicles used to put the satellite into small low Earth orbits can comprise a single rocket stage. In addition to the rocket stages, launch vehicles also comprise boosters that are used to aid the rockets during main orbital manoeuvres or to provide small orbital corrections. Figure 3.23 shows the launch sequence of the GSLV-FO4 launch vehicle developed by the Indian Space Research Organization (ISRO). It is a three stage vehicle comprising a solid rocket engine based first stage (GS-1) and liquid rocket engine based second and third stages (GS-2 and GS-3).

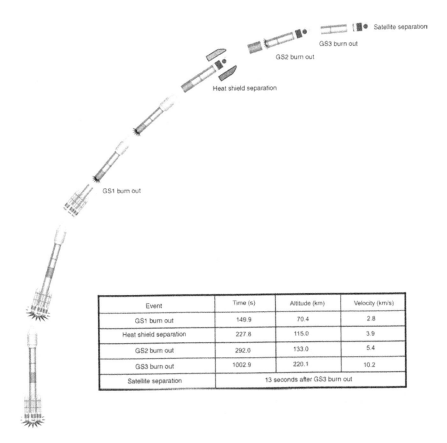

Event	Time (s)	Altitude (km)	Velocity (km/s)
GS1 burn out	149.9	70.4	2.8
Heat shield separation	227.8	115.0	3.9
GS2 burn out	292.0	133.0	5.4
GS3 burn out	1002.9	220.1	10.2
Satellite separation	13 seconds after GS3 burn out		

Figure 3.23 Launch sequence of GSLV-FO4

Ariane (Europe), Atlas (USA), Delta (USA), GSLV (India), PSLV (India), Long March (China) and Proton (Russia) are some common expendable launch vehicles being used internationally to launch satellites.

Reusable launch vehicles (RLV) are designed to be recovered intact and used again for subsequent launches. The Space Shuttle (Figure 3.24) from the United States is one example. It is generally used for human spaceflight missions.

On the basis of launch capacity, which is the most common method of classification, launch vehicles are classified as heavy lift launch vehicles (HLLVs), large launch vehicles (LLVs), medium launch vehicles (MLVs) and small launch vehicles (SLVs).

Figure 3.24 Space Shuttle (Courtesy: NASA)

A *heavy lift launch vehicle* is capable of putting more than 10 000 kg of payload into low Earth orbit and more than 5000 kg of payload into a geostationary transfer orbit. Some well-known launch vehicles belonging to this category include the Space Shuttle, Titan-III and IV, Proton-D1 and Ariane-5. As an example, the Ariane-5 ECA (Enhanced Capability-A; Figure 3.25) launch vehicle of the Ariane-5 series has the capacity of launching 12 000 kg of payload to geostationary transfer orbit. As another example, Titan-IV can put a payload of up to 21 680 kg in low Earth orbit, 17 600 kg in polar low Earth orbit, 5760 kg in geostationary orbit and 5660 kg in a heliocentric orbit.

A *large launch vehicle* is capable of putting 5000–10 000 kg of payload in to a low Earth orbit and 2000–5000 kg of payload into a geostationary transfer orbit. Some common launch vehicles in this category are the GSLV, Long March LM-3A/3B (Figure 3.26), Atlas-IIAS and Ariane-44L. As an example, Long March LM-3A has a maximum payload launch capacity of 8500 kg for low Earth orbit and 2600 kg for geostationary transfer orbit. As another example, Ariane-44L can launch a 5000–7600 kg payload in low Earth orbit and a 2000–4300 kg payload in geostationary transfer orbit. GSLV is yet another type of large launch vehicle. The GSLV launch vehicle (Figure 3.27), developed by the ISRO can launch a 2000–2500 kg satellite into geostationary transfer orbit.

A *medium launch vehicle* is capable of putting 2000–5000 kg of payload in to a low Earth orbit and 1000–2000 kg of payload into a geostationary transfer orbit. Some common launch vehicles in this category are PSLV, Long March LM-3, Molniya and Delta II-6925/7925. As an example, Long March LM-3 has a maximum payload launch capacity of 5000 kg for low

Launch Vehicles 103

Figure 3.25 Ariane-5 ECA (Reproduced by permission of © ESA-D.DUCROS)

Figure 3.26 Long March LM-3A/3B

Figure 3.27 GSLV (Courtesy: ISRO)

Earth orbit and 1340 kg for geostationary transfer orbit. As another example, Delta II can launch up to 5000 kg payload in low Earth orbit and a payload up to 1800 kg in geostationary transfer orbit.

A *small launch vehicle* is capable of putting less than 2000 kg of payload into low Earth orbit and <1000 kg of payload into geostationary transfer orbit. Some common launch vehicles in this category are ASLV (Augmented Satellite Launch Vehicle), Taurus, Athena-1 and 2, Cosmos, and Pegasus-XL. As an example, Taurus (Figure 3.28) has a maximum payload launch capacity of 1320 kg for low Earth orbit. As another example, Athgena-1 can launch a 794–1896 kg payload in low Earth orbit.

Launch vehicles are also frequently identified by the company or consortium that manufactures the launch vehicle and the country or space agency responsible for the launch. For example, the Ariane series of launch vehicles are associated with the ESA and United Launch Alliance manufactures and launches the Delta-IV and Atlas-V series of launch vehicles.

3.3.3 Anatomy of a Launch Vehicle

This section discusses the anatomy of a typical expendable launch vehicle, briefly describing its different components. A typical launch vehicle is composed of a number of separate sections called stages. Most expendable launch vehicles have two or three stages, although launch vehicles with as many as five stages have been successfully built. The ISRO's ASLV is an example of a five-stage launch vehicle. The first stage has the task of imparting the initial thrust needed to overcome Earth's gravity and lift the total weight of the vehicle along with its payload up off the surface of the Earth. As a result, it is the heaviest part of the vehicle and has the largest rocket engines, and the largest fuel and oxidizer tanks. As each stage exhausts its fuel, which constitutes the largest part of its mass, it is detached, thereby reducing the

Launch Vehicles 105

Figure 3.28 Taurus

mass that the next stage must propel. This also reduces the quantum of energy required to lift the remaining vehicle mass. As a consequence of this, the subsequent stages are increasingly lighter. The detached stages usually fall back towards the Earth's surface, disintegrating and evaporating on encountering atmospheric heating on their fall back towards the Earth.

Each stage of a vehicle is made up of four basic subsystems: propulsion, structure, fuel and oxidizer tanks, and guidance and control. There are two basic types of propulsion systems, namely solid fuelled and liquid fuelled systems. Solid fuelled rocket engines are relatively cheaper to design and build, are more robust and the fuel can be stored for longer periods of time. Liquid fuelled rocket engines on the other hand are fragile, but they offer better control and higher energy. A combination of solid and liquid fuelled rocket motors is generally used in a multi-stage launch vehicle. While solid fuelled rocket engines are generally used by smaller vehicles and as strap-on boosters by larger launch vehicles, liquid fuelled engines are used both for upper stages as well as strap-on boosters. It may be mentioned here that strap-on booster rocket motors can be used to increase the payload capacity of the launch vehicle without the need to redesign the entire launch vehicle. The Ariane-4 family of launch vehicles is an example where a single core vehicle can be tailored to meet the requirements of different payloads by using a combination of solid or liquid fuelled strap-on booster motors. Different structural configurations have been used to build launch vehicle stages in different launch vehicles. These include using a separate skin and structure surrounding the fuel tanks and engines, as is the case in Soyuz rockets, and outside of fuel tanks serving as the skin like in Delta-2 rockets. In the case of some launch vehicles, the fuel pressure on the tank provides the required strength to prevent the skin from buckling under the weight of the launch vehicle and its payload. Liquid fuel is stored in vessels and suitable plumbing is used to direct it. Major functions of the guidance and control system of the launch vehicle include control of the time of initiation and duration of engine firing, initiation of stage separation, initiate mission abort and self destruct sequence in case of a fatal problem during launch, and in some cases initiate an action to compensate for a failed system by using another part of the launch vehicle.

3.3.4 *Principal Parameters*

The principal parameters of launch vehicles include:

1. Initial launch mass
2. Final mass
3. Propellant mass
4. Mass fraction
5. Propellant mass fraction
6. Thrust
7. Thrust-to-weight ratio
8. Specific impulse
9. Type of propellant
10. Number of stages

The terms *initial mass, final mass* and *propellant mass* are self explanatory. The initial mass is the mass of the launch vehicle at the time of lift-off and it includes the propellant mass and the dry mass. Also, initial mass equals the sum of the propellant mass and the final mass.

The *mass fraction* of a launch vehicle is defined as the fraction of the initial mass of the launch vehicle that does not reach the destination. The *propellant mass fraction* is the fraction of initial mass of the launch vehicle that is used by the propellant and therefore is given by the ratio of propellant mass to initial mass. It may be mentioned here that in the case of single stage to orbit launch vehicles that do not drop any part of the vehicle during flight, the mass fraction equals the propellant mass fraction. Also, the mass fraction of a multi-stage launch vehicle is larger than the propellant mass fraction because parts of the launch vehicle are dropped en route to the destined orbit. A high mass fraction implies that a larger quantity of propellant is driving a relatively smaller mass of launch vehicle and payload. This enables the higher velocities desirable for putting the payload into intended orbit to be achieved.

Thrust is the force generated by the propulsion system of the launch vehicle through the application of Newton's third law of motion and is measured in Newtons or kg-f. It is the sum of two components: momentum thrust and pressure thrust. Momentum thrust is the force generated due to the reaction to accelerating mass of a fluid and is given by the product of propellant mass flow rate and exhaust velocity. Pressure thrust is generated by a difference in exhaust velocity pressure and atmospheric pressure acting over an area of the exit plane of the exhaust. The two components of the thrust are combined into a single component by defining it as the product of propellant mass flow rate and effective exhaust velocity. It may be mentioned here that the initial thrust generated at the time of lift-off must exceed the total weight of the launch vehicle. That is why it is particularly important to have high thrust level in the first stage of the launch vehicle. This is what led to the preference for solid rocket motors in the first stage of the launch vehicle. In the case of thrust being less than the launch vehicle weight the vehicle will continue to rest on the launch pad until loss of weight due to propellant burning brings the total weight to less than the thrust. After lift-off, thrust is continuously required to accelerate the launch vehicle against the force of the Earth's gravity and to achieve the minimum required altitude and velocity to put the payload into the intended orbit.

Thrust-to-weight ratio, as the name suggests, is the ratio of the thrust generated by the propulsion system to the weight of the vehicle. It is a dimensionless quantity and is an indicator of the acceleration capability of the launch vehicle. It is expressed as a number that is a multiple of gravitational acceleration. Since the launch vehicles are supposed to travel through a changing gravitational environment and also since there is a weight loss due to continuous burning of the propellant, the thrust-to-weight ratio is generally expressed as the thrust-to-Earth weight ratio, which is the ratio of thrust to weight of the launch vehicle at sea level on Earth at the time of lift-off. It is also referred to as the g-force capability of the launch vehicle and is a maximum just before the propellant is fully consumed. A thrust-to-weight ratio of 1.2 is desirable. A figure less than unity is acceptable in upper stages.

Specific impulse is defined as the ratio of thrust to the rate of consumption of propellant and is a measure of the efficiency of the propulsion system. It is measured in seconds. A specific impulse of 1.0 s means that 1.0 kg/s of propellant weight flow rate produces 1.0 kg force of thrust. In other words, 1.0 kg of propellant produces a thrust of 1.0 kg-f for a period of 1.0 s. Higher specific impulse is desirable. As an example, in the two-stage Ariane-5 launch vehicle, the first and second stages have specific impulse specifications of 475 s and 324 s, respectively. As another example, the two-stage liquid fuelled core of the Titan-IV launch vehicle had specific impulse specifications of 316 s and 302 s for the first and second stages respectively. The three-stage rocket of ISRO's GSLV has specific impulse specifications of 460 s, 295 s and 166 s for the first, second and third stages respectively.

There are three basic types of propellants used in rocket engines: liquid propellants, solid propellants, and hybrid propellants. A propellant usually consists of a fuel and an oxidizer, and a chemical reaction between them produces the thrust. The chemical reaction is enhanced in some cases by adding a catalyst. Some common liquid propellants include kerosene/liquid oxygen (LOX), also known as RP-1 (Refined Petroleum-1 or Rocket Propellant-1), liquid hydrogen/liquid oxygen (LH_2/LOX), unsymmetrical dimethyl hydrazine/nitrogen tetroxide (UDMH/N_2O_4), and hydrazine (N_2H_4)/Aerozine-50. RP-1 is a highly refined form of kerosene. The propellant is stable at room temperature and is relatively much less toxic and carcinogenic. However, it has relatively lower vacuum specific impulse of about 355 s. It has been used in the first stage of launch vehicles such as Titan-I, Saturn-I, Saturn-V and Energia, and in first-stage boosters of Soyuz, Delta-I and III, and the Atlas series of launch vehicles. LH_2/LOX has a vacuum specific impulse of about 455 s and has been used in different stages of the Space Shuttle, Delta-IV, Ariane-5 and Centaur launch vehicles. UDMH/N_2O_4 is hypergolic and can be stored for long periods of time. It has a vacuum specific impulse specification of about 345 s. It has been used in the lower stages of Russian and Chinese boosters. N_2H_4/Aerozine-50, like UDMH/N_2O_4, is also hypergolic and can be stored for long periods of time. It can also be used as monopropellant with a suitable catalyst. A bipropellant combination of cryogenic LOX and hydrogen is the preferred choice for upper stages.

Different families of solid propellants have been used over the years for a variety of applications. These include *black powder (BP) propellants* composed of charcoal as fuel, potassium nitrate as oxidizer and sulfur as additive, *zinc–sulfur (ZS) propellants* composed of zinc metal as fuel and powdered sulfur as oxidizer, *candy propellants* using sugar (usually dextrose, sorbitol or sucrose) as fuel and potassium nitrate as oxidizer, *double-base (DB) propellants* composed of two monopropellant fuel components, typically nitroglycerine dissolved in nitrocellulose with additives, *composite propellants* that are either ammonium nitrate based (ANCP) or ammonium perchlorate based (APCP) and comprising a mixture of a powdered metal fuel and a powdered oxidizer with an added rubbery binder also acting as a fuel, *high energy composite (HEC) propellants* composed of a standard composite propellant mixture such as APCP with added high energy explosive, *composite modified double-base propellants* composed of nitrocellulose/nitroglycerine DB propellant as a binder with added solids generally used in composite propellants and *minimum signature* or *smokeless propellants*.

Hybrid propellants comprise a mixture of a solid propellant, usually the fuel, and a liquid propellant, usually the oxidizer. In a rocket engine using hybrid propellant, the liquid part is injected into the solid part and the storage chamber of the solid propellant acts as the combustion chamber. One common hybrid propellant comprises pressurized liquefied nitrous oxide as the oxidizer and cellulose as the fuel. Compared to solid and liquid rocket engines, these offer significant advantage in terms of simplicity, safety and operational cost.

Launch vehicles are also characterized by the *number of stages* used to build the vehicle. Most present-day expendable launch vehicles have two or three stages. In the past, launch vehicles having up to five stages have also been built. In a multi-stage launch vehicle the first stage is the heaviest, with the largest rocket engines generating the highest thrust to be able to lift the total weight of the vehicle along with payload against the force of the Earth's gravity. When the propellants of the first stage are used up, it becomes detached from the rest of the launch vehicle and falls back to Earth. Each successive stage is lighter than the immediately preceding stage. Each of the upper stages, having completed its intended mission, either falls back towards Earth, usually disintegrating on the way due to atmospheric heating, or enters an orbit itself.

3.3.5 Major Launch Vehicles

The evolution of launch vehicles was briefly discussed in section 1.4 of Chapter 1. Of all the launch vehicles outlined in section 1.4, the prominent ones are discussed in more detail in the present section. These include the Ariane series from Europe, the Atlas, Delta and Titan series from the United States, the Long March series from China, GSLV and PSLV from India, and the Proton, Soyuz and Energia series from Russia in the category of expendable launch vehicles. In addition, in the category of reusable launch vehicles, the Space Shuttle from the United States and Buran from Russia are also covered.

3.3.5.1 Ariane Series

The Ariane series of launch vehicles was developed by the ESA. Beginning with Ariane-1, the first in the series, Ariane launch vehicles have seen five generations through to Ariane-5. Ariane-1 was the basic launch vehicle derived from missile technology. The three subsequent versions of Ariane-2, Ariane-3, and Ariane-4 were enhancements of the basic vehicle in terms of engine technology, enabling larger payload capacity and multiple satellite launch capability. Ariane-3 and Ariane-4 were equipped with booster stages. For Ariane-1 to Ariane-4, a total of 144 launches were carried out between 1979 and 2003, of which 137 were successful. Ariane-5 is a completely new design. The single core cryogenic stage that replaces the first two stages of its predecessor vehicles serves as one of the launcher's key propulsion systems. Two solid rocket boosters, each capable of providing a thrust of 6.2 MN (equivalent of 630 tonnes-force), are attached to the sides. Ariane-5 is capable of delivering up to 10 tonnes of payload into geostationary transfer orbit and up to 20 tonnes of payload into low Earth orbit. The launcher has been operational since 1996 and had 68 launches up to 2013, of which 64 were successful. Figure 3.29 illustrates the Ariane series of launch vehicles. Major performance parameters are summarized in Table 3.1.

Table 3.1 Performance parameters of the Ariane series of launch vehicles

Launch vehicle	Ariane-1	Ariane-2	Ariane-3	Ariane-4	Ariane-5
Launch mass (tons)	210	219	234	243–280	750–780
Height (m)	47.4	49	49	59	47–57
Payload capacity to GTO (tons)	1.75	2.20	2.60	2.0–4.8	6.9–10
Payload capacity to LEO (tons)	1.40	–	–	5.0–7.6	16–21
Thrust (MN)	2.5	2.7	2.7	2.7–5.4	12–13
Number of stages	4	3	3/4	2	2
Launches	11	6	11	116	68*
Successes	9	5	10	113	64
Design life	1981–1986	1986–1989	1984–1989	1988–2003	1996 to present

*Number of launches to March 2013.
GTO, geosynchronous transfer orbit; LEO, low Earth orbit.

3.3.5.2 Proton Series

Proton is a series of Russian launch vehicles manufactured by the Khrunichev Space Centre. It is one of the most successful launch vehicles of the Soviet Union and now Russia, and carries

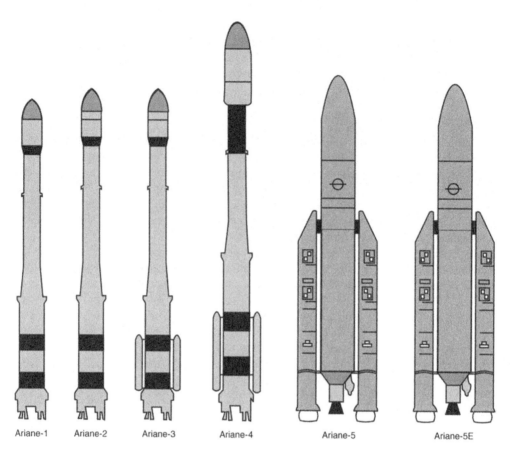

Figure 3.29 The Ariane series of launch vehicles, Ariane-1 to Ariane-5

out both government and commercial launches. It has been successfully used in the past to launch satellites in the low Earth and geostationary orbits, spacecraft to Mars and Venus, and elements of Salyut, Mir and the ISS. The Proton rocket was originally conceived as a super intercontinental ballistic missile (ICBM) capable of delivering a 100 megaton or larger nuclear warhead over distances of 13 000 km but was considered highly oversized to be deployed in that role. Subsequently, it became the largest expendable launch vehicle until the first launch of Russian Energia in 1987 and Titan-IV of the United States in 1989. Proton launch vehicles are made in two-, three- and four-stage variants with the first three stages fuelled by a UDMH and nitrous oxide combination. The fourth stage used a combination of LOX and LH_2 in earlier versions, but this was replaced by a nitrous tetroxide and UDMH combination in later versions. All Proton launches are carried out from the Baikonur Cosmodrome. The Proton launch vehicle can launch a payload of up to 22 tonnes into low Earth orbit, up to 6.7 tonnes into geosynchronous transfer orbit and up to 3.5 tonnes into geosynchronous orbit. Figure 3.30 shows a photograph of the Proton launch vehicle.

Figure 3.30 Proton launch vehicle (Courtesy: NASA)

There are two major categories of launch vehicles in the Proton series: Proton-K, originally designated as UR-500, and Proton-M. Proton-K was a three- or four-stage launch vehicle. The baseline version of the launch vehicle had three stages. A large number of launches used a fourth booster stage to deliver payloads into higher orbits. These included the Block D-1, D-2, DM, DM-2, DM-3, DM-4, DM-5 and DM-2M upper stages. The first launch of Proton-K took place in 1965 and over the years it has been successfully used to launch a variety of payloads, including all of the Salyut space stations and modules of the Mir space station and the ISS. Proton-K remained in service for the period 1965–2012, during which time it had 311 launches. The Proton-M launch vehicle is a modified version of the baseline Proton with reduced structural mass, increased thrust and a closed loop guidance system on the first stage that ensures complete utilization of propellant, thereby enhancing its performance and also minimizing residual toxic chemicals. The Briz-M, also called Breeze-M, upper stage has been used with Proton-M launch vehicles to inject larger payloads into low, medium and high altitude geosynchronous orbits. Block DM-2 and DM-3 upper stages have also been used with Proton-M launch vehicles. Up to May 2013, there had been 68 Proton-M launches.

3.3.5.3 Soyuz Series

The Soyuz series of launch vehicles is derived from the R-7 ICBM first successfully tested in August 1957. The R-7 ICBM had the capability to deliver much larger and heavier nuclear

Figure 3.31 Soyuz-U launch vehicle (Courtesy: NASA)

warheads compared to the ICBMs of US origin at that time, which gave the erstwhile Soviet Union a significant advantage in terms of payload capacity if R-7 design were to be used in a space launch vehicle. In fact, on 4 October 1957, the R-7 ICBM was used to launch the first Soviet satellite, Sputnik-1. Subsequently, a number of R-7 variants was produced, each having an upper stage designed to best suit the intended mission. These variants include Vostok, Voshdok, Molniya and Soyuz. The Soyuz family of launch vehicles also has several variants, including Soyuz, Soyuz-L, Soyuz-M, Soyuz-U (Figure 3.31), Soyuz-U2, Soyuz-FG and Soyuz-2. Soyuz had its first launch in 1966 and is still in service today. It is the most widely used launch vehicle. All of the Soyuz launch vehicles, with the exception of Soyuz-U2, use RP-1 and liquid oxygen propellant. Soyuz-U2 uses an RP-1 variant called Syntin with LOX. Over the years the Soyuz family of launch vehicles has been used to launch both manned missions, such as those to the moon and other planets, and unmanned missions, such as the Progress spacecraft for transfer of supplies and crew changeover to the ISS. Major specifications of the Soyuz series of launch vehicles are summarized in Table 3.2.

Table 3.2 Major specifications of the Soyuz series of launch vehicles

Launch vehicle	Launch mass (tons)	Height (m)	Thrust (MN)	Payload capacity to LEO (tons)	Payload capacity to GTO (tons)	Number of stages	Launches	Successes
Soyuz	308	45.6	1.972*	6.45	–	2	30	28
Soyuz-L	300	50	1.972*	5.50	–	2	3	3
Soyuz-M	300	50	–	6.6	–	2	8	8
Soyuz-U	313	51.1	–	6.9	–	2	745	724
Soyuz-U2	297.8	34.54	–	7.05	–	2	66–92	66–90
Soyuz-FG	305	49.5	1.828*	7.1	–	2 or 3	43	43
Soyuz-2	305	46.1	2.021	7.8	3.25	2 or 3	23**	21

*Near surface thrust.
**Number of launches to April 2013.
GTO, geosynchronous transfer orbit; LEO, low Earth orbit.

3.3.5.4 Energia

Energia was conceived as an HLLV in 1976 with its primary mission being the launch of the reusable Soviet shuttle Buran, which unlike its US counterpart did not have its own propulsion system. The Energia launch vehicle, named after its developer, NPO Energia, produces a maximum thrust of 29 MN and is capable of placing up to 200 tonnes of payload in low Earth orbit. The flexible design of Energia allows for two, four, six or eight booster rockets and the giant core booster can also be a fly back booster. The original configuration employed four strap-on booster stages around a central core stage. Each of the booster stages was powered by a liquid fuelled (kerosene/LOX) RD170 engine. The central core stage was powered by four LH_2/LOX fuelled RD0120 engines. Energia was first test launched in May 1987 with the Polyus spacecraft as its payload. Its second flight in November 1988 successfully launched the Buran shuttle. Production of Energia launch vehicles has reportedly been discontinued since the fall of the Soviet Union and the end of the Soviet shuttle Buran project.

Three main variants of Energia were conceived: the smallest Energia-M, Energia-2 and Vulcan-Hercules. Energia-M was derived from original Energia designed to launch Buran and used only two instead of four booster rockets in the original configuration. The central core also used only one RD0120 engine. Energia-M was intended to launch relatively smaller payloads and could also be used to launch the new Russian mini-shuttle MAKS, which could sit atop the core booster as illustrated in Figure 3.32. It was intended to replace the Russian Proton rocket but lost the competition in favour of the Angara rocket, which was developed by the Russian Space Agency.

Energia-2 (Figure 3.33) is an enhanced version of Energia with a fly back capable core booster thus making it a reusable launch vehicle with the ability to take off and land like a space shuttle. Vulcan-Hercules was the largest of all the variants. It employed eight booster rockets and a central core based on Energia-M. With a launch mass of 4500 tonnes, it was designed to deliver up to 200 tonnes of payload into low Earth orbit. With its gigantic lifting power, it was considered capable of launching lunar and Mars missions in a single shot without

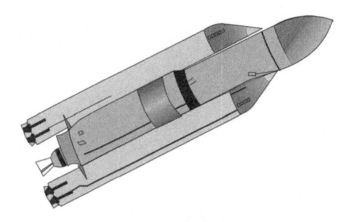

Figure 3.32 Energia-M launch vehicle with Russian mini-shuttle MAKS as payload

Figure 3.33 Energia-2

the requirement of any sub-orbital assembly. However, it was never flown and the project has been reportedly abandoned.

3.3.5.5 Delta Series

The Delta series of expendable launch vehicles has its origin in the Thor Intermediate Range Ballistic Missiles (IRBMs) and Atlas and Titan Inter Continental Ballistic Missiles (ICBMs) of the 1950s. Delta launch vehicles have undergone various stages of evolution over the decades. In early days different launch vehicles took letter designations, namely Delta-A, Delta-B, Delta-C, Delta-D, Delta-E, Delta-F, Delta-G, Delta-J, Delta-K, Delta-L, Delta-M and Delta-N. In order to better represent different technological advances and enhancements the Delta launch vehicles were undergoing a new number based designation system was introduced. The four digits specified the first stage/boosters (digit 1), the number of boosters (digit 2), the second stage (digit 3) and the third stage (digit 4). A suffixed letter is used for heavy configuration. Delta-0100, Delta-1000, Delta-2000, Delta-3000, Delta-4000, Delta-5000, Delta-6000, Delta-7000, Delta-8000 and Delta-9000 are the different series of Delta launch vehicles in this new number designation. Delta-6000 and Delta-7000 series and the two variants of Delta-7000 series called 'Light' and 'Heavy' are collectively known as the Delta-II series. Delta-8000 and Delta-9000 along with its 'Heavy' variant are collectively known as the Delta-IV series. The Delta-II and Delta-IV series of launch vehicles are currently in use and are briefly described in the following paragraphs.

The Delta-II series was originally designed and manufactured by McDonnell Douglas and, after its merger with Boeing, by Boeing Integrated Defence Systems. Since December 2006, production of Delta launch vehicles has been the responsibility of United Launch Alliance. The Delta-II series vehicles use a four digit designation system. The first digit indicates the series, digit 6 for 6000-series and 7 for 7000-series. The second digit indicates the number of boosters (the launch vehicle employs up to nine boosters). In the case of nine boosters, six are fired at the time of lift-off and the remaining three are fired approximately one minute into flight. In the case of three or four boosters, all are fired at lift-off. The third digit is 2, indicating a second stage with Aerojet10 engine. The fourth digit specifies third stage with 0 indicating no third stage, 5 indicating a payload assist stage with Star 48B solid motor and 6 indicating a Star 37FM solid motor. The first stage uses a Rocketdyne RS-27 engine fueled by RP-1/LOX. Solid rocket boosters are used to provide the extra thrust during the initial two minutes of the flight. A spacer called inter-stage separates the first and second stages. The second stage is powered by a re-startable Aerojet AJ10-118K hypergolic engine. This stage also houses the inertial platform and the guidance system. Stage three provides most of the velocity change needed to inject the spacecraft on a trajectory beyond Earth's orbit. It has no active guidance control and is imparted proper orientation by the second stage before the second and third stages separate. A thin metal or composite payload fairing is used to protect the spacecraft during its flight through the Earth's atmosphere.

Delta-IV launch vehicles are made in five different variants, namely Medium, Medium+ (4, 2), Medium+ (5, 2), Medium+ (5, 4) and Heavy (Figure 3.34), to specifically suit different payload sizes and weights. The first stage of a Delta-IV launch vehicle comprises one Medium class and three Heavy class common booster cores (CBC) powered by an LH_2/LOX fuelled Rocketdyne RS-68 engine. The second stage of Delta-IV is a cryogenic LH_2/LOX fuelled Pratt & Whitney RL-10B2 engine. The inter-stage used to connect first and second stages in 4 M variants is a tapering inter-stage narrowing in diameter from 5 m to 4 m. In the case of 5 M variants, a cylindrical inter-stage is used. Second stage, as is the case for the Delta-II launch vehicle, also contains the guidance, control and communications systems. A composite payload fairing is used to protect the payload. The major specifications of Delta-II and Delta-IV series launch vehicles are summarized in Table 3.3.

3.3.5.6 Atlas Series

The Atlas series of launch vehicles also has its origin in the first successful launch of the ICBM in 1957 and was initially designed to carry nuclear warheads. This ICBM used three liquid fuelled engines burning RP-1 and LOX in the core and two outboard boosters. The boosters were jettisoned during ascent. A year later, in 1958, it successfully launched the first communication satellite, named SCORE (Signal Communication by Orbiting Relay Equipment). The satellite was used to broadcast US President Eisenhower's pre-recorded Christmas address around the world. Subsequently, a large number of Atlas variants were produced to launch

Table 3.3 Major specifications of Delta-II and Delta-IV series launch vehicles

Launch vehicle	Height (m)	Launch mass (tons)	Stages	Payload capacity to LEO (tons)	Payload capacity to GTO (tons)	No. of boosters	Total launches	Successes
Delta-II 6000-series	38.2–39	151.7–231.87	2 or 3	2.7–6.1	0.9–2.17	9	17	17
Delta-II 7000-series	38.2–39	151.7–231.87	2 or 3	2.7–6.1	0.9–2.17	3, 4 or 9	128	126
Delta-II 7000H-series	38.2–39	151.7–231.87	2 or 3	2.7–6.1	0.9–2.17	9	6	6
Delta-III 8000-series	35	301.45	2	8.29	3.81	9	3	0
Delta-IV Medium	63–72	249.5–733.4	2	8.6–22.56	3.9–12.98	0, 2, 4	16	6
Delta-IV Heavy	63–72	249.5–733.4	2	8.6–22.56	3.9–12.98	2	6	5

GTO, geosynchronous transfer orbit; LEO, low Earth orbit.

Figure 3.34 Delta-IV Heavy launch vehicle

many scientific and exploratory missions for better understanding of space and the Earth, including the launch of the Friendship-7 Mercury spacecraft by the Atlas launch vehicle in 1962 that carried John Glenn, the first American to orbit the planet Earth, to join another iconic figure, Yuri Gagarin of the erstwhile Soviet Union.

As outlined above, evolution of the Atlas series of launch vehicles began in 1957 with the successful launch of the Atlas ICBM, the SM-65 Atlas missile. These missiles were taken off service in 1965 and many of them were converted into launch vehicles. These converted launch vehicles were used to launch early Block-I GPS satellites, signal communication satellites, and manned missions such as Project Mercury. Subsequently, beginning in 1960 and 1963 respectively, hypergolic propellant fuelled Agena and liquid hydrogen fuelled Centaur upper stages were used extensively on Atlas launch vehicles. Atlas launch vehicles with Agena upper stages were used to launch Sigint and Mariner-2 and also for the space rendezvous practice missions of Gemini. Atlas rockets with Centaur upper stages were used for launching lunar lander spacecraft under the Surveyor programme and Mars bound spacecraft under the Mariner programme. During the period 1958–1990, a number of Atlas variants were developed. These included Atlas-Vega, Atlas-Able, Atlas LV-3A, Atlas LV-3B, Atlas-SLV-3, Atlas SLV-3A, Atlas LV-3C, Atlas LV-3D, Atlas-D, Atlas-E, Atlas-F, Atlas-G and Atlas-H. This was followed by development of Atlas-I, Atlas-II, Atlas-IIA, Atlas-IIAS, Atlas-IIIA, Atlas-IIIB, Atlas-V 400, Atlas-V 500 and Atlas-V HLV. Atlas-II, Atlas-III and Atlas-V are the more recent versions of Atlas launch vehicles. There is no Atlas-IV launch vehicle. Atlas-V Heavy is comparable to Delta-IV in terms of lifting capability. Major performance parameters of the Atlas-II, Atlas-III and Atlas-V series of launch vehicles are summarized in Table 3.4. Figure 3.35 shows the Atlas-II, Atlas-III and Atlas-V series of launch vehicles.

3.3.5.7 Titan Series

The Titan series of expendable launch vehicles also has its origin in the ICBM development programme. Titan series was successfully used from 1960 to 1985 to launch a number of intelligence gathering satellites and inter-planetary missions to Mars, Jupiter, Saturn, Uranus

Table 3.4 Major performance parameters of the Atlas series rockets/launch vehicles

Launch vehicle	Height (m)	Launch mass (tons)	Stages	Payload capacity to LEO (tons)	Payload capacity to GTO (tons)	No. of boosters	Total launches	Successes
Atlas-II	47.54	204.3	3.5	6.58	2.81	1	10	10
Atlas-IIA	47.54	204.3	3.5	6.58	2.81	1	23	23
Atlas-IIAS	47.54	204.3	3.5	6.58	2.81	5	30	30
Atlas-IIIA	52.8	214.338	2	8.64	4.055	–	2	2
Atlas-IIIB	52.8	214.338	2	10.218	4.5	–	4	4
Atlas-V 400	58.3	334.5	2	9.37–15.718	4.75–7.7	1 to 4	26	25
Atlas-V 500	58.3	334.5	2	8.123–20.52	3.775–8.9	1 to 6	12	12
Atlas-V Heavy		941.45	2	29.4	13.605	2		

GTO, geosynchronous transfer orbit; LEO, low Earth orbit.

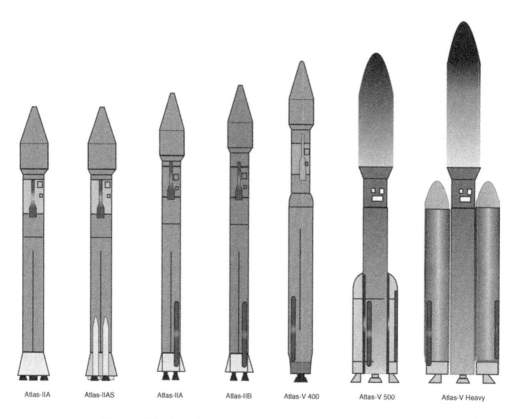

Figure 3.35 Atlas-II, Atlas-III and Atlas-V series launch vehicles

and Neptune. The Titan series of launch vehicles were phased out in 2005 in favour of extending the Atlas series and joint ventures with Delta-IV medium and heavy lift vehicles. The primary reasons for discontinuation of the Titan series (Titan-II, Titan-III and Titan-IV) were high cost and toxicity issues of hydrazine and nitrogen tetroxide as fuel as compared to LH_2 or RP-1-LOX oxidizer-fuel combination.

Different variants of Titan family of launch vehicles include Titan-I, Titan-II, Titan-III and Titan-IV. Titan-I, the first in the family of Titan launch vehicles, like Atlas was the outcome of the ICBM project. It comprised a two-stage rocket powered by an LR-87 liquid fuelled engine with RP-1 and LOX combination as the fuel. The rocket used radio command guidance instead of the inertial guidance originally planned for the vehicle. It remained operational from 1962 to 1965. Titan-II had a modified LR-87 liquid fuelled engine that used a hypergolic combination of Aerozine-50 (50/50 mixture of hydrazine and UDMH) and nitrous tetroxide. The rocket used inertial guidance. It remained in operation from 1963 to 1987. Some of the deactivated Titan-II rockets were converted into space launch vehicles, which remained in operation until 2003. Titan-III is a modified variant of Titan-II with optional solid rocket boosters. It was mainly developed as a HLLV to be used to launch intelligence gathering, military communications, and observation and reconnaissance satellites. Titan-III has different variants in Titan-IIIA, Titan-IIIB, Titan-IIIC, Titan-IIID, Titan-IIIE and Titan-34D. Titan-IV is the last in the family of Titan rockets. It used a two-stage liquid fuelled core with a hypergolic combination of Aerozine-50 and nitrous tetroxide and two solid fuel rocket boosters. It was also intended to be used for carrying military and intelligence gathering payloads. It had two variants in Titan-IVA and Titan-IVB, and remained in operation until 2005. Depending on mission requirement, Titan-IV can be launched in several configurations: no upper stage, inertial upper stage (IUS) and Centaur upper stage. Figure 3.36 shows the Titan-IVB rocket. The major performance parameters of the Titan series of rockets/launch vehicles are summarized in Table 3.5.

Figure 3.36 Titan-IVB rocket (Courtesy: NASA)

3.3.5.8 Long March

The Long March series of launch vehicles, called Chang Zheng in Chinese and designed by the China Academy of Launch Vehicle Technology, is the primary family of expendable launch vehicles from the People's Republic of China. The Long March series of launch vehicles has evolved through various phases of design and development. Beginning with Long March-1, there have been a number of different rocket families: Long March-2, Long March-3, Long March-4, Long March-5, Long March-6, Long March-7, Long March-9 and Long March-11. Long March-1 is derived from the Chinese two-stage intermediate range ballistic missile Dong Feng-2 (DF-2), Long March-2, Long March-3 and Long March-4 are the derivatives of the Chinese two-stage inter-continental ballistic missile Dong Feng-5 (DF-5). The Long March-5 family of launch vehicles adopts a different design and is considered to be the new generation of launch vehicles in the Long March series. Long March-6 and Long March-7 are variants of Long March-5. Long March-9 is planned to be a super HLLV with a payload capacity up to 100 tonnes to low Earth orbit and Long March-11 is intended to be a solid fuel rocket

Table 3.5 Major performance parameters of the Titan series of rockets/launch vehicles

Launch vehicle	Height (m)	Launch mass (tons)	Stages	Payload capacity to LEO (tons)	Payload capacity to GTO (tons)	No. of boosters	Total launches	Successes
Titan-I	31	105.14	2	1.8	–	–	70*	53
Titan-II	31.394	154.00	2	3.6	–	–	106**	101
Titan-IIIA	42	161.73	3	3.1	–	–	4	3
Titan-IIIB	45	156.54	3	3	–	–	68	62
Titan-IIIC	42	626.19	2/3	13.1	3	2	36	31
Titan-IIID	36	612.99	2	12.3	–	2	22	22
Titan-IIIE	48	632.97	3/4	15.4	–	2	7	6
Titan-34D	50	723.49	4	14.515	5	2	15	12
Titan-IVA	44	943.05	3 to 5	21.68	–	2	22	17
Titan-IVB	44	943.05	3 to 5	21.68	–	2	20	15

*Used as a missile.
**Used as both missile and launch vehicle.
GTO, geosynchronous transfer orbit; LEO, low Earth orbit.

engine driven launch vehicle with the ability to rapidly launch satellites during emergencies or disasters.

The propellants used or planned to be used in different rocket families in the Long March series of launch vehicles are as follows. Long March-1 uses nitric acid and UDMH in the first two stages and a solid rocket engine in upper stages. Long March-2, Long March-3 and Long March-4 use UDMH as the fuel and dinitrogen tetroxide as the oxidizer in the main stages and associated booster rockets. Long March-2 is a two-stage rocket while Long March-3 and Long March-4 are three-stage rockets. The upper stages of Long March-3 use engines driven by LH_2 as fuel and LOX as oxidizer. Long March-4 was initially designed to be the back-up option for Long March-3 for launching communication satellites but after the successful maiden launch of Long March-3, it was subsequently employed for launching sun-synchronous satellites. Long March-5 and its derivatives Long March-6 and Long March-7 are designed to use an LOX/kerosene combination in the core stage and the boosters, and an LOX/LH_2 combination in the upper stages. Figure 3.37 shows different variants of the Long March series of launch vehicles.

One or more variants of Long March-2, Long March-3 and Long March-4 are currently in service. Long March-5 is the next generation HLLV, with payload capacity of up to 25 tonnes to low Earth orbit and up to 14 tonnes to geostationary transfer orbit. Different variants of this rocket family, namely CZ-5-200, CZ-5-320, CZ-5-504, CZ-5-522 and CZ-5-540, intended for a variety of mission requirements, are currently under development. These have capabilities broadly matching those of the US Delta-IV and Atlas-V launch vehicles. These will replace the in-service Long March-2, Long March-3 and Long March-4 series launch vehicles. Figure 3.37 shows images of different variants of Long March-2, Long March-3, Long March-4 and Long March-5 launch vehicles. The major performance parameters of the Long March series of rockets/launch vehicles are summarized in Table 3.6.

Launch Vehicles 121

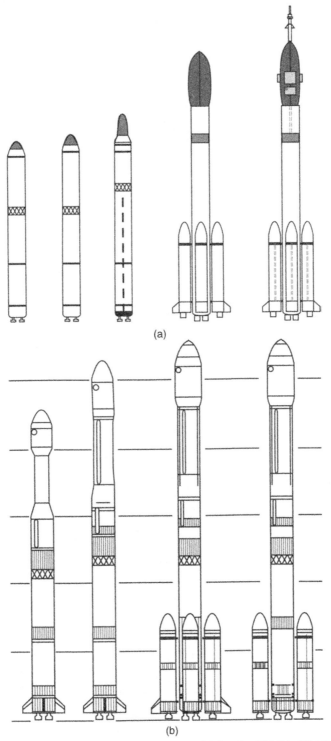

Figure 3.37 Long March series of launch vehicles: (a) CZ-2 series (CZ-2A, CZ-2C, CZ-2D, CZ-2E, CZ-2F); (b) CZ-3 series (CZ-3, CZ-3A, CZ-3B, CZ-3C); (c) CZ-4 series (CZ-4A, CZ-4B, CZ-4C); (d) CZ-5 series (CZ-5-200, CZ-5-300, CZ-5-340, CZ-5-500 core, CZ-5-522)

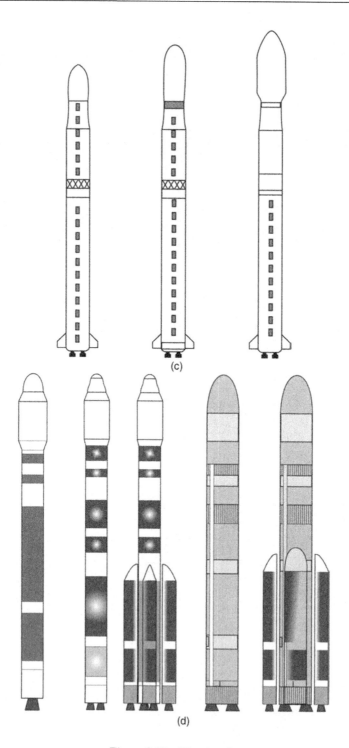

Figure 3.37 (*Continued*)

Table 3.6 Major performance parameters of the Long March series of rockets/launch vehicles

Launch vehicle	Height (m)	Launch mass (tons)	Number of stages	Payload capacity to LEO (tons)	Payload capacity to GTO (tons)	Number of boosters	Total launches	Successes
Long March-1	29.86	81.6	3	0.3	–	–	2	2
Long March-1D	28.22	81.1	3	0.93	–	–	1	1
Long March-2A	31.17	190	2	1.8	–	–	1	0
Long March-2C	35.15	192	2	2.4	–	–	39	38
Long March-2D	33.67	232	2	3.1	–	–	19	19
Long March-2E	49.68	462	2	9.5	3.5	4	7	5
Long March-2F	58.34	480	2	8.4	3.37	4	10	10
Long March-3	43.8	202	3	5	1.5	–	13	10
Long March-3A	52.52	241	3	8.5	2.6	–	23	23
Long March-3B	54.84		3	12	5.1	4	10	8
Long March-3C	55.64	345	3	–	3.8	2	10	10
Long March-4A	41.9	249	3	4	1.5	–	2	2
Long March-4B	44.1	254	3	4.2	2.2	–	18	18
Long March-4C			3	4.2	2.8	–	10	10
Long March-5	–	–	3	25	14	–	–	–
Long March-6	–	–	3	0.5 (SSO)	–	–	–	–
Long March-7	57	–	2	–	–	–	–	–
Long March-9	–	–	–	–	–	–	–	–
Long March-11	–	–	3	–	–	–	–	–

GTO, geosynchronous transfer orbit; LEO, low Earth orbit.

3.3.5.9 Polar Satellite Launch Vehicle

The polar satellite launch vehicle (PSLV) is a medium launch vehicle that is the first operational expendable launch vehicle developed and operated by ISRO. It was developed by ISRO to initially launch Indian remote sensing satellites (IRS) into sun-synchronous orbits. PSLV-standard and its variants PSLV-CA (PSLV-Core Alone) and PSLV-XL (PSLV-Extended) have proved their multi-payload and multi-mission capability over the years by successfully launching payloads into sun-synchronous polar orbits and geosynchronous transfer orbits.

The salient features and capabilities of the PSLV include a height of 44.4 m with a lift-off weight of 295 tonnes. It is a four-stage launch vehicle that alternately uses solid and liquid propulsion systems. The first stage is a solid fuel propelled stage whose thrust is augmented by six strap-on boosters. The second stage employs a liquid propelled engine using a UDMH/N_2O_4 fuel/oxidizer combination. The third stage is propelled by a hydroxyl terminated polybutadiene (HTPB) based solid propellant while the fourth and terminal stage uses a liquid fuel engine with mono-methyl hydrazine (MMH) as fuel and mixed oxides of nitrogen (MON) as oxidizer. The PSLV launch record includes 24 total launches, 11 of PSLV-standard, 9 of PSLV-CA and 4 of PSLV-Extended, to date out of which 22 have been successful. Recent PSLV launches include flight C-25 (PSLV-XL) in October 2013 and flight C-24 (PSLV-CA) in December 2013. The PSLV series of launch vehicles have a payload capacity of up to 3.25 tonnes to low Earth orbit, 1.6 tonnes to heliocentric orbit and 1.41 tonnes to geosynchronous transfer orbit. While PSLV and its variants have been extensively used to launch a variety of satellites of both Indian and foreign origin, one of the notable payloads has been Chandrayaan-I aboard PSLV-XL (flight C-11) on 22 October 2008.

3.3.5.10 Geosynchronous satellite launch vehicle

The geosynchronous satellite launch vehicle (GSLV) is, like the PSLV, an expendable medium lift launch vehicle developed and operated by ISRO. It was developed by ISRO to have indigenous capability to launch the INSAT-series of satellites into geostationary orbit. GSLV has so far had eight launches (six of GSLV Mk-I and two of GSLV Mk-II), out of which three have been successful, four have been failures and one was a partial failure. GSLV Mk-III, a HLLV, is considered to be a technological successor to GSLV Mk-I and Mk-II. GSLV Mk-III is currently under development. GSLV Mk-III, once completed and flight tested, is intended to make India self reliant in launching heavier communication satellites of INSAT-4 class weighing 4.5–5 tonnes. The launch vehicle is designed for multi-mission launch capability for geosynchronous transfer orbit, low Earth orbit, and polar and intermediate circular orbits. Figure 3.38 shows a comparison of PSLV, GSLV and GSLV-III.

The salient features and capabilities of GSLV Mk-I and Mk-II include a height of 49 m with a lift-off weight of 414 tonnes. They are three-stage launch vehicles using solid, liquid and cryogenic stages. The first stage uses HTPB based propellant. The four strap-on boosters surrounding the first stage and the second stage use UDMH/N_2O_4 propelled liquid engines. The third stage is a cryogenic one using LOX as fuel and LH_2 as oxidizer. GSLV Mk-I and Mk-II series of launch vehicles have a payload capacity of up to 5 tonnes to low Earth orbit and 2–2.5 tonnes to geosynchronous transfer orbit.

GSLV Mk-III is 42.4 m tall and has a lift-off weight of 630 tonnes. It is a three-stage launch vehicle. The first stage is a solid propellant driven engine surrounded by two strap-on solid boosters. The second stage is a UDMH/N_2O_4 propelled liquid stage comprising two liquid engines. The third stage is a cryogenic stage fuelled by an LH_2/LOX combination. GSLV Mk-III has a payload capacity of up to 10 tonnes to low Earth orbit and up to 4–5 tonnes to geosynchronous transfer orbit. The vehicle is currently under development and its maiden flight is scheduled for 2014.

3.3.5.11 Space Shuttle

The Space Shuttle is classified as a reusable or recoverable launch vehicle. In contrast to the expendable launch vehicles discussed in the preceding sections, in the case of a reusable launch vehicle most of the expensive hardware comprising the spacecraft is recovered and reused for subsequent missions. One of the main objectives of the development of such a reusable launcher was to bring back to Earth safely and economically crews of astronauts, scientific equipment and the expensive components of the launch vehicles. It may be mentioned here that the economic advantages obtained through the use of reusable launchers are partially offset by the increased size of the launcher, which substantially increases its manufacturing and operational costs. As an example, the ratio of payload mass put into the orbit by the US

Launch Vehicles 125

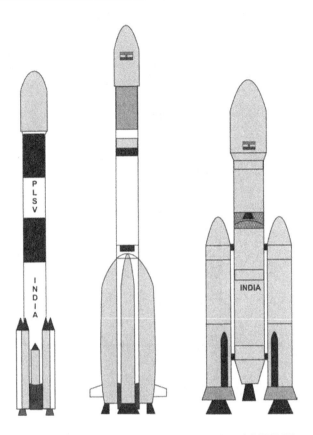

Figure 3.38 Comparison of PSLV, GSLV and GSLV-III

Space Shuttle to its total mass at lift-off is about 1.35% as compared to a figure of 2.8% for the Ariane launcher. The Space Shuttle is a reusable, partially reusable to be more precise, low Earth orbital spacecraft manufactured by United Space Alliance, Thiokol/Alliant Techsystems, Lockheed Martin and Boeing/Rockwell, and operated by NASA. The Space Shuttle was the world's first reusable spacecraft and was conceived during the years of the Apollo lunar program. Major objectives included providing service to space stations and reducing the cost of space travel. The test flights of the Space Shuttle began in 1981 leading to operational flights from 1982 onwards. A total of 135 missions were carried out between April 1981 and July 2011, out of which there were 134 successful launches and one launch failure. There were 133 successful re-entries. Of the two failed re-entries, one is attributed to the sole launch failure and the other to a re-entry failure. The failures included the one launch failure of Challenger and one re-entry failure of Columbia.

The Space Shuttle comprises three main components, including the *orbiter vehicle*, an expendable *external tank* and the two recoverable *solid rocket boosters* (Figure 3.39). The orbiter vehicle is an aeroplane-like crew and cargo carrying craft. The large external tank carries LOX oxidizer and LH_2 fuel. The pair of solid rocket boosters are packed with powdered aluminium and rubber fuel providing 6 million pounds of thrust at lift-off. The solid rocket

Figure 3.39 Components of the Space Shuttle (Courtesy: NASA)

boosters are jettisoned to parachute into the ocean before the vehicle reaches orbit for reuse. The external tank is also jettisoned soon after the solid rocket boosters. The external tank burns up during re-entry into the atmosphere. The Space Shuttle is launched vertically, like the conventional launchers discussed in the preceding sections. At the end of the mission, the orbital manoeuvring system of the orbiter vehicle is fired, forcing the orbiter to drop out of orbit and re-enter the atmosphere. Subsequent to this, the orbiter is glided to a runway landing.

The shuttle is 56.1 m tall and has a lift-off mass of 2030 tonnes. Its payload capacity is as follows: 24.4 tonnes to low Earth orbit, 3.81 tonnes to geosynchronous transfer orbit and 12.7 tonnes to polar orbit. Payload capacity to landing is 14.4 tonnes. As outlined earlier, it employs a pair of solid rocket boosters. The first stage is powered by three main engines fuelled by LOX/LH_2 and the second stage is powered by two orbit manoeuvring engines fuelled by MMH/N_2O_4. A total of six variants of orbiter vehicle were built. These included Enterprise, Columbia, Challenger, Discovery, Atlantis and Endeavour. Enterprise did not have capability to fly into orbit and was built exclusively for carrying out approach and landing tests. The remaining five were operational orbiters. Endeavour was developed to replace Challenger after the later was lost in a mission accident in 1986. Columbia, too was lost in a mission failure in 2003. Some of the notable payloads include tracking and data relay satellites, Spacelab, Hubble telescope, Galileo, Magellan, ISS components and the Mir docking module.

The Space Shuttle began with first orbital test flight (STS-1) of Columbia on 12 April 1981 ended on 8 July 2011 with the last Atlantis flight, which was also the last space shuttle flight (STS-135).

3.3.5.12 Buran

Buran is the Soviet space shuttle, which is similar to the American Space Shuttle except for the fact that the former does not have its own launch system. Buran depends on the Energia rocket launcher discussed in section 3.3.5.4 for propulsion. The Buran shuttle is 36.4 m long, with a lift-off mass of 105 tonnes and a maximum touchdown mass of 82 tonnes. It is capable of carrying up to 30 tonnes of payload at launch. A maximum of 20 tonnes is returned to Earth. The Buran shuttle contains two rocket engines each of 8.8 tonnes used to put the spacecraft into orbit. In addition, it has 38 380 kg attitude control systems and eight 20 kg precision thrusters for carrying out orbital manoeuvres.

One of the salient features of Buran shuttle is its auto pilot system. The shuttle can fly unmanned in the vicinity of a space station for as many as seven days in the exploitation phase and subsequently up to 30 days. In addition, it can be piloted by a crew from a pressurized cabin. The first and the last orbital launch of the Buran shuttle occurred on 15 November 1988 from Baikonur Cosmodrome. It was an unmanned flight that orbited Earth twice and spent three hours in space. The flight made history by being the first ever test flight of its kind when a spacecraft of this size and complexity was launched, completed orbit manoeuvres and re-entry into atmosphere, and subsequently landed safely, all in a fully automatic mode. Although the second flight of Buran was planned to take place in 1993 and was projected to stay in orbit for 15–20 days this never happened because the Buran programme was cancelled after the dissolution of the Soviet Union. It has been reported that the Buran programme may be reviewed with the hope of restarting the test launches of a manned spacecraft design similar to that of Buran as early as 2015.

3.4 Space Centres

A space centre is a centre used for launching space vehicles such as satellites, space probes, space stations and so on and subsequently monitoring their operation throughout the mission. The following sections present an overview of a space centre with particular reference to considerations for selection of the right location for the centre, major constituents of a space centre and a brief on features, facilities and operations of major international space centres.

3.4.1 Location Considerations

The choice of space centre location is mainly governed by safety considerations, accessibility, geological stability, infrastructure and economic factors. Launch safety is a primary consideration. Preferred locations are (a) sand deserts or ice plains, (b) coastal areas and (c) islands. The choice is driven by the fact that the desired space centre needs necessarily to be away from populated areas to fully mitigate the serious consequences of an accident during launch. The merits and demerits of the above mentioned types of locations are as follows. Deserts and ice plains encounter extreme temperatures and therefore suffer from uncomfortable working conditions, but finding recoverable parts is easy in such locations. The Jiuquan satellite launch centre in China is an example of a desert location 1600 km from Beijing. Both islands and coastal areas have the advantage of vast open areas, which is a primary requirement from the viewpoint of safety, but these locations have the associated problems of high humidity levels, tropical insects and violent cyclonic conditions. A large number of space launch centres are located on sea coasts. These include the Guiana Space Centre in Kourou, French Guiana and the Satish Dhawan Space Centre in Sriharikota, India. The Tanegashima Space Centre in Japan and the John F. Kennedy Space Centre in Cape Canaveral are located on islands.

Good accessibility via air and sea routes is another important consideration. This is because of the need to have sufficient human resource and logistic infrastructure at the space centre location. The lack of such resources in an isolated remote area would involve huge investments to be made to establish them. Geological stability in terms of making the space launch centre immune to seismic shocks due to earthquakes is another important choice criterion.

Economic considerations that ultimately decide the likely profitability must also be taken into account. Launches from a space centre close to the equator, such as the one at Kourou in French Guiana (5°, 14′ North), are helped by the Earth's rotation, which allows a given rocket to launch a relatively heavier satellite as compared to the payload capacity of the same rocket if the launch were carried out from another launch centre such as the John F. Kennedy Space Centre in Cape Canaveral (28°, 30′ North) or Xichang satellite launch centre in China (27°, 58′), which are located farther away from equator. In other words, a launch centre close to equator when used to launch a given satellite payload with a given rocket will allow the satellite to carry much more fuel, thereby extending its operational life.

3.4.2 *Constituent Parts of a Space Centre*

A space centre essentially requires the services of a technical centre, a launch complex and a control centre. The *technical centre* is where different parts of the launch vehicle and its payloads are assembled, tested and then prepared for launch. It is equipped with the required instrumentation and sophisticated computer systems for assembly and evaluation tasks. The launch vehicle is then configured for launch at the *launch complex* followed by initiation of countdown operations, including filling of fuel tanks and their pressurization. This is then followed up by a synchronized sequence of checks before the rocket engines are finally ignited. There are two broad categories of launch pads: fixed and mobile.

For a fixed launch pad, the launch vehicle is integrated, checked and launched from the same spot. In this case the infrastructure required for assembly integration and testing of the launch vehicle has to move to and from the launch pad at different times thereby allowing accessibility to the vehicle at different heights. The infrastructure is in the form of mobile towers with platforms equipped with the required instrumentation. The advantage of such a launch pad is its simplicity of design but the disadvantage is that only a limited number of launches, usually five or six, can be carried out in a year.

On the other hand, a mobile launch pad allows simultaneous assembly and integration of one launch vehicle and final launch preparation of another. This almost doubles the number of launches executable in a year. After the space vehicle has been successfully launched from the launch site, the control centres monitor the launch vehicle and subsequently the payload throughout the mission life. This is done with the help of a network of tracking, telemetry and command (TT&C) stations. Tracking, telemetry and command stations are discussed in section 4.7 of Chapter 4.

3.4.3 Major Space Centres

This section gives an overview of major space launch centres located in different parts of the world covering all major countries and the space agencies responsible for their operation and maintenance. The space centres covered in the following section include the following.

1. John F. Kennedy Space Centre at Cape Canaveral, United States
2. Baikonur Cosmodrome, Kazakhstan
3. Guiana Space Centre at Kourou, French Guiana
4. Yuri Gagarin Cosmonaut Training Centre (GCTC), Russia
5. Xichang Satellite Launch Centre, China
6. Jiuquan Satellite Launch Centre, China
7. Uchinoura Space Centre, Japan
8. Tanegashima Space Centre, Japan
9. German Aerospace Centre, Germany
10. Satish Dhawan Space Centre, Sriharikota (SHAR), India

3.4.3.1 John F. Kennedy Space Centre

The John F. Kennedy Space Centre, located on CityMerritt Island, is north–northwest of Cape Canaveral on Florida's coastline. It is the largest space launch centre and is approximately 55 km long and 10 km wide, spread over 570 km^2. The launch site's origin dates back to 1958 and was authorized by the administration of US president Dwight D. Eisenhower and known by the name of Launch Operations Directorate at that time. Expansion plans for the centre had taken shape from the early 1960s, with the then US President John F. Kennedy's declaration in 1961 of the US goal of landing on the moon before 1970. The site was renamed the Launch Operations Centre on 1 July 1962 and subsequently rechristened the John F. Kennedy Space Centre on 23 November 1963 following the death of John F. Kennedy. Since 1968, the centre has been used for each and every manned space flight carried out by the United States.

The John F. Kennedy Space Centre is equipped with comprehensive assembly, integration, test and evaluation, launch and tracking facilities. These facilities are classified under the two broad headings of *launch complex facilities* and *industrial area facilities*. Key facilities at the launch complex include the following:

1. *Shuttle landing facility*: This facility includes a 150 × 168 m parking apron and a 3.2 km long tow-way connecting it to the orbiter processing facility. The facility is equipped with various navigation and landing aids.

2. *Orbiter processing facility*: In this facility the orbiter is subjected to safing procedures, which are followed by removal of previous mission payloads, inspection, testing and refurbishment of the orbiter for the next mission.
3. *Vehicle assembly building*: The vehicle assembly building is 218 m (L) × 158 m (W) × 160 m (H) and covers an area of 3.25 hectares. Different bays in the building are used for integration and stacking of launch vehicle, external tank check out and storage, payload canister operations, contingency handling, and storage of solid rocket boosters and orbiters. Figure 3.40 shows a view of vehicle assembly building and the surrounding area at the John F. Kennedy Space Centre.

Figure 3.40 Vehicle assembly building and surrounding area at the John F. Kennedy Space Centre (Courtesy: NASA)

4. *Launch control centre*: The launch control centre is 115.2 m (L) × 55.1 m (W) × 23.5 m (H). It houses telemetry, tracking, instrumentation, data reduction and evaluation equipment, computers of the central data subsystem that perform most checkout and launch functions, and firing rooms with checkout, control and monitor subsystems.
5. *Crawler transporter*: Crawler transporters are used to carry the assembled launch vehicles from the vehicle assembly building to the launch pad. The centre is equipped with two such crawler transporters. Figure 3.41 shows an image of a crawler transporter carrying the assembled space shuttle to the launch pad.
6. *Mobile launch platform*: There are three mobile launch platforms. Each of them is a two-storey steel structure measuring 49 m (L) × 41 m (W) × 7.6 m (H) with an empty weight of 4190 tonnes.
7. *Operations support building*: This building houses a technical documentation centre, library and photo analysis area.
8. *Processing control centre building*: This facility is used for shuttle orbiter testing and launch team training. It is also used for launch processing system maintenance.
9. *Logistics facility*: This is used for storage of space shuttle hardware parts and is equipped with automated handling equipment for retrieval of different parts.
10. *Launch complex 39A and 39B*: Launch complex 39 consists of two pads, Pad A and Pad B, that are 2647 m apart. The two pads are more or less identical. The launch pads had been originally designed for Apollo programme and were subsequently modified to support Space Shuttle launch operations. Major facilities at the pads include the fixed support

Figure 3.41 Crawler transporter carrying the Space Shuttle to the launch pad (Courtesy: NASA)

structure, the rotating service structure, the hypergolic umbilical system, the gaseous oxygen vent arm, the orbiter access arm and the mobile launch platform supported by six permanent and four extensible pedestals.

Key industrial area facilities include the following:

1. *Operations and checkout building*: This is used for assembly and integration of horizontally integrated payloads.
2. *Central instrumentation facility*: This facility is equipped with comprehensive instrumentation to receive, monitor, process, display and record information received during test, launch, flight and landing phases of the space vehicle. It also houses calibration laboratories.
3. *Vertical processing facility*: This facility is used for assembly and integration of vertically processed payloads.
4. *Space station processing facility*: This facility was built for processing of flight hardware for the ISS. The technical facilities included two processing bays, operational control rooms, an airlock and laboratories.
5. *Hypergolic maintenance and checkout facility*: This facility is used for processing and storage of hypergolic-fueled modules.
6. *Spacecraft assembly and encapsulation facility*: This facility is primarily used for spacecraft assembly, test and encapsulation.
7. *Payload hazardous servicing facility*: This is used as part of the payload processing facility by supporting integration of solid rocket boosters with the payloads and also ordnance and hazardous fuel servicing.
8. *Launch equipment test facility*: This facility is used for testing launch critical ground support systems and equipment.
9. *Parachute refurbishment facility*: This facility is used for refurbishment of recovered parachutes. After refurbishment, they are stored in this facility.
10. *Merritt Island spaceflight tracking and data network stations*: The antennas and the associated communication equipment at the facility are used to provide a communication link between the space vehicle and its control centre.

Major projects/programmes accomplished from the John F. Kennedy Space Centre over the last five decades (1960–2010) are as follows:

1. The Apollo mission of accomplishing a manned lunar landing was the first major programme executed from the John F. Kennedy Space Centre. It may be mentioned here that the lunar landing mission was accomplished in three stages. The first stage was the Mercury mission, which had the objective of investigating issues related to human performance and recovery of crew and spacecraft through the launch of a manned spacecraft in near Earth orbit. The mission was executed from the US Air Force Base at Cape Canaveral in the early 1960s.

    Launching the Gemini spacecraft was the second step. Gemini spacecraft were launched with the objective of carrying out rendezvous, docking and extra-vehicular activity missions. A total of 12 Gemini flights (Gemini-1 to Gemini-12) were carried out, again from Cape Canaveral's base. The first two flights were unmanned. The 10 manned flights took place between March 1965 and November 1966.

    The next important milestone was the Apollo mission of lunar landing. The mission was executed with a three-stage Saturn-V rocket. There were three unmanned flights beginning with Apollo-4 in November 1967, which was also the first launch from the John F. Kennedy Space Centre. Apollo-8's lunar orbiting mission aboard Saturn-V was the first manned mission and was successfully carried out in December 1968. Apollo-9 and Apollo-10 flights were used to test the lunar module in Earth orbit and lunar orbit, respectively. The historic moon landing took place on 20 July 1969 through the launch of Apollo-11 from Launch Pad 39A on 16 July 1969. Apollo-12 repeated the feat four months later. Apollo-13 to Apollo-17 missions were launched during the period 1970–1972. The final Apollo spacecraft was launched from Launch Pad 39B in 1975 for Apollo–Soyuz test project.
2. Saturn-V's last launch was used to put the Skylab space station into orbit. The launch took place from Launch Pad 39A in 1975. This was followed by three manned missions to Skylab aboard Saturn-IBs from modified Launch Pad 39B during the same year.
3. The Space Shuttle programme dominated launch operations from the John F. Kennedy Space Centre during the two decade period of 1980–2000. It began with launch of Columbia on 12 April 1981. A total of 135 shuttle flights were carried out between April 1981 and July 2011. There were two failures: first a launch failure of Challenger (the 25th shuttle flight) on 28 January 1986 and second a re-entry failure of Columbia (the 113th shuttle flight) on 1 February 2003. Space Shuttle missions were successfully used to launch satellites and interplanetary probes, construction and servicing of the ISS, for visits to the Russian Mir Space Station, deployment of the Hubble telescope and so on.

3.4.3.2 Baikonur Cosmodrome

Baikonur Cosmodrome, also known as Tyuratam, was built by the Soviet Union in 1955 in the desert steppes of Kazakhstan. It is one of the two major space launch complexes of the Russian federation, and is located approximately 2100 km south-east of Kazakhstan and spread over an elliptical area measuring 90 × 85 km. It is currently leased to Russia till 2050 by the Kazakh government. Baikonur Cosmodrome offers launch facilities to a large number of commercial, military and scientific missions every year. It is also the launch base for all Russian manned space missions.

Baikonur Cosmodrome has all the infrastructural facilities required for launching manned and unmanned spacecraft. Over the last five decades and more, different generations of Russian spacecraft, which prominently include Soyuz, Proton, Zenit, Tsyklon and Buran, have been launched and supported by the cosmodrome. Figure 3.42 shows an image of a Soyuz rocket being erected into position at the launch pad of Baikonur Cosmodrome. The launch complex was even used to provide supply services to the ISS through Soyuz and Progress spacecraft during a temporary disruption in the US space shuttle programme following the Columbia shuttle disaster in 2003. The launch complex has 16 launch pads, of which five are either inactive or destroyed. These are designated Pad 1/5, Pad 31/6, Pad 41/15, Pad 45/1, Pad 45/2 (destroyed in 1990), Pad 81/23 or 81L, Pad 81/24 or 81P, Pad 90/19 or 90L (inactive since 1989), Pad 90/20 or 90R, Pad 109/95, Pad 110/37 or 110L, Pad 110/38 or 110R (inactive since 1969), Pad 175/59, Pad 200/39 or 200L, Pad 200/40 or 200R (inactive since 1991) and Pad 250 (inactive since 1987).

Figure 3.42 Proton lift-off from site 200 of Baikonur Cosmodrome (Courtesy: NASA)

Baikonur Cosmodrome has been the centre of activity for many historic missions, including the launch of the first operational ICBM R-7 on 27 February 1961, the launch of Sputnik-1, the first man-made satellite, on 4 October 1957, the first manned orbital flight by Yuri Gagarin aboard Vostok-1 on 12 April 1961 and the first spacecraft (Luna-1 and Luna-2) that travelled close to moon on 2 January 1959. Valentina Tereshkova, first woman to go to space, in 1963, also took off from Baikonur Cosmodrome.

3.4.3.3 Guiana Space Centre

The Guiana Space Centre also known as the Centre Spatial Guyanais (CSG) is the space launch complex for launching a variety of French space agency Centre National d'Études Spatiales (CNES), European Space Agency (ESA) and Arianespace space missions. Many non-European companies carry out their launches from here. The Guiana Space Centre is situated in Kourou, French Guiana, north-east of South America. The space centre was founded in 1968. Subsequently, the French government offered to share the facilities at the centre with

ESA after it came into existence in 1975. Since then, ESA has provided two-thirds of the funds required for its operations in addition to funds provided for the creation of new facilities.

Located close to the equator (latitude 5°, 3′) and offering a launch angle of 102°, it is geographically placed to allow execution of all possible space missions. The fact that the site is also sparsely populated, with 90% of the country covered with equatorial forests, and that there is no risk of earthquakes or cyclones, means Kourou scores very highly with regard to safety considerations. An eastward boost provided by the spinning motion of the Earth at Kourou due to its closer proximity to the equator compares to the 406 m/s provided by Cape Canaveral or the John F. Kennedy Space Centre. Another advantage of being closer to equator is that manoeuvring satellites for geosynchronous orbits is simpler.

The Guiana Space Centre is equipped with several launch sites, including ELA or ELA-1, ELA-2, ELA-3 and ELS. ELA-1 (l'ensemble de lancement Ariane-1) was built after the demolition of an earlier launch pad named ELV (l'ensemble de lancement Vega), which was designed to launch the Europa-II rocket. ELA-1 was used for launches of Ariane-1, Ariane-2 and Ariane-3 launch vehicles until 1989, when it was retired from service. Subsequently, it was refurbished for the Vega (Vettore Europeo Generazione Avanzata) rocket.

The first launch from ELA-1 of the Europa-II rocket occurred on 5 November 1971. The first Ariane-1 launch occurred on 24 December 1979. The first launch of the Vega rocket from the refurbished ELA-1 pad occurred on 13 February 2012. The launch pad is currently active.

The ELA-2 launch pad is currently inactive. It was used for two Ariane-3 launches, V17 in 1986 and V25 in 1988, second Ariane-2 launch in 1987 and all the 116 launches of Ariane-4 between 1988 and 2003. In September 2011, the launch pad's mobile tower was demolished. Figure 3.43 shows an overview of launch sites ELA-1 and ELA-2. ELA-3 was built for Ariane-5 launches. The first Ariane-5 launch was carried out on 4 June 1996. A total of 56 launches have been carried out up to 16 February 2011. The launch pad is currently active.

Figure 3.43 Overview of ELA-1 and ELA-2 at the Guiana Space Centre (Courtesy: CNES)

ELS (l'ensemble de lancement Soyuz) was built for launching Russian built Soyuz-2 rockets. It is located 27 km from Kourou harbour. In contrast to the way Soyuz is traditionally fully assembled in a horizontal position and then transported to the launch pad, where it is made vertical for launch, at ELS the rocket alone is assembled in a horizontal position and it is then transported to the launch pad. The spacecraft is transported separately to the launch pad and then attached to the rocket. The launch pad is equipped with a closed gantry, which can be moved away from the launch pad during the launch. Soyuz launches from the ELS launch complex began with launch of two Galileo IOV-1 and IOV-2 satellites using the Soyuz-ST rocket on 21 October 2011. This was followed by launches of the Pleiades-1A earth imaging satellite, four ELISA electronic intelligence satellites and one SSOT (Sistema Satelital para Observacion de la Tierra) remote sensing satellite on 17 December 2011 and the Pleiades-1B satellite on 1 December 2012.

Another important facility at the Guiana Space Centre is the *launcher integration building*. This facility is used for the mating of different stages of the launch vehicle. The 58 m tall building is linked to the launch table via an umbilical mast. The launch table, along with assembled launch vehicle, is then transported to the *final assembly building* via a dual track rail line.

The assembled rocket is delivered to the 90 m tall final assembly building for payload integration. It is 2.8 km from launch pad ELA-3. After payload integration the launch vehicle is transported to the launch pad by the mobile launch table. The Guiana Space Centre is at present one of the most modern and best-equipped launch complexes in the world. A large number of countries around the world, including those in Europe, North and South America, as well as Japan and India, use the base for launching their satellites.

3.4.3.4 Yuri Gagarin Cosmonaut Training Centre

The Yuri Gagarin Cosmonaut Training Centre (GCTC) is located in Star city on the outskirts of Moscow. The centre is used to train and prepare cosmonauts for space missions. It was founded in January 1960 and subsequently named after Yuri Gagarin in 1969. It took its present form in 1995 after the merger of the Cosmonaut Training Centre and the Air Force Test and Training Regiment. The centre underwent another transformation when the Russian president, through a presidential decree, freed the centre from the joint ownership and operational management of the Russian Ministry of Defence and the Russian Federal Space Agency and transferred it to the Russian Federal Space Agency.

Under the Intercosmos programme, a space programme of the Soviet Union offering access to manned and unmanned space missions to nations friendly to the Soviet Union, the training centre also offered its training facilities to trainee cosmonauts from other countries, including the east European countries of the Warsaw pact and other socialist/communist nations such as Cuba, Mongolia and Vietnam. The training centre later had staff members from part-time NATO countries, such as France, and non-aligned nations, such as India. The infrastructural facilities available at the centre include the following:

1. For all training exercises, models of different space ships and space stations are available in original sizes. GCTC is equipped with mock-ups of major space craft, including Soyuz and Buran, space stations MIR and the ISS, TKS modules and orbital stations of the Salyut programme. These are used to familiarize trainees with the different functions and operations of various instruments they will use as cosmonauts. MIR is a modular space station assembled in orbit during 1986–1996 and owned initially by Soviet Union and then by Russia. Figure 3.44 shows a view of the training module of the MIR space station at GCTC. ISS, like MIR, also has a modular structure. It is currently the largest body

Figure 3.44 Training module of MIR space station (Courtesy: NASA)

in orbit and is used as a microgravity and space environment research laboratory. The Salyut programme was the first space station programme of the former Soviet Union and comprised a series of four manned scientific research space stations and two manned military reconnaissance space stations. The TKS space craft comprising two spacecraft mated together was designed for both manned and unmanned cargo resupply flights.
2. Zero gravity training aircraft to simulate conditions of weightlessness. These include the MiG-15 UTI, Tupolev Tu-104 and IL-76 MDK. Training aircraft are based at the Russian Air Force base at Chkalovskiy airfield.
3. A planetarium capable of projecting 9000 stars.
4. A medical observation and testing facility.

GCTC has been the centre of cosmonaut space training for more than 50 years. The main training tasks include simulation of space flights, g-force training, mission specific training for extra-vehicular activity, medical investigation of trainee cosmonauts and monitoring of them before and after flights, survival training for out-of-control landings and training in the field of celestial navigation.

3.4.3.5 Xichang Satellite Launch Centre

The Xichang Satellite Launch Centre is situated about 60 km north of Xichang city in Sichuan province. The choice of site is significant because it has high altitude, low latitude, canyon topography and a May–October launch window. The centre is under the jurisdiction of the PLA's 27th test and training base, which is a subsidiary of the General Armament Department. It is also known as Base-27. Although work on the construction of the launch centre began in 1970, the site only became operational in 1984. It is mainly used to launch satellite launch vehicles and geostationary and weather satellites. The fact that the centre is almost exclusively used for China's geostationary transfer orbit and geostationary earth orbit launches means it experiences the highest number of space launches in China.

The centre is equipped with multiple launch towers, command centres, telemetry stations and logistics infrastructure. Launch area-1 is presently used as a viewing area. It was initially planned to support China's project 714 and the launch of the Shuguang One spacecraft, but the launch area was never completed because the programme was cancelled. Launch area-2 is used for HLLVs, namely Long March-2E (LM-2E or CZ-2E), Long March-3A (LM-3A or CZ-3A), Long March-3B (LM-3B or CZ-3B) and Long March-3C (LM-3C or CZ-3C). Launch area-3 has launch pads for Long March-2C (LM-2C or CZ-2C), Long March-3 (LM-3 or CZ-3) and Long March-3A (LM-3A or CZ-3A). It was upgraded between 2005 and 2006 to support the lunar exploration programme.

The technical centre is equipped with state-of-the-art infrastructure for integration and testing of launch vehicle and payload. The mission control and command centre is 7 km away from the launch pad and is connected to Xichang Qingshou airport, which is 50 km from the launch site and connected by a dedicated railway and highway.

Some of the important missions executed from the centre include the following:

1. Launch of the first LM-2E rocket on 16 July 1990. The rocket was used to send into orbit Pakistan's first indigenously developed digital communications and experimental satellite Badr-1 and the US communications satellite HS-601.
2. There was a launch failure and subsequent crash of the first launch of the new LM-3B rocket carrying INTELSAT-708 on 15 February 1996. The crash reportedly destroyed 80 homes in a nearby village.
3. The first successful Anti-satellite (ASAT) test was carried out by China from this centre on 11 January 2007. The SC-19 ASAT weapon was used to destroy a Chinese weather satellite (FY-1C of the Fengyun series) at an altitude of 865 km using a kinetic kill vehicle.
4. An unmanned moon orbiter Chang'e-1 of the Chang'e lunar exploration programme was successfully launched on 24 October 2007.
5. The first LM-3C rocket was launched on 25 April 2008. The rocket sent into orbit China's first data relay satellite, called Tianlian-I. Figure 3.45 shows a view of the LM-3C rocket carrying data relay satellite Tianlian-I on the launch pad at Xichang Satellite Launch Centre.

Figure 3.45 Aerial view of the Xichang Satellite Launch Centre

6. The EUTELSAT W3C communications satellite was successfully launched by LM-3B on 7 October 2011.
7. Chinasat-1A was successfully launched by LM-3B on 17 September 2011.
8. China's second lunar orbiter Chang'e-2 was launched on 1 October 2010. The mission was completed on 9 June 2011.
9. Chinasat-12 was successfully launched by LM-3B on 27 November 2012.
10. Chang'e-3, China's first lunar rover, was launched on December 1, 2013 by LM-3B. The landing rover descended on the lunar surface on 14 December 2013.

It is reported that after the Wenchang Space Centre on Hainan Island becomes operational in 2014, the Xichang centre will mainly be used for military launches.

3.4.3.6 Jiuquan Satellite Launch Centre

The Jiuquan Satellite Launch Centre was the first of the three space vehicle launch facilities in the People's Republic of China and is situated in Jiuquan city in Gansu province and spread over 2800 km^2. The other two space centres are the Taiyuan Satellite Launch Centre in Xinzhou in Shanxi province and the Xichang Satellite Launch Centre in Xichang city in Sichuan province. The Jiuquan centre is located in the Gobi desert about 1600 km from Beijing, south of the China–Mongolia border. It was founded in 1958 and was operational from 1960 onwards. It is part of Dong Feng space city, which is also known as the Dong Feng Base or Base-10. This site, in addition to the launch centre, also houses the PLA Air Force test flight facility and a space museum.

The launch centre is equipped with state-of-the art infrastructural facilities and the instrumentation and control hardware required for different stages of a satellite launch campaign. Major facilities include the technical centre, a launch complex comprising several launch areas (Launch area-2, Launch area-3 and Launch area-4) each with multiple launch pads, the launch control centre, the mission control and command centre, a propellant fuelling system, a tracking system, a gas supply system, a weather forecast system and a logistic support system. Launch area-2 and Launch area-3 are currently inactive. The only currently active launch area (Launch area-4) has two launch pads. The launch centre is equipped with a 75 m high umbilical tower with an explosion proof elevator. The technical centre, 1.5 km away from launch centre, includes the vertical processing building, which is world's tallest single floor concrete building.

The centre was initially intended as a range facility for flight testing of ballistic, surface-to-air and air-to-air missiles. Flight testing of surface-to-air and air-to-air missiles was handed over to the PLA Air Force in the 1970s. The centre continued to be used for testing ballistic missiles. It also began to be used for launching a variety of artificial satellites, beginning with the launch of China's first artificial satellite Dong Fang Hong-1 (DFH-1) on 24 April 1970 using Long March-1. Subsequent to this, over the last 40 years, a number of manned and unmanned space missions have been executed from this centre.

Some significant milestone missions include the following:

1. Launch of the first artificial satellite Dong Fang Hong-1 (DFH-1) on 24 April 1970. The satellite carried a radio transmitter that was used to broadcast the song 'The East is Red' for 26 days.

2. The first manned space mission Shenzhou-5 was launched on 15 October 2003 by the Long March-2F launcher. With this launch China became the third country in the world after the Soviet Union and the United States to have independent human space flight capability.
3. The second manned space flight Shenzhou-6 was launched on 12 October 2005, again by the Long March-2F rocket. The two crew members stayed in low Earth orbit for five days.
4. The third manned space flight mission Shenzhou-7 was launched on 25 September 2008 by the Long March-2F rocket. Extra vehicular activity (EVA) was carried out for the first time by China. The Shenzhou-7 mission made China's space programme the third after the Soviet Union and the United States to have conducted EVA.
5. An unmanned space mission Shenzhou-8 was launched on 31 October 2011 by the Long March-2F rocket. This was China's first experiment with unmanned docking. Shenzhou-8 automatically docked with the Tiangong-1 space module on 3 November 2011 and again on 14 November 2011. Such a rendezvous and docking feat had been accomplished only by the Soviet Union, Japan and the ESA before this.
6. The fourth manned space flight Shenzhou-9 was launched on 16 June 2012 and carried as crew China's first woman astronaut. This was China's first manned spacecraft to dock with the Tiangng-1 space module.
7. China's fifth manned space mission Shenzhou-10 was launched on 11 June 2013. It docked with Tiangong-1 on 13 June 2013 and returned safely to Earth on 26 June 2013 after a series of successful docking tests and science and technology experiments. Figure 3.46 shows Shenzhou-10 blasting off from the Jiuquan Satellite Launch Centre.

Figure 3.46 Shenzhou-10 blasting off from the Jiuquan Satellite Launch Centre (Courtesy: NASA)

3.4.3.7 Uchinoura Space Centre

The Uchinoura Space Centre was established in 1962 as a part of the Institute of Industrial Science, University of Tokyo. It became a part of the Institute of Space and Aeronautical Science (ISAS), University of Tokyo, in 1964. It became an independent research facility, the Kagoshima Space Centre, but still attached to ISAS, in 1981. After the formation of the Japan Aerospace Exploration Agency (JAXA) on 1 October 2003, it was renamed the Uchinoura Space Centre. It is situated in the Japanese town of Kimotsuki on the Pacific coast of Kagoshima prefecture and is shown in Figure 3.47. The centre is used for launching sounding rockets and scientific satellites. As many as 390 rockets have been launched from it since 1962. In addition, 30 scientific satellites and probes have been launched from the centre since the launch of first Japanese satellite Osumi in 1970.

The major technical facilities at the Uchinoura Space Centre include the following:

1. The M rocket launch site is spread over 25 000 m^2 and comprises a launching pad, a rocket assembly building, a satellite integration centre and a launch control centre.
2. The KS centre is used for launching sounding rockets such as SS-520, S-520 and S-310. Japan's first satellite, Osumi, was launched from this site in 1970 using the L-4S-5 rocket.
3. The control centre infrastructure is used to manage the firing sequence, flight path and flight safety of rockets launched from KS centre. It also has a balloon-data acquisition system for launch-angle correction, a marine-monitoring radar system, a meteorological data receiver for weather forecasting and a thunder detecting/forecasting system.
4. The telemetry centre is equipped with 10 m, 20 m and 34 m antennas capable of tracking both Earth orbiting satellites and deep space probes.
5. The radar centre is equipped with a rocket-tracking radar system.

Figure 3.47 Uchinoura Space Centre

Some of the prominent launches from the Uchinoura Space Centre are as follows:

1. Launch of first scientific satellite Osumi on 11 February 1970 using the Lambda-4S-5 rocket.
2. The first orbital launch attempt of the Mu family of rockets (Mu-4S) was conducted on 25 September 1970.
3. The launch of the Tansei-1 scientific satellite on 16 February 1971 using the Mu-4S rocket.
4. The launch of the Tansei-3 scientific satellite on 19 February 1977 using the M-3H rocket.
5. The launch of the HALCA (MUSES-B) satellite on 12 February 1997 using the Mu-5 rocket.
6. The launch of HAYABUSA (MUSES-C) on 9 May 2003 using the Mu-5 rocket.
7. The launch of the SPRINT-A (Spectroscopic Planet Observatory for Recognition of Interaction of Atmosphere) satellite on 27 August 2013 using the Epsilon-1 rocket.

Figure 3.48 Tanegashima Space Centre

3.4.3.8 Tanegashima Space Centre

The Tanegashima Space Centre was founded in 1969 at the time of the formation of the original National Space Development Agency (NASDA) of Japan prior to the establishment of JAXA in 2003. It is located on Tanegashima Island, the easternmost of the Osumi Islands, south of the island of Kyushu. The space centre is extensively used for launching satellites and also for carrying out firing tests of rocket engines. It is the largest space development centre in Japan and is spread over an area of 9 700 000 m^2.

Major technical facilities at the centre include the two launch complexes, named Yoshinobu and Osaki, the test and assembly buildings, the centre for radar and optical tracking of launch vehicles and spacecraft, and the spacecraft and fairing building. The Yoshinobu launch complex, with its two launch pads, is the launch site for large rockets such as H-2A and H-2B. The test and assembly buildings are used for testing, assembly and integration of launch vehicle stages, satellites and other payloads, such as explorers. The tracking and control centre includes the Masuda tracking and communication station in the north and the Uchugaoka radar station and optical observation facility in the west.

One of the most recent and significant launches was the launch of the H-2B rocket carrying the Konotori cargo transporter to the ISS. The world's first talking humanoid robot astronaut, named Kirobo, was packed inside the space ship HTV-4 and launched by H-2B rocket on 4 August 2013.

3.4.3.9 German Aerospace Centre

The German Aerospace Centre was established in 1969 as a national centre for research in aerospace, energy and transportation through the merger of several institutes. It has 29 institutes located in 16 different locations, 13 in Germany and three out of Germany in Brussels, Paris and Washington, DC. The centre has its headquarters in Cologne, Germany. In 1989, it was renamed DLR, an abbreviation of a German phrase meaning German Research Institute for

Aviation and Spaceflight. Another institution named DARA (German Agency for Spaceflight Affairs) was also created around the same time. These two institutions merged on 1 October 1997 and the space centre got its present name DLR, which literally means German Centre for Aviation and Spaceflight. It is known as the German Aerospace Centre in English language publications. Figure 3.49 shows an aerial view of it.

Figure 3.49 German Aerospace Centre (Attribution: DLR, CC-BY 3.0)

The German Aerospace Centre has diverse areas of activity. As part of the German space agency, the centre is responsible for planning and implementing the German space programme. Some of its important space missions and related activities are as follows.

1. Launch of the high resolution stereo camera (HRSC) originally designed for the Russian Mars 96 mission but subsequently redeveloped for ESA's Mars Express mission. HRSC was developed at the Institute of Planetary Studies at the German Aerospace Centre and was launched aboard Mars Express on 2 June 2003 to record high resolution images of the Martian surface and help in studying the geology, mineralogy and atmosphere of Mars.
2. The launch of TerraSAR-X, the German Earth observation satellite, on 15 June 2007 and its twin satellite TanDEM-X on 21 June 2010, with the objective of providing radar remote sensing data to scientific and commercial users.
3. The launch of a number of scientific missions, including AZUR, first scientific mission, in November 1969, the Aeronomy satellite missions AEROS-A and AEROS-B during 1970–1974, the HELIOS-A and HELIOS-B interplanetary space flight missions during 1974–1976, AMPTE (Active Magnetosphere Particle Explorer) during 1984–1986, the

X-ray satellite ROSAT between 1990 and 1999 and, of course, as outlined above, the TerraSAR-X and TanDEM-X missions in 2007 and 2010, respectively.
4. The centre has also carried out research to build a reusable sub-orbital space plane that can lift-off vertically and land like a glider, making fast intercontinental passenger transport possible.
5. The space centre has developed research aircraft for diverse research missions, for example VFW 614 ATTAS, HALO (high altitude and long range research aircraft), ATRA (advanced technology research aircraft), and SOFIA (stratospheric observatory for infrared astronomy), which is carried on a Boeing 747SP aircraft with modified fuselage.
6. The German Aerospace Centre participates in ESA's manned spaceflights and also in space missions with the US Space Shuttle and Russian Mir and Soyuz spacecraft.

3.4.3.10 Satish Dhawan Space Centre

The Satish Dhawan Space Centre (SDSC), previously known as the Sriharikota High Altitude Range (SHAR) became operational in 1971. It is the primary launch facility of ISRO and is used to launch a variety of satellites and other unmanned space missions. It is also used to launch the Rohini series of sounding rockets. Earlier sounding rockets were launched from the Thumba Equatorial Rocket Launch Station (TERLS) located on the west coast of India. SDSC has also been identified as the launch centre for manned space missions, for which the existing infrastructure facilities at the space centre are being augmented and new facilities are being created.

The space centre is located in Sriharikota Island on the eastern coast of Andhra Pradesh, about 80 km north of Chennai. It is spread over an area of 145 km^2, which is largely inhabited. This is an advantage from the viewpoint of safety considerations. Its proximity to the equator (latitude 13°, 43' N) also helps eastward launches. It was renamed the Satish Dhawan Space Centre in September 2002 in memory of Professor Satish Dhawan, the former chairman of the Space Commission. Figure 3.50 shows a view of the space centre.

Figure 3.50 Satish Dhawan Space Centre (Courtesy: ISRO)

Major technical facilities at the centre include the vehicle assembly, static test and evaluation complex (VAST), the solid propellant space booster plant (SPROB), telemetry, tracking and control (TT&C) systems, a liquid propellant storage and servicing facility (LSSF), the PSLV launch complex, equipped with a 76 m tall mobile service tower providing a SP-3 payload clean room, and a closed centre equipped with computers, CCTV, real time tracking system and meteorological observation equipment. The closed centre is linked to the three Radar SYSTEMS at SHAR and also to the five stations of ISRO's telemetry, tracking and command (ISTRAC) network.

The space centre has two operational launch pads. A third launch pad, likely to be ready by 2017, is being prepared. The first launch pad was built in 1990 and is used for both PSLV and GSLV launches. The second launch pad is configured as a universal launch pad. It became operational in 2005 and is capable of handling both existing launch vehicles and futuristic ISRO launch vehicles. The third launch pad is being specifically built for carrying out manned space missions in the future.

Important missions carried out from Satish Dhawan Space Centre are as follows.

1. The centre became operational with the launch of three small sounding rockets of the Rohini-125 series on 9 and 10 October 1971.
2. The first test launch of the satellite launch vehicle 3 (SLV-3) rocket was carried out on 10 August 1979 with partial success. It was followed by a successful launch on 18 July 1980. Out of a total of four launches of SLV-3, two were successful. The last SLV-3 flight took place on 17 April 1983.
3. The first test of the augmented satellite launch vehicle (ASLV) was carried out on 24 March 1987 and met with failure. Of the four ASLV launches during the period 1987–1994, only the fourth and last were successful, launching the SROSS-C2 satellite.
4. The first launch of the PSLV occurred on 20 September 1993. Out of 23 launches so far, 22 have been successful. The most recent launches of PSLV have been those of PSLA-XL on 1 July 2013 and 5 November 2013. PSLV has been repeatedly used to successfully launch Indian remote sensing satellites (IRS) in addition to providing launch services to other nations.
5. The PSLV-XL launch vehicle was used to successfully carry out India's first mission to the moon, Chandrayaan-1. The launch took place on 22 October 2008.
6. The first launch of the GSLV took place on 18 April 2001 and was a failure. Subsequent to this seven more launches of different versions of the GSLV have been carried out. Only three flights have been successful. The most recent flight of the GSLV (GSLV-D5/GSAT-14 mission) took place on 19 August 2013.

3.5 Orbital Perturbations

The satellite, once placed in its orbit, experiences various perturbing torques that cause variations in its orbital parameters with time. These include gravitational forces from other bodies like solar and lunar attraction, magnetic field interaction, solar radiation pressure, asymmetry of Earth's gravitational field and so on. Due to these factors, the satellite orbit tends to drift and its orientation also changes and hence the true orbit of the satellite is different from that defined using Kepler's laws. The satellite's position thus needs to be controlled both in the east–west as well as the north–south directions. The east–west location needs to be maintained to prevent radio frequency (RF) interference from neighbouring satellites. It may be mentioned here that in the case of a geostationary satellite, a 1° drift in the east or west direction is equivalent to

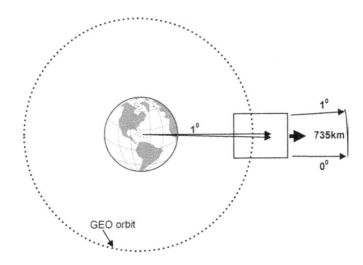

Figure 3.51 Drift of a geostationary satellite

a drift of about 735 km along the orbit (Figure 3.51). The north–south orientation has to be maintained to have proper satellite inclination.

The Earth is not a perfect sphere and is flattened at the poles. The equatorial diameter is about 20–40 km more than the average polar diameter. Also, the equatorial radius of the Earth is not constant. In addition, the average density of Earth is not uniform. All of this results in a non-uniform gravitational field around the Earth which in turn results in variation in gravitational force acting on the satellite due to the Earth. The effect of variation in the gravitational field of the Earth on the satellite is more predominant for geostationary satellites than for satellites orbiting in low Earth orbits as in the case of these satellites the rapid change in the position of the satellite with respect to the Earth's surface will lead to the averaging out of the perturbing forces. In the case of a geostationary satellite, these forces result in an acceleration or deceleration component that varies with the longitudinal location of the satellite.

In addition to the variation in the gravitational field of the Earth, the satellite is also subjected to the gravitational pulls of the sun and the moon. The Earth's orbit around the sun is an ellipse whose plane is inclined at an angle of 7° with respect to the equatorial plane of the sun. The Earth is tilted around 23° away from the normal to the ecliptic. The moon revolves around the Earth with an inclination of around 5° to the equatorial plane of the Earth. Hence, the satellite in orbit is subjected to a variety of out-of-plane forces which change the inclination on the satellite's orbit. The gravitational pulls of Earth, sun and moon have negligible effect of the satellites orbiting in LEO orbits, where the effect of atmospheric drag is more predominant.

As the perturbed orbit is not an ellipse anymore, the satellite does not return to the same point in space after one revolution. The time elapsed between the successive perigee passages is referred to as anomalistic period. The anomalistic period (T_A) is given by equation 3.17.

$$t_A = \frac{2\pi}{\omega_{\text{mod}}} \qquad (3.17)$$

Where,

$$\omega_{\text{mod}} = \omega_0 \left[1 + \frac{K(1 - 1.5 \sin^2 i)}{a^2 (1 - e^2)^{3/2}} \right]$$

ω_0 is the angular velocity for spherical Earth, $K = 66\,063.1704 \text{ km}^2$, a is the semi-major axis, e is the eccentricity and $i = \cos^{-1} W_Z$, W_Z is the Z axis component of the orbit normal.

The attitude and orbit control system maintains the satellite's position and its orientation and keeps the antenna pointed correctly in the desired direction (bore-sighted to the centre of the coverage area of the satellite). The orbit control is performed by firing thrusters in the desired direction or by releasing jets of gas. It is also referred to as station keeping. Thrusters and gas jets are used to correct the longitudinal drifts (in-plane changes) and the inclination changes (out-of-plane changes). It may be mentioned that the manoeuvres required for correcting longitudinal drifts (referred to as the north-south manoeuvre) require a much larger velocity increment as compared to the manoeuvres required for correcting the inclination changes (referred to as the east-west manoeuvre). Hence, generally a different set of thrusters or gas jets is used for north-south and east-west manoeuvres.

3.6 Satellite Stabilization

Commonly employed techniques for satellite attitude control include:

1. Spin stabilization
2. Three-axis or body stabilization

3.6.1 Spin Stabilization

In a spin-stabilized satellite, the satellite body is spun at a rate between 30 and 100 rpm about an axis perpendicular to the orbital plane (Figure 3.52). Like a spinning top, the rotating body offers inertial stiffness, which prevents the satellite from drifting from its desired orientation. Spin-stabilized satellites are generally cylindrical in shape. For stability, the satellite should be spun about its major axis, having a maximum moment of inertia. To maintain stability, the moment of inertia about the desired spin axis should at least be 10 % greater than the moment of inertia about the transverse axis.

There are two types of spinning configurations employed in spin-stabilized satellites. These include the simple spinner configuration and the dual spinner configuration. In the simple spinner configuration, the satellite payload and other subsystems are placed in the spinning section, while the antenna and the feed are placed in the de-spun platform. The de-spun platform is spun in a direction opposite to that of the spinning satellite body. In the dual spinner configuration, the entire payload along with the antenna and the feed is placed on the de-spun platform and the other subsystems are located on the spinning body. Modern spin-stabilized satellites almost invariably employ the dual spinner configuration. It may be mentioned here that mounting of the antennae system on the de-spun platform in both the configurations ensures a constant pointing direction of the antennae. In both configurations, solar cells are mounted on the cylindrical body of the satellite. Intelsat-1 to Intelsat-4, Intelsat-6 and

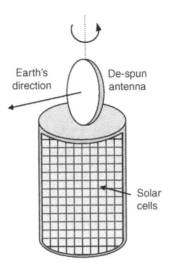

Figure 3.52 Spin stabilized satellite

Figure 3.53 Spin-stabilized satellite (Intelsat-4) (Reproduced by permission of © Intelsat)

TIROS-1 are some of the popular spin stabilized satellites. Figure 3.53 shows the photograph of Intelsat-4 satellite.

3.6.2 Three-axis or Body Stabilization

In the case of three-axis stabilization, also known as body stabilization, the stabilization is achieved by controlling the movement of the satellite along the three axes, that is yaw, pitch and roll, with respect to a reference (Figure 3.54). The system uses reaction wheels or momentum wheels to correct orbit perturbations. The stability of the three-axis system is provided by

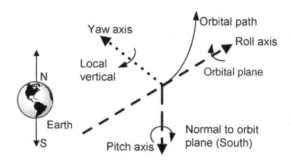

Figure 3.54 Three-axis stabilization

the active control system, which applies small corrective forces on the wheels to correct the undesirable changes in the satellite orbit.

Most three-axis stabilized satellites use momentum wheels. The basic control technique used here is to speed up or slow down the momentum wheel depending upon the direction in which the satellite is perturbed. The satellite rotates in a direction opposite to that of speed change of the wheel. For example, an increase in speed of the wheel in the clockwise direction will make the satellite rotate in a counterclockwise direction. The momentum wheels rotate in one direction and can be twisted by a gimbal motor to provide the required dynamic force on the satellite.

An alternative approach is to use reaction wheels. Three reaction wheels are used, one for each axis. They can be rotated in either direction depending upon the active correction force. The satellite body is generally box shaped for three-axis stabilized satellites. Antennae are mounted on the Earth-facing side and on the lateral sides adjacent to it. These satellites use flat solar panels mounted above and below the satellite body in such a way that they always point towards the sun, which is an obvious requirement.

Some popular satellites belonging to the category of three-axis stabilized satellites include Intelsat-5, Intelsat-7, Intelsat-8, GOES-8, GOES-9, TIROS-N and the INSAT series of satellites. Figure 3.55 is a photograph of the Intelsat-5 satellite.

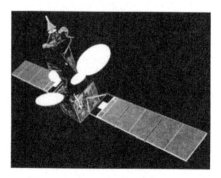

Figure 3.55 Three-axis stabilized satellite (Intelsat-5) (Reproduced by permission of © Intelsat)

3.6.3 Comparison between Spin-stabilized and Three-axis Stabilized Satellites

1. In comparison to spin-stabilized satellites, three-axis stabilized satellites have more power generation capability and more additional mounting area available for complex antennae structures.
2. Spin-stabilized satellites are simpler in design and less expensive than three-axis stabilized satellites.
3. Three-axis stabilized satellites have the disadvantage that the extendible solar array used in these satellites are unable to provide power when the satellite is in the transfer orbit, as the array is still stored inside the satellite during this time.

3.6.4 Station Keeping

Station keeping is the process of maintenance of the satellite's orbit against different factors that cause temporal drift. Satellites need to have their orbits adjusted from time to time because the satellite, even though initially placed in the correct orbit, can undergo a progressive drift due to some natural forces such as minor gravitational perturbations due to the sun and moon, solar radiation pressure, Earth being an imperfect sphere, and so on. The orbital adjustments are usually made by releasing jets of gas or by firing small rockets tied to the body of the satellite.

In the case of spin-stabilized satellites, station keeping in the north–south direction is maintained by firing thrusters parallel to the spin axis in a continuous mode. The east–west station keeping is obtained by firing thrusters perpendicular to the spin axis. In the case of three-axis stabilization, station keeping is achieved by firing thrusters in the east–west or the north–south directions in a continuous mode. The amplitude and direction of thrusts required to carry out a variety of orbital adjustments have already been discussed at length in Section 3.1.

3.7 Orbital Effects on Satellite's Performance

As we know the satellite is revolving constantly around the Earth. The motion of the satellite has significant effects on its performance. These include the Doppler shift, effect due to variation in the orbital distance, effect of solar eclipse and sun's transit outrage.

3.7.1 Doppler Shift

The geostationary satellites appear stationary with respect to an Earth station terminal whereas in the case of satellites orbiting in low Earth orbits, the satellite is in relative motion with respect to the terminal. However, in the case of geostationary satellites also there are some variations between the satellite and the Earth station terminal. As the satellite is moving with respect to the Earth station terminal, the frequency of the satellite transmitter also varies with respect to the receiver on the Earth station terminal. If the frequency transmitted by the satellite is f_T, then the received frequency f_R is given by equation 3.18.

$$\left(\frac{f_R - f_T}{f_T}\right) = \left(\frac{\Delta f}{f_T}\right) = \left(\frac{v_T}{v_P}\right) \tag{3.18}$$

Where,
v_T is the component of the satellite transmitter velocity vector directed towards the Earth station receiver
v_P is the phase velocity of light in free space (3×10^8 m/s)

3.7.2 Variation in the Orbital Distance

Variation in the orbital distance results in variation in the range between the satellite and the Earth station terminal. If a Time Division Multiple Access (TDMA) scheme is employed by the satellite, the timing of the frames within the TDMA bursts should be worked out carefully so that the user terminals receive the correct data at the correct time. Range variations are more predominant in low and medium Earth orbiting satellites as compared to the geostationary satellites.

3.7.3 Solar Eclipse

There are times when the satellites do not receive solar radiation due to obstruction from a celestial body. During these periods the satellites operate using onboard batteries. The design of the battery is such so as to provide continuous power during the period of the eclipse. Ground control stations perform battery conditioning routines prior to the occurrence of an eclipse to ensure best performance during the eclipse. These include discharging the batteries close to their maximum depth of discharge and then fully recharging them just before the eclipse occurs. Also, the rapidity with which the satellite enters and exits the shadow of the celestial body creates sudden temperature stress situations. The satellite is designed in such a manner so as to cope with these thermal stresses.

3.7.4 Sun Transit Outrage

There are times when the satellite passes directly between the sun and the Earth as shown in Figure 3.56. The Earth station antenna will receive signals from the satellite as well as the microwave radiation emitted by the sun (the sun is a source of radiation with an equivalent temperature varying between 6000 K to 11000 K depending upon the time of the 11-year sunspot cycle). This might cause temporary outrage if the magnitude of the solar radiation exceeds the fade margin of the receiver. The traffic of the satellite may be shifted to other satellites during such periods.

3.8 Eclipses

With reference to satellites, an eclipse is said to occur when the sunlight fails to reach the satellite's solar panel due to an obstruction from a celestial body. The major and most frequent source of an eclipse is due to the satellite coming in the shadow of the Earth (Figure 3.57). This is known as a solar eclipse. The eclipse is total; that is the satellite fails to receive any light whatsoever if it passes through the umbra, which is the dark central region of the shadow, and receives very little light if it passes through the penumbra, which is the less

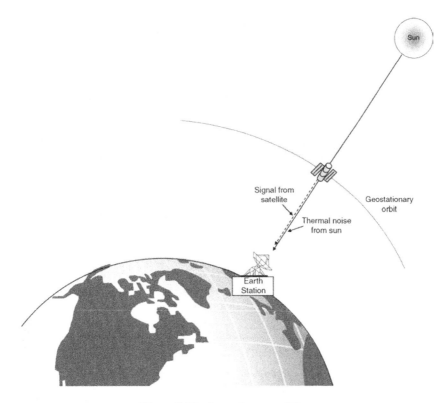

Figure 3.56 Sun outrage conditions

dark region surrounding the umbra (Figure 3.58). The eclipse occurs as the Earth's equatorial plane is inclined at a constant angle of about 23.5° to its ecliptic plane, which is the plane of the Earth's orbit extended to infinity. The eclipse is seen on 42 nights during the spring and an equal number of nights during the autumn by the geostationary satellite. The effect is the worst during the equinoxes and lasts for about 72 minutes. The equinox, as explained earlier, is the point in time when the sun crosses the equator, making the day and night equal in length. The spring and autumn equinoxes respectively occur on 20–21 March and 22–23 September. During the equinoxes in March and September, the satellite, the Earth and the sun are aligned at midnight local time and the satellite spends about 72 minutes in total darkness. From 21 days before and 21 days after the equinoxes, the satellite crosses the umbral cone

Figure 3.57 Solar eclipse

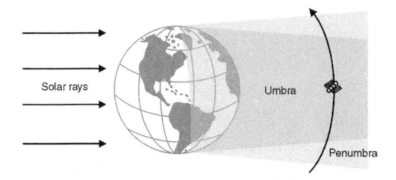

Figure 3.58 Umbra and penumbra

each day for some time, thereby receiving only a part of solar light for that time. During the rest of the year, the geostationary satellite orbit passes either above or below the umbral cone. It is at the maximum distance at the time of the solstices, above the umbral cone at the time of the summer solstice (20–21 June) and below it at the time of the winter solstice (21–22 December). Figure 3.59 further illustrates the phenomenon.

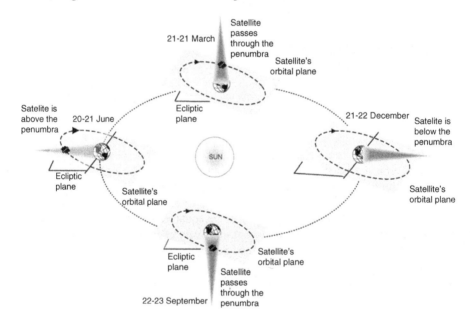

Figure 3.59 Positions of the geostationary satellite during the equinoxes and solstices

Hence, the duration of an eclipse increases from zero to about 72 minutes starting 21 days before the equinox and then decreases from 72 minutes to zero during 21 days following the equinox. The duration of an eclipse on a given day around the equinox can be seen from the graph in Figure 3.60. Another type of eclipse known as the lunar eclipse occurs when the moon's shadow passes across the satellite (Figure 3.61). This is much less common and

Eclipses

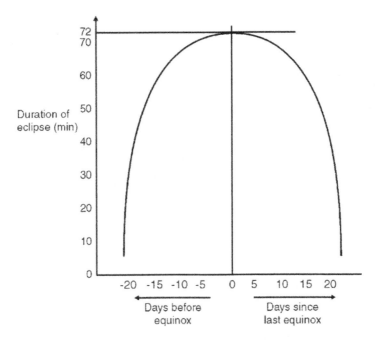

Figure 3.60 Duration of the eclipse before and after the equinox

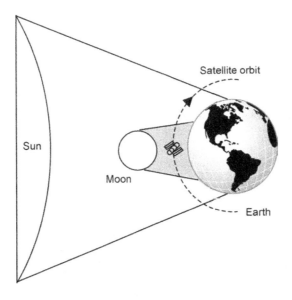

Figure 3.61 Lunar eclipse

occurs once in 29 years. In fact, for all practical purposes, when an eclipse is mentioned with respect to satellites, it is a solar eclipse that is referred to.

While a solar eclipse takes place, the failure of sunlight to reach the satellite interrupts the battery recharging process. The satellite is depleted of its electrical power capacity. It does not significantly affect low power satellites, which can usually continue their operation with back-up power. The high power satellites, however, shut down for all but essential services.

3.9 Look Angles of a Satellite

The look angles of a satellite refer to the coordinates to which an Earth station must be pointed in order to communicate with the satellite and are expressed in terms of azimuth and elevation angles. In the case where an Earth station is within the footprint or coverage area of a geostationary satellite, it can communicate with the satellite by simply pointing its antenna towards it. The process of pointing the Earth station antenna accurately towards the satellite can be accomplished if the azimuth and elevation angles of the Earth station location are known. Also, the elevation angle, as we shall see in the following paragraphs, affects the slant range, that is line of sight distance between the Earth station and the satellite.

In order to determine the look angles of a satellite, its precise location should be known. The location of a satellite is very often determined by the position of the sub-satellite point. The sub-satellite point is the location on the surface of the Earth that lies directly between the satellite and the centre of the Earth. To an observer on the sub-satellite point, the satellite will appear to be directly overhead (Figure 3.62).

3.9.1 Azimuth Angle

The azimuth angle A of an Earth station is defined as the angle produced by the line of intersection of the local horizontal plane and the plane passing through the Earth station, the satellite and the centre of the Earth with the true north (Figure 3.63). We can visualize that this line of intersection between the two above-mentioned planes would be one of the many possible tangents that can be drawn at the point of location of the Earth station. Depending upon the location of the Earth station and the sub-satellite point, the azimuth angle can be computed as follows:

Earth station in the northern hemisphere:

$A = 180° - A'$.   when the Earth station is to the west of the satellite   (3.19)

$A = 180° + A'$.   when the Earth station is to the east of the satellite   (3.20)

Earth station in the southern hemisphere:

$A = A'$ ...   when the Earth station is to the west of the satellite   (3.21)

$A = 360° - A'$ ...   when the Earth station is to the east of the satellite   (3.22)

Look Angles of a Satellite

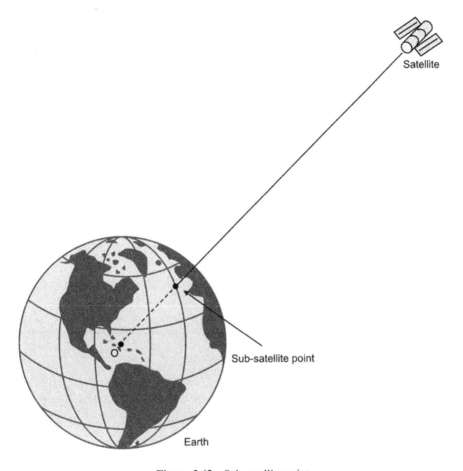

Figure 3.62 Sub-satellite point

where A' can be computed from

$$A' = \tan^{-1}\left(\frac{\tan|\theta_s - \theta_L|}{\sin\theta_l}\right) \tag{3.23}$$

where
θ_s = satellite longitude
θ_L = Earth station longitude
θ_l = Earth station latitude

3.9.2 Elevation Angle

The Earth station elevation angle E is the angle between the line of intersection of the local horizontal plane and the plane passing through the Earth station, the satellite and the centre of

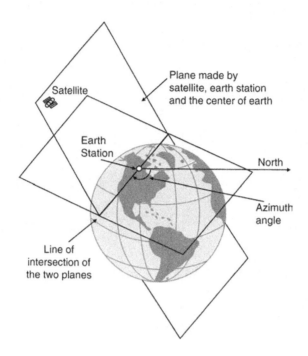

Figure 3.63 Azimuth angle

the Earth with the line joining the Earth station and the satellite. Figures 3.64 (a) and (b) show the elevation angles for two different satellite and Earth station positions. It can be computed from

$$E = \tan^{-1}\left[\frac{r - R\cos\theta_l \cos|\theta_s - \theta_L|}{R\sin\{\cos^{-1}(\cos\theta_l \cos|\theta_s - \theta_L|)\}}\right] - \cos^{-1}(\cos\theta_l \cos|\theta_s - \theta_L|) \quad (3.24)$$

where
r = orbital radius, R = Earth's radius
θ_s = Satellite longitude, θ_L = Earth station longitude, θ_l = Earth station latitude

3.9.3 Computing the Slant Range

Slant range of a satellite is defined as the range or the distance of the satellite from the Earth station. The elevation angle E, as mentioned earlier, has a direct bearing on the slant range. The smaller the elevation angle of the Earth station, the larger is the slant range and the coverage angle. Refer to Figure 3.65.

The slant range can be computed from

$$\text{Slant range } D = \sqrt{R^2 + (R+H)^2 - 2R(R+H)\sin\left[E + \sin^{-1}\left\{\left(\frac{R}{R+H}\right)\cos E\right\}\right]}$$

$$(3.25)$$

Look Angles of a Satellite 157

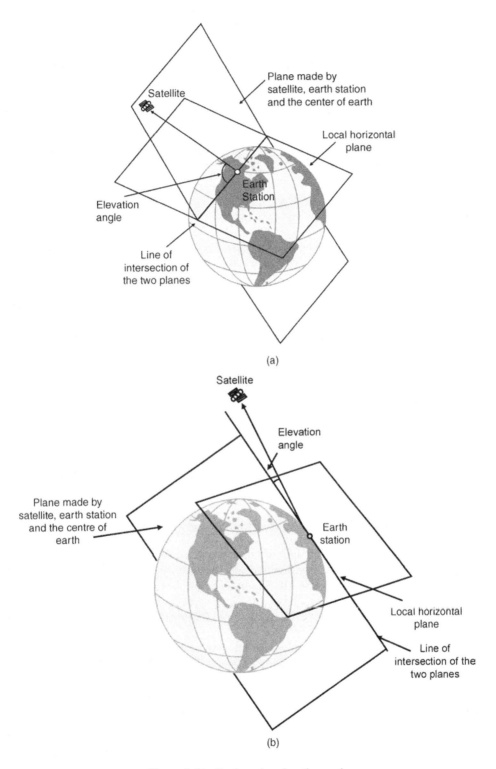

Figure 3.64 Earth station elevation angle

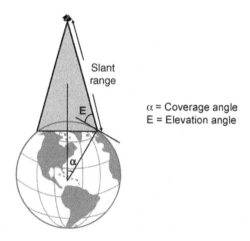

Figure 3.65 Elevation angle, slant range and coverage angle

$$\text{Coverage angle } \alpha = \sin^{-1}\left\{\left(\frac{R}{R+H}\right)\cos E\right\} \quad (3.26)$$

Where

R = radius of the Earth
E = angle of elevation
H = height of the satellite above the surface of the Earth

It is evident from the above expression that a zero angle of elevation leads to the maximum coverage angle. A larger slant range means a longer propagation delay time and a greater impairment of signal quality, as the signal has to travel a greater distance through the Earth's atmosphere.

3.9.4 Computing the Line-of-Sight Distance between Two Satellites

Refer to Figure 3.66. The line-of-sight distance between two satellites placed in the same circular orbit can be computed from triangle ABC formed by the points of location of two satellites and the centre of the Earth. The line-of-sight distance AB in this case is given by

$$\text{AB} = \sqrt{(\text{AC}^2 + \text{BC}^2 - 2\ \text{AC}\ \text{BC}\ \cos\theta)} \quad (3.27)$$

Note also that angle θ will be the angular separation of the longitudes of the two satellites. For example, if the two satellites are located at 30°E and 60°E, θ would be equal to 30°. If the two locations are 30°W and 60°E, then in that case θ would be 90°. The maximum line-of-sight distance between these two satellites occurs when the satellites are placed so that the line joining the two becomes tangent to the Earth's surface, as shown in Figure 3.67.

In this the case, the maximum line-of-sight distance (AB) equals OA + OB, which further equals 2OA or 2OB as OA = OB. If R is the radius of the Earth and H is the height of satellites

Look Angles of a Satellite

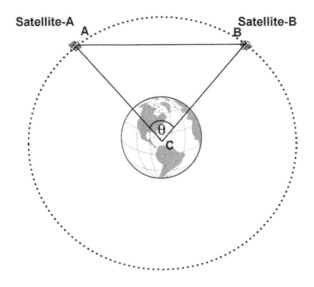

Figure 3.66 Line-of-sight distance between two satellites

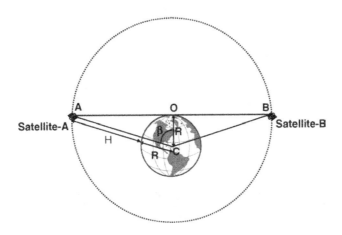

Figure 3.67 Maximum line-of-sight distance between two satellites

above the surface of the Earth, then

$$OA = AC \sin \beta = (R + H) \sin \beta \qquad (3.28)$$

Now

$$\beta = \cos^{-1}\left(\frac{R}{R + H}\right) \qquad (3.29)$$

Therefore

$$OA = (R+H)\sin\left[\cos^{-1}\left(\frac{R}{R+H}\right)\right] \quad (3.30)$$

and

$$\text{Maximum line-of-sight distance} = 2(R+H)\sin\left[\cos^{-1}\left(\frac{R}{R+H}\right)\right] \quad (3.31)$$

Problem 3.7
Determine the maximum possible line-of-sight distance between two geostationary satellites orbiting the Earth at a height of 36 000 km above the surface of the Earth. Assume the radius of the Earth to be 6370 km.

Solution:
Maximum line-of-sight distance can be computed from

$$2(R+H)\sin\left[\cos^{-1}\left\{\frac{R}{(R+H)}\right\}\right]$$

where

R = Radius of Earth
H = Height of satellite above surface of Earth
Therefore, Maximum line-of-sight distance is given by:

$2(6370 + 36\,000)\sin[\cos^{-1}\{6370/(6370+36\,000)\}]$

$= 2(42\,370)\sin[\cos^{-1}\{6370/42\,370\}] = 84\,740 \times 0.989 = 83\,807.86\,\text{km}$

Problem 3.8
A satellite in the Intelsat-VI series is located at 37°W and another belonging to the Intelsat-VII series is located at 74°E (Figure 3.68). If both these satellites are in a circular equatorial geostationary orbit with an orbital radius of 42 164 km, determine the inter-satellite distance.

Solution: The inter-satellite distance can be computed from

$$\sqrt{(D_1^2 + D_2^2 - 2D_1 D_2 \cos\theta)}$$

where

D_1 = orbital radius of the first satellite
D_2 = orbital radius of the second satellite
θ = angle formed by two radii

$D_1 = D_2 = 42\,164\,\text{km}$ and $\theta = 37° + 74° = 111°$

Look Angles of a Satellite

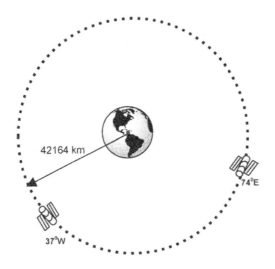

Figure 3.68 Figure for Problem 3.8

Inter-satellite distance $= \sqrt{[(42\,164)^2 + (42\,164)^2 - (2 \times 42\,164 \times 42\,164 \times \cos 111°)]}$

$= \sqrt{[2 \times (42\,164)^2(1 - \cos 111°)]}$

$= 69\,486.27$ km

Problem 3.9

An Earth station is located at 30°W longitude and 60°N latitude. Determine the Earth station's azimuth and elevation angles with respect to a geostationary satellite located at 50°W longitude. The orbital radius is 42 164 km. (Assume the radius of the Earth to be 6378 km.)

Solution: Since the Earth station is in the northern hemisphere and is located towards east of the satellite, the azimuth angle A is given by (180° + A'), where A' can be computed from

$$A' = \tan^{-1}\left(\frac{\tan|\theta_s - \theta_L|}{\sin \theta_l}\right)$$

where

θ_s = satellite longitude = 50°W
θ_L = Earth station longitude = 30°W
θ_l = Earth station latitude = 60°N

Therefore

$$A' = \tan^{-1}\left(\frac{\tan 20°}{\sin 60°}\right) = \tan^{-1}\left(\frac{0.364}{0.866}\right) = \tan^{-1}(0.42) = 22.8°$$

and

$$A = 180° + 22.8° = 202.8°$$

The Earth station elevation angle is given by

$$E = \tan^{-1}\left[\frac{r - R\cos\theta_l \cos|\theta_s - \theta_L|}{R\sin\{\cos^{-1}(\cos\theta_l \cos|\theta_s - \theta_L|)\}}\right] - \cos^{-1}(\cos\theta_l \cos|\theta_s - \theta_L|)$$

where

r = Satellite orbital radius
R = Earth's radius

Substituting the values of various parameters gives

$$E = \tan^{-1}\left[\frac{42\,164 - 6378 \cos 60° \cos 20°}{6378 \sin\{\cos^{-1}(\cos 60° \cos 20°)\}}\right] - \cos^{-1}(\cos 60° \cos 20°)$$

$$= \tan^{-1}\left[\frac{42\,164 - 2998}{6378 \sin(\cos^{-1} 0.47)}\right] - \cos^{-1} 0.47$$

$$= \tan^{-1}\left(\frac{39\,166}{5631}\right) - 62°$$

$$= 81.8° - 62° = 19.8°$$

Therefore,

$$\text{Azimuth} = 202.8° \text{ and Elevation} = 19.8°$$

Problem 3.10
Consider two Earth stations, X and Y, with longitudes at 60°W and 90°W respectively and latitudes at 30°N and 45°N respectively. They are communicating with each other via a geostationary satellite located at 105°W. Find the total delay in sending 500 kilo bits of information if the transmission speed is 10 Mbps. Assume the orbital radius to be 42 164 km and the radius of the Earth to be 6378 km.

Solution: In the first step, the elevation angles of the two Earth stations are determined:

$$\text{Earth station X latitude, } \theta_{lX} = 30° \text{ N}$$
$$\text{Earth station X longitude, } \theta_{LX} = 60° \text{ W}$$
$$\text{Satellite longitude, } \theta_s = 105° \text{ W}$$

The elevation angle E_X of the Earth station X is given by

$$E_X = \tan^{-1}\left[\frac{r - R\cos\theta_{lX} \cos|\theta_s - \theta_{LX}|}{R\sin\{\cos^{-1}(\cos\theta_{lX} \cos|\theta_s - \theta_{LX}|)\}}\right] - \cos^{-1}(\cos\theta_{lX} \cos|\theta_s - \theta_{LX}|)$$

where
r = Satellite orbital radius
R = Earth's radius

$$E_X = \tan^{-1}\left[\frac{42\,164 - 6378\cos 30°\cos 45°}{6378\sin\{\cos^{-1}(\cos 30°\cos 45°)\}}\right] - \cos^{-1}(\cos 30°\cos 45°)$$

$$= \tan^{-1}\left(\frac{42\,164 - 6378 \times 0.612}{6378\sin 52.3°}\right) - \cos^{-1}(0.612)$$

$$= \tan^{-1}\left(\frac{38\,260.66}{5044.998}\right) - 52.3° = 82.6° - 52.3° = 30.3°$$

Earth station Y latitude, $\theta_{lY} = 45°$ N

Earth station Y longitude, $\theta_{LY} = 90°$ W

Satellite longitude, $\theta_s = 105°$ W

The elevation angle E_Y of the Earth station Y is given by

$$E_Y = \tan^{-1}\left(\frac{42\,164 - 6378\cos 45°\cos 15°}{6378\sin\{\cos^{-1}(\cos 45°\cos 15°)\}}\right) - \cos^{-1}(\cos 45°\cos 15°)$$

$$= \tan^{-1}\left(\frac{37\,507.826}{4656.174}\right) - \cos^{-1} 0.683$$

$$= 82.92° - 46.92° = 36°$$

In the next step, the slant range of the two Earth stations will be determined. Refer to Figure 3.69. The slant range (d_X) of the Earth station X can be computed from

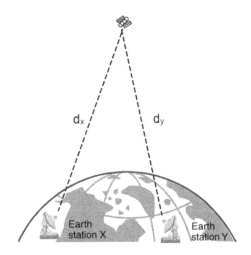

Figure 3.69 Figure for Problem 3.10

$$(d_X)^2 = R^2 + (R+H)^2 - 2R(R+H)\sin\left[E + \sin^{-1}\left\{\left(\frac{R}{R+H}\right)\cos E_X\right\}\right]$$

$$= (6378 \times 10^3)^2 + (42\,164 \times 10^3)^2 - 2 \times 6378 \times 10^3 \times 42\,164 \times 10^3$$

$$\times \sin\left[30.3° + \sin^{-1}\left\{\left(\frac{6378 \times 10^3}{42\,164 \times 10^3}\right)\cos 30.3°\right\}\right]$$

$$= 40\,678\,884 \times 10^6 + 1\,777\,802\,896 \times 10^6 - 537\,843\,984 \times 10^6 \times \sin 37.8°$$

$$= 1\,488\,783\,418 \times 10^6$$

This gives

$$d_X = \sqrt{1\,488\,783\,418 \times 10^6} = 38\,584.76 \text{ km}$$

The Slant range d_Y for the Earth station Y can computed from

$$(d_Y)^2 = R^2 + (R+H)^2 - 2R(R+H)\sin\left[E + \sin^{-1}\left\{\left(\frac{R}{R+H}\right)\cos E_Y\right\}\right]$$

$$= (6378 \times 10^3)^2 + (42\,164 \times 10^3)^2 - 2 \times 6378 \times 10^3 \times 42\,164 \times 10^3$$

$$\times \sin\left[36° + \sin^{-1}\left\{\left(\frac{6378 \times 10^3}{42\,164 \times 10^3}\right)\cos 36°\right\}\right]$$

$$= 40\,678\,884 \times 10^6 + 1\,777\,802\,896 \times 10^6 - 537\,843\,984 \times 10^6 \times \sin 43°$$

$$= 1\,451\,672\,183 \times 10^6$$

This gives

$$d_Y = \sqrt{1\,451\,672\,183 \times 10^6} = 38\,100.8 \text{ km}$$

Total range to be covered $= 38\,584.76 + 38\,100.8 = 76\,685.56$ km

Propagation delay $= (76\,685.56 \times 10^3 / 3 \times 10^8) = 255.62$ ms

The time required to transmit 500 kilo bits of information at a transmission speed of 10 Mbps is given by

$$\frac{500\,000}{10^7} = 50 \text{ ms}$$

Total delay $= 255.62 + 50 = 305.62$ ms

Problem 3.11
Two geostationary satellites, A and B, moving in an orbit of 42 164 km radius are stationed at 85°W and 25°W longitudes, as shown in Figure 3.70. The two satellites have slant ranges of 38 000 km and 36 000 km from a common Earth station. Determine the angular

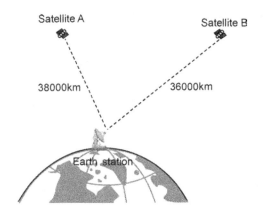

Figure 3.70 Figure for Problem 3.11

separation of the two satellites as viewed by the Earth station. Also find the intersatellite separation in orbit.

Solution: If d is the inter-satellite distance between the two satellites, then

$$d^2 = d_A^2 + d_B^2 - 2d_A d_B \cos\theta$$

where

θ = Angular separation between the two satellites
d_A = Slant range of Satellite A
d_B = Slant range of Satellite B

Also,

$$d^2 = r^2 + r^2 - 2r^2 \cos\beta = 2r^2(1 - \cos\beta)$$

where

r = Satellite orbital radius
β = Longitudinal difference between the two satellites

This gives

$$d_A^2 + d_B^2 - 2d_A d_B \cos\theta = 2r^2(1 - \cos\beta)$$

or

$$\theta = \cos^{-1}\left[\frac{d_A^2 + d_B^2 - 2r^2(1 - \cos\beta)}{2d_A d_B}\right]$$

Substituting these values gives

$$\theta = \cos^{-1}\left[\frac{(38\,000)^2 + (36\,000)^2 - 2 \times (42\,164)^2(1 - \cos 60°)}{2 \times 38\,000 \times 36\,000}\right]$$

$$= \cos^{-1}\left(\frac{1444 \times 10^6 + 1296 \times 10^6 - 1777.8 \times 10^6}{2736 \times 10^6}\right)$$

$$= \cos^{-1}\left(\frac{962.2}{2736}\right) = 69.4°$$

The intersatellite distance can be computed from

$$d = \sqrt{[2r^2(1 - \cos \beta)]} = 1.414 \times r\sqrt{(1 - \cos 60°)} = 1.414 \times 42\,164\sqrt{0.5} = 42\,164 \text{ km}$$

Angular separation = 69.4°
Inter-satellite distance = 42 164 km

3.10 Earth Coverage and Ground Tracks

Earth coverage, also known as the 'footprint', is the surface area of the Earth that can possibly be covered by a given satellite. In the discussion to follow, the effect of satellite altitude on Earth coverage provided by the satellite will be examined. Ground track is the path followed by the sub-satellite point. The effects of altitude of the satellite and its latitude on ground track are also discussed.

3.10.1 Satellite Altitude and the Earth Coverage Area

Refer to Figure 3.71. It is evident that the coverage area increases with the height of the satellite above the surface of the Earth. It varies from something like 1.5 % of the Earth's surface area for a low Earth satellite orbit at 200 km to about 43 % of the Earth's surface area for a satellite at a geostationary height of 36 000 km. Table 3.7 shows the variation of coverage area as a function of the satellite altitude. It can be seen from the table that the increase in coverage area with an increase in altitude is steeper in the beginning than it is as the altitude increases beyond 10 000 km. The coverage angle, from Equation (3.26), can be computed from

$$\text{Coverage angle } \alpha = \sin^{-1}\left\{\left(\frac{R}{R+H}\right)\cos E\right\}$$

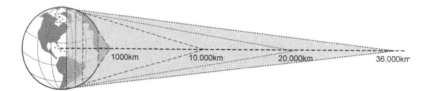

Figure 3.71 Satellite altitude and Earth coverage area

Table 3.7 Variation of the coverage area as a function of the satellite altitude

Satellite altitude (km)	Coverage area (% of Earth's surface area)
200	1.5
300	2.0
400	2.5
500	3.0
600	3.5
700	4.5
800	5.5
900	6.0
1 000	7.0
2 000	12.0
4 000	18.5
5 000	21.5
6 000	24.0
7 000	26.0
8 000	27.5
9 000	29.0
10 000	30.0
15 000	35.0
20 000	37.5
25 000	40.0
30 000	41.5
36 000	43.0

For maximum possible coverage, $E = 0°$. The expression reduces to

$$\text{Coverage angle } \alpha = \sin^{-1}\left(\frac{R}{R+H}\right) \quad (3.32)$$

The full coverage angle α can be computed to be approximately 150° for a satellite at 200 km and 17° for a satellite at a geostationary height of 36 000 km. Table 3.8 shows the variation of the full coverage angle as a function of the satellite altitude

3.10.2 Satellite Ground Tracks

The ground track of an orbiting satellite is the path followed by the sub-satellite point, that is the point formed by the projection of the line joining the orbiting satellite with the centre of the Earth on the surface of the Earth (Figure 3.72). If the Earth were not rotating, the ground track would simply be the circumference of the great circle formed by the bisection of the Earth with the orbital plane of the satellite. In reality, however, the Earth does rotate, with the result that the ground track gets modified from what it would be in the hypothetical case of a non-rotating Earth.

Two factors that influence the ground track due to Earth's rotation include the altitude of the satellite, which in turn determines the satellite's angular velocity, and the latitude at which the satellite is located, which determines the component of the Earth's rotation applicable at that point. In simple words, if we know the ground track that would have been there had the Earth

Table 3.8 Variation of the full coverage angle as a function of the satellite altitude

Satellite altitude (km)	Full Coverage angle (deg)
200	150
300	144
400	138
500	134
600	130
700	126
800	124
900	120
1 000	118
2 000	100
4 000	76
5 000	68
6 000	60
7 000	56
8 000	52
9 000	48
10 000	44
15 000	32
20 000	26
25 000	22
30 000	18
36 000	17

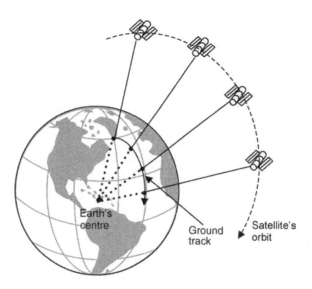

Figure 3.72 Satellite ground tracks

Earth Coverage and Ground Tracks

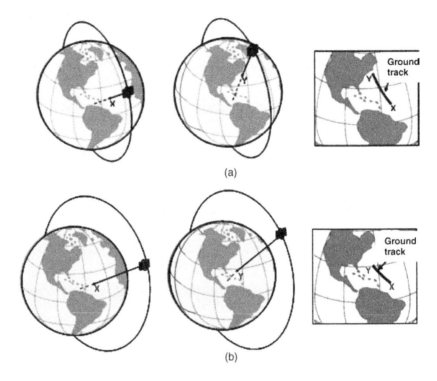

Figure 3.73 Effect of satellite altitude on the ground track

been static, modification to this track at any given point in the satellite orbit would depend on the satellite altitude at that point and also on the latitude of that point.

3.10.2.1 Effect of Altitude

The higher the altitude of the satellite, the smaller is the angular velocity and the greater will be the displacement of the ground track towards the west due to the Earth's rotation (Figure 3.73). In the case of circular orbits, the ground track of the satellite at a higher altitude will shift more than that of the satellite in a lower altitude orbit. In the case of eccentric orbits, the shift in the ground track is much less around the perigee point for the same reasons.

3.10.2.2 Effect of Latitude

The Earth's relative rotation rate decreases with an increase in latitude, becoming zero at the poles. As a consequence, the shift in the ground track reduces as the latitudes over which the satellite moves increase. In fact, the shift in the ground track is zero at the poles. Another point worth mentioning here is that in the case of a satellite in a prograde orbit, the ground track intersects increasingly westerly meridians. In the case of satellites in a prograde orbit, the ground track intersects increasingly westerly meridians when the satellite's angular speed is less than the rotational rate of Earth and increasingly easterly meridians when it is more than the Earth's rotation rate.

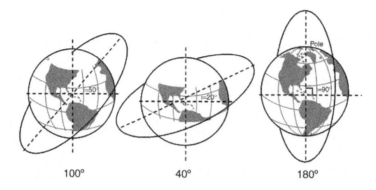

Figure 3.74 Effect of satellite orbital inclination on the latitude coverage

3.10.3 Orbit Inclination and Latitude Coverage

The northern and southern latitudes of the terrestrial segment covered by the satellite's ground track depend upon the satellite orbit inclination. The zone from the extreme northern latitude to the extreme southern latitude, which is symmetrical about the equator, is called the latitude coverage. Figure 3.74 illustrates the extent of latitude coverage for different inclinations.

It can be seen that the latitude coverage is 100 % only in the case of polar orbits. The higher the orbit inclination, the greater is the latitude coverage. This also explains why an equatorial orbit is not useful for higher latitude regions and also why a highly inclined Molniya orbit is more suitable for the territories of Russia and other republics of the former USSR.

Problem 3.12
Determine the theoretical maximum area of the Earth's surface that would be in view from a geostationary satellite orbiting at a height of 35 786 km from the surface of the Earth. Also determine the area in view for a minimum elevation angle of 10°. (Assume that the radius of the Earth is 6378 km.)

Solution: Refer to Figure 3.75. For the maximum possible coverage angle, the elevation angle E must be 0°. In that case the coverage angle α is given by

$$\text{Coverage angle } \alpha = \sin^{-1}\left[\left(\frac{R}{R+H}\right)\cos E\right] = \sin^{-1}\left(\frac{R}{R+H}\right)$$

where
$R =$ Earth's radius
$H =$ Height of the satellite above the Earth's surface
Thus

$$\alpha = \sin^{-1}\left(\frac{6378}{42\,164}\right) = 8.7°$$

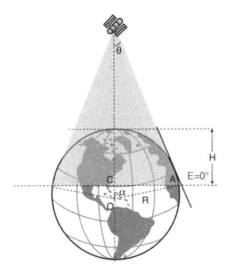

Figure 3.75 Figure for Problem 3.12

This gives

$$\theta = 90° - \alpha - E = 90° - 8.7° = 81.3°$$

In the right-angled triangle OAC, OC = OA $\times \sin 8.7°$, (Angle OAC = 8.7°) and OC = 6378 × 0.151 = 963.1 km. From the geometry, the covered surface area is given by

$$2\pi R(6378 - 963.1) = 2\pi \times 6378 \times 5414.9 = 216\,997\,546.7 \text{ km}^2$$

For $E = 10°$,

$$\alpha = \sin^{-1}\left[\left(\frac{R}{R+H}\right)\cos E\right] = \sin^{-1}\left[\left(\frac{6378}{42\,164}\right)\cos 10°\right] = 8.57°$$

This gives

$$\theta = 90° - 8.57° - 10° = 71.43°$$

The new value of OC is

6378 sin 18.57° = 2028.2 km

Covered area = $2\pi R(6378 - 2028.2) = 2\pi \times 6378 \times 4349.8 = 174\,314\,563.3 \text{ km}^2$

Problem 3.13
A satellite is orbiting the Earth in a low Earth orbit inclined at 30° to the equatorial plane. Determine the extreme northern and southern latitudes swept by its ground track.

Solution:
The extreme latitudes covered in northern and southern hemispheres are the same as orbit inclination.
Therefore, extreme northern latitude covered = 30°N
Extreme southern latitude covered = 30°S
In fact, the ground track would sweep all latitudes between 30°N and 30°S.

Further Readings

Bander, R. (1998) *Launching and Operating Satellites: Legal Issues*, Kluwer Law International for Martinus Nijhoff Publishers.

Capderou, M. and Lyle, S. (translator) (2005) *Satellites: Orbits and Missions*, Springer-Verlag, France.

Elbert, R.B. (2001) *Satellite Communication Ground Segment and Earth Station Handbook*, Artech House, Boston, Massachusetts.

Gatland, K. (1990) *Illustrated Encyclopedia of Space Technology*, Crown, New York.

Logsdon, T. (1998) *Orbital Mechanics: Theory and Applications*, John Wiley & Sons, Inc., New York.

Montenbruck, O. and Gill, E. (2000) *Satellite Orbits: Models, Methods, Applications*, Springer-Verlag, Berlin, Heidelberg, New York.

Pattan, B. (1993) *Satellite Systems: Principles and Technologies*, Van Nostrand Reinhold, New York.

Richharia, M. (1999) *Satellite Communication Systems*, Macmillan Press Ltd.

Verger, F., Sourbes-Verger, I., Ghirardi, R., Pasco, X., Lyle, S. and Reilly, P. (2003) *The Cambridge Encyclopedia of Space*, Cambridge University Press.

Internet Sites

1. http://www.centennialofflight.gov/essay/Dictionary/STABILIZATION/DI172.htm
2. http://spaceinfo.jaxa.jp/note/eisei/e/eis12a_e.html
3. www.fas.org/spp//guide/launcher.htm
4. www.abyss.uoregon.edu
5. www.britannica.com/ebchecked/topic/332323/launch-vehicle
6. www.nasa.gov/externalflash/issrg/pdfs/launchvehicles.pdf
7. www.en.wikipedia.org/wiki/Space_center

Glossary

Ariane: Launch vehicle series by European Space Agency (ESA)
Atlas: Series of expendable launch vehicles from the USA
Azimuth angle – Earth station: The azimuth angle of an Earth station is the angle produced by the line of intersection of the local horizontal plane and the plane passing through the satellite, Earth station and centre of the Earth with the true north

Buran: Russian counterpart of US Space Shuttle

Delta: Series of launch vehicles from the USA derived from Thor IRBM

De-spun Antenna: An antenna system placed on a platform that is spun in a direction opposite to the direction of spin of the satellite body. This ensures a constant pointing direction for the satellite antenna system

Earth coverage: Surface area of the Earth that can possibly be covered by a satellite

Eclipse: An eclipse is said to occur when sunlight fails to reach the satellite's solar panel due to an obstruction from a celestial body. The major and most frequent source of an eclipse is due to the satellite coming in the shadow of Earth, known as the solar eclipse. Another type of eclipse known as the lunar eclipse occurs when the moon's shadow passes across the satellite

Elevation angle – Earth station: The elevation angle of an Earth station is the angle produced by the line of intersection of the local horizontal plane and the plane passing through the satellite, Earth station and centre of the Earth with the line joining the Earth station and the satellite

Energia: Heavy lift launch vehicle from Russia and counterpart of US Space Shuttle Expendable launch vehicle. It is designed to be used only once

Footprint: Same as Earth coverage

Ground track: This is an imaginary line formed by the locus of the lowest point on the surface of the Earth. The lowest point is the point formed by the projection of the line joining the satellite with the centre of the Earth on the surface of the Earth

GSLV: Geosynchronous satellite launch vehicle from ISRO, India

Long March: Primary family of launch vehicles from the People's Republic of China

Proton: Heavy lift launch vehicle from Russia manufactured by the Khurnichev Space Centre

PSLV: Polar satellite launch vehicle from ISRO, India

Reusable launch vehicle: Launch vehicle designed to be recovered and used again for subsequent launches

Slant range: The line-of-sight distance between the satellite and the Earth station

Soyuz: Series of launch vehicles from Russia derived from R-7 ICBM

Space Shuttle: Reusable launch vehicle from the USA

Spin stabilization: A technique for stabilizing the attitude of a satellite in which the satellite body is spun around an axis perpendicular to the orbital plane. Like a spinning top, the spinning satellite body offers inertial stiffness, thus preventing the satellite from drifting from its desired orientation

Station keeping: Station keeping is the process of maintenance of the satellite's attitude against different factors that cause temporal drift

Three-axis stabilization: Also known as body stabilization, a technique for stabilizing the attitude of a satellite in which stabilization is achieved by controlling the movement of the satellite along the three axes, that is yaw, pitch and roll, with respect to a reference

Titan: Series of launch vehicles from the USA having its origin in the ICBM programme

4

Satellite Hardware

In the two previous chapters, i.e. Chapters 2 and 3, the fundamental issues relating to the operational principle of satellites, the dynamics of the satellite orbits, the launch procedures and various in-orbit operations have been addressed. Having studied at length how a satellite functions and before getting on to application-related aspects of it, in the present chapter a closer look will be taken at what a typical satellite comprises, irrespective of its intended application. Different subsystems making up a typical satellite will be briefly discussed and issues like the major function performed by each one of these subsystems along with a brief consideration of their operational aspects, will be addressed.

4.1 Satellite Subsystems

Irrespective of the intended application, be it a communications satellite or a weather forecasting satellite or even a remote sensing satellite, different subsystems comprising a typical satellite include the following:

1. Mechanical structure
2. Propulsion subsystem
3. Thermal control subsystem
4. Power supply subsystem
5. Telemetry, tracking and command (TT&C) subsystem
6. Attitude and orbit control subsystem
7. Payload subsystem
8. Antenna subsystem

1. The *mechanical structural subsystem* provides the framework for mounting other subsystems of the satellite and also an interface between the satellite and the launch vehicle.
2. The *propulsion subsystem* is used to provide the thrusts required to impart the necessary velocity changes to execute all the manoeuvres during the lifetime of the satellite. This would include major manoeuvres required to move the satellite from its transfer orbit to the

geostationary orbit in the case of geostationary satellites and also the smaller manoeuvres needed throughout the lifespan of the satellite, such as those required for station keeping.

3. The *thermal control subsystem* is essential to maintain the satellite platform within its operating temperature limits for the type of equipment on board the satellite. It also ensures the desirable temperature distribution throughout the satellite structure, which is essential to retain dimensional stability and maintain the alignment of certain critical equipment.

4. The primary function of the *power supply subsystem* is to collect the solar energy, transform it to electrical power with the help of arrays of solar cells and distribute electrical power to other components and subsystems of the satellite. In addition, the satellite also has batteries, which provide standby electrical power during eclipse periods, during other emergency situations and also during the launch phase of the satellite when the solar arrays are not yet functional.

5. The *telemetry, tracking and command (TT&C) subsystem* monitors and controls the satellite right from the lift-off stage to the end of its operational life in space. The tracking part of the subsystem determines the position of the spacecraft and follows its travel using angle, range and velocity information. The telemetry part gathers information on the health of various subsystems of the satellite, encodes this information and then transmits the same. The command element receives and executes remote control commands to effect changes to the platform functions, configuration, position and velocity.

6. The *attitude and orbit control subsystem* performs two primary functions. It controls the orbital path, which is required to ensure that the satellite is in the correct location in space to provide the intended services. It also provides attitude control, which is essential to prevent the satellite from tumbling in space and also to ensure that the antennas remain pointed at a fixed point on the Earth's surface.

7. *The payload subsystem* is that part of the satellite that carries the desired instrumentation required for performing its intended function and is therefore the most important subsystem of any satellite. The nature of the payload on any satellite depends upon its mission. The basic payload in the case of a communication satellite is the transponder, which acts as a receiver, an amplifier and a transmitter. In the case of a weather forecasting satellite, a radiometer is the most important payload. High resolution cameras, multispectral scanners and thematic mappers are the main payloads on board a remote sensing satellite. Scientific satellites have a variety of payloads depending upon the mission. These include telescopes, spectrographs, plasma detectors, magnetometers, spectrometers and so on.

8. *Antennas* are used for both receiving signals from ground stations as well as for transmitting signals towards them. There are a variety of antennas available for use on board a satellite. The final choice depends mainly upon the frequency of operation and required gain. Typical antenna types used on satellites include horn antennas, centre-fed and offset-fed parabolic reflectors and lens antennas.

4.2 Mechanical Structure

The mechanical structure weighs between 7 and 10% of the total mass of the satellite at launch. It performs three main functions namely:

1. It links the satellite to the launcher and thus acts as an interface between the two.
2. It acts as a support for all the electronic equipments carried by the satellite.

3. It serves as a protective screen against energetic radiation, dust and micrometeorites in space.

The mechanical structure that holds the satellite and links it to the launcher is very important. Some of the important design considerations that need to be addressed while designing the mechanical structure of the satellite are briefly described in the following paragraphs.

4.2.1 Design Considerations

1. The cost of launching a satellite is a function of its mass. As a result of this, the cost of launching one is very high, more so in the case of a geostationary satellite. One of the most basic requirements is therefore lightness of its mechanical structure. All efforts are therefore made to reduce the structural mass of the satellite to the minimum. This is achieved by using materials that are light yet very strong. Some of the materials used in the structure include aluminium alloys, magnesium, titanium, beryllium, Kevlar fibres and more commonly the composite materials. All these materials are characterized by high strength and stiffness and yet low weight and low density. In addition to the lightness of the structure, the choice of material is also governed by many other material properties. The design of the structural subsystem relies heavily on the results of a large number of computer simulations where the structural design is subjected to stresses and strains similar to those likely to be encountered by the satellite during the mission.
2. The structural subsystem design should be such that it can withstand mechanical accelerations and vibrations, which are particularly severe during the launch phase. Therefore the material should be such that it can dampen vibrations.
3. The satellite structure is subjected to thermal cycles throughout its lifetime. It is subjected to large differences in temperature as the sun is periodically eclipsed by Earth. The temperatures are typically several hundred degrees Celsius on the side facing the sun and several tens of degrees below zero degrees Celsius on the shaded side. Designers keep this in mind while choosing material for the structural subsystem.
4. The space environment generates many other potentially dangerous effects. The satellite must be protected from collision with micrometeorites, space junk and charged particles floating in space. The material used to cover the outside of a satellite should also be resistant to puncture by these fast travelling particles.
5. The structural subsystem also plays an important role in ensuring reliable operation in space of certain processes such as separation of the satellite from the launcher, deployment and orientation of solar panels, precise pointing of satellite antennas, operation of rotating parts and so on.

4.2.2 Typical Structure

Figure 4.1 shows a photograph of Intelsat-5 telecommunications satellite. The structure itself weighs 140 kg only, though the total mass of the satellite is greater than 1000 kg. It consists of carbon fibre tubing and honeycomb panels, as illustrated in Figure 4.2. The cellular structure made up of aluminium is sandwiched between two layers of carbon.

Figure 4.1 Photograph of Intelsat-5 satellite (Reproduced by permission of © Intelsat)

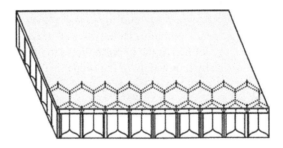

Figure 4.2 Carbon fibre tubing and honeycomb panels used in the Intelsat-5 satellite structure

4.3 Propulsion Subsystem

As briefly mentioned earlier, the propulsion subsystem is used to provide the thrusts required to impart the necessary velocity changes to execute all the manoeuvres during the lifetime of the satellite. This would include major manoeuvres required to move the satellite from its transfer orbit to the geostationary orbit in the case of geostationary satellites and also the smaller station-keeping manoeuvres needed throughout the lifespan of the satellite.

It will be seen in the following paragraphs that most of the onboard fuel, about 95 %, is required for east–west or longitudinal station-keeping manoeuvres and only about 5 % of the fuel is required for north–south or latitudinal manoeuvres. A small quantity of fuel is retained for the end of the satellite's life so that it can be moved out of the orbit by a few kilometres at the end of its lifespan.

4.3.1 Basic Principle

The propulsion system works on the principle of Newton's third law, according to which 'for every action, there is an equal and opposite reaction'. The propulsion system uses the principle of expelling mass at some velocity in one direction to produce thrust in the opposite direction. In the case of solid and liquid propulsion systems, ejection of mass at a high speed involves the generation of a high pressure gas by high temperature decomposition of propellants. The high pressure gas is then accelerated to supersonic velocities in a diverging–converging nozzle. In the case of ion propulsion, thrust is produced by accelerating charged plasma of an ionized elemental gas such as xenon in a highly intense electrical field.

4.3.2 Types of Propulsion System

Depending upon the type of propellant used and the mechanism used to produce the required thrust, there are three types of propulsion systems in use. These are:

1. Solid fuel propulsion
2. Liquid fuel propulsion
3. Electric and ion propulsion

Irrespective of the type, the performance of a propulsion system is measured by two parameters, namely the thrust force and the specific impulse. Thrust force is measured in Newton or pounds (force). Specific impulse (I_{sp}) is defined as the impulse that is the product of thrust force and time, imparted during a time (dt) by unit weight of the propellant consumed during this time. In mathematical terms,

$$I_{sp} = \frac{F \, dt}{g \, dM} = \frac{F}{g \, (dM/dt)} \tag{4.1}$$

where

F = thrust force
dM = mass of the propellant consumed in time dt
g = acceleration due to gravity = $9.807 \, m/s^2$

The term $g(dM/dt)$ in the above expression is simply the rate at which the propellant weight is consumed. Specific impulse can thus also be defined as the thrust force produced per unit weight of the propellant consumed per second.

Thrust force F produced when a mass dM is ejected at a velocity v with respect to the satellite can be computed from the velocity increment dv it imparts to the satellite of mass M using the law of conservation of momentum. According to the law of conservation of momentum,

$$M \, dv = v \, dM \tag{4.2}$$

which can also be written as

$$\frac{M \, dv}{dt} = \frac{v \, dM}{dt} = F \tag{4.3}$$

Substituting the value of F given in equation 4.3 in equation 4.1, we get

$$I_{sp} = \frac{v}{g} \qquad (4.4)$$

which confirms how the specific impulse is expressed in seconds.

Specific impulse is also sometimes expressed in Newton-seconds/kg or lb (force)-seconds/lb (mass). The three are interrelated as

$$I_{sp}(\text{in seconds}) = I_{sp}[\text{in lb (force)-seconds/lb (mass)}] = \frac{I_{sp}(\text{Newton-seconds/kg})}{9.807} \qquad (4.5)$$

The significance of specific impulse lies in the fact that it describes the mass of the propellant necessary to provide a certain velocity increment to the satellite of a known initial mass. Either of the following expressions can be used for the purpose:

$$m = M_i \left[1 - \exp\left(\frac{-\Delta v}{g I_{sp}}\right) \right] \qquad (4.6)$$

$$m = M_f \left[\exp\left(\frac{\Delta v}{g I_{sp}}\right) - 1 \right] \qquad (4.7)$$

where $M_i = M_f + m$. Here, M_i is the initial mass of the propellant, M_f the final mass of the propellant and m the mass of the propellant necessary to provide a velocity increment Δv. Also,

$$\Delta v = v \log\left(\frac{M_i}{M_f}\right) \qquad (4.8)$$

Another relevant parameter of interest is the time of operation T of the thrust force F. Quite obviously, from equation 4.1, we can write

$$T = \frac{g m I_{sp}}{F} \qquad (4.9)$$

which means that when a mass m of a propellant with a specific impulse specification I_{sp} is consumed to produce a thrust force F, then the operational time is given by the ratio of the product of specific impulse and weight of propellant consumed to the thrust force produced.

4.3.2.1 Solid Fuel Propulsion

Figure 4.3 shows the cross-section of a typical solid fuel rocket motor. The system comprises a case usually made of titanium, at the exit of which is attached a nozzle assembly made of carbon composites. High strength fibres have also been used as the case material. The case is filled with a relatively hard, rubbery, combustible mixture of fuel, oxidizer and binder. The combustible mixture is ignited by a pyrotechnic device in the motor case referred to as an igniter, as shown in the figure. Two igniters are usually employed in order to have redundancy. The combustible mixture when ignited burns very rapidly, producing an intense thrust, which could be as high as 10^6 N. This type of rocket motor produces a specific impulse of about 300 seconds. Because of high magnitudes of thrust force and specific impulse, a solid fuel rocket

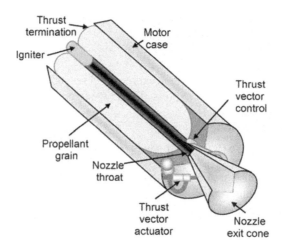

Figure 4.3 Cross-section of a typical solid fuel rocket motor

motor is particularly suitable for major orbital manoeuvres such as apogee or perigee kick operations. Either a single solid motor can be integrated into the spacecraft, in which case the empty case remains within the spacecraft throughout its lifetime, or it can be attached to the bottom of the spacecraft and discarded after use.

Although a bipropellant liquid fuel rocket motor, to be described in the following paragraphs, also produces thrust force and specific impulse of the same order as the solid fuel rocket motor and is widely used on geostationary satellites for executing major orbital manoeuvres, the solid rocket motor continues to be used for the same purpose due to its simplicity and efficiency.

4.3.2.2 Liquid Propulsion

Liquid propellant motors can be further classified as monopropellant motors and bipropellant motors. Monopropellant motors use a single combustible propellant like hydrazine, which on contact with a catalyst decomposes into its constituents. The decomposition process releases energy, resulting in a high pressure gas at the nozzle. Figure 4.4 shows the cross-section of a typical monopropellant liquid fuel motor.

The specific impulse of a typical monopropellant motor is 200 seconds and the thrust generated is in the range of 0.05 to 0.25 N. These motors are used in relatively smaller orbital

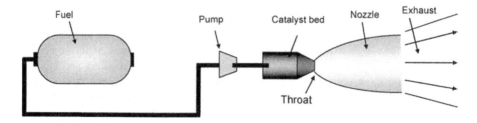

Figure 4.4 Cross-section of a typical monopropellant liquid fuel motor

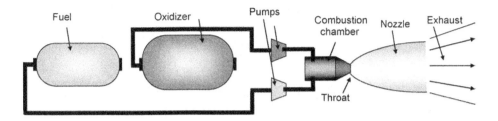

Figure 4.5 Cross-sectional view of a bipropellant liquid fuel motor

manoeuvres such as station-keeping manoeuvres, where low levels of thrust are required. Propulsion systems of this type have been used on the GEOS, SPOT, ERS and Helios families of satellites.

The performance, in terms of thrust produced, of a monopropellant can be achieved by using an electrically heated thruster (EHT) in which the propellant is heated using an electrical winding. A bipropellant liquid fuel propulsion system uses a separate fuel and oxidizer. The fuel and the oxidizer are stored in separate tanks and are brought together only in the combustion chamber. Figure 4.5 shows a simplified cross-sectional view of a bipropellant thruster. Having separate fuel and oxidizer tanks produces a greater thrust for the same weight of fuel. This allows a bipropellant system to yield a longer life. Conversely, it saves on fuel for a given life. Some of the commonly used fuel–oxidizer combinations include kerosene–liquid oxygen, liquid hydrogen–oxygen and hydrazine or its derivatives–nitrogen tetraoxide. The specific impulse can be as high as 300 to 400 seconds and the thrust produced can go up to 10^6 N. These systems are particularly suited to major orbital changes requiring large amounts of thrust.

Figure 4.6 shows the bipropellant propulsion system used on board the ETS-VIII (engineering test satellite) satellite. The fuel–oxidiser combination used is MMH (mono-methyl hydrazine) and MON-3 (mixed oxides of nitrogen). The system is capable of producing both the apogee impulse (a thrust of 500 N) and the orbital and attitude control impulses (thrust levels of 20 N).

Earlier satellites had independent propulsion systems for various thrust requirements such as one for apogee injection, another for orbit control and yet another for attitude control. A recent trend is to have a common propellant tank system for producing multiple thrusts. Such a system is called a unified propulsion system (UPS). One such UPS from EADS space company is used on the Meteosat satellite. It feeds two liquid apogee engines of 400 N thrust each and six relatively smaller thrusters of 10 N each.

4.3.2.3 Electric and Ion Propulsion

All electrical propulsion systems use electrical power derived from the solar panels to provide most of the thrust. Some of the common names include ARCJET propulsion, pulsed plasma thruster (PPT), Hall thruster and Ion propulsion.

The ARCJET thruster produces a high energy exhaust plume by electrical heating (using electrical power in the range of 1 to 20 kW) of ammonia used as fuel. It uses a nozzle to control the plume. It is capable of producing specific impulse in the range 500 to 800 seconds and a

Figure 4.6 Bipropellant propulsion system used on board ETS-VIII Satellite (Provided by the Japan Aerospace Exploration Agency (JAXA))

thrust that is an order of magnitude smaller than that produced by a monopropellant hydrazine liquid thruster.

The pulsed plasma thruster (PPT) uses a Teflon propellant and an electrical power in the range of 100 to 200 watts. The magnitude of thrust produced in this case is much smaller than even the ARCJET thruster and is three orders of magnitude smaller than that produced by a monopropellant hydrazine liquid thruster. The magnitude of thrust can be adjusted by varying the pulse rate. This type of thruster is capable of producing a specific impulse in excess of 1000 seconds.

The Hall thruster supports an electric discharge between two electrodes placed in a low pressure propellant gas. A radial magnetic field generates an electric current due to the Hall effect. The electric current interacts with the magnetic field to produce a force on the propellant gas in the downstream direction. This type of thruster was developed in Russia and is capable of producing a thrust magnitude of the order of 1600 seconds.

In the case of ion propulsion, thrust is produced by accelerating charged plasma of an ionized elemental gas such as xenon in a highly intense electrical field. It carries very little fuel and in turn relies on acceleration of charged plasma to a high velocity. The ion thruster is capable of producing a specific impulse of the order of 3000 seconds at an electrical power of about 1 kW. Thrust magnitude is quite low, which necessitates that the thruster is operated over extended periods of time to achieve the required velocity increments.

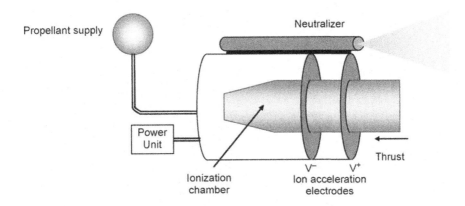

Figure 4.7 Basic arrangement of an ion thruster

Figure 4.7 shows the basic arrangement of an ion thruster, which is self-explanatory. A high value of specific impulse coupled with low thrust magnitude makes it ideally suited to attitude control and station-keeping operations. As a result, this type of thruster is increasingly popular with geostationary satellites where station-keeping requirements remain throughout the lifetime of the satellite. Such a thruster can lead to large savings of the station-keeping fuel.

The ion thruster was first flown aboard PanAmSat-5 (Pan American satellite), launched successfully in 1997. Another popular and proven ion propulsion system is the RITA (radio frequency ion thruster assembly) propulsion system (Figure 4.8). Some of the salient features of the RITA propulsion system include a specific impulse in the range of 3000 to 5000 seconds, adjustable thrust from 15 to 135 %, operating life of greater than 20 000 hours and 85 % less propellant requirement than bipropellant liquid thrusters. These attributes enable the RITA thruster to provide a significant saving in the satellite's propulsion mass and volume, which in turn can be used for more payload and/or a reduced launch cost. For instance, a 4100 kg spacecraft in a geostationary orbit and having a lifetime of 15 years would save about 574 kg in propellant mass by using RITA-type ion thruster rather than using solid and liquid fuel propellants.

The RITA ion thruster has been used successfully onboard ARTEMIS (advanced relay technology mission satellite), a European Space Agency's geostationary telecommunications satellite launched in July 2001 with the primary objective of testing new technologies. Although RITA was used onboard ARTEMIS (Figure 4.9) as an experimental propulsion system to control the orbital drift perpendicular to the satellite's orbital plane, it was also successfully used to take the satellite from a circular orbit having an altitude of 31 000 km to the final geostationary orbit having an altitude of around 36 000 km. This 5000 km increase in orbit height was achieved with a thrust magnitude of 0.015 N, realizing an orbital height increment of 15 km per day.

Ion thrusters particularly suit station-keeping and attitude control operations though they can be used for orbital transfers between the LEO, MEO and GEO. Compared to classical propellants, ion thrusters do not pollute the space environment as they are driven by

Figure 4.8 RITA ion propulsion system (Reproduced by permission of © ESA)

environmentally friendly xenon gas. Minimal fuel consumption and long life make them ideally suited for research flights as well as deep space missions.

Problem 4.1
The propulsion system of a certain satellite uses a propellant with a specific impulse of 250 seconds. Compute the ejection velocity of the propellant mass.

Solution:
Specific impulse, $I_{sp} = v/g$
Therefore, $v = I_{sp} \times g = 250 \times 9.807 = 2451.75$ m/s

Problem 4.2
Given that an Intelsat-6 series satellite with an initial mass of 4330 kg uses a bipropellant liquid propulsion system having a specific impulse specification of 290 seconds. Calculate the mass of the propellant necessary to be burnt to provide a velocity increment of 100 m/s to carry out a certain orbit inclination correction manoeuvre.

Figure 4.9 ARTEMIS satellite (Reproduced by permission of © ESA-J. Huart)

Solution:

$$m = M_i \left[1 - \exp\left(\frac{-\Delta v}{gI_{sp}}\right)\right]$$

where the terms have their usual meaning. Substituting for various parameters gives

$$m = 4330 \times \left[1 - \exp\left(\frac{-100}{9.807 \times 290}\right)\right]$$
$$= 150 \, \text{kg}$$

Problem 4.3

A satellite has an initial mass of 2950 kg. Calculate the mass of the propellant that would be consumed to produce a thrust of 450 N for a time period of 10 seconds, given that the propellant used has a specific impulse parameter of 300 seconds.

Solution:

From the definition of specific impulse, $I_{sp} = \dfrac{F \times T}{m \times g}$

Therefore, $m = \dfrac{F \times T}{I_{sp} \times g} = (450 \times 10)/(300 \times 9.807) = 1.53 \, \text{kg}$

4.4 Thermal Control Subsystem

The primary objective of the thermal control subsystem is to ensure that each and every subsystem on board the satellite is not subjected to a temperature that falls outside its safe operating temperature range. Different pieces of equipment may have different normal operating temperature ranges. As an illustration, the majority of electronic equipment has an operating temperature range of $-20\,°C$ to $+55\,°C$, the batteries used on board the satellite usually have an operating temperature range of $0\,°C$ to $+30$–$40\,°C$, the solar cells have a relatively much wider permissible operating temperature range of $-190\,°C$ to $+60\,°C$ and energy dissipating components such as power amplifiers have an operational temperature limits of $-10\,°C$ to $+80\,°C$. The thermal control system maintains the satellite platform within its operating temperature limits for the type of equipment on board. The thermal control subsystem also ensures the desired temperature distribution throughout the satellite structure, which is essential to retain dimensional stability and to maintain the alignment of certain critical equipment.

4.4.1 Sources of Thermal Inequilibrium

In the following paragraphs, the sources that produce temperature variations on the satellite platform will be discussed. There are both internal and external sources that cause changes in the temperature.

There are fundamentally three sources of radiation external to the spacecraft. The first and the foremost is the radiation from the sun. The sun is equivalent to a perfect black body radiating at an absolute temperature of 5760 K. About 40 % of this radiant energy is in the visible spectrum and about 50 % of it is in the infrared (IR). This radiant energy produces a flux of about $1370\,W/m^2$ at the Earth's orbit. Earth and its atmosphere constitute the second source of radiation. Earth along with its atmosphere also acts like a black body radiating at 250 K. This radiation is predominantly in the infrared and produces a radiant flux of about $150\,W/m^2$ in the low Earth satellite orbits. However, flux due to this radiation at geostationary orbit is negligible. The third source is the space itself, which acts like a thermal sink at 0 K.

Internal to the satellite platform, there are a large number of sources of heat generation as no piece of equipment is 100 % efficient. For example, a TWT (travelling wave tube) amplifier with a typical power rating of 200 to 250 watts may have an efficiency of about 40 %, thus dissipating heat power to the tune of 150 watts per amplifier.

4.4.2 Mechanism of Heat Transfer

In order to ensure that each and every subsystem on board the satellite operates within its prescribed temperature limits, there is a need to have some mechanism of heat transfer to and from these subsystems. There are three modes of heat transfer that can be used to remove heat from or add heat to a system. These are conduction (mechanism of heat transfer through a solid), convection (mechanism of heat transfer through a fluid) and radiation (mechanism of heat transfer through vacuum).

On Earth, convection is the dominant mode of heat transfer. Here, heat transfer through radiation is often not significant. Radiation is, however, the dominant mode of heat transfer in space. The convection mode of heat transfer does exist on the satellite platform to a lesser extent, where it is used to redistribute heat rather then remove or add any. On a satellite platform, all heat removal or addition must therefore be done through radiation.

4.4.3 Types of Thermal Control

Thermal control systems are either passive or active. Passive techniques include having multi-layer insulation surfaces, which either absorb or reflect radiation that is produced internally or generated by an external source. They have no moving parts or electrical power input. These techniques include a good layout plan for the equipment, careful selection of materials for structure, use of thermal blankets, coatings, reflectors, insulators, heat sinks and so on.

It may be mentioned here that the external conditions are widely different when the satellite is facing the sun than they would be during the eclipse periods. Also, some satellites will always be in an orbit where one side of their body is always facing the sun and the other side is facing the colder side of space. In order to achieve thermal regulation, the satellite is shaded as much as possible from changes of radiation from the sun by using highly reflective coatings and other forms of thermal insulation called thermal blanketing. Thermal blankets are usually golden in colour (gold is a good IR reflector) and are used to shade the satellite from excessive heating due to sunlight or to retain internal satellite heat to prevent too much cooling. The photograph shown in Figure 4.10 shows thermal blanketing on the FUSE (far

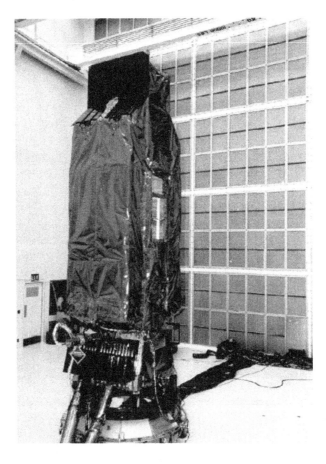

Figure 4.10 Thermal blanketing on FUSE satellite (Courtesy: NASA / FUSE Project at JHU)

ultraviolet spectroscopic explorer) satellite. Optical solar reflectors are also used on some satellites for the same purpose.

Active techniques are usually employed to cope with sudden changes in temperature of relatively larger magnitude such as those encountered during an eclipse when the temperature falls considerably. Active systems include remote heat pipes, controlled heaters and mechanical refrigerators. The heaters and refrigerators are controlled either by sensors on board or activated by ground commands.

Heat pipes are highly effective in transferring heat from one location to another. A typical heat pipe of spacecraft quality has an effective thermal conductivity several thousand times that of copper. Heat pipes in fact are a fundamental aspect of satellite thermal and structural subsystem design, for in the space environment radiation and conduction are the sole means of heat transfer. Wherever possible, heat producing components such as power amplifiers are mounted on the inner side of the outside wall. The excess heat is transferred to the outside through the thermally conducting heat pipes.

A heat pipe consists of a hermetically sealed tube whose inner surface has a wicking profile. The tube can be of any size or length and can be made to go around corners and bends. The tube is filled with a liquid with a relatively low boiling point. Ammonia is the commonly used liquid. Heat enters the heat pipe at the hot end from where the heat is to be transferred. This heat causes the liquid at that end to boil (Figure 4.11). The resulting vapours expand into the pipe carrying the heat. On reaching the cold end, they condense back to the liquid form, releasing heat in the process. This heat then flows out of the pipe to warm that part of the satellite. The condensed fluid then travels back to the starting point of the pipe along the wick. The cycle of vaporization and condensation is repeated. It may be mentioned here that operation of the heat pipe is passively driven only by the heat that is transferred. This continuous cycle of vaporization and condensation transfers large quantities of heat with very low thermal gradients. Figure 4.12 shows a view of a heat pipe thermal control system onboard the third Space Shuttle mission [STS (shuttle transportation system)-3] being prepared for testing at the Goddard Space Flight Center, NASA.

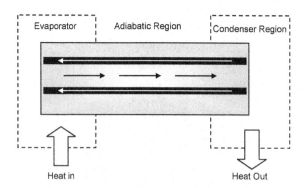

Figure 4.11 Principle of operation of a heat pipe

Thermal designs of spin-stabilized and three-axis stabilized satellites are different from each other. In a spin-stabilized satellite, the main equipments are mounted within the rotating drum. The rotation of the drum enables every equipment to receive some solar energy and

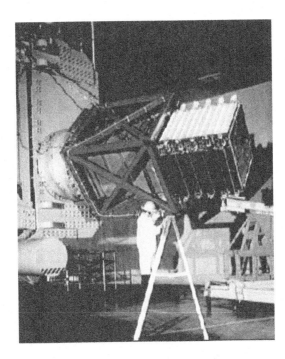

Figure 4.12 Heat pipe thermal control system onboard the third space shuttle mission (STS-3) (Courtesy: NASA)

to maintain the temperature at around 20–25 °C for most of the time, except during eclipse periods. To minimize the effect of the eclipse, the drum is isolated from the equipment by heat blankets. The temperature fluctuation is more in the case of three-axis stabilized satellites, as their orientation remains fixed with respect to Earth. Therefore, an insulation blanket is placed around the satellite to maintain the temperature of the equipment inside within desirable limits.

4.5 Power Supply Subsystem

The power supply subsystem generates, stores, controls and distributes electrical power to other subsystems on board the satellite platform. The electrical power needs of a satellite depend upon the intended mission of the spacecraft and the payloads that it carries along with it in order to carry out the mission objectives. The power requirement can vary from a few hundreds of watts to tens of kilowatts.

4.5.1 Types of Power System

Although power systems for satellite applications have been developed based on the use of solar energy, chemical energy and nuclear energy, the solar energy driven power systems are undoubtedly the favourite and are the most commonly used ones. This is due to abundance of mostly uninterrupted solar energy available in the space environment. Here reference is being made to the use of photon energy in solar radiation. The radiant flux available at the Earth's orbit is about $1370 \, W/m^2$.

There are power systems known as heat generators that make use of heat energy in solar radiation to generate electricity. A parabolic dish of mirrors reflects heat energy of solar radiation through a boiler, which in turn feeds a generator, thus converting solar energy into electrical power. This mode of generating power is completely renewable and efficient if the satellite remains exposed to solar radiation. It can also be used in conjunction with rechargeable batteries. Heat generators, however, are very large and heavy and are thus appropriate only for large satellites.

Batteries store electricity in the form of chemical energy and are invariably used together with solar energy driven electrical power generators to meet the uninterrupted electrical power requirements of the satellite. They are never used as the sole medium of supplying the electrical power needs of the satellite. The batteries used here are rechargeable batteries that are charged during the period when solar radiation is falling on the satellite. During the periods of eclipse when solar radiation fails to reach the satellite, the batteries supply electrical power to the satellite.

Nuclear fission is currently the commonly used technique of generating nuclear energy and eventually may be replaced by nuclear fusion technology when the latter is perfected. In nuclear fission, the heavy nucleus of an atom is made to split into two fragments of roughly equivalent masses, releasing large amounts of energy in the process. On satellites, nuclear power is generated in radio isotopic thermoelectric generators (RTGs). The advantage of nuclear power *vis-à-vis* its use on satellites is that it is practically limitless and will not run out before the satellite becomes useless for other reasons. The disadvantage is the danger of radioactive spread over Earth in the event of the rocket used to launch the satellite exploding before it escapes the Earth's atmosphere. Nuclear power is not used in Earth orbiting satellites because when its orbit decays after the completion of the mission life, the satellite falls back to Earth and burns up in the atmosphere, and would therefore spread radioactive particles over Earth. Nuclear power is effective in the case of satellites intended for space exploration as these satellites go deep into space too far from the sun for any solar energy driven power system to be effective. Therefore, nuclear power, though seen to be advantageous for interplanetary spacecraft, is not exploited for commercial satellites because of the cost and possible environmental hazards.

4.5.2 Solar Energy Driven Power Systems

In the paragraphs to follow, a solar energy driven power system for a satellite will be discussed at length. Solar energy will mean the photon energy of the solar radiation unless otherwise specified.

The major components of a solar power system are the solar panels (of which the solar cell is the basic element), rechargeable batteries, battery chargers with inbuilt controllers, regulators and inverters to generate various d.c and a.c voltages required by various subsystems. Figure 4.13 shows the basic block schematic arrangement of a regulated bus power supply system. The diagram is self-explanatory. Major components like the solar panels and the batteries are briefly described in the following paragraphs. During the sunlight condition, the voltage of the solar generator and also the bus is maintained at a constant amplitude with the voltage regulator connected across the solar generator. The battery is decoupled from the

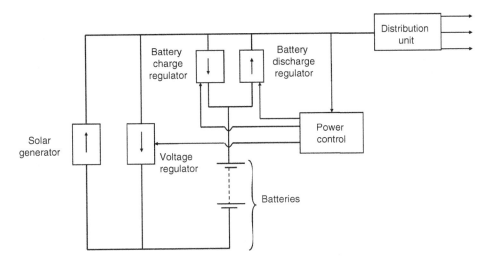

Figure 4.13 Basic block schematic arrangement of a regulated bus power supply system

bus during this time by means of a battery discharge regulator (BDR) and is also charged using the battery charge regulator (BCR) as shown in the figure. During the eclipse periods, the battery provides power to the bus and the voltage is maintained constant by means of the BDR.

4.5.2.1 Solar Panels

The solar panel is nothing but a series and parallel connection of a large number of solar cells. Figure 4.14 (a) shows this series–parallel arrangement of solar cells and Figure 4.14 (b) shows the image of a solar panel. The voltage output and the current delivering capability of an individual solar cell are very small for it to be of any use as an electrical power input to any satellite subsystem. The series–parallel arrangement is employed to get the desired output voltage with the required power delivery capability. A large surface area is therefore needed in order to produce the required amount of power. The need for large solar panels must, however, be balanced against the need for the entire satellite to be as small and light weight as possible.

The three-axis body stabilized satellites use flat solar panels (Figure 4.15) whereas spin-stabilized satellites use cylindrical solar panels (Figure 4.16). Both types have their own advantages and disadvantages. In the case of three-axis stabilized satellites, the flat solar panels can be rotated to intercept maximum solar energy to produce maximum electric power. For example, 15 foot long solar panels on Intelsat-V series satellites produce in excess of 1.2 kW of power. However, as the solar panels always face the sun, they operate at relatively higher temperatures and thus reduced efficiency as compared to solar panels on spin-stabilized satellites, where the cells can cool down when in shadow.

On the other hand, in the case of spin-stabilized satellites, such as Intelsat-VI series satellites, only one-third of the solar cells face the sun at a time and hence greater numbers of cells are needed to get the desired power, which in turn leads to an increase in the mass of the satellite.

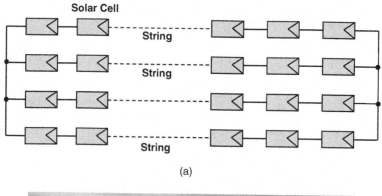

Figure 4.14 (a) Series–parallel arrangement of solar cells and (b) solar panel ((b) Reproduced by permission of © NREL/PIX, Credit, Warren Gretz)

This disadvantage is, however, partially offset by reduction in the satellite mass due to use of a relatively simpler thermal control system and attitude control system in the case of spin-stabilized satellites. It may be mentioned here that in the case of newer satellites requiring more power, the balance may tilt in favour of three-axis stabilized satellites.

4.5.2.2 Principle of Operation of a Solar Cell

The operational principle of the basic solar cell is based on the photovoltaic effect. According to the photovoltaic effect, there is generation of an open circuit voltage across a P–N junction when it is exposed to light, which is the solar radiation in the case of a solar cell. This open

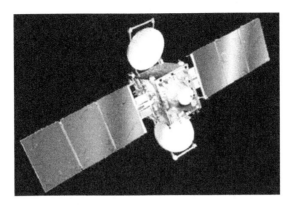

Figure 4.15 Flat solar panels used on three-axis stabilized satellites (Courtesy: ISRO)

Figure 4.16 Cylindrical solar panels used on spin-stabilized satellites (Reproduced by permission of © EADS SPACE)

circuit voltage leads to flow of electric current through a load resistance connected across it, as shown in Figure 4.17.

It is evident from the figure that the impinging photon energy leads to the generation of electron–hole pairs. The electron–hole pairs either recombine and vanish or start to drift in the opposite directions, with electrons moving towards the N-layer and holes moving towards the P-layer. This accumulation of positive and negative charge carriers constitutes the open circuit voltage. As mentioned before, this voltage can cause a current to flow through an external load. When the junction is shorted, the result is a short circuit current whose magnitude is proportional to the incident light intensity.

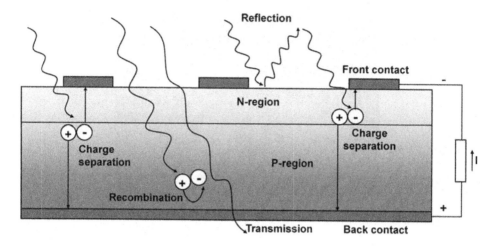

Figure 4.17 Principle of operation of a solar cell

Figure 4.18 shows the current–voltage and power–voltage characteristics of a solar cell. It is evident from the figure that the solar cell generates its maximum power at a certain voltage. The power–voltage curve has a point of maximum power, called the maximum power point (MPP). The cell voltage and the corresponding current at the maximum power point are less than the open circuit voltage and the short circuit current respectively.

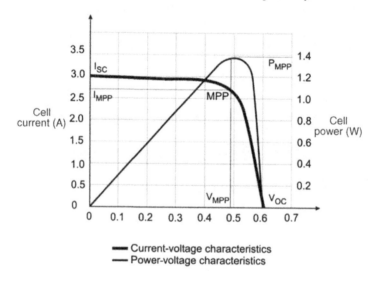

Figure 4.18 Current–voltage and power–voltage characteristics of a solar cell

Solar efficiency is the ratio of the maximum electrical solar cell power to the radiant power incident on the solar cell area. The efficiency figure for some crystalline solar cells is in excess of 20%. The most commonly used semiconductor material for making solar cells is silicon. Both crystalline and amorphous silicon are used for the purpose. Another promising material for making solar cells is gallium arsenide. Gallium arsenide solar cells, when perfected, will be lightweight and more efficient.

4.5.3 Batteries

Batteries are used on board the satellite to meet the power requirements when the same cannot be provided by solar panels, as is the case during eclipse periods. Rechargeable batteries are almost invariably used for the purpose. These are charged during the period when the solar radiation is available to the satellite's solar panels and then employed during eclipse periods or to meet a short term peak power requirements. Batteries are also used during the launch phase, before the solar panels are deployed.

The choice of the right battery technology for a given satellite mission is governed by various factors. These include the frequency of use, magnitude of load and depth of discharge. Generally, fewer cycles of use and less charge demanded on each cycle lead to a longer battery life. The choice of battery technology is closely related to the satellite orbit. Batteries used on board low Earth orbit satellites encounter much larger number of charge/discharge cycles as compared to batteries onboard geostationary satellites. LEO satellites have an orbital period of the order of 100 min and the eclipse period is 30–40 min per orbit. For GEO satellites, the orbital period is 24 hours and the eclipse duration varies from 0 to a maximum of 72 min during equinoxes. Batteries on LEO satellites are therefore subjected to a lower depth of discharge. On the other hand, batteries on geostationary satellites are subjected to a greater depth of discharge. The batteries for LEO satellites have typical DoD of 40% whereas those for GEO satellites have a typical DoD of 80%. One of the major points during battery design is that their capacity is highly dependent on the temperature. As an example, the nickel metal hydride (NiMH) battery has a maximum capacity between the operating temperature of 10 to 15 °C and its capacity decreases at a rate of 1 Ah/°C outside this range.

Commonly used batteries onboard satellites are the nickel–cadmium (NiCd), nickel metal hydride (NiMH) and nickel–hydrogen (NiH$_2$) batteries. These have specific energy specifications of 20–30 W h/kg (in the case of NiCd batteries), 35–55 W h/kg (in the case of NiMH and NiH$_2$ batteries) and 70–110 Wh/kg (in case of Li Ion batteries). Small satellites in low Earth orbits mostly employ nickel–cadmium batteries. Nickel–hydrogen batteries are slowly replacing these because of their higher specific energy and longer life expectancy. Currently, GEO satellites mostly employ nickel–hydrogen batteries. Lithium ion batteries is the battery of the future and can be used on LEO, MEO and GEO satellites. Each of these battery types is briefly described in the following paragraphs.

4.5.3.1 Nickel–cadmium Batteries

The nickel–cadmium battery is the most commonly used rechargeable battery, particularly for household appliances. The basic galvanic cell in a nickel–cadmium battery uses a cadmium anode, a nickel hydroxide cathode and an alkaline electrolyte. It may be mentioned here that the anode is the electrode at which oxidation takes place and the cathode is the electrode that is reduced. In the case of a rechargeable battery, the negative electrode is the anode and the positive electrode is the cathode while discharging. They can offer high currents at a constant voltage of 1.2 V. However, they are highly prone to what is called the 'memory effect'. Memory effect means that if a battery is only partially discharged before recharging repeatedly, it can forget that it can be further discharged. If not prevented, it can reduce the battery's lifetime. The best way to prevent this situation is to fully charge and discharge the battery on a regular basis. The other problem with this battery is the toxicity of cadmium, as a result of which it needs to be recycled or disposed of properly. Also, nickel–cadmium batteries have a lower energy per mass ratio as compared to nickel metal hydride and nickel–hydrogen batteries. This means that, for a given battery capacity, nickel-cadmium batteries are relatively heavier as compared to nickel metal hydride and nickel–hydrogen batteries. Nickel metal hydride and nickel hydrogen batteries have more or less completely replaced nickel cadmium batteries in most applications.

Nickel–cadmium batteries are mostly used on LEO satellites (having an orbital period of 100 min and an eclipse period of 30–40 min per orbit) due to their robustness versus cycle numbers. They were used on GEO satellites in the 1960s and 1970s, but now have been replaced by nickel-hydrogen batteries. Some of the satellites employing nickel–cadmium batteries include SPOT satellites. Figure 4.19 shows the battery pack on board the SPOT-4 satellite. It comprises four batteries each consisting of 24 nickel–cadmium accumulators, storing 40 A h and weighing almost 45 kg. The average power consumption of SPOT-4 satellite is 1 kW.

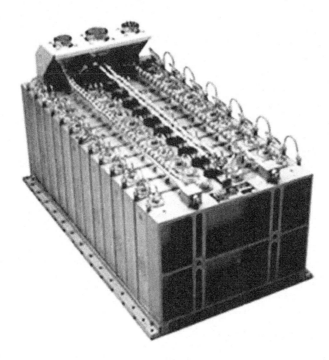

Figure 4.19 Nickel-cadmium battery pack used on SPOT-4 satellite (Reproduced by permission of © Saft)

4.5.3.2 Nickel Metal Hydride Battery

The nickel metal hydride batteries are cadmium-free replacements of nickel–cadmium batteries. The anode of the battery is made of a hydrogen storage metal alloy and the cathode is made of nickel oxide. The basic cell produces a voltage of 1.2 V. These batteries offer relatively higher energy density as compared to nickel–cadmium batteries, which implies an increased run time for the battery with no additional weight implications. Also, these batteries are less affected by the memory effect as compared to nickel–cadmium batteries. However, these batteries have problems at very high and low temperatures and therefore are not considered suitable for space applications.

4.5.3.3 Nickel–hydrogen Batteries

The nickel–hydrogen battery combines the technologies of batteries and fuel cells. The battery uses nickel hydroxide as the cathode as in the nickel–cadmium cell. Like the hydrogen–oxygen fuel cell, the battery uses hydrogen as the active element in the anode. The battery is characterized by a high specific energy (in excess of 50 W h/kg), high power density and high cyclic stability (greater than 5000 cycles). Its resistance to repeated deep discharge and tolerance for overcharge makes it the chosen battery in many aerospace applications, especially for geosynchronous and low Earth orbit satellites. Its disadvantages include its high cost and low volumetric energy density. Nickel-hydrogen batteries are being used on both LEO and GEO satellites. The batteries for LEOs have a typical depth-of-discharge (DoD) of 40 % whereas those for GEO applications have a typical DoD of 80 %. Some satellites using these batteries include Arabsat-2, Arabsat-3, Hispasat-1C, INSAT-3, Intelsat-7, Intelsat-7A, MT-Sat (multifunctional transport satellite), NStar, Superbird, Thaicom, etc. Figure 4.20 shows nickel–hydrogen batteries that have been flown on board many satellites.

Figure 4.20 Nickel–hydrogen batteries (Reproduced by permission of © Saft)

4.5.3.4 Lithium Ion Battery

Lithium ion batteries produce the same energy as nickel metal hydride batteries but weigh approximately 30 % less. These batteries do not suffer from the memory effect unlike their nickel–cadmium and nickel metal hydride counterparts. These batteries however, require special handling as lithium ignites very easily. They can be used for LEO, MEO as well as GEO satellites.

Problem 4.4
Prove that a spin-stabilized satellite with its solar panels mounted around its cylindrical body of diameter D and length or height L will need π times more solar cells than those needed by a three-axis stabilized satellite with a flat solar panel of width D and length L.

Solution: Area of the cylindrical body = πDL
The projected area towards the sun will be maximum when the sun-rays arrive perpendicular to the solar cells and is given by DL. In the case of the flat panel of width D and length L, the projected area will be DL. Therefore,

$$\text{Ratio of the two areas} = \frac{\pi DL}{DL} = \pi$$

Thus, the solar panel area required in the case of a spin-stabilized satellite using - cylindrical solar panels will be π times the solar panel area required in the case of a three-axis stabilized satellite using flat solar panels.

Problem 4.5
Consider a case where a spin-stabilized satellite has to generate 2000 watts of electrical power from the solar panels. Assuming that the solar flux falling normal to the solar cells in the worst case is 1250 W/m², the area of each solar cell is 4 cm² and the conversion efficiency of the solar cells including the losses due to cabling, etc., is 15 %, determine the number of solar cells needed to generate the desired power. What would be the number of cells required if the sun rays fell obliquely, making an angle of 10° with the normal?

Solution: The expression for the power P can be written as

$$P = \phi n s \eta$$

where

P = power to be generated (W)
ϕ = solar flux arriving normal to the solar array (W/m²)
n = number of solar cells
s = surface area of each cell (m²)
η = conversion efficiency of the solar cell

Therefore,

$$n = \frac{P}{\phi s \eta}$$

Substituting the values gives

$$n = \frac{2000}{1250 \times 4 \times 10^{-4} \times 0.15} = 26\,666.67$$

The power generated would be 2000 W provided 26 667 solar cells face the sun with solar radiation falling normal to the solar cells. Since it is a spin-stabilized satellite, the number of cells facing the sun is equal to the total number of cells divided by π. Therefore,

$$\text{Required number of cells} = 26\,667 \times \pi = 83\,777 \text{ cells}$$

If the sun rays are making an angle of 10° with the normal, then the actual solar flux falling normal to the cells would reduce by a factor of cosine of 10°. The number of cells required to get the same power would therefore increase by a factor of 1/cos 10°. Therefore, number of cells required when sunrays are making an angle of 10° are

$$\text{Number of cells} = \frac{83\,777}{\cos 10°} = \frac{83\,777}{0.9848} = 85\,070 \text{ cells}$$

Problem 4.6
It is desired that the battery system on board the satellite is capable of meeting the full power requirement of 3600 watts for the worst case eclipse period of 72 minutes. If the satellite uses nickel–hydrogen cells of 1.3 volts, 90 A h capacity each with an allowable depth of discharge of 80 %, and discharge efficiency of 95 %, find (a) the number of cells required and (b) the total mass of the battery system. Given that the specific energy specification for the battery technology used is 60 W h/kg.

Solution: (a) Power required, P = 3600 watts and the worst case eclipse period = 72 minutes = 1.2 hour. Therefore,

$$\text{Required energy} = 3600 \times 1.2 = 4320 \text{ W h}$$

The, energy stored by n cells is given by

Capacity of each cell (in Ah) × voltage of each cell (in volts) × depth of discharge

× discharge efficiency × n

Substituting different values gives

$$\text{Energy stored by } n \text{ cells} = 90 \times 1.3 \times 0.8 \times 0.95 \times n = 88.92 \times n$$

This gives $88.92 \times n = 4320$ or $n = 49$ cells.

(b) Energy required to be stored in the battery system = $\dfrac{4320}{0.8 \times 0.95}$ = 5684.2 Wh

Therefore,

$$\text{Mass of the battery system} = \frac{5684.2}{60} = 94.74 \text{ kg}$$

4.6 Attitude and Orbit Control

As briefly mentioned during introduction to various subsystems in the earlier part of the chapter, the attitude and orbit control subsystem performs twin functions of controlling the orbital path, which is required to ensure that the satellite is in the correct location in space to provide the intended services and to provide attitude control, which is essential to prevent the satellite from tumbling in space. In addition, it also ensures that the antennas remain pointed at a fixed point on the Earth's surface. The requirements on the attitude and orbit control subsystem differ during the launch phase and the operational phase of the satellite.

4.6.1 Attitude Control

Attitude of a satellite, or for that matter any space vehicle, is its orientation as determined by the relationship between its axes (yaw, pitch and roll) and some reference plane. The attitude control subsystem is used to maintain a certain attitude of the satellite, both when it is moving in its orbit and also during its launch phase. As mentioned in the previous chapter, two types of attitude control systems are in common use, namely spin stabilization and three-axis stabilization. During the launch phase, the attitude control system maintains the correct attitude of the satellite so that it is able to maintain link with the ground Earth station and controls its orientation such that the satellite is in the correct direction for an orbital manoeuvre. When the satellite is in orbit, the attitude control system maintains the antenna of the satellite pointed accurately in the desired direction. The precision with which the attitude needs to be controlled depends on the satellite antenna beam width. Spot beams and shaped beams require more precise attitude control as compared to Earth coverage or regional coverage antennas.

Attitude control in spin stabilized satellites requires pitch correction only on the de-spun antenna system and can be obtained by varying the speed of the spin motor. Yaw and roll are controlled by pulsing radially mounted jets at appropriate intervals of time. In the case of three-axis stabilized satellites, the speed of the inertia wheel needs to be controlled.

For satellites orbiting in low Earth and medium Earth orbits, the gravitational pull from the Earth is very strong. These satellites often use a long pole referred to as the gravity gradient boom, pointing towards the centre of the Earth. This pole dampens the oscillations in the direction towards the centre of the Earth from the satellite by virtue of the difference in the gravitational field between the top and the bottom of the pole.

Attitude control systems can be either passive or active. Passive systems maintain the satellite attitude by obtaining equilibrium at the desired orientation. There is no feedback mechanism to check the orientation of the satellite. Active control maintains the satellite attitude by sensing its orientation along the three axes and making corrections based on these measurements. The basic active attitude control system has three components: one that senses the current attitude of the platform, second that computes the deviations in the current attitude from the desired attitude and third that controls and corrects the computed errors.

Sensors are used to determine the position of the satellite axis with respect to specified reference directions (commonly used reference directions are Earth, sun or a star). Earth sensors sense infrared emissions from Earth and are used for maintaining the roll and the pitch axis. Sun and star sensors are generally used to measure the error in the yaw axis. The error between the current attitude and the desired attitude is computed and a correction torque is generated in proportion to the sensed error.

4.6.2 Orbit Control

Orbit control is required in order to correct for the effects of perturbation forces. These perturbation forces may alter one or more of the orbital parameters. The orbit control subsystem provides correction of these undesired changes. This is usually done by firing thrusters. During the launch phase, the orbit control system is used to affect some of the major orbit manoeuvres and to move the satellite to the desired location.

In the case of geostationary satellites, the inclination of the orbit increases at an average rate of about 0.85° per year. In general, the geostationary satellites have to remain within a block of ±0.05° or so. The east-west and north-south station keeping manoeuvres are carried out at intervals of two weeks each. North-south manoeuvres require more fuel to be expended than any other orbital correction.

In the case of non-circular orbits, the velocity of the satellite needs to be increased or decreased on a continuous basis. This is done by imparting corrections in the direction tangential to the axis lying in the orbital plane. In a spin stabilized satellite, radial jets are fired in this direction whereas in the case of three-axis stabilized satellites, two pairs of X-axis jets acting in opposite directions are used.

4.7 Tracking, Telemetry and Command Subsystem

The tracking, telemetry and command (TT&C) subsystem monitors and controls the satellite right from the lift-off stage to the end of its operational life in space. The tracking part of the subsystem determines the position of the spacecraft and follows its travel using angle, range and velocity information. The telemetry part gathers information on the health of various subsystems of the satellite. It encodes this information and then transmits the same towards the Earth control centre. The command element receives and executes remote control commands from the control centre on Earth to effect changes to the platform functions, configuration, position and velocity. The TT&C subsystem is therefore very important, not only during orbital injection and the positioning phase but also throughout the operational life of the satellite.

Figure 4.21 shows the block schematic arrangement of the basic TT&C subsystem. Tracking, as mentioned earlier, is used to determine the orbital parameters of the satellite on a regular basis. This helps in maintaining the satellite in the desired orbit and in providing look-angle information to the Earth stations. Angle tracking can, for instance, be used to determine the azimuth and elevation angles from the Earth station. The time interval measurement technique can be used for the purpose of ranging by sending a signal via the command link and getting a return via the telemetry link. The rate of change of range can be determined either by measuring the phase shift of the return signal as compared to that of the transmitted signal or by using a pseudorandom code modulation and the correlation between the transmitted and the received signals.

During the orbital injection and positioning phase, the telemetry link is primarily used by the tracking system to establish a satellite-to-Earth control centre communications channel. After the satellite is put into the desired slot in its intended orbit, its primary function is to monitor the health of various subsystems on board the satellite. It gathers data from a variety of sensors and then transmits that data to the Earth control centre. The data include a variety of electrical and non-electrical parameters. The sensor output could be analogue or digital. Wherever necessary, the analogue output is digitized. With the modulation signal as digital, various signals are multiplexed using the time division multiplexing (TDM) technique. Since the bit rates involved in telemetry signals are low, it allows a smaller receiver bandwidth to be used at the Earth control centre with good signal-to-noise ratio.

The command element is used to receive, verify and execute remote control commands from the satellite control centre. The functions performed by the command element include

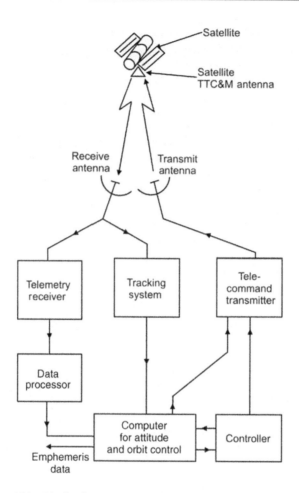

Figure 4.21 Block schematic arrangement of the basic TT&C subsystem

controlling certain functions during the orbital injection and positioning phase, including firing the apogee boost motor and extending solar panels, during the launch phase. When in orbit, it is used to control certain onboard equipment status including transponder switching, antenna pointing control, battery reconditioning, etc. The control commands received by the command element on the satellite are first stored on the satellite and then retransmitted back to the Earth control station via a telemetry link for verification. After the commands are verified on the ground, a command execution signal is then sent to the satellite to initiate intended action.

Two well-established and better-known integrated TT&C networks used worldwide for telemetry, tracking and command operations of satellites include the ESTRACK (European space tracking) network of the ESA (European Space Agency) and the ISTRAC (ISRO telemetry, tracking and command) network of the ISRO (Indian Space Research Organization).

The European Space Agency operates the ESTRACK network of ground stations used for telemetry, tracking and command in support of spacecraft operations. The ESTRACK network stations are connected to the ESAs mission control centre in Darmstadt, Germany, via OPSNET (operations network), which is ESA's ground communications network. OPSNET has permanent links with the NASCOM (NASA astronomical satellite communications) network of NASA (national aeronautics and space administration) and also temporary links with CNES (Centre National d'Études Spatiales) of France, DLR/GSOC (Deutschen Zentrums für huft- und Raumfahrt/German Space Operations Center) of Germany and NASDA (National Space Development Agency) of Japan.

The ISTRAC network of ISRO with its headquarters at Bangalore provides TT&C and mission control support to launch vehicle missions and near Earth orbiting satellites through an integrated network of ground stations located at Bangalore, Lucknow, Sriharikota, Port Blair, Thiruvanthapuram (all in India), Mauritius, Bearslake (Russia), Brunei and Biak (Indonesia), with a multimission spacecraft control centre at Bangalore. ISTRAC has also established the SPACENET, which connects various ISRO centres. ISTRAC network provides the following major functions:

1. TT&C support to satellites launched from Sriharikota right from the lift-off stage till the satellite injection stage. This includes range tracking support for satellite injection, monitoring and preliminary orbit determination.
2. TT&C support including housekeeping data acquisition throughout the mission lifetime for low Earth orbiting satellites, their health monitoring operations and control operations.
3. Data reception and processing from scientific payloads for payload analysts.
4. TT&C support to international space agencies under commercial arrangements through the Antrix Corporation.

4.8 Payload

Payload is the most important subsystem of any satellite. Payload can be considered as the brain of the satellite that performs its intended function. The payload carried by a satellite depends upon the mission requirements. The basic payload in the case of a communication satellite, for instance, is a transponder, which acts as a receiver/amplifier/transmitter. A transponder can be considered to be a microwave relay channel that also performs the function of frequency translation from the uplink frequency to the relatively lower downlink frequency. Thus, a transponder (Figure 4.22) is a combination of elements like sensitive high gain antennas for transmit–receive functions, a subsystem of repeaters, filters, frequency shifters, low noise

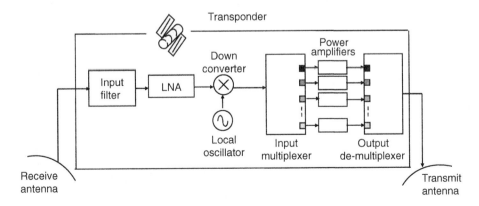

Figure 4.22 Transponder

amplifiers (LNAs), frequency mixers and power amplifiers. Satellites employ the L, S, C, X, Ku and Ka microwave frequency bands for communication purposes, with the Ka band being the latest entry into the satellite communication bands. Due to the low atmospheric absorption at the L (2 GHz/1 GHz), S (4 GHz/2 GHz) and C (6 GHz/4 GHz) bands, they were first to be employed for satellite broadcasting applications. The C band is the most popular band and is being used for providing domestic and international telephone services. Due to advances in the technology of microwave devices, high frequency Ku (12–18 GHz) and Ka (27–40 GHz) bands are also being extensively used. These high frequency bands have advantages of higher bandwidth and reduced antenna size. This has led to the revolutionary development in the field of DTH (direct-to-home) services enabling the individual home users to receive TV and broadcast services using antenna sizes as small as 30–50 cm. Transponders are covered in detail in Chapter 9 on Communication Satellites.

In the case of weather forecasting satellites, the radiometer is the most important payload. The radiometer is used as a camera and has a set of detectors sensitive to the radiation in the visible, near-IR and far-IR bands. Visible images show the amount of sunlight being reflected from Earth or clouds whereas the IR images provide information on the temperature of the cloud cover or the Earth's surface. The meteorological payload on board INSAT-3 series satellites, for instance, includes a very high resolution radiometer (VHRR) with 2 km resolution in visible and 8 km resolution in IR and water vapour channels and a CCD (change coupled device) camera in the visible (0.63–0.69 µm), near-IR (0.77–0.86 µm) and short-wave IR (1.55–1.70 µm) bands with 1 km resolution. The payload on board the Meteosat weather forecasting satellite includes a very high resolution radiometer. Depending upon the mode of operation, radiometers are classified as imagers and sounders. More details on meteorological payloads follow in Chapter 11 on Weather Satellites.

High resolution visible (HRV) cameras, multispectral scanners and thematic mapper are the main payloads on board an Earth observation satellite. Light and heat reflected and emitted from land and oceans, which contain specific information of the various living and nonliving things, are picked up by these sensors. The images produced are then digitized and transmitted to the Earth stations, where they are processed to give the required information.

The Indian remote sensing satellite IRS-P4 launched in 1999 contains an ocean colour monitor (OCM), operating in eight very near IR bands with a resolution of 360 m, and a multifrequency scanning microwave radiometer. CARTOSAT-1 (IRS-P5), launched in 2005 contains two panchromatic cameras with a spatial resolution of 2.5 m and a solid state recorder for storing payload data. RESOURCESAT-1 (IRS-P6), launched in 2003 carries multispectral cameras LISS-III (linear imaging self-scanner) and LISS-IV, having a spatial resolution of 23.5 m and 5.8 m respectively.

The US remote sensing satellite Landsat-7 consists of an enhanced thematic mapper plus (ETM+) payload which observes Earth in eight spectral bands ranging from the visible to the thermal IR region. There are three visible bands, a near-IR band and a middle IR band with a resolution of 30 m, two thermal IR bands with resolutions of 120 m and 15 m respectively. Remote sensing satellite payloads are covered in more detail in Chapter 10 on Remote Sensing Satellites.

Scientific satellites have varied payloads depending on their mission. Satellites observing the stars carry telescopes to collect light from stars and spectrographs operating over a wide range of ultraviolet (UV) wavelengths from 120 to 320 nm. Satellites for planetary exploration have varied equipment, like the plasma detector to study solar winds and radiation belts, the magnetometer to investigate the possible magnetic field around the planet, the gamma spectrometer to determine the radioactivity of surface rocks, the neutral mass spectroscope, the ion mass spectroscope, etc. Payloads on board the scientific satellites will be covered in a little more detail in Chapter 13 on Scientific Satellites.

4.9 Antenna Subsystem

The antenna subsystem is one of the most critical components of the spacecraft design because of several well-founded reasons. Some of these are the following:

1. The antenna or antennas (as there are invariably more than one antenna) on board the spacecraft cannot be prohibitively large as large antennas are difficult to mount.
2. Large antennas also cause structural problems as they need to be folded inside the launch vehicle during the launch and orbital injection phase and are deployed only subsequent to the satellite reaching the desired orbit.
3. The need for having a large antenna arises from the relationship between antenna size and its gain. If the antenna could be as large as desired, there would not be a need to generate so much power on board the satellite to achieve the required power density at the Earth station antenna.
4. All satellites need a variety of antennas. These include an omnidirectional antenna, which is an isotropic radiator, a global or Earth coverage antenna, a zone coverage antenna and antennas that produce spot beams. In addition, antennas producing spot beams may have a fixed orientation with respect to Earth or may be designed to be steered by remote commands.

The omnidirectional antenna is used for TT&C operations during the phase when the satellite has been injected into its parking orbit until it reaches its final position. Unless the high gain directional antennas are fully deployed and oriented properly, the omnidirectional antenna is the only practicable means of establishing a communication channel for tracking, telemetry and command operations.

The global or Earth coverage antenna has a beam width of 17.34°, which is the angle subtended by Earth at a geostationary satellite, as shown in Figure 4.23. Any beam width lower than that would have a smaller coverage area while a beam width larger than that would lead to loss of power.

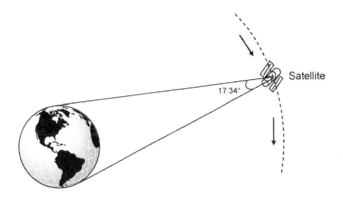

Figure 4.23 Beam width of a global antenna

The zone antenna offers a coverage that is smaller than that of an Earth coverage antenna. Such an antenna ensures that the desired areas on Earth are within the satellite's footprint.

Spot beam antennas concentrate power into a much narrower beam by using large sized reflectors and thus illuminate a much smaller area on Earth.

Figure 4.24, which shows a photograph of the Intelsat-4 satellite, gives a fairly good idea of the variety of antennas used on board a satellite. As is evident from the photograph, the satellite carries an omnidirectional antenna for TT&C operations, a horn antenna for Earth coverage and two reflector antennas for producing spot beams. Figures 4.25 to 4.27 show photographs of some of the antennas to illustrate the variety of antennas found on board a satellite.

Figure 4.24 Intelsat-4 (Reproduced by permission of © Intelsat)

Figure 4.25 Parabolic reflector antenna (Reproduced by permission of © 3amSystems)

Figure 4.26 Horn antenna (Reproduced by permission of © A.H. Systems, Inc.)

Figure 4.27 Omnidirectional antenna (Reproduced by permission of © Dudley lab)

In the following paragraphs, the important antenna parameters are discussed, followed by the operational basics of different antenna types relevant to satellites.

4.9.1 Antenna Parameters

The undermentioned performance parameters are briefly described in the following paragraphs:

1. Gain
2. Effective isotropic radiated power (EIRP)
3. Beam width
4. Bandwidth
5. Polarization
6. Aperture

The *gain* of an antenna is simply its ability to concentrate the radiated energy in a given direction, with the result that the power density in that specific direction has to be greater than it would be had the antenna been an isotropic radiator. The power gain of an antenna is defined as the ratio of the power density at a given distance in the direction of maximum radiation intensity to the power density at the same distance due to an isotropic radiator for the same input power fed to the two antennas.

The antenna gain tells us about the amplifying or more appropriately the directivity characteristics of the antenna without taking into consideration the actual transmitter power delivered to it. The *effective isotropic radiated power* (EIRP) is the more appropriate figure-of-merit of the antenna. It is given by the product of the transmitter power and the antenna gain. An antenna with a power gain of 40 dB and a transmitter power of 1000 W would mean an EIRP of 10 MW. This means that 10 MW of transmitter power when fed to an isotropic radiator would be as effective in the desired direction as 1000 W of power fed to a directional antenna having a power gain of 40 dB in the desired direction. The EIRP is of fundamental importance to the designers and operators of satellite communication systems as magnitude and distribution of a satellite's EIRP over its coverage area are major determinants of Earth station design.

The *beam width* gives angular characteristics of the radiation pattern of the antenna. It is taken as the angular separation either between the half power points on its power density radiation pattern, as shown in Figure 4.28 (a), or between −3 dB points on the field intensity radiation pattern, as shown in Figure 4.28 (b). The beam width is related to the power gain G by

$$G(\theta, \Phi) = \frac{4\pi}{\Omega} \qquad (4.10)$$

where

Ω = solid angle (in steradians) = $\Delta\theta \, \Delta\Phi$
$\Delta\theta$ = beam width in the azimuth direction (in radians)
$\Delta\Phi$ = beam width in the elevation direction (in radians)

Antenna *bandwidth* in general is the operating frequency range over which the antenna gives a certain specified performance. It is always defined with reference to a certain parameter such as gain, in which case it is taken as the frequency range around the nominal centre frequency over which the power gain falls to half of its maximum value.

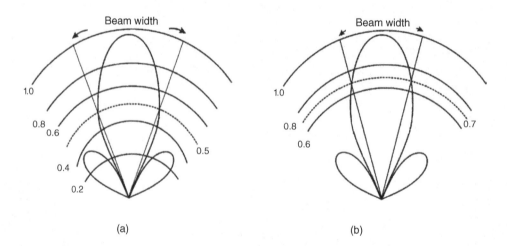

Figure 4.28 Definition of the beam width

Polarization is the direction of the electric field vector with respect to ground in the radiated electromagnetic wave while transmitting and orientation of the electromagnetic wave again in terms of the direction of the electric field vector at which the response of the antenna is maximum, while receiving. The polarization of an antenna can be classified into two broad categories of linear polarization and elliptical polarization. In the case of linear polarization, the electric field vector lies in a plane. It is called linear polarization because the direction of resultant electric field vector is constant with respect to time. In the generalized case of a linearly polarized wave, the two mutually perpendicular components of the electric field vector are in phase. Linear polarization is either horizontal or vertical. If the plane is horizontal, it is horizontally polarized and if the plane is vertical, it is vertically polarized. An inclined plane leads to what is called slant polarization. Slant polarization has both horizontal and vertical components.

When the two components of the electric field vector are not in phase, it can be verified that the tip of the resultant electric field vector traverses an ellipse as the RF signal goes through one complete cycle. This is called elliptical polarization. Elliptical polarization has two orthogonal linearly polarized components. When the magnitudes of these orthogonal components become equal, a circularly polarized wave results. Therefore, circular polarization is a special case of elliptical polarization. Figure 4.29 shows an electrical vector representation in the case of linearly polarized [Figure 4.29 (a)], elliptically polarized [Figure 4.29 (b)] and circularly polarized [Figure 4.29 (c)] waves.

Cross polarization is the component that is orthogonal to the desired polarization. A well-designed antenna should have a cross-polarized component at least 20 dB below the desired polarization in the direction of the main lobe and 5 to 10 dB below the desired polarization in the direction of side lobes.

If the received electromagnetic wave is of a polarization different from the one the antenna is designed for, a polarization loss results. In the case of a linearly polarized wave, polarization loss can be computed from

$$\text{Polarization loss} = 20 \log \left(\frac{1}{\cos \Phi} \right) \qquad (4.11)$$

where Φ = angle between the polarization of the received wave and that of the antenna.

The *antenna aperture* is the physical area of the antenna projected onto a plane perpendicular to the direction of the main beam or the main lobe. In the case of the main beam axis being parallel to the principle axis of the antenna, it is the same as the physical aperture of the antenna itself. For a given antenna aperture A, the directive gain of the antenna is given by

$$\text{Gain} = \frac{4\pi A}{\lambda^2} \qquad (4.12)$$

The above expression is valid only when the aperture is uniformly illuminated. If the illumination is non-uniform, the gain would be less than what would be given by the above expression. It is then given by

$$\text{Gain} = \frac{4\pi A_e}{\lambda^2} \qquad (4.13)$$

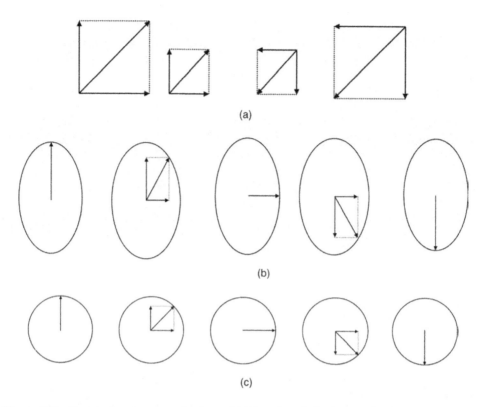

Figure 4.29 Electrical vector representations of (a) linearly polarized wave, (b) elliptically polarized wave and (c) circularly polarized wave

where

$$A_e = \eta A$$

and

A_e = effective aperture   and η = aperture efficiency

4.9.2 Types of Antennas

There is a large variety of antennas having varied features and characteristics. In the present section, those types that are relevant to satellite applications will be described. These include reflector antennas, horn antennas, helical antennas, lens antennas and phased array antennas.

4.9.2.1 Reflector Antennas

A reflector antenna, made in different types, shapes and configurations depending upon the shape of the reflector and type of feed mechanism, is by far the most commonly used antenna

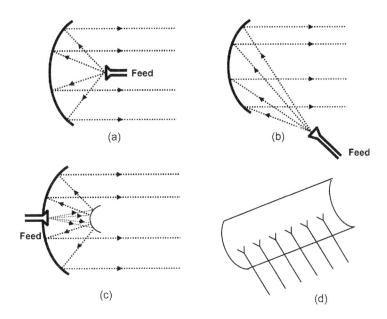

Figure 4.30 (a) Focal point fed parabolic reflector, (b) offset fed sectioned parabolic reflector, (c) cassegrain fed reflector and (d) array fed cylindrical reflector

type in all those applications that require high gain and directivity. A reflector antenna in essence comprises a reflector and a feed antenna. Depending upon the shape of the reflector and feed mechanism, different types of reflector antenna configurations are available. The reflector is usually a paraboloid, also called a parabolic reflector, or a section of a paraboloid or a cylinder. A cylindrical reflector has a parabolic surface in one direction only. The feed mechanisms include the feed antenna placed at the focal point of the paraboloid or the feed antenna located off the focal point. Another common feed mechanism is the cassegrain feed. Cylindrical reflectors are fed by an array of feed antennas. The feed antenna is usually a dipole or a horn antenna. Some of the more commonly employed reflector antenna configurations include:

1. Focal point fed parabolic reflector [Figure 4.30 (a)]
2. Offset fed sectioned parabolic reflector [Figure 4.30 (b)]
3. Cassegrain fed reflector [Figure 4.30 (c)]
4. Array fed cylindrical reflector [Figure 4.30 (d)]

A parabolic reflector has a very important property: for any point on its surface, the sum of distances of this point from the focal point and the directrix [shown dotted in Figure 4.31 (a)] is constant. If the source of radiation is placed at its focal point, the waves travelling after reflection from different points on the reflector surface will reach the directrix in phase due to equal path lengths involved with the result that the emitted beam is highly concentrated along the axis of the antenna. Similarly, on reception, the waves approaching the antenna parallel to the axis get focused on the feed antenna, whereas the waves arriving from an off-axis direction will focus on a different point rather than the focal point and thus get diffused. Such a phenomenon makes this type of antenna inherently a highly directional one. What is expressed by the ray diagram of Figure 4.31 (a) can also be expressed by the wave diagram of Figure 4.31 (b). In the wave diagram, spherical waves from the focus are reflected from the reflector surface and become planar and approaching planar waves after reflection become spherical.

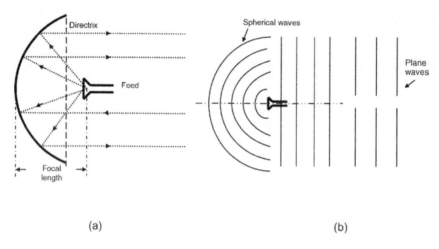

Figure 4.31 Parabolic reflector

The basic design parameters of the focal point fed reflector antenna are the size of the reflector, focal length, the feed antenna's beam pattern and the feed blockage of the reflector surface. The size of the reflector and its illumination pattern determine the antenna gain, the beam width and to some extent the side lobe pattern as well. The gain of such an antenna can be computed from

$$G = \frac{4\pi\eta A}{\lambda^2} \qquad (4.14)$$

In terms of the mouth diameter D, the expression can be written as

$$G = \eta\pi^2 \left(\frac{D}{\lambda}\right)^2 \qquad (4.15)$$

The 3 dB beam width of such an antenna is given by $70\left(\frac{\lambda}{D}\right)$ (4.16)

If the feed antenna beam width is excessive, it causes a spillover, producing an undesired antenna response in that direction. On the other hand, if it is too small then only a portion of the reflector is illuminated, with the result that the antenna produces a wider beam and consequently a lower gain.

The focal length is another important design parameter. A long focal length reflector antenna would produce more error signals at the feed than that produced by a short focal length reflector antenna. However, the focal length cannot be increased arbitrarily as long focal length reflectors need a larger support structure for the feed resulting in greater aperture blockage.

The directional pattern of the feed determines the illumination of the reflector. The angle subtended by the feed antenna at the edges of the reflector is given by $4\tan^{-1}[1/(4f/D)]$. According to a rule of thumb, the 3 dB beam width should be equal to 0.9 times the subtended angle.

Feed, together with its support, is one of the major causes of aperture blockage, which is further one of the major causes of side lobes. In applications where the feed antenna is so large that it blocks a substantial portion of the reflector aperture, resulting in significant effects on the radiated beam in terms of increased side lobe content, the offset fed parabolic reflector antenna is one of the solutions. Figure 4.32 shows the arrangement.

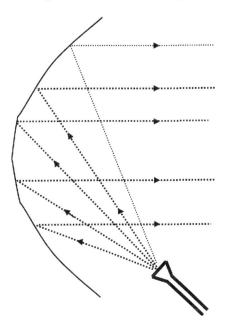

Figure 4.32 Offset fed parabolic reflector antennas

Some of the shortcomings of the focal point fed parabolic reflector antenna, such as aperture, blockage and lack of control over the main reflector illumination, can be overcome by adding a secondary reflector. The contour of the secondary reflector determines the distribution of power along the main reflector, thereby giving control over both amplitude and phase in the aperture. The cassegrain antenna derived from telescope designs is the most commonly used antenna using multiple reflectors. The feed antenna illuminates the secondary reflector, which is a hyperboloid. One of the foci of the secondary reflector and the focus of the main reflector are coincident. The feed antenna is placed on the other focus of the secondary reflector. The reflection from the secondary reflector illuminates the main reflector. Figure 4.33 (a) shows the arrangement.

Symmetrical cassegrain systems usually produce large aperture blockage, which can be minimized by choosing the diameter of the secondary reflector equal to that of the feed. Blockage can be completely eliminated by offsetting both the feed as well as the secondary reflector, as shown in Figure 4.33 (b). Such an antenna is capable of providing a very low side lobe level.

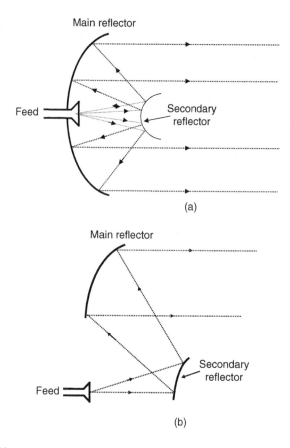

Figure 4.33 (a) Symmetrical cassegrain antenna and (b) offset cassegrain antenna

A cylindrical paraboloid antenna uses a reflector that is a parabolic surface only in one direction and is not curved in the other. It is fed from an array of feed antennas, which gives it much better control over reflector illumination. Electronic steering of the output beam is also more convenient in an array fed cylindrical antenna. However, symmetrical parabolic cylindrical reflectors suffer from a large aperture blockage. A cylindrical reflector fed from an offset placed multiple element line source offers excellent performance.

4.9.2.2 Horn Antennas

Just as in the case of a transmission line with open-circuit load end, not all the electromagnetic energy is reflected and some of it does escape to the surrounding atmosphere; the same is also true for waveguides. This radiation of electromagnetic energy is, however, inefficient due to a combination of factors, the most prominent being the impedance mismatch between the transmission line (or the waveguide) and the atmosphere.

It is observed that the energy coupling of the transmission line to the atmosphere could be enhanced and the radiation efficiency significantly improved by opening out the open end of

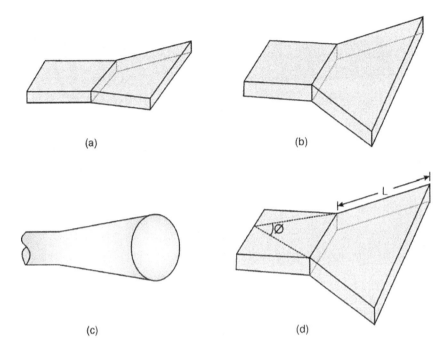

Figure 4.34 (a) Sectoral horn antenna, (b) rectangular pyramidal horn antenna and (c) conical horn antenna. (d) important design parameters of a horn antenna

the line and straightening the conductors so as to take the shape of the dipole. If the same principles are applied in the case of waveguides to improve the coupling of electromagnetic energy to the atmosphere, the waveguide's abrupt discontinuity is transformed into a more gradual one. What is obtained is a horn antenna, a sectoral horn [Figure 4.34 (a)], where the flare is only on one side, a rectangular pyramidal horn [Figure 4.34 (b)], where the flare is on both sides, and a conical horn [Figure 4.34 (c)], which is a natural extension of a circular waveguide.

Important design parameters of a horn antenna include the flare length (L) and flare angle (ϕ) [Figure 4.34 (d)]. The flare angle cannot be either too large or too small. If it is too small, the antenna has low directivity and also the emitted waves are spherical and not planar. Too large a flare angle also leads to loss of directivity due to diffraction effects.

Horns can have simple straight flares or exponential flares. Horn antennas are commonly used as the feed antennas for reflector antennas. In case more demanding antenna performance is desired in terms of polarization diversity, low side lobe level, high radiation efficiency, etc., the feeds are also more complex. Segmented [Figure 4.35 (a)], finned [Figure 4.35 (b)] and multimode [Figure 4.35 (c)] horns may be used.

4.9.2.3 Helical Antenna

The Helical antenna is a broadband VHF (very high frequency) and UHF (ultra high frequency) antenna. In addition to its broadband capability, it has found most of its applications as a result

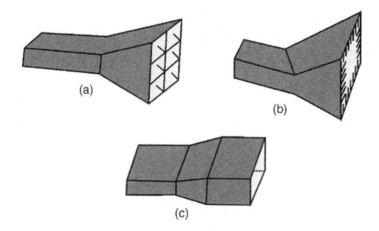

Figure 4.35 (a) Segmented horn antenna, (b) finned horn antenna and (c) multimode horn antenna

of the circularly polarized waves that it produces. VHF and UHF propagation undergoes a random change in its polarization as it propagates through the atmosphere due to various factors, like the Earth's magnetic field, ionization of different regions of the atmosphere, with Faraday's rotation being the main cause. The propagation becomes more severely affected in the case of trans-ionospheric communications, such as those involving satellites. Circular polarization is to a large extent immune to these polarization changes. On the other hand, a horizontally polarized wave will not be received at all if its polarization is rotated by 90° and it becomes vertically polarized.

Figure 4.36 shows a typical helical antenna. The ground plane is a wire mesh. This antenna has two modes of operation, with one producing a circularly or elliptically polarized broadside

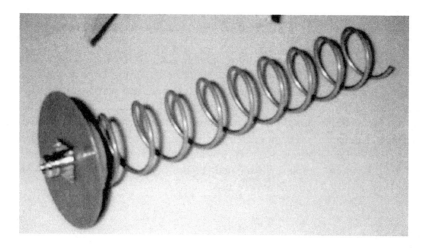

Figure 4.36 Typical helical antenna (Fabricated by High Gain Antenna Co. Ltd. Reproduced by permission of Hak-Keun-Choi)

Figure 4.37 Helical fed reflector antenna (Reproduced by permission of © Dr. Robert Suding/ultimatecharger.com)

pattern with the emitted wave perpendicular to the helical axis and the other producing a circularly polarized end-fire pattern with the emitted wave along the helical axis.

For the first mode, the helix circumference is much smaller than the operating wavelength, whereas for the second mode, which is the more common of the two, the helix circumference is equal to the operating wavelength. Figure 4.37 shows the photograph of a helical fed reflector antenna.

4.9.2.4 Lens Antenna

Like reflector antennas, lens antennas are another example of the application of rules of optics to microwave antennas. While reflector antennas rely on the laws of reflection, the lens

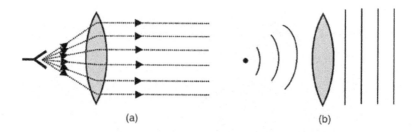

Figure 4.38 (a) Principle of operation of a lens antenna. (b) Wave diagram of a lens antenna

antennas depend for their operation on the refraction phenomenon. Lens antennas are made of a dielectric material. Figure 4.38(a) illustrates the principle of operation of such an antenna. A point source of radiation is placed at the focus of the lens. The rays arriving at the lens closer to the edges of the lens encounter a larger curvature as compared to those arriving at the centre portion of the lens. The rays closer to the edges are therefore refracted more than the rays closer to the centre. This explains the collimation of the rays. Similarly, on reception, the rays arriving parallel to the lens axis are focused onto the focal point where the feed antenna is placed.

Another way of explaining this is to look at the wave diagram shown in Figure 4.38 (b). Spherical waves emitted by the point source get transformed into plane waves during transmission. The reason for this is that those portions of the wave front closer to the centre are slowed down relatively more than those portions that are closer to the edges, with the result that outgoing waves are planar. By the same reasoning, planar waves incident on the lens antenna during reception emerge as spherical waves travelling towards the feed.

The precision with which these transformations take place depends upon the thickness of the lens in terms of the operating wavelength. In fact, the thickness of the lens at the centre should be much larger than the operating wavelength. This makes lens antennas less attractive at lower microwave frequencies. For an operating frequency of 3 GHz, the required lens thickness at the centre may be as much as 1 metre if the thickness is to be ten times the wavelength. It is because of this that lens antennas are not the favoured ones for frequencies less than 10 GHz. Even at frequencies around 10 GHz, the problems of thickness and weight are present. These problems can be overcome in what is called a Fresnel or zoned lens. Two types of zoned lenses are shown in Figure 4.39. Zoning not only overcomes the weight problem but it also absorbs less percentage of the radiation. A thicker lens would absorb a higher proportion of the radiation. The thickness t of each step in a zoned lens is related to the wavelength in order to ensure that the phase difference between the rays passing through the centre and those passing through the adjacent section is 2π radians or an integral multiple of it. A zone lens has a small operational frequency range because its thickness is related to the operating wavelength.

4.9.2.5 Phased Array Antenna

A phased array antenna is the one where the radiated beam (the axis of the main lobe of the radiated beam) can be steered electronically without any physical movement of the antenna structure. This is done by feeding the elements of the array with signals having a certain fixed phase difference between adjacent elements of the array during transmission. On reception,

Figure 4.39 Zoned lenses

they work in exactly the same way and instead of splitting the signals among elements, the elemental signals are summed up. The receive steering uses the same phase angles as the transmit steering due to the antenna reciprocity principle.

The elements used in the array are usually either horns or microstrip antennas and the array can have any one of a large number of available configurations. A linear array is a one-dimensional array with multiple elements along its length lying on a single line [Figure 4.40(a)]. This type of array, quite understandably, would be capable of steering the beam only in one direction, depending on the orientation of the array. A planar array is a two-dimensional array with multiple elements in both dimensions, with all its elements lying on the same plane [Figure 4.40(b)]. Such an array is capable of steering the beam in both azimuth and elevation. The steering angle θ between the antenna axis and the observation axis, the phase difference between adjacent elements of the array $\Delta\Phi$, operating wavelength (λ) and the spacing between adjacent elements S are interrelated by

$$\theta = \sin^{-1}\left(\lambda \frac{\Delta\Phi}{360S}\right) \qquad (4.17)$$

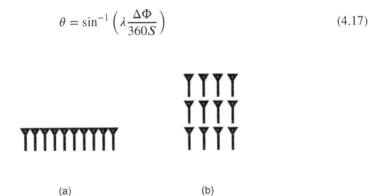

(a)   (b)

Figure 4.40 (a) Linear and (b) planar phased array antennas

Figure 4.41 Photograph of a typical phased array antenna structure (Courtesy: NASA)

Based on the feeding methodology, these electronically steered arrays are configured in one of the following ways:

1. Conventional arrays
2. Lens arrays
3. Reflector arrays

Conventional arrays have their elements fed from a common source by using power dividers and combiners. *Lens arrays* have their elements fed from a common source which radiates spherical waves to one side of the elements during transmission. These elements receive, process and re-radiate a plane wave from the other side. The *reflector array* is also a space fed array. In this case, elements receive, process and re-radiate on the transmission side. Figure 4.41 shows a photograph of a typical phase array antenna structure.

Problem 4.7

An antenna designed for tracking applications produces a pencil-like beam with both azimuth and elevation beam widths equal to 0.5° each. Determine the gain of the antenna in dB. Also determine the antenna aperture, if the operating frequency is 6 GHz.

Solution:

$$\text{Azimuth beam width } \Delta\theta = \text{elevation beam width } \Delta\Phi$$

$$= 0.5° = \left(\frac{0.5\pi}{180}\right)$$

$$= 0.00873 \text{ rad}$$

Antenna Subsystem

$$\text{Antenna gain} = \frac{4\pi}{\Delta\theta\Delta\Phi} = \frac{4\pi}{0.00873 \times 0.00873} = 164\,885.09$$

$$\text{Gain in dB} = 10\log 164\,885.09 = 52.17\,\text{dB}$$

Operating frequency $f = 6\,\text{GHz}$

Therefore,

$$\text{Operating wavelength in cm } (\lambda) = \frac{3 \times 10^{10}}{6 \times 10^9} = 5\,\text{cm}$$

The antenna gain expressed in terms of the antenna aperture A is given by

$$G = \frac{4\pi A}{\lambda^2}$$

which gives

$$A = \frac{G\lambda^2}{4\pi} = \frac{164\,885.09 \times 0.05 \times 0.05}{4\pi} = 32.80\,\text{m}^2$$

Problem 4.8
A certain receiving antenna has an actual projected area equal to the received beam of 10 square metres. The main lobe of its directional pattern has a length efficiency of 0.7 in the elevation and 0.5 in the azimuth direction. Determine the effective aperture of the antenna.

Solution:
Length efficiency in azimuth direction = 0.5
Length efficiency in elevation direction = 0.7
Therefore, Aperture efficiency, $\eta = 0.5 \times 0.7 = 0.35$
Effective aperture, $A_e = \eta \times A = 0.35 \times 10 = 3.5\,\text{m}^2$

Problem 4.9
A certain antenna when fed with a total power of 100 watts produces a power density of 10 mW/m² at a distance of 1 km from the antenna in the direction of its maximum radiation. It is also observed that same power density can be produced at the same point when the antenna is replaced by an isotropic radiator fed with a total power of 10 kW. Determine the directivity of the antenna.

Solution:
Directivity (in dB) = 10 log [10 000/100] = 10 log [100] = 20 dB

Problem 4.10
Determine the beam width between nulls of a paraboloid reflector antenna having a 3-dB beam width of 0.4° and an effective aperture of 5 m².

Solution:
The null-to-null beam width of a paraboloid reflector antenna is twice its 3-dB beam width. Therefore, null-to-null beam width = $2 \times 0.4° = 0.8°$

Problem 4.11
The received signal strength in a certain horizontally polarized antenna is 20 dB when receiving a right-hand circularly polarized electromagnetic wave. Compute the received signal strength when:

(a) the incident wave is horizontally polarized;
(b) the incident wave is vertically polarized;
(c) the incident wave is left-hand circularly polarized;
(d) the received wave polarization makes an angle of 60° with the horizontal.

Solution: When the incident polarization is circularly polarized and the antenna is linearly polarized, there is a polarization loss of 3 dB. Therefore,

$$\text{Incident signal strength} = 20 + 3 = 23 \text{ dB}$$

(a) When the received polarization is the same as antenna polarization, the polarization loss is zero. Therefore, the received signal strength = 23 dB.
(b) When the incident wave is vertically polarized, the angle between the incident polarization and antenna polarization is 90°. Therefore, polarization loss = $20\log[1/\cos \Phi]$ = infinity and the received signal strength = 0.
(c) When the incident wave is left-hand circularly polarized and the antenna polarization is linear, there will be a polarization loss of 3 dB and the received signal strength will be 20 dB.
(d) In this case, polarization loss is given by

$$20 \log \left(\frac{1}{\cos 60°} \right) = 20 \log 2 = 20 \times 0.30 = 6 \text{ dB}$$

Therefore the received signal strength = 23 − 6 = 17 dB.

Problem 4.12
A dish antenna meant for satellite down link reception has an effective aperture of 1.0 m. Compute the gain (in dB) and 3-dB beam width (in degrees) for a down link operating frequency of 11.95 GHz.

Solution:

Effective aperture = 1.0 m
Therefore, aperture area, $A_e = (\pi \times 1 \times 1)/4 = 0.785$ m^2
Operating frequency = 11.95 GHz
Therefore, operating wavelength = $(3 \times 10^8/11.95 \times 10^9) = 0.025$ m
Now, antenna gain = $4\pi A_e/\lambda^2 = (4 \times \pi \times 0.785)/(0.025 \times 0.025) = 15\,783.36$
Antenna gain in dB = $10 \log(15\,783.36) = 41.98$ dB
3-dB beam width = $70(\lambda/D) = (70 \times 0.025)/1 = 1.75°$

Problem 4.13
A focal point fed parabolic reflector antenna has the following characteristics of the reflector: mouth diameter = 2.0 m and focal length = 2.0 m. If the 3 dB beam width of the antenna has been chosen to be 90 % of the angle subtended by the feed at the edges of the reflector, determine the 3 dB beam width and null-to-null beam width of the feed antenna.

Solution: The angle θ subtended by the focal point feed at the edges of the reflector is given by

$$\theta = 4 \tan^{-1}\left(\frac{1}{4f/D}\right)$$

where f = focal length and D = mouth diameter. This gives

$$\theta = 4 \tan^{-1}\left(\frac{1}{4 \times 2/2}\right) = 4 \tan^{-1} 0.25 = 56.16°$$

Therefore, the 3 dB beam width = $0.9 \times 56.16° = 50.54°$ and the null-to-null beam width = $2 \times (3\,\text{dB beam width}) = 101.08°$.

Problem 4.14
A linear periodic array of five elements has an inter-element spacing of 10 cm (Figure 4.42). If the operating frequency is 2.5 GHz, determine the desired phase angles of all the elements, if the beam is to be steered by 10° towards the right side of the array axis. The phase for element-1 can be taken as zero.

Figure 4.42 Figure for Problem 4.14

Solution: The phase difference angle between adjacent elements $\Delta\Phi = (360S/\lambda) \sin \theta$. Then

$$\lambda = \frac{3 \times 10^8}{2.5 \times 10^9} = 0.12\,\text{m}$$

Therefore,

$$\Delta\Phi = \left(\frac{360 \times 0.1}{0.12}\right)\sin 10° = 52.2°$$

Thus the phase angles for elements 1, 2, 3, 4 and 5 are respectively 0°, 52.2°, 104.4°, 156.6° and 208.8°.

Problem 4.15

An Earth station antenna having a maximum gain of 60 dB at the operational frequency is fed from a power amplifier generating 10 kW. If the feed system has a loss of 2 dB, determine the Earth station EIRP.

Solution:

Power fed to the antenna = 10 kW = 40 dB
Antenna gain = 60 dB
Power loss in the feed system = 2 dB
Therefore EIRP = 40 + 60 − 2 = 98 dB

4.10 Space Qualification and Equipment Reliability

Satellites operate in a harsh environment and it is necessary that the components onboard them are space qualified so that they perform their intended function with a high reliability in space. Also, redundancy is built into the system so that if some component fails, the operation of the satellite is not affected. In this section, we discuss in brief the aspects of space qualification and system reliability and redundancy designed to provide continued satellite operation.

4.10.1 Space Qualification

Geostationary satellites are subjected to a more hostile environment as compared to satellites orbiting in low Earth orbits. Geostationary satellites operate in a total vacuum and the sun's irradiance falling on the surface of the satellite is of the order of 1.5 kW/m^2. During the period of the eclipse, space acts as an infinite heat sink and the surface temperatures of the satellite will fall towards absolute zero. As we have studied earlier, in the case of geostationary satellites, the eclipse occurs once in a period of 24 hours and lasts for a maximum of 72 minutes. LEO satellites also face severe thermal problems as they move from sunlight to shadow every 100 minutes or so.

As we have seen in Section 4.4, satellites employ a thermal control system so that the components and equipment on board the satellite do not have to face such large temperature variations. The thermal control system ensures that the temperature variation experienced by the components and equipment inside the satellite is of the order of 0°C to 70°C or so.

The selection of components to be used on board a satellite is a multi-stage process. The first stage is selection and screening of components. Components that are space worthy are selected. In the second stage, each component is tested individually or as a subsystem so as to ensure its proper functioning. After each component and subsystem is tested individually, the complete satellite is tested as a system. This process is referred to as Quality Control or Quality Assurance.

When a satellite is designed, three models, namely the mechanical model, the thermal model and the electrical model are usually built and tested. The mechanical model contains all of the structural and the mechanical parts of the satellite. The model is tested over a wide temperature range under vacuum conditions and is also subjected to vibration and shock testing. The thermal model contains all of the electronic packages and other components which must be maintained at the correct temperature. The electrical model contains all of the electronic parts of the satellite and is tested for correct electrical performance over the entire operating temperature range under vacuum conditions. The testing of these prototypes is carried out to overstress the system beyond extreme operating conditions. The prototype models used in these tests are usually not flown. Flight models are built and are subjected to same tests as the prototype but are not subjected to extreme conditions.

4.10.2 Reliability

The reliability of different satellite parts and subsystems is calculated in order to ascertain the probability that the part or the subsystem will still be working after a given time period and to add redundancy where the probability of failure is large. The reliability of a part or a component is expressed in terms of the probability of its failure after time t. The probability of failure of a part or a component follows a bathtub curve, as shown in Figure 4.43, wherein the probability of failure is higher in the beginning and later near the end of its life than in the middle. Satellite components are selected after extensive testing under the worst case conditions which they may encounter. The initial period of reduced reliability is generally eliminated by a burn-in period before the component is installed in the satellite. Semiconductor components and integrated circuits are generally subjected to burn-in periods of 100 to 1000 hours, often under over-stressed conditions in order to induce failures and get beyond the initial low reliability part of the bathtub curve.

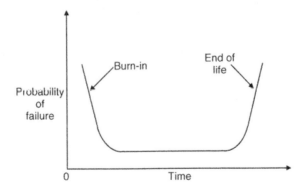

Figure 4.43 Probability of failure curve

Further Reading

Inglis, A.F (1997) *Satellite Technology: An Introduction*, Butterworth Heinemann, Massachusetts.

Karam, R.D. (1998) *Satellite Thermal Control for System Engineers*, American Institute of Aeronautics and Astronautics, Inc., Virginia.

Levitan, B. and Harte, L. (2003) *Introduction to Satellite Systems: Technology Basics, Market Growth, Systems, and Services*, Excerpted from *Wireless Communications Basics*, Althos Publishing.

Luther, A.C. and Inglis, A.F. (1997) *Satellite Technology: Introduction*, Focal Press, Boston, Massachusetts.

Pattan, B. (1993) *Satellite Systems: Principles and Technologies*, Van Nostrand Reinhold, New York.

Sarafin, T.P. (1995) *Spacecraft Structures and Mechanisms: From Concept to Launch*, Microcosm, Inc., USA and Kluwer Academic Publishers, The Netherlands.

Tajmar, M. (2002) *Advanced Space Propulsion*, Springer, New York.

Turner, M.J.L. (2000) *Rocket and Spacecraft Propulsion: Principles, Practice and New Developments*, Praxis Publishing Ltd, Chichester.

Internet Sites

1. www.ahsystems.com
2. http://spaceinfo.jaxa.jp/note/eisei/e/eis9910_strctrplan_a_e.html
3. http://esamultimedia.esa.int/docs/industry/SME/2005-Training/3-Structures-course/02-The-structural-subsystem-in-the-frame-of-the-spacecraft-system_Configuration-and-functional-relationships_Case-studies.pdf
4. http://en.wikipedia.org/wiki/Spacecraft_propulsion
5. http://www.grc.nasa.gov/WWW/K-12/airplane/bgp.html
6. http://www.vectorsite.net/tarokt.html
7. http://www.esa.int/esaMI/Launchers_Technology/SEMVLQ2PGQD_0.html
8. http://homepage.mac.com/sjbradshaw/msc/propuls.html
9. http://nmp.nasa.gov/ds1/tech/ionpropfaq.html
10. http://en.wikipedia.org/wiki/Ion_thruster
11. http://www.cheresources.com/htpipes.shtml
12. http://spot4.cnes.fr/spot4_gb/thermic.htm
13. http://collections.ic.gc.ca/satellites/english/anatomy/power/solar.html
14. http://www.howstuffworks.com/solar-cell.htm/printable
15. http://en.wikipedia.org/wiki/Solar_cell
16. http://www.saftbatteries.com
17. www.seaveyantenna.com

Glossary

Antenna: Antennas are used for both receiving signals from ground stations as well as for transmitting signals towards them. There are a variety of antennas available for use on board a satellite. The ultimate choice depends upon the frequency of operation and required gain

Antenna aperture: This is the physical area of the antenna projected on a plane perpendicular to the direction of the main beam or the main lobe. In the case of main beam axis being parallel to the principle axis of the antenna, this is the same as the physical aperture of the antenna itself

Attitude and orbit control: This subsystem performs two primary functions. It controls the orbital path, which is required to ensure that the satellite is in the correct location in space to provide the intended services. It also provides attitude control, which is essential to prevent the satellite from tumbling in space and also to ensure that the antennas remain pointed at a fixed point on the Earth's surface

Beam width: Defined with respect to the antenna, it is the angular separation between the half power points on the power density radiation pattern

Cross-polarization: This is the component that is orthogonal to the desired polarization

Effective isotropic radiated power (EIRP): This is given by the product of the transmitter power and the antenna gain. An antenna with a power gain of 40 dB and a transmitter power of 1000 W would have an EIRP of 10 MW; i.e. 10 MW of transmitter power when fed to an isotropic radiator would be as effective in the desired direction as 1000 W of power fed to a directional antenna having a power gain of 40 dB in the desired direction

Heat pipe: Heat pipe consists of a hermetically sealed tube whose inner surface has a wicking profile and it is filled with a liquid with a relatively low boiling point. Heat pipes transfer heat based on repeated cycles of vaporization and condensation

Horn antenna: A type of microwave antenna constructed from a section of a rectangular or circular waveguide

Ion propulsion: In the case of ion propulsion, the thrust is produced by accelerating charged plasma of an ionized elemental gas such as xenon in a highly intense electrical field

Lens antenna: An antenna made from dielectric material and depending upon refraction phenomenon for operation

Payload: This is that part of the satellite that carries the desired instrumentation required for performing its intended function and is therefore the most important subsystem of any satellite. The nature of the payload on any satellite depends upon its mission. The basic payload in the case of a communication satellite is the transponder, a radiometer in the case of a weather forecasting satellite, high resolution cameras, multi spectral scanners, etc., in the case of a remote sensing satellite and equipment like spectrographs, telescopes, plasma detectors, magnetometers, etc., in the case of scientific satellites

Phased array antenna: An antenna array in which the radiated beam axis can be electronically steered by having a certain phase difference between the signals fed to adjacent elements

Polarization: This is the direction of the electric field vector with respect to the ground in the radiated electromagnetic wave while transmitting and orientation of the electromagnetic wave again in terms of the direction of the electric field vector that the antenna responds to best while receiving

Polarization loss: Polarization loss results if the received electromagnetic wave is of a polarization different from the one the antenna is designed for

Power supply subsystem: This is used to collect the solar energy, transform it to electrical power with the help of arrays of solar cells and distribute the electrical power to other components and subsystems of the satellite. In addition, a satellite also has batteries, which provide standby electrical power during eclipse periods, other emergency situations and also during the launch phase of the satellite when the solar arrays are not yet functional

Propulsion subsystem: This is the satellite subsystem used to provide the thrusts required to impart the necessary velocity changes to execute all the manoeuvres during the lifetime of the satellite. This would include major manoeuvres required to move the satellite from its transfer orbit to the geostationary orbit in the case of geostationary satellites and also the smaller manoeuvres needed throughout the lifespan of the satellite, such as those required for station keeping

Reflector antenna: This comprises a reflector and a feed antenna and is capable of offering a very high gain. Reflector antennas are made in a variety of shapes, sizes and configurations depending upon the type of reflector and feed antenna used

Solar panel: This is simply a series and parallel connection of a large number of solar cells to get the desired output voltage and power delivery capability

Specific impulse: A parameter of the propulsion system, it is the ratio of the thrust force to the mass expelled to produce the desired thrust. It is measured in seconds. A specific impulse indicates how much mass is to be ejected to produce a given orbit velocity increment

Structural subsystem: This is the satellite subsystem that provides the framework for mounting other subsystems of the satellite and also an interface between the satellite and the launch vehicle

Telemetry, tracking and command (TT&C) subsystem: This is the satellite subsystem that monitors and controls the satellite from the lift-off stage to the end of its operational life in space. The tracking part of the subsystem determines the position of the spacecraft and follows its travel using angle, range and velocity information. The telemetry part gathers information on the health of various subsystems of the satellite, encodes this information and then transmits the same. The command element receives and executes remote control commands to effect changes to the platform functions, configuration, position and velocity

Thermal control subsystem: This is the satellite subsystem that is used to maintain the satellite platform within its operating temperature limits for the type of equipment on board the satellite. It also ensures a reasonable temperature distribution throughout the satellite structure, which is essential to retain dimensional stability and maintain the alignment of certain critical equipments

5

Communication Techniques

In Chapters 1 to 4, after an introduction to the evolution of satellites and satellite technology, the principles of satellite launch, in-orbit operations and its functioning have been covered in detail followed by a detailed account of satellite hardware including various subsystems that combine to make the spacecraft. In Chapters 5, 6 and 7, the focus shifts to topics that relate mainly to communication satellites, which account for more than 80 % of satellites present in space, more than 80 % of applications configured around satellites and more than 80 % of the budget spent on development of satellite technology as a whole. The topics covered in the first of the three chapters, that is the present chapter, mainly include modulation and demodulation techniques (both analogue and digital) and multiplexing techniques, followed by multiple access techniques in the next chapter. Chapter 7 focuses on satellite link design related aspects. All three chapters are amply illustrated with a large number of solved problems.

5.1 Types of Information Signals

When it comes to transmitting information over an RF communication link, be it a terrestrial link or a satellite link, it is essentially voice or data or video. A communication link therefore handles three types of signals, namely voice signals like those generated in telephony, radio broadcast and the audio portion of a television broadcast; data signals produced in computer-to-computer communications; and video signals like those generated in a television broadcast or video conferencing. Each of these signals is referred to as a base band signal. The base band signal is subjected to some kind of processing, known as base band processing, to convert the signal to a form suitable for transmission. Band limiting of speech signals to 3000 Hz in telephony and use of coding techniques in the case of digital signal transmission are examples of base band processing. The transformed base band signal then modulates a high frequency carrier so that it is suitable for propagation over the chosen transmission link. The demodulator on the receiver end recovers the base band signal from the received modulated signal. Modulation, demodulation and other relevant techniques will be discussed in the following pages of this chapter. However, before that, a brief outline on the three types of information signals mentioned above follows.

Satellite Technology: Principles and Applications, Third Edition. Anil K. Maini and Varsha Agrawal.
© 2014 John Wiley & Sons, Ltd. Published 2014 by John Wiley & Sons, Ltd. Companion Website: www.wiley.com/go/maini3

5.1.1 Voice Signals

Though the human ear is sensitive to a frequency range of 20 to 20 kHz, the frequency range of a speech signal is less than this. For the purpose of telephony, the speech signal is band limited to an upper limit of 3400 Hz during transmission. The quality of a received analogue voice signal has been specified by CCITT (Comité Consultatif International de Télégraphique et Téléphonique) to give a worst case base band signal-to-noise ratio of 50 dB. Here the signal is considered to be a standard test tone and the maximum allowable base band signal noise power is 10 nW. Apart from the signal bandwidth and signal-to-noise ratio, another important parameter that characterizes the voice signal is its dynamic range. A speech or voice signal is characterized to have a large dynamic range of 50 dB.

In the case of digital transmission, the quality of the recovered speech signal depends upon the number of bits transmitted per second and the bit error rate (BER). The BER required to give a good quality speech is considered to be 10^{-4}; i.e. 1 bit error in 10 kB though a BER of 10^{-5} or better is common.

5.1.2 Data Signals

Data signals refer to a digitized version of a large variety of information services, including voice telephony and a video and computer generated information exchange. It is indeed the most commonly used vehicle for information transfer due to its ability to combine on a single transmission support the data generated by a number of individual services, which is of great significance when it comes to transmitting multimedia traffic integrating voice, video and application data.

Again it is the system bandwidth that determines how fast the data can be sent in a given period of time, expressed in bits per second (bps). Obviously, the bigger the size of the file to be transferred in a given time, faster is the required data transfer rate or larger is the required bandwidth. Transmission of a video signal requires a much larger data transmission rate (or bandwidth) than that required by transmission of a graphics file. A graphics file requires a much larger data transfer rate than that required by a text file. The desired data rate may vary from a few tens of kbps to hundreds of Mbps for various information services. However, data compression techniques allow transmission of signals at a rate much lower than that theoretically needed to do so.

5.1.3 Video Signals

The frequency range or bandwidth of a video signal produced as a result of television quality picture information depends upon the size of the smallest picture information, referred to as a pixel. The larger the number of pixels, the higher is the signal bandwidth. As an example, in the 625 line, 50 Hz television standard where each picture frame having 625 lines is split into two fields of 312 $1/2$ lines and the video signal is produced as a result of scanning 50 fields per second in an interlaced scanning mode. Assuming a worst case picture pattern where pixels alternate from black to white to generate one cycle of video output, the highest video frequency is given by:

$$f = \frac{\frac{a\,N}{2}}{t_h} \tag{5.1}$$

where

a = aspect ratio = 4/3
N = number of lines per frame
t_h = time period for scanning one horizontal line

For the 625 line, 50 Hz system, it turns out to be 6.5 MHz. The above calculation does not, however, take into account the lines suppressed during line and frame synchronization. For actual picture transmission, the chosen bandwidth is 5 MHz for the 625 line, 50 Hz system and 4.2 MHz for the 525 line, 60 Hz system. The reduced bandwidth does not seem to have any detrimental effect on picture quality.

5.2 Amplitude Modulation

Amplitude modulation (AM) is not used as such in any of the satellite link systems. A brief overview of it is given as it may be used to modulate individual voice channels, which then can be multiplexed using frequency division multiplexing before the composite signal finally modulates another carrier.

In amplitude modulation, the instantaneous amplitude of the modulated signal varies directly as the instantaneous amplitude of the modulating signal. The frequency of the modulated signal remains the same as the carrier signal frequency. Figure 5.1 shows the modulating signal, the carrier signal and the modulated signal in the case of a single tone modulating signal.

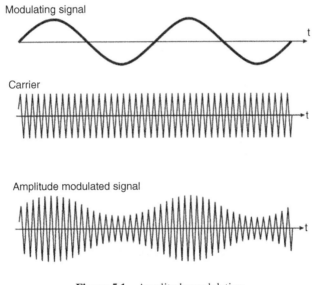

Figure 5.1 Amplitude modulation

If the modulating signal and the carrier signal are expressed respectively by $v_m(t) = V_m \cos \omega_m t$ and $v_c(t) = V_c \cos \omega_c t$, then the amplitude modulated signal ($v_{AM}(t)$) can be expressed mathematically by

$$v_{AM}(t) = V_c(1 + m \cos \omega_m t) \cos \omega_c t \qquad (5.2)$$

where m = modulation index = V_m/V_c. When more than one sinusoidal or cosinosoidal signals with different amplitudes modulate a carrier, the overall modulation index in that case is given by

$$m = \sqrt{(m_1^2 + m_2^2 + m_3^2 + \cdots)} \tag{5.3}$$

where m_1, m_2 and m_3 are modulation indices corresponding to the individual signals. The percentage of modulation or depth of modulation is given by $(m \times 100)$ and for a depth of modulation equal to 100 %, $m = 1$ or $V_m = V_c$.

5.2.1 Frequency Spectrum of the AM Signal

Expanding the expression for the modulated signal given above, we get

$$v_{AM}(t) = V_c \cos \omega_c t + \frac{mV_c}{2} \cos(\omega_c - \omega_m)t + \frac{mV_c}{2} \cos(\omega_c + \omega_m)t \tag{5.4}$$

The frequency spectrum of an amplitude modulated signal in the case of a single frequency modulating signal thus contains three frequency components, namely the carrier frequency component (ω_c), the sum frequency component ($\omega_c + \omega_m$) and the difference frequency component ($\omega_c - \omega_m$). The sum component represents the upper side band and the difference component the lower side band. Figure 5.2 shows the frequency spectrum.

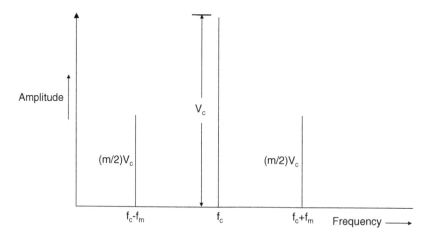

Figure 5.2 Frequency spectrum of the AM signal for a single frequency modulating signal

It may be mentioned here that in actual practice, the modulating signal is not a single frequency tone. In fact, it is a complex signal. This complex signal can always be represented mathematically in terms of sinusoidal and cosinosoidal components. Thus if a given modulating signal is equivalently represented as a sum of, say, three components (ω_{m1}, ω_{m2} and ω_{m3}), then the frequency spectrum of the AM signal, when such a complex signal modulates a carrier, contains the frequency components ω_c, $\omega_c + \omega_{m1}$, $\omega_c - \omega_{m1}$, $\omega_c + \omega_{m2}$, $\omega_c - \omega_{m2}$, $\omega_c + \omega_{m3}$ and $\omega_c - \omega_{m3}$.

5.2.2 Power in the AM Signal

The total power P_t in an AM signal is related to the unmodulated carrier power P_c by

$$P_t = P_c \left(1 + \frac{m^2}{2}\right) \qquad (5.5)$$

This can be interpreted as

$$P_t = P_c + \frac{P_c m^2}{4} + \frac{P_c m^2}{4} \qquad (5.6)$$

where $P_c m^2/4$ is the power in either of the two side bands, i.e. the upper and lower side bands. For 100 % depth of modulation for which $m = 1$, the total power in an AM signal is $3P_c/2$ and the power in each of the two side bands is $P_c/4$ with the total side band power equal to $P_c/2$. These expressions indicate that even for 100 % depth of modulation, the power contained in the side bands, which contain actual information to be transmitted, is only one-third of the total power in the AM signal.

The power content of different parts of the AM signal can also be expressed in terms of the peak amplitude of an unmodulated carrier signal (V_c) as

$$\text{Total power in AM signal } P_t = \frac{V_c^2}{2R} + \frac{mV_c^2}{8R} + \frac{mV_c^2}{8R} \qquad (5.7)$$

$$\text{Power in either of two side bands } P_s = \frac{mV_c^2}{8R} \qquad (5.8)$$

where, R is the resistance in which the power is dissipated (e.g. antenna resistance).

5.2.3 Noise in the AM Signal

The noise performance when an AM signal is contaminated with noise will now be examined. S, C and N are assumed to be the signal, carrier and noise power levels respectively. It is also assumed that the receiver has a bandwidth B, which in the case of a conventional double side band system equals $2f_m$, where f_m is the highest modulating frequency. If N_b is the noise power at the output of the demodulator, then

$$N_b = AN \qquad (5.9)$$

where A is the scaling factor for the demodulator. Signal power in each of the side band frequencies at maximum is equal to one-quarter of the carrier power, as explained in the earlier paragraphs; i.e.

$$S_L = S_U = \frac{C}{4} \qquad (5.10)$$

and

$$S_{bL} = S_{bU} = \frac{AC}{4} \qquad (5.11)$$

where

S_L = signal power in the lower side band frequency before demodulation
S_U = signal power in the upper side band frequency before demodulation
S_{bL} = signal power in the lower side band frequency after demodulation
S_{bU} = signal power in the upper side band frequency after demodulation

Since both lower and upper side band frequencies are identical before and after demodulation, they will add coherently in the demodulator to produce a total base band power S_b given by

$$S_b = 2(S_{bL} + S_{bU}) = 2\left(\frac{AC}{4} + \frac{AC}{4}\right) = AC \qquad (5.12)$$

Combining the expressions for S_b and N_b gives the following relationship between S_b/N_b and C/N:

$$\frac{S_b}{N_b} = \frac{C}{N} \qquad (5.13)$$

where $N = N_o B$, with N_o being the noise power spectral density in W/Hz and B being the receiver bandwidth.

This relationship is, however, valid only for a modulation index of unity. The generalized expression for the modulation index of m will be

$$\frac{S_b}{N_b} = m^2 \left(\frac{C}{N}\right) \qquad (5.14)$$

So far, a single frequency modulating signal has been discussed. In the case where the modulating signal is a band of frequencies, the frequency spectrum of the AM signal would comprise of lower and upper frequency bands such as the one shown in Figure 5.3. Incidentally, the spectrum shown represents a case where the modulating signal is the base band telephony signal ranging from 300 Hz to 3400 Hz.

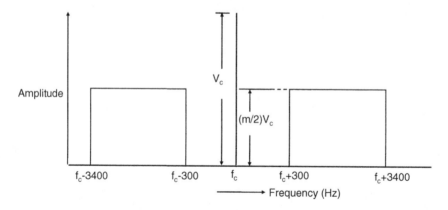

Figure 5.3 Frequency spectrum of the AM signal for a multifrequency modulating signal

5.2.4 Different Forms of Amplitude Modulation

In the preceding paragraphs it was seen that the process of amplitude modulation produces two side bands, each of which contains the complete base band signal information. Also, the carrier contains no base band signal information. Therefore, if one of the side bands was suppressed and only one side band transmitted, it would make no difference to the information content of the modulated signal. In addition, it would have the advantage of requiring only one half of the bandwidth as compared to the conventional double side band signal. If the carrier was also suppressed before transmission, it would lead to a significant saving in the required transmitted power for a given power in the information carrying signal. That is why, the single side band suppressed carrier mode of amplitude modulation is very popular. In the following paragraphs, some of the practical forms of amplitude modulation systems will be briefly outlined.

5.2.4.1 A3E System

This is the standard AM system used for broadcasting. It uses a double side band with a full carrier. The standard AM signal can be generated by adding a large carrier signal to the double side band suppressed carrier (DSBSC) or simply the double side band (DSB) signal. The DSBSC signal in turn can be generated by multiplying the modulating signal $m(t)$ and the carrier ($\cos \omega_c t$). Figure 5.4 shows the arrangement of generating the DSBSC signal.

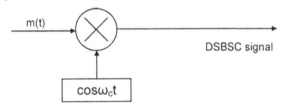

Figure 5.4 Generation of the DSBSC signal

Demodulation of the standard AM signal is very simple and is implemented by using what is known as the envelope detection technique. In a standard AM signal, when the amplitude of the unmodulated carrier signal is very large, the amplitude of the modulated signal is proportional to the modulating signal. Demodulation in this case simply reduces to detection of the envelope of the modulated signal regardless of the exact frequency or phase of the carrier. Figure 5.5 shows the envelope detector circuit used for demodulating the standard AM signal. The capacitor (C) filters out the high frequency carrier variations.

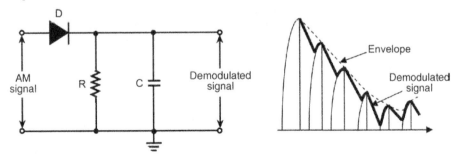

Figure 5.5 Envelope detector for demodulating standard AM signal

Demodulation of the DSBSC signal is carried out by multiplying the modulated signal by a locally generated carrier signal and then passing the product signal through a lowpass filter (LPF), as shown in Figure 5.6.

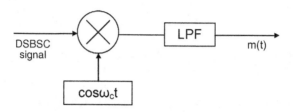

Figure 5.6   Demodulation of the DSBSC signal

5.2.4.2 H3E System

This is the single side band full carrier system (SSBFC). H3E transmission could be used with A3E receivers with distortion not exceeding 5 %. One method to generate a single side band (SSB) signal is first to generate a DSB signal first and then suppress one of the side bands by the process of filtering using a bandpass filter (BPF). This method, known as the frequency discrimination method, is illustrated in Figure 5.7. In practice, this approach poses some difficulty because the filter needs to have sharp cut-off characteristics.

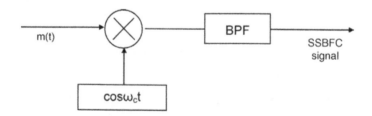

Figure 5.7   Frequency discrimination method for generating the SSBFC signal

Another method for generating an SSB signal is the phase shift method. Figure 5.8 shows the basic block schematic arrangement. The blocks labelled $-\pi/2$ are phase shifters that add a lagging phase shift of $\pi/2$ to every frequency component of the signal applied at the input to the block. The output block can either be an adder or a subtractor. If $m(t)$ is the modulating signal and $m'(t)$ is the modulating signal delayed in phase by $\pi/2$, then the SSB signal produced at the output can be represented by

$$x_{SSB}(t) = m(t)\cos \omega_c t \pm m'(t) \sin \omega_c t \qquad (5.15)$$

The output with the plus sign is produced when the output block is an adder and with the minus sign when the output block is a subtractor.

Amplitude Modulation

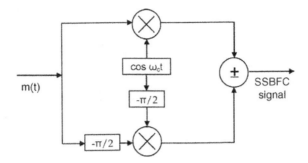

Figure 5.8 Phase shift method for generating the SSBFC signal

The difference signal represents the upper side band SSB signal while the sum represents the lower side band SSB signal. For instance, if $m(t)$ is taken as $\cos \omega_m t$, then $m'(t)$ would be $\sin \omega_m t$. The SSB signal in the case of the minus sign would then be

$$\cos \omega_m t \cos \omega_c t - \sin \omega_m t \sin \omega_c t = \cos(\omega_m t + \omega_c t) = \cos(\omega_m + \omega_c) t \quad (5.16)$$

and in the case of the plus sign, it would be

$$\cos \omega_m t \cos \omega_c t + \sin \omega_m t \sin \omega_c t = \cos(\omega_c t - \omega_m t) = \cos(\omega_c - \omega_m) t \quad (5.17)$$

5.2.4.3 R3E System

This is the single side band reduced carrier (SSBRC) system, also called the pilot carrier system. Re-insertion of the carrier with much reduced amplitude before transmission is aimed at facilitating receiver tuning and demodulation. This reduced carrier amplitude is 16 dB or 26 dB below the value it would have, had it not been suppressed in the first place. This attenuated carrier signal, while retaining the advantage of saving in power, provides a reference signal to help demodulation in the receiver.

5.2.4.4 J3E System

This is the single side band suppressed carrier (SSBSC) system and is usually referred to as the SSB, in which the carrier is suppressed by at least 45 dB in the transmitter. It was not popular initially due to the requirement for high receiver stability. However, with the advent of synthesizer driven receivers, it has now become the standard form of radio communication.

Generation of SSB signals was briefly described in earlier paragraphs. Suppression of the carrier in an AM signal is achieved in the building block known as the balanced modulator (BM). Figure 5.9 shows the typical circuit implemented using field effect transistors (FETs). The modulating signal is applied in push–pull to a pair of identical FETs as shown and, as a result, the modulating signals appearing at the gates of the two FETs are 180° out of phase. The carrier signal, as is evident from the circuit, is applied to the two gates in phase. The modulated output currents of the two FETs produced as a result of their respective gate signals are combined in the centre tapped primary of the output transformer. If the two halves of the

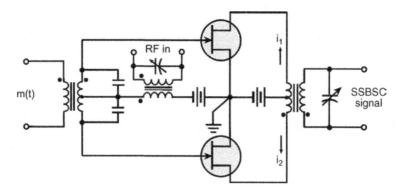

Figure 5.9 Balanced modulator

circuit are perfectly symmetrical, it can be proved with the help of simple mathematics that the carrier signal frequency will be completely cancelled in the modulated output and the output would contain only the modulating frequency, the sum frequency and the difference frequency components. The modulating frequency component can be removed from the output by tuning the output transformer.

Demodulation of SSBSC signals can be implemented by using a coherent detector scheme, as outlined in case of demodulation of the DSBSC signal in earlier paragraphs. Figure 5.10 shows the arrangement.

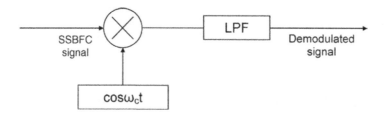

Figure 5.10 Coherent detector for demodulation of the SSBSC signal

5.2.4.5 B8E System

This system uses two independent side bands with the carrier either attenuated or suppressed. This form of amplitude modulation is also known as independent side band (ISB) transmission and is usually employed for point-to-point radio telephony.

5.2.4.6 C3F System

Vestigial side band (VSB) transmission is the other name for this system. It is used for transmission of video signals in commercial television broadcasting. It is a compromise between SSB and DSB modulation systems in which a vestige or part of the unwanted side band is also transmitted, usually with a full carrier along with the other side band. The typical bandwidth required to transmit a VSB signal is about 1.25 times that of an SSB signal. VSB transmission is used in commercial television broadcasting to conserve bandwidth. Figure 5.11 shows the spectrum of transmitted signals in the case of NTSC (National Television System Committee) TV standards followed in the United States, Canada and Japan Figure 5.11 (a) and PAL (Phase Alternation Line) TV standards followed in Europe, Australia and elsewhere [Figure 5.11 (b)]. As can be seen from the two figures, if the channel width is from, say, A to B MHz, the picture carrier is at $(A + 1.25)$ MHz and the sound carrier is at $(B - 0.25)$ MHz.

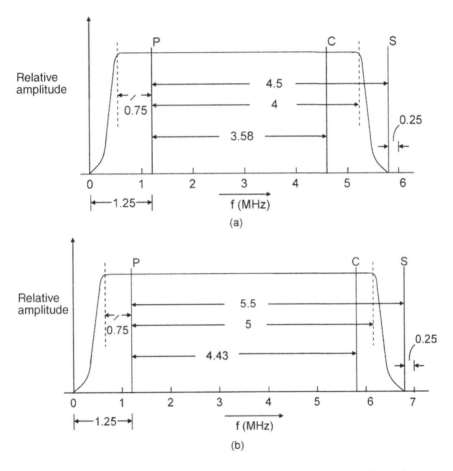

Figure 5.11 (a) NTSC TV standard signal and (b) PAL TV standard signal

The VSB signal can be generated by passing a DSB signal through an appropriate side band shaping filter, as shown in Figure 5.12. The demodulation scheme for the VSB signal is shown in Figure 5.13.

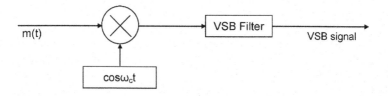

Figure 5.12 Generation of the VSB Signal

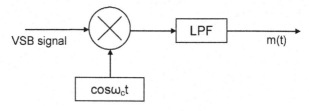

Figure 5.13 Demodulation of the VSB signal

Problem 5.1
Compute the saving in power in case of an SSBSC signal as compared to a standard AM signal for (a) modulation index, $m = 0.5$ and (b) modulation index, $m = 1$.

Solution: Total power in case of standard AM signal is given by

$$P_t = P_c \left(1 + \frac{m^2}{2}\right)$$

and total power in case of SSBSC signal is given by

$$P_t = P_c \left(\frac{m^2}{4}\right)$$

(a) For $m = 0.5$, in case of standard AM signal, $P_t = 1.125 \times P_c$ and for SSBSC signal $P_t = 0.0625 \times P_c$
Therefore, percentage saving in power = $[(1.125P_c - 0.0625P_c)/1.125P_c] \times 100\%$ = 94.4%

(b) For $m = 1$, in case of standard AM signal, $P_t = 1.5 \times P_c$ and for SSBSC signal $P_t = 0.25 \times P_c$
Therefore, percentage saving in power = $[(1.5P_c - 0.25P_c)/1.5P_c] \times 100\%$ = 83.3%

Problem 5.2
A 500 watt carrier signal is amplitude modulated with a modulation percentage of 60 %. Compute the total power in the modulated signal if the form of amplitude modulation used is (a) A3E and (b) J3E.

Solution: (a) A3E is the double side band AM with a full carrier. The total power in the modulated signal in this case is given by

$$P_t = P_c \left(1 + \frac{m^2}{2}\right)$$

Frequency Modulation

Here $P_c = 500$ watt and $m = 0.6$. Therefore, $P_t = 500 \times [1 + (0.6 \times 0.6/2)] = 500 \times 1.18 = 590$ watt.

(b) J3E is a single side band suppressed carrier system. The total power in this case is given by

$$P_t = P_c \left(\frac{m^2}{4} \right)$$

Therefore, $P_t = P_c(m^2/4) = 500 \times (0.6 \times 0.6/4) = 45$ watt.

Problem 5.3

The standard AM signal (A3E form of AM) broadcast from a station has an average percent of modulation of 60 %. If it is decided to shift to J3E form of transmission, what would be average power saving if the signal strength in the reception area is to remain unaltered?

Solution: A3E is the double side band AM with full carrier. Total power in the modulated signal would then be

$$P_t = P_c \left(1 + \frac{m^2}{2} \right)$$

Substituting the value of (m), we get

$$P_t = 1.18 \times P_c$$

In case of J3E transmission,

$$P_t = P_c \left(\frac{m^2}{4} \right)$$

Substituting the value of (m), we get

$$P_t = 0.09 \times P_c$$

Percentage power saving $= [(1.18 P_c - 0.09 P_c)/1.18 P_c] \times 100\% = 1.09/1.18 \times 100\% = 92.4\%$

Problem 5.4

A message signal $m(t)$ is band-limited to ω_M. It is frequency translated by multiplying it by a signal $A \cos \omega_c t$. What should be the value of ω_c, if the bandwidth of the resultant signal is 0.5 % of the carrier frequency ω_c.

Solution: Multiplication of the two signals gives an amplitude modulated signal with sum $(\omega_c + \omega_M)$ and difference $(\omega_c - \omega_M)$ frequency components. The bandwidth of the resultant signal is $2\omega_M$. Now, $0.5 \omega_c/100 = 2\omega_M$. Therefore, $\omega_c = 400 \omega_M$.

5.3 Frequency Modulation

In frequency modulation (FM), the instantaneous frequency of the modulated signal varies directly as the instantaneous amplitude of the modulating or the base band signal. The rate at which these frequency variations take place is of course proportional to the modulating frequency. If the modulating signal is expressed by $v_m = V_m \cos \omega_m t$, then the instantaneous frequency f of an FM signal is mathematically expressed by

$$f = f_c(1 + KV_m \cos \omega_m t) \tag{5.18}$$

where

f_c = unmodulated carrier frequency
$V_m \cos \omega_m t$ = instantaneous modulating voltage
V_m = peak amplitude of the modulating signal
ω_m = modulating frequency
K = constant of proportionality

The instantaneous frequency is maximum when $\cos \omega_m t = 1$ and minimum when $\cos \omega_m t = -1$. This gives

$$f_{max} = f_c(1 + KV_m) \quad \text{and} \quad f_{min} = f_c(1 - KV_m)$$

where, f_{max} is the maximum instantaneous frequency and f_{min} is the minimum instantaneous frequency.

Frequency deviation (δ) is one of the important parameters of an FM signal and is given by $(f_{max} - f_c)$ or $(f_c - f_{min})$. This gives

$$\text{Frequency deviation } \delta = f_{max} - f_c = f_c - f_{min} = KV_m f_c \tag{5.19}$$

Figures 5.14 (a) to (c) show the modulating signal (taken as a single tone signal in this case), the unmodulated carrier and the modulated signal respectively. An FM signal can be

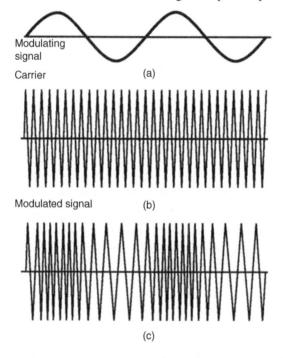

Figure 5.14 Frequency modulation

mathematically represented by

$$v_{FM}(t) = A \sin[\omega_c t + (\delta/f_m) \sin \omega_m t] = A \sin[\omega_c t + m_f \sin \omega_m t] \quad (5.20)$$

where m_f = modulation index = δ/f_m. A is the amplitude of the modulated signal which in turn is equal to the amplitude of the carrier signal.

The depth of modulation in the case of an FM signal is defined as the ratio of the frequency deviation (δ) to the maximum allowable frequency deviation. The maximum allowable frequency deviation is different for different services and is also different for different standards, even for a given type of service using this form of modulation. For instance, the maximum allowable frequency deviation for a commercial FM radio broadcast is 75 kHz. It is 50 kHz for an FM signal of television sound in CCIR (Consultative Committee on International Radio) standards and 25 kHz for an FM signal of television sound in FCC (Federal communications commission) standards. Therefore,

Depth of modulation for commercial FM radio broadcast = δ(in kHz)/75
Depth of modulation for TV FM sound in CCIR standards = δ(in kHz)/50
Depth of modulation for TV FM sound in FCC standards = δ(in kHz)/25

5.3.1 Frequency Spectrum of the FM Signal

We have seen that an FM signal involves a sine of a sine. The solution of this expression involves the use of Bessel functions. The expression for the FM signal can be rewritten as

$$\begin{aligned} v_{FM}(t) = A\{ & J_0(m_f) \sin \omega_c t + J_1(m_f)[\sin(\omega_c + \omega_m)t - \sin(\omega_c - \omega_m)t] \\ & + J_2(m_f)[\sin(\omega_c + 2\omega_m)t - \sin(\omega_c - 2\omega_m)t] \\ & + J_3(m_f)[\sin(\omega_c + 3\omega_m)t - \sin(\omega_c - 3\omega_m)t] + \cdots \} \end{aligned} \quad (5.21)$$

Thus the spectrum of an FM signal contains the carrier frequency component and apparently an infinite number of side bands. In general, $J_n(m_f)$ is the Bessel function of the first kind and nth order. It is evident from this expression that it is the value of m_f and the value of the Bessel functions that will ultimately decide the number of side bands having significant amplitude and therefore bandwidth. Figure 5.15 shows how the carrier and side band amplitudes vary as a function of the modulation index. In fact, the curves shown in Figure 5.15 are nothing but plots of $J_0(m_f), J_1(m_f), J_2(m_f), J_3(m_f), \ldots$ as a function of m_f. Also, $J_0(m_f), J_1(m_f), J_2(m_f), J_3(m_f), \ldots$ respectively represent the amplitudes of the carrier, the first side band, the second side band, the third side band and so on.

The following observations can be made from this expression:

1. For every modulating frequency, the FM signal contains an infinite number of side bands in addition to the carrier frequency. In an AM signal, there are only three frequency components, i.e. the carrier frequency, the lower side band frequency and the upper side band frequency.
2. The modulation index (m_f) determines the number of significant side bands. The higher the modulation index, the more the number of significant side bands. Figure 5.16 shows

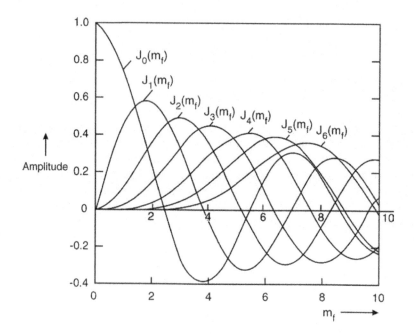

Figure 5.15 Variation of carrier and side band amplitudes as a function of the modulation index

the spectra of FM signals for a given sinusoidal modulating signal for different values of the modulation index. As is evident from the figure, a higher modulation index leads to a larger number of side band frequency components having significant amplitude, i.e. side band frequency components having an appreciable relative amplitude.

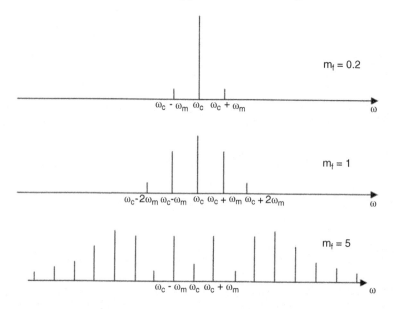

Figure 5.16 Spectra of FM signals for different values of the modulation index

3. The side band distribution is symmetrical about the carrier frequency. The pair of side band frequencies for which the Bessel function has a negative value signifies a 180° phase change for that pair.
4. In the case of an AM signal, when the modulation index increases so does the side band power and hence the transmitted power. In the case of an FM signal also, as the modulation index increases, the side band power increases. However, it does so only at the cost of carrier power so that the total transmitted power remains constant.
5. In FM, the carrier component can disappear completely for certain specific values of m_f for which $J_0(m_f)$ becomes zero. These values are 2.4, 5.5, 8.6, 11.8 and so on.

5.3.2 Narrow Band and Wide Band FM

An FM signal, whether it is a narrow band FM signal or a wide band FM signal, is decided by its bandwidth and in turn by its modulation index. For a modulation index m_f much less than 1, the signal is considered as the narrow band FM signal. It can be shown that for m_f less than 0.2, 98 % of the normalized total signal power is contained within the bandwidth as given by:

$$\text{Bandwidth} = 2(m_f + 1)\omega_m \approx 2\omega_m \qquad \text{for } m_f \ll 1 \tag{5.22}$$

where ω_m is the sinusoidal modulating frequency.

In the case of an FM signal with an arbitrary modulating signal m(t) band limited to ω_M, another parameter called the deviation ratio (D) can be defined as

$$D = \frac{\text{maximum frequency deviation}}{\text{bandwidth of } m(t)} \tag{5.23}$$

The deviation ratio D has the same significance for arbitrary modulation as the modulation index m_f for sinusoidal modulation. The bandwidth in this case is given by

$$\text{Bandwidth} = 2(D + 1)\omega_M \tag{5.24}$$

This expression for bandwidth is generally referred to as Carson's rule.

In the case of $D \ll 1$, the FM signal is considered as narrow band signal and the bandwidth is given by the expression

$$\text{Bandwidth} = 2(D + 1)\omega_M \approx 2\omega_M \tag{5.25}$$

In the case $m_f \gg 1$ (for sinusoidal modulation) or $D \gg 1$ (for an arbitrary modulation signal band limited to ω_M), the FM signal is termed as the wide band FM. The bandwidth of a wide band FM signal in the case of a sinusoidal modulating signal is given by

$$\text{Bandwidth} \approx 2m_f \omega_m \tag{5.26a}$$

The bandwidth of a wide band FM signal in case of an arbitrary modulating signal band limited to ω_M is given by

$$\text{Bandwidth} \approx 2D\omega_M \qquad (5.26b)$$

5.3.3 Noise in the FM Signal

It will be seen in the following paragraphs that frequency modulation is much less affected by the presence of noise as compared to the effect of noise on an amplitude modulated signal. Whenever a noise voltage with peak amplitude V_n is present along with a carrier voltage of peak amplitude V_c, the noise voltage amplitude modulates the carrier with a modulation index equal to V_n/V_c. It also phase modulates the carrier with a phase deviation equal to $\sin^{-1}(V_n/V_c)$. This expression for phase deviation results when a single frequency noise voltage is considered vectorially and the noise voltage vector is superimposed on the carrier voltage vector, as shown in Figure 5.17. An FM receiver is not affected by the amplitude change as it can be removed in the limiter circuit inside the receiver. Also, an AM receiver will not be affected by the phase change. It is therefore the effect of phase change on the FM receiver and the effect of amplitude change on the AM receiver that can be used as the yardstick for determining the noise performance of the two modulation techniques. Two very important aspects that need to be addressed when the two communication techniques are compared *vis-à-vis* their noise performance are the effects of the modulation index and the signal-to-noise ratio at the receiver input. These are considered in the following paragraphs and are suitably illustrated with the help of examples.

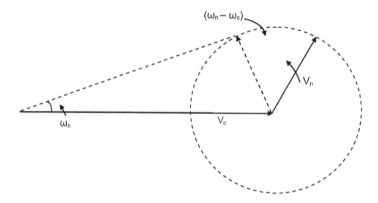

Figure 5.17 Effect of noise on signal

5.3.3.1 Effect of the Modulation Index

Let us take the case of noise voltage amplitude being one-fourth of the carrier amplitude. The carrier-to-noise voltage ratio of 4 is equivalent to the carrier-to-noise power ratio of 16 or 12 dB. The reasons for taking these figures will be obvious in the latter part of this section.

A modulating frequency of 15 kHz is assumed, which is the highest possible frequency for voice communication. A modulation index of unity for both AM and FM is considered. In the case of an AM receiver, the noise-to-signal ratio will be 0.25 or 25 % as the noise voltage amplitude is one-fourth of the carrier voltage amplitude. In the case of an FM receiver, the noise-to-signal ratio will be 14.5°/57.3°(1 radian = 57.3°). This equals 0.253 or 25.3 %. Thus, when a signal-to-noise ratio of 12 dB and a unity modulation index for AM and FM systems is assumed, the noise performance of the two systems is more or less the same, slightly better in the case of the AM system.

The effect of change in modulating noise frequencies will now be examined. It should be remembered that the noise frequency will interfere with the desired signals only when the noise difference frequency, that is the frequency produced by the mixing action of the carrier and noise frequencies, lies within the pass band of the receiver. A change in modulating and noise difference frequencies does not have any effect on the noise modulation index and signal modulation index in the case of AM. As a result, the noise-to-signal ratio in the case of AM remains unaltered. In the case of FM, when the noise difference frequency is lowered, there again is no effect of this change on the noise modulation index as a constant noise-to-carrier voltage means a constant phase modulation due to noise and hence a constant noise modulation index. However, a reduction in modulating frequency implies an increase in the signal modulation index in the same proportion. This leads to a reduction in the noise-to-signal ratio in the same proportion. For instance, in the example considered earlier, if the modulating frequency is decreased from 15 kHz to 30 Hz, the noise-to-signal ratio in the case of FM will reduce to $(0.253 \times 30/15000) = 0.0005$ or 0.05 %, while the noise-to-signal ratio in the case of AM remains the same at 0.25 or 25 %.

Figure 5.18 shows how noise at the receiver output varies with the noise difference frequency or noise side band frequency assuming that noise frequencies are evenly spread over the entire pass band of the receiver. Figure 5.18 (a) is for the case where $m_f = 1$ at the highest modulating frequency and Figure 5.18 (b) depicts the case where $m_f = 5$ at the highest modulating frequency. A rectangular distribution in the case of AM is obvious.

5.3.3.2 Effect of the Signal-to-Noise Ratio

An FM receiver uses a limiter circuit that precedes the FM demodulator. The idea behind the use of a limiter circuit is the fact that any amplitude variations in an FM signal are spurious and contain no intelligence information. Since FM demodulator circuits to some extent respond to amplitude variations, removing these amplitude variations results in a better noise performance in an FM receiver. The amplitude limiter acts on the stronger signals and tends to reject the weaker signals. Thus when the signal-to-noise ratio at the limiter input is very low, that is when the signal is weak, the FM system offers a poorer performance as compared to an AM system. The FM system offers a better performance with respect to an AM system only when the signal-to-noise ratio is above a certain threshold value, which is 8 dB (or 9 dB). This is depicted in Figure 5.19. It is clear from the curves shown in Figure 5.19, the FM system offers full improvement over the AM system when the signal-to-noise ratio is about 3 dB greater than this threshold of 9 dB. It is evident from Figure 5.19 that the AM system has a definite advantage as compared to the FM system for an input signal-to-noise

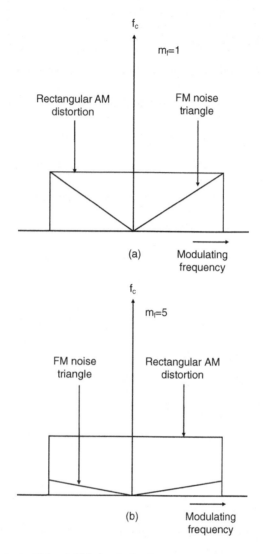

Figure 5.18 (a) Noise in the FM and AM signals for the modulation index equal to 1. (b) Noise in FM and AM signals for the modulation index equal to 5

ratio less than 9 dB. Improvement of the FM system over the AM system is visible for an input signal-to-noise ratio greater than 9dB. The quantum of improvement increases with an increase in the signal-to-noise ratio until it reaches its maximum value at the signal-to-noise ratio of 12 dB. From then onwards, FM offers this maximum improvement over AM. It may be mentioned that the first threshold (point X) represents a modulation index (m_f) of unity and the second threshold (point Y) corresponds to a modulation index (m_f) equal to the deviation ratio, which is the ratio of maximum frequency deviation to maximum modulating frequency.

Frequency Modulation

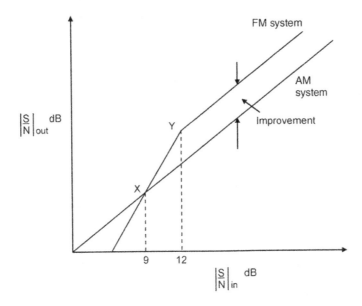

Figure 5.19 Signal-to-noise ratio in FM and AM systems

In summary, it can be stated that an FM system offers a better performance than an AM system provided:

1. The modulation index is greater than unity.
2. The amplitude of the carrier is greater than the maximum noise peak amplitudes.
3. The receiver is insensitive to amplitude variations.

5.3.3.3 Pre-emphasis and De-emphasis

In the case of FM, noise has a greater effect on higher modulating frequencies than it has on lower ones. This is because of the fact that FM results in smaller values of phase deviation at the higher modulating frequencies whereas the phase deviation due to white noise is constant for all frequencies. Due to this, the signal-to-noise ratio deteriorates at higher modulating frequencies. If the higher modulating frequencies above a certain cut-off frequency were boosted at the transmitter prior to modulation according to a certain known curve and then reduced at the receiver in the same fashion after the demodulator, a definite improvement in noise immunity would result. The process of boosting the higher modulating frequencies at the transmitter and then reducing them in the receiver are respectively known as pre-emphasis and de-emphasis. Figure 5.20 shows the pre-emphasis and de-emphasis curves.

Having discussed various aspects of noise performance of an FM system, it would be worthwhile presenting the mathematical expression that could be used to compute the base band signal-to-noise ratio at the output of the demodulator. Without getting into intricate

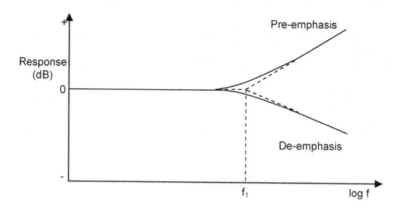

Figure 5.20 Pre-emphasis and de-emphasis curves

mathematics, the following expression can be written for the base band signal-to-noise ratio (S_b/N_b):

$$\frac{S_b}{N_b} = 3 \left(\frac{f_d}{f_m}\right)^2 \left(\frac{B}{2f_m}\right)\left(\frac{C}{N}\right) \tag{5.27}$$

where

f_d = frequency deviation
f_m = highest modulating frequency
B = receiver bandwidth
C = carrier power at the receiver input
N = noise power (kTB) in bandwidth (B)

The above expression does not take into account the improvement due to the use of pre-emphasis and de-emphasis. In that case the expression is modified to

$$\frac{S_b}{N_b} = \left(\frac{f_d}{f_1}\right)^2 \left(\frac{B}{f_m}\right)\left(\frac{C}{N}\right) \tag{5.28}$$

where f_1 = cut-off frequency for the pre-emphasis/de-emphasis curve.

5.3.4 Generation of FM Signals

In the case of an FM signal, the instantaneous frequency of the modulated signal varies directly as the instantaneous amplitude of the modulating or base band signal. The rate at which these frequency variations take place is of course proportional to the modulating frequency. Though there are many possible schemes that can be used to generate the signal characterized above, all of them depend simply on varying the frequency of an oscillator circuit in accordance with the modulating signal input.

One of the possible methods is based on the use of a varactor (a voltage variable capacitor) as a part of the tuned circuit of an L–C oscillator. The resonant frequency of this oscillator will not vary directly with the amplitude of the modulating frequency as it is inversely proportional to the square root of the capacitance. However, if the frequency deviation is kept

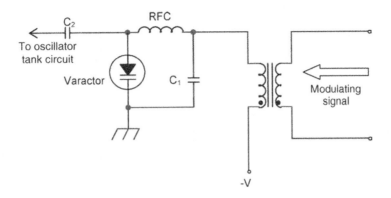

Figure 5.21 L–C oscillator based direct method of FM signal generation

small, the resulting FM signal is quite linear. Figure 5.21 shows the typical arrangement when the modulating signal is an audio signal. This is also known as the direct method of generating an FM signal, as in this case the modulating signal directly controls the carrier frequency.

Another direct method scheme that can be used for generation of an FM signal is the reactance modulator. In this, the reactance offered by a three-terminal active device such as an field effect transistor (FET) or a bipolar junction transistor (BJT) forms a part of the tuned circuit of the oscillator. The reactance in this case is made to vary in accordance with the modulating signal applied to the relevant terminal of the active device. For example, in the case of FET, the drain source reactance can be shown to be proportional to the transconductance of the device, which in turn can be made to depend upon the bias voltage at its gate terminal. The main advantage of using the reactance modulator is that large frequency deviations are possible and thus less frequency multiplication is required. One of the major disadvantages of both these direct method schemes is that carrier frequency tends to drift and therefore additional circuitry is required for frequency stabilization. The problem of frequency drift is overcome in crystal controlled oscillator schemes.

Although it is known that crystal control provides a very stable operating frequency, the exact frequency of oscillation in this case mainly depends upon the crystal characteristics and to a very small extent on the external circuit. For example, a capacitor connected across the crystal can be used to change its frequency, typically from 0.001 % to 0.005 %. The frequency change may be linear only up to a change of 0.001 %. Thus a crystal oscillator can be frequency modulated over a very small range by a parallel varactor. The frequency deviation possible with such a scheme is usually too small to be used directly. The frequency deviation in this case is then increased by using frequency multipliers as shown in Figure 5.22.

Another approach that eliminates the requirement of extensive chains of frequency multipliers is an indirect method where frequency deviation is not introduced at the source of the RF carrier signal, that is the oscillator. The oscillator in this case is crystal controlled to get the desired stability of the unmodulated carrier frequency and the frequency deviation is introduced at a later stage. The modulating signal phase modulates the RF carrier signal produced by the crystal controlled oscillator. Since, frequency is simply the rate of change of phase, phase modulation of the carrier has an associated frequency modulation. Introduction of a leading phase shift would lead to an increase in the RF carrier frequency and a lagging

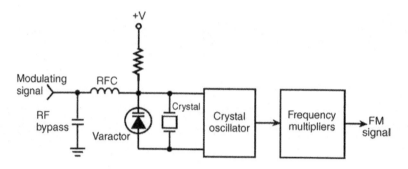

Figure 5.22 Crystal oscillator based scheme for FM signal generation

phase shift results in a reduced RF carrier frequency. Thus, if the phase of the RF carrier is shifted by the modulating signal in a proper way, the result is a frequency modulated signal. Since phase modulation also produces little frequency deviation, a frequency multiplier chain is required in this case too. Figure 5.23 shows the typical schematic arrangement for generating an FM signal via the phase modulation route.

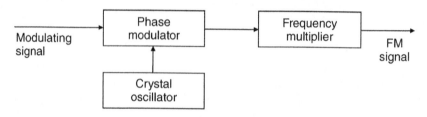

Figure 5.23 Indirect method of generating an FM signal by employing phase modulation

5.3.5 Detection of FM Signals

Detection of an FM signal involves the use of some kind of frequency discriminator circuit that can generate an electrical output directly proportional to the frequency deviation from the unmodulated RF carrier frequency. The simplest of the possible circuits would be the balanced slope detector, which makes use of two resonant circuits, one off-tuned to one side of the unmodulated RF carrier frequency and the other off-tuned to the other side of it. Figure 5.24 shows the basic circuit. When the input to this circuit is at the unmodulated carrier frequency, the two off-tuned slope detectors (or the resonant circuits) produce equal amplitude but out-of-phase outputs across them. The two signals, after passing through their respective diodes, produce equal amplitude opposing DC outputs, which combine together to produce a zero or near-zero output. When the received signal frequency is towards either side of the centre frequency, one output has a higher amplitude than the other to produce a net DC output across the load. The polarity of the output produced depends on which side of the centre frequency the received signal is. Figure 5.25 explains all of this. Such a detector circuit, however, does not find application for voice communication because of its poor response linearity.

Frequency Modulation

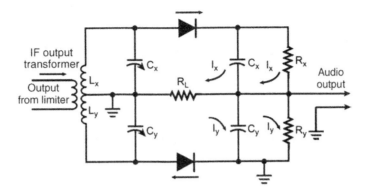

Figure 5.24 Basic circuit of the balanced slope detector (IF, intermediate frequency)

Another class of FM detectors, known as quadrature detectors, use a combination of two quadrature signals, i.e. two signals 90° out of phase, to obtain the frequency discrimination property. One of the two signals is the FM signal to be detected and its quadrature counterpart is generated by using either a capacitor or an inductor, as shown in Figure 5.26. The two signals here have been labelled E_a and E_b.

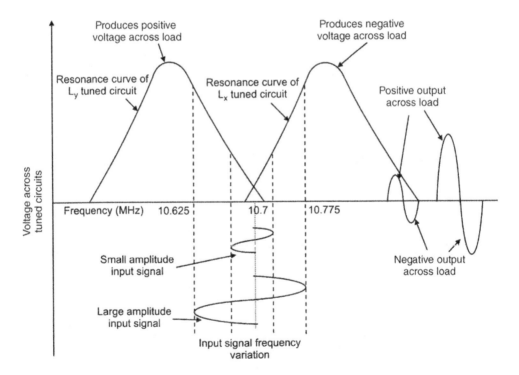

Figure 5.25 Output of the balanced slope detector

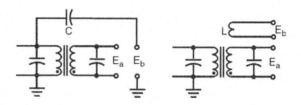

Figure 5.26 Generation of the quadrature signal

If the secondary of the transformer in the arrangements of Figure 5.26 is tuned to the unmodulated carrier frequency, then at this frequency, E_a and E_b are 90° out of phase, as shown in Figure 5.27 (a). The phase difference between E_a and E_b is greater than 90° when the input frequency is less than the unmodulated carrier frequency [Figure 5.27 (b)] and less than 90°, in case the input frequency is greater than the unmodulated carrier frequency [Figure 5.27 (c)]. The resultant is a signal that is proportional to $E_a E_b \cos \phi$, where ϕ is the phase deviation from 90°. This in fact forms the basis of two of the most commonly used FM detectors, namely the Foster–Seeley frequency discriminator and the ratio detector.

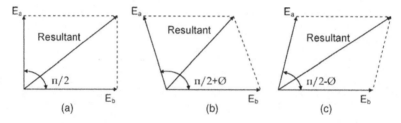

Figure 5.27 Phase diagrams of FM signals

In the Foster–Seeley frequency discriminator circuit of Figure 5.28 (a), the two quadrature signals are provided by the primary signal (E_p) as appearing at the centre tap of secondary and E_b. It should be appreciated that E_a and E_b are 180° out of phase and also that E_p, available at the centre tap of the secondary, is 90° out of phase with the total secondary signal. Signals E_1 and E_2 appearing across the two halves of the secondary have equal amplitudes when the received signal is at the unmodulated carrier frequency as shown in the phasor diagrams Figure 5.28 (b). E_1 and E_2 cause rectified currents I_1 and I_2 to flow in the opposite directions, with the result that voltage across R_1 and R_2 are equal and opposite. The detected voltage is zero for $R_1 = R_2$. The conditions when the received signal frequency deviates from the unmodulated carrier frequency value are also shown in the phasor diagrams. In the case of frequency deviation, there is a net output voltage whose amplitude and polarity depends upon the amplitude and sense of frequency deviation. The frequency response curve for this type of FM discriminator circuit is also given in the Figure 5.28 (c).

Another commonly used FM detector circuit is the ratio detector. This circuit has the advantage that it is insensitive to short term amplitude fluctuations in the carrier and therefore does not require an additional limiter circuit. The circuit configuration, as shown in Figure 5.29, is similar to the one given in the Foster–Seeley frequency discriminator circuit except for a couple of changes. These are a reversal of diode connections and the addition of a large capacitor (C_5). The time constant ($R_1 + R_2)C_5$ is much larger than the time period of even the

Frequency Modulation

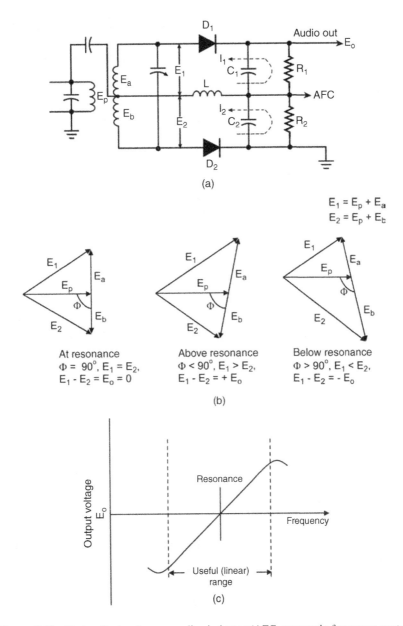

Figure 5.28 Foster–Seeley frequency discriminator (AFC, automatic frequency control)

lowest modulating frequency of interest. The detected signal in this case appears across the C_3–C_4 junction. The sum output across R_1–R_2 and hence across C_3–C_4 remains constant for a given carrier level and is also insensitive to rapid fluctuations in the carrier level. However, if the carrier level changes very slowly, C_5 charges or discharges to the new carrier level. The detected signal is therefore not only proportional to the frequency deviation but also depends upon the average carrier level.

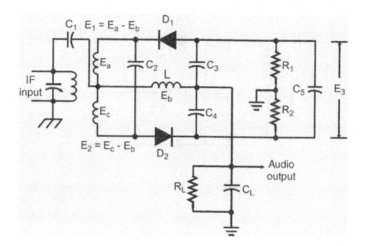

Figure 5.29 Ratio detector

Yet another form of FM detector is the one implemented using a phase locked loop (PLL). A PLL has a phase detector (usually a double balanced mixer), a lowpass filter and an error amplifier in the forward path and a voltage controlled oscillator (VCO) in the feedback path. The detected output appears at the output of the error amplifier, as shown in Figure 5.30. A PLL-based FM detector functions as follows.

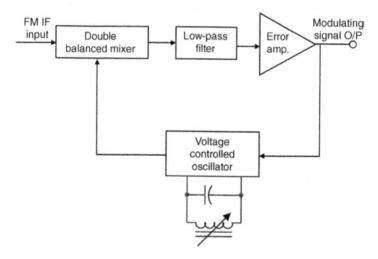

Figure 5.30 PLL based FM detector

The FM signal is applied to the input of the phase detector. The VCO is tuned to a nominal frequency equal to the unmodulated carrier frequency. The phase detector produces an error voltage depending upon the frequency and phase difference between the VCO output and the instantaneous frequency of the input FM signal. As the input frequency deviates from the centre frequency, the error voltage produced as a result of frequency difference after passing

Frequency Modulation

through the lowpass filter and error amplifier drives the control input of the VCO to keep its output frequency always in lock with the instantaneous frequency of the input FM signal. As a result, the error amplifier always represents the detected output. The double balanced mixer nature of the phase detector suppresses any carrier level changes and therefore the PLL-based FM detector requires no additional limiter circuit.

A comparison of the three types of FM detector reveals that the Foster–Seeley type frequency discriminator offers excellent linearity of response, is easy to balance and the detected output depends only on the frequency deviation. However, it needs high gain RF and IF stages to ensure limiting action. The ratio detector circuit, on the other hand, requires no additional limiter circuit; detected output depends both on the frequency deviation as well as on the average carrier level. However, it is difficult to balance. The PLL-based FM detector offers excellent reproduction of the modulating signal, is easy to balance and has low cost and high reliability.

Problem 5.5
An FM signal is represented by the equation $v(t) = 10 \sin(7.8 \times 10^8 t + 6 \sin 1450t)$. Determine the unmodulated carrier frequency, the modulating frequency and the modulation index.

Solution:
Comparing the given equation with standard form of equation for FM signal given by $v(t) = V_m \sin(\omega_c t + m \sin \omega_m t)$.
Unmodulated carrier frequency, $\omega_c = 7.8 \times 10^8$ or $f_c = (7.8 \times 10^8 / 2\pi)$Hz = 124.14 MHz
Modulation index, $m = 6$
Modulating frequency, $\omega_m = 1450$ or $f_m = 1450/2\pi = 230.77$ Hz

Problem 5.6
An FM signal is represented by $v(t) = 15 \cos[10^8 \pi t + 6 \sin(2 \times 10^3 \pi t)]$. Determine (a) Maximum phase deviation and (b) Maximum frequency deviation

Solution:
Comparing the given equation with the standard form of FM signal expression, we get, $(\omega_c t + \phi) = [(10^8 \pi)t + 6 \sin(2 \times 10^3 \pi t)]$. Therefore, phase $(\phi) = 6 \sin 2\pi(10^3)t$. This gives maximum phase deviation $\phi_{max} = 6$ radian.
Frequency is nothing but rate of change of phase. Therefore, maximum frequency deviation is nothing but maximum value of $(d\phi/dt)$.
Now, $(d\phi/dt) = 6 \times 2\pi(10^3) \cos 2\pi(10^3)t = 12\pi(10^3) \cos 2\pi(10^3)t$

Maximum frequency deviation = $12\pi(10^3)$ radian/sec = 6 kHz

Problem 5.7
An FM signal is represented by

$$v(t) = 15 \cos[2\pi(10^8)t + 150 \cos 2\pi(10^3)t]$$

Determine the bandwidth of the signal. Also write the expression for the instantaneous frequency of the signal.

Solution: Comparing the given expression with the standard expression for the FM signal, we get the value of modulation index (m_f) as

$$m_f = 150$$

The modulating frequency $f_m = 2\pi(10^3)/(2\pi) = 1$ kHz. Therefore, the frequency deviation $\delta = 150 \times 1$ kHz $= 150$ kHz. The bandwidth can now be computed from:

$$\text{Bandwidth} = 2(m_f + 1)f_m = (2 \times 151 \times 1)\text{kHz} = 302 \text{ kHz}$$

Expression for instantaneous frequency can be written by taking the derivative of the expression for instantaneous phase ϕ, where

$$\phi = 2\pi(10^8)t + 150 \cos 2\pi(10^3)t$$

$$\text{Instantaneous frequency } \omega = \frac{d\phi}{dt} = 2\pi(10^8) - 150 \times 2\pi(10^3) \sin 2\pi(10^3)t$$

$$= 2\pi(10^8) - 300\pi(10^3) \sin 2\pi(10^3)t$$

$$f = 10^8 - 150(10^3) \sin 2\pi(10^3)t$$

Problem 5.8
A 95 MHz carrier is frequency modulated by a sinusoidal signal and the modulated signal is such that maximum frequency deviation achieved is 50 kHz. Determine the modulation index and bandwidth of the modulated signal, if the modulating signal frequency is (a) 1 kHz and (b) 100 kHz.

Solution:
In the first case, frequency deviation = 50 kHz and modulating frequency = 1 kHz
Therefore, modulation index = $(50 \times 10^3)/(1 \times 10^3) = 50$
Bandwidth = $[2 \times (50 + 1) \times 1]$ kHz = 102 kHz
In the second case, modulating frequency = 100 kHz
Therefore, modulation index = $(50 \times 10^3)/(100 \times 10^3) = 0.5$
Bandwidth = $[2 \times (0.5 + 1) \times 100]$ kHz = 300 kHz

Problem 5.9
Find the bandwidth of an FM signal produced in a commercial FM broadcast with modulating signal frequency being in the range of 50 Hz to 15 kHz and maximum allowable frequency deviation being 75 kHz.

Solution:
Maximum allowed frequency deviation = 75 kHz and highest modulating frequency = 15 KHz
Therefore, deviation ratio, $D = (75 \times 10^3)/(15 \times 10^3) = 5$
Bandwidth = $2(D+1)f_m = [2 \times (5+1) \times 15]$ kHz = 180 kHz

Problem 5.10

A carrier when frequency modulated by a certain sinusoidal signal of 1 kHz produces a modulated signal with a bandwidth of 20 kHz. If the same carrier signal is frequency modulated by another modulating signal whose peak amplitude is 3 times that of the previous signal and the frequency is one-half of the previous signal, determine the bandwidth of the new modulated signal.

Solution: The bandwidth $= 2(m_f + 1)f_m$. This gives $2(m_f + 1) \times 1 = 20$ or $m_f = 9$. Since the amplitude of the new modulating signal is 3 times that of the previous signal, it will produce a frequency deviation that is 3 times that produced by the previous signal. Also, new modulating frequency is one-half of the previous signal frequency. Therefore, the new modulation index, which is the ratio of the frequency deviation to the modulating frequency, will be 6 times that produced by the previous signal. Therefore,

New modulation index $= 9 \times 6 = 54$

New bandwidth $= 2(m_f + 1)f_m = [2 \times (54 + 1) \times 0.5]\text{kHz} = 55 \text{ kHz}$

5.4 Pulse Communication Systems

Pulse communication systems differ from continuous wave communication systems in the sense that the message signal or intelligence to be transmitted is not supplied continuously as in the case of AM or FM. In this case, it is sampled at regular intervals and it is the sampled data that is transmitted. All pulse communication systems fall into either of two categories, namely analogue pulse communication systems and digital pulse communication systems. Analogue and digital pulse communication systems differ in the mode of transmission of sampled information. In the case of analogue pulse communication systems, representation of the sampled amplitude may be infinitely variable whereas in the case of digital pulse communication systems, a code representing the sampled amplitude to the nearest predetermined level is transmitted.

5.4.1 Analogue Pulse Communication Systems

Important techniques that fall into the category of analogue pulse communication systems include:

1. Pulse amplitude modulation
2. Pulse width (or duration) modulation
3. Pulse position modulation

5.4.1.1 Pulse Amplitude Modulation

In the case of pulse amplitude modulation (PAM), the signal is sampled at regular intervals and the amplitude of each sample, which is a pulse, is proportional to the amplitude of the modulating signal at the time instant of sampling. The samples, shown in Figure 5.31, can have either a positive or negative polarity. In a single polarity PAM, a fixed DC level can be added to the signal, as shown in Figure 5.31 (c). These samples can then be transmitted either by a cable or used to modulate a carrier for wireless transmission. Frequency modulation is usually employed for the purpose and the system is known as PAM-FM.

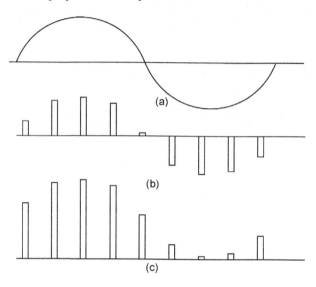

Figure 5.31 Pulse amplitude modulation

5.4.1.2 Pulse Width Modulation

In the case of pulse width modulation (PWM), as shown in Figure 5.32, the starting time of the sampled pulses and their amplitude is fixed. The width of each pulse is proportional to the amplitude of the modulating signal at the sampling time instant.

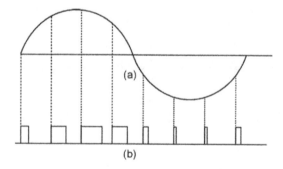

Figure 5.32 Pulse width modulation

5.4.1.3 Pulse Position Modulation

In the case of pulse position modulation (PPM), the amplitude and width of the sampled pulses is maintained as constant and the position of each pulse with respect to the position of a recurrent reference pulse varies as a function of the instantaneous sampled amplitude of the modulating signal. In this case, the transmitter sends synchronizing pulses to operate timing circuits in the receiver.

A pulse position modulated signal can be generated from a pulse width modulated signal. In a PWM signal, the position of the leading edges is fixed whereas that of the trailing edges depends upon the width of the pulse, which in turn is proportional to the amplitude of the modulating signal at the time instant of sampling. Quite obviously, the trailing edges constitute the pulse position modulated signal. The sequence of trailing edges can be obtained by differentiating the PWM signal and then clipping the leading edges as shown in Figure 5.33. Pulse width modulation and pulse position modulation both fall into the category of pulse time modulation (PTM).

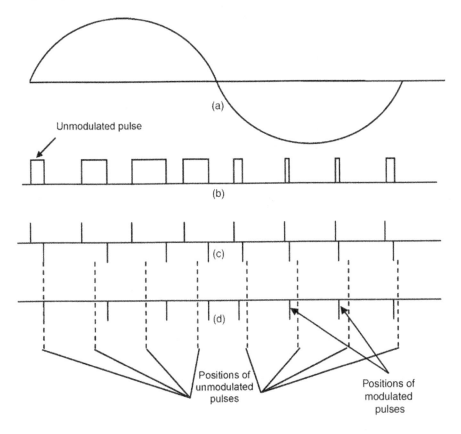

Figure 5.33 Pulse position modulation

5.4.2 Digital Pulse Communication Systems

Digital pulse communication techniques differ from analogue pulse communication techniques described in the previous paragraphs in the sense that in the case of analogue pulse modulation the sampling process transforms the modulating signal into a train of pulses, with each pulse in the pulse train representing the sampled amplitude at that instant of time. It is one of the characteristic features of the pulse, such as amplitude in the case of PAM, width in the case of PWM and position of leading or trailing edges in the case of PPM, that is varied in accordance with the amplitude of the modulating signal. What is important to note here is that the characteristic parameter of the pulse, which is amplitude or width or position, is infinitely variable. As an illustration, if in the case of pulse width modulation, every volt of modulating signal amplitude corresponded to 1 μs of pulse width, then 5.23 volt and 5.24 volt amplitudes would be represented by 5.23μs and 5.24 μs respectively. Further, there could be any number of amplitudes between 5.23 volts and 5.24 volts. It is not the same in the case of digital pulse communication techniques, to be discussed in the paragraphs to follow, where each sampled amplitude is transmitted by a digital code representing the nearest predetermined level.

Important techniques that fall into the category of digital pulse communication systems include:

1. Pulse code modulation (PCM)
2. Differential PCM
3. Delta modulation
4. Adaptive delta modulation

5.4.2.1 Pulse Code Modulation

In pulse code modulation (PCM), the peak-to-peak amplitude range of the modulating signal is divided into a number of standard levels, which in the case of a binary system is an integral power of 2. The amplitude of the signal to be sent at any sampling instant is the nearest standard level. For example, if at a particular sampling instant, the signal amplitude is 3.2 volts, it will not be sent as a 3.2 volt pulse, as might have been the case with PAM, or a 3.2 μs wide pulse, as for the case with PWM; instead it will be sent as the digit 3, if 3 volts is the nearest standard amplitude. If the signal range has been divided into 128 levels, it will be transmitted as 0000011. The coded waveform would be like that shown in Figure 5.34 (a). This process is known as quantizing. In fact, a supervisory pulse is also added with each code group to facilitate reception. Thus, the number of bits for 2^n chosen standard levels per code group is $n + 1$. Figure 5.34 (b) illustrates the quantizing process in PCM.

It is evident from Figure 5.34 (b) that the quantizing process distorts the signal. This distortion is referred to as quantization noise, which is random in nature as the error in the signal's amplitude and that actually sent after quantization is random. The maximum error can be as high as half of the sampling interval, which means that if the number of levels used were 16, it would be 1/32 of the total signal amplitude range. It should be mentioned here that it would be unfair to say that a PCM system with 16 standard levels will necessarily have a signal-to-quantizing noise ratio of 32:1, as neither the signal nor the quantizing noise will always have its maximum value. The signal-to-noise ratio also depends upon many other factors and also its dependence on the number of quantizing levels is statistical in nature. Nevertheless, an increase in the number of standard levels leads to an increase in the signal-to-noise ratio. In practice, for speech signals, 128 levels are considered as adequate. In addition, the more the number of levels, the larger is the number of bits to be transmitted and therefore higher is the required bandwidth.

Pulse Communication Systems

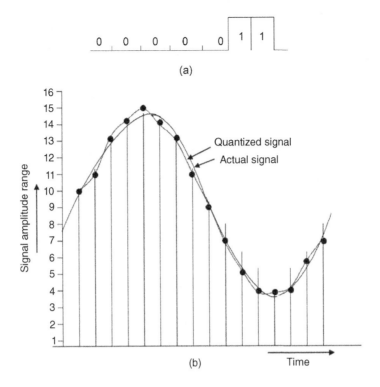

Figure 5.34 Quantizing process

In binary PCM, where the binary system of representation is used for encoding various sampled amplitudes, the number of bits to be transmitted per second would be given by nf_s, where

$n = log_2 L$
L = number of standard levels

and

$$f_s \geq f_m \text{ where } f_m = \text{message signal bandwidth}$$

Assuming that the PCM signal is a lowpass signal of bandwidth f_{PCM}, then the required minimum sampling rate would be $2f_{PCM}$. Therefore,

$$2f_{PCM} = nf_s \quad \text{or} \quad f_{PCM} = \left(\frac{n}{2}\right) f_s$$

Generating a PCM signal is a complex process. The message signal is usually sampled and first converted into a PAM signal, which is then quantized and encoded. The encoded signal can then be transmitted either directly via a cable or used to modulate a carrier using analogue or digital modulation techniques. PCM-AM is quite common.

5.4.2.2 Differential PCM

Differential PCM is similar to conventional PCM. The difference between the two lies in the fact that in differential PCM, each word or code group indicates a difference in amplitude (positive or negative) between the current sample and the immediately preceding one. Thus, it is not the absolute but the relative value that is indicated. As a consequence, the bandwidth required is less as compared to the one required in case of normal PCM.

5.4.2.3 Delta Modulation

Delta modulation (DM) has various forms. In one of the simplest forms, only one bit is transmitted per sample just to indicate whether the amplitude of the current sample is greater or smaller than the amplitude of the immediately preceding sample. It has extremely simple encoding and decoding processes but then it may result in tremendous quantizing noise in case of rapidly varying signals.

Figure 5.35 (a) shows a simple delta modulator system. The message signal $m(t)$ is added to a reference signal with the polarity shown. The reference signal is an integral part of the delta modulated signal. The error signal $e(t)$ so produced is fed to a comparator. The output of

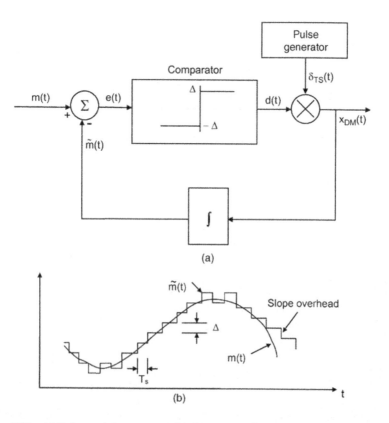

Figure 5.35 (a) Delta modulator system. (b) Output waveform of a delta modulator system

the comparator is $(+\Delta)$ for $e(t) > 0$ and $(-\Delta)$ for $e(t) < 0$. The output of the delta modulator is a series of impulses with the polarity of each impulse depending upon the sign of $e(t)$ at the sampling instants of time. Integration of the delta modulated output $x_{DM}(t)$ is a staircase approximation of the message signal $m(t)$, as shown in Figure 5.35 (b).

A delta modulated signal can be demodulated by integrating the modulated signal to obtain the staircase approximation and then passing it through a lowpass filter. The smaller the step size (Δ), the better is the reproduction of the message signal. However, a small step size must be accompanied by a higher sampling rate if the slope overload phenomenon is to be avoided. In fact, to avoid slope overload and associated signal distortion, the following condition should be satisfied:

$$\frac{\Delta}{T_s} \geq \left|\frac{dm(t)}{dt}\right|_{max} \tag{5.29}$$

where T_s = time between successive sampling time instants.

5.4.2.4 Adaptive Delta Modulator

This is a type of delta modulator. In delta modulation, the dynamic range of amplitude of the message signal $m(t)$ is very small due to threshold and overload effects. This problem is overcome in an adaptive delta modulator. In adaptive delta modulation, the step size (Δ) is varied according to the level of the message signal. The step size is increased as the slope of the message signal increases to avoid overload. The step size is reduced to reduce the threshold level and hence the quantizing noise when the message signal slope is small. In the case of adaptive delta modulation, however, the receiver also needs to be adaptive. The step size at the receiver should also be made to change to match the changes in step size at the transmitter.

5.5 Sampling Theorem

During the discussion on digital pulse communication techniques such as pulse code modulation, delta modulation, etc., it was noted that the three essential processes of such a system are sampling, quantizing and encoding. Sampling is the process in which a continuous time signal is sampled at discrete instants of time and its amplitudes at those discrete instants of time are measured. Quantization is the process by which the sampled amplitudes are represented in the form of a finite set of levels. The encoding process designates each quantized level by a code.

Digital transmission of analogue signals has been made possible by sampling the continuous time signal at a certain minimum rate, which is dictated by what is called the sampling theorem. The sampling theorem states that a band limited signal with the highest frequency component as f_M Hz, can be recovered completely from a set of samples taken at a rate of f_s samples per second, provided that $f_s \geq 2f_M$. This theorem is also known as the uniform sampling theorem for base band or lowpass signals. The minimum sampling rate of $2f_M$ samples per second is called the Nyquist rate and its reciprocal the Nyquist interval. For sampling bandpass signals, lower sampling rates can sometimes be used.

The sampling theorem for bandpass signals states that if a bandpass message signal has a bandwidth of f_B and an upper frequency limit of f_u, then the signal can be recovered from the sampled signal by bandpass filtering if $f_s = 2f_u/k$, where f_s is the sampling rate and k is the largest integer not exceeding f_u/f_B.

5.6 Shannon–Hartley Theorem

The Shannon–Hartley theorem describes the capacity of a noisy channel (assuming that the noise is random). According to this theorem,

$$C = B \log_2[1 + (S/N)] \tag{5.30}$$

where

C = channel capacity in bps
B = channel bandwidth in Hz
S/N = signal-to-noise ratio at the channel output or receiver input

The Shannon–Hartley theorem underlines the fundamental importance of bandwidth and signal-to-noise ratio in communication. It also shows that for a given channel capacity, increased bandwidth can be exchanged for decreased signal power. It should be mentioned that increasing the channel bandwidth by a certain factor does not increase the channel capacity by the same factor in a noisy channel, as would apparently be suggested by the Shannon–Hartley theorem. This is because increasing the bandwidth also increases noise, thus decreasing the signal-to-noise ratio. However, channel capacity does increase with an increase in bandwidth; the increase will not be in the same proportion.

Problem 5.11
A binary channel with a capacity of 48 kb/second is used for PCM voice transmission. If the highest frequency component in the message signal is taken as 3.2 kHz, find appropriate values for sampling rate (f_s), number of quantizing levels (L) and number of bits (n) used per sample.

Solution:
As per Nyquist criterion, $f_s \geq 2f_M$. This gives $f_s \geq 6400$ samples/sec
Also, $nf_s \leq$ Bit transmission rate; this gives $nf_s \leq 48000$ or $n \leq (48000/6400) = 7.5$
This gives $n = 7$ and $L = 2^7 = 128$ and $f_s = 48000/7$ Hz $= 6.857$ kHz

Problem 5.12
An analog signal is sampled at 10 kHz. If the number of quantizing levels is 128, find the time duration of one bit of binary encoded signal.

Solution:
Number of bits per sample $(n) = \log_2 L = \log_2 128 = 7$
Where L = Number of quantizing levels
Now f_s = 10 kHz. Therefore time duration of one bit of binary encoded signal = $1/(nf_s)$ = $1/(7 \times 10000) = 14.286$ μs

Problem 5.13
Find the Nyquist rate for the message signal represented by

$$m(t) = 10 (\cos 1000\pi t)(\cos 4000\pi t)$$

Solution:

$$10(\cos 1000\pi t)(\cos 4000\pi t) = 5 \times [2(\cos 1000\pi t)(\cos 4000\pi t)]$$
$$= 5[(\cos 5000\pi t) + (\cos 3000\pi t)]$$

This is a band limited signal with the highest frequency component equal to (5000π) radians/second, or 2500 Hz. Therefore,

$$\text{Nyquist rate} = 2 \times 2500 = 5000 \text{ Hz} = 5 \text{ kHz}$$

Problem 5.14

A message signal given by $m(t) = A \sin \omega_m t$ is applied to a delta modulator having a step size of Δ. Show that slope overload distortion will occur if $A > \left(\dfrac{\Delta}{2\pi}\right)\left(\dfrac{f_s}{f_m}\right)$ where f_s is the sampling frequency.

Solution:

$$m(t) = A \sin \omega_m t$$

$$\frac{dm(t)}{dt} = A\omega_m \cos \omega_m t$$

The condition for avoiding slope overload is given by

$$\frac{\Delta}{T_s} \geq \left|\frac{dm(t)}{dt}\right|_{max} = A\omega_m$$

or

$$A \leq \frac{\Delta}{(T_s \omega_m)} = \frac{\Delta f_s}{2\pi f_m} = \frac{\Delta}{2\pi}\left(\frac{f_s}{f_m}\right)$$

This is the condition for avoiding slope overload. Thus slope overload will occur when

$$A > \left(\frac{\Delta}{2\pi}\right)\left(\frac{f_s}{f_m}\right) \tag{5.31}$$

5.7 Digital Modulation Techniques

Base band digital signals have significant power content in the lower part of the frequency spectrum. Because of this, these signals can be conveniently transmitted over a pair of wires or coaxial cables. At the same time, for the same reason, it is not possible to have efficient wireless transmission of base band signals as it would require prohibitively large antennas, which would not be a practical or a feasible proposition. Therefore, if base band digital signals are to be transmitted over a wireless communication link, they should first modulate a continuous wave (CW) high frequency carrier. Three well-known techniques available for the purpose include:

1. Amplitude shift keying (ASK)
2. Frequency shift keying (FSK)
3. Phase shift keying (PSK)

Each one of these is described in the following paragraphs. Of the three techniques used for digital carrier modulation, PSK and its various derivatives, like differential PSK (DPSK), quadrature PSK (QPSK) and offset quadrature PSK (O-QPSK), are the most commonly used ones, more so for satellite communications, because of certain advantages they offer over others. PSK is therefore described in a little more detail.

5.7.1 Amplitude Shift Keying (ASK)

In the simplest form of amplitude shift keying (ASK), the carrier signal is switched ON and OFF depending on whether a '1' or '0' is to be transmitted (Figure 5.36). For obvious reasons, this form of ASK is also known as ON–OFF keying (OOK). The signal in this case is represented by

$$x(t) = A \sin \omega_c t \quad \text{for bit '1'} \quad (5.32)$$
$$= 0 \quad \text{for bit '0'}$$

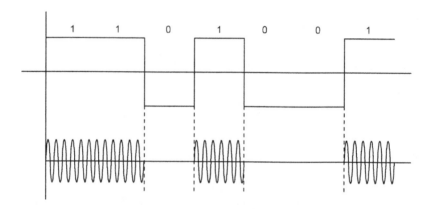

Figure 5.36 Amplitude shift keying

The ON–OFF keying has the disadvantage that appearance of any noise during transmission of bit '0' can be misinterpreted as data. This problem can be overcome by switching the amplitude of the carrier between two amplitudes, one representing a '1' and the other representing a '0', as shown in Figure 5.37. Again, the carrier can be suppressed to have maximum power in information carrying signals and also one of the side bands can be filtered out to conserve the bandwidth.

5.7.2 Frequency Shift Keying (FSK)

In frequency shift keying (FSK), it is the frequency of the carrier signal that is switched between two values, one representing bit '1' and the other representing bit '0', as shown in

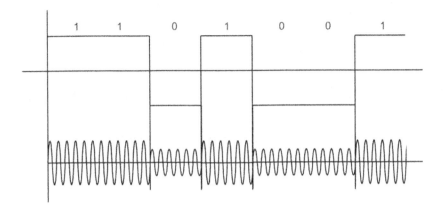

Figure 5.37 Two level amplitude shift keying

Figure 5.38. The modulated signal in this case is represented by

$$x(t) = A \sin \omega_{c1} t \quad \text{for bit '1'}$$
$$= A \sin \omega_{c2} t \quad \text{for bit '0'} \quad (5.33)$$

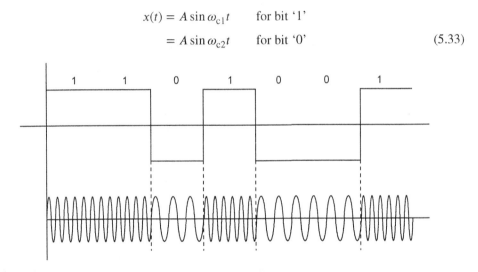

Figure 5.38 Frequency shift keying

In the case of FSK, when the modulation rate increases, the difference between the two chosen frequencies to represent a '1' and a '0' also needs to be higher. Keeping in mind the restriction in available bandwidth, it would not be possible to achieve a bit transmission rate beyond a certain value.

5.7.3 *Phase Shift Keying (PSK)*

In phase shift keying (PSK), the phase of the carrier is discretely varied with respect to either a reference phase or to the phase of the immediately preceding signal element in accordance

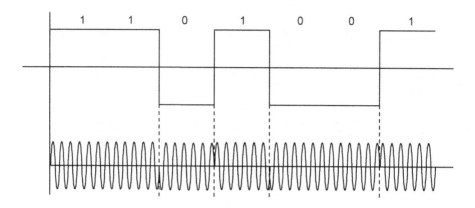

Figure 5.39 Binary phase shift keying

with the data being transmitted. For example, when encoding bits, the phase shift could be 0° for encoding a bit '0' and 180° for encoding a bit '1', as shown in Figure 5.39. The phase shift could have been −90° for encoding a bit '0' and +90° for encoding a bit '1'. The essence is that representations for '0' and '1' are a total of 180° apart. Such PSK systems in which the carrier can assume only two different phase angles are known as binary phase shift keying (BPSK) systems. In the BPSK system, each phase change carries one bit of information. This, in other words, means that the bit rate equals the modulation rate. Now, if the number of recognizable phase angles is increased to 4, then two bits of information could be encoded into each signal element.

Returning to the subject of BPSK, the carrier signals used to represent '0' and '1' bits could be expressed by equations 5.34(a) and 5.34(b) respectively.

$$x_{c0}(t) = A \cos(\omega_c t + \theta_0) \quad (5.34a)$$

$$x_{c1}(t) = A \cos(\omega_c t + \theta_1) \quad (5.34b)$$

Since the phase difference between two carrier signals is 180°, i.e. $\theta_1 = \theta_0 + 180°$, then

$$x_{c0}(t) = A \cos(\omega_c t + \theta_0) \quad \text{and} \quad x_{c1}(t) = -A \cos(\omega_c t + \theta_0)$$

5.7.4 Differential Phase Shift Keying (DPSK)

Another form of PSK is differential PSK (DPSK). In DPSK instead of instantaneous phase of the modulated signal determining which bit is transmitted, it is the change in phase that carries message intelligence. In this system, one logic level (say '1') represents a change in phase of the modulated signal and the other logic level (i.e. '0') represents no change in phase. In other words, if a digit changes in the bit stream from 0 to 1 or from 1 to 0, a '1' is transmitted in the form of change in phase of the modulated signal. In the case where there is no change, a '0' is transmitted in the form of no phase change in the modulated signal.

The BPSK signal is detected using a coherent demodulator where a locally generated carrier component is extracted from the received signal by a PLL circuit. This locally generated carrier assists in the product demodulation process where the product of the carrier and the received

Digital Modulation Techniques

modulated signal generates the demodulated output. There could be a difficulty in successfully identifying the correct phase of the regenerated signal for demodulation. Differential PSK takes care of this ambiguity to a large extent.

5.7.5 Quadrature Phase Shift Keying (QPSK)

Quadrature phase shift keying (QPSK) is the most commonly used form of PSK. A QPSK modulator is nothing but two BPSK modulators operating in quadrature. The input bit stream $(d_0, d_1, d_2, d_3, d_4, ...)$ representing the message signal is split into two bit streams, one having, say, even numbered bits $(d_0, d_2, d_4, ...)$ and the other having odd numbered bits $(d_1, d_3, d_5, ...)$. Also, in QPSK, if each pulse in the input bit stream has a duration of T seconds, then each pulse in the even/odd numbered bit streams has a pulse duration of $2T$ seconds, as shown in Figure 5.40.

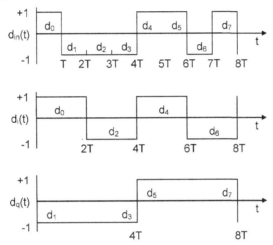

Figure 5.40 Quadrature phase shift keying

Figure 5.41 shows the block schematic arrangement of a typical QPSK modulator. One of the bit streams $[d_i(t)]$ feeds the in-phase modulator while the other bit stream $[d_q(t)]$ feeds the

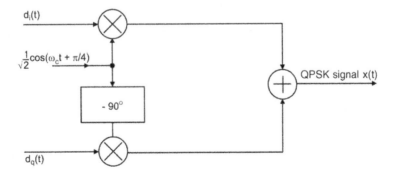

Figure 5.41 Block schematic arrangement of a typical QPSK modulator

quadrature modulator. The modulator output can be written as

$$x(t) = \left(\frac{1}{\sqrt{2}}\right)\left\{d_i(t)\cos\left(\omega_c t + \frac{\pi}{4}\right)\right\} + \left(\frac{1}{\sqrt{2}}\right)\left\{d_q(t)\sin\left(\omega_c t + \frac{\pi}{4}\right)\right\} \quad (5.35)$$

This expression can also be written in a simplified form as

$$x(t) = \cos[\omega_c t + \theta(t)] \quad (5.36)$$

In the input bit stream shown in Figure 5.40, an amplitude of +1 represents a bit '1' and an amplitude of −1 represents a bit '0'. The in-phase bit stream represented by $d_i(t)$ modulates the cosine function and has the effect of shifting the phase of the function by 0 or π radians. This is equivalent to BPSK. The other pulse stream represented by $d_q(t)$ modulates the sine function, thus producing another BPSK-like output that is orthogonal to the one produced by $d_i(t)$. The vector sum of the two produces a QPSK signal, given by

$$x(t) = \cos[\omega_c t + \theta(t)]$$

Figure 5.42 illustrates this further.

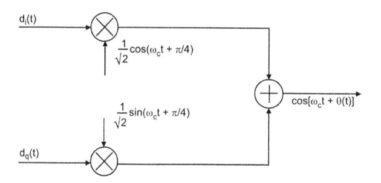

Figure 5.42 Conceptual diagram of QPSK

Depending on the status of the pair of bits having one bit from the $d_i(t)$ bit stream and the other from the $d_q(t)$ bit stream, $\theta(t)$ will have any of the four values of 0°, 90°, 180° and 270°. Four possible combinations are 00, 01, 10 and 11. Figure 5.43 illustrates the process further. It shows all four possible phase states. The in-phase bits operate on the vertical axis at phase states of 90° and 270° whereas the quadrature phase channel operates on the horizontal axis at phase states of 0° and 180°. The vector sum of the two produces each of the four phase states as shown. As mentioned earlier, the phase state of the QPSK modulator output depends on a pair of bits. The phase states in the present case would be 0°, 90°, 180° and 270° for input combinations 11, 10, 00 and 01 respectively.

Since each symbol in QPSK comprises two bits, the symbol transmission rate is half of the bit transmission rate of BPSK and the bandwidth requirement is halved. The power spectrum for QPSK is the same as that for BPSK.

Digital Modulation Techniques

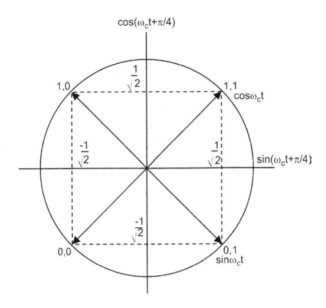

Figure 5.43 QPSK phase diagram

5.7.6 Offset QPSK

The offset QPSK is similar to QPSK with the difference that the alignment of the odd/even streams is shifted by an offset equal to T seconds, as shown in Figure 5.44. In the case of QPSK, as explained earlier, a carrier phase change can occur every $2T$ seconds. If neither of the two streams changes sign, the carrier phase remains unaltered. If only one of them changes sign, the carrier phase undergoes a change of $+90°$ or $-90°$ and if both change sign, the carrier phase undergoes a change of $180°$. In such a situation, the QPSK signal no longer

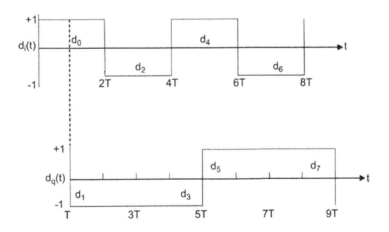

Figure 5.44 Offset QPSK waveform

has a constant envelope if it is filtered to remove the spectral side lobes. If such a QPSK signal is passed through a nonlinear amplifier, the amplitude variations could cause spectral spreading to restore unwanted side lobes, which in turn could lead to interference problems. Offset QPSK overcomes this problem. Due to staggering in-phase and quadrature-phase bit streams, the possibility of the carrier phase changing state by 180° is eliminated as only one bit stream can change state at any time instant of transition. A phase change of +90° or −90° does cause a small drop in the envelope, but it does not fall to near zero as in the case of QPSK for a 180° phase change.

5.7.7 8PSK and 16PSK

8PSK and 16PSK digital modulation techniques are an extension of QPSK. While in QPSK there are four different defined phase positions allowing each symbol to be represented by two bits, in the case of 8PSK and 16PSK there are 8 and 16 defined phase changes, respectively. Each phase change is represented by three bits for 8PSK while for 16PSK each phase change is represented by four bits. In other words, 8PSK employs eight symbols with constant carrier amplitude and a phase shift of 45° between adjacent phase states and 16PSK employs 16 symbols with constant amplitude carrier signal and a phase shift of 22.5°. While 8PSK transmits three bits per symbol, 16PSK transmits four bits per symbol. Consequently, for the same bandwidth specification or symbol rate, data bit rates for 8PSK and 16PSK are, respectively, three and four times conventional BPSK. When compared to QPSK, these rates are 50 % higher for 8PSK and 100 % higher for 16PSK. Figures 5.45 (a) and (b) show I-Q phase diagrams for 8PSK and 16PSK, respectively.

Multiple phase shift keying (M-PSK) digital modulation techniques such as 8PSK and 16PSK have the advantage of being spectrally very efficient. Another advantage is that a constant amplitude carrier allows the use of power efficient non-linear amplifiers such as class-C amplifiers. For a given bandwidth, 8PSK and 16PSK offer increased data capacity at the same bit error rate. The disadvantage is that a greater number of relatively smaller phase shifts makes the demodulation process very complex, more so in the presence of noise and interference. When the distance between different positions on the I-Q phase diagram is shorter there is increased risk of the symbols being misinterpreted. This requires additional error coding bits, which effectively reduces data throughput of the required data.

One common example of the use of the 8PSK technique is in enhanced data rate for global evolution (EDGE) technology. EDGE is a technology that can be used to enhance the data rate of existing global systems for mobile communications (GSMs) and digital-advanced mobile phone system (D-AMPS) communication networks. Both GSM and D-AMPS are second generation mobile communication systems. EDGE is an official member of the IMT-2000 family and in Europe it is seen as an intermediate technology for transition of 2G mobile communication systems like GSM to 3G systems like the universal mobile telecommunication system (UMTS).

5.7.8 Quadrature Amplitude Modulation (QAM)

M-PSK modulation techniques such as 8PSK and 16PSK achieve a higher bit rate for a given symbol rate by having larger number of phase positions of the constant amplitude carrier signal. As the number of phase positions increases further beyond 8 or 16 in an M-PSK

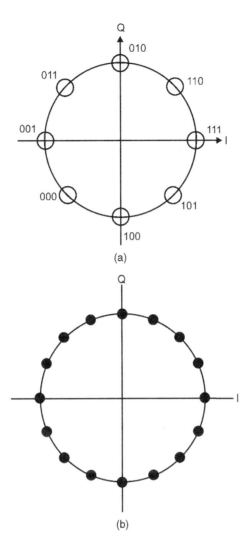

Figure 5.45 I-Q phase diagram: (a) 8PSK and (b) 16PSK

system, the distance between adjacent phase positions reduces to a point where the advantage of increased bit rate is outweighed by increased demodulation complexity and bit error rate. In the case of 32PSK and 64PSK, the phase shifts between two adjacent positions are 11.25° and 5.625°. Quadrature amplitude modulation (QAM) is a method of increasing the bit rate for a given symbol rate without causing a significant reduction in phase shift between adjacent phase positions.

QAM is a combination of amplitude and phase shift keying modulation techniques, that is, it makes use of both phase shifts as well as amplitude variations to increase the bit rate. For example, 8QAM makes use of four carrier signal phases and two different amplitude levels. Other variations of QAM are 16QAM, 32QAM, 64QAM, 128QAM and 256QAM. Figures 5.46 (a) and (b) show I-Q phase diagrams of 16QAM in circular and rectangular

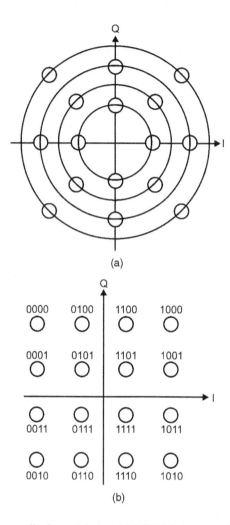

Figure 5.46 Quadrature amplitude modulation: (a) 16QAM-circular and (b) 16QAM-rectangular

configurations respectively. The phase diagram of Figure 5.46 (a) has eight different phases and four amplitude levels. The rectangular configuration of Figure 5.46 (b) has 12 different phases and four amplitude levels.

As the number of points or distinct states increases, there is an increase in the number of amplitude levels, which does not allow use of high efficiency non-linear amplifiers. Though spectrally very efficient, power output and efficiency suffer due to the requirement to use linear amplifiers. QAM is widely used in TV, Wi-Fi wireless LANs, satellites and cellular telephone systems.

5.7.9 Amplitude Phase Shift Keying (APSK)

As outlined in case of QAM, higher levels of QAM of 16 and above have many amplitude levels and phase shifts. An increased number of amplitude levels makes noise and interference more likely. Amplitude phase shift keying (APSK), which is a combination of amplitude and phase shift keying, is a variant of M-PSK and QAM that incorporates the advantages of both M-PSK and QAM. It has relatively fewer amplitude levels compared to QAM, thus allowing the use of non-linear power amplifiers, and also has relatively greater phase shifts between adjacent phase positions compared to M-PSK. While the former feature produces higher power output and efficiency; latter feature produces a lower bit error rate. Fewer amplitude levels and a smaller difference in amplitude between different levels make it possible to operate in the non-linear region of the power amplifier to boost output power level.

APSK uses fewer amplitude levels. In this case, symbols are arranged into two or more concentric rings with a constant phase shift offset. Figure 5.47 (a) shows the I-Q phase diagram of 16APSK using a double ring PSK format called 4+12 APSK with four symbols in the centre ring and 12 symbols in the outer ring. Figure 5.47 (b) shows the phase diagram of a 32APSK. APSK is primarily used in satellites due to its suitability for use with travelling wave tube (TWT) amplifiers.

5.8 Multiplexing Techniques

Multiplexing techniques are used to combine several message signals into a single composite signal so that they can be transmitted over a common channel. Multiplexing ensures that the different message signals in the composite signal do not interfere with each other and that they can be conveniently separated out at the receiver end. The two basic multiplexing techniques in use include:

1. Frequency division multiplexing (FDM)
2. Time division multiplexing (TDM)

While frequency division multiplexing is used with signals that employ analogue modulation techniques, time division multiplexing is used with digital modulation techniques where the signals to be transmitted are in the form of a bit stream. The two techniques are briefly described in the following paragraphs.

5.8.1 Frequency Division Multiplexing

In case of frequency division multiplexing (FDM), different message signals are separated from each other in the frequency domain. Figure 5.48 illustrates the concept of FDM showing simultaneous transmission of three message signals over a common communication channel. It is clear from the block schematic arrangement shown that each of the three message signals modulates a different carrier. The most commonly used modulation technique is single side band (SSB) modulation. Any type of modulation can be used as long as it is ensured that the carrier spacing is sufficient to avoid a spectral overlap. On the receiving side, bandpass filters separate out the signals, which are then coherently demodulated as shown. The composite signal formed by combining different message signals after they have modulated their respective carrier signals may be used to modulate another high frequency carrier before it is transmitted over the common link. In that case, these individual carrier signals are known as subcarrier signals.

FDM is used in telephony, commercial radio broadcast (both AM and FM), television broadcast, communication networks and telemetry. In the case of a commercial AM broadcast, the carrier frequencies for different signals are spaced 10 kHz apart. This separation is definitely

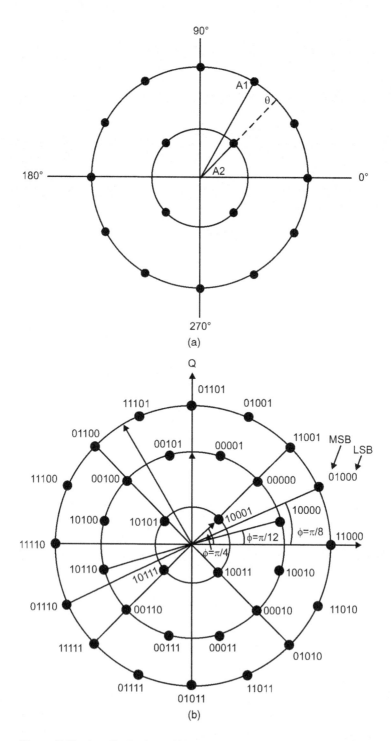

Figure 5.47 Amplitude phase shift keying: (a) 16APSK and (b) 32APSK

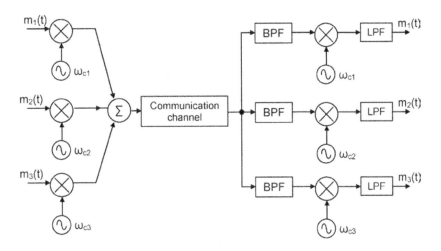

Figure 5.48 Frequency division multiplexing

not adequate if we consider a high fidelity voice signal with a spectral coverage of 50 Hz to 15 kHz. For this reason, AM broadcast stations using adjacent carrier frequencies are usually geographically far apart to minimize interference. In the case of an FM broadcast, the carrier frequencies are spaced apart at 200 kHz or more. In the case of long distance telephony, 600 or more voice channels, each with a spectral band of 200 Hz to 3.2 kHz, can be transmitted over a coaxial or microwave link using SSB modulation and a carrier frequency separation of 4 kHz.

5.8.2 Time Division Multiplexing

Time division multiplexing (TDM) is used for simultaneous transmission of more than one pulsed signal over a common communication channel. Figure 5.49 illustrates the concept. Multiple pulsed signals are fed to a type of electronic switching circuitry, called a commutator in the figure. All the message signals, which have been sampled at least at the Nyquist rate

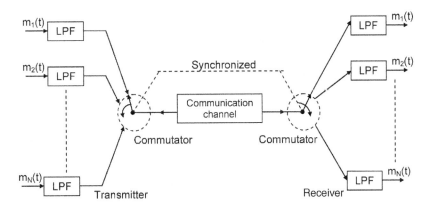

Figure 5.49 Time division multiplexing

(the sampling is usually done at 1.1 times the Nyquist rate to avoid aliasing problems), are fed to the commutator. The commutator interleaves different samples from different sampled message signals in order to form a composite interleaved signal. This composite signal is then transmitted over the link. In case all message signals have the same bandwidth, one commutation cycle will contain one sample from each of the messages. Where signals have different bandwidths, more samples would need to be transmitted per second of the signals having larger bandwidths. As an illustration, if there are three message signals with respective sampling rates of 2.4 kHz, 2.4 kHz and 4.8 kHz, then each cycle of commutation will have one sample each from the first two messages and two samples from the third message.

At the receiving end, the composite signal is de-multiplexed using a similar electronic switching circuitry that is synchronized with the one used at the transmitter. TDM is widely used in telephony, telemetry, radio broadcasting and data processing applications.

If T is the sampling time interval of the time multiplexed signal of n different signals, each having a sampling interval of T_s, then

$$T = \frac{T_s}{n} \qquad (5.37)$$

Also, if the time multiplexed signal is considered as a lowpass signal having a bandwidth of f_{TDM} and f_m is the bandwidth of individual signals, then

$$f_{TDM} = n f_m \qquad (5.38)$$

Problem 5.15
Three message signals $m_1(t)$, $m_2(t)$ and $m_3(t)$ with bandwidths of 2.4 kHz, 3.2 kHz and 3.4 kHz respectively are to be transmitted over a common channel in a time multiplexed manner. Determine the minimum sampling rate for each of the three signals if a uniform sampling rate is to be chosen. Also determine the sampling interval of the composite signal.

Solution: Since the sampling has to be uniform for the three signals, the minimum sampling rate for each of the signals would be twice the highest frequency component, i.e. 2×3.4 kHz $= 6.8$ kHz. The sampling rate of the composite signal $= 3 \times 6.8$ kHz $= 20.4$ kHz and therefore the sampling interval of the composite signal $= (10^6)/(20.4 \times 10^3) = 49 \,\mu s$.

Problem 5.16
In a certain digital telephony system comprising 24 voice channels, with each voice channel band limited to 3.2 kHz and using an 8-bit PCM, is transmitted over a common communication channel using the TDM approach. If the signal is sampled at 1.2 times the Nyquist rate and a single synchronization bit is added at the end of each frame, determine:

(a) duration of each bit and
(b) transmission rate

Solution: (a) Sampling rate $= (2 \times 3.2 \times 1.2)$kHz $= 7.68$ kHz. Therefore, the time period of each multiplexed frame $= (10^6/7.68 \times 10^3) = 130.2 \,\mu s$. Number of bits in each frame $= 24 \times 8 + 1 = 193$, therefore the bit duration $= 130.2/193 = 0.675 \,\mu s$.
(b) The transmission rate $= 1/0.675 = 1.482$ Mbps.

5.8.3 Code Division Multiplexing

Code division multiplexing (CDM) allows message signals from multiple independent signal sources to be transmitted simultaneously over a common frequency band. This is unlike either FDM or TDM. In the case of FDM, as discussed in section 5.8.1, multiple message signals are transmitted simultaneously but are separated in frequency domain. In the case of TDM, as discussed in section 5.8.2, multiple message signals are transmitted over a common frequency band and are separated from each other by transmitting them in different time slots. While FDM is used in the case of analogue signals, the latter is applicable to transmission of digital bit streams.

CDM is accomplished by using orthogonal codes called spreading codes, which spread each of the message signals over a common large frequency band, larger than the minimum bandwidth that would otherwise be required to transmit these signals. After different signals are spread over a large common frequency band using one of the spread spectrum techniques, such as direct sequence spread spectrum, they modulate a common carrier and are transmitted over a common channel. At the receiver end, appropriate orthogonal codes are used to recover the corresponding message signals. In other words, at the receiver relevant orthogonal code is used to recover the intended signal from the knowledge of the type of spreading code used for that signal at the transmitter. Figures 5.50 (a) and (b) show the block schematic arrangements used for the direct sequence spread spectrum for spreading and de-spreading operations, respectively, at the transmitter and receiver.

To summarize, the spread spectrum technique of CDM is a three-step process of (a) spreading the bandwidth of each message signal by using a spreading code that is independent of

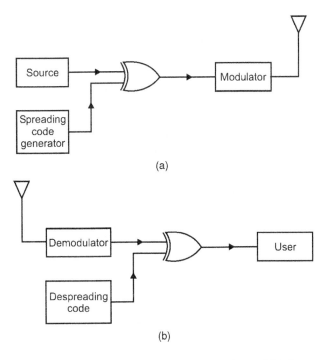

Figure 5.50 Direct sequence spectrum for (a) spreading operation and (b) de-spreading operation

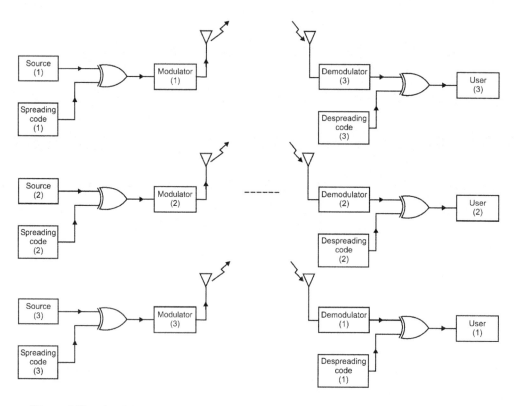

Figure 5.51 Simplified arrangement of communication system using code division multiplexing

message information, (b) modulating a carrier with the spread spectrum signal and transmitting it over a common transmission channel and (c) using the relevant de-spreading codes at the receivers to recover the intended message signals. Figure 5.51 shows a simplified arrangement of a communication system using CDM. It may be mentioned here that CDM techniques inherently provide reliability, noise immunity and security features to the communication and therefore are extensively used in a number of commercial cellular telephone systems. The technique is used in many contemporary mobile telephone standards including (a) cdmaOne, the first cellular standard to use CDM and implement the code division multiple access (CDMA) scheme and also known as IS-95, (b) CDMA2000, a family of 3G mobile technology standards also known as IMT multi-carrier (IMT-MC) employing CDM and CDMA access to send voice, data and signalling data between mobile phones and cell sites and (c) the universal mobile telecommunications system (UMTS), a 3G mobile cellular system employing wide band code division multiple access (W-CDMA) technology and used for networks based on the GSM standard.

Further Readings

Calcutt, D. and Tetley, L. (1994) *Satellite Communications: Principles and Application*, Edward Arnold, a member of the Hodder Headline Group, London.

Elbert, B.R. *Introduction to Satellite Communication*, Altech House, Boston, Massachusetts.

Gedney, R.T., Schertler, R. and Gargione, F. (2000) *The Advanced Communication Technology*, SciTech., New Jersey.

Kadish, J.E. (2000) *Satellite Communications Fundamentals*, Artech House, Boston, Massachusetts.

Kennedy, G. and Davis, B. (1992) *Electronic Communication Systems*, McGraw-Hill Education Europe.

Maral, G. and Bousquet, M. (2002) *Satellite Communication Systems: Systems, Techniques and Technology*, John Wiley & Sons, Ltd, Chichester.

Richharia, M. (1999) *Satellite Communication Systems*, Macmillan Press Ltd.

Stallings, W. (2004) *Wireless Communications and Networks*, Prentice-Hall, Englewood Cliffs, New Jersey.

Tomasi, W. (1997) *Advanced Electronic Communications Systems*, Prentice-Hall, Englewood Cliffs, New Jersey.

Tomasi, W. (2003) *Electronic Communications System: Fundamentals Through Advanced*, Prentice-Hall, Englewood Cliffs, New Jersey.

Internet Sites

1. http://en.wikipedia.org/wiki/Amplitude_modulation
2. http://en.wikipedia.org/wiki/Frequency_modulation
3. http://www.rfcafe.com/references/electrical/frequency_modulation.htm
4. http://www.tm.agilent.com/data/static/downloads/eng/Notes/interactive/an-150-1/hp-am-fm.pdf
5. http://en.wikipedia.org/wiki/Pulse-code_modulation
6. http://www2.uic.edu/stud_orgs/prof/pesc/part_3_rev_F.pdf
7. http://www.cisco.com/warp/public/788/signalling/waveform_coding.html
8. http://www.web-ee.com/primers/files/AN-236.pdf
9. http://www.dip.ee.uct.ac.za/~nicolls/lectures/eee482f/04_chancap_2up.pdf
10. http://en.wikipedia.org/wiki/Shannon-Hartley_theorem
11. http://www.complextoreal.com/chapters/mod1.pdf
12. http://users.pandora.be/educypedia/electronics/rfdigmod.htm
13. http://www.maxim-ic.com/appnotes.cfm/appnote_number/686
14. http://en.wikipedia.org/wiki/QPSK

Glossary

16PSK: The 16PSK digital modulation technique is an extension of quadrature phase shift keying. It has 16 defined phase changes and each phase change is represented by four bits. It employs 16 symbols with constant amplitude carrier signal and a phase shift of 22.5°

8PSK: 8PSK digital modulation technique is an extension of quadrature phase shift keying. It has eight defined phase changes and each phase change is represented by three bits. It employs eight symbols with constant amplitude carrier signal and a phase shift of 45°

A3E system: This is the standard AM system used for broadcasting. It uses a double side band with a full carrier

Adaptive delta modulation: This is a type of delta modulator. In delta modulation, the dynamic range of the amplitude of the message signal $m(t)$ is very small due to threshold and overload effects. In adaptive delta modulation, the step size (Δ) is varied according to the level of the message signal. The step size is increased as the slope of the message signal increases to avoid overload. The step size is reduced to reduce the threshold level and hence the quantizing noise is reduced when the message signal slope is small

Amplitude modulation: This is the analogue modulation technique in which the instantaneous amplitude of the modulated signal varies directly as the instantaneous amplitude of the modulating signal. The frequency of the carrier signal remains constant

Amplitude phase shift keying: Amplitude phase shift keying (APSK) is a combination of amplitude and phase shift keying. It is a variant of M-PSK and QAM that incorporates the advantages of M-PSK and QAM

Amplitude shift keying: This is a digital modulation technique. In the simplest form of amplitude shift keying (ASK), the carrier signal is switched ON and OFF depending upon whether a '1' or '0' is to be transmitted. This is also known as ON–OFF keying. In another form of ASK, the amplitude of the carrier is switched between two different amplitudes

B8E system: This system uses two independent side bands with the carrier either attenuated or suppressed. This form of amplitude modulation is also known as independent side band (ISB) transmission and is usually employed for point-to-point radio telephony

C3F system: Vestigial side band (VSB) transmission is the other name for this system. It is used for transmission of video signals in commercial television broadcasting. It is a compromise between SSB and DSB modulation systems in which a vestige or a part of the unwanted side band is also transmitted, usually with a full carrier along with the other side band. The typical bandwidth required to transmit a VSB signal is about 1.25 times that of an SSB signal

Delta modulation: Delta modulation has various forms. In one of the simplest forms, only one bit is transmitted per sample just to indicate whether the amplitude of the current sample is greater or smaller than the amplitude of the immediately preceding sample

Differential PCM: In differential PCM, each word or code group indicates a difference in amplitude (positive or negative) between the current sample and the immediately preceding one

Differential phase shift keying: This is another form of PSK. In this, instead of the instantaneous phase of the modulated signal determining which bit is transmitted, it is the change in phase that carries message intelligence. In this system, one logic level (say '1') represents a change in phase of the carrier and the other logic level (i.e. '0') represents no change in phase

Frequency division multiplexing: In the case of frequency division multiplexing (FDM), different message signals are separated from each other in frequency. FDM is used in telephony, commercial radio broadcast (both AM and FM), television broadcast, communication networks and telemetry

Frequency modulation: This is the analogue modulation technique in which the instantaneous frequency of the modulated signal varies directly as the instantaneous amplitude of the modulating or base band signal. The rate at which these frequency variations take place is proportional to the modulating frequency

Frequency shift keying: In frequency shift keying (FSK), it is the frequency of the carrier signal that is switched between two values, one representing bit '1' and the other representing bit '0'

H3E system: This is the single side band, full carrier AM system

J3E system: This is the single side band suppressed carrier AM system. It is the system usually referred to as SSBSC, in which the carrier is suppressed by at least 45 dB in the transmitter

Multiple phase shift keying: Multiple phase shift keying (M-PSK) modulation techniques such as 8PSK and 16PSK have a larger number of phase positions of the constant amplitude carrier signal, thereby achieving a higher bit rate for a given symbol rate

Offset QPSK: Offset QPSK is similar to QPSK with the difference that the alignment of the odd/even streams is shifted in case of offset QPSK by an offset equal to T seconds, which is the pulse duration of the input bit stream

Phase shift keying: In phase shift keying (PSK), the phase of the carrier is discretely varied with respect to either a reference phase or to the phase of the immediately preceding signal element in accordance with the data being transmitted. For example, when encoding bits, the phase shift could be $0°$ for encoding a bit '0' and $180°$ for encoding a bit '1'. The phase shift could have been $-90°$ for encoding a bit '0' and $+90°$ for encoding a bit '1'

Pulse amplitude modulation: In the case of pulse amplitude modulation (PAM), the signal is sampled at regular intervals and the amplitude of each sample, which is a pulse, is proportional to the amplitude of the modulating signal at the time instant of sampling

Pulse code modulation: In pulse code modulation, the peak-to-peak amplitude range of the modulating signal is divided into a number of standard levels, which in the case of a binary system is an integral power of 2. The amplitude of the signal to be sent at any sampling instant is the nearest standard level. This nearest standard level is then encoded into a group of pulses. The number of pulses (n) used to encode a sample in binary PCM equals $\log_2 L$, where L is the number of standard levels

Pulse position modulation: In the case of pulse position modulation (PPM), the amplitude and width of the sampled pulses are maintained as constant and the position of each pulse with respect to the position of a recurrent reference pulse varies as a function of the instantaneous sampled amplitude of the modulating signal

Pulse width modulation: In the case of pulse width modulation (PWM) the starting time of the sampled pulses and their amplitude is fixed. The width of each pulse is made proportional to the amplitude of the signal at the sampling time instant

Quadrature amplitude modulation: Quadrature amplitude modulation (QAM) is a combination of amplitude and phase shift keying modulation techniques. It makes use of both phase shifts and amplitude variations to increase the bit rate

Quadrature phase shift keying: Quadrature phase shift keying (QPSK) is the most commonly used form of PSK. A QPSK modulator is simply two BPSK modulators operating in quadrature. The input bit stream ($d_0, d_1, d_2, d_3, d_4, \ldots$) representing the message signal is split into two bit streams, one having, say, even numbered bits ($d_0, d_2, d_4, \ldots$) and the other having odd numbered bits ($d_1, d_3, d_5, \ldots$). The two bit streams modulate the carrier signals, which have a phase difference of $90°$ between them. The vector sum of the output of two modulators constitutes the QPSK output

R3E system: This is the single side band reduced carrier type AM system, also called the pilot carrier system

Sampling theorem: The sampling theorem states that a band limited signal with the highest frequency component as f_M Hz can be recovered completely from a set of samples taken at the rate of f_s samples per second provided that $f_s \geq 2f_M$

Shannon–Hartley theorem: The Shannon–Hartley theorem describes the capacity of a noisy channel (assuming that the noise is random). According to this theorem,

$$C = B \log_2 (1 + S/N) \quad \text{bps}$$

The Shannon–Hartley theorem underlines the fundamental importance of the bandwidth and the signal-to-noise ratio in communication. It also shows that for a given channel capacity, increased bandwidth can be exchanged for decreased signal power

Time division multiplexing: Time division multiplexing is used for simultaneous transmission of more than one pulsed signal over a common communication channel and different message signals are separated from each other in time. TDM is widely used in telephony, telemetry, radio broadcast and data processing

6

Multiple Access Techniques

Multiple access means access to a given facility or a resource by multiple users. In the context of satellite communication, the facility is the transponder and the multiple users are various terrestrial terminals under the footprint of the satellite. The transponder provides the communication channel(s) that receives the signals beamed at it via the uplink and then retransmits the same back to Earth for intended users via the downlink. Multiple users are geographically dispersed and certain specific techniques, to be discussed in this chapter, are used to allow them a simultaneous access to the satellite's transponder. The text matter is suitably illustrated with the help of a large number of problems.

6.1 Introduction to Multiple Access Techniques

Commonly used multiple access techniques include the following:

1. Frequency division multiple access (FDMA)
2. Time division multiple access (TDMA)
3. Code division multiple access (CDMA)
4. Space domain multiple access (SDMA)

In the case of frequency division multiple access (FDMA), different Earth stations are able to access the total available bandwidth in the satellite transponder(s) by virtue of their different carrier frequencies, thus avoiding interference amongst multiple signals. The term should not be confused with frequency division multiplexing (FDM), which is the process of grouping multiple base band signals into a single signal so that it could be transmitted over a single communication channel without the multiple base band signals interfering with each other. Here, multiple base band signals modulate different carrier frequencies called subcarrier frequencies and the multiplexed signal then modulates a common relatively higher frequency carrier, which then becomes the signal to be transmitted from the Earth station. Similarly, other stations may also have similar frequency division multiplexed signals with a different final carrier frequency. These multiplexed signals, by virtue of their different final carrier frequencies, are able to access the satellite simultaneously.

In the case of time division multiple access (TDMA), different Earth stations in the satellite's footprint make use of the transponder by using a single carrier on a time division basis. Again it should not be confused with time division multiplexing (TDM), which is the technique used at a given Earth station to simultaneously transmit digitized versions of multiple base band signals over a common communication channel by virtue of their separation on the timescale. The composite time multiplexed signal modulates a high frequency carrier using any of the digital carrier modulation techniques. Multiple time multiplexed signals from other stations having the same carrier frequency are then able to access the satellite by allowing each station to transmit during its allotted time slot.

In the case of code division multiple access (CDMA), the entire bandwidth of the transponder is used simultaneously by multiple Earth stations at all times. Each transmitter spreads its signal over the entire bandwidth, which is much wider than that required by the signal otherwise. One of the ways of doing this is by multiplying the information signal by a pseudorandom bit sequence. Interference is avoided as each transmitter uses a unique code sequence. Receiving stations recover the desired information by using a matched decoder that works on the same unique code sequence used during transmission.

Space domain multiple access (SDMA) uses spatial separation where different antenna beam polarizations can be used to avoid interference between multiple transmissions. Beams with horizontal and vertical or right-hand circular and left-hand circular polarizations may be used for the purpose. Use of the SDMA technique on board a single satellite platform to cover the same Earth surface area with multiple beams having different polarizations allows for frequency re-use. In the overall satellite link, SDMA is usually achieved in conjunction with other types of multiple access techniques such as FDMA, TDMA and CDMA.

6.1.1 Transponder Assignment Modes

In addition to the multiple access techniques outlined in the preceding paragraphs, there are also certain transponder assignment modes. The commonly used ones include:

1. Preassigned multiple access (PAMA)
2. Demand assigned multiple access (DAMA)
3. Random multiple access (RMA)

In the case of preassigned multiple access (PAMA), the transponder is assigned to the individual user either permanently for the satellite's full lifetime or at least for long durations. The preassignment may be that of a certain frequency band, time slot or a code. When it is used infrequently, a link set-up with preassigned channels is not only costly to the user but the link utilization is also not optimum.

Demand assigned multiple access (DAMA) allows multiple users to share a common link wherein each user is only required to put up a request to the control station or agency when it requires the link to be used. The channel link is only completed as required and a channel frequency is assigned from the available frequencies within the transponder bandwidth. It is very cost effective for small users who have to pay for using the transponder capacity only for the time it was actually used.

In the case of random multiple access (RMA), access to the link or the transponder is by contention. A user transmits the messages without knowing the status of messages from other users. Due to the random nature of transmissions, data from multiple users may collide. In case a collision occurs, it is detected and the data are retransmitted. Retransmission is carried out with random time delays and sometimes may have to be done several times. In such a situation, when all the stations are entirely independent, there is every likelihood that the messages that collided would be separated out in time on retransmission.

6.2 Frequency Division Multiple Access (FDMA)

It is the earliest and still one of the most commonly employed forms of multiple access techniques for communications via satellite. In the case of frequency division multiple access (FDMA), as outlined earlier, different Earth stations are able to access the total available bandwidth of satellite transponder by virtue of their different carrier frequencies, thus avoiding interference among multiple signals. Figure 6.1 shows the typical arrangement for carrier frequencies for a C band transponder for both uplink and downlink. The transponder receives transmissions at around 6 GHz and retransmits them at around 4 GHz. Figure 6.1 shows the case of a satellite with 12 transponders, with each transponder having a bandwidth of 36 MHz and a guard band of 4 MHz between adjacent transponders to avoid interference.

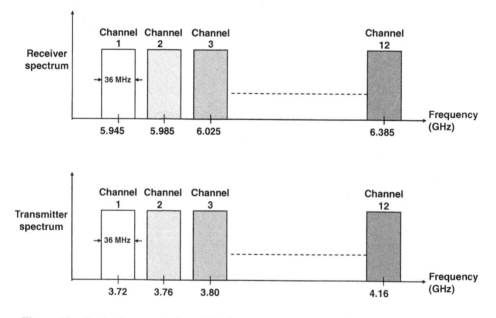

Figure 6.1 Carrier frequencies for a C band transponder for both uplink and downlink channels

Each of the Earth stations within the satellite's footprint transmits one or more message signals at different carrier frequencies. Each carrier is assigned a small guard band, as

Frequency Division Multiple Access (FDMA)

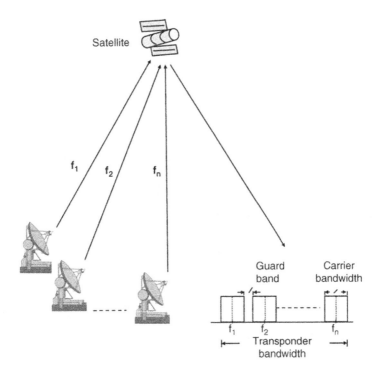

Figure 6.2 Basic concept of FDMA

mentioned above, to avoid overlapping of adjacent carriers. The satellite transponder receives all carrier frequencies within its bandwidth, does the necessary frequency translation and amplification and then retransmits them back towards Earth. Figure 6.2 illustrates the basic concept of FDMA in satellite communications. Different Earth stations are capable of selecting the carrier frequency containing messages of their interest. Two FDMA techniques are in operation today. One of them is the multichannel per carrier (MCPC) technique, where the Earth station frequency multiplexes several channels into one carrier base band assembly, which then frequency modulates an RF carrier and transmits it to an FDMA satellite transponder. In the other technique, called the single channel per carrier (SCPC), each signal channel modulates a separate RF carrier, which is then transmitted to the FDMA transponder. The modulation technique used here could either be frequency modulation (FM) in case of analogue transmission or phase shift keying (PSK) for digital transmission.

Major advantages of FDMA include simplicity of Earth station equipment and the fact that no complex timing and synchronizing techniques are required. Disadvantages include the likelihood of intermodulation problems with its adverse effect on the signal-to-noise ratio. The intermodulation products result mainly from the non-linear characteristics of the travelling wave tube amplifier (TWTA) of the transponder, which is required to amplify a large number of carrier frequencies. The problem is further compounded when the TWTA is made to operate near saturation so as to be able to supply certain minimum carrier power in order to reduce downlink noise and by the fact that the TWTA when operated near saturation exhibits higher non-linearity.

It may be mentioned here that an FDMA system can either be power-limited or bandwidth limited in terms of the number of carriers that can access the satellite transponder. The maximum number of carriers that can access the transponder is given by ($n = B_{TR}/B_C$), where B_{TR} is the total transponder bandwidth and B_C is the carrier bandwidth. If the EIRP is sufficient to meet the (C/N) requirements, then the system can support (n) carriers and is said to be bandwidth-limited. In case the EIRP is insufficient to meet the (C/N) requirements, the number of carriers that can access the satellite is less than (n). The system in this case is power-limited.

6.2.1 Demand Assigned FDMA

In a demand assigned FDMA system, the transponder frequency is subdivided into a number of channels and the Earth station is assigned a channel depending upon its request to the control station. Demand assignment may be carried out either by using the polling method or by using the random access method. In the polling method, the master Earth station continuously polls all of the Earth stations in sequence and if the request is encountered, frequency slots are assigned to that Earth station which had made the request. The polling method introduces delays more so when the number of Earth stations is large. In the random access method, the problem of delays does not exist. The random access method can be of two types namely the centrally controlled random access method and the distributed control random access method. In the case of centrally controlled random access, the Earth stations make requests through the master Earth station as the need arises. In the case of distributed control random access, the control is exercised at each Earth station.

6.2.2 Pre-assigned FDMA

In a preassigned FDMA system, the frequency slots are pre-assigned to the Earth stations. The slot allocations are pre-determined and do not offer flexibility. Hence, some slots may be facing the problem of over-traffic, while other slots are sitting idle.

6.2.3 Calculation of C/N Ratio

The overall noise-to-carrier ratio for a satellite link is given by equation 6.1.

$$\left[\frac{N}{C}\right]_{OV} = \left[\frac{N}{C}\right]_{U} + \left[\frac{N}{C}\right]_{D} + \left[\frac{N}{C}\right]_{IM} \tag{6.1}$$

Where,
$[N/C]_{OV}$ = Overall noise-to-carrier ratio
$[N/C]_{U}$ = Uplink noise-to-carrier ratio
$[N/C]_{D}$ = Downlink noise-to-carrier ratio
$[N/C]_{IM}$ = Inter-modulation noise-to-carrier ratio

The value of noise-to-carrier ratio given by equation 6.1 should be less than the required design value of noise-to-carrier ratio $[(N/C)_{REQ}]$, that is

$$\left[\frac{N}{C}\right]_{OV} \leq \left[\frac{N}{C}\right]_{REQ} \tag{6.2}$$

Combining equations 6.1 and 6.2 we get,

$$\left[\frac{N}{C}\right]_{REQ} \geq \left[\frac{N}{C}\right]_{U} + \left[\frac{N}{C}\right]_{D} + \left[\frac{N}{C}\right]_{IM} \tag{6.3}$$

In an FDMA system, the up-link noise is usually negligible and the inter-modulation noise is brought to an acceptable level by employing back-off in power amplifiers. It may be mentioned here that the operating point of the power amplifiers mostly TWTA is shifted closer to the linear portion of the curve in order to reduce the inter-modulation distortion. The reduction in the input power is referred to as input back-off and is the difference in dB between the carrier input at the operating point and the saturation input that is required for single carrier operation. The output back-off is the corresponding drop in the output power in dB. Therefore, equation 6.3 can be rewritten as

$$\left[\frac{N}{C}\right]_{REQ} \geq \left[\frac{N}{C}\right]_{D} \tag{6.4}$$

Equation 6.4 can again be rewritten as

$$\left[\frac{C}{N}\right]_{REQ} \leq \left[\frac{C}{N}\right]_{D} \tag{6.5}$$

The downlink carrier-to-noise ($[C/N]_D$) is expressed by equation 6.6.

$$\left[\frac{C}{N}\right]_{D} = [EIRP]_D + \left[\frac{G}{T}\right]_D - [LOSSES]_D - [k] - [B] \tag{6.6}$$

Where,

$[EIRP]_D$ = Satellite Equivalent Isotropic Radiated Power
$[G/T]_D$ = Earth-station receiver G/T
$[LOSSES]_D$ = Free space and other losses at the downlink frequency
$[k]$ = Boltzmann's constant in dB
$[B]$ = Signal bandwidth and is equal to the noise bandwidth

Therefore,

$$\left[\frac{C}{N}\right]_{REQ} \leq [EIRP]_D + \left[\frac{G}{T}\right]_D - [LOSSES]_D - [k] - [B] \tag{6.7}$$

In the case of single carrier per channel (SCPC) systems, the satellite will have saturation value of EIRP ($[EIRP]_{sat}$) and transponder bandwidth of B_{TR}, both of which are fixed. In this case, there is no back-off and equality sign of equation 6.7 applies. Therefore,

$$\left[\frac{C}{N}\right]_{REQ} = [EIRP]_{SAT} + \left[\frac{G}{T}\right]_D - [LOSSES]_D - [k] - [B_{TR}] \tag{6.8}$$

Equation 6.8 can be rewritten as

$$\left[\frac{C}{N}\right]_{REQ} - [EIRP]_{SAT} - \left[\frac{G}{T}\right]_D + [LOSSES]_D + [k] + [B_{TR}] = 0 \qquad (6.9)$$

Let us now consider the case of multiple carriers per channel (MCPC) systems having N carriers with each carrier sharing the output power equally and having a bandwidth of B. The output back-off is given by $[BO]_O$. The output power for each of the FDMA carriers is given by equation 6.10.

$$[EIRP]_D = [EIRP]_{SAT} - [BO]_O - [N] \qquad (6.10)$$

The transponder bandwidth (B_{TR}) is shared by all the carriers but due to power limitation imposed by the need of back-off, the whole bandwidth is not utilized. Let us assume that the fraction of the bandwidth utilized is alpha (α). Therefore,

$$NB = \alpha B_{TR} \qquad (6.11)$$

Expressing in terms of decilogs

$$[B] = [\alpha] + [B_{TR}] - [N] \qquad (6.12)$$

Substituting the values of $[EIRP]_D$ and $[B]$ given by equations 6.10 and 6.12 respectively in equation 6.7, we get

$$\left[\frac{C}{N}\right]_{REQ} \leq [EIRP]_{SAT} - [BO]_O + \left[\frac{G}{T}\right]_D - [LOSSES]_D - [k] - [\alpha] - [B_{TR}] \qquad (6.13)$$

Equation 6.13 can be rearranged as

$$\left[\frac{C}{N}\right]_{REQ} - [EIRP]_{SAT} - \left[\frac{G}{T}\right]_D + [LOSSES]_D + [k] + [B_{TR}] \leq -[BO]_O - [\alpha] \qquad (6.14)$$

As we can see from equation 6.9, in the case of SCPC system LHS of equation 6.14 is zero. For MCPC systems, it is less than zero.
Therefore,

$$0 \leq -[BO]_O - [\alpha] \quad or \quad [\alpha] \leq -[BO]_O. \qquad (6.15)$$

The best that can be achieved in a MCPC system is to make $[\alpha] \leq -[BO]_O$.

6.3 Single Channel Per Carrier (SCPC) Systems

In the paragraphs to follow, we shall discuss two common forms of SCPC systems namely:

1. SCPC/FM/FDMA system
2. SCPC/PSK/FDMA system

Each one of them is briefly described below.

6.3.1 SCPC/FM/FDMA System

As outlined earlier, in this form of SCPC system, each signal channel modulates a separate RF carrier and the modulation system used here is frequency modulation. The modulated signal is then transmitted to the FDMA transponder. The transponder bandwidth is subdivided in such a way that each base band signal channel is allocated a separate transponder subdivision and an individual carrier. This type of SCPC system is particularly used on thin route satellite communication networks. Though it suffers from the problem of power limitation resulting from the use of multiple carriers and the associated intermodulation problems, it does enable a larger number of Earth stations to access and share the capacity of the transponder using smaller and more economic units as compared to multiple channels per carrier systems. Another advantage of the SCPC/FM/FDMA system is that it facilitates the use of voice activated carriers. This means that the carriers are switched off during the periods when there is no speech activity, thus reducing power consumption. This in turn leads to availability of more transponder power and hence higher channel capacity. This type of SCPC system also has the advantage that the power of the individual transmitted carriers can be adjusted to the optimum value for given link conditions. Some channels may operate at higher power levels than others, depending on the requirement of back-off for the transponder output power device. It may be mentioned here that the output back-off or simply the back-off of the transponder output power device is the ratio of the saturated output power to the desired output power. However, this type of SCPC system requires automatic frequency control to maintain spectrum centering for individual channels, which is usually achieved by transmitting a pilot tone in the centre of the transponder bandwidth.

Figure 6.3 shows the transmission path for an SCPC/FM/FDMA system. The diagram is self-explanatory. Different base band signals frequency-modulate their respective allocated carriers, which are combined and then transmitted to the satellite over the uplink.

The signal-to-noise power ratio (S/N) at the output of the demodulator for the SCPC/FM/FDMA system can be computed from

$$\frac{S}{N} = \left(\frac{C}{N}\right) \times 3B \times \left(\frac{f_d^2}{f_2^2 - f_1^2}\right) \tag{6.16}$$

where

C = carrier power at the receiver input (in W)
N = noise power (in W) in bandwidth B (in Hz)
B = RF bandwidth (in Hz)
f_d = test tone frequency deviation (in Hz)
f_2 = upper base band frequency (in Hz)
f_1 = lower base band frequency (in Hz)

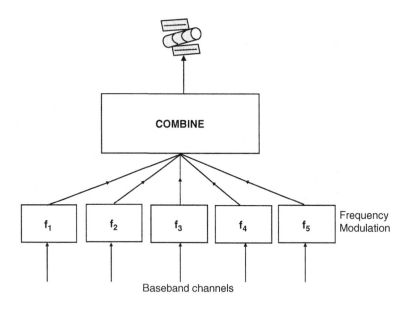

Figure 6.3 Transmission path for the SCPC/FM/FDMA system

6.3.2 SCPC/PSK/FDMA System

This is the digital form of the SCPC system in which the modulation technique used is phase shift keying (PSK). SPADE (single channel per carrier PCM multiple access demand assignment equipment) was the first operational SCPC/PSK/FDMA system. It was designed for use on Intelsat-4 and subsequent Intelsat satellites. This system employs PCM for base band signal encoding and QPSK as the carrier modulation technique. With this, it is possible to accommodate a 64 kbps voice channel in a bandwidth of 38.4 kHz as compared to the requirement of a full 45 kHz in the case of frequency modulation. With the use of 45 kHz per channel in QPSK, the guard band is effectively included in this bandwidth, which enables the SPADE system to handle 800 voice channels within a 36 MHz transponder bandwidth. The SPADE system offers the advantage of voice activation described earlier. The ECS-2 (European communications satellite) business service is another example of SCPC/PSK/FDMA system.

The channel capacity can be determined from the carrier-to-noise density ratio. The carrier-to-noise ratio necessary to support each carrier can be computed from

$$\left[\frac{C}{N}\right]_{\text{th}} = \left[\frac{E_b}{N_0}\right]_{\text{th}} - [B] + [R] + [M] \qquad (6.17)$$

where

$[C/N]_{\text{th}}$ = carrier-to-noise ratio (in dB) at the threshold error rate
$[E_b/N_0]_{\text{th}}$ = bit energy-to-noise density ratio (in dB) at the threshold error rate
$[B]$ = noise bandwidth (in Hz)
$[R]$ = data rate (in bps)
$[M]$ = system margin to allow for impairments (in dB)

The carrier-to-noise density ratio $[C/N_0]$ can be computed from

$$\left[\frac{C}{N_0}\right] = \left[\frac{C}{N}\right]_{th} + 10 \log B \tag{6.18}$$

where $[C/N_0]$ is the total available carrier-to-noise density (in dB) for the transponder

6.4 Multiple Channels Per Carrier (MCPC) Systems

As the name suggests, in this type of multiple access arrangement, multiple signal channels are first grouped together to form a single base band signal assembly. These grouped base band signals modulate preassigned carriers which are then transmitted to the FDMA transponder. Based on the multiplexing technique used to form base band assemblies and the carrier modulation technique used for onward transmission to the satellite transponder, there are two common forms of MCPC systems in use. These are:

1. MCPC/FDM/FM/FDMA system
2. MCPC/PCM-TDM/PSK/FDMA system

Each of them is briefly described in the following paragraphs.

6.4.1 MCPC/FDM/FM/FDMA System

In this arrangement, multiple base band signals are grouped together by using frequency division multiplexing to form FDM base band signals. The FDM base band assemblies frequency modulate pre-assigned carriers and are then transmitted to the satellite. The FDMA transponder receives multiple carriers, carries out frequency translation and then separates out individual carriers with the help of appropriate filters. Multiple carriers are then multiplexed and transmitted back to Earth over the downlink. The receiving station extracts the channels assigned to that station. Figure 6.4 shows the typical block schematic arrangement of such a system, which is suitable only for limited access use. The channel capacity falls with an increase in the number of carriers. Larger number of carriers causes more intermodulation products, with the result that intermodulation-prone frequency ranges cannot be used for traffic.

The signal-to-noise ratio at the demodulator output for such a system is given by:

$$\frac{S_b}{N_b} = \left(\frac{f_d}{f_m}\right)^2 \times \left(\frac{B}{b}\right) \times \left(\frac{C}{N}\right) \tag{6.19}$$

where

f_d = RMS (root mean square) test tone deviation (in Hz)
f_m = highest modulation frequency (in Hz)
B = bandwidth of the modulated signal (in Hz)
b = base band signal bandwidth (in Hz)
C = carrier power at the receiver input (in W)
N = noise power (= kTB) in bandwidth B (in W)

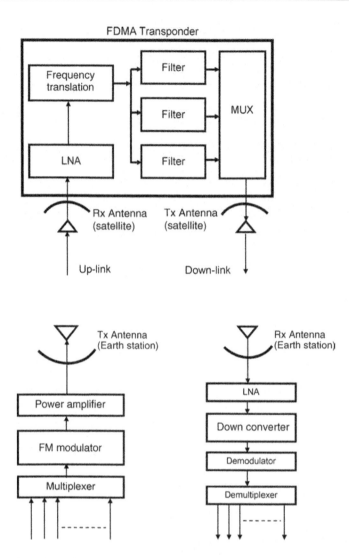

Figure 6.4 Typical block diagram of the MCPC/FDM/FM/FDMA system (MUX, multiplexer; Rx, receiver; Tx, transmitter)

6.4.2 MCPC/PCM-TDM/PSK/FDMA System

In this arrangement, multiple base band signals are first digitally encoded using the PCM technique and then grouped together to form a common base band assembly using time division multiplexing. This time division multiplexed bit stream then modulates a common RF carrier using phase shift keying as the carrier modulation technique. The modulated signal is then transmitted to the satellite, which uses FDMA to handle multiple carriers.

6.5 Time Division Multiple Access (TDMA)

As outlined earlier, time division multiple access (TDMA) is a technique in which different Earth stations in the satellite footprint having a common satellite transponder use a single carrier on a time division basis. Different Earth stations transmit traffic bursts in a period time-frame called the TDMA frame. Over the length of a burst, each Earth station has the entire transponder bandwidth at its disposal. The traffic bursts from different Earth stations are synchronized so that all bursts arriving at the transponder are closely spaced but do not overlap. The transponder works on a single burst at a time and retransmits back to Earth a sequence of bursts. All Earth stations can receive the entire sequence and extract the signal of their interest. Figure 6.5 illustrates the basic concept of TDMA. The disadvantages of TDMA include a requirement for complex and expensive Earth station equipment and stringent timing and synchronization requirements. TDMA is suitable for digital transmission only.

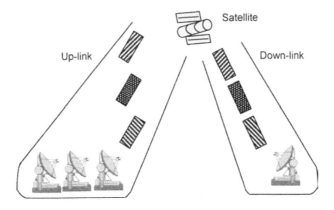

Figure 6.5 Basic concept of TDMA

TDMA systems can be classified as preassigned TDMA systems, demand assigned TDMA systems, satellite switched TDMA and limited preassigned TDMA systems. In preassigned TDMA systems, every Earth station is allotted a specific time slot. In a demand assigned TDMA system, the time slots are allotted to the Earth stations on request from the control station. In satellite switched TDMA systems, several antenna spot beams are utilized to provide services to different regions on the Earths surface. Limited preassigned TDMA is a technique that allows the traffic to be handled during busy hours by demand.

6.6 TDMA Frame Structure

As mentioned above, in a TDMA network, each of the multiple Earth stations accessing a given satellite transponder transmits one or more data bursts. The satellite thus receives at its input a set of bursts from a large number of Earth stations. This set of bursts from various Earth stations is called the TDMA frame. Figure 6.6 shows a typical TDMA frame structure.

It is evident from the frame structure that the frame starts with a reference burst transmitted from a reference station in the network. The reference burst is followed by

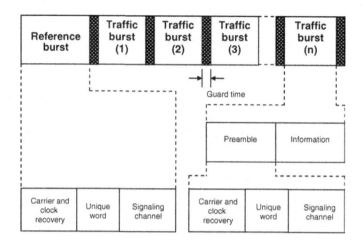

Figure 6.6 Typical TDMA frame structure

traffic bursts from various Earth stations with a guard time between various traffic bursts from different stations. The traffic bursts are synchronized to the reference burst to fix their timing reference. Different parts of the TDMA frame structure are briefly described in the following paragraphs.

6.6.1 Reference Burst

The reference burst is usually a combination of two reference bursts (RB-1 and RB-2). The primary reference burst, which can be either RB-1 or RB-2, is transmitted by one of the stations, called the primary reference station, in the network. The secondary reference burst, which is RB-1 if the primary reference burst is RB-2 and RB-2 if the primary reference burst is RB-1, is transmitted by another station, called the secondary reference station, in the network. The reference burst automatically switches over to the secondary reference burst in the event of primary reference station's failure to provide reference burst to the TDMA network. The reference burst does not carry any traffic information and is used to provide timing references to various stations accessing the TDMA transponder.

6.6.2 Traffic Burst

Different stations accessing the satellite transponder may transmit one or more traffic bursts per TDMA frame and position them anywhere in the frame according to a burst time plan that coordinates traffic between various stations. The timing reference for the location of the traffic burst is taken from the time of occurrence of the primary reference burst. With this reference, a station can locate and then extract the traffic burst or portions of traffic bursts intended for it. The reference burst also provides timing references to the stations for transmitting their traffic bursts so as to ensure that they arrive at the satellite transponder within their designated positions in the TDMA frame.

6.6.3 Guard Time

Different bursts are separated from each other by a short guard time, which ensures that the bursts from different stations accessing the satellite transponder do not overlap. This guard time should be long enough to allow for differences in transmit timing inaccuracies and also for differences in range rate variations of the satellite.

6.7 TDMA Burst Structure

As outlined above, a TDMA frame consists of reference and traffic bursts separated by the guard time. The traffic burst has two main parts, namely the information carrying portion and another sequence of bits preceding the information data, called preamble. The purpose of preamble sequence of bits is to synchronize the burst and to carry management and control information. The preamble usually consists of three adjacent parts, namely (a) the carrier and clock recovery sequence, (b) the unique word and (c) the signalling channel. The reference burst carries no traffic data and contains only the preamble.

6.7.1 Carrier and Clock Recovery Sequence

Different Earth stations have slight differences in frequency and bit rate. Therefore, the receiving stations must be able to establish accurately the frequency and bit rate of each burst. This is achieved with the help of carrier and clock recovery sequence bits. The length of this sequence usually depends on the carrier-to-noise ratio at the input of the demodulator and the carrier frequency uncertainty. A higher carrier-to-noise ratio and a lower carrier frequency uncertainty require a smaller bit sequence for carrier and clock recovery and vice versa.

6.7.2 Unique Word

Unique word is again a sequence of bits that follows the carrier and clock recovery sequence of bits in the preamble. In the reference burst, this bit sequence allows the Earth station to locate the position of the received TDMA frame. The unique word bit sequence in the traffic burst provides a timing reference on the occurrence of the traffic burst and also provides a timing marker to allow the Earth stations to extract their part of the traffic burst. The timing marker allows the identification of the start and finish of a message in the burst and helps to correct decoding. For obvious reasons, the unique word should have a high probability of detection. For instance, when the unique word of a traffic burst is missed, the entire traffic burst is lost. To achieve this, the unique word is a sequence of 1's and 0's selected to exhibit good correlation properties to enhance probability of detection.

Figure 6.7 shows a type of digital correlator circuit that can be used to detect the unique word bit sequence. Here, the unique word has N bits and is correlated with a stored pattern of itself. The received data is input to one of the N-bit shift registers, as shown in the figure, in synchronization with the data clock rate. The other N-bit shift register has a stored pattern of the unique word. Each stage of the shift register feeds a 2-bit adder, whose output is a '0' if the bits are in agreement and a '1' if they are in disagreement. The outputs of N

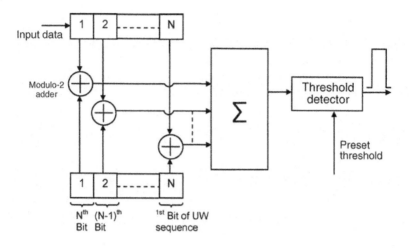

Figure 6.7 Digital correlator circuit

adders are summed up. The output of the summer is a step function that depends upon the number of agreements or disagreements between the received unique word bit pattern and the stored pattern. The output of summer then feeds a threshold detector that specifies the acceptable number of disagreements. If the number of mismatches is less than or equal to the preset threshold value, the unique word is considered to have been detected. Remember that a declaration of the detection of the unique word occurs at the time instant of reception of the last bit or symbol of the unique word. As mentioned earlier, if the unique word belongs to the reference burst, it marks the TDMA receive frame timing. In the case of traffic burst, it marks the receive traffic burst timing.

6.7.3 Signalling Channel

The signalling channel is used to carry out system management and control functions. The signalling channel of the reference burst has three channels, namely (a) an order wire channel used to pass instructions to and from Earth stations, (b) a management channel transmitted by reference stations to all traffic stations carrying frame management instructions, such as changes in the burst time plan that coordinates traffic between different stations, and (c) a transmit timing channel that carries acquisition and synchronization information to different traffic stations, enabling them to adjust their transmit burst timing (TBT) so that similar bursts from different stations reach the satellite transponder within the correct time slot in the TDMA frame.

The signalling channel of the traffic burst also has an order wire channel, which performs the same function as it does in the case of reference burst. It also has a service channel, which performs functions like carrying traffic station's status to the reference station, carrying information such as the high bit error rate or unique word loss alarms, etc., to other traffic stations.

6.7.4 Traffic Information

Traffic bursts follow the reference burst in the TDMA frame structure. Each station in the TDMA network can transmit and receive many traffic bursts and sub-bursts per frame. The length of each sub-burst, which represents information on a certain channel, depends upon the type of service and the number of channels being supported in the traffic burst. For instance, while transmitting a PCM voice channel that is equivalent to a data rate of 64 kbps, each sub-burst for this channel would be 64 bits long if the frame time available for the purpose was 1 ms.

6.8 Computing Unique Word Detection Probability

As outlined in earlier paragraphs, accurate detection of the unique word is of the utmost importance in a TDMA network. Without going into mathematical details, in this part, an outline will be given of the important mathematical expressions that can be used to compute probabilities of miss detection and false detection. In general, for a given length (in bits or symbols) of unique word, increasing the detection threshold, i.e. the maximum number of acceptable correlation errors, reduces the probability of miss detection but at the same time increases the probability of false detection. On the other hand, lowering the detection threshold to improve the false detection probability increases the probability of miss detection.

In view of this, if the detection threshold is designated as ϵ, the probability of miss detection is nothing but the probability of it having $\epsilon + 1$ or more errors. The probability P_i that i bits or symbols out of N will be in error is given by the binomial distribution as follows:

$$P_i = \left[\frac{N!}{i!(N-i)!} \right] \times p^i \times (1-p)^{N-i} \tag{6.20}$$

where p = average probability of error for receive data.

The probability of correct detection P_C, which is the sum of probabilities of $0, 1, 2, 3, \ldots, \epsilon$ errors, and the probability of miss detection P_M, which equals $(1 - P_C)$, are then given respectively by the following expressions:

$$P_C = \sum_{i=0}^{\epsilon} \left[\left\{ \frac{N!}{i!(N-i)!} \right\} \times p^i \times (1-p)^{N-i} \right] \tag{6.21}$$

$$P_M = \sum_{i=(\epsilon+1)}^{N} \left[\left\{ \frac{N!}{i!(N-i)!} \right\} \times p^i \times (1-p)^{N-i} \right] \tag{6.22}$$

The probability of false detection P_F is another parameter of interest and can be computed on the basis of the logic that it is the probability of a random string of N bits or symbols accidentally corresponding to the stored unique word pattern to an extent that the number of bits or symbols in disagreement does not exceed the detection threshold. It is only then that there would be a false detection of the unique word. Since there are 2^N possible combinations in which random data can occur, the probability of occurrence of the combination matching the stored unique word pattern would be $1/2^N$. If the detection threshold is ϵ, then $1/2^N$ is also the probability of false detection for $\epsilon = 0$, i.e. random data completely matching the unique word pattern with no correlation errors. In the case where the detection threshold is ϵ, the probability of false detection can be computed from

$$P_F = \frac{\sum_{i=0}^{\epsilon}\left[\frac{N!}{i!(N-i)!}\right]}{2^N} \qquad (6.23)$$

The above expression is obvious as the total number of possible combinations in which correlation errors are less than equal to ϵ will be

$$\sum_{i=0}^{\epsilon}\left[\frac{N!}{i!(N-i)!}\right] \qquad (6.24)$$

By analysing the expressions for probabilities of miss detection and false detection, the following observations can be made:

1. For a given value of error probability p of the link, the probability of miss detection can be reduced by decreasing the length N of the unique word or by increasing the detection threshold ϵ. These values are so chosen as to yield $P_M \ll p$.
2. The probability of false detection can be reduced by increasing the length N of the unique word or by decreasing the detection threshold ϵ.

6.9 TDMA Frame Efficiency

TDMA frame efficiency is defined as the percentage of total frame length allocated for transmission of traffic data. It is expressed as

$$\eta = 1 - \left(\frac{T_x}{T_f}\right) \qquad (6.25)$$

where

T_x = overhead portion of the frame (guard times, preambles)
T_f = frame length

In the case where the frame has n bursts, T_x can further be expressed as

$$T_x = n \times T_g + \sum_{i=0}^{n} T_{p,i} \qquad (6.26)$$

where

T_g = guard time between bursts
$T_{p,i}$ = preamble of the ith burst

Frame efficiency should be as high as possible. One of the methods to achieve this is to reduce the overhead portion of the frame. This cannot be done arbitrarily. For instance, the carrier and clock recovery sequence must be long enough to provide stable acquisition of the carrier and to minimize the ill effects of interburst interference. Also, the guard time in between the bursts should be long enough to allow for differences in transmit timing inaccuracies and also for differences in range rate variations of the satellite. All these factors need to be considered carefully in any TDMA system design.

Frame efficiency can also be increased by increasing the frame length. However, higher the frame length, larger is the amount of memory needed to perform functions like storing the incoming terrestrial data at a continuous rate for one frame, transmitting the data at a much higher burst bit rate to the satellite, storing the receive traffic bursts and then converting the received data to lower continuous outgoing terrestrial data. Another reason that puts an upper limit on the frame length is that it needs to be kept at a small fraction of the maximum satellite round trip delay of about 274 ms in order to avoid adding a significant delay to the transmission of voice traffic, for which it is usually selected to be less than 20 ms.

6.10 Control and Coordination of Traffic

Two of the most important functions in a TDMA network are controlling the position of the burst in the frame and coordination of traffic between different stations. Both functions are intended to ensure that any change in position and length of bursts does not lead to any burst overlapping or service disruption. Both these functions are performed by the reference station by using the reference burst through the transmit timing channel to control the burst position and the management channel to coordinate traffic between stations.

To do the twin tasks mentioned above, the reference station makes use of what is called a superframe structure. It should be appreciated that to perform control and coordination functions, the reference station has to address all traffic stations in the network. In order to do this in the conventional way, it is observed that it needs a large amount of additional time slots in the reference burst, leading to a reduction in frame efficiency. For instance, if there are N traffic stations, the reference station will need to send N messages to N stations in the transmit timing channel and also another N messages to N stations in the management channel of the reference burst per frame. In order to make transmission and reception of these messages reliable, which is essential, some form of coding is usually employed. One such coding is the 8:1 redundancy coding algorithm where an information bit is repeated eight times in a predetermined pattern and then decoded by using majority decision logic at the receive end. This further increases the time slot required for the purpose eight times. The same is true for the service channel of traffic bursts.

This problem is overcome if the reference station sends one message to one station per frame and the process of sending N messages is then completed in N frames. In other words, station 1 may be addressed by the reference station in frame 1, station 2 in frame 2 and so on, with station N being addressed in frame N. The same procedure can be followed by traffic stations if they have to send a status report to the reference station or some other information to another traffic station. This reduces the lengths of reference burst and traffic burst preambles and increases frame efficiency.

Figure 6.8 shows a typical superframe structure with N frames. There are various ways by which frames can be identified in a superframe. It could be an identification number carried by the management channel in the reference burst with the identification number of frame 1 serving as the superframe marker. Another method is to use different unique words by the reference bursts and traffic bursts to distinguish superframe markers from frame markers.

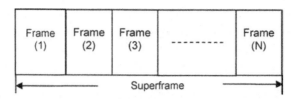

Figure 6.8 Typical superframe structure with N frames

Having discussed the superframe structure and its significance in performing control and coordination functions, yet another aspect of control and coordination that needs to be addressed is the situation where the number of traffic stations in the network is growing. If the number is fixed or its maximum is known, it is relatively easy for the service channel of traffic bursts to transmit its message over N frames. However, when the number of stations in the network is variable, for instance when the network is growing and the demand assignment mode has been employed for the transponder assignment, it may be appropriate to transmit the messages in the service channel of traffic bursts and demand assignment messages in a separate superframe short burst (SSB) at the superframe rate. In this case, each of the stations in the network transmits a superframe short burst once per superframe; i.e. if station 1 transmits an SSB during superframe 1, station 2 would transmit an SSB during superframe 2 and so on. The advantage of having the service channel in the superframe short burst lies particularly in a situation where a station transmits more than one traffic bursts per frame. Since the messages to be transmitted from a given station through the service channels of traffic bursts in the same frame are likely to be identical, redundancy of messages leads to a reduction in frame efficiency. Figure 6.9 shows the position of the superframe short burst in the superframe structure. The superframe short burst structure is shown in Figure 6.10.

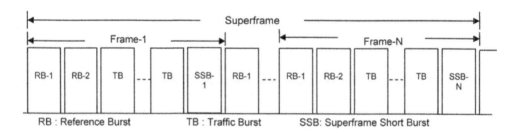

Figure 6.9 Position of superframe short burst in the superframe structure

Carrier and Clock Recovery	Unique Word	Service Channel	Others (Demand Assignment)

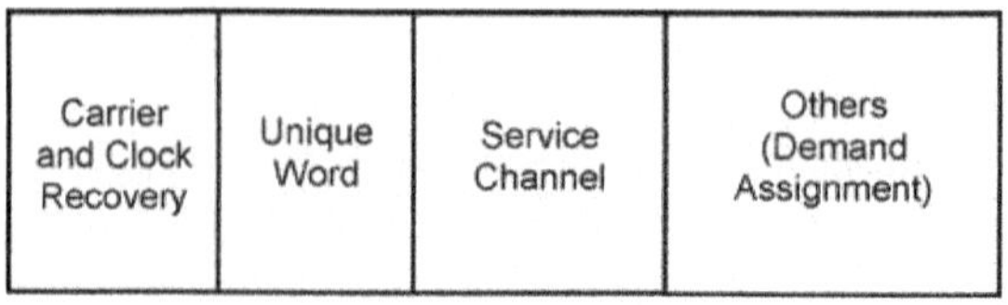

Figure 6.10 Superframe short burst structure

6.11 Frame Acquisition and Synchronization

Frame acquisition and synchronization, both during the receive and transmit phases, are vital to proper functioning of a TDMA network. While receiving, the station should be able to receive the traffic bursts addressed to it by the satellite transponder(s) in every frame. Similarly, while transmitting, the station should be able to transmit the traffic bursts in such a way that the bursts arrive at the transponder(s) at the correct position within the frame without any overlap with the bursts from other traffic stations.

6.11.1 Extraction of Traffic Bursts from Receive Frames

To ensure a proper receive operation, the traffic station establishes the receive frame timing (RFT), which is defined as the time instant of occurrence of the last bit or symbol of the unique word of the primary reference burst. This sets the time marker from which the location of the traffic burst intended for a given station can be fixed. This is achieved by identifying the receive burst timing (RBT), which is determined by knowing the offset between the receive frame timing reference and the transmit burst timing. The amount offset in bits or symbols is contained in a receive burst time plan, which is stored in the foreground memory of the traffic station. Thus with the help of a receive burst time plan, the traffic station can extract all traffic bursts addressed to it in different frames. This whole process is called the receive frame acquisition.

6.11.2 Transmission of Traffic Bursts

A prerequisite for proper transmission of traffic burst is that it should reach the satellite transponder within the allocated position in the TDMA frame so as not to cause any overlap with traffic bursts transmitted to the transponder by other traffic stations. This can be ensured again by establishing what is called transmit frame timing (TFT) and transmit burst timing (TBT). While the former marks the start of the station's time frame, the latter marks the start of traffic burst. Again, the position of the traffic burst in a transmitted frame is determined by the offset between the transmit frame timing and the transmit burst timing. The information on the offset is contained in the transmit burst time plan stored in the foreground memory of the traffic station. The traffic burst that is transmitted at its transmit burst timing will fall into its appropriate position in the TDMA frame at the transponder. Similarly, traffic bursts from other stations accessing a particular transponder fall into their preassigned or designated positions in the TDMA frame without causing any burst overlapping. This whole process is called transmit frame acquisition.

6.11.3 Frame Synchronization

What has been discussed in the preceding paragraphs is the acquisition of receive frame and transmit frame timings. The acquisition process is needed when a station enters or re-enters operation. The process of maintaining the acquired timing references is synchronization. Synchronization becomes a necessity due to small changes in the satellite orbit caused by a variety of factors. A geostationary satellite orbit may be specified in terms of its inclination angle with the equatorial plane, its eccentricity and its east–west drift. A variation in the inclination angle causes north–south drift and a variation in eccentricity causes a variation in altitude. In the case of a geostationary orbit, the peak-to-peak variation in altitude is 0.2 % of the orbit radius (=42 164 km), which amounts to about 85 km (0.2 × 42164/100). Peak-to-peak north–south drift and peak-to-peak east–west drift are about 0.2°, which amounts to about 150 km in terms of distance. These errors introduce a maximum range variation of 172 km [=$\sqrt{(85^2 + 150^2)}$]. This is equivalent to a one-way propagation delay of 0.575 ms and a round trip maximum delay variation of 1.15 ms. Assuming that the satellite takes about 8 hours to move from its nominal position to a position where maximum delay variation occurs, this leads to a maximum Doppler shift of 40 ns/s. This Doppler shift causes errors in the burst positions as they arrive at the transponder. Thus, frame synchronization is necessary for proper transmission and reception of traffic bursts.

Problem 6.1
A TDMA frame and burst structure has the following parameters:

1. TDMA frame length = 20 ms
2. Length of carrier and clock recovery sequence = 352 bits
3. Length of unique word = 48 bits
4. Length of order wire channel = 510 bits
5. Length of management channel = 256 bits
6. Length of transmit timing channel = 320 bits
7. Length of service channel = 24 bits
8. Guard time = 64 bits

Also each of the 10 stations in the network transmits two traffic bursts in each frame and each frame contains two reference bursts in addition.

Determine the following:

(a) Length of the reference burst preamble (in bits)
(b) Length of the traffic burst preamble (in bits)
(c) Total number of overhead bits

Solution:

(a) It is known that the reference burst contains only the preamble with the carrier and clock recovery sequence, unique word, order wire channel, management channel and transmit timing channel as various component parts. Therefore, the length of the reference burst preamble (from given data) = 352 + 48 + 510 + 256 + 320 = 1486 bits.
(b) The preamble of the traffic burst contains the carrier and clock recovery sequence, unique word, order wire channel and service channel. Therefore, the length of the traffic burst preamble (from given data) = 352 + 48 + 510 + 24 = 934 bits.
(c) Since there is a total of 22 bursts per frame (2 reference bursts and 20 traffic bursts), the total guard time per frame is given by 22 × 64 = 1408 bits. Therefore, the total number of overhead bits is 1486 × 2 + 934 × 20 + 1408 = 2972 + 18 680 + 1408 = 23 060 bits.

Problem 6.2
If in Problem 6.1, the TDMA burst bit rate is 90 Mbps, determine the TDMA frame efficiency.

Solution: The total number of bits in the frame = $90 \times 10^6 \times 20 \times 10^{-3} = 1\,800\,000$ bits. Therefore, the frame efficiency = $[(1\,800\,000 - 23\,060)/1\,800\,000] \times 100\% = 98.72\,\%$.

Problem 6.3
Considering the TDMA frame whose data are given in Problems 6.1 and 6.2, if the traffic data is assumed to be voice and voice is PCM encoded, with each voice channel data rate being 64 kbps, determine the maximum number of PCM voice channels carried in a frame.

Solution: The length of the TDMA frame = 20 ms. Therefore, the number of bits in a frame for a voice sub-burst = $64 \times 10^3 \times 20 \times 10^{-3} = 1280$ bits. Now, the total number of bits available in a frame for carrying traffic is given by $1\,800\,000 \times 0.9872 = 1\,776\,960$. Therefore, the maximum number of PCM voice channels in a frame = $1\,776\,960/1280 = 1388$ channels.

Problem 6.4
A geostationary satellite has an orbital radius of 42 150 km. If the station-keeping accuracy of the satellite is $\pm 0.15°$ and the altitude variation caused by the orbit inclination is 0.25 %, determine the maximum Doppler shift caused by these variations in terms of variation in the round trip propagation delay as a function of time. Assume that it takes the satellite about 8 hours to move from its nominal position to a position of maximum delay variation.

Solution:
Variation in altitude caused by orbit inclination = $0.0025 \times 42\,150 = 105.4$ km

Variation due to a station-keeping error of $0.3° = 42\,150 \times 2 \times \pi \times 0.3/360 = 220.7$ km

Both these errors will introduce a maximum range variation of

$$\sqrt{(220.7)^2 + (105.4)^2} = 244.6 \text{ km}$$

This causes a one-way propagation delay of $(244.6 \times 10^3 / 3 \times 10^8)\text{s} = 0.815$ ms
Therefore the round-trip propagation delay = 2×0.815 ms = 1.63 ms
Doppler shift = 1.63 ms in 8 hours = 56.25 ns/s

6.12 FDMA vs. TDMA

TDMA and FDMA systems offer some advantages and disadvantages with respect to each other. In this section we present the comparison between the two systems.

6.12.1 Advantages of TDMA over FDMA

1. The bit capacity of the TDMA system is independent of the number of accesses.
2. The power amplifiers in the transponder of a TDMA system can be operated in the saturation mode as opposed to that of an FDMA system. This increases the capacity of multiple accesses.
3. The duty cycle of the Earth station in a TDMA system is low.
4. TDMA systems allow the use of digital techniques like digital speech interpolation, satellite on-board switching and so on.
5. The TDMA systems offer more flexibility due to use of high-speed logic circuits and processors which offer high data rates.
6. TDMA systems are more economical as compared to FDMA systems as they are easy to multiplex, independent of distance and can be easily interfaced with terrestrial services.
7. TDMA systems can tolerate higher levels of interference noise.

6.12.2 Disadvantages of TDMA over FDMA

1. The peak power of the amplifier in a TDMA system is always large.
2. The TDMA system is more complex as compared to the FDMA system. The Earth station of the TDMA system requires ADC, clock recovery, synchronization, burst control and data processing before transmission.
3. TDMA systems use high-speed PSK/FSK circuits.

6.13 Code Division Multiple Access (CDMA)

In case of code division multiple access (CDMA), the entire bandwidth of the transponder is used simultaneously by multiple Earth stations at all times. Code division multiple access, therefore allows multiple Earth stations to access the same carrier frequency and bandwidth at the same time. Each transmitter spreads its signal over the entire bandwidth, which is much wider than that required by the signal otherwise. One of the techniques to do this is to multiply the information signal, which has a relatively lower bit rate, by a pseudorandom bit sequence with a much higher bit rate. Interference between multiple channels is avoided as each transmitter uses a unique pseudorandom code sequence. Receiving stations recover the desired information by using a matched decoder that works on the same unique code sequence used during transmission. In the following paragraphs, it will be seen how such a system works.

It is assumed that the message signal is a PCM bit stream. Each message bit is combined with a predetermined code bit sequence. This predetermined code sequence of bits is usually a pseudorandom noise (PN) signal. The bit rate of the PN sequence is kept much higher than the bit rate of the message signal. This spreads the message signal over the entire available bandwidth of the transponder. It is because of this reason that this technique of multiple access is often referred to as spread spectrum multiple access (SSMA). The spread spectrum operation enables the signal to be transmitted across a frequency band that is much wider than the minimum bandwidth required for the transmission of the message signal. The PN sequence bits are often referred to as 'chips' and their transmission rate as the 'chip rate'. The receiver is able to retrieve the message addressed to it by using a replica of the PN sequence used at the transmitter, which is synchronized with the transmitted PN sequence.

CDMA uses direct sequence (DS) techniques to achieve the multiple access capability. In this, each of the N users is allocated its own PN code sequence. PN code sequences fall into the category of orthogonal codes. Cross-correlation of two orthogonal codes is zero, while their auto-correlation is unity. This forms the basis of each of the N stations being able to extract its intended message signal from a bit sequence that looks like white noise.

6.13.1 DS-CDMA Transmission and Reception

Figure 6.11 shows the basic DS-CDMA transmitter. The diagram is self-explanatory. The transmitter generates a bit stream by multiplying in the time domain the message bit stream $m_i(t)$ and the code information $a_i(t)$. Multiplication in the time domain is convolution in the frequency domain. Therefore, the product of $m_i(t)$ and $a_i(t)$ produces a signal whose spectrum is nothing but convolution of the spectrum of $m_i(t)$ and the spectrum of $a_i(t)$. Also, if the bandwidth of the message signal is much smaller than the bandwidth of the code signal, the product signal has a bandwidth approaching that of the code signal.

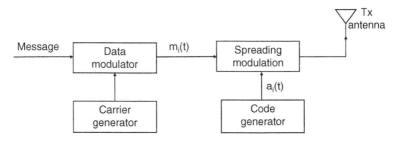

Figure 6.11 Basic block schematic arrangement of the DS-CDMA transmitter

The message signal is either an analogue or a digital signal. In the majority of cases, it is a digital signal. In that case, message signal modulation as shown in the diagram is omitted and the message signal is directly multiplied by the code signal. The resulting signal then modulates a wideband carrier using a digital modulation technique, which is usually some form of phase shift keying.

Figure 6.12 shows the basic block schematic arrangement of the DS-CDMA receiver. The diagram is self-explanatory. It is assumed that the receiver is configured to receive the

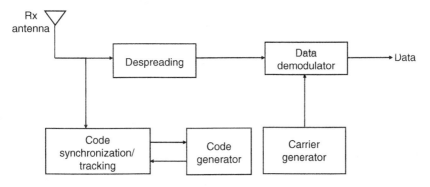

Figure 6.12 Basic block schematic arrangement of the DS-CDMA receiver

message signal $m_i(t)$. The receiver in this case generates a code signal $a_i(t)$ synchronized with the received message. If the signals represented by suffix j constitute undesired signals, i.e. noise, then the bit stream present at the output of the first stage of the receiver and at the input of the demodulator is given by

$$\overline{[a_i^2(t)][m_i(t)]} + \sum_{\substack{i=1 \\ j \neq i}}^{N} \overline{[a_i(t)][a_j(t)][m_j(t)]} \tag{6.27}$$

In the case of orthogonal codes,

$$\int_0^T a_i^2(t) = 1 \tag{6.28}$$

$$\int_0^T [a_i(t)][a_j(t)] = 0 \quad \text{for } i \neq j \tag{6.29}$$

By substituting these values, it is found that only the first part of the expression for the received bit sequence is the desired one; the remaining part is just noise.

A term called processing gain (G) is defined in such systems. It is given by

$$G = 10 \log \text{(Number of chips per bit)} = 10 \log \text{(chip rate/message information rate)} \tag{6.30}$$

A chip rate of 2.5 Mbps and a message bit rate of 25 kbps would give a processing gain of 20 dB. Since the undesired component or the noise is spread over the entire bandwidth and the receiver responds only to the $1/G$ part of this bandwidth, this has the effect of reducing the noise by the same factor. The thermal noise and intermodulation noise are also reduced by the factor G. To sum up, the input to the DS-CDMA receiver is wide band and contains both desired and undesired components. When this received bit stream is applied to the correlator, the output of the correlator is the desired message signal centred on the intermediate frequency. The undesired signals remain spread over the entire bandwidth and only that portion within the receiver bandwidth causes interference. This is further illustrated in Figure 6.13.

In practice, the cross-correlation function is not equal to zero, which leads to some performance degradation. The level of interference is directly determined by the ratio of the peak cross-correlation value to the peak auto-correlation value of the pseudorandom code sequences.

DS-CDMA is further divided into two types, namely:

1. Sequence synchronous DS-CDMA
2. Sequence asynchronous DS-CDMA

6.13.1.1 Sequence Synchronous DS-CDMA

In the case of the sequence synchronous DS-CDMA system, the message bit duration is chosen to be a certain N times the PN sequence bit duration. This implies that the ratio of the chip rate to message bit rate is N, which is also the processing gain of the system. Also, the system is so synchronized that the PN sequence period of each carrier is time-aligned at the satellite. This is similar to what is done in TDMA, but the synchronization requirements in this case are far less stringent. Here, synchronization in time should be of the order of one-fifth of the

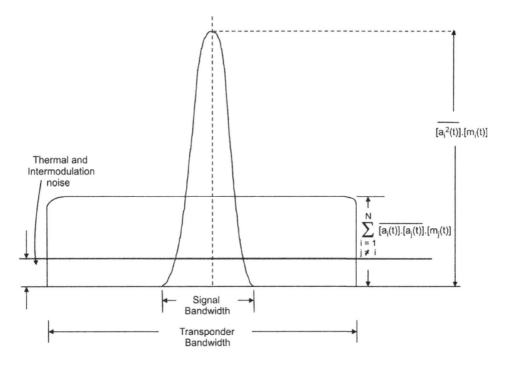

Figure 6.13 Signal output of the correlator

PN sequence bit duration. Knowing the timing error caused by the Doppler effect, this sets an upper limit on the maximum usable chip rate.

6.13.1.2 Sequence Asynchronous DS-CDMA

In the case of the sequence asynchronous DS-CDMA system, the message bit duration is chosen to be a certain N times the PN sequence bit duration. The difference between the two systems lies in the fact that here no attempt is made to align the PN sequence period at the satellite. Therefore, there will be time shifts between different carriers. It can be verified that the sequence synchronous DS-CDMA system offers no definite advantage over the sequence asynchronous DS-CDMA system when the user population is much larger than the number of simultaneous users.

6.13.2 Frequency Hopping CDMA (FH-CDMA) System

CDMA is also referred to as spread spectrum multiple access because of the reason that the carrier spectrum is spread over a much larger bandwidth as compared to the information rate. The spread spectrum signal is inherently immune to jamming as it forces the jammer to deploy its transmitted jamming power over a much wider bandwidth than would have been necessary for a conventional system. In other words, for a given power of the jammer, the power spectral density produced by the jammer becomes reduced by a factor of increase in the bandwidth due to spreading. The direct-sequence CDMA discussed in earlier paragraphs is one way to spread the carrier spectrum where the message bit sequence is multiplied by a pseudorandom code sequence. Frequency hopping (FH), to be discussed in the following paragraphs, is another method used to do the same.

In the case of a frequency hopping CDMA (FH-CDMA) system, the carrier is sequentially hopped into a series of frequency slots spread over the entire bandwidth of the satellite transponder. The transmitter operates in synchronization with the receiver, which remains always tuned to the frequency of the transmitter. The transmitter transmits a short burst of data on a narrowband, then tunes to another frequency and transmits again. The transmitter thus hops its frequency over a given bandwidth several times per second, transmitting one frequency for a certain period of time, then hopping to another frequency and transmitting again. This is achieved by using a frequency synthesizer whose output is controlled by a pseudorandom code sequence. The pseudorandom code sequence decides the instantaneous transmission frequency. On the receiver side, the data can be recovered by using an identical frequency synthesizer controlled by an identical pseudorandom sequence. Figures 6.14 and 6.15 show the block schematic arrangements of typical FH-CDMA transmitter and receiver respectively. The diagrams are self-explanatory.

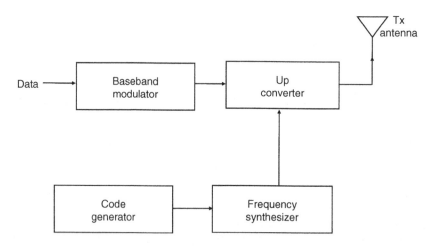

Figure 6.14 Block schematic arrangement of a typical FH-CDMA transmitter

Due to the random hopping pattern as governed by the pseudorandom sequence, to an observer, the carrier appears to use the entire transponder bandwidth over the pseudorandom sequence period. At a given instant of time, however, it uses a particular frequency slot. The hopping rate of the carrier may equal the coded symbol rate in the case of slow hop systems or be several times that of the coded symbol rate in the case of fast hop systems. In the case of FH-CDMA, each Earth station is assigned a unique hop pattern. Commonly used modulation schemes in frequency hop CDMA systems include noncoherent M-ary FSK (FSK techniques having M frequency levels) and differential QPSK. Also, non-coherent demodulation is used as it is very difficult to maintain phase coherence between the hops.

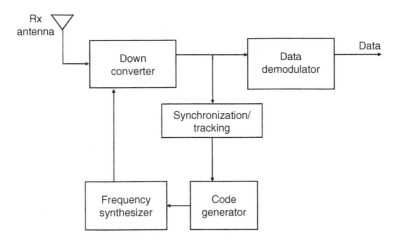

Figure 6.15 Block schematic arrangement of a typical FH-CDMA receiver

6.13.3 Time Hopping CDMA (TH-CDMA) System

In the case of the time hopping CDMA (TH-CDMA) system, the pseudorandom bit sequence determines the time instant of transmission of information. In fact, the signal is transmitted by the user in rapid bursts during time intervals determined by the pseudorandom code assigned to the user. A given user transmits only during one of the M time slots each frame has been divided into. However, the time slot used by a given user for transmission of data in successive frames depends upon the code assigned to it. Since each user transmits its data only during one of the M time slots in each frame, the bandwidth available to it increases by a factor of M. Figure 6.16 shows the block schematic arrangement of a typical TH-CDMA transmitter. The typical receiver block schematic is shown in Figure 6.17.

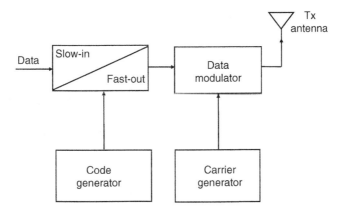

Figure 6.16 Block schematic arrangement of a typical TH-CDMA transmitter

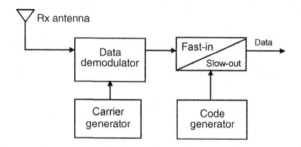

Figure 6.17 Block schematic arrangement of a typical TH-CDMA receiver

6.13.4 Comparison of DS-CDMA, FH-CDMA and TH-CDMA Systems

The operational principle of the three systems can be illustrated with the help of frequency–time graphs, as shown in Figures 6.18 (a), (b) and (c). These graphs depict the frequency usage of the three systems as a function of time. It is evident from the graph shown in Figure 6.18 (a) that a DS-CDMA system occupies the whole of the available bandwidth when it transmits, whereas an FH-CDMA system uses only a small part of the bandwidth at a given instant of time when it transmits, as shown in Figure 6.18 (b). However, the location of this part differs with time. The bandwidth occupied by an FH-CMDA signal for a given hop frequency depends not only on the bandwidth of the message signal but also on the shape of the hopping signal and the hopping frequency. In the case of a slow hop system, when the hopping frequency is much smaller than the message signal bandwidth, the occupied bandwidth is mainly decided by the message signal bandwidth. In case of a fast hop system, when the hopping frequency

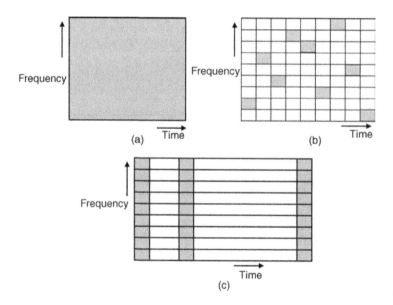

Figure 6.18 Frequency–time graphs of (a) DS-CDMA system (b) FH-CDMA system and (c) TH-CDMA system

Code Division Multiple Access (CDMA)

is much larger than the message signal bandwidth, the occupied bandwidth is mainly decided by the shape of the hopping signal at a given hopping frequency.

In the case of time hopping CDMA (TH-CDMA), depicted in Figure 6.18 (c), as compared to FH-CDMA, the whole of the available bandwidth is used for short time periods instead of parts of the bandwidth being used all the time. To sum up, DS-CDMA uses an entire bandwidth all the time, FH-CDMA uses a small part of the bandwidth at a given time instant but the chosen frequency slot varies with time in order to cover the entire bandwidth and TH-CDMA uses the entire bandwidth for short periods of time.

Problem 6.5

A geostationary satellite has a round-trip propagation delay variation of 40 ns/s due to station-keeping errors. If the time synchronization of DS-CDMA signals from different Earth stations is not to exceed 20 % of the chip duration, determine the maximum allowable chip rate so that a station can make a correction once per satellite round - trip delay. Assume the satellite round-trip delay to be equal to 280 ms.

Solution:

Doppler effect variation due to station-keeping errors = 40 ns/s

Satellite round trip delay = 280 ms
Therefore,

Time error due to the Doppler effect in one satellite round trip = $40 \times 10^{-9} \times 280 \times 10^{-3}$

$$= 11.2 \text{ ns}$$

Let T_c = chip duration. Then

$$0.2 \times T_c = 11.2 \text{ ns} \quad \text{or} \quad T_c = 56 \text{ ns}$$

which gives maximum chip rate = (1/56) Gbps = 1000/56 Mbps = 17.857 Mbps

Problem 6.6

If, in Problem 6.5, the maximum chip rate is to be 25 Mbps, what should be the maximum permissible Doppler effect variation due to station-keeping errors?

Solution:

$$\text{Chip rate} = 25 \text{ Mbps}$$

Therefore

$$\text{Chip duration} = 1/25 \; \mu s = 40 \text{ ns}$$

Maximum allowable timing error per satellite round trip = 0.2×40 ns = 8 ns
This 8 ns error is to occur in 280 ms. Therefore,
Maximum permissible Doppler effect variation = $(8/280 \times 10^{-3})$ ns/s = 28.57 ns/s

Problem 6.7

In a DS-CDMA system, the information bit rate and chip rate are respectively 20 kbps and 20 Mbps. Determine the processing gain in dB and also determine the noise reduction (in dB) achievable in this system.

Solution:

Chip rate = 20 Mbps
Information bit rate = 20 kbps
Processing gain = 10 log (chip rate/information bit rate)
$= 10 \log(20 \times 10^6 / 20 \times 10^3) = 10 \log(1000) = 30$ dB
Noise reduction achievable = processing gain = 30 dB

6.14 Space Domain Multiple Access (SDMA)

So far, multiple access techniques have been discussed that allow multiple Earth stations to access a given transponder(s) capacity without causing any interference among them. In the case of the frequency division multiple access (FDMA) technique, different Earth stations are able to access the total available bandwidth in satellite transponder(s) by virtue of their different carrier frequencies, thus avoiding interference amongst multiple signals. Here each Earth station is allocated only a part of the total available transponder bandwidth. In the case of the time division multiple access (TDMA) technique, different Earth stations in the satellite's footprint make use of the transponder by using a single carrier frequency on a time division basis. In this case, the transponder's entire bandwidth is available to each Earth station on a time-shared basis. In the case of the code division multiple access technique (CDMA), the entire bandwidth of the transponder is used simultaneously by multiple Earth stations at all times. Each transmitter spreads its signal over the entire transponder bandwidth. One of the methods to do so is by multiplying the information signal by a unique pseudorandom bit sequence. Others include frequency hopping and time hopping techniques. Interference is avoided as each transmitter uses a unique code sequence. Receiving stations recover the desired information by using a matched decoder that works on the same unique code sequence as used during transmission.

Space domain multiple access (SDMA), as outlined in the beginning of the chapter, is a technique that primarily allows frequency re-use where adjacent Earth stations within the footprint of the satellite can use the same carrier transmission frequency and still avoid co-channel interference by using orthogonal antenna beam polarization. Also, transmissions from/to a satellite to/from multiple Earth stations can use the same carrier frequency by using narrow antenna beam patterns. As also mentioned earlier, in an overall satellite link, SDMA is usually achieved in conjunction with other types of multiple access techniques such as FDMA, TDMA and CDMA. That is why these techniques were briefly mentioned here again before going into SDMA in a little more detail.

6.14.1 Frequency Re-use in SDMA

Frequency re-use, as outlined above, is the key feature and the underlying concept of space domain multiple access (SDMA). In the face of continually increasing demands on the frequency spectrum, it becomes important that frequency bands assigned to satellite communications are efficiently utilized. One of the ways of achieving this is to re-use all or part of the frequency band available for the purpose. Another way could be employment of efficient user access methods. Yet another approach could be the use of efficient modulation, encoding and compression techniques in order to pack more information into available bandwidths.

Restricting the discussion to frequency re-use, which is the present topic, the two methods in common use today for the purpose are beam separation and beam polarization. Beam separation is based on the fact that if two beams are so shaped that they illuminate two different regions on the surface of the Earth without overlapping, then the same frequency band could be used for the two without causing any mutual interference. One could do so by using two different antennas [Figure 6.19(a)] or a single antenna with two feeds [Figure 6.19(b)].

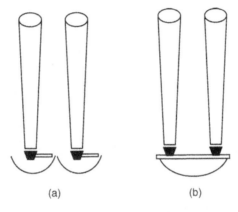

(a)   (b)

Figure 6.19 Frequency re-use using beam separation employing (a) Two antennas and (b) Single antenna

Beam polarization, on the other hand, relies on the principle of using two orthogonally polarized electromagnetic waves to transmit and receive using the same frequency band with no mutual interference between the two. Orthogonal polarizations used commonly include horizontal and vertical polarizations or right-hand circular and left-hand circular polarizations.

Both techniques have the capability of doubling the transmission capacity individually, and when used in tandem can increase the capacity four times. SDMA is seldom used in isolation. It is usually used in conjunction with other types of multiple access techniques discussed earlier, including FDMA, TDMA and CDMA. In the following paragraphs, the employment of SDMA separately with each one of these multiple access techniques will be briefly discussed.

6.14.2 SDMA/FDMA System

Figure 6.20 shows a typical block schematic arrangement of the SDMA/FDMA system in which the satellite uses fixed links to route an incoming uplink signal as received by a receiving antenna to a particular downlink transmitter antenna. It is clear from the diagram that the satellite uses multiple antennas to produce multiple beams. The transmitting antenna--receiving antenna combination defines the source and destination Earth stations.

The desired fixed links can be set on board the satellite by using some form of a switch, which can be selected only occasionally when the satellite needs to be reconfigured. The links

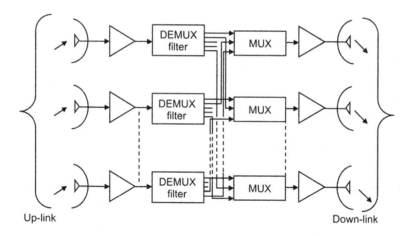

Figure 6.20 Typical block schematic arrangement of an SDMA/FDMA system (DEMUX, de-multiplexer)

could also be configured alternatively by switching the filters with a switch matrix operated by a command link. It may once again be mentioned here that satellite switches are changed only occasionally when the satellite is to be reconfigured.

6.14.3 SDMA/TDMA System

This system uses a switching matrix to form uplink/downlink beam pairs. In conjunction with TDMA, the system allows TDMA traffic from the uplink beams to be switched to the downlink beams during the course of a TDMA frame. The link between a certain source–destination combination exists at a specified time for the burst duration within the TDMA frame. As an example, the signal on beam 1 may be routed to beam 3 during, say, the first 40 μs of a 2 ms TDMA frame and then routed to beam n during the next 40 μs slot. The process continues until every connection for the traffic pattern has been completed. Figure 6.21 shows a

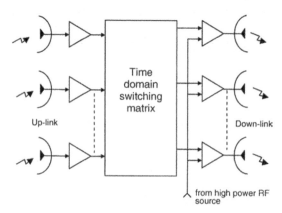

Figure 6.21 Typical transponder arrangement for an SDMA/SS/TDMA system

typical transponder arrangement for an SDMA/SS/TDMA system. (SS here stands for 'satellite switched').

6.14.4 SDMA/CDMA System

CDMA provides multiple access to the satellite. The satellite receives an uplink CDMA bit stream, decodes it to determine the destination address and then routes it to the desired downlink. The bit stream is usually re-timed, regenerated and stored in onboard processors before it is retransmitted. This implies that a downlink configuration does not need to be the same as the uplink configuration, thus allowing each link to be optimized.

Further Readings

Calcutt, D. and Tetley, L. (1994) *Satellite Communications: Principles and Application*, Edward Arnold, a member of the Hodder Headline Group, London.

Chartant, R.M. (2004) *Satellite Communications for the Nonspecialist*, SPIE, Washington.

Elbert, B.R. (1997) *The Satellite Communication Applications Handbook*, Artech House, Boston, Massachusetts.

Elbert, B.R. (1999) *Introduction to Satellite Communication*, Altech House, Boston, Massachusetts.

Gedney, R.T., Schertler, R. and Gargione, F. (2000) *Advanced Communications Technology Satellite: Insider's Account of the Emergence of Interactive Broadband Services in Space*, SciTech, New Jersey.

Gedney, R.T., Schertler, R. and Gargione, F. (2000) *The Advanced Communication Technology*, SciTech, New Jersey.

Glibson, J.D. (2002) *The Communications Handbook (The Electrical Engineering Handbook Series)*, CRC Press, Boca Raton, Florida.

Glisic, S.G. and Leppanen, P.A. (1997) *A Handbook for Design, Installation and Service Engineers*, Kluwer Academic Publishers.

Harte, L. (2004) *Introduction to CDMA: Network, Services, Technologies, and Operation*, Althos.

Kadish, J.E. (2000) *Satellite Communications Fundamentals*, Artech House, Boston, Massachusetts.

Lamb, G. (1998) *The TDMA Book*, Cordero Consulting Inc.

Lewis, E.G. (1992) *Communication Services via Satellite: A Handbook for Design, Installation and Service Engineers*, Butterworth-Heinemann.

Maral, G. and Bousquet, M. (2002) *Satellite Communication Systems: Systems, Techniques and Technology*, John Wiley & Sons, Ltd, Chichester.

Richharia, M. (1999) *Satellite Communication Systems*, Macmillan Press Ltd.

Internet Sites

1. http://www.bee.net/mhendry/vrml/library/cdma/cdma.htm
2. http://www.iec.org/online/tutorials/tdma/index.html
3. www.sergiochacon.com/Communication/Wireless/Satellite TDMAFDMA.ppt
4. http://www.mlab.t.u-tokyo.ac.jp/ mori/courses/WirelessMobileComm2000/CDMA.pdf
5. http://www.eas.asu.edu/ junshan/pub/net_capacity.pdf
6. http://www.umtsworld.com/technology/cdmabasics.htm
7. http://en.wikipedia.org/wiki/Space-division_multiple_access

Glossary

Code division multiple access (CDMA): A multiple access technique in which the entire bandwidth of the transponder is used simultaneously by multiple Earth stations at all times

Demand assigned multiple access (DAMA): A transponder assignment mode that allows multiple users to share a common link, with each user only required to put up a request to the control station or agency for the same as and when it requires to use the link

Direct sequence CDMA: In this form of CDMA, the information signal, which has a relatively lower bit rate, is multiplied by a pseudorandom bit sequence with a much higher bit rate, with the result that the carrier frequency spectrum is spread over a much larger bandwidth. Interference between multiple channels is avoided as each transmitter uses a unique pseudorandom code sequence

Frequency division multiple access (FDMA): A multiple access technique in which different Earth stations are able to access the total available bandwidth in satellite transponder/s by virtue of their different carrier frequencies, thus avoiding interference among multiple signals

Frequency hopping CDMA: In the case of a frequency hopping CDMA system, the carrier is sequentially hopped into a series of frequency slots spread over the entire bandwidth of the satellite transponder. The transmitter hops its frequency over a given bandwidth several times per second, transmitting on one frequency for a certain period of time, and then hopping to another frequency and transmitting again

Guard time: Different bursts are separated from each other by a short time period, known as the guard time, which ensures that bursts from different stations accessing the satellite transponder do not overlap

Multichannel per carrier FDMA: This is a type of FDMA where the Earth station frequency multiplexes several channels into one carrier base band assembly, which then frequency-modulates an RF carrier and transmits it to the FDMA satellite transponder

Multiple access: Multiple access means access to a given facility or resource by multiple users. In the context of satellite communication, the facility is the transponder and the multiple users are various terrestrial terminals under the footprint of the satellite

Preassigned multiple access: A transponder assignment mode in which the transponder is assigned to the user either permanently for the satellite's full lifetime or at least for long durations. The preassignment may be that of a certain frequency band, time slot or a code

Random multiple access: A transponder assignment mode in which access to the link or the transponder is by contention. A user transmits the messages without knowing the status of messages from other users. Due to the random nature of transmissions, data from multiple users may collide. In case a collision occurs, it is detected and the data are retransmitted. Retransmission is carried out with random time delays and sometimes may have to be done several times

Reference burst: The reference burst is used to provide timing references to various stations accessing the TDMA transponder. It does not carry any traffic information

Signalling channel: The signalling channel is used to carry out system management and control functions

Single channel per carrier FDMA: This is a type of FDMA in which each signal channel modulates a separate RF carrier which is then transmitted to the FDMA transponder. The modulation technique used here could either be frequency modulation (FM) in the case of analogue transmission or phase shift keying (PSK) for digital transmission

Space domain multiple access (SDMA): SDMA is a technique that primarily allows frequency re-use where adjacent Earth stations within the footprint of the satellite can use the same carrier transmission frequency and still avoid co-channel interference by using either orthogonal antenna beam polarization or narrow antenna beam patterns

Spread spectrum communications: A technique in which the carrier frequency spectrum is spread over a much larger bandwidth as compared to the information rate. This not only makes the system immune to interception by an enemy but also gives it an anti-jamming capability. CDMA and its different variants employ the spread spectrum technique

TDMA frame: In a TDMA network, each of the multiple Earth stations accessing a given satellite transponder transmits one or more data bursts. The satellite thus receives at its input a set of bursts from a large number of Earth stations. This set of bursts from various Earth stations is called the TDMA frame

TDMA frame efficiency: TDMA frame efficiency is defined as the percentage of total frame length allocated for transmission of traffic data

Time division multiple access (TDMA): A multiple access technique in which different Earth stations in the satellite's footprint make use of a transponder by using a single carrier on a time division basis

Time hopping CDMA: In this form of CDMA, the data signal is transmitted in rapid bursts at time intervals determined by a pseudorandom code sequence assigned to the user. Time hopping CDMA uses a wideband spectrum for short periods of time instead of parts of the spectrum all the time

Unique word: The function of the unique word is to establish the existence of burst and to enable determination of a timing marker, which can be used to establish the position of each bit in remainder of the burst

7

Satellite Link Design Fundamentals

In this chapter, as the title of the chapter itself suggests, a comprehensive look will be taken at the important parameters that govern the design of a satellite communication link. The significance of each one of these parameters will be discussed *vis-à-vis* the overall link performance in terms of both quantity and quality of services provided by the link, without of course losing sight of the system complexity of both the Earth station and the space segment and the associated costs involved therein. What is implied by the previous sentence is that for a given link performance, a better system performance could perhaps be provided by making the Earth station and spacecraft instrumentation more complex, thereby increasing overall costs, sometimes to the extent of making it an unviable proposition.

The designer must therefore attempt to optimize the overall link, giving due attention to each element of the link and the factors associated with its performance. The chapter begins with a brief introduction to various parameters that characterize a satellite link or influence its design. Each one of these parameters is then dealt with in greater detail. This is followed by the basics of link design and the associated mathematical treatment, which is suitably illustrated with a large number of illustrations and design examples.

7.1 Transmission Equation

The transmission equation relates the received power level at the destination, which could be the Earth station or the satellite in the case of a satellite communication link, to the transmitted RF power, the operating frequency and the transmitter–receiver distance. It is fundamental to the design of not only a satellite communication link but also any radio communication link because the quality of the information delivered to the destination is governed by the level of the signal power received. The reason for this is that it is the received carrier-to-noise ratio that is going to decide the quality of information delivered, and for a given noise contribution from various sources, both internal and external to the system, the level of received power is vital to the design of the communication link. An estimation of received power level in a satellite communication link is made in the following paragraphs.

It is assumed that the transmitter radiates a power P_T watts with an antenna having a gain G_T as compared to the isotropic radiation level. The power flux density (P_{RD} in W/m^2) due to the radiated power in the direction of the antenna bore sight at a distance d metres is given by

$$P_{RD} = \frac{P_T G_T}{4\pi d^2} \quad (7.1)$$

The product $P_T G_T$ is the effective isotropic radiated power (EIRP). Also, if the radiating aperture A_T of the transmitting antenna is large as compared to λ^2, where λ is the operating wavelength, then G_T equals $(4\pi A_T/\lambda^2)$. If A_R is the aperture of the receiving antenna, then the received power P_R at the receiver at a distance d from the transmitter can be expressed as

$$P_R = \left(\frac{P_T G_T}{4\pi d^2}\right) A_R \quad (7.2)$$

where A_R is related to the receiver antenna gain by $G_R = 4\pi A_R/\lambda^2$. The expression for the received power is modified to

$$P_R = \frac{P_T G_T G_R \lambda^2}{(4\pi d)^2} \quad (7.3)$$

or

$$P_R = \frac{P_T G_T G_R}{(4\pi d/\lambda)^2} = \frac{P_T G_T G_R}{L_P} \quad (7.4)$$

The term $(4\pi d/\lambda)^2$ represents the free space path loss L_P. The above expression is also known as the Friis transmission equation. The received power can be expressed in decibels as

$$10 \log P_R = 10 \log P_T + 10 \log G_T + 10 \log G_R - 10 \log L_P$$
$$P_R(\text{in dBW}) = \text{EIRP}(\text{in dBW}) + G_R(\text{in dB}) - L_P(\text{in dB}) \quad (7.5)$$

The above equation can be modified to include other losses, if any, such as losses due to atmospheric attenuation, antenna losses, etc. For example, if L_A, L_{TX} and L_{RX} are the losses due to atmospheric attenuation, transmitting antenna and receiving antenna respectively, then the above equation can be rewritten as

$$P_R = \text{EIRP} + G_R - L_P - L_A - L_{TX} - L_{RX} \quad (7.6)$$

Problem 7.1
A geostationary satellite at a distance of 36 000 km from the surface of the Earth radiates a power of 10 watts in the desired direction through an antenna having a gain of 20 dB. What would be the power density at a receiving site on the surface of the Earth and also the power received by an antenna having an effective aperture of 10 m^2?

Solution: The power density can be computed from

$$\text{Power flux density} = \frac{P_T G_T}{4\pi d^2}$$

where the terms have their usual meaning. Here,

$$G_T = 20\,\text{dB} = 100, \qquad P_T = 10\,\text{watts}, \qquad d = 36\,000\,\text{km} = 36 \times 10^6\,\text{m}$$

This gives

Power flux density = $(10 \times 100)/[4 \times \pi \times (36 \times 10^6)^2] = 0.0614 \times 10^{-12}\,\text{W/m}^2$

Power received by the receiving antenna = $0.0614 \times 10^{-12} \times 10 = 0.614\,\text{pW}$

7.2 Satellite Link Parameters

Important parameters that influence the design of a satellite communication link include the following:

1. Choice of operating frequency
2. Propagation considerations
3. Noise considerations
4. Interference-related problems

7.2.1 Choice of Operating Frequency

The choice of frequency band from those allocated by the International Telecommunications Union (ITU) for satellite communication services such as the fixed satellite service (FSS), the broadcast satellite service (BSS) and the mobile satellite service (MSS) is mostly governed by factors like propagation considerations, coexistence with other services, interference-related issues, technology status, economic considerations and so on. While it may be more economic to use lower frequency bands, there would be interference-related problems as a large number of terrestrial microwave links use frequencies within these bands. Also, lower frequency bands would offer lower bandwidths and hence a reduced transmission capacity. Higher frequency bands offer higher bandwidths but suffer from the disadvantage of severe rain-induced attenuation, particularly above 10 GHz. Also, above 10 GHz, rain can have the effect of reducing isolation between orthogonally polarized signals in a frequency re-use system. It may be mentioned here that for frequencies less than 10 GHz and elevation angles greater than 5°, atmospheric attenuation is more or less insignificant.

7.2.2 Propagation Considerations

The nature of propagation of electromagnetic waves or signals through the atmospheric portion of an Earth station–satellite link has a significant bearing on the link design. From the viewpoint of a transmitted or received signal, it is mainly the operating frequency and to a lesser extent the polarization that would decide how severe the effect of atmosphere is going to be. From the viewpoint of atmosphere, it is the first few tens of kilometres constituting the troposphere and then the ionosphere extending from about 80 km to 1000 km that do the damage. The effect of atmosphere on the signal is mainly in the form of attenuation caused by atmospheric scattering and scintillation and depolarization caused by rain in the troposphere and Faraday rotation in the ionosphere. While rain-induced attenuation is very severe for frequencies above 10 GHz, polarization changes due to Faraday rotation are severe at lower frequencies and are almost insignificant beyond 10 GHz. In fact, atmospheric attenuation is the least in the 3 to 10 GHz window. That is why it is the preferred and most widely used one for satellite communications.

7.2.3 Noise Considerations

In both analogue and digital satellite communication systems, the quality of signal received at the Earth station is strongly dependent on the carrier-to-noise ratio of the satellite link. The satellite link comprises an uplink, the satellite channel and a downlink. The quality of the signal received on the uplink therefore depends upon how strong the signal is, as it leaves the originating Earth station and how the satellite receives it. On the downlink, it depends upon how strongly the satellite can retransmit the signal and then how the destination Earth station receives it. Because of the large distances involved, the signals received by the satellite over the uplink and received by the Earth station over the downlink are very weak. Satellite communication systems, moreso the geostationary satellite communication systems are therefore particularly susceptible to noise because of their inherent low received power levels. In fact, neither the absolute value of the signal nor that of the noise should be seen in isolation for gauging the effectiveness of the satellite communication link. If the received signal is sufficiently weak as compared to the noise level, it may become impossible to detect the signal. Even if the signal is detectable, steps should be taken within the system to reduce the noise to an acceptable level lest it impairs the quality of the signal received.

The sources of noise include natural and man-made sources, as well as the noise generated in the Earth station and satellite equipment. While the man-made noise mainly arises from electrical equipment and is almost insignificant above 1 GHz, the natural sources of noise include solar radiation, sky noise, background noise contributed by Earth, galactic noise due to electromagnetic waves emanating from radio stars in the galaxy and the atmospheric noise caused by lightening flashes and absorption by oxygen and water vapour molecules followed by re-emission of radiation. Sky noise and solar noise can be avoided by proper orientation and directionality of antennas. Galactic noise is insignificant above 1 GHz. Noise due to lightning flashes is also negligible at satellite frequencies.

7.2.4 Interference-related Problems

Major sources of interference include interference between satellite links and terrestrial microwave links sharing the same operational frequency band, interference between two satellites sharing the same frequency band, interference between two Earth stations accessing different satellites operating in the same frequency band, interference arising out of cross-polarization in frequency re-use systems, adjacent channel interference inherent to FDMA systems and interference due to intermodulation phenomenon.

Interference between satellite links and terrestrial links could further be of two types: first where terrestrial link transmission interferes with reception at an Earth station and the second where transmission from an Earth station interferes with terrestrial link reception. The level of inter-satellite and inter-Earth station interference is mainly governed by factors like the pointing accuracy of antennas, the width of transmit and receive beams, intersatellite spacing in the orbit of two co-located satellites, and so on. Cross-polarization interference is caused by coupling of energy from one polarization state to another polarization state when a frequency re-use system employs orthogonal linear polarizations (horizontal and vertical polarization) or orthogonal circular polarization (left-hand and right-hand circular polarization). This coupling of energy occurs due to a finite value of cross-polarization discrimination of the Earth station and satellite antennas and also to depolarization caused by rain. Adjacent channel interference arises out of overlapping amplitude characteristics of channel filters. Intermodulation interference is caused by the intermodulation products produced in the satellite transponder when multiple carriers are amplified in the high power amplifier that has both amplitude as well as phase nonlinearity.

7.3 Frequency Considerations

The choice of operating frequency for a satellite communication service is mainly governed by factors like propagation considerations, coexistence with other services, noise considerations and interference-related issues. These have been briefly discussed in the preceding paragraphs and will be dealt with at length in the sections to follow. The requirement for co-existence with other services and the fact that there will always be competition to use the optimum frequency band by various agencies with an eye to getting the best link performance show that there is always a need for a mechanism for allocation and coordination of the frequency spectrum at the international level. A brief on frequency allocation and the coordination mechanism is given in the following paragraphs.

7.3.1 Frequency Allocation and Coordination

Satellite communication employs electromagnetic waves for transmission of information between Earth and space. The bands of interest for satellite communications lie above 100 MHz and include the VHF, UHF, L, S, C, X, Ku and Ka bands. (Various microwave frequency bands are listed in Table 7.1). Higher frequencies are employed for satellite communication, as the frequencies below 100 MHz are either reflected by the ionosphere or they suffer varying degrees of bending from their original paths due to refraction by the ionosphere. Initially, the satellite communication was mainly concentrated in the C band (6/4 GHz) as it offered fewest propagation as well as attenuation problems. However, due to overcrowding in the C band and the advances made in the field of satellite technology, which enabled it to deal with the propagation problems in the higher frequency bands, newer bands like the X, Ku and Ka bands are now being employed for commercial as well as military satellite applications. Moreover, use of these higher frequencies gives satellites an edge over terrestrial networks in terms of the bandwidth offered by a satellite communication system.

As the frequency spectrum is limited, it is evident the frequency bands are allocated in such a manner as to ensure their rational and efficient use. The International Telecommunication

Table 7.1 Microwave frequency bands for satellite communication

Band	Frequency (GHz)
L band	1–2
S band	2–4
C band	4–8
X band	8–12
Ku band	12–18
K band	18–27
Ka band	27–40
V band	40–75
W band	75–110

Union (ITU), formed in the year 1865, is a specialized institution that ensures the proper allocation of frequency bands as well as the orbital positions of the satellite in the GEO orbit. ITU carries out these regulatory activities through its four permanent organs, namely the General Secretariat, the International Frequency Registration Board (IFRB), the International Radio Consultative Committee (CCIR) and the International Telegraph and Telephone Consultative Committee (CCITT). The ITU organizes the international radio conferences such as World Administrative Radio Conferences (WARC) and other regional conferences to issue guidelines for frequency allocation for various services. These allocations are made either on an exclusive or shared basis and can be put into effect worldwide or limited to a region. The frequency bands that are allocated internationally are in turn reallocated by the national government bodies of the individual countries. Each country is assigned frequencies according to its requirements. The main purpose is to ensure that the technical parameters are retained for a period of at least 10 to 20 years and to guarantee that all countries can access the satellite service when needed. The ITU has divided the frequency allocations for different services into various categories: primary, secondary, planned, shared, etc. In primary frequency allocation, a service has an exclusive right of operation. In secondary frequency allocation, a service is not guaranteed interference protection and neither is it permitted to cause interference to services with primary status. WARC divided the globe into three regions for the purpose of frequency allocations. They are:

Region 1. Including Europe, Africa, USSR and Mongolia
Region 2. Including North and South America, Hawaii and Greenland
Region 3. Including Australia, New Zealand and those parts of Asia and the Pacific not included in regions 1 and 2.

It has also classified the various satellite services into the following categories, including fixed satellite services (FSS), intersatellite links (ISL), mobile satellite services (MSS), broadcasting satellite services (BSS). Earth exploration services, space research activities, meteorological activities, space operation, amateur radio services, radio determination, radio navigation, aeronautical radio navigation and maritime radio navigation. Among these applications the FSS, MSS and BSS are the principle communication-related applications of satellites.

The fixed satellite service (FSS) refers to the two-way communication between two Earth stations at fixed locations via a satellite. It supports the majority of commercial applications including satellite telephony, satellite television and data transmission services. The FSS primarily uses two frequency bands: the C band (6/4 GHz), which provides lower power transmission over a wide geographic area requiring large antennas for reception, and the Ku band (14/11 GHz, 14/12 GHz), which offers higher transmission power over a smaller geographical area, enabling reception by small receiving antennas. All of the C band and much of the Ku band have been allocated for international and domestic FSS applications. The X band (8/7 GHz) is used to provide fixed services to government and military users. EHF (extremely high frequency), UHF (ultra high frequency) and SHF (super high frequency) band transponders are also used in military satellites.

The broadcast satellite service (BSS) refers to the satellite services that can be received at many unspecified locations by relatively simple receive-only Earth stations. These Earth stations can either be community Earth stations serving various distribution networks or located in homes for direct-to-home transmission. The Ku band (18/12 GHz) and the Ka band (30/20 GHz) are mainly used for BSS applications, like television broadcasting and DTH applications.

The mobile satellite service (MSS) refers to the reception by receivers that are in motion, like ships, cars, lorries, and so on. It comprises land mobile services, maritime mobile services and aeronautical mobile services. Increasingly, MSS networks are providing relay communication services to portable handheld terminals. The L band (2/1 GHz) and the S Band (4/2.5 GHz) are mainly employed for MSS services because, at these lower microwave frequencies, broader beams are transmitted from the satellite, enabling the reception by antennas even if they are not pointed towards the satellite. This in turn makes these frequencies attractive for mobile and personal communications. The lower frequencies in the VHF and UHF bands are mainly employed for messaging and positioning applications, as these applications require smaller bandwidths. Some LEO satellite constellations are using these frequencies to provide the above-mentioned applications. Table 7.2 enumerates a partial list of frequency allocations for satellite communications and the satellites using them.

Current trends in satellite communication indicate the opening of new higher frequency bands for various applications, like the Q band (33–46 GHz) and the V band (46–56 GHz), which are being considered for use in FSS, BSS and intersatellite communication applications. Another important concept is the sharing of the same frequency bands between GEO and new systems for personal satellite communications using non-GEO orbits.

When an organization intends to add a new satellite system, frequencies close to those used in the existing system are preferred in order to minimize the impact of the planned expansion on the existing customers and also changes to the space segment. For many applications like the DTH service, frequency and other parameter considerations have already been assigned by the ITU for each country. In these cases the new frequency assignment becomes easy and straight forward. When such a frequency plan does not exist, the system designer selects frequencies from the ITU allocations based on the trade-off analysis taking into account the technical (propagation, state of the technology) and economic factors. A notification is sent to the ITU about the details of the new system (planned satellite location). After selecting the frequency band, the next step is to resolve all the interference related problems that might occur by coordinating with all the system operators. After resolving all the issues with other operators, ITU is again notified and the ITU enters the allocations in its frequency register. The new system is then considered operational.

Table 7.2 Frequency allocations for satellite communication

Frequency band	Satellite service	Examples
VHF (<200 MHz)	Messaging	Starsys constellation of 24 satellites in the LEO orbit, having an uplink frequency of 148–150 MHz and a downlink frequency of 137–138 MHz
UHF (200 MHz–1 GHz)	Messaging and positioning, voice and fax	Gonets is a Russian messaging satellite constellation comprising 18 satellites in the LEO. Uplink frequencies are 312–315 MHz and 1624.5–1643.4 MHz and downlink frequencies are 387–390 MHz and 1523–1541.9 MHz
L band (2/1 GHz)	MSS and positioning	Inmarsat-II satellite having 4 L band transponders supporting mobile-to-mobile services with an uplink frequency of 1.6 GHz and a downlink frequency of 1.5 GHz
S band (4/2.5 GHz)	MSS	NStar-C satellite having 20 S band transponders provides S band mobile communication services to Japan
C band Uplink 5.925–6.425 GHz Downlink 3.7–4.2 GHz	FSS	Palapa-B, Indonesia's domestic satellite comprises 20 C band transponders each capable of carrying 1000 two-way voice circuits or a colour television transmission channel. It has uplink frequencies of 5.925–6.425 GHz and downlink frequencies of 3.7–4.2 GHz
X band Uplink 7.9–8.4 GHz Downlink 7.25–7.75 GHz	Military applications	Hispasat-1B satellite operated by the Spanish government carries four X band transponders for military applications

Ku band
Europe
  FSS
    Uplink 14–14.8 GHz
    Downlink 10.7–11.7 GHz
  BSS
    Uplink 17.3–18.1 GHz
    Downlink 11.7–12.5 GHz
  Telecom
    Uplink 14–14.8 GHz
    Downlink 12.2–12.75 GHz
USA
  FSS
    Uplink 14–14.5 GHz
    Downlink 11.7–12.2 GHz
  BSS
    Uplink 17.3–17.8 GHz
    Downlink 12.2–12.7 GHz
	FSS, BSS and Telecom	Hot Bird-6 satellite consists of 28 Ku band transponders, which provide television broadcast and multimedia services over the entire European continent, North Africa and the Middle East

Ka band
FSS
  Uplink 27–31 GHz
  Downlink 17–21 GHz
	FSS, BSS	Superbird-6 satellite, having 23 Ku and 6 Ka band transponders, provides business communication services to Japan in Ku and Ka bands and additional Ka band services using a steerable spot beam

7.4 Propagation Considerations

As outlined earlier, the nature of propagation of electromagnetic waves through the atmosphere has a significant bearing on the satellite link design. As will be seen in the paragraphs to follow, it is the first few tens of kilometres constituting the troposphere and then the ionosphere extending from about 80 km to 1000 km that do the major damage.

The effect of the atmosphere on the signal is mainly in the form of atmospheric gaseous absorption, cloud attenuation, topospheric scintillation causing refraction, Faraday rotation in the ionosphere, ionospheric scintillation, rain attenuation and depolarization. Both attenuation and depolarization come from interactions between the propagating electromagnetic waves and different sections of the atmosphere.

Attenuation is defined as the difference between the power that would have been received under ideal conditions and the actual power received at a given time.

$$A(t) = P_{rideal}(t) - P_{ractual}(t) \qquad (7.7)$$

Where,

$A(t)$ is the attenuation at any given time t
$P_{rideal}(t)$ is the received power under ideal conditions at time t
$P_{ractual}(t)$ is the actual received power at time t

Depolarization refers to the conversion of energy from the wanted channel to the unwanted channel. Here, wanted channel refers to the co-polarized channel and the unwanted channel refers to the cross-polarized channel. Due to de-polarization, co-channel interference and cross-talk occurs between dual-polarized satellite links.

Propagation losses are further of two types, namely those that are more or less constant and therefore predictable and those that are random in nature and therefore unpredictable. While free-space loss belongs to the first category, attenuation caused due to rain is unpredictable to a large extent. The second category of losses can only be estimated statistically. The combined effect of these two types of propagation losses is to reduce the received signal strength. Due to the random nature of some types of losses, the received signal strength fluctuates with time and may even reduce to a level below the minimum acceptable limit for as long a period as an hour in 24 hours during the period of severe fading. This is amply illustrated in the graph of Figure 7.1.

7.4.1 Free-space Loss

Free-space loss is the loss of signal strength only due to distance from the transmitter. While free space is a theoretical concept of space devoid of all matter, in the present context it implies remoteness from all material objects or forms of matter that could influence propagation of electromagnetic waves. In the absence of any material source of attenuation of electromagnetic signals, therefore, the radiated electromagnetic power diminishes as the inverse square of the distance from the transmitter, which implies that the power received by an antenna of 1 m² cross-section will be $P_t/(4\pi R^2)$ where P_t is the transmitted power and R is the distance of the receiving antenna from the transmitter. In the case of uplink, the Earth station antenna

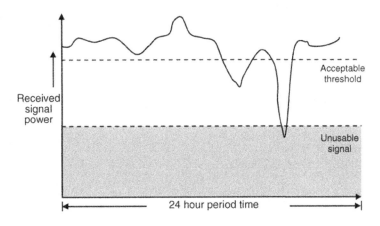

Figure 7.1 Fading phenomenon

becomes the transmitter and the satellite transponder is the receiver. It is the opposite in the case of downlink.

The free-space path loss component can be computed from

$$L_{FS} = \left(\frac{4\pi R}{\lambda}\right)^2 = 20 \log\left(\frac{4\pi R}{\lambda}\right) \quad \text{dB} \tag{7.8}$$

where L_{FS} is the free space loss and λ = operating wavelength. Also, $\lambda = c/f$, where

c = velocity of electromagnetic waves in free space
f = operating frequency

If c is taken in km/s and f in MHz, then the free-space path loss can also be computed from

$$L_{FS}(\text{dB}) = (32.4 + 20 \log R + 20 \log f) \tag{7.9}$$

7.4.2 Gaseous Absorption

Electromagnetic energy gets absorbed and converted into heat due to gaseous absorption as it passes through the troposphere. The absorption is primarily due to the presence of molecular oxygen and uncondensed water vapour and has been observed to be not so significant as to cause problems in the frequency range of 1 to 15 GHz. Of the other gases, atmospheric nitrogen does not have a peak while carbon dioxide has shown a peak at around 300 GHz. The presence of free electrons in the atmosphere also causes absorption due to collision of electromagnetic waves with these electrons. However, electron absorption is significant only at frequencies less than 500 MHz. Figure 7.2 shows the plot of total absorption (in dB) towards the zenith, that is for an elevation angle of 90° as a function of frequency. It can be seen from this plot that there are specific frequency bands where the absorption is maximum, near total. The first absorption band is caused due to the resonance phenomenon in water vapour and occurs at

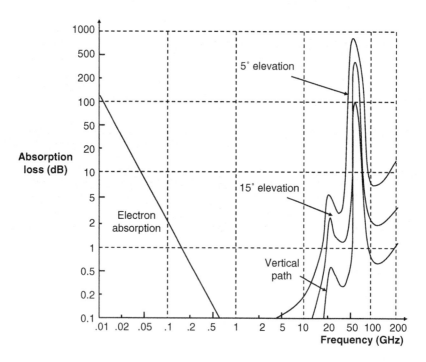

Figure 7.2 Gaseous absorption as a function of frequency and elevation angle

22.2 GHz. The second band is caused by a similar phenomenon in oxygen and occurs around 60 GHz. These bands are therefore not employed for either uplinks or downlinks. However, they can be used for intersatellite links. It should also be mentioned that absorption at any frequency is a function of temperature, pressure, relative humidity (RH) and elevation angle.

Absorption increases with a decrease in elevation angle E due to an increase in the transmission path. Absorption at any elevation angle E less than 90° can be computed by multiplying the absorption figure for 90° degree elevation by $1/\sin E$. Applying this correction, it is observed that the one-way absorption figure range in the 1–15 GHz frequency band increases from (0.03–0.2 dB) for 90° elevation to (0.35–2.3 dB) for 5° elevation. Absorption is also observed to increase with humidity. In the case of resonance absorption of water vapours around 22.2 GHz, it varies from as low as 0.05 dB (at 0 % RH) to about 1.8 dB (at 100 % RH). It may also be mentioned here that the data given in Figure 7.2 applies to the Earth station at sea level and the losses would reduce with an increase in height of the Earth station.

It is evident from the plots of Figure 7.2 that there are two transmission windows in which absorption is either insignificant or has a local minimum. The first window is in the frequency range of 500 MHz to 10 GHz and the second is around 30 GHz. This explains the wide use of the 6/4 GHz band. The increasing interest in the 30/20 GHz band is due to the second window, which shows a local minimum around 30 GHz. Losses at the 14/11 GHz satellite band are within acceptable limits with values of about 0.8 dB for 5° elevation and 0.2 dB for 15° elevation.

7.4.3 Attenuation due to Rain

After the free-space path loss, rain is the next major factor contributing to loss of electromagnetic energy caused by absorption and scattering of electromagnetic energy by rain drops. It may be mentioned here that the loss of electromagnetic energy due to gaseous absorption discussed in the earlier paragraphs tends to remain reasonably constant and predictable. On the other hand, the losses due to precipitation in the form of rain, fog, clouds, snow, and so on, are variable and far less predictable. However, losses can be estimated in order to allow the satellite links to be designed with an adequate link margin wherever necessary.

Losses due to rain increases with an increase in frequency and reduction in the elevation angle. Figure 7.3 shows rain attenuation in dB as a function of frequency for a set of different elevation angles. It is evident from the family of curves that there is not much to worry about from rain attenuation for C band satellite links. Attenuation becomes significant above 10 GHz and therefore when a satellite link is planned to operate beyond 10 GHz, an estimate of rain-caused attenuation is made by making extensive measurements at several locations in the coverage area of the satellite system.

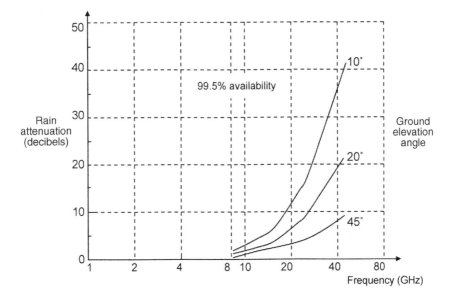

Figure 7.3 Rain-caused attenuation as a function of frequency and elevation angle

Attenuation of electromagnetic waves due to rain (A_{rain}) extended over a path length of L can be computed from

$$A_{rain} = \int_0^L \alpha \, d\alpha \qquad (7.10)$$

where α = specific attenuation of rain in dB/km. Specific attenuation again depends upon various factors like rain drop size, drop size distribution, operating wavelength and the refractive index. In practice, rain attenuation is estimated from

$$\alpha = aR^b \tag{7.11}$$

where a and b are frequency and temperature-dependent constants and R is the surface rain rate at the location of interest.

In addition to attenuation of electromagnetic energy, another detrimental effect caused by rain is depolarization. This reduces the cross-polarization isolation. This effect has been observed to be more severe on circularly polarized waves than linearly polarized waves. Depolarization of electromagnetic waves due to rain is caused by flattening of supposedly spherical rain drops affected by atmospheric drag. Depolarization is not particularly harmful at the C band and lower parts of the Ku band. It is severe at frequencies beyond 15 GHz.

7.4.4 Cloud Attenuation

Attenuation due to clouds is more or less irrelevant for lower frequency bands (L, S, C and Ku bands), but is largely relevant for satellite systems employing Ka and V band frequencies. The attenuation figure for water-filled clouds is much larger than the attenuation figure for clouds made from ice crystals. It may be mentioned here that the attenuation figure for ice clouds is negligible for the frequency bands of interest. The typical figure of attenuation for water filled clouds is of the order of 1 to 3 dB for frequency bands around 30 GHz at elevation angle of 30° in temperate latitude locations. However, the attenuation figures increase with increase in the thickness of the clouds and its probability of occurrence. In addition, the cloud attenuation increases for lower elevation angles.

7.4.5 Signal Fading due to Refraction

Refraction is the phenomenon of bending of electromagnetic waves as they pass through the different layers of the atmosphere. Refraction of the satellite beams occur in the troposphere (lower layer of the atmosphere from the Earth's surface to a height of 15 km approximately) due to the variations in the refractive index of the air column. The variations in the refractive index is the result of the turbulent mixing of the different columns of the air due to the agitative convective activity in the troposphere caused by heating of the Earth's surface by the sun. The variations in the refractive index lead to bending of electromagnetic waves resulting in fluctuations in the received signal levels, also referred to as scintillations. The result of bending of electromagnetic waves is depicted in Figure 7.4. It leads to a virtual position for the satellite slightly above the true position of the satellite. The random nature of bending due to discontinuities and fluctuations caused by unstable atmospheric conditions like temperature inversions, clouds and fog produces signal fading, which leads to loss of signal strength. Fading is the phenomenon wherein the Earth station receiving antenna receives the signal transmitted by the satellite via different paths with different phase shifts. The fading phenomenon is more adverse at lower elevation angles.

It may be mentioned here that the topospheric scintillation does not cause depolarization. The scintillation effects increase for higher frequencies, lower path elevation angles and for warmer and humid climate. The effect is more severe for terrestrial links and not too worrisome

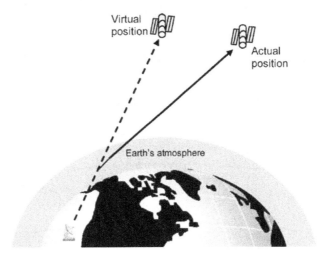

Figure 7.4 Bending of electromagnetic waves caused by refraction in the atmosphere

for satellite links, as the amount of bending caused in a satellite link is small relative to the beam width of the satellite and Earth station antennas.

7.4.6 Ionosphere-related Effects

The ionosphere is an ionized region in space, extending from about 80–90 km to 1000 km (Figure 7.5) formed by interaction of solar radiation with different constituent gases of the atmosphere. Electromagnetic waves travelling through the ionosphere are affected in more than one way, some more predominant than the other from the viewpoint of satellite communications. The effects that are of concern and need attention include polarization rotation, also called the Faraday effect, and scintillation, which is simply rapid fluctuation of the signal amplitude, phase, polarization or angle of arrival. The ionosphere also affects the propagating electromagnetic waves in many other ways, such as absorption, causing propagation delay, dispersion, etc. However, these effects are negligible at the frequencies of main interest for satellite communications, except for very small time periods during intense solar activity such as solar flares. Also, most ionospheric effects including those of primary interest, like polarization rotation and scintillation, decrease with an increase in frequency having a $1/f^2$ dependence. The two major effects are briefly described in the following paragraphs.

7.4.6.1 Polarization Rotation – Faraday Effect

When an electromagnetic wave passes through a region of high electron content like the ionosphere, the plane of polarization of the wave gets rotated due to interaction of the electromagnetic wave with the Earth's magnetic field. The angle through which the plane of polarization rotates is directly proportional to the total electron content of the ionized region and inversely proportional to the square of the operating frequency. It also depends upon the

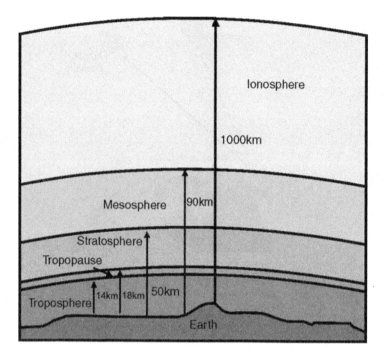

Figure 7.5 Different layers of atmosphere

state of the ionosphere, time of the day, solar activity, the direction of the incident wave, and so on. Directions of polarization rotation are opposite for transmit and receive signals.

Due to its $1/f^2$ dependence, the effect is observed to be pronounced only at frequencies below 2 GHz. The worst case polarization rotation angle may be as large as 150° in certain conditions at 1 GHz. Applying $1/f^2$ dependence, the rotation angle may be as small as 9° at 4 GHz and 4° at 6 GHz. It would reduce to a fraction of a degree in the Ku band frequency range. Except for a small time period during unusual atmospheric conditions caused by intense solar activity, magnetic storms, and so on, the Faraday effect is more or less predictable and therefore can be compensated for by adjusting the polarization of the receiving antenna. Circular polarization is virtually unaffected by Faraday effect and therefore its impact can be minimized by using circular polarization.

The polarization rotation angle ($\Delta\Psi$) for a path length through the ionosphere of Z metres is given by

$$\Delta\Psi = \int \left(\frac{2.36 \times 10^4}{f^2}\right) ZNB_o \cos\theta \, dz \qquad (7.12)$$

Where,

$\Delta\Psi$ is the rotation angle (radians)
θ is the angle between the geomagnetic field and the direction of propagation of the wave
N is the electron density (electrons/cm^3)
B_o is the geomagnetic flux density (Tesla)
f is the operating frequency (Hz)

For a polarization rotation angle (also referred to as the polarization mismatch angle) of $\Delta\Psi$, the attenuation of the co-polar signal given by

$$A_{PR} = -20 \log(\cos \Delta\Psi) \qquad (7.13)$$

where, A_{PR} is the attenuation due to polarization rotation in dB.
The mismatch also produces a cross-polarized component, which reduces the cross-polarization discrimination (X_{PD}), given by

$$X_{PD} = -20 \log(\tan \Delta\Psi) \qquad (7.14)$$

where, X_{PD} is the cross-polarization discrimination in dB.
The magnitudes of attenuation and X_{PD} due to polarization mismatch will be 0.1 dB and 16 dB respectively at 4 GHz for which $\Delta\Psi = 9°$.

7.4.6.2 Ionospheric Scintillation

As mentioned above, scintillation is nothing but the rapid fluctuations of the signal amplitude, phase, polarization or angle-of-arrival. In the ionosphere, scintillation occurs due to small scale refractive index variations caused by local electron concentration fluctuations. The total electron concentration (total number of electrons existing in a vertical column of 1m^2 area) of the ionosphere increases by two orders of magnitude during the day as compared to night due to the energy received from the sun. This rapid change in the value of total electron concentration from the daytime value to the nighttime value gives rise to irregularities in the ionosphere. It mainly occurs in the F-region of the ionosphere due to the highest electron concentration in that region.

The irregularities cause refraction resulting in rapid variations in the signal amplitude and phase, which leads to rapid signal fluctuations that are referred to as ionospheric scintillations. As a result, the signal reaches the receiving antenna via two paths, the direct path and the refracted path, as illustrated in Figure 7.6. Multipath signals can lead to both signal enhancement as well as signal cancellation depending upon the phase relationship with which they arrive at the receiving antenna. The resultant signal is a vector addition of the direct and the refracted signal. In the extreme case, when the strength of the refracted signal is comparable to that of the direct signal, cancellation can occur when the relative phase difference between the two is 180°. On the other hand, an instantaneous recombination of the two signals in phase can lead to signal amplification up to 6 dB.

The scintillation effect is inversely proportional to the square of the operating frequency and is predominant at lower microwave frequencies, typically below 4 GHz. Scintillation, however, increases during periods of high solar activity and other extreme conditions such as the occurrence of magnetic storms. Scintillation has also been observed to be maximum in the region that is $\pm 25°$ around the equator. Under such adverse conditions, scintillation can cause problems at 6/4 GHz band too. At the Ku band and beyond, however, the effect is negligible. Also, unlike scintillation caused by the troposphere, ionospheric scintillation is independent of the elevation angle.

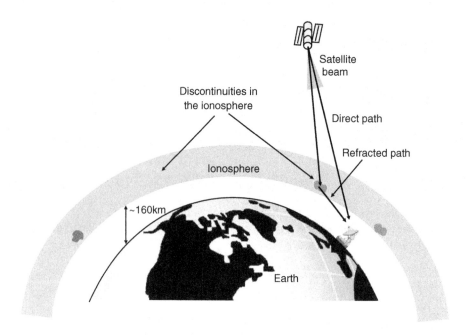

Figure 7.6 Ionospheric scintillation

In regions where scintillation is expected to cause problems, an adequate link margin in the form of additional transmitted power should be catered for to take care of the effect and maintain link reliability.

7.4.7 Fading due to Multipath Signals

Ionospheric scintillation, as illustrated above, results in a multipath phenomenon where the indirect signal is produced as a result of refraction caused by pockets of ion concentration in the F-region of the ionosphere. Multipath signals also result from reflection and scattering from obstacles such as buildings, trees, hills and other man-made and natural objects. In the case of fixed satellite terminals, the situation remains more or less the same and does not change with time as long as the satellite remains in the same position with respect to the satellite terminal (Figure 7.7). However, in the case of mobile satellite terminals, the situation keeps changing with time. In a typical case, a mobile terminal could receive the direct signal and another signal reflected off the highway, buildings, neighbouring hills or trees. The relative phase difference between the two signals could produce either a signal enhancement or fading (Figure 7.8). Moreover, the fading signal varies with time as the satellite moves with respect to the points of reflection.

Thus, the situation in the case of mobile satellite service (MSS) terminals is far more severe and uncertain than it is in the case of fixed receiving terminals. The mobile station–satellite path profile keeps changing continuously with the movement of the mobile terminal. Also, mobile terminals usually employ broadband receiving antennas, which do not provide adequate discrimination against signals received via indirect paths.

Propagation Considerations

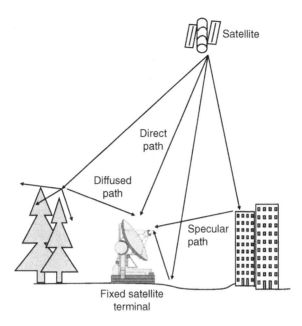

Figure 7.7 Fading due to multipath signals for a fixed satellite terminal

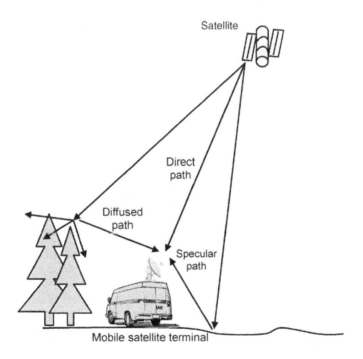

Figure 7.8 Mobile satellite terminal receiving multipath signals

Problem 7.2
Compute the free-space path loss in decibels for the following conditions:

1. For a path length of 10 km at 4 GHz operating frequency
2. Earth station transmitting antenna EIRP = 50 dBW, satellite receiving antenna gain = 20 dB and received power at satellite = −120 dBW

Solution:

(a) Path length, $R = 10$ km, operating frequency, $f = 4$ GHz

$$\text{Operating wavelength, } \lambda = c/f = (3 \times 10^8 / 4 \times 10^9)\,\text{m} = 0.075\,\text{m}$$

$$\text{Path loss (in dB)} = 20\log(4\pi R/\lambda)$$

$$= 20\log(4\pi \times 10\,000/0.075)$$

$$= 124.48\,\text{dB}$$

(b) Path loss can be computed from:
Received power = EIRP + receiving antenna gain − path loss

Therefore, path loss = EIRP + receiving antenna gain − received power

$$= 50 + 20 - (-120) = 50 + 20 + 120 = 190\,\text{dB}$$

Problem 7.3
Under certain atmospheric conditions, a 2 GHz linearly polarized signal experiences a rotation of its plane of polarization by 75°. How much polarization rotation would have been experienced by a 10 GHz signal under similar atmospheric conditions? Also determine the attenuation (in dB) experienced by the co-polar component due to polarization rotation if it is not corrected for at the receiving antenna.

Solution: Polarization rotation is inversely proportional to the square of the operating frequency. Here, the frequency has increased by a factor of 5. Therefore, the polarization rotation angle will decrease by a factor of 25; i.e.
Polarization rotation experienced = 75/25 = 3°
Now, the attenuation due to the polarization loss can be computed from

$$\text{Attenuation (in dB)} = -20\log(\cos\Delta\Psi)$$

where $\Delta\Psi$ = polarization mismatch angle. In the first case, $\Delta\Psi = 75°$. Therefore,

$$\text{Attenuation} = -20\log(\cos 75°) = -20\log(0.2588)$$

$$= -20 \times (-0.587) = 11.74\,\text{dB}$$

In the second case, $\Delta\Psi = 3°$. Therefore,

$$\text{Attenuation} = -20\log(\cos 3°) = -20\log(0.9986)$$

$$= -20 \times (-0.0006) = 0.012\,\text{dB}$$

7.5 Techniques to Counter Propagation Effects

As mentioned before, the propagation effects can be broadly classified as attenuation effects and depolarization effects. The attenuation effects can be countered by employing the following techniques:

1. Power Control
2. Signal Processing
3. Diversity

The depolarization effects can be negated by employing depolarization compensation. In this section all of these techniques are briefly explained.

7.5.1 Attenuation Compensation Techniques

The attenuation compensation techniques are described in the following paragraphs.

1. **Power Control:** Power control refers to varying the EIRP of the signal to enhance the C/N ratio. Adaptive power control is applied wherein the transmitter power is adjusted to compensate for the changes in the signal attenuation along the transmission path. This is referred to as up-link power control (ULPC) and it results in enhancement of the overall availability of the connection. ULPC can be done either in open-loop mode or in closed loop mode. In the open-loop mode, the fade level on the downlink signal is used to predict the likely fade level occurring on the uplink signal and accordingly the signal power is increased. In the closed-loop mode, the signal power is detected at the satellite and it sends a control signal to the Earth station to adjust the transmitted power in accordance with the attenuation in real-time. As is evident, closed loop control is more accurate that the open loop control. However, it is more expensive than open loop control.
2. **Signal Processing:** Signal processing refers to onboard processing techniques on the satellite to translate the signals coming from the Earth station via the uplink to baseband levels for processing and onward transmission back to the Earth. The satellite transponder demodulates the incoming up-link signal from the Earth station, demultiplexes it and decodes it. Each traffic packet is handled at the baseband level and therefore most of the bit errors can be removed. In addition, the signal level of each packet is detected and the transmitting Earth station is alerted in case the energy level of the received packet has fallen beyond a certain threshold. The packets are then coded, multiplexed and modulated for retransmission back to the Earth.
3. **Diversity:** Many diversity schemes exist that can be used to enhance the signal levels but are not implemented due to the costs involved. Diversity schemes include time diversity, frequency diversity and site diversity. Site diversity scheme is most potential candidate to be employed as it offers the maximum gain. In time diversity, additional slots are assigned to the rain-affected link in the TDMA frame so that the same signal can be sent at a smaller bit rate, thereby reducing the bandwidth and increasing the C/N ratio. In frequency diversity, the frequency band of the signal is changed from the band having large attenuation to the band having less attenuation. As an example, a rain affected Ku band link can be switched

to a C band which is not attenuated significantly by rain. This requires additional C band transmitting capability and the receiving capability at the transmitter and receiver Earth stations respectively. Site diversity is a technique wherein two or more Earth stations are located sufficiently apart so that the rain and other attenuation impairments observed by signals from each of the Earth stations to the satellite are not correlated. The Earth stations are connected to each other so that any Earth station can be used to support the traffic from other Earth station/s suffering from attenuation.

7.5.2 *Depolarization Compensation Techniques*

Depolarization compensation is a technique used to compensate for the depolarization effects caused to the signal during its propagation through the atmosphere. Here the feed system of the antenna is adjusted in such a way to correct for depolarization in the path. Another compensation technique is to cross-couple the orthogonal channels at the receiver and determine the depolarized signal. The depolarized signal is then removed by subtracting it from the received signal. Only a few Earth stations have implemented depolarization compensation as it is an expensive proposition.

7.6 Noise Considerations

Satellite communication systems are particularly susceptible to noise because of their inherent low received power levels, as the signals received by the satellite over the uplink and received by the Earth station over the downlink are very weak due to involvement of large distances. Sources of noise include natural and man-made sources and also the noise generated inside the Earth station and satellite equipment, as outlined during the introductory paragraphs at the beginning of the chapter. From the viewpoint of satellite communications, the natural and man-made sources of noise can either be taken care of or are negligible. It is mainly the noise generated in the equipment where attention primarily needs to be paid. In the paragraphs to follow, various parameters that can be used to describe the noise performance of various building blocks individually and also as a system, which is a cascaded arrangement of those building blocks, will be briefly discussed.

7.6.1 *Thermal Noise*

Thermal noise is generated in any resistor or resistive component of any impedance due to random motion of molecules, atoms and electrons. It is called thermal noise as the temperature of a body is the statistical RMS value of the velocity of motion of these particles. It is also called 'white' noise because, due to randomness of the motion of particles, the noise power is evenly spread over the entire frequency spectrum. It is also known as Johnson noise.

According to kinetic theory, the motion of these particles ceases at absolute zero temperature, that is zero degrees kelvin. Therefore, the noise power generated in a resistor or resistive component is directly proportional to its absolute temperature, in addition to the bandwidth

Noise Considerations

over which it is measured; that is

$$P_n \propto TB = kTB \tag{7.15}$$

where

T = absolute temperature (in K)
B = bandwidth of interest (in Hz)
k = Boltzmann's constant = 1.38×10^{-23} J/K
P_n = noise power output of a resistor (in W)

Thermal noise power at room temperature (T = 290 K) in dBm (decibels relative to a power level of 1mW) can also be computed from

$$P_n = -174 + 10 \log B \tag{7.16}$$

If the resistor is considered as a noise generator with an equivalent noise voltage equal to V_n, then this noise generator will transfer maximum noise power P_n to a matched load that is given by

$$P_n = \frac{V_n^2}{4R} \tag{7.17}$$

which gives expression for noise voltage (V_n) as

$$V_n = \sqrt{(4kTRB)} \tag{7.18}$$

Expression for noise current (I_n) can be deduced from the expression for noise voltage and is given by

$$I_n = \sqrt{(4kTB/R)} \tag{7.19}$$

Another term that is usually defined in this context is the noise power spectral density given by

$$P_{no} = kT \tag{7.20}$$

where, P_{no} is the noise power spectral density in W/Hz.
This implies that the noise power spectral density increases with physical temperature of the device. Also, thermal noise generated in a device can be reduced by reducing its physical temperature or the bandwidth over which the noise is measured or both.

7.6.2 Noise Figure

The noise figure F of a device can be defined as the ratio of the signal-to-noise power at its input to the signal-to-noise power at its output; i.e.

$$F = \frac{S_i/N_i}{S_o/N_o} = \frac{N_o}{(S_o/S_i)N_i} = \left(\frac{N_o}{N_i}\right)\left(\frac{1}{G}\right) \tag{7.21}$$

where

S_i = available signal power at the input
N_i = available noise power at the input
S_o = available signal power at the output
N_o = available noise power at the output (in a noiseless device)
G = power gain over the specified bandwidth = S_o/S_i

Now, $N_i = kT_iB$, where T_i is the ambient temperature in kelvin. Therefore, the noise figure (F) is expressed as

$$F = \frac{N_o}{GkT_iB} \tag{7.22}$$

The actual amplifier, however, introduces some noise, which is added to the output noise power. If the noise power introduced is ΔN, then

$$N_o = GkT_iB + \Delta N$$

$$F = \frac{GkT_iB + \Delta N}{GkT_iB} = 1 + \frac{\Delta N}{GkT_iB} \tag{7.23}$$

The noise figure is thus the ratio of the actual output noise to that which would remain if the device itself did not introduce any noise. In the case of a noiseless device, $\Delta N = 0$, which gives $F = 1$. Thus the noise figure in the case of an ideal device is unity. Any value of the noise figure greater than unity means a noisy device.

7.6.3 Noise Temperature

Yet another way of expressing noise performance of a device is in terms of its equivalent noise temperature T_e. It is the temperature of a resistance that would generate the same noise power at the output of an ideal (i.e. noiseless) device as that produced at the output by an actual device when terminated at its input by a noiseless resistance, i.e. a resistance at absolute zero temperature.

Now, noise generated by the device $\Delta N = GkT_eB$, which when substituted in the expression for noise figure mentioned above gives

$$F = 1 + \frac{T_e}{T_i} \quad \text{or} \quad T_e = T_i(F - 1) \tag{7.24}$$

In the preceding paragraphs, noise figure and noise temperature specifications of devices have been discussed and a relationship between the two parameters commonly used for measuring a device's noise performance has been established. It will now be shown how the expression for the effective noise temperature is modified in the case where the device is purely a resistive attenuator with a specified attenuation or loss factor.

If L is the loss factor, then the gain G for this attenuator can be expressed as $G = 1/L$. The expression for the total noise power at the output of the attenuator (N_o) can be written as

$$N_o = GkT_iB + GkT_eB = \frac{kT_iB + kT_eB}{L} \tag{7.25}$$

where T_e = effective noise temperature of the attenuator. If the attenuator is considered to be at the same temperature (T_i) as that of the source resistance from which it is fed, then the value of output noise (N_o) is given by

$$N_o = kT_i B \tag{7.26}$$

This gives

$$\frac{kT_i B + kT_e B}{L} = kT_i B$$

or

$$\frac{T_i + T_e}{L} = T_i \quad \text{which gives } T_e = T_i(L - 1) \tag{7.27}$$

This expression gives the effective noise temperature of a noise source at temperature T_i, followed by a resistive attenuator having a loss factor L.

7.6.4 Noise Figure and Noise Temperature of Cascaded Stages

So far the noise figure and noise temperature specifications of individual building blocks have been discussed. A system such as a receiver would have a large number of individual building blocks connected in series and it is important to determine the overall noise performance of this cascaded arrangement. In the following paragraphs, expressions will be derived for the noise figure and effective noise temperature for a cascaded arrangement of multiple stages.

Consider a cascaded arrangement of three stages with individual gains given as G_1, G_2 and G_3 and input noise temperature parameters as T_1, T_2 and T_3, as shown in Figure 7.9 (a). In the case of the cascaded arrangement, the total noise power at the output (N_{TO}) can be computed from

$$N_{TO} = G_3 kT_3 B + G_3 G_2 kT_2 B + G_3 G_2 G_1 kT_1 B$$
$$= G_3 G_2 G_1 kB \left(T_1 + \frac{T_2}{G_1} + \frac{T_3}{G_1 G_2} \right) \tag{7.28}$$

Now if T_e is the effective input noise temperature of the cascaded arrangement of the three stages, which would have a gain of $G_3 G_2 G_1$ as shown in Figure 7.9 (b), then the total noise

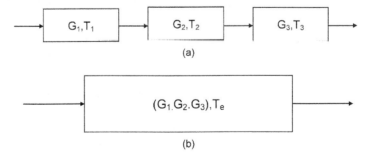

Figure 7.9 (a) Cascaded arrangement of three stages. (b) Equivalent single stage

power at the output (N_{TO}) can also be computed from

$$N_{TO} = G_3 G_2 G_1 k T_e B \qquad (7.29)$$

Equating the two expressions for N_{TO} gives

$$T_e = T_1 + \frac{T_2}{G_1} + \frac{T_3}{G_1 G_2} \qquad (7.30)$$

Generalizing the expression for n stages gives

$$T_e = T_1 + \frac{T_2}{G_1} + \frac{T_3}{G_1 G_2} + \frac{T_4}{G_1 G_2 G_3} + \cdots + \frac{T_n}{G_1 G_2 G_3 \ldots G_{n-1}} \qquad (7.31)$$

The same expression can also be written in terms of noise figure specifications of individual stages as

$$F = F_1 + \frac{F_2 - 1}{G_1} + \frac{F_3 - 1}{G_1 G_2} + \frac{F_4 - 1}{G_1 G_2 G_3} + \frac{F_n - 1}{G_1 G_2 G_3 \ldots G_{n-1}} \qquad (7.32)$$

where, $F_1, F_2, F_3 \ldots F_n$ are the noise figures for stages 1,2,3 ... n respectively and $G_1, G_2, G_3 \ldots G_n$ are the gains for stages 1,2,3 ... n respectively. It may be mentioned here that the gain values in the expression are not in decibels.

The expressions for the noise figure and effective noise temperature derived above highlight the significance of the first stage. As is evident from these expressions, the noise performance of the overall system is largely governed by the noise performance of the first stage. That is why it is important to have the first stage with as low noise as possible.

7.6.5 Antenna Noise Temperature

The antenna noise temperature is a measure of the noise entering the receiver via the antenna. The antenna picks up noise radiated by various man-made and natural sources within its directional pattern. Various sources of noise, include noise generated by electrical equipment, noise emanating from natural sources including solar radiation, sky noise, background noise contributed by Earth, galactic noise due to electromagnetic waves emanating from radio stars in the galaxy and atmospheric noise caused by lightning flashes and absorption by oxygen and water vapour molecules followed by the re-emission of radiation.

Noise from these sources could enter the receiver both through the main lobe as well as through the side lobes of the directional pattern of the receiving antenna. Thus, the noise output from a receiving antenna is a function of the direction in which it is pointing, its directional pattern and the state of its environment. The noise performance of an antenna, as mentioned before, can be expressed in terms of a noise temperature called the antenna noise temperature. If the antenna noise temperature is T_A K, it implies that the noise power output of the antenna is equal to the thermal noise power generated in a resistor at a temperature of T_A K.

The noise temperature of the antenna can be computed by integrating the contributions of all the radiating bodies whose radiation lies within its directional pattern. It is given by

$$T_A = \left(\frac{1}{4\pi}\right) \iint G(\theta, \phi)\, T_b(\theta, \phi)\, \sin\theta\, d\theta\, d\phi \tag{7.33}$$

where

θ = azimuth angle
ϕ = elevation angle
$G(\theta, \phi)$ = antenna gain in the θ and ϕ directions
$T_b(\theta, \phi)$ = brightness temperature in the θ and ϕ directions

There are two possible situations to be considered here. One is that of the satellite antenna when the uplink is referred to and the other is that of the Earth station antenna when the downlink is referred to. These two cases need to be considered separately because of the different conditions that prevail both in terms of the antenna's directional pattern and the significant sources of noise.

In the case of satellite antenna (uplink scenario), the main sources of noise are Earth and outer space. Again, the noise contribution of the Earth depends upon the orbital position of the satellite and the antenna beam width. In case, the satellite antenna's beam width is more or less equal to the angle-of-view of Earth from the satellite, which is 17.5° for a geostationary satellite, the antenna noise temperature depends upon the frequency of operation and orbital position. In case the beam width is smaller as in case of a spot beam, the antenna noise temperature depends upon the frequency of operation and the area being covered on Earth. Different areas radiate different noise levels, as is obvious from Figure 7.10, which shows the brightness temperature model of the Earth in the Ku band.

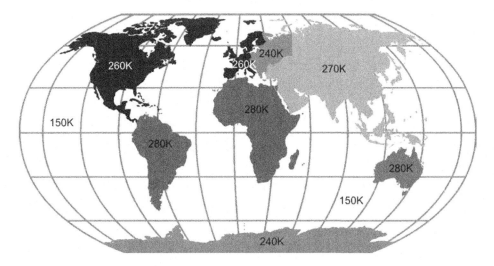

Figure 7.10 Earth's brightness temperature model in the Ku band

In the case of the Earth station antenna (downlink scenario), the main sources of noise that contribute to the antenna noise temperature include the sky noise and the ground noise.

The sky noise is primarily due to sources such as radiation from the sun and the moon and the absorption by oxygen and water vapour in the atmosphere accompanied by re-emission. The noise from other sources such as cosmic noise originating from hot gases of stars and interstellar matter, galactic noise due to electromagnetic waves emanating from radio stars in the galaxy is negligible at frequencies above 1 GHz. Here again there are two distinctly different conditions, one of clear sky devoid of any meteorological formations and the other that of sky with meteorological formations such as clouds, rain, etc.

In the clear sky conditions, the noise contribution is from sky noise and ground noise. The sky noise enters the system mainly through the main lobe of the antenna's directional pattern and the ground noise enters the system mainly through the side lobes and only partly through the main lobe, particularly at low elevation angles [Figure 7.11 (a)]. With reference to Figure 7.11 (b), if the attenuation and noise temperature figures for the meteorological formation are A_m and T_m respectively, then the noise contribution due to sky noise is T_{sky}/A_m and that due to the meteorological formation is $T_m(1 - 1/A_m)$.

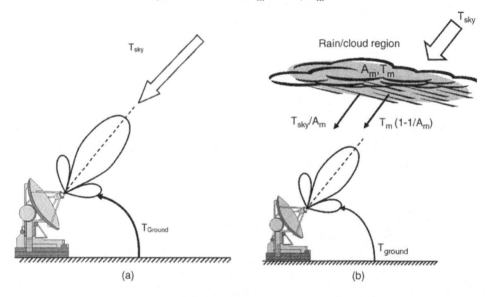

Figure 7.11 Constituents of noise temperature of an Earth station antenna in (a) the clear sky condition and (b) the sky with meteorological formations

The sky noise contribution can be computed knowing the brightness temperature of the sky in the direction of the antenna within its beam width. Figure 7.12 illustrates a family of curves showing the brightness temperature of clear sky as a function of operating frequency for various practical elevation angles ranging from 5° to 60°. The contribution of ground noise can also be similarly computed knowing the brightness temperature of the ground. The brightness temperature of the ground may be as large as 290 K for a side lobe whose elevation angle is less than −10°, as shown in Figure 7.13 (a), and as small as 10 K for an elevation angle greater than 10°, as shown in Figure 7.13 (b).

In addition to the sky noise component discussed above, there are also a number of high intensity point sources in the sky noise. These sources subtend an angle that is only a few arcs

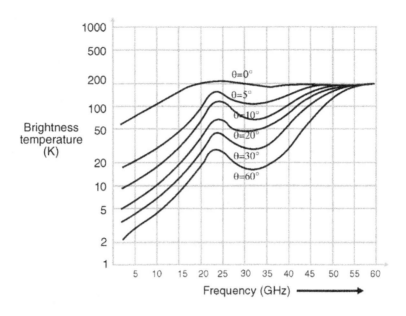

Figure 7.12 Brightness temperature of the clear sky

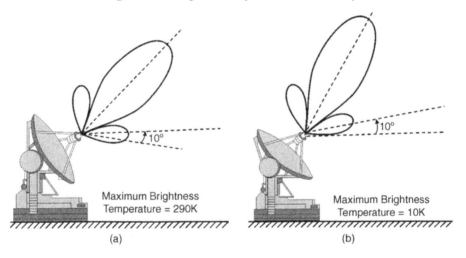

Figure 7.13 Ground noise for elevation angles (a) less than −10° (b) Greater than 10°

of a minute and therefore they matter only if the Earth station antenna is highly directional and the radiation source lies within the antenna beam width.

The sky noise contribution increases significantly when heavenly bodies like the sun and the moon become aligned with the Earth station-satellite path. The increase in noise temperature is a function of operating frequency and the size of the antenna. The average noise temperature contribution due to the sun can be approximated as $12\,000 f^{-0.75}$, where f is the operating frequency in GHz. It may increase by a factor of 100 to 10 000 during the periods of high solar activity.

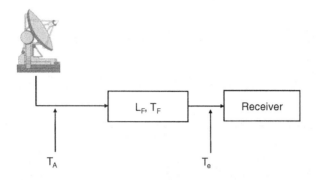

Figure 7.14 Computation of overall system noise temperature

7.6.6 Overall System Noise Temperature

The overall system is a cascaded arrangement of the antenna, the feeder connecting the antenna output to the receiver input and the receiver as shown in Figure 7.14. Expressions can be written for the noise temperature at two points, one at the output of the antenna, i.e. input of the feeder, and second at the input of the receiver. The expression for the system noise temperature with reference to the output of the antenna (T_{SAO}) can be written as

$$T_{SAO} = T_A + T_F(L_F - 1) + T_e L_F \tag{7.34}$$

where

T_A = antenna noise temperature
T_F = thermodynamic temperature of the feeder, often taken as the ambient temperature
L_F = attenuation factor of the feeder
T_e = effective input noise temperature of the receiver

The expression for the system noise temperature when referred to the receiver input (T_{SRI}) can be written as

$$T_{SRI} = \frac{T_A}{L_F} + T_F\left(\frac{L_F - 1}{L_F}\right) + T_e \tag{7.35}$$

The above expression for the noise temperature takes into account the noise generated by the antenna and the feeder together with the receiver noise. The two expressions for the noise temperature given above highlight another very important point that the noise temperature at the antenna output is larger than the noise temperature at the receiver input by a factor of L_F. This underlines the importance of having minimum losses before the first RF stage of the receiver.

Problem 7.4
A 12 GHz receiver consists of an RF stage with gain G_1 = 30 dB and noise temperature T_1 = 20 K, a down converter with gain G_2 = 10 dB and noise temperature T_2 = 360 K and an IF amplifier stage with gain G_3 = 15 dB and noise temperature T_3 = 1000 K.

Noise Considerations

Calculate the effective noise temperature and noise figure of the system. Take the reference temperature to be 290 K.

Solution: The effective noise temperature T_e can be computed from

$$T_e = T_1 + \frac{T_2}{G_1} + \frac{T_3}{G_1 G_2}$$

Now,

$T_1 = 20\,\text{K}, \quad T_2 = 360\,\text{K} \quad \text{and} \quad T_3 = 1000\,\text{K}$

$G_1 = 30\,\text{dB} = 1000, \quad G_2 = 10\,\text{dB} = 10$

Therefore,

$T_e = 20 + 360/1000 + 1000/(1000 \times 10) = 20 + 0.36 + 0.1 = 20.46\,\text{K}$

The system noise figure F can be computed from

$$F = 1 + \frac{T_e}{T_i} \quad \text{where } T_i = 290\,\text{K}$$

$$= 1 + \frac{20.46}{290} = 1.07$$

Problem 7.5

For the receiver of Problem 7.4, compute the noise figure specifications of the three stages and then compute the overall noise figure from the individual noise figure specifications.

Solution: In problem 7.4, $T_1 = 20\,\text{K}$, $T_2 = 360\,\text{K}$, $T_3 = 1000\,\text{K}$, $T_i = 290\,\text{K}$, $G_1 = 30\,\text{dB} = 1000$, $G_2 = 10\,\text{dB} = 10$. Let F_1, F_2 and F_3 be the noise figure specifications of the three stages. Then

$$F_1 = 1 + \frac{T_1}{T_i} = 1 + \frac{20}{290} = 1.069$$

$$F_2 = 1 + \frac{T_2}{T_i} = 1 + \frac{360}{290} = 2.24$$

$$F_3 = 1 + \frac{T_3}{T_i} = 1 + \frac{1000}{290} = 4.45$$

The overall noise figure can be computed from

$$F = F_1 + \frac{F_2 - 1}{G_1} + \frac{F_3 - 1}{G_1 G_2}$$

$$= 1.069 + \frac{2.24 - 1}{1000} + \frac{4.45 - 1}{10\,000}$$

$$= 1.069 + 0.001\,24 + 0.000\,345 = 1.07$$

Problem 7.6

The effective input noise temperature of a satellite receiver is 30 K when the effect of noise contributions from the antenna and feeder are not taken into consideration. If the receiver is fed from an antenna having a noise temperature of 50 K via a feeder with a loss factor of 2.5 dB, determine the effective input noise temperature of the receiver considering the effect of the antenna and the feeder noise contributions. Assume $T_i = 290$ K and also that the feeder is at a temperature T_i. Also compute the noise figure in the two cases in decibels.

Solution: Loss factor L of the feeder = 2.5 dB = 1.778
The contribution of the antenna noise temperature when referred to the input of the receiver is given by

$$\left(\frac{T_A}{L}\right) = \frac{50}{1.778} = 28.1 \text{ K}$$

The contribution of the feeder noise when referred to the input of the receiver is given by

$$\frac{T_F(L-1)}{L} = \frac{T_i(L-1)}{L} = \frac{290(1.778-1)}{1.778} = 290\left(\frac{0.778}{1.778}\right) = 126.9 \text{ K}$$

Therefore, the effective input noise temperature of the receiver taking into account the effect of noise contributions from the antenna and feeder is given by

$$28.1 + 126.9 + 30 = 185 \text{ K}$$

Noise figure in the first case = $1 + \frac{30}{290} = 1.103 = 0.426$ dB

Noise figure in the second case = $1 + \frac{185}{290} = 1.638 = 2.14$ dB

Problem 7.7

An otherwise ideal receiver when fed from a non-ideal antenna feeder combination had its effective input noise temperature increased by 50 K. If $T_A = 40$ K and $T_F = T_i = 290$ K, determine the loss factor of the feeder in decibels.

Solution: If T_e is the effective input noise temperature of the ideal receiver without considering the effect of antenna and feeder, then $T_e = 0$ K. Considering the effect of the antenna and feeder noise contributions, the effective input noise temperature of the receiver becomes $(0 + 50) = 50$ K, or

$$\frac{T_A}{L} + \frac{T_F(L-1) + T_e}{L} = 50$$

As $T_e = 0\,K$, therefore

$$\left(\frac{1}{L}\right)[T_A + T_F(L-1)] = 50$$

$$\left(\frac{1}{L}\right)[40 + 290(L-1)] = 50$$

$$240L = 250$$

which gives

$$L = \frac{250}{240} = 1.0417 = 0.177\,dB$$

Therefore, loss factor in dB is 0.177 dB.

Problem 7.8
Consider a satellite receiver with the following data:

1. Antenna noise temperature $T_A = 50\,K$
2. Thermodynamic temperature of the feeder $T_F = 300\,K$
3. Effective input noise temperature of the receiver $T_e = 50\,K$

Compute the system noise temperature at the receiver input for (a) no feeder loss and (b) a feeder loss of 1.5 dB.

Solution: The system noise temperature at the receiver input can be computed from

$$\frac{T_A}{L_F} + T_F\left(\frac{L_F - 1}{L_F}\right) + T_e$$

where the terms have their usual meaning.

(a) $L_F = 0\,dB = 1$. Therefore,

$$\text{System noise temperature} = T_A + T_e = 50 + 50 = 100\,K$$

(b) $L_F = 1.5\,dB = 1.413$. Therefore,

$$\text{System noise temperature} = \frac{50}{1.413} + 300\left(\frac{0.413}{1.413}\right) + 50$$

$$= 173.072\,K$$

7.7 Interference-related Problems

The major sources of interference were outlined in the introductory discussion on the topic in the earlier part of the chapter. These included:

1. Intermodulation distortion
2. Interference between the satellite and the terrestrial link sharing the same frequency band
3. Interference between two satellites sharing the same frequency band
4. Interference arising out of cross-polarization in frequency re-use systems
5. Adjacent channel interference inherent to FDMA systems

Each of them will be briefly described in the following paragraphs.

7.7.1 Intermodulation Distortion

Intermodulation distortion is caused as a result of the generation of intermodulation products within the satellite transponder as a result of amplification of multiple carriers in the power amplifier, which is invariably a TWTA. The generation of intermodulation products is due to both amplitude nonlinearity and phase nonlinearity. Figure 7.15 shows the transfer characteristics (output power versus input power) of a typical TWTA. It is evident from the figure, that the characteristics are linear only up to a certain low input drive level and become increasingly nonlinear as the output power approaches saturation. Intermodulation products can be avoided by operating the amplifier in the linear region by reducing or backing off the input drive. A reduced input drive leads to a reduced output power. This results in a downlink power-limited system that is forced to operate at a reduced capacity.

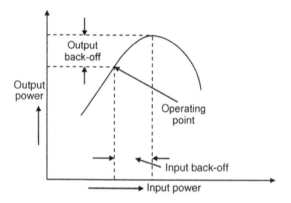

Figure 7.15 Transfer characteristics of TWTA

Intermodulation products are generated whenever more than one signal is to be amplified by the amplifier with non-linear characteristics. Filtering helps to remove the intermodulation products but when these products are within the bandwidth of the amplifier, filtering is not of much use. The transfer characteristics of an amplifier can be written as

$$V_{out} = AV_{in} + B\left(V_{in}\right)^2 + C\left(V_{in}\right)^3 \tag{7.36}$$

Where, $A \gg B \gg C$

Interference-related Problems

It may be mentioned here that the intermodulation products are mainly generated by the third-order component in the equation as the third-order intermodulation products have frequencies close to the input frequencies and hence lie within the transponder bandwidth.

Let us consider that the signal applied to the input of the amplifier is given by

$$V_{in} = V_1 \cos \omega_1 t + V_2 \cos \omega_2 t \tag{7.37}$$

In other words, two unmodulated carriers at frequencies f_1 and f_2 are applied to the input of the amplifier.

The output of the amplifier is given by

$$V_{out} = A\left[V_1 \cos \omega_1 t + V_2 \cos \omega_2 t\right] + B\left[V_1 \cos \omega_1 t + V_2 \cos \omega_2 t\right]^2$$
$$+ C\left[V_1 \cos \omega_1 t + V_2 \cos \omega_2 t\right]^3 \tag{7.38}$$

The first term is the linear term and it amplifies the input signal by A and represents the desired output of the amplifier (V_{desout}).

$$V_{desout} = A\left[V_1 \cos \omega_1 t + V_2 \cos \omega_2 t\right]$$

The total desired power output from the amplifier, referenced to a one ohm load is, therefore given by

$$P_{desout} = \frac{1}{2}A^2 V_1^2 + \frac{1}{2}A^2 V_2^2 = A^2 (P_1 + P_2)$$

Where,

$P_1 = 1/2(V_1^2)$
$P_2 = 1/2(V_2^2)$

The second term in equation 7.38 is the second order term and can be expanded as

$$V_{2out} = B\left[V_1 \cos \omega_1 t + V_2 \cos \omega_2 t\right]^2$$
$$= B\left[V_1^2 \cos^2 \omega_1 t + V_2^2 \cos^2 \omega_2 t + 2V_1 V_2 \cos \omega_1 t \cos \omega_2 t\right]$$
$$= B\left[V_1^2 \{(\cos 2\omega_1 t + 1)/2\} + V_2^2 \{(\cos 2\omega_2 t + 1)/2\} + V_1 V_2 \{\cos(\omega_1 + \omega_2)t \right.$$
$$\left. + \cos(\omega_1 - \omega_2)t\}\right] \tag{7.39}$$

The term V_{2out} contains frequency components $2f_1$, $2f_2$, $(f_1 + f_2)$ and $(f_1 - f_2)$. All these components can be removed from the amplifier output with the help of band pass filters.

The third term in the equation 7.38 is the third order term and can be expanded as

$$\begin{aligned}
V_{3out} &= C\left[V_1 \cos\omega_1 t + V_2 \cos\omega_2 t\right]^3 \\
&= C\left[V_1^3 \cos^3\omega_1 t + V_2^3 \cos^3\omega_2 t + 2\left(V_1^2 \cos^2\omega_1 t\right)\left(V_2 \cos\omega_2 t\right)\right. \\
&\quad \left. + 2\left(V_2^2 \cos^2\omega_2 t\right)\left(V_1 \cos\omega_1 t\right)\right] \\
&= C\left[V_1^3 \cos^3\omega_1 t + V_2^3 \cos^3\omega_2 t + V_1^2 V_2 \left(1+\cos 2\omega_1 t\right)\left(\cos\omega_2 t\right)\right. \\
&\quad \left. + V_2^2 V_1 \left(1+\cos 2\omega_2 t\right)\left(\cos\omega_1 t\right)\right] \\
&= C\begin{bmatrix} V_1^3 \cos^3\omega_1 t + V_2^3 \cos^3\omega_2 t + V_1^2 V_2 \cos\omega_2 t + V_1^2 V_2/2\left\{\cos\left(2\omega_1-\omega_2\right)t\right. \\ \left. + \cos\left(2\omega_1+\omega_2\right)t\right\} \\ + V_2^2 V_1 \cos\omega_1 t + V_2^2 V_1/2\left\{\cos\left(2\omega_2-\omega_1\right)t + \cos\left(2\omega_2+\omega_1\right)t\right\} \end{bmatrix}
\end{aligned}$$

(7.40)

The first two terms contain frequency components f_1, f_2, $3f_1$ and $3f_2$. The triple frequency component can be removed from the amplifier output with the help of band pass filters. The fifth and the last terms contain the frequency components $(2f_1 + f_2)$ and $(2f_2 + f_1)$ which can again be removed by the band pass filters. The frequency components $(2f_1 - f_2)$ and $(2f_2 - f_1)$ in the fourth and seventh terms can fall within the bandwidth of the transponder and are referred to as third-order intermodulation products of the amplifier. Therefore, the intermodulation products of concern are

$$V_{3IM} = C\left[V_1^2 V_2 \cos\left(2\omega_1 - \omega_2\right)t + V_2^2 V_1 \cos\left(2\omega_2 - \omega_1\right)t\right] \tag{7.41}$$

The power of the intermodulation components is given by

$$P_{IM} = \frac{1}{2}C^2 V_1^4 V_2^2 + \frac{1}{2}C^2 V_2^4 V_1^2 = 4C^2(P_1^2 P_2 + P_2^2 P_1) \tag{7.42}$$

It is clear from the equation 7.42 that the ratio of intermodulation power to the desired power increase in proportion to the cubes of the signal power and also to square of (C/A). The greater the non-linearity of the amplifier, larger is the value of (C/A) and larger the value of intermodulation products. Also, the intermodulation terms increase rapidly as the amplifier operates near its saturation region.

The easiest way to reduce the intermodulation problems is to reduce the levels of input signals to the amplifier. The output backoff is defined as the difference in decibels between the saturated output power of the amplifier and its actual power. When the transponder is operated with output backoff, the power level at the input is reduced by input backoff. As the characteristics of the amplifier are non-linear, the value of input backoff is greater than the value of output backoff.

The overall C/N ratio taking into account the contribution of intermodulation products is given by

$$\left(\frac{C}{N}\right)_o = 1 \Big/ \left[\left\{1\Big/\left(\frac{C}{N}\right)_{up}\right\} + \left\{1\Big/\left(\frac{C}{N}\right)_{down}\right\} + \left\{1\Big/\left(\frac{C}{N}\right)_{IM}\right\}\right] \tag{7.43}$$

Intermodulation distortion is a serious problem when the transponder is made to handle two or more carrier signals. That is why satellite links that use frequency division multiple access technique are particularly prone to this type of interference. On the other hand, a single carrier per transponder TDMA system is becoming increasingly popular as in this case the satellite TWTA can be operated at or close to the saturation level without any risk of generating intermodulation products. This maximizes the EIRP for the downlink.

Another intermodulation interference-related problem associated with the FDMA system is that the Earth station needs to exercise a greater control over the transmitted power in order to minimize the overdrive of the satellite transponder and the consequent increase in intermodulation interference. Intermodulation considerations also apply to Earth stations transmitting multiple carriers, which forces the amplifiers at the Earth station to remain underutilized.

7.7.2 Interference between the Satellite and Terrestrial Links

Satellite and terrestrial microwave communication links cause interference to each other when they share a common frequency band. The 6/4 GHz frequency band is allocated to both the satellite as well as terrestrial microwave links. An Earth station receiving at 4 GHz is susceptible to interference from terrestrial stations transmitting at 4 GHz. Similarly, an Earth station transmitting at 6 GHz is a source of interference for a terrestrial station receiving at 6 GHz. The level of mutual interference between the two is a function of a number of parameters including carrier power, carrier power spectral density and the frequency offset between the two carriers.

The level of interference caused by a terrestrial transmission to a satellite signal reception would depend upon the spectral density of the terrestrial interfering signal and the bandwidth of the satellite signal received by the Earth station. As an example, for a broadband satellite signal, the whole of the interfering carrier power may be applicable whereas for a narrowband satellite signal, the interfering carrier power is reduced by a factor equal to the ratio of the total carrier power and the carrier power included in the narrow bandwidth.

Interference from a narrowband satellite transmission to a terrestrial microwave system can be reduced by using a frequency offset between the satellite and terrestrial carriers. The amount of interference depends upon the frequency difference between the interfering satellite carrier frequency and the terrestrial carrier frequency. The interference reduction factor in this case can be determined by convolving the power spectral densities of the interfering satellite carrier signal and the terrestrial carrier signal.

7.7.3 Interference due to Adjacent Satellites

This type of interference is caused by the presence of side lobes in addition to the desired main lobe in the radiation pattern of the Earth station antenna. If the angular separation between two adjacent satellite systems is not too large, it is quite possible that the power radiated through the side lobes of the antenna's radiation pattern, whose main lobe is directed towards the intended satellite, interferes with the received signal of the adjacent satellite system. Similarly, transmission from an adjacent satellite can interfere with the reception of an Earth station through the side lobes of its receiving antenna's radiation pattern.

This type of interference phenomenon is illustrated in Figure 7.16. Satellite A and satellite B are two adjacent satellites. The transmitting Earth station of satellite A on its uplink, in addition to directing its radiated power towards the intended satellite through the main lobe of its transmitting antenna's radiation pattern, also sends some power, though unintentionally, towards satellite B through the side lobe. The desired and undesired paths are shown by solid

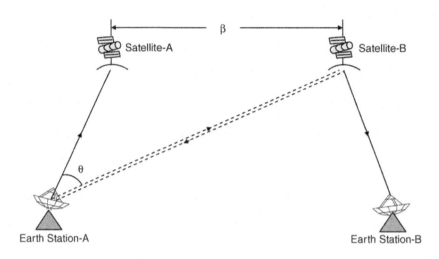

Figure 7.16 Interference due to adjacent satellites

and dotted lines respectively. In the figure, θ is the angular separation between two satellites as viewed by the Earth stations and β is the angular separation between the satellites as viewed from the centre of the Earth; i.e. β is simply the difference in longitudinal positions of the two satellites. Coming back to the problem of interference, transmission from satellite B on its downlink, in addition to being received by its intended Earth station shown by a solid line again, also finds its way to the receiving antenna of the undesired Earth station through the side lobe shown by the dotted line. Quite obviously, this would happen if the off-axis angle of the radiation pattern of the Earth station antenna is equal to or more than the angular separation θ between the adjacent satellites. θ and β are interrelated by the following expression:

$$\theta = \cos^{-1}\left[\frac{d_A^2 + d_B^2 - 2r^2(1 - \cos\beta)}{2d_A d_B}\right] \qquad (7.44)$$

where

d_A = slant range of satellite A
d_B = slant range of satellite B
r = geostationary orbit radius

For a known value of θ, the worst case acceptable value of the off-axis angle of the antenna's radiation pattern can be computed. Similarly, for a given radiation pattern and known off-axis angle, it is possible to find the minimum required angular separation between the two adjacent satellites for them to coexist without causing interference to each other. Let us take the case of downlink interference and determine the expression for carrier-to-interference (C/I) ratio.

The desired carrier power C_D for the downlink channel in dBW can be expressed as

$$C_D = \text{EIRP} - L_D + G \qquad (7.45)$$

where

EIRP = desired EIRP (in dBW, or decibels relative to a power level of 1 W)
L_D = downlink path loss for the beam from the desired satellite (in dB)
G = Earth station antenna gain in the direction of the desired satellite (in dB)

The interfering carrier power for the downlink channel (I_D) in dBW is given by

$$I_D = \text{EIRP}' - L_{D'} + G' \tag{7.46}$$

where

EIRP' = interfering EIRP (in dBW)
$L_{D'}$ = downlink path loss for the beam from interfering satellite (in dB)
G' = Earth station antenna gain in the direction of the interfering satellite (in dB)

The expression for (C/I) in the case of downlink can then be written as

$$(C/I)_D = (\text{EIRP} - L_D + G) - (\text{EIRP}' - L_{D'} + G')$$
$$= (\text{EIRP} - \text{EIRP}') - (L_D - L_{D'} + (G - G')) \tag{7.47}$$

where $(C/I)_D$ is the C/I for the downlink channel in dB.
If the path losses are considered as identical, then

$$(C/I)_D = (\text{EIRP} - \text{EIRP}') + (G - G') \tag{7.48}$$

Also, the term $(G - G')$ is the receive Earth station antenna discrimination, which is defined as the antenna gain in the direction of the desired satellite minus the antenna gain in the direction of the interfering satellite. According to CCIR standards, in cases where the ratio of the antenna diameter to the operating wavelength is greater than 100, G' as a function of the off-axis angle θ should at the most be equal to $(32 - 25 \log \theta)$ dB (forward gain of an antenna compared to an idealized isotropic antenna), where θ is in degrees. The requirement of FCC standards for the same is $(29 - 25 \log \theta)$ dB. Figure 7.17 shows the typical Earth station antenna pattern, which is a plot of the gain versus the off-axis angle along with CCIR requirements. This gives

$$(C/I)_D = (\text{EIRP} - \text{EIRP}') + (G - 32 + 25 \log \theta) \tag{7.49}$$

A similar calculation can be made for the uplink interference, where a satellite may receive an unwanted signal from an interfering Earth station. In the case of uplink, the expression for C/I can be written as

$$(C/I)_U = (\text{EIRP} - \text{EIRP}') + (G - G') \tag{7.50}$$

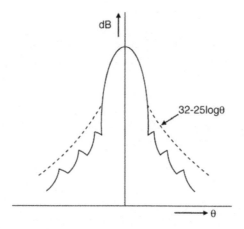

Figure 7.17 Typical Earth station antenna pattern

where

$(C/I)_U$ = C/I for the uplink channel in dB
EIRP = EIRP of the desired Earth station in dBW
EIRP′ = EIRP of the interfering Earth station in the direction of the satellite in dBW
  G = gain of the satellite receiving antenna in the direction of the desired Earth station in dB
  G′ = gain of the satellite receiving antenna in the direction of the interfering Earth station in dB

EIRP′ is further equal to

$$\text{EIRP}' = \text{EIRP}^* - G_I + (32 - 25\log\theta) \qquad (7.51)$$

where

EIRP* = EIRP of the interfering Earth station in dBW
  G_I = on-axis transmit antenna gain of the interfering Earth station in dB
  θ = viewing angle of the satellite from the desired and interfering Earth Stations

The overall carrier-to-interference ratio (C/I) for adjacent satellite interference is given by

$$\frac{C}{I} = \left[\left(\frac{C}{I}\right)_U^{-1} + \left(\frac{C}{I}\right)_D^{-1}\right]^{-1} \qquad (7.52)$$

where the subscripts U and D imply uplink and downlink respectively. Where the interference is noise like, it is possible to combine the effects of noise and interference. The combined carrier-to-noise ratio (C/NI) is given by

$$(C/NI) = [(C/N)^{-1} + (C/I)^{-1}]^{-1} \qquad (7.53)$$

It may be mentioned here that the various terms in equations 7.52 and 7.53 are not in decibels.

7.7.4 Cross-polarization Interference

Cross-polarization interference occurs in frequency re-use satellite systems. It occurs due to coupling of energy from one polarization state to the other orthogonally polarized state in communications systems that employ orthogonal linear polarizations (horizontal and vertical polarization) and orthogonal circular polarizations (right-hand circular and left-hand circular). The coupling of energy from one polarization state to the other polarization state takes place due to finite cross-polarization discrimination of the Earth station and satellite antennas and also by depolarization caused by rain, particularly at frequencies above 10 GHz.

Cross-polarization discrimination is defined as the ratio of power received by the antenna in principal polarization to that received in orthogonal polarization from the same incident signal. A cross-polarization discrimination figure of 30 to 40 dB along the antenna axis is considered very good. The combined effect of finite values of cross-polarization discrimination for the Earth station and satellite antennas can be expressed in the form of net minimum cross-polarization discrimination for the overall link as follows:

$$X = \frac{1}{2}[(X_e)^{-1} + (X_s)^{-1}]^{-1} \tag{7.54}$$

where X is the worst case carrier-to-cross polarization interference ratio. X_e and X_s are the cross-polarization discrimination for the Earth station antenna and for the satellite antenna respectively. It can be taken as an additional source of interference while computing the overall carrier-to-noise plus interference ratio.

7.7.5 Adjacent Channel Interference

Adjacent channel interference occurs when the transponder bandwidth is simultaneously shared by multiple carriers having closely spaced centre frequencies within the transponder bandwidth. When the satellite transmits to Earth stations lying within its footprint, different

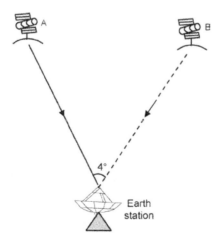

Figure 7.18 Figure for Problem 7.9

carriers are filtered by the receiver so that each Earth station receives its intended signal. Filtering would have been easier to realize had there been a large guard band between adjacent channels, which is not practically feasible as that would lead to the inefficient use of the transponder bandwidth. The net result is that a part of the power of the carrier in the channel adjacent to the desired one is also captured by the receiver due to overlapping amplitude characteristics of the channel filters. This becomes a source of noise.

Problem 7.9

Refer to Figure 7.18. Satellite A radiates an EIRP of 35 dBW on the downlink to an Earth station whose receive antenna gain is 50 dB. Emission from satellite B located in the vicinity of the first satellite produces interference to the desired downlink. If the EIRP of the interfering satellite is 30 dBW, determine the carrier-to-interference (C/I) ratio assuming that the path loss on the downlink channel for the interfered and interfering satellites are the same and that the angle between the line-of-sight between the Earth station and the desired satellite and the line-of-sight between the Earth station and the interfering satellite is 4°.

Solution: The carrier-to-interference ratio can be computed from

$$C/I = (EIRP - EIRP') + (G - 32 + 25\log\theta)$$

where
EIRP = EIRP from the desired satellite = 35 dBW
EIRP' = EIRP from the interfering satellite = 30 dBW
  G = gain of the Earth station receiving antenna = 50 dB
  θ = viewing angle of the two satellites from the Earth station = 4°

Therefore,

$$C/I = (EIRP - EIRP') + (G - 32 + 25\log\theta)$$
$$= (35 - 30) + (50 - 32 + 25\log 4)$$
$$= 23 + 25\log 4 = 38.05 \text{ dB}$$

Problem 7.10

Refer to Figure 7.19. The EIRP values of Earth stations A and B are 80 dBW and 75 dBW respectively. The transmit antenna gains in the two cases are 50 dB each. If the gain of the receiving antenna of the satellite uplinked from Earth station A is 20 dB in the direction of Earth station A and 15 dB in the direction of Earth station B, determine the carrier-to-interference ratio at the satellite due to interference caused by Earth station B. Assume that the viewing angle of the satellite from the two Earth stations is 4°.

Solution: $(C/I)_U$ can be computed from

$$(C/I)_U = (EIRP - EIRP') + (G - G') \quad \text{where} \quad EIRP' = EIRP^* - G_1 + (32 - 25\log\theta)$$

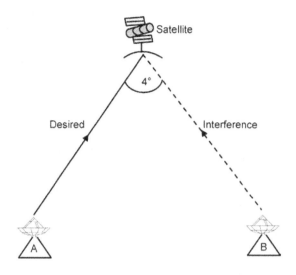

Figure 7.19 Figure for Problem 7.10

Different terms used above have their usual meaning, as explained in the text:

EIRP = 80 dBW,   EIRP* = 75 dBW,   G = 20 dB,   G' = 15 dB, G_1 = 50 dB

EIRP' = 75 − 50 + 32 − 25 log 4° = 42 dBW

Therefore,
$$(C/I)_U = 80 - 42 + 20 - 15 = 43 \text{ dB}$$

Problem 7.11
In a point-to-point satellite communication system, the carrier signal strength at the satellite as received over the uplink is 40 dB more than the strength of the interference signal from an interfering Earth station. Also, the strength of the signal power received at the desired Earth station over the downlink is 35 dB more than the strength of the interference signal power due to an interfering satellite. Determine the total carrier-to-interference ratio of the satellite link.

Solution: It is given that $(C/I)_U$ = 40 dB = 10 000 and $(C/I)_D$ = 35 dB = 3162.28. The total carrier-to-interference (C/I) ratio can be computed from

$$C/I = [(C/I)_U^{-1} + (C/I)_D^{-1}]^{-1}$$
$$= [(10\,000)^{-1} + (3162.28)^{-1}]^{-1} = 2402.53$$

$$(C/I)(\text{in dB}) = 10 \log 2402.53 = 33.8 \text{ dB}$$

Problem 7.12
The angle formed by the slant ranges of two geostationary satellites from a certain Earth station as shown in Figure 7.20 is 5°. Determine the longitudinal location of the two

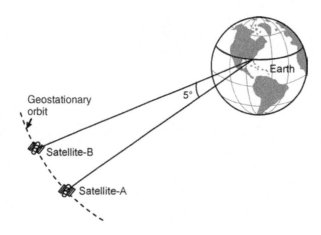

Figure 7.20 Figure for Problem 7.12

satellites given that the two slant ranges of satellites A and B are 42 100 km and 42 000 km respectively. Given that the radius of the geostationary orbit is equal to 42 164 km.

Solution: If β is the longitudinal separation of the two satellites, it can be computed from

$$\cos\theta = \left[\frac{d_A^2 + d_B^2 - 2r^2(1-\cos\beta)}{2d_A d_B}\right]$$

where

d_A = slant range of satellite A = 42 100 km
d_B = slant range of satellite B = 42 000 km
r = geostationary orbit radius = 42 164 km
θ = angular separation of the two satellites as viewed from the Earth station = 5°

Therefore,

$$\cos 5° = \frac{(42\,100 \times 10^3)^2 + (42\,000 \times 10^3)^2 - 2 \times (42\,164 \times 10^3)^2 \times (1-\cos\beta)}{2 \times (42\,100 \times 10^3) \times (42\,000 \times 10^3)}$$

or

$$0.996 = \frac{(177\,241 \times 10^{10}) + (176\,400 \times 10^{10}) - 2 \times (177\,780.3 \times 10^{10}) \times (1-\cos\beta)}{2 \times (176\,820 \times 10^{10})}$$

or

$$\cos\beta = 1 - 0.004 = 0.996$$

or

$$\beta = 5.126°$$

Therefore, the longitudinal separation between the two satellites = 5.126°.

7.8 Antenna Gain-to-Noise Temperature (G/T) Ratio

The antenna gain-to-noise temperature (G/T) ratio, usually defined with respect to the Earth station receiving antenna, is an indicator of the sensitivity of the antenna to the downlink carrier signal from the satellite. It is a figure-of-merit used to indicate the combined performance of the Earth station antenna and low noise amplifier combined in receiving weak carrier signals. G is the receive antenna gain, usually referred to the input of the low noise amplifier. It equals the receive antenna gain as computed, for instance, from $G = \eta(4\pi A_e/\lambda^2)$ minus the power loss in the waveguide connecting the output of the antenna to the input of the low noise amplifier. T is the Earth station effective noise temperature, also referred to as the noise temperature at the input of the low noise amplifier. It may be mentioned here that the value of the G/T ratio is invariant irrespective of the reference point chosen to measure its value. The input of the low noise amplifier is chosen because it is the point where its contribution is clearly shown.

During the discussion on system noise temperature earlier in the chapter, expressions were derived for the overall system noise temperature as referred to the output of the receive antenna as well as the input of the low noise amplifier. The system noise temperature at the output of the antenna (T_{SAO}) was derived as

$$T_{SAO} = T_A + T_F(L_F - 1) + T_e L_F \qquad (7.55)$$

where

T_A = antenna noise temperature
T_F = thermodynamic temperature of the feeder, often taken as the ambient temperature
L_F = attenuation factor of the feeder
T_e = effective input noise temperature of the receiver

The system noise temperature when referred to the input of the low noise amplifier (T_{SRI}) was derived as

$$T_{SRI} = \frac{T_A}{L_F} + T_F\left(\frac{L_F - 1}{L_F}\right) + T_e \qquad (7.56)$$

where T_e can further be expressed as

$$T_e = T_1 + \frac{T_2}{G_1} + \frac{T_3}{G_1 G_2} \qquad (7.57)$$

Here, T_1, T_2 and T_3 are the noise temperatures of different stages in the receiver, beginning with the low noise amplifier, and G_1, G_2 and G_3 are the corresponding gain values.

From the known values for G and T, the G/T ratio can be computed. To conclude the discussion on the G/T ratio, the following observations can be made:

1. The higher the antenna gain and the lower the loss of the feeder connecting the output of the antenna to the input of the low noise amplifier, the higher will be the G/T ratio.
2. The lower the noise temperature of the low noise amplifier, the higher will be the G/T ratio.
3. The higher the gain of the low noise amplifier, the lower will be the noise contribution of successive stages in the receiver and the higher will be the G/T ratio.

Problem 7.13

Refer to the block diagram of the receiver side of the satellite link, as shown in Figure 7.21. It is given that $G_A = 60$ dB, $T_A = 60$ K, $L_1 = 0.5$ dB, $T_1 = 290$ K, $G_2 = 60$ dB, $T_2 = 140$ K, $T_3 = 10\,000$ K. Determine the G/T ratio in dB/K referred to the input of the low noise amplifier.

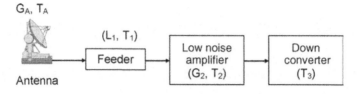

Figure 7.21 Figure for Problem 7.13

Solution: The gain as referred to the input of the low noise amplifier = $60 - 0.5 = 59.5$ dB. The system noise temperature as referred to the input of the low noise amplifier can be computed from

$$T_{SRI} = \frac{T_A}{L_1} + T_1 \left(\frac{L_1 - 1}{L_1}\right) + T_2 + \frac{T_3}{G_2}$$

$$L_1 = 0.5 \text{ dB} = 1.12$$

$$G_2 = 60 \text{ dB} = 10^6$$

$$T_A = 60 \text{ K}, T_1 = 290 \text{ K}, T_2 = 140 \text{ K}, T_3 = 10\,000 \text{ K}$$

Therefore,

$$T_{SRI} = \frac{60}{1.12} + 290 \left(\frac{1.12 - 1}{1.12}\right) + 140 + \frac{10\,000}{1\,000\,000}$$

$$= 53.57 + 31.07 + 140 + 0.01 = 224.65 \text{ K}$$

Therefore,

$$G/T(\text{in dB/K}) = G(\text{in dB}) - 10 \log T$$

$$= 59.5 - 10 \log 224.65$$

$$= 59.5 - 23.5 = 36 \text{ dB/K}$$

Problem 7.14

Determine the G/T ratio for the data given in Problem 7.13 as referred to the output of the antenna. What do you deduce from the result obtained?

Solution: G in this case is equal to 60 dB. The system noise temperature as referred to the output of the antenna is

$$T_{SAO} = T_A + T_1(L_1 - 1) + L_1 \left(T_2 + \frac{T_3}{G_2}\right)$$

This gives

$$T_{SAO} = 60 + 290(1.12 - 1) + 1.12(140 + 0.01)$$
$$= 60 + 34.8 + 156.8 = 251.6 \, K$$

Therefore,

$$G/T(\text{in dB/K}) = G(\text{in dB}) - 10 \log T$$
$$= 60 - 10 \log 251.6$$
$$= 60 - 24 = 36 \, dB/K$$

It is evident from solutions to Problems 7.13 and 7.14 that the G/T ratio is invariant regardless of the reference point, in agreement with a statement made earlier in the text.

7.9 Link Design

The design of any satellite based communication system is based on the two objectives of meeting a minimum C/N ratio for a specified time period and carrying maximum traffic at minimum cost. Both of these objectives are contradictory and the art of system design is to reach the best compromise so as to meet all the system parameters at minimum cost.

There are many parameters that affect the design of a satellite based communication system. These parameters may be categorized according the sub-system to which they are of prime importance (Earth station, satellite and satellite-Earth station channel). The parameters related to the Earth station design are geographical location of the Earth station (which provides an estimate of rain fades, satellite look angle, satellite EIRP and path loss), transmit antenna gain and transmitter power, receive antenna gain, system noise temperature and characteristics of the different modules of the Earth station (demodulator characteristics, filter characteristics, cross-polarization discrimination). Satellite related parameters are location of the satellite (determines the coverage region and Earth station look angle), transmit antenna gain, radiation pattern and transmitted power, receive antenna gain and radiation pattern and transponder, type, gain and its noise characteristics. The parameters of the transmission channel that affect the link design are operating frequency (path loss and link margin) and propagation characteristics (decide the link margin and choice of modulation and coding). A one way satellite link consists of two separate paths; an up-link path from the Earth station to the satellite and a down-link path from the satellite to the Earth station. A typical two way satellite link comprises of four separate paths; an uplink path from the first Earth station to the satellite, downlink path from the satellite to the second Earth station, uplink from the second Earth station to the satellite and a downlink from the satellite to the first Earth station. The design procedure for a one way satellite link is highlighted in the following paragraphs. The design for a two way link can also be carried out on similar lines.

7.9.1 Link Design Procedure

The satellite communication link design procedure is as follows:

1. Determine the frequency band in which the system will operate.
2. Determine the communication parameters of the satellite.
3. Calculate the signal-to-noise ratio and the bit error rate for the baseband channel.
4. Determine the parameters of the transmitting and the receiving Earth stations.
5. The design starts with the transmitting Earth station. Determine the carrier-to-noise ratio for the uplink using the uplink budget and the transponder noise power budget.
6. Determine the output power of the transponder based on the value of the transponder gain.
7. Determine the carrier-to-noise ratio for the downlink and establish the noise budget for the receiving Earth station at the edge of the coverage zone.
8. Determine the propagation conditions under which the system operates and calculate the value of atmospheric attenuation and other losses caused due to the atmospheric conditions.
9. Determine the link margin by calculating the link budget. Compare the result with the desired specifications. Change the system parameters to obtain the desired value of the link margin.

7.9.2 Link Budget

The link budget is a way of analysing and predicting the performance of a microwave communication link for given values of vital link parameters that contribute to either signal gain or signal loss. It is the algebraic sum of all gains and losses expressed in decibels as we move from the transmitter to the receiver. The final value thus obtained provides us with the means of knowing the available signal strength at the receiver and therefore also knowing how strong the received signal is with respect to the minimum acceptable level, called the threshold level. The difference between the actual value and the threshold is known as the link margin. The higher the value of the link margin, the better is the quality of the microwave link. The link budget is thus a tool that can be used for optimizing various link parameters in order to get the desired performance.

The concept of the link budget can be illustrated further with the help of a one-way microwave communication link schematic shown in Figure 7.22. Various parameters of interest in this case would be:

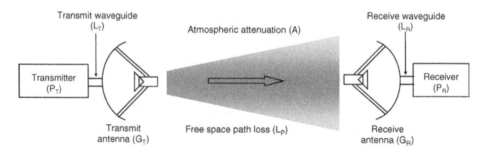

Figure 7.22 Link budget

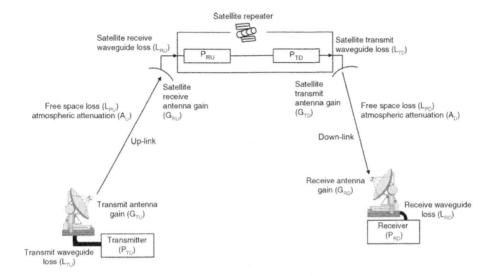

Figure 7.23 Satellite link budget analysis

1. Transmitter power P_T
2. Power loss in the waveguide connecting the transmitter output to the antenna input L_T
3. Transmitting antenna gain G_T
4. Free-space path loss L_P
5. Attenuation due to rain, clouds, fog, etc., A
6. Receive antenna gain G_R
7. Power loss in the waveguide connecting the receive antenna output to the receiver input L_R
8. Received signal power P_R

The power balance equation describing the link budget in this case would be given by

$$P_T - L_T + G_T - L_P - A + G_R - L_R = P_R \qquad (7.58)$$

It may be mentioned here that all the power levels in the above expression are in dBW and the gain, attenuation and loss terms are in dB.

With reference to a satellite link, such an equation can be written for both the uplink as well as the downlink. Figure 7.23 shows the schematic arrangement of a satellite link indicating various parameters that typically contribute to the link budget. For the satellite link shown in Figure 7.23, the uplink and downlink power balance equations can be written as

$$P_{TU} - L_{TU} + G_{TU} - L_{PU} - A_U + G_{RU} - L_{RU} = P_{RU} \dots \qquad \text{Uplink} \qquad (7.59)$$

$$P_{TD} - L_{TD} + G_{TD} - L_{PD} - A_D + G_{RD} - L_{RD} = P_{RD} \dots \qquad \text{Downlink} \qquad (7.60)$$

Design of the uplink channel is often easier than the downlink channel as higher power transmitters can be placed on the Earth stations than on the satellites. However, there are some systems like the VSATs which have limitations on the size of the antennas that can be used on the Earth station. The design of the uplink channel is done with the aim of producing a certain flux density or a certain power level at the input of the satellite transponder. The desired EIRP of the transmitting Earth station is calculated based on this requirement.

The design of the downlink channel is done by taking into consideration the back-off requirements of the satellite transponder amplifier to reduce the intermodulation problems. This is made possible either by reducing the uplink transmitter power or by reducing the satellite transponder amplification.

As an example, let us look at the link budget of a typical Ku band satellite-to-DTH receiver downlink. Typical values for the various parameters are:

1. Transmit power $P_{TD} = 25$ dBW
2. Transmit waveguide loss $L_{TD} = 1$ dB
3. Transmit antenna gain $G_{TD} = 30$ dB
4. Free-space path loss $L_{PD} = 205$ dB
5. Receive antenna gain G_{RD} (for a 50 cm diameter dish) = 39.35 dB
6. Receive waveguide loss $L_{RD} = 0.5$ dB

The received signal power can be computed from the above data to be equal to -112.15 dBW. If the receive system noise temperature is taken to be 140 K and the receive bandwidth to be 27 MHz, which are typical values for the link under consideration, then the receiver noise power would be

$$N = kTB = 1.38 \times 10^{-23} \times 140 \times 27 \times 10^6 \text{ watts} = 5216.4 \times 10^{-17} \text{ watts} = -132.83 \text{ dBW}$$

Therefore, the received carrier-to-noise C/N ratio for this link would be 20.68 dB. This figure can be used to determine the link margin and hence the quality of service provided by the link under clear sky conditions. It can also be used to assess the deterioration in the quality of service in case of hostile atmospheric conditions.

Problem 7.15
A certain 6/4 GHz satellite uplink has the following data on various gains and losses:

1. Earth station EIRP = 80 dBW
2. Earth station satellite distance = 35 780 km
3. Attenuation due to atmospheric factors = 2 dB
4. Satellite antenna's aperture efficiency = 0.8
5. Satellite antenna's aperture area = 0.5 m²
6. Satellite receiver's effective noise temperature = 190 K
7. Satellite receiver's bandwidth = 20 MHz

Determine the link margin for a satisfactory quality of service if the threshold value of the received carrier-to-noise ratio is 25 dB.

Solution:

$$\lambda = 3 \times 10^8 / 6 \times 10^9 = 0.05 \text{ m}$$

$$\text{Satellite antenna's gain} = \eta \left(\frac{4\pi A_e}{\lambda^2}\right)$$

$$= \frac{0.8 \times 4 \times \pi \times 0.5}{0.05 \times 0.05}$$

$$= 2010.62 = 33.03\,\text{dB}$$

$$\text{Receiver's noise power} = 10\log(kTB) = 10\log(1.38 \times 10^{-23} \times 190 \times 20 \times 10^6)$$

$$= -132.8\,\text{dB}$$

$$\text{Free-space path loss} = 20\log\left(\frac{4\pi R}{\lambda}\right)\,\text{dB} = 20\log\left[\frac{(4 \times \pi \times 35780 \times 10^3)}{0.05}\right]\,\text{dB}$$

$$= 199.08\,\text{dB}$$

From the given data,

$$\text{Received power at the satellite} = 80 - 2 - 199.08 + 33.03$$

$$= -88.05\,\text{dBW}$$

Thus, the received carrier is 44.75 dB [−88.05 − (−132.8)] stronger than the noise. It is 19.75 dB (44.75 − 25) more than the required threshold value. Therefore, the link margin = 19.75 dB.

7.10 Multiple Spot Beam Technology

Multiple spot beam technology is a concept in broadband satellite communications that allows frequency reuse, thereby maximizing the bandwidth capacity and minimizing the required frequency allocation. Inherent to the operational concept of multiple spot beam technology and the related frequency reuse feature is the capability of the satellite employing the technology to transmit multiple streams of information using the same frequency without causing unwanted interference. With increase in demand for satellite capacity, the traditional broadcast satellite technology operating on Ku-band becomes a limiting factor. Due to a large footprint at Ku-band, which would be much larger than what it would be at Ka-band for the same antenna size, the traditional Ku-band satellites do not support flexible distribution of the bandwidth required in the case of broadband internet.

While Ku-band is from 12 to 18 GHz, Ka-band extends from 26.5 to 40 GHz, including the upper part of the super high frequency (SHF) band of 3–30 GHz and the lower part of the extremely high frequency (EHF) band of 30–300 GHz. Typically, 27.5–30 GHz is used for uplink transmissions and 17.7–20.2 GHz is used for downlink transmissions. For a given antenna size, Ka-band operation provides higher gain and narrower beam width, leading to a much smaller footprint on Earth. This allows deployment of multiple beams to cover a given area. Figure 7.24 shows a comparison between use of a single large beam and multiple spot beams to provide the same coverage.

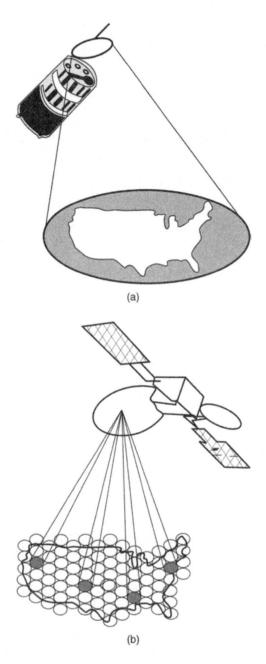

Figure 7.24 Comparison between (a) single large beam and (b) multiple spot beams to cover a given area

While a traditional Ku-band satellite may use a single downlink beam to cover a given area on Earth, a Ka-band satellite uses multiple spot beams to provide the same coverage. The multiple spot beam concept of covering a larger area is analogous to a cellular telephone network with each of the spot beams corresponding to a cell. It may be mentioned here that if one uses a single wide beam to cover the whole of a very large geographical area, then one can transmit the same information to every receiver in that area tuned to receive that information at the same time. The limitation is that only one stream of information can be transmitted at a time, which is fine with applications such as live television broadcasts. This would not be an efficient mode of communication for internet transmissions.

Use of multiple spot beams with each beam covering a small well-defined area in the larger area to be covered implies that multiple streams of information can be transmitted using the same frequency. This allows services to be provided to many more customers with one satellite.

There are a large number of communication satellites that employ multiple spot beam technology. It may also be mentioned here that not all spot beam services are equivalent and that different satellite operators are using spot beam technology at both Ku-band and Ka-band for various purposes. Ka-band operation of course allows a much smaller footprint with the result that multiple spot beam Ka-band satellites have increased capacity as compared to Ku-band satellites. Ka-band satellites employing multiple spot beam technology are also sometimes called high throughput satellites (HTS). HTSs are defined as a class of communication satellites that provide at least twice the total throughput of a classical fixed satellite service (FSS) satellite for the same allocated spectrum. The throughput of an HTS is usually greater by a factor of 20 or more.

Some of the well-known communication Ka-band satellites employing multiple spot beam technology include Spaceway-3 and EchoStar-17, also known as Jupiter-1 and Spaceway-4, from the EchoStar Corporation, Anik-F2 from Telesat Canada, KA-SAT from Eutelsat, Yahsat Y1B from AlYahsat, Thaicom-4 from Thaicom Public Company Ltd, ViaSat-1 from ViaSat, Astra-2E from Astra Communications and HYLAS-2 from Avanti Communications.

Spaceway-3 from Hughes Network Systems, a subsidiary of EchoStar Corporation, is one of the largest operational Ka-band satellite networks in the world and is also the first with onboard switching and routing. With 10 Gbps capacity, it enables communications directly between customer locations in a single hop without requiring a central hub. EchoStar-17 is also a geostationary communications satellite operated by Hughes Network Systems and used for satellite broadband. It has 60 Ka-band transponders that produce 60 narrow downlink beams to deliver broadband satellite services in North America.

Anik-F2 from Telesat, Canada is a communications satellite providing Ka-band multimedia services across North America in addition to offering an FSS service including internet access. It has a total of 114 transponders, including 50 operating in Ka-band, 40 in Ku-band and 24 in C-band.

Eutelsat's *KA-SAT* is Europe's HTS providing 82 Ka-band spot beams connected to a network of 10 ground stations. The configuration allows frequency reuse, thereby increasing the throughput beyond 90 Gbps. Figure 7.25 shows use of multiple spot beams by KA-SAT to cover Europe.

Yahsat-Y1B from AlYahsat is a communications satellite equipped with 25 110 MHz transponders to provide a high speed internet service to south-west Asia, the Middle East, Africa and Eastern Europe using multiple spot beam technology.

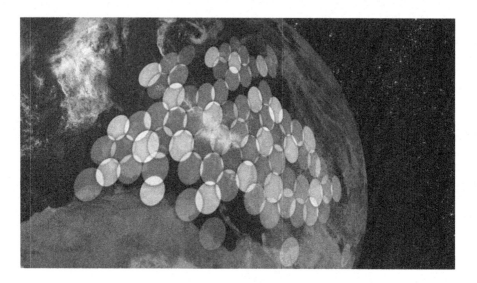

Figure 7.25 KA-SAT employing multiple spot beam technology covering Europe

Thaicom-4, also called IPSTAR, is a broadband communications satellite designed for high speed internet/multimedia services. The satellite has a capacity of 45 Gbps. Its multiple spot beam technology offers 20 times higher bandwidth compared to traditional Ku-band satellites.

ViaSat-1 from ViaSat is a high capacity communications satellite with a capacity exceeding 140 Gbps. It is equipped with 56 Ka-band transponders that provide broadband internet services to North America. *ViaSat-2* is scheduled to be launched during 2016 and will further augment capacity and coverage area.

Astra-2E from Astra Communications is a communications satellite designed to provide free-to-air and encrypted direct-to-home (DTH) digital television and satellite broadband services for Europe and the Middle East. The satellite has three Ku-band downlink beams with two of them covering Europe and the third covering the Middle East. The Ka-band footprint spreads across Europe.

HYLAS-2 (Highly Adaptable Satellite) from Avanti Communications is a high throughput communications satellite equipped with 24 active Ka-band transponders producing 24 Ka-band downlink beams, which along with four gateway beams provide two-way communications services such as corporate networking, broadband internet access, business continuity services and video distribution. The satellite provides coverage to northern and southern Africa, Eastern Europe and the Middle East, and is also equipped with steerable spot beam to provide coverage anywhere on Earth within the spacecraft's visibility.

Further Readings

Calcutt, D. and Tetley, L. (1994) *Satellite Communications: Principles and Application*, Edward Arnold, a member of the Hodder Headline Group, London.

Chartant, R.M. (2004) *Satellite Communications for the Nonspecialist*, SPIE, Washington.

Elbert, B.R. (1997) *The Satellite Communication Applications Handbook*, Artech House, Boston, Massachusetts.

Elbert, B.R. (1999) *Introduction to Satellite Communication*, Altech House, Boston, Massachusetts.

Elbert, R.B. (2001) Satellite *Communication Ground Segment and Earth Station Handbook*, Artech House, Boston, Massachusetts.

Gedney, R.T., Schertler, R. and Gargione, F. (2000) *Advanced Communications Technology Satellite: Insider's Account of the Emergence of Interactive Broadband Services in Space*, SciTech, New Jersey.

Gedney, R.T., Schertler, R. and Gargione, F. (2000) *Advanced Communication Technology*, SciTech, New Jersey.

Geoffrey, L.E. *Communication Services via Satellite*, Butterworth Heinemann, Oxford.

Inglis, A.F. (1997) *Satellite Technology: An Introduction*, Butterworth Heinemann, Massachusetts.

Kadish, J.E. (2000) *Satellite Communications Fundamentals*, Artech House, Boston, Massachusetts.

Maral, G. and Bousquet, M. (2002) *Satellite Communication Systems: Systems, Techniques and Technology*, John Wiley & Sons, Ltd, Chichester.

Pattan, B. (1993) *Satellite Systems: Principles and Technologies*, Van Nostrand Reinhold, New York.

Perez, R. (1998) *Wireless Communications Design Handbook: Space Interference*, Academic Press, London.

Richharia, M. (1999) *Satellite Communication Systems*, Macmillan Press Ltd.

Richharia, M. (2001) *Mobile Satellite Communications: Principles and Trends*, Addison Wesley.

Sherrif, R.E. and Hu, Y.F. *Mobile Satellite Communication Networks*, John Wiley & Sons, Ltd, Chichester.

Internet Sites

1. http://www.satsig.net/linkbugt.htm
2. http://engr.nmsu.edu/~etti/spring97/communications/nsn/linkbudget.html
3. http://en.wikipedia.org/wiki/Thermal_noise
4. http://www.satsig.net/noise.htm
5. http://homes.esat.kuleuven.be/~cuypers/satellite_noise.pdf
6. http://www.tutorialsweb.com/satcom/link-power-budget/uplink-noise-ratio.htm

Glossary

Anik-F2: A high throughput communications satellite from Telesat Canada

Antenna gain-to-noise temperature (G/T) ratio: This is usually defined with respect to the Earth station receiving antenna and is an indicator of the sensitivity of the antenna to the downlink carrier signal from the satellite. It is the figure-of-merit used to indicate the performance of the Earth station antenna and low noise amplifier combined to receive weak carrier signals

Antenna noise temperature: This is a measure of noise entering the receiver via the antenna. The noise temperature of the antenna can be computed by integrating contributions of all the radiating bodies whose radiation lies within the directional pattern of the antenna

Astra-2E: Communications satellite from Astra Communications

Broadcast Satellite Services (BSS): This refers to the satellite services that can be received at many unspecified locations by relatively simple receive-only Earth stations

Cross-polarization discrimination: This is defined as the ratio of power received by the antenna in principal polarization to that received in orthogonal polarization from the same incident signal

Cross-polarization interference: This occurs in frequency re-use satellite systems. It occurs due to coupling of energy from one polarization state to the other orthogonally polarized state in communications systems that employ orthogonal linear polarizations (horizontal and vertical polarization) and orthogonal circular polarizations (right-hand circular and left hand circular). The coupling of energy from one polarization state to the other polarization state takes place due to finite values of cross-polarization discrimination of the Earth station and satellite antennas and also by depolarization caused by rain, particularly at frequencies above 10 GHz

Fixed Satellite Services (FSS): This refers to the two-way communication between two Earth stations at fixed locations via a satellite

Free-space loss: This is the loss of signal strength due to the distance from the transmitter. While free space is a theoretical concept of space devoid of all matter, in the present context it implies remoteness from all material objects or forms of matter that could influence propagation of electromagnetic waves

High throughput satellite (HTS): A class of communication satellites that provide at least twice the total throughput of a classical fixed satellite service satellite from the same allocated spectrum

HYLAS-2: High throughput communications satellite from Avanti Communications

Intermodulation interference: This is caused due to generation of intermodulation products within the satellite transponder as a result of amplification of multiple carriers in the power amplifier, which is invariably a TWTA

Johnson noise: Another name for thermal noise or white noise

KA-SAT: Eutelsat's high throughput satellite

Link budget: This is a way of analysing and predicting the performance of a microwave communication link for given values of vital link parameters that contribute to either signal gain or signal loss. It is the algebraic sum of all gains and losses expressed in decibels when travelling from the transmitter to the receiver

Mobile Satellite Services (MSS): This refers to the reception by receivers that are in motion, like ships, cars, lorries, and so on.

Multiple spot beam technology: A concept in broadband satellite communications that allows frequency reuse to maximize bandwidth capacity and minimize required frequency allocation

Noise figure: It is defined as the proportion of the signal-to-noise ratio at the input of the amplifier to the signal-to-noise ratio at its output

Noise temperature: This is just another way of expressing the noise performance of a device in terms of its equivalent noise temperature. It is the temperature of a resistance that would generate the same noise power at the output of an ideal (i.e. noiseless) device as that produced at its output by an actual device when terminated at its input by a noiseless resistance, that is a resistance at absolute zero temperature

Polarization rotation: When an electromagnetic wave passes through a region of high electron content, such as ionosphere, the plane of polarization of the wave is rotated due to interaction of the electromagnetic wave and the Earth's magnetic field. The angle through which the plane of polarization rotates is directly proportional to the total electron content of the ionosphere and inversely proportional to the square of the operating frequency. It also depends upon the state of the ionosphere, time of the day, solar activity, direction of the incident wave, etc. The directions of polarization rotation are opposite for transmit and receive signals

Scintillation: This is simply the rapid fluctuation of the signal amplitude, phase, polarization or angle of arrival. In the ionosphere, scintillation occurs due to small scale refractive index variations caused by local ion concentration. The scintillation effect is inversely proportional to the square of the operating frequency and is predominant at lower microwave frequencies, typically below 4 GHz

Spaceway-3: A high throughput satellite from EchoStar Corporation

Thaicom-4: Broadband communications satellite designed for high speed internet/multimedia services

Thermal noise: Thermal noise is generated in any resistor or resistive component of any impedance due to the random motion of molecules, atoms and electrons. It is called thermal noise as the temperature of a body is the statistical RMS value of the velocity of motion of these particles. It is also called white noise as, due to the randomness of the motion of particles, the noise power is evenly spread over the entire frequency spectrum

ViaSat-1: High capacity communications satellite from ViaSat

White noise: Another name for thermal noise or Johnson noise

Yahsat-Y1B: Communications satellite from AlYahsat

8
Earth Station

Three essential elements of any satellite communication network or system include the Earth segment, the space segment and the up/down link between the space segment and the Earth segment. The space segment, mainly in terms of the different hardware components that constitute the satellite and related topics, was discussed in detail earlier in Chapter 4 on *Satellite Hardware*. Chapter 7 titled *Satellite Link Design Fundamentals* addressed different topics relevant to the design of an optimum satellite link. This chapter comprehensively covers different subsystems that make up a typical satellite Earth station and the key factors governing its design. Beginning with a brief introduction to the Earth station in terms of its role and significance in the overall satellite communication network, the chapter goes on to discuss different types of Earth stations along with their architecture and different subsystems constituting an Earth station. Key performance parameters and other factors governing the design of an Earth station are also discussed. The chapter concludes in overview of some of the major Earth stations around the world in terms of available infrastructure, range of services offered and so on. After reading the chapter you will learn the following:

- The role of an Earth station in overall satellite communication set-up
- Types of Earth station with reference to size and complexity and type of service
- Earth station architecture
- Design considerations for an Earth station
- Earth station subsystems and function of each subsystem
- Earth station figure-of-merit
- Satellite tracking methodologies
- Services offered by some of the major Earth stations around the world

8.1 Earth Station

An Earth station is a terrestrial terminal station mainly located on the Earth's surface. It could even be airborne or maritime. Those located on the Earth's surface could either be

Satellite Technology: Principles and Applications, Third Edition. Anil K. Maini and Varsha Agrawal.
© 2014 John Wiley & Sons, Ltd. Published 2014 by John Wiley & Sons, Ltd. Companion Website: www.wiley.com/go/maini3

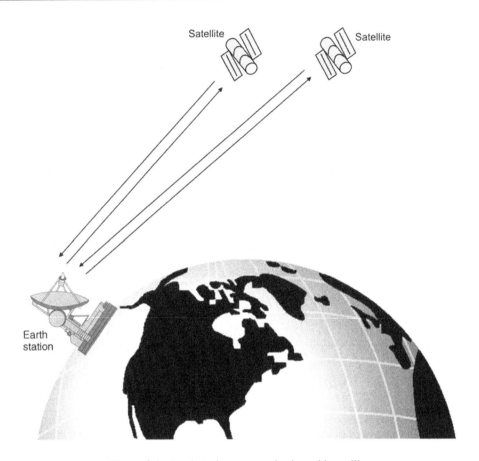

Figure 8.1 Earth station communicating with satellites

fixed or mobile. The Earth station is intended for communication with one or more manned or unmanned space stations as shown in Figure 8.1 or with one or more terrestrial stations of the same type via one or more reflecting satellites or other objects in space as depicted in Figure 8.2. In most of the applications related to communication satellites, Earth stations transmit to and receive from satellites. In some special applications, the Earth stations only transmit to or receive from satellites. Receive-only Earth station terminals are mainly of relevance in the case of broadcast transmissions. Transmit-only Earth station terminals are relevant to data gathering applications.

Major subsystems comprising an Earth station include (a) *transmitter system* whose complexity depends upon the number of different carrier frequencies and satellites simultaneously handled by the Earth station; (b) *receiver system* whose complexity again depends upon the number of frequencies and satellites handled by the Earth station; (c) *antenna system* that is usually a single antenna used for both transmission and reception with a multiplex arrangement to allow simultaneous connection to multiple transmit and receive chains; (d) *tracking system* to ensure that the antenna points to the satellite; (e) *terrestrial interface equipment*; (f) *primary power* to run the Earth station and (g) *test equipment* required for routine maintenance of the Earth station and terrestrial interface.

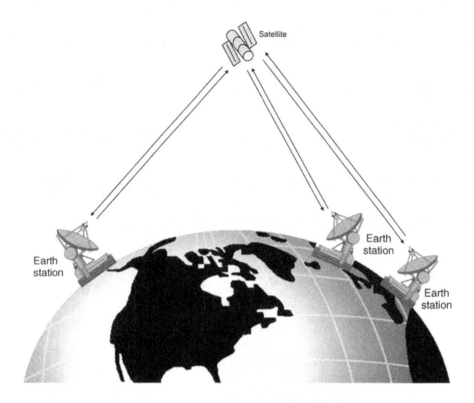

Figure 8.2 Earth station communicating with another Earth station

Earth station design is mainly governed by the type of service to be provided such as fixed satellite service (FSS), broadcast satellite service (BSS), mobile satellite service (MSS) etc.; quality of service to be provided mainly dictated by Earth station G/T; type of communication requirements such as telephony, data, television etc.; international regulations; cost considerations and site constraints. The Earth station is characterized by frequency band (6/4 GHz, 14/12 GHz etc.), polarization (linear, circular etc.), antenna diameter, effective isotropic radiated power ($EIRP$), G/T, receive antenna gain, modulation type, access method (FDMA, TDMA etc.) and so on.

8.2 Types of Earth Station

Earth stations are generally categorized on the basis of type of services or functions provided by them though they may sometimes be classified according to the size of the dish antenna. Based on the type of service provided by the Earth station, they are classified into the following three broad categories.

1. Fixed Satellite Service (FSS) Earth Stations
2. Broadcast Satellite Service (BSS) Earth Stations
3. Mobile Satellite Service (MSS) Earth Stations

Earth stations are also sometimes conveniently categorized into three major functional groups depending upon their usage. These categories are the following.

1. Single function stations
2. Gateway stations
3. Teleports

Each of the above mentioned types is briefly described in the following paragraphs.

8.2.1 Fixed Satellite Service (FSS) Earth Station

Under the group of FSS Earth stations, we have the large Earth stations ($G/T \cong 40$ dB/K) (Figure 8.3), medium Earth stations ($G/T \cong 30$ dB/K), small Earth stations ($G/T \cong 25$ dB/K), very small terminals with transmit/receive functions ($G/T \cong 20$ dB/K) (Figure 8.4) and very small terminals with receive only functions ($G/T \cong 12$ dB/K) (Figure 8.5).

Figure 8.3 Large Earth station

Fixed satellite service (FSS) is a term that is mainly used in North America. The service involves the use of geostationary communication satellites for telephony, data communications and radio and television broadcast feeds. FSS satellites operate in either the C band (3.7 GHz to 4.2 GHz) or the Ku band (11.45 GHz to 11.7 GHz and 12.5 GHz to 12.75 GHz in Europe, and 11.7 GHz to 12.2 GHz in the USA).

FSS satellites operate at relatively lower power levels as compared to Broadcast Satellite Service (BSS) satellites and therefore consequently require a much larger dish. Also, FSS satellite transponders use linear polarization as compared to circular polarization employed by BSS satellite transponders.

Figure 8.4 Very Small terminal (Transmit/Receive)

8.2.2 Broadcast Satellite Service (BSS) Earth Stations

Under the group of BSS Earth stations, we have large Earth stations ($G/T \cong 15$ dB/K) used for community reception and small Earth stations ($G/T \cong 8$ dB/K) used for individual reception. Technically, broadcast satellite service or BSS as it is known by the International Telecommunications Union (ITU) refers only to the services offered by satellites in specific frequency bands. These frequency bands for different ITU regions include 10.7 GHz to 12.75 GHz in ITU region-1 (Europe, Russia, Africa), 12.2 GHz to 12.7 GHz in ITU region-2 (North and South America) and 11.7 GHz to 12.2 GHz in ITU region-3 (Asia, Australia). ITU adopted an international BSS plan in the year 1977. Under this plan, each country was allotted specific frequencies for use at specific orbital locations for domestic services. It is also known by the name of Direct Broadcast Service or DBS or more commonly as Direct-to-Home or DTH. The term DBS is often used interchangeably with DTH to cover both analog and digital video and audio services received by relatively small dishes.

Figure 8.5 Very small terminal (Receive only)

8.2.3 Mobile Satellite Service (MSS) Earth Stations

Under the group of MSS Earth stations, we have the large Earth stations ($G/T \cong -4$ dB/K), medium Earth stations ($G/T \cong -12$ dB/K) and small Earth stations ($G/T \cong -24$ dB/K). While both large and medium Earth stations require tracking, small MSS Earth stations are without tracking equipment.

Satellite phone is the most commonly used mobile satellite service. It is a type of mobile that connects to satellites instead of terrestrial cellular sites. Mobile satellite services are provided both by the geostationary as well as low Earth orbit satellites. In the case of the former, three or four satellites can maintain near continuous global coverage. These satellites are very heavy and therefore very expensive to build and launch. Geostationary satellite based mobile services also suffer from noticeable delay while making a telephone call or using data services.

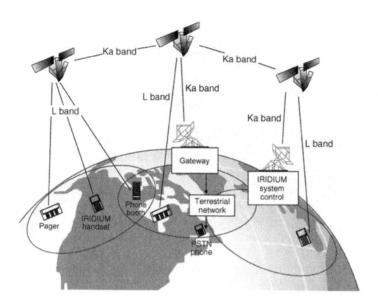

Figure 8.6 Iridium system

Yet another disadvantage of geostationary satellite system is frequent absence of line-of-sight between the satellite and the phone due to obstacles present in-between the two.

The disadvantages of the geostationary satellite system are overcome in Low Earth Orbit (LEO) satellite systems. In the case of LEO satellite systems, an obstacle would block the satellite access only for a short time until another satellite passes overhead. The major advantage of LEO-satellites-based communication systems is worldwide wireless coverage with no gaps. However, a constellation of LEO satellites would be required to maintain uninterrupted coverage. Iridium (Figure 8.6) and Globalstar are the two major LEO satellite systems offering mobile satellite services. Globalstar uses 44 satellites with the orbital inclination of the satellites being 52°. It may be mentioned here that the polar regions are not covered by the Globalstar constellation. Iridium operates 66 satellites orbiting in polar orbits. Radio links are used between the satellites in order to relay data to the nearest satellite connected to the Earth station.

8.2.4 Single Function Stations

Single function stations are characterized by a single type of link to a satellite or a satellite constellation. These stations may be transmit-only, receive-only or both. Some common examples of single function stations include television receive-only (TVRO) terminals used for TV reception by an individual (Figure 8.7), satellite radio terminals, receive-only terminals used at a television broadcast station to pick up contribution feeds, two-way VSAT terminals used at retail stores for point-of-sale communications with the corporate hub, hand-held satellite telephone terminals designed to work with a single satellite constellation and many more.

Figure 8.7 TVRO terminal

8.2.5 Gateway Stations

Gateway stations serve as an interface between the satellites and the terrestrial networks and also serve as transit points between satellites. These stations are connected to terrestrial networks by various transmission technologies, both wired such as coaxial cables, optical fibres and so on, and wireless such as microwave towers. Unlike single function Earth stations where it is just up-linking and down-linking operations that comprise the core activity, in the case of gateway stations, signal processing is the major activity.

A gateway station receives a large variety of terrestrial signals at any given time. These include telephone signals, television signals, and data streams and so on. These signals come in different formats; use various levels of multiplexing and telecommunication standards. A lot of signal manipulation activities therefore need to be carried out on these signals before they are routed to the intended satellite. There are both independent as well as satellite system owner's gateway stations. Antennas used at gateway stations working with a specific satellite system need to be designed and manufactured in accordance with the standards promulgated by the satellite fleet owner. Type approved Earth station equipment that is a particular satellite system specific is available from many manufacturers.

8.2.6 Teleports

Teleport is a type of gateway station operated by firms that are usually not a part of a specific satellite system. Teleports are useful for those companies whose not-too-high requirement of satellite connectivity does not justify having their own dishes. They are also useful for business houses located in crowded places inhibiting line-of-sight to the satellite of interest due to the close proximity of another tall building or some other obstacle. Teleports are usually located on the outskirts of the city and the connectivity from the subscriber company to the teleport station is usually provided through a hub. All subscribers are linked to the hub and the hub in turn is connected to the teleport through a fibre-optic or a microwave link.

Modern teleport stations are versatile and often have a wide range of dishes conforming to the standards of many satellite operators so as to be able to offer a wide range of services to the subscribers. The services offered by teleport stations typically include format conversion, encryption, production and post production, turn-around services and even leasing transportable uplinks for temporary events.

8.3 Earth Station Architecture

The major components of an Earth station include the *RF section*, the *baseband equipment* and the *terrestrial interface*. In addition, every Earth station has certain support facilities such as power supply unit with adequate back-up, monitoring and control equipment and thermal and environment conditioning unit (heating, air-conditioning and so on). Though the actual architecture of an Earth station depends on the application; the block schematic arrangement of Figure 8.8 is representative of a generalized Earth station.

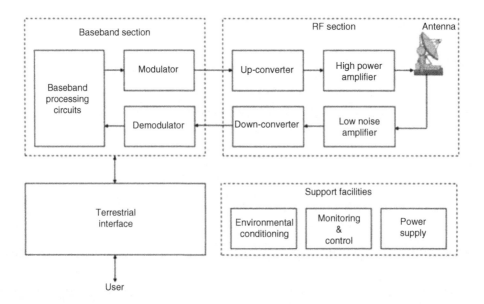

Figure 8.8 Block schematic arrangement of a generalized Earth station

The *RF section* as shown in the block schematic arrangement of Figure 8.8 mainly comprises the antenna subsystem, the up-converter and the high power amplifier (HPA) in the up-link channel and the antenna subsystem, low noise amplifier (LNA) and the down-converter in the down-link channel. In the case of an Earth station being a major hub of a network or if service reliability were a major concern, equipment redundancy is used in the RF section. RF section interfaces with the modem subsystem of the baseband section. The job of up-converter in the up-link channel is to up-convert the baseband signal to the desired frequency. The up-converted signal is then amplified to the desired level before it is fed to the feed system for subsequent transmission to the intended satellite. Similarly, a low noise amplifier amplifies the weak signals received by the antenna. The amplified signal is then down-converted to the intermediate frequency level before it is fed to the modem in the baseband section. The antenna feed system provides the necessary aperture illumination, introduces the desired polarization and also provides isolation between the transmitted and the received signals by connecting HPA output and LNA input to the cross-polarized ports of the feed.

The *baseband section* performs the modulation/demodulation function with the specific equipment required depending upon the modulation technique and the multiple access method employed. For example, in the case of a two-way digital communication link, the baseband section would comprise of a digital modem and a time division multiplexer. The baseband section input/output is connected to the terrestrial network through a suitable interface known as *terrestrial interface*. It may be connected directly to the user in some applications. The terrestrial network could be a fibre optic cable link or a microwave link or even a combination of the two. In addition to the three abovementioned components of an Earth station, every Earth station has support facilities such as tracking, control and monitoring equipment, power supply with back-up and environmental conditioning unit.

The complexity of Earth station architecture depends upon the application. For example, a TVRO Earth station would be far less complex than a FSS Earth station interconnecting large traffic nodes. Figure 8.9 shows the detailed block schematic of a typical large FSS Earth station. Redundancy of equipment as outlined earlier is evident in the RF and the baseband sections. The diagram shown is typical of the Earth station used in the INTELSAT network.

Figure 8.10 shows the block schematic of a typical VSAT remote terminal showing both the outdoor and the indoor units along with the dish antenna. The outdoor unit is typically of the size of a shoe box or even smaller and contains different subsystems of the RF section. The dish antenna is typically 0.55 to 2.4 metre in diameter. The indoor unit, typically of the size of a domestic video recorder, contains different subsystems of the baseband section. These include modulator and demodulator, multiplexer and demultiplexer and user interfaces.

8.4 Earth Station Design Considerations

Design of an Earth station is generally a two-step process. The first step involves identification of Earth station requirement specifications, which in turn govern the choice of system parameters. The second step is about identifying the most cost effective architecture that achieves the desired specifications.

Requirement specifications affecting the design of an Earth station include type of service offered (Fixed satellite service, Broadcast satellite service or Mobile satellite service), communication requirements (telephony, data, television and so on), required base band quality at the destination, system capacity and reliability. Major system parameters relevant to Earth station

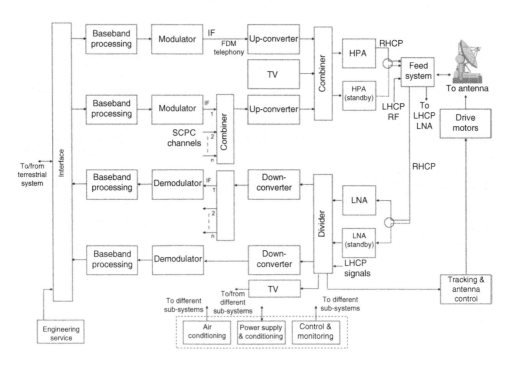

Figure 8.9 Block schematic of a typical large FSS Earth station

design include transmitter *EIRP* (Effective Isotropic Radiated Power), receiver figure-of-merit (G/T), system noise and interference and allowable tracking error.

When it comes to designing a satellite communication system, it is always advisable to minimize the overall system costs including both development as well as recurring costs of the Earth and space segments. A trade-off is always possible between the two where the cost of one segment can be reduced at the cost of the other. That is, cost incurred on the Earth station could be reduced by having a more expensive space segment. According to the most fundamental economic rule of satellite telecommunications, every dollar spent on the space segment gets divided by the number of potential users on the ground whereas every dollar spent on the user terminal gets multiplied by the same number. This leads to the practice of designing less expensive user terminals and more expensive satellites, a trend that started with advent of geostationary satellites way back in 1960s and continued for more than three decades. Several trade-offs are possible in Earth station design optimization, which are discussed in detail in Section 8.4.2. However, as we shall see, these trade-offs are subjected to some technical and regulatory constraints, which are also briefly outlined during the discussion.

8.4.1 Key Performance Parameters

Key performance parameters governing Earth station design include the *EIRP* (Effective or Equivalent Isotropic Radiated Power) and the figure-of-merit (G/T). While the former is a transmitter parameter, the latter is indicative of receiver performance in terms of sensitivity and the quality of the received signal.

Earth Station Design Considerations

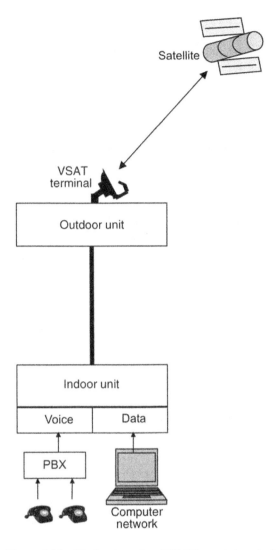

Figure 8.10 Block schematic of VSAT remote terminal

Effective (or Equivalent) Isotropic Radiated Power (*EIRP*). *EIRP* gives the combined performance of the high power amplifier (HPA) and the transmitting antenna. It is given by the product of the power output of HPA at the antenna and the gain of the transmitting antenna. Expressed in decibels, *EIRP* is the sum of the power output of HPA in dB and the gain of transmitting antenna in dB. If a particular HPA-transmitting antenna combine had an *EIRP* of 60 dBW, it would imply that the RF power radiated by the antenna is the same as that radiated by an isotropic radiator in that direction when fed with million times more power at its input.

EIRP is defined for both Earth station transmitting antenna as well as satellite transmitting antenna. It is important to note that *EIRP* is always measured at the antenna. When we see a footprint map with *EIRP* numbers for a given transponder on a satellite, these numbers are indicative of the amount of power sent down to the Earth station and measured as it left the satellite's down-link dish. Some satellite operators have the practice of taking space loss in account while publishing the satellite footprint maps for their users. They prefer to give the signal strength as is received on ground thus correcting for the space loss at the frequency of operation. This number is known as Illumination Level and is given by (*EIRP* − Space loss). Some operators prefer to specify received power per unit bandwidth. The unit of bandwidth is typically taken as 4 kHz, which is the bandwidth of a typical analog telephony channel. In that case, the new value called Power Flux Density (PFD) is given by (*EIRP* − Space loss − Bandwidth). It may be mentioned here that the PFD is specified in the decibels scale.

Receiver Figure-of-merit (G/T). Receiver figure-of-merit is indicative of how the receiving antenna performs together with the receiving electronics to produce a useful signal. While the *EIRP* gives the performance of the transmitting antenna and HPA combination; receiver figure-of-merit, tells us about the sensitivity of the receiving antenna and the Low Noise Amplifier (LNA) combine to weak received signals. As it is effectively a measurement of the sensitivity of the receiving antenna to weak signals, the larger the value of receiver figure-of-merit, the better it is. The response of the receiving system to weak signals is largely governed by the receiving system gain and the overall system noise. The figure-of-merit is therefore defined by a parameter called G/T ratio, which is the ratio of receiving antenna gain to system noise temperature. G/T is expressed in dB/K. G/T of the Earth station may be enhanced by increasing the receiving antenna gain or lowering the noise temperature or both. For any practical communication link, *EIRP* of the satellite transmitting antenna and the G/T of the Earth station receiving antenna and the *EIRP* of the Earth station transmitting antenna and G/T of the satellite receiving antenna have to work together to get the desired results. A poorer G/T necessitates a higher *EIRP* and vice versa. Both *EIRP* and G/T were discussed at length along with illustrative examples in Chapter 7.

8.4.2 Earth Station Design Optimization

As outlined earlier, the transmitter *EIRP* and receiver G/T together dictate the performance of the communication system and therefore one can be traded off against the other during the design optimization process. In the early days of development of satellite technology, available *EIRP* from satellites was pretty low, which made complex and expensive Earth stations a necessity. In those days, Earth station antennas were several tens of metres in diameter and cost a few million US dollars apiece. Current trend is to minimize Earth station complexity at the cost of a complex space segment. It is more so for applications that involve a large user population such as direct broadcast, business use, mobile communication and so on.

Possible trade-offs can be best understood by resorting to expression for Earth station G/T. The generalized expression for G/T is given by equation 8.1.

$$G/T = C/N_o - EIRP + \left(L_p + L_m\right) + k \tag{8.1}$$

Where C/N_o, *EIRP*, L_p, L_m and k are carrier-to-total noise power spectral density, satellite's effective isotropic radiated power, path loss, link margin and Boltzmann constant (in dBs) respectively. For a minimal cost Earth station, G/T should be minimized. This can be possible by either using relatively higher *EIRP* in the satellite or being able to afford a lower carrier-to-noise ratio or both. For desired base band quality at the receiver, this can be achieved by using modulation schemes that are more immune to noise. In the case of digital base band, coding allows a further reduction in G/T.

Other factors governing Earth station complexity and hence its cost include the Earth station *EIRP*, antenna tracking requirements, traffic handling capacity and terrestrial interface requirements. In addition, there are international regulatory issues and technical constraints that drive the optimization process.

In the early days, International Telecommunications Union (ITU) had put certain limitations on the transmitted *EIRP* of the FSS satellites sharing their frequency bands with terrestrial systems in order to allow them to co-exist. For applications such as direct broadcast, mobile communications etc. where a small size terminal is a requirement, limiting the satellite *EIRP* would put a lower limit on the diameter of the dish antenna. This implies that G/T cannot be reduced below a certain value. Even if G/T were reduced by using a smaller antenna, reduction in size would increase antenna side lobes to undesired levels, which would further lead to more interference to and from adjacent satellite systems. This has been overcome by having exclusive frequency allocations for these services, thus permitting relatively much higher *EIRP* for the satellites.

The satellite *EIRP* is also limited by the DC power available on the satellite, maximum power that can be generated by the high power amplifiers on board the satellite and the practical constraints imposed on the satellite antenna diameter limiting the antenna gain. Also, for a given antenna size, gain reduces with decrease in operational frequency. That is why, satellite *EIRP* limitation is more acute in L-band used for mobile communications.

Having decided on the *EIRP* and G/T values, the next obvious step is to choose an optimum configuration of the antenna, high power amplifier and the low noise amplifier to achieve the desired values. Specified *EIRP* and G/T may be obtained by any of the possible options. A small size antenna, which would be low cost, and a relatively low noise LNA, which would be expensive, is one option. Another would be the use of large size antenna and LNA with a higher noise figure. Antenna size also affects the *EIRP* as a small size antenna may require a prohibitively large HPA.

8.4.3 Environmental and Site Considerations

It is important to consider a number of environmental and locational factors while making a decision on the site of an Earth station. Environmental parameters of interest include external temperature and humidity, rainfall and snow, wind conditions, likelihood of earthquakes, corrosive conditions of the atmosphere and so on. Careful site selection can take care of the ill effects of some but not all of these factors.

Minimizing radio frequency interference (RFI) and electromagnetic interference (EMI) is another requirement. RFI and EMI produced by the Earth station can cause interference to other RF installations. Also, RFI and EMI from external sources can adversely affect the Earth station performance. It is usually necessary to carry out a radio frequency survey at various possible sites before a final choice is made on the Earth station location.

An essential requirement is to have a clear line-of-sight to the satellites of interest. Availability of sufficient space for the Earth station equipment, easy transportation to the Earth station and reliable electrical power are the other requirements.

Though all efforts are made to take into account the abovementioned factors while choosing a suitable site for the Earth station; it is important that the satellite operators specify all possible environmental factors and site constraints to potential manufacturers of the Earth station equipment. Also, the manufacturers should build into the design of Earth station equipment the ability to operate reliably under specified environmental and interference conditions.

8.5 Earth Station Testing

Having chosen the Earth station equipment, it is important to ensure that the equipment would not only meet the specified requirements of the intended Earth station; it is also necessary to ensure that the Earth station would not cause any problems either to other users of the satellite or to any adjacent satellites. This is achieved by performing different levels of testing, which begins with testing at component or unit level followed up by subsystem level testing. These two levels of testing form part of Earth station hardware and software commissioning process and therefore precede any integrated testing of the overall Earth station. Overall Earth station testing also includes what is called line-up testing, which involves checking the performance of the Earth station in conjunction with the other Earth stations, which the newly commissioned Earth station is intended to work with.

8.5.1 Unit and Subsystem Level Testing

Unit or *component level testing* is usually done at the manufacturer's premises and the test data is made available to the subsystem designer making use of the components. The user may choose to witness the tests, if the component happens to be of a new design.

In *subsystem* or *equipment level testing*, different subsystems are comprehensively tested for their electrical, mechanical and environmental specifications. The critical tests are witnessed by the user. Test data generated as a part of comprehensive testing is usually supplied to the user before some selected tests are repeated in the presence of the user. The selected tests carried out in the presence of the user are repeated once the equipment or subsystem is installed on the site.

8.5.2 System Level Testing

System level testing is carried out after subsystem testing and integration has been completed. In cases where the complete system has been ordered on a single supplier, as many subsystems as possible are integrated at the premises of the supplier and the performance of the integrated set verified. The rest of the integration job is carried out on site followed up by full system testing for a wide range of parameters. These tests are also called acceptance tests. Tests are carried out to verify that the system meets all the performance specifications and also all the mandatory requirements of the satellite system to be used. The system is also tested for its adherence to international regulatory standards and fulfillment of desired base band signal quality requirement. A wide range of transmit and receive tests are carried out to meet the abovementioned requirements. These tests fall into two broad categories namely the mandatory tests and the additional tests. These tests are briefly described below.

8.5.2.1 Mandatory Tests

Mandatory tests include measurements of (a) Transmit cross-polarization isolation (b) Receiver figure-of-merit (c) *EIRP* stability and (d) Spectral shape. Each one of these is briefly covered in the following paragraphs.

Transmit Cross-polarization Isolation Measurement. Transmit cross-polarization isolation measurement is performed to guarantee that the power level of the cross-polarized component is either nil or within the tolerance limit so as not to cause any significant interference to other users. This test is typically performed only on-axis. If required, full transmit cross-polarization isolation measurement may be performed.

Figure 8.11 shows the simplified schematic arrangement used for measuring transmit cross-polarized isolation for both linearly as well as circularly polarized antennas. As outlined earlier, the test is usually performed on-axis. Under the control of the monitoring station, the antenna under test (AUT) is initially driven to transmit a relatively lower level carrier at the test frequency. This is used for the purpose of boresighting by observing the change in the co-polarized carrier power level as the AUT is driven off boresight in azimuth and elevation. Once the AUT has been boresighted, the next step is optimization of the polarization angle of the antenna under test by observing the power level of the cross-polarized component as the AUT rotates the feed. This is done to determine the polarization angle corresponding to the minimum power level of the cross-polarized component.

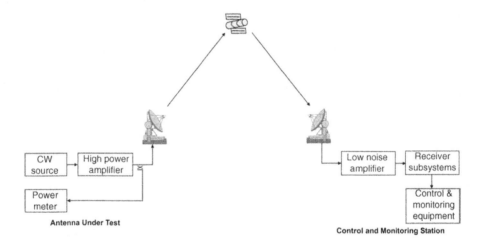

Figure 8.11 Schematic arrangement of transmit cross-polarization isolation measurement

The on-axis transmit cross-polarization isolation is then computed to determine whether or not the measured value of cross-polarization isolation meets the required on-axis transmit cross-polarization isolation specification. The above procedure is for linearly polarized antennas. The test set-up and measurement procedure for circularly polarized antennas is identical to that of linearly polarized antennas except that the polarization angle optimization is not required in the case of circularly polarized antennas.

Receiver Figure-of-merit Measurement. Figure 8.12 shows the schematic arrangement of a possible test set-up that can used for measurement of receiver figure-of-merit

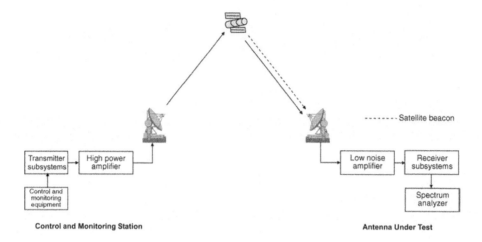

Figure 8.12 Schematic arrangement of test set-up for measurement of receiver G/T

that is G/T. The underlying principle for measurement of G/T is as follows. The downlink carrier to noise spectral density C/N_o and the corresponding satellite *EIRP* are measured. G/T is obtained by rearranging the downlink equation. Though the method is simple; it is prone to in accuracies due to its susceptibility to variations the in atmospheric loss.

The antenna under test (AUT) measures the received power level of either an unmodulated beacon, or a test carrier. The downlink *EIRP* of the unmodulated beacon or the test carrier is also measured. The receive system noise contribution is then measured by steering the antenna off the spacecraft and measuring the noise floor. The difference between the two measurements is the downlink (C + N)/N ratio. From this the downlink, C/N_o is calculated by taking into account corrections required for the effects of the system thermal noise, spectrum analyser detection non-linearity and noise bandwidth. G/T is then computed by solving the downlink equation.

Another method used for measurement of receiver G/T is the *Gain and System Temperature Method*. This test procedure requires measurement of receiver gain and system noise temperature to obtain G/T. Figure 8.13 shows the test set-up for measurement of receiver gain [Figure 8.13(a)] and system noise temperature [Figure 8.13(b)]. This method leads to more accurate results than are obtained by the spectrum analyser method described above. Receiver gain can be measured either by pattern integration technique or by determination of 3 dB and 10 dB beamwidths. In the case of pattern integration technique, azimuth and elevation narrowband (±5° corrected) patterns are measured. Directive gain of the antenna is then measured through integration of the sidelobe patterns. The receiver gain is then determined by reducing the directive gain by the antenna inefficiencies.

In order to measure the receiver gain using the beamwidth method, the AUT measures the corrected azimuth and elevation 3 dB/10 dB beamwidths. Receiving gain of the antenna can then be computed from known values of the two corrected beamwidths by using formula given in equation 8.2. The equation does not account for gain losses due to insertion loss of the feed

Earth Station Testing

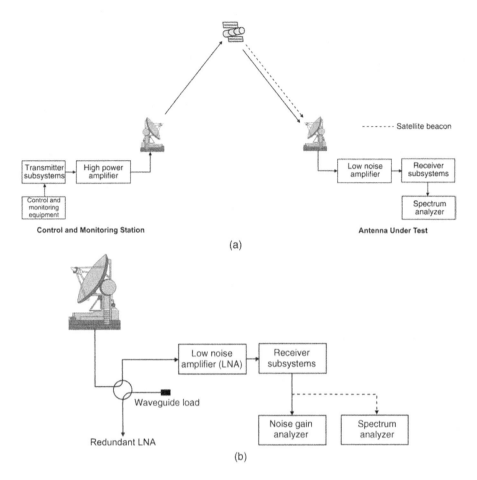

Figure 8.13 Receiver gain and system temperature method (a) Measurement of receiver gain (b) Measurement of system temperature

mechanism and antenna reflector inaccuracies.

$$G = 10\log_{10}\left[1/2\left\{\frac{31000}{(\theta_{3dB} \times \phi_{3dB})}\right\} + \left\{\frac{91000}{(\theta_{10dB} \times \phi_{10dB})}\right\}\right] \quad (8.2)$$

Where,

G = Receiving gain of the antenna (in dBi)
θ_{3dB} = Corrected azimuth 3° beamwidth (in deg)
ϕ_{3dB} = Corrected elevation 3° beamwidth (in deg)
θ_{10dB} = Corrected azimuth 10° beamwidth (in deg)
ϕ_{10dB} = Corrected elevation 10° beamwidth (in deg)

The system noise temperature can be calculated from what is known as Y-factor measurement where Y represents the difference in the noise power when the input to the receive system is terminated in hot and cold loads. Figure 8.13 (b) shows the test set-up. In the above test configuration, a spectrum analyser could be used as a less accurate alternative to the noise gain analyser.

In order to measure the system noise temperature, the noise power is measured for hot and cold load conditions. The antenna under test is pointed towards the clear sky to simulate the cold load condition. An input waveguide load at the ambient temperature provides the hot load. This waveguide load typically forms an integral part of the LNA redundancy switching system. It can be switched in or out using the LNA controller. The difference between the noise power levels measured with hot and cold loads is the Y-factor. The system noise temperature can then be computed with the help of expression given in equation 8.3.

$$T_{SYS} = \left[\frac{(T_{LOAD} + T_{LNA})}{Y} \right] \quad (8.3)$$

Where,

T_{SYS} = System noise temperature (in K)
T_{LOAD} = Noise temperature of waveguide/test load (in K)
T_{LNA} = Noise temperature of LNA (in K)
Y = Y-factor expressed as a ratio

Receiver G/T is then computed using equation 8.4.

$$G/T = G_R - 10 \log_{10} T_{SYS} \quad (8.4)$$

Where,

G/T = Receiver G/T in dB/K
G_R = Receive gain of antenna (in dBi)
T_{SYS} = Receive noise temperature (in K)

Yet another method to determine receiver G/T is to use radio stars. This method too is very accurate but it can be used to measure G/T of antennas having G/T values typically greater than 36 dB/K.

EIRP Stability. The quality of service provided by the system is dependent on the stability of *EIRP*. It is particularly so in the case of high usage transponders and high power uplinks. As the atmospheric effects also contribute to power flux density (PFD) variations at the satellite, the allowable *EIRP* stability figure should be such that the PFD variations caused by *EIRP* instability and atmospheric effects together are within allowable PFD variations at the satellite. The *EIRP* stability of a station transmitting digital services should be better than ± 0.5 dB. The *EIRP* stability, although a mandatory requirement, is usually not measured as part of the Earth station verification tests. It may however be monitored as a part of initial service line-up testing or when there is a reason to believe that *EIRP* instability limit has been exceeded.

Spectral Shape. The spectral shape of the modulated carrier is initially measured during carrier line-up testing and is also measured subsequently on a regular basis. Though there are no specific measurements to be made by the users; it is important for the users to note that control on spectral shape and hence the bandwidth is required to avoid undesired interference to other system users.

Earth Station Testing 397

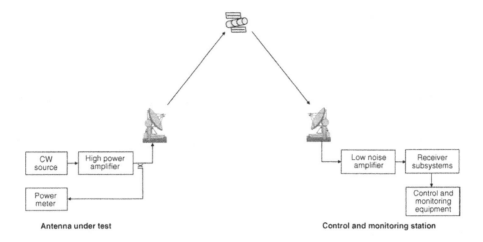

Figure 8.14 Test set-up for measurement of transmit sidelobe pattern

8.5.2.2 Additional Tests

In addition to the mandatory tests described in the previous pages, transmitting and receiving antenna patterns including both co-polarized and cross-polarized patterns may also be measured. These measurements are briefly described in the following paragraphs.

Figure 8.14 shows the test set-up for measurement of transmit sidelobe pattern. The test configuration is the same for measuring both co-polarized and cross-polarized patterns. For measurement of transmit sidelobe pattern, the antenna under test is made to transmit an unmodulated carrier. Co-polarization and cross-polarization components are then measured. Initially, the AUT transmits a low level test carrier usually 15 dB below what is defined in the test plan at the test frequency. The power level is subsequently increased until the nominal power level for the test plan is reached. The AUT is swept over angular displacements, typically ±15° corrected. For antennas employing elevation over azimuth mounts, the azimuth angle needs to be corrected for the elevation angle. Corrected azimuth angle (A_Z') in terms of the azimuth angle from boresight (A_Z) as measured from encoders and the elevation angle (A_{EL}) is given by equation 8.5.

$$A_Z' = 2\sin^{-1}\left[\sin\left(A_Z/2\right) \times \cos\left(A_{EL}\right)\right] \qquad (8.5)$$

It may be mentioned here that all the angles in equation 8.5 are specified in degrees. Figure 8.15 shows the test set-up for measurement of receive sidelobe pattern. In this case, the antenna under test (AUT) receives an unmodulated beacon, or alternatively an unmodulated carrier transmitted from the payload operations centre. Co-polarized and cross-polarized receive azimuth and elevation sidelobe patterns are then measured at the AUT site. Prior to the commencement of the test, the azimuth and elevation tracking velocities of the AUT are measured possibly over the full angular displacement range of the measurement. The antenna is then swept over the required angular displacement (typically ±15° corrected) to measure the receiver sidelobe patterns.

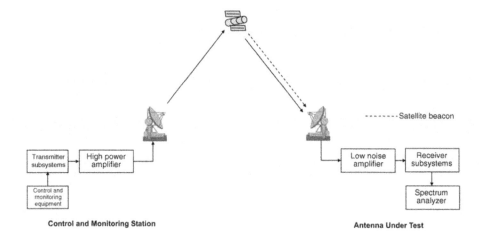

Figure 8.15 Test set-up for receive sidelobe pattern measurement

For antennas employing elevation over azimuth mounts, the azimuth angle needs to be corrected for the elevation angle. Equation 8.5 is used to determine the actual azimuth angle from measured values of azimuth and elevation angles.

8.5.2.3 Line-up test

Line-up testing involves checking the performance of the newly commissioned Earth station *vis-à-vis* other Earth stations it is intended to operate with. Other stations also include the control and monitoring Earth station of the satellite operator, which provides the assistance to carry out these tests. During these tests, the carrier *EIRP* is set to provide the desired carrier-to-noise ratio at the receiving end. The link is cleared for traffic only on completion of line-up tests.

8.6 Earth Station Hardware

Most Earth station hardware can be categorized into one of the three groups namely *RF equipment, IF and baseband equipment* and *terrestrial interface equipment*. Basic functions performed by each one of these equipment classes were briefly outlined in Section 8.3 on Earth station architecture. In the present section, these are described in more detail with focus on individual building blocks constituting these three groups.

8.6.1 RF Equipment

The RF equipment comprises of up-converters, high power amplifiers (HPA) and the transmit antenna in the transmit channel, and the receive antenna, low noise amplifiers (LNA) and down-converters in the receive channel. While the output of HPA feeds the transmit antenna; the receive antenna is connected to the input of the LNA. Transmit and receive antenna

Earth Station Hardware

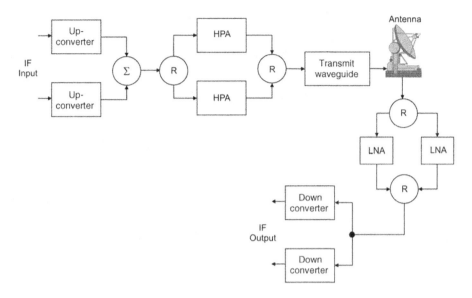

Figure 8.16 Block schematic of the RF portion of the Earth station

functions are almost invariably performed by the same antenna. Figure 8.16 shows the block schematic arrangement of the RF portion of the Earth station equipment.

From the viewpoint of *EIRP* and also Earth station G/T, it is always desirable to have minimal losses in the waveguide/cable connecting the antenna and the HPA output or LNA input. To achieve this, one option is to house the RF section in a separate shelter or cabinet adjacent to the antenna. Another option is to package the uplink and downlink equipment separately. The uplink equipment mainly comprises of the modulator and the up-converter and the downlink equipment has down-converter and the demodulator. Yet another practice prevalent in the case of VSAT and TVRO terminals is to combine the LNA and first stage of the down-converter into a single block known as low noise block (LNB). This has the distinct advantage of offering low noise amplification and down conversion to L-band, which allow them the use of inexpensive coaxial cable to further carry the signal.

8.6.1.1 Antenna

Different types of antenna and their performance parametres of relevance to satellite communications have been discussed earlier in Chapter 4 on satellite hardware. A brief description of antennas of relevance to Earth stations is given in this section.

Different variants of reflector antenna are commonly used as Earth station antenna. These mainly include the prime focus fed parabolic reflector antenna, offset fed sectioned parabolic reflector antenna and cassegrain fed reflector antenna. The prime focus fed parabolic reflector antenna as shown in Figure 8.17 is used for an antenna diameter of less than 4.5 m, more so for receive only Earth stations. An offset fed sectioned parabolic reflector antenna (Figure 8.18) is used for antenna diameters of less than 2 m. Offset feed configuration eliminates the blockage of the main beam due to feed and its mechanical support system and thus improves antenna efficiency and reduces side lobe levels.

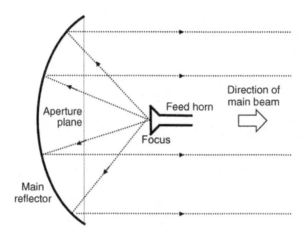

Figure 8.17 Prime focus fed parabolic reflector antenna

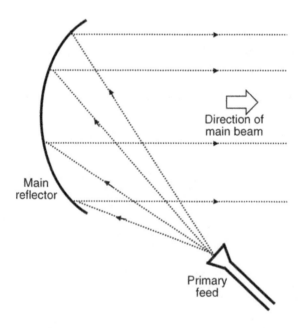

Figure 8.18 Offset fed sectioned parabolic reflector antenna

In a variation of the prime focus fed parabolic reflector antenna, a piece of hook shaped waveguide extending from the vertex of the parabolic reflector is connected to the feed horn. In this case, the low noise block (LNB) is connected to the waveguide behind the parabolic reflector. This allows placement of electronics without causing any obstruction to the main beam in addition to allowing an easy access to it.

Earth Station Hardware

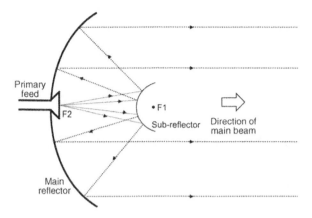

Figure 8.19 Cassegrain antenna

Cassegrain antennas overcome most of the shortcomings of the prime focus fed parabolic reflector antennas. The cassegrain antenna uses a hyperbolic reflector placed in front of the main reflector, closer to the dish than the focus as shown in Figure 8.19. This hyperbolic reflector receives the waves from the feed placed at the centre of the main reflector and bounces them back towards the main reflector. In the case of cassegrain antenna, the front end electronics instead of being located at the prime focus is positioned on or even behind the dish. Offset feed configuration is also possible in case of Cassegrain antenna (Figure 8.20).

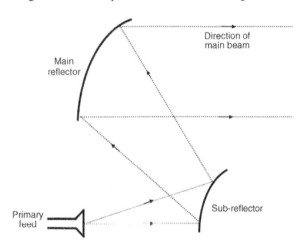

Figure 8.20 Offset fed Cassegrain antenna

Yet another common reflector antenna configuration is the Gregorian antenna [Figure 8.21(a)]. This configuration uses a concave secondary reflector just behind the prime focus. The purpose of this reflector is also to bounce the waves back towards the dish. The front end in this case is located between the secondary reflector and the main reflector. Offset feed configuration is also possible in case of Gregorian antenna [Figure 8.21(b)].

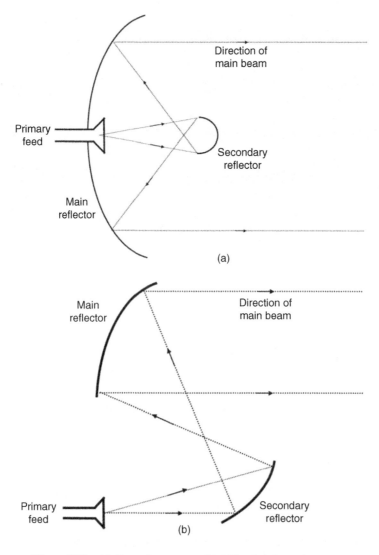

Figure 8.21 (a) Gregorian antenna (b) Offset fed Gregorian antenna

8.6.1.2 High Power Amplifier

EIRP, which is the product of the power output of the high power amplifier (HPA) minus the waveguide losses and gain of transmit antenna, is an important parameter in deciding the uplink performance of the Earth station. To achieve the specified *EIRP* of the Earth station, one could have a combination of moderate output power HPA and a high gain antenna. The other option is to have a relatively higher power output HPA feeding a moderate sized antenna. One could always draw a family of curves for different frequency bands (C, Ku, Ka) showing a variation of HPA power output against antenna diameter for desired value of *EIRP*. Figure 8.22 shows one such family of curves drawn for an *EIRP* of 80 dB. As is evident from the curves, one

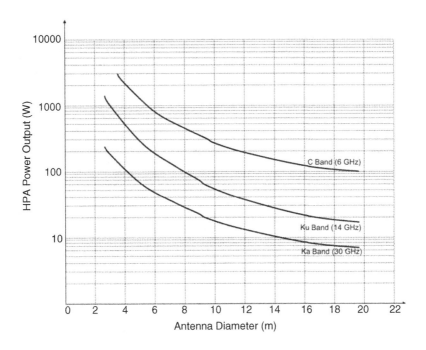

Figure 8.22 HPA power output versus antenna diametre for a given *EIRP*

would need to have a 800 watt HPA for a C-band transponder, if the antenna diameter were to be around 6 m or so. A 10 m antenna on the other hand would need only a 300 watt HPA.

Different types of power amplifiers used in Earth stations include (a) Travelling wave tube (TWT) amplifiers (b) Klystron amplifiers and (c) Solid state power amplifiers (SSPA). SSPA are used for relatively lower power applications while tube based amplifiers are used when the required power levels are high. Klystrons are narrow band devices providing a bandwidth of the order of 40 to 80 MHz that is tunable over a range of 500 MHz or more. Power levels offered are from several hundred watts to few kilowatts. On the other hand, TWTA is a wideband amplifier offering a bandwidth as large as 500 MHz or more and a power level from a few watts to a few kilowatts. However, klystrons are less expensive, simple to operate and easy to maintain. Solid state power amplifiers are comparatively cheaper and more reliable though the power level offered by them is limited as compared to klystrons and TWTAs.

Apart from frequency, power level, linearity and bandwidth, other important characteristics of high power amplifiers include gain, variation of group delay with frequency, noise performance and AM/PM conversion. While variation of group delay with frequency is also a cause of intermodulation components, AM/PM conversion produces intelligible crosstalk and intermodulation noise.

Commonly used amplifier configurations for multi-carrier operation include the single amplifier and multiple amplifier configurations. In the case of single amplifier configuration (Figure 8.23), different carriers are combined before the amplifier and the composite signal is fed to the input of the amplifier. The amplifier is operated in the linear region of its operating characteristics to minimize the inter-modulation noise. In the figure shown, redundant HPA is used to improve the system reliability. It is terminated in a matched load. In the case of

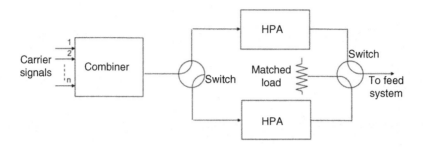

Figure 8.23 Single amplifier HPA configuration

multiple amplifier configuration, each HPA amplifies either a single or a group of carriers as shown in Figure 8.24. Amplified signals are then combined at the output of HPAs. This configuration allows the HPA to be operated near its full power rating, which increases the overall efficiency of the Earth station. However this comes at the cost of additional HPAs.

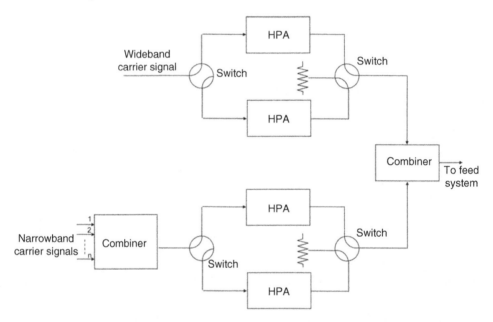

Figure 8.24 Multiple amplifier HPA configuration

8.6.1.3 Up-converters/Down-converters

Up-converters and down-converters are frequency translators that convert the IF used in the modems and baseband equipment to the operating RF frequency bands (C, Ku and Ka) and vice versa. The up-converter translates the IF signal at 70 MHz (or 140 MHz) from the modulator to the operating RF frequency in C or Ku or Ka band as the case may be. The down-converter translates the received RF signal in C or Ku or Ka band into IF signal, which is subsequently

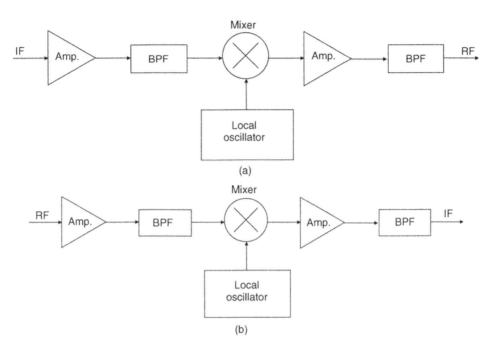

Figure 8.25 Simplified block diagram of single frequency conversion frequency converters (a) up-converter (b) down-converter

fed to the demodulator. Either single or double frequency conversion topologies are used for up-converters and down-converters.

Figures 8.25(a) and (b) respectively show the schematic diagrams of up-converters and down-converters employing single frequency conversion topology. A typical up-converter uses a stage of amplification before the mixer stage. Mixer along with local oscillator (LO) provides frequency conversion. A frequency synthesizer is used for LO so as to be able to generate any frequency within the satellite up-link band. The signal is further amplified after frequency conversion before it is fed to the high power amplifier. A band pass filter at the output of the mixer eliminates LO frequency and its harmonics from reaching the up-link path. Insertion loss in the filter causes a reduction of the effective isotropic radiated power (*EIRP*). The operation of down-converter can be explained on similar lines. Amplification stage provides gain and reduces the noise contribution of mixer and the IF equipment. The frequency synthesizer provides frequency agility in the receive frequency operation.

Double frequency conversion topology employs a two mixer conversion stage. In the case of an up-converter using double conversion, the IF frequency is first up-converted to another intermediate frequency usually in the L-band. The signal is then amplified and fed to the second mixer stage where it is up-converted to the final operational RF frequency band. As outlined in the case of single stage converters, an amplifier precedes the mixer and a band pass filter follows the same. Figure 8.26(a) shows block schematic of a C-band up-converter employing double frequency conversion topology. Figure 8.26(b) shows the arrangement of double frequency conversion topology based down-converter for C-band operation. The diagrams are self explanatory.

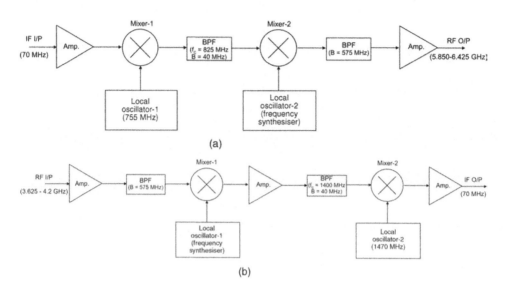

Figure 8.26 Simplified block diagram of double frequency conversion frequency converters (a) up-converter (b) down-converter

8.6.1.4 Low Noise Amplifier (LNA)

While the high power amplifier (HPA) is an important element of the up-link path that together with the transmit antenna gain decides the *EIRP* of the Earth station; the low noise amplifier (LNA) is one of the key components deciding the system noise temperature and hence the figure-of-merit G/T of the Earth station. The design of LNA and the active devices around which the design of a LNA is configured have undergone many changes since the advent of satellite communication. Design of LNA in the early days used to be configured around masers and subsequently parametric amplifiers. Requirements on LNA in those days used to be far more stringent than they are today. This has been made possible due to improvements in antenna efficiency and feed techniques and also increase in the transmit power capability of satellites.

Present day LNAs are configured around either Gallium Arsenide FET (GaAs FET) or High Electron Mobility Transistors (HEMT). These designs are far more compact and reliable than their parametric amplifier counterparts. The uncooled GaAs FET or HEMT based LNAs offer a noise temperature of about 75–170 K and compared with cryogenically cooled parametric amplifiers of early days giving noise temperature of 30–90 K. Table 8.1 gives a performance comparison of different LNA technologies.

There are two variations of low noise amplifier (LNA). In one of the variants, particularly where small size antennas are used such as those for TVRO or small business applications, the low noise amplifier section feeds a single stage down-converter in a single block called low noise block (LNB). Note that the frequency converters integrated into LNAs are all down-converters shifting the received frequency to some lower frequency. This allows use of coaxial cables to transport the signal from the antenna to inside the premises. The output of LNB is a standard IF signal of around 1 GHz frequency. LNB is usually placed on the antenna structure itself and is connected to the feed directly. Figure 8.27 shows a photograph of a DTH dish and a co-located LNB.

Earth Station Hardware

Table 8.1 Performance comparison of LNA technologies

Type of Amplifier	Frequency Range (GHz)	Typical Noise Temperature (K)
Parametric Amplifier (Cooled)	3.7–4.2	30
	11–12	90
Parametric Amplifier (Uncooled)	3.7–4.2	40
	11–12	100
GaAs FET (Cooled)	3.7–4.2	50
	11–12	125
GaAs FET (Uncooled)	3.7–4.2	75
	11–12	170

Figure 8.27 DTH dish and co-located LNB

Another variation of LNA is the LNC ('C' stands for converter). In LNC, the amplifier can typically be tuned to amplify over the entire bandwidth of a single transponder, whatever that bandwidth may be, before it down converts. The basic difference between LNB and LNC lies in the conversion bandwidth. LNB uses a block converter and is capable of handling block of frequencies from different transponders on the satellite. LNC uses the signal from a single transponder.

8.6.2 IF and Baseband Equipment

The nature and complexity of baseband equipment in an Earth station is mainly governed by the range of services offered by it and the requirement specifications that would be needed to provide those services. In the case of large Earth stations such as gateways, this portion of Earth station hardware also involves the largest investment. Important building blocks of IF and baseband equipment of the Earth station hardware include baseband processing circuits, modulator/demodulator (MODEM), multiplexer/ demultiplexer etc.

The architecture of the IF and baseband section depends upon parametres like the modulation/demodulation scheme, multiple access method and so on. For example, in the case of an FDMA station, there must be one modem for each frequency resulting in use of a large number of such units. On the other hand, a TDMA Earth station needs to have only one modem for obvious reasons. However, the bandwidth requirement of the modem in the case of a TDMA station would be much larger than what it would be in the case of an FDMA station.

Figure 8.28 shows the block schematic arrangement of FDMA Earth station capable of providing full duplex digital transmission for multiple carriers. In the arrangement shown, each carrier has its own dedicated modem tuned to a separate frequency in the transponder. The modems interface with the terrestrial network through a TDM multiplexer. Individual channels are combined into a single higher bandwidth channel. Though the arrangement does not depict redundancy; full or partial redundancy is almost invariably provided to maintain high reliability.

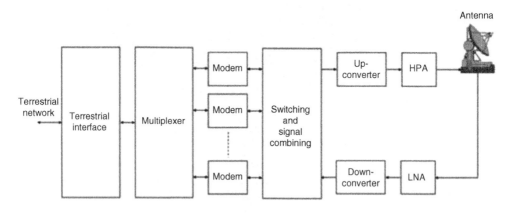

Figure 8.28 Block schematic of a full duplex FDMA digital communication Earth station

Figure 8.29 shows the simplified block schematic arrangement of a typical TDM/ TDMA interactive VSAT terminal showing both the hub site and the remote locations. One can see the use of a single modem. In the case of TDMA, the frequency band occupied by the carrier is shared by several Earth stations on time basis. This implies that there needs to be only a single modem per Earth station. The modem receives bursts of data from different Earth stations in a manner that they do not overlap in time. In the case of CDMA however, different stations transmit on the same frequency simultaneously. Different multiple access techniques are discussed at length in Chapter 6. Modulation, demodulation and multiplexing techniques are also discussed at length in Chapter 5.

Earth Station Hardware

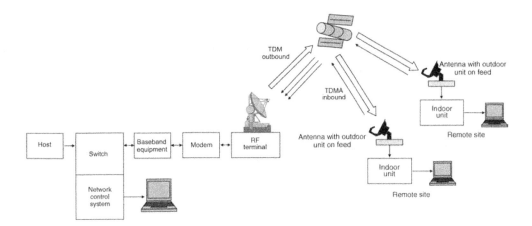

Figure 8.29 Block schematic arrangement of a typical TDM/ TDMA interactive VSAT terminal

8.6.3 Terrestrial Interface

Terrestrial interface is that part of the Earth station that connects the Earth station to the users. Its importance lies in the fact that an improperly designed interface can significantly degrade the quality of service. The nature and complexity of the terrestrial interface depends upon the range of services or functions provided by the Earth station. The interface requirement varies from practically no interface in the case of portable user terminals such as satellite phones to a simple interface in the case of VSAT or TVRO terminals where the Earth station provides the services by directly feeding the consumer equipment, which could be a TV set or a personal computer. In the case of large Earth stations, depending upon service provided by it, terrestrial interface may even look like a telephone exchange or a broadcast studio.

Two major components of terrestrial interface include the terrestrial tail and the interface. Terrestrial tail links are needed to connect the main Earth station to one or more remote user locations with line-of-sight microwave and fibre optic cable being the two principle options. Common interfaces needed in satellite links and terrestrial networks include telephone interface (voice), data transmission interface (data) and television interface (video).

8.6.3.1 Terrestrial Tail Options

The length of terrestrial tail may vary from few tens of metres to hundreds of kilometres. C-band satellite systems suffer from problems of radio frequency interference (RFI). This necessitates that the Earth station be located far away from the city leading to use of an elaborate tail. On the other hand, Ku and Ka band systems do not have to worry much about interference related issues and therefore have relatively shorter tails. In the case of Ka band, site diversity may be used to maintain reliable service in the event of adverse weather conditions. In such situations, one would also need to have a tail link between diverse sites. This tail link would also need to handle large bandwidth and thus involve large investment. In addition, there may also be some short links connecting various facilities within the Earth station complex.

Figure 8.30 shows a typical set up depicting tail links connecting various centres. The diagram is self explanatory. It shows a fibre optic cable link connecting the RF terminal

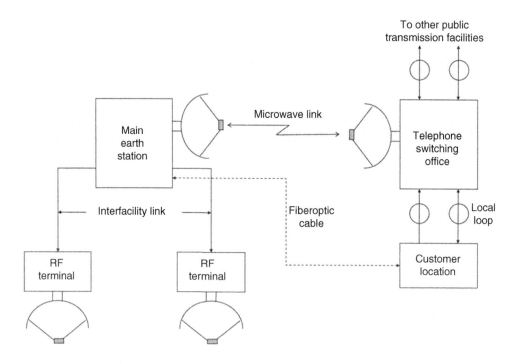

Figure 8.30 Typical Earth station set-up with terrestrial tail links

and the Earth station's main building, a microwave tail connecting the Earth station and a switching office, which in turn connects to user locations through public or private loops and an alternative fibre optic link between the Earth station and the customer location.

Both fibre optic cable and microwave links are effective and reliable technologies. Fibre optic cable may be the preferred choice in the case of short tails such as those connecting the Earth station with other facilities or a VSAT terminal to customer. It is low noise and is immune to electromagnetic interference (EMI). Single hop microwave is also a good alternative for short tails.

For long and elaborate tails, microwave link is a better choice. Fibre optic cable in that case turns to be relatively more expensive option, more so in a metropolitan area. Fibre optic cable may retain the edge in terms of cost for tail lengths shorter than 20 km, beyond which the microwave link certainly entails relatively lower cost.

8.6.3.2 Interface

As outlined earlier, terrestrial interface equipment need could vary from practically no requirement as is the case with receive-only or satellite phone terminals to very elaborate interface equipment in the case of a large commercial satellite Earth station. Such stations are required to handle massive traffic comprising of hundreds of telephone channels together with data and video reaching the station through microwave and fibre optic systems using time division or frequency division terrestrial multiplex methods. Signals received from the terrestrial network therefore need to be de-multiplexed and then changed from the existing terrestrial formats to

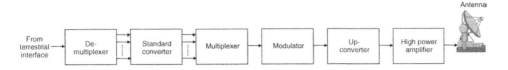

Figure 8.31 Terrestrial interface – up-link

formats suitable for satellite transmission. After this format/standards conversion, the signals are processed further in the up-link chain of the Earth station as shown in Figure 8.31. On the down-link side, the signals received from satellite/s are processed in the down-link chain before they are sent to standard converter. After reformatting, the signals are multiplexed and put on the terrestrial network as shown in Figure 8.32.

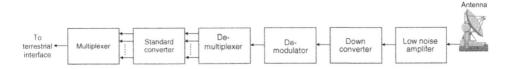

Figure 8.32 Terrestrial interface – down-link

Another interface related issue that is of particular relevance to handling digital signals arises out of the variation in the data rate at the receiving station over the period of the sidereal day caused by path length variation due to slight inclination and eccentricity of the orbit. As an example, at a nominal data rate of 9.6 kbps, a delay of 1.1 ms in the path length produces a peak-to-peak data rate variation of about 10.56 bits/s. It increases to 1.6984 kbps for a nominal data rate of 1.544 Mbps. This causes problems while interfacing with terrestrial networks that use synchronous transmission. Since these terrestrial networks cannot accommodate data rate variations of this magnitude, an elastic buffer that can absorb the expected peak-to-peak data rate variations is used between the satellite facilities and the terrestrial network as shown in Figure 8.33. An elastic buffer is nothing but a FIFO (First-in First-out) random access memory. The chosen elastic buffer should be large enough to absorb peak-to-peak data rate variations. For example, for peak-to-peak data rate variation of 10.56 bits/s, 16-bit buffer may be used. As the buffer is also going to add to the delay to the satellite link, smallest memory meeting the requirement should be used.

Figure 8.33 Use of elastic buffer to absorb data rate variations

8.7 Satellite Tracking

The Earth station antenna needs to track the satellite when the beam width of the antenna is only marginally wider than the satellite drift seen by it. Given the fact that satellite drift is typically in the range of 0.5–3° per day, antennas with large beamwidths such as DBS receivers do not require to track the satellite. On the other hand, large Earth stations do need some form of tracking with tracking accuracy depending upon the intended application. The tasks performed by the Earth station's satellite tracking system include some or all of the following.

1. Satellite acquisition
2. Manual tracking
3. Automatic tracking
4. Programme tracking

The *acquisition system* acquires the desired satellite by either moving the antenna manually around the expected position of the satellite or by programming the antenna to perform a scan around the anticipated position of the satellite. Automatic tracking is initiated only after the received signal strength due to the beacon signal transmitted by the satellite is above a certain threshold value, which allows the tracking receiver to lock to the beacon. *Manual track* option is used in the event of total failure of auto track system. *Automatic tracking* ensures continuous tracking of the satellite. Commonly used tracking techniques are described in the latter part of this section. In the case of *programme tracking*, the antenna is driven to the anticipated position of the satellite usually predicted by the satellite operator. Unlike automatic tracking, which is a closed loop system; programme track is an open loop system and therefore its accuracy is relatively much lower than that of auto track mode of operation.

8.7.1 Satellite Tracking System -- Block Diagram

Figure 8.34 shows the generalized block schematic arrangement of the satellite tracking system. The Earth station antenna makes use of the beacon signal to track itself to the desired positions in both azimuth and elevation. The auto track receiver derives the tracking correction data or in some cases the estimated position of the satellite. The estimated position is compared with the measured position in the control subsystem whose output feeds the servomechanism. In the case of manual and programme track modes, the desired positions of the satellite in the two orthogonal axes are respectively set by the operator and the computer. The difference in actual and desired antenna positions constitutes the error signal that is used to drive the antenna.

8.7.2 Tracking Techniques

Tracking techniques are classified on the basis of the methodology used to generate angular errors. Commonly used tracking techniques include the following.

1. Lobe switching
2. Sequential lobing
3. Conical scan

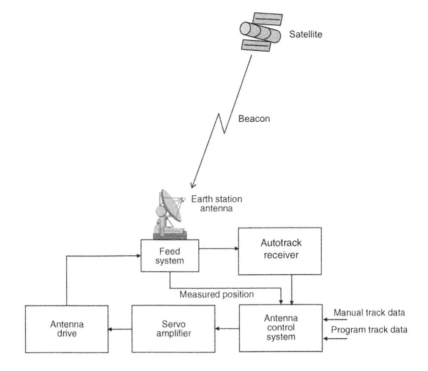

Figure 8.34 Block schematic arrangement of satellite tracking system

4. Monopulse track
5. Step track
6. Intelligent tracking

Of all the abovementioned techniques, the last four are more common in the case of satellite tracking. Sequential lobing with the rapid switching of a single beam has also been tried in some cases. Each of the abovementioned concepts with relative merits and demerits is briefly described in the following paragraphs.

8.7.2.1 Lobe Switching

In the case of lobe switching tracking methodology, the antenna beam is rapidly switched between two positions around the antenna axis in a single plane as shown in Figure 8.35. The amplitudes of the echo from the object to be tracked are compared for the two lobe positions. The difference between the two amplitudes is indicative of the location of the target with respect to the antenna axis. When the object to be tracked is on the axis, the echo amplitudes for the two positions of the beam are equal and the difference between the two is zero. When the object is on one side of the antenna axis, the amplitude and sense of the difference signal tells how much and what side of the antenna axis the object is located. The difference signal can then be used to generate correction signal, which with the help of servo control loop can

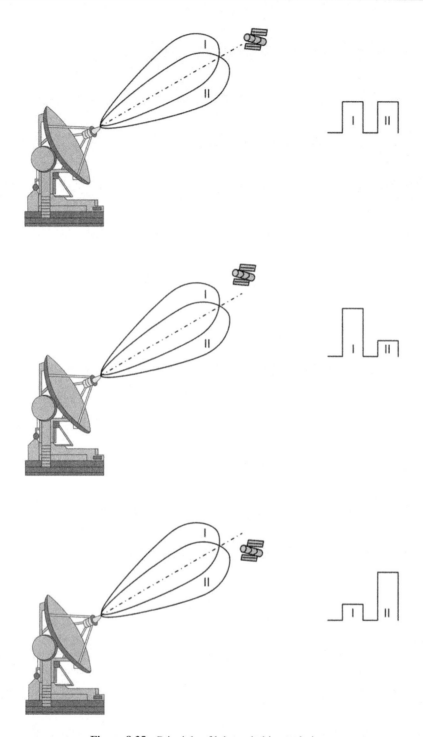

Figure 8.35 Principle of lobe switching technique

Satellite Tracking

be used to drive the antenna to bring the object on to the antenna axis. The lobe switching technique is prone to inaccuracies if the object cross-section as seen by the antenna changes between different returns in one scan.

8.7.2.2 Sequential Lobing

In sequential lobing, the beam axis is slightly shifted off the antenna axis. This squinted beam is sequentially placed in discrete angular positions, usually four, around the antenna axis (Figure 8.36). The angular information about the object to be tracked is determined by processing several echo signals. The track error information is contained in the echo signal amplitude variations. The squinting and beam switching is done with the help of electronically controlled feed and therefore can be done very rapidly practically simulating simultaneous lobing.

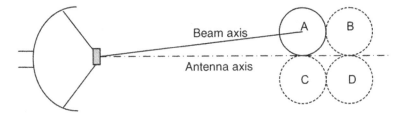

Figure 8.36   Principle of sequential lobing

8.7.2.3 Conical Scan

This is similar to sequential lobing except that in the case of conical scan, the squinted beam is scanned rapidly and continuously in a circular path around the axis as shown in Figure 8.37. If the object to be tracked is off the antenna axis, the amplitude of the echo signal varies with antenna's scan position. The tracking system senses the amplitude variations and the phase delay as function of scan position to determine the angular co-ordinates. The amplitude variation provides information on the amplitude of the angular error and the phase delay indicates direction. The angular error information is then used to steer the antenna axis to make it to coincide with the object location. The technique offers good tracking accuracy and an average response time. It is however not in common use now.

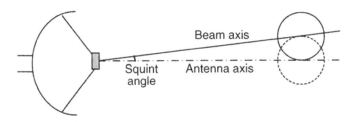

Figure 8.37   Principle of conical scan

8.7.2.4 Monopulse Tracking

One of the major disadvantages of sequential techniques including lobe switching, sequential lobing and conical scan is that the tracking accuracy is severely affected if the cross-section of the object to be tracked changes during the time the beam was being switched or scanned to get the desired number of samples. Monopulse tracking overcomes these shortcomings by generating the required information on the angular error by simultaneous lobing of the received beacon. There are two techniques of monopulse tracking namely amplitude comparison monopulse tracking and phase comparison monopulse tracking.

In the case of amplitude comparison monopulse tracking, the antenna uses four feeds placed symmetrically around the focal point. The wavefronts of the received signal in the case of to-be-tracked satellite being on antenna axis and off antenna axis are shown in Figures 8.38(a) and (b) respectively. In the on-axis case, the wavefront will be focused onto a spot on the antenna

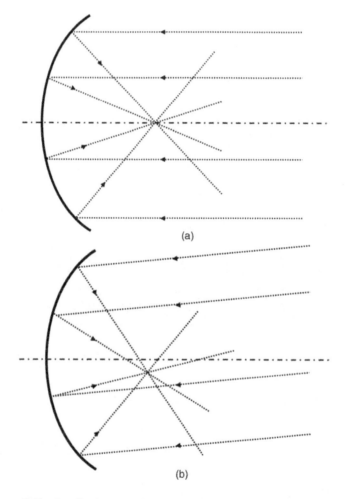

Figure 8.38 Amplitude comparison monopulse tracking – Received wavefront

axis as shown in Figure 8.38(a). For off-axis location of satellite, the focus spot will also be off the antenna axis. As a consequence, in the case of satellite being on-axis, the amount of energy falling on the four feeds representing four quadrants (A, B, C and D in Figure 8.39) will be the same. When the satellite is located off-axis, the amount of energy falling on the four feeds will be different depending upon which quadrant around the antenna axis, the satellite is located. Figures 8.39(a) to (e) show five different cases with satellite on-axis [Figure 8.39(a)], satellite located above antenna axis with same azimuth location [Figure 8.39(b)], satellite located below antenna axis with same azimuth location [Figure 8.39(c)], satellite located towards right of antenna axis with same elevation location [Figure 8.39(d)] and satellite located towards left of antenna axis with same elevation location [Figure 8.39(e)].

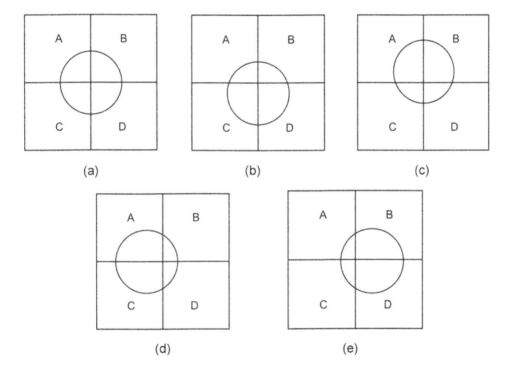

Figure 8.39 Amplitude comparison monopulse tracking – Spot for different angular positions

The amplitudes of the received pulse at the output of the four feeds are appropriately processed to determine azimuth and elevation errors required for tracking. In the amplitude comparison monopulse tracking technique, it is important that the signals arriving at different feeds are in phase. This is not a problem when using reflector antennas with feeds that are physically small, usually a few wavelengths across. In the case of arrays, where antenna surface is very large, signals arriving from different off-axis angles present different phases to the different segments into which the array has been divided. These phases need to be equalized before the error signals are derived.

In the case of phase comparison monopulse tracking, it is the phase difference between the received signals in different antenna elements that contains information on angular errors. At least two antenna elements are required for both azimuth and elevation error detection. The magnitude and sense of the phase difference determines the magnitude and direction of the off-axis angle. When the satellite is on axis as shown in Figure 8.40(a), the magnitude of phase difference is zero. Figure 8.40(b) depicts the case when the satellite is off-axis.

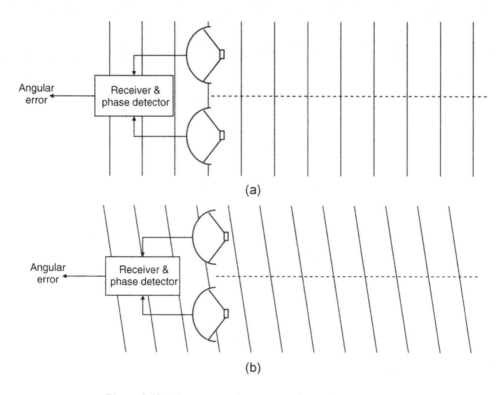

Figure 8.40 Phase comparison monopulse tracking technique

The sensitivity of this technique, that is the phase difference produced per unit angular error increases with increase in spacing between different the antenna elements. However, if they were too far apart, an off-axis signal may produce identical phases at the antenna elements. This gives rise to ambiguity. A practical system (Figure 8.41) could have two pairs of elements each for azimuth and elevation with the outer pair giving the desired sensitivity and the inner pair resolving ambiguity.

Monopulse tracking technique offers very high tracking accuracy and fast response time. Due to absence of any mechanical parts, the feed system requires very little maintenance. The disadvantages include high cost, large and complex feed system and need to have at least two-channel coherent receivers and good RF phase stability. It is commonly employed in large Earth stations and also in those Earth stations that require accurate tracking of non-geostationary satellites.

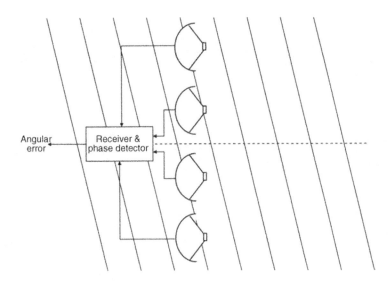

Figure 8.41 System to resolve ambiguity in phase comparison monopulse tracking technique

8.7.2.5 Step Track

In the case of step track, antenna axes are moved in small incremental steps in an effort to maximize the received signal strength. Amplitude sensing is the basis of this tracking methodology. It is simple and low cost and RF phase stability is not important. It is best suited to small and medium Earth stations. As expected, the technique is susceptible to amplitude perturbations caused by scintillation, signal fading and so on. Tracking accuracy is primarily determined by the step size and signal to noise ratio. For a high signal-to-noise ratio, tracking error approaches the step size. Accuracy is sensitive to amplitude interference.

8.7.2.6 Intelligent Tracking

In the case of intelligent tracking, the satellite position is obtained by optimally combining antenna position estimate data obtained from a gradient tracking algorithm with the prediction data on satellite position obtained from a satellite position model. In the case of signal amplitude fluctuations, the antenna position may be updated by using prediction data from satellite position model. Intelligent tracking offers all advantages of step track. It is however susceptible to amplitude fluctuations during initial acquisition. Full accuracy is achieved several hours after acquisition. Intelligent track may be used in small, medium and large Earth stations, particularly those susceptible to scintillation and signal fades.

8.8 Some Representative Earth Stations

8.8.1 Goonhilly Satellite Earth Station

Goonhilly satellite Earth station (Figure 8.42) is a large telecommunications site located on Goonhilly Downs near Helston on the Lizard peninsula in Cornwall, UK. It once was the largest satellite Earth station in the world with over 60 dishes. The site also links into undersea

Figure 8.42 A view of Goonhilly Earth station (Courtesy: Loz Flowers)

cable lines. British telecommunications in the year 2006 announced the shutting down of all operations from this Earth station from 2008 onwards.

The first dish antenna at Goonhilly Earth station called 'Arthur' (Figure 8.43) employed the open parabolic design. It was built in 1962 to link with Telstar-1 satellite (Telstar-1 was

Figure 8.43 First dish antenna called 'Arthur' commissioned at Goonhilly Earth station (Courtesy: Madnzany)

the first true commercial communications satellite) and measured 25.9 metres in diameter. The 'Arthur' dish received the first live transatlantic television broadcasts from the United States via the Telstar-1 satellite on 11 July 1962.

The Earth station had the infrastructure to transmit to every corner of the world via satellite and through a network of undersea fibre optic cables. It simultaneously handled millions of international telephone calls, e-mails and TV broadcasts. The site's largest dish, named 'Merlin', had a diameter of 32 metres. Other dishes included 'Guinevere', 'Tristan' and 'Isolde'. The Earth station was powered by the national grid and had a power back of 20 seconds provided by four one-megawatt diesel generators.

8.8.2 Madley Communications Centre

Madley communications centre (Figure 8.44) is British Telecom's satellite tracking Earth station located between Madley and Kingstone, Herefordshire, England. The Earth station has been in use for international telephone, fax and television transmission and reception since September 1978. Madley was the first Earth station of UK to transmit a fully digital transmission employing time division multiple access (TDMA) methodology.

Figure 8.44 Madley communications centre (Copyright Philip Halling)

8.8.3 Madrid Deep Space Communications Complex

The Madrid Deep Space Communications Complex (MDSCC) is a ground station located in Robledo de Chavela, Spain. Figure 8.45 gives a view of the station. The facility is part of the NASA's Deep Space Network run by the Jet Propulsion Laboratory. The NASA Deep Space Network (DSN) is an international network of antennas intended primarily to support interplanetary spacecraft missions and radio and radar astronomy observations for exploration of the universe and the solar system and also selected Earth-orbiting missions. The three Deep

Figure 8.45 Madrid deep space communications complex (Courtesy NASA/JPL-Caltech)

Space Network complexes are the Madrid Deep Space Communications Complex, Goldstone Deep Space Communications Complex (in USA) and Canberra Deep Space Communications Complex (in Australia). The strategic placement of the three complexes (120° apart around the world) allows constant observation of spacecraft as the Earth rotates. All DSN antennas are steerable, high-gain, parabolic reflector antennas.

The facility at Madrid provides the vital two-way communications link that guides and controls the remotely controlled drones. It also receives the images and other scientific data collected by them. The facility is equipped with five antennas designated as DSS-54, DSS-55, DSS-63, DSS-65 and DSS-66. DSS-54 and DSS-55 are 34-metre beam waveguide (BWG) antennas. DSS-63 was initially built as a 64 metre antenna in 1974, which was later upgraded to a 70 metre antenna in 1980s. It can transmit in S and X bands and receive in L, S and X bands. DSS-65, 34 metre diameter high efficiency antenna can transmit in X-band with a maximum power of 20 kW. It receives in S and X bands. DSS-66 has a diameter of 26 metres and is primarily intended to support near Earth missions and early orbital phase of deep space missions. All these antennas along with other associated hardware are used to perform spacecraft position and velocity tracking functions and acquire telemetry data from spacecraft. The facility is also used for astronomical observations and monitor and control the performance of deep space network.

8.8.4 Canberra Deep Space Communications Complex

The Canberra deep space communications complex (CDSCC), commonly referred to as Tidbinbilla deep space tracking station, is a ground station located in Canberra, Australia.

Some Representative Earth Stations 423

Figure 8.46 Canberra deep space communications complex (Courtesy NASA/JPL-Caltech)

The complex (Figure 8.46) is a part of the deep space network run by NASA's Jet Propulsion Laboratory (JPL). As outlined earlier, the complex is one of the three deep space network complexes of NASA. The other two are the Madrid deep space communications complex located in Spain and the Goldstone deep space communications complex in the United States. Since March 2003, Raytheon Australia has managed the Canberra complex on behalf of CSIRO and NASA.

Antennas equipping the complex include a 34 metre dish designated DSS-34 built in 1997 and utilizing a waveguide to place the receiving and transmitting hardware underground, a 70 metre dish designated DSS-43 built in the year 1976 and extended in 1987 being the largest steerable parabolic antenna in the southern hemisphere, a 34 metre dish designated DSS-45 built in the year 1985, and a 64 metre dish designated DSS-49 located at Parkes. There are plans to build two additional 34 metre beam waveguide antennas by the year 2016.

8.8.5 *Goldstone Deep Space Communications Complex*

The Goldstone Deep Space Communications Complex (GDSCC), commonly known as the Goldstone Observatory, is located in California's Mojave Desert (USA). The primary function of the observatory is to track and communicate with space missions. As outlined earlier, it is a part of NASA's deep space network and is one of the three such deep space communications complexes, the other two being the Madrid deep space communications complex and the Canberra deep space communications complex. It includes the Pioneer deep space

Figure 8.47 Pioneer deep space station (Courtesy NASA/JPL-Caltech)

station (Figure 8.47). It is operated by the ITT Corporation for the Jet Propulsion Laboratory, NASA.

The Goldstone deep space communications complex has a number of antennas. Figure 8.48 shows a photograph of one such antenna, a 70 metre diameter dish. These antennas have also been used as sensitive radio telescopes for a wide range of scientific experiments and investigations, which include mapping of celestial radio sources; radar mapping of planets, comets and asteroids and spotting comets and asteroids with the potential to impact Earth. Yet another application is the use of large aperture radio antennas to search for ultra-high energy neutrino interactions in the moon.

8.8.6 Honeysuckle Creek Tracking Station

Honeysuckle Creek Tracking Station (Figure 8.49) was established in 1967 near Canberra in Australia. The tracking station played a vital role during the Apollo 11 mission. It was instrumental in providing the first pictures of the Moonwalk on 21 July 1969. In addition to the television pictures of the Moonwalk, the station also had voice and telemetry contact with the lunar and command modules. After the completion of the Apollo mission in 1972, Honeysuckle Creek Tracking Station began to support regular Skylab passes and the Apollo scientific stations left on the Moon by astronauts. It also provided assistance to the deep space network with interplanetary tracking commitments. After the completion of the Skylab programmeme in the year 1974, the station became a part of the Deep Space Network as Deep Space Station 44. After its closure in 1981, its 26 metre antenna was relocated to the Canberra Deep Space Communications Complex. It was renamed Deep Space Station 46 and was in use until 2009.

Figure 8.48 70 metre dish antenna at Goldstone deep space communications complex (Courtesy NASA/JPL-Caltech)

Figure 8.49 Honeysuckle Creek Tracking Station (Reproduced by permission of Honeysuckle Creek. net © 2010)

8.8.7 Kaena Point Satellite Tracking Station

The Kaena Point Satellite Tracking Station (Figure 8.50) is a military installation of the United States Air Force (USAF) located at Kaena point on the island of Oahu in Hawaii. The station was originally established in the year 1959 to support the highly classified Corona satellite programme. It is a part of Air Force satellite control network responsible for tracking satellites in orbit, many of which belong to the United States Department of Defense. The facility originally had one large exposed antenna. The present day set up has a large number of smaller radomes.

Figure 8.50 Kaena Point Satellite Tracking Station (Author: Xpda)

8.8.8 Bukit Timah Satellite Earth Station

The Bukit Timah Satellite Earth Station (Figure 8.51) is Singapore's first satellite Earth station and is managed and owned by Singapore Telecommunications Limited. It started operations in 1986.

8.8.9 INTELSAT Teleport Earth Stations

Intelsat offers a network of a large number of teleport facilities to complement their space assets. These include the following facilities.

1. Located at Atlanta-Ellenwood, Georgia, USA; it is Intelsat's primary AOR gateway equipped with more than 53 antennas, RF and fibre interconnect, primary and backup TT&C, etc.

Figure 8.51 Bukit Timah Satellite Earth Station (Author: mailer_diablo)

2. Located at Clarksburg, Maryland, USA; it is a prime C and Ku band launch support and IOT site, equipped with primary and backup TT&C, 26 antennas, fibre interconnect and offers up-link/down-link and carrier monitoring services.
3. Located at Mountain-Hagerstown, Maryland, USA; it houses the Intelsat mission critical operations centre and disaster recovery facilities and is equipped with 17 antennas and fibre interconnect. It offers up-link/down-link and carrier monitoring services. It also houses customer co-located equipment for over 20 customers.
4. Located at Fillmore, California, USA; it is Intelsat's prime C-band launch and support site, IOT site and TT&C site. It is equipped with 20 antennas and fibre interconnect. It offers emergency restoration services, transfer orbit support and carrier monitoring services.
5. Located at Napa, California, USA; it is Intelsat's primary POR gateway and has a more than 50 simultaneous transponder up-link capability. It is equipped with 20 antennas, primary and backup TT&C and fibre interconnect.
6. Located at Riverside, California, USA; it is Intelsat's prime C-band launch support site and also primary and backup TT&C site. It is equipped with 19 antennas and fibre interconnect. It offers up-link/down-link and carrier monitoring services. It also houses customer co-located equipment for over 25 customers.
7. Located at Castle Rock, Colorado, USA; it is Intelsat's prime Ku-band launch site. It is also an IOT site and TT&C site. It is equipped with 20 antennas and fibre interconnect. It offers emergency restoration, transfer orbit support and carrier monitoring services.
8. Located at Paumalu, Hawaii, USA; it is Intelsat's prime C-band launch and support site. It is also a primary TT&C site. It is equipped with seven antennas and offers up-link/down-link and carrier monitoring services.
9. Located at Fuchsstadt, Germany; it is Intelsat's secondary C-band launch support site and also a primary and backup TT&C site. It is equipped with 45 antennas and fibre

interconnect and offers up-link/down-link and carrier monitoring services. It also houses customer co-located equipment for over 30 customers.

8.8.10 SUPARCO Satellite Ground Station

SUPARCO is an abbreviation for Space and Upper Atmosphere Research Commission. As the executive and bureaucratic space agency of the government of Pakistan, SUPARCO is responsible for research in aeronautics and aerospace science, and also for Pakistan's public and civil space programme. The SUPARCO Satellite Ground Station is located at Rawat near Islamabad in the Rawalpindi district of Punjab, Pakistan. The ground station is the control centre for Earth observation and remote sensing satellites. It has a data acquisition zone of approximately 2500 km radius covering the whole of Pakistan and wholly or partially 25 other countries in the South, Central, Western and Middle East Asian regions. It is equipped to acquire and store satellite data from different Earth resources and remote sensing satellite missions. It provides data and related services to different agencies both within and outside Pakistan.

8.8.11 Makarios Satellite Earth Station

Makarios Satellite Earth Station is located in Cyprus and is operated by the Cyprus Telecommunication Authority (CYTA), a semi-government organization and a leading provider of integrated electronic communications services in Cyprus, including fixed line telecommunications, mobile telecommunications and internet access. The Earth station has been in operation since 1980. The geographical location of Cyprus at the eastern end of the Mediterranean Sea offers full visibility of the geostationary satellites stationed anywhere between 33.5° West and 100.5° East with elevation angles of greater than 10°. Also, the climate of Cyprus coupled with access to almost any point on the globe makes it an ideal choice for satellite operations. Makarios Satellite Earth Station is connected to CYTA's national and international fibre networks, providing capacity and dedicated direct links with major international nodes to support a wide range of satellite communication systems and services, such as satellite television, internet connectivity and hosting services. The station provides a multitude of satellite links in C, Ku and DBS frequency bands. Figure 8.52 shows an aerial view of Makarios Satellite Earth Station.

8.8.12 Raisting Earth Station

Raisting Earth Station is located in the German foothills of the Alps in a shallow valley at the southern end of Ammersee Lake in Bavaria, Germany. The terrain here is particularly favourable for building an Earth station as the foreground is free and the Alps and the rolling hills in the other directions offer protection against any interference. Raisting Earth Station is one of the three largest teleport facilities in the world. It is equipped with 19 parabolic dish antennas and about 12000 m^2 of data centre facilities. Its antennas include two 32 metre, two 28.5 metre, one 18 metre, one 15 metre, one 13 metre, seven 11 metre and four other antennas of different sizes to provide satellite communications services in various bands. Figure 8.53 shows a photograph of different antennas at Raisting Earth Station.

Figure 8.52 Makarios Satellite Earth Station (Courtesy: CytaGlobal)

Figure 8.53 Raisting Earth Station (Attribution: Michael Lucan, (CC-BY-SA 3.0))

8.8.13 Indian Deep Space Network

The Indian Deep Space Network (IDSN) is located at Byalalu, approximately 40 km from Bangalore, India. It was inaugurated in October 2008. It is a network of large antennas and communication facilities to support India's interplanetary spacecraft. The network consists of 11 metre, 18 metre and 32 metre antennas. The 18 metre and 32 metre antennas are co-located at the Byalalu site.

The 18 metre antenna is a fully steerable antenna capable of receiving two right circularly polarized (RCP) and left circularly polarized (LCP) downlink carriers in S-band and two in X-band. The uplink is either LCP or RCP. The 32 metre antenna is of wheel-and-track design providing uplink in both S-band and X-band either through LCP or RCP and downlink in S-band and X-band through simultaneous LCP and RCP. The IDSN is connected to the spacecraft control centre (SCC) and network control centre (NCC) through a fibre optic link. The station is equipped for remote control from the ISTRAC network control centre. The 11 metre terminal antenna became operational in 2009. It was built for the ASTROSAT mission, India's first dedicated satellite for applications in astronomy, scheduled for launch on board PSLV during 2014.

The IDSN also houses the ISRO Navigation Centre (INC). The centre became operational in June 2013 with the launch of the Indian Regional Navigational Satellite System (IRNSS-1). The centre is also equipped with a high stability atomic clock used to coordinate the activities of 21 Earth stations across India.

Glossary

Baseband equipment: The baseband equipment performs the modulation/demodulation function with the specific equipment required depending upon the modulation technique and the multiple access method employed.

Broadcast Satellite Service (BSS) Earth Station: Under the group of BSS Earth stations, we have large Earth stations ($G/T \cong 15$ dB/K) used for community reception and small Earth stations ($G/T \cong 8$ dB/K) used for individual reception.

Bukit Timah Satellite Earth Station: The Bukit Timah Satellite Earth Station is Singapore's first satellite Earth station and is managed and owned by Singapore Telecommunications Limited.

Canberra Deep Space Communications Complex: Canberra Deep Space Communications Complex (CDSCC), commonly referred to as Tidbinbilla deep space tracking station, is a ground station located in Canberra, Australia. The complex is a part of the deep space network run by NASA's Jet Propulsion Laboratory (JPL).

Cassegrain fed reflector antenna: The cassegrain antenna uses a hyperbolic reflector placed in front of the main reflector, closer to the dish than the focus low noise block (LNB)

Conical scan: In the case of conical scan, the squinted beam is scanned rapidly and continuously in a circular path around the axis. If the object to be tracked is off the antenna axis, the amplitude of the echo signal varies with the antenna's scan position.

Down-converter: Down-converter is a frequency translator that converts the signals in the operating frequency bands (C, Ku and Ka) to the IF frequency used in the modems and the baseband equipment.

Earth station: An Earth station is a terrestrial terminal station mainly located on the Earth's surface. It could even be airborne or maritime.

Effective Isotropic Radiated Power (*EIRP*): It is given by the product of the power output of HPA at the antenna and the gain of the transmitting antenna.

Fixed Satellite Service (FSS) Earth Station: Under the group of FSS Earth stations, there are large Earth stations ($G/T \cong 40$ dB/K), medium Earth stations ($G/T \cong 30$ dB/K), small Earth stations ($G/T \cong 25$ dB/K), very small terminals with transmit/receive functions ($G/T \cong 20$ dB/K) and very small terminals with receive only functions ($G/T \cong 12$ dB/K)

Gateway station: Gateway stations serve as an interface between the satellites and the terrestrial networks and also serve as transit points between satellites.

Goldstone Deep Space Communications Complex: The Goldstone Deep Space Communications Complex (GDSCC), commonly known as the Goldstone Observatory, is located in California's Mojave

Desert (USA) and is a part of NASA's deep space network. The primary function of the observatory is to track and communicate with space missions.

Goonhilly Satellite Earth Station: Goonhilly Satellite Earth Station is a large telecommunications site located on Goonhilly Downs near Helston on the Lizard peninsula in Cornwall, UK.

Gregorian antenna: This configuration uses a concave secondary reflector just behind the prime focus. The purpose of this reflector is to bounce the waves back towards the dish. The front end in this case is located between the secondary reflector and the main reflector.

Honeysuckle Creek Tracking Station: Honeysuckle Creek Tracking Station was established in the year 1967 near Canberra in Australia. The tracking station played a vital role during Apollo 11 mission. It is currently a part of Deep Space Network of NASA.

Indian Deep Space Network: A network of large antennas and communication facilities to support India's interplanetary spacecraft. It is located in Byalalu, near Bengaluru, India

Intelligent tracking: In intelligent tracking, the satellite's position is obtained by optimally combining antenna position estimate data obtained from a gradient tracking algorithm with the prediction data on satellite position obtained from a satellite position model.

Intelsat teleport Earth stations: Intelsat offers a network of large number of teleport facilities to compliment their space assets.

Kaena Point Satellite Tracking Station: The Kaena Point Satellite Tracking Station is a military installation of the United States Air Force (USAF) located at Kaena Point on the island of Oahu in Hawaii.

Lobe switching: In the case of lobe switching tracking methodology, the antenna beam is rapidly switched between two positions around the antenna axis in a single plane. The amplitudes of the echo from the object to be tracked are compared for the two lobe positions. The difference between the two amplitudes is indicative of the location of the target with respect to the antenna axis.

Madley Communications Centre: Madley Communications Centre is British Telecom's satellite tracking Earth station located between Madley and Kingstone, Herefordshire, England.

Madrid Deep Space Communications Complex: The Madrid Deep Space Communications Complex (MDSCC) is a ground station located in Robledo de Chavela, Spain.

Makarios Earth Station: Satellite Earth station located in Cyprus and operated by the Cyprus Telecommunication Authority (CYTA)

Mobile Satellite Service (MSS) Earth Station: Under the group of MSS Earth stations, we have the large Earth stations ($G/T \cong -4$ dB/K), medium Earth stations ($G/T \cong -12$ dB/K) and small Earth stations ($G/T \cong -24$ dB/K). While both large and medium Earth stations require tracking, small MSS Earth stations are without tracking equipment.

Monopulse tracking: There are two techniques of monopulse tracking namely amplitude comparison monopulse tracking and phase comparison monopulse tracking.

Programme tracking: In the case of programme tracking, the antenna is driven to the anticipated position of the satellite usually predicted by the satellite operator.

Raisting Earth Station: Satellite earth station located at the Southern end of Ammersee lake in Bavaria, Germany

Receiver figure-of-merit (G/T): Receiver figure-of-merit is indicative of how the receiving antenna performs together with the receiving electronics to produce a useful signal. The receiver figure-of-merit tells us about the sensitivity of the receiving antenna and the Low Noise Amplifier (LNA) combined to weak received signals.

RF section: The RF section mainly comprises of antenna subsystem, the up-converter and the high power amplifier (HPA) in the up-link channel and the antenna subsystem, low noise amplifier (LNA) and the down-converter in the down-link channel.

Satellite acquisition system: Satellite acquisition system acquires the desired satellite by either moving the antenna manually around the expected position of the satellite or by programming the antenna to perform a scan around the anticipated position of the satellite.

Satellite tracking: Satellite tracking refers to the tracking of the satellite by the Earth station antenna.

Sequential lobing: In sequential lobing, the beam axis is slightly shifted off the antenna axis. This squinted beam is sequentially placed in discrete angular positions, usually four, around the antenna axis. The angular information about the object to be tracked is determined by processing several echo signals. The track error information is contained in the echo signal amplitude variations.

Single function station: Single function stations are characterized by a single type of link to a satellite or a satellite constellation. These stations may be transmit-only, receive-only or both.

Step track: In step track, antenna axes are moved in small incremental steps in an effort to maximize the received signal strength.

SUPARCO Ground Station: Satellite ground station located in Rawat, near Islamabad in Pakistan

Teleport: Teleport is a type of gateway station operated by firms that are usually not a part of a specific satellite system.

Terrestrial Tail: Terrestrial tail is that part of the Earth station that connects the Earth station to the users

Up-converter: Up-converter is a frequency translator that converts the IF used in the modems and baseband equipment to the operating RF frequency bands (C, Ku and Ka).

9
Networking Concepts

9.1 Introduction

Satellites today provide a wide range of applications and services, which include traditional telephony, radio and television broadcast services, and broadband and internet services. The ultimate objective of satellite networks is to provide these services, which are continuously evolving both in terms of type of services as well as the quality of service, to users irrespective of their geographical location and particularly to those in remote locations that are not conveniently accessible to terrestrial networks. Given the enormity and requirements of satellite applications and services in a modern satellite network, one is likely to find use of a single satellite and multiple satellites with intersatellite links or multi-hop transmission, bench top user terminals and handheld terminals, fixed Earth stations and transportable Earth stations, satellites-only networks and satellite networks integrated with terrestrial networks. With the integration of audio, video and broadband internet services and due to the need to support a wide range of applications and services to users wherever they are, internetworking with terrestrial networks and protocols is also an important part of satellite networks. Some specific aspects of satellite networking such as satellite link designing, satellite communication services, VSAT networks, satellite constellations and so on have been discussed in other chapters. The core topics of satellite networking, including networking technologies (circuit switching, packet switching, internet-working with terrestrial networks), networking protocols (OSI reference model, internet protocol, transmission control protocol, asynchronous transmission mode), network topologies (star, bus, ring, mesh, tree), internet-working with terrestrial networks, intersatellite links and network security, are discussed in this chapter. The chapter begins with an introduction to satellite network characteristics, applications and services, and then goes on to briefly discuss the other core topics outlined above in the subsequent sections.

9.2 Network Characteristics

The fundamental characteristics that can be used for the purpose of comparing various available networking options and solutions include the following:

- availability
- reliability
- security
- speed or throughput
- scalability
- topology
- cost

9.2.1 Availability

Network availability refers to the ability of the network to respond to the users' requests to access it. Availability is the probability that the network will be available to users as and when required. It is typically measured as a percentage of the total time the network is required to be available. One way of computing availability, known as the *availability ratio*, is to take the ratio of the time in minutes the network was actually available during a year to the total number of minutes in a year. The time period considered for quantifying availability is very important. A given percentage availability when measured over a one year period is indicative of a much higher degree of availability than when it is measured for a period of one month. Another method of expressing availability, particularly with reference to satellite networks, is the *mean time between outages* (MTBO), which is defined as the average duration of a time interval during which the connection is available from the service perspective. Consecutive intervals of available time during which the user attempts to use the network are concatenated.

Availability is generally expressed as different classes of availability. The standard method of defining classes of availability is to express availability on a scale of nines, such as two nines, three nines and so on. Two nines, three nines and five nines mean availability of 99%, 99.9% and 99.999%, respectively. As before, 99.9% availability measured over a one year period implies a much higher availability level than 99.9% availability measured over one month. The typical availability class for internet service providers and mainstream business systems is three nines (99.9%) and for a defence system it is six nines (99.9999%). Use of redundant hardware and intelligent software can minimize the run down time and increase availability. This becomes particularly important when achieving availability of four nines or better.

With reference to a satellite network, the total availability of the satellite network (A) may be measured as the combined effect of three different components, namely the availability of the satellite A_1, the availability of the satellite link A_2 and the availability of the satellite resources A_3, and is given by the product of thses components: $A = A_1 A_2 A_3$. One way to improve the availability of a satellite link is to employ Earth-to-space diversity and in-orbit network diversity. Earth-to-space diversity uses more than one satellite, which allows an improvement in physical availability by providing redundancy at the physical or data link level or by decreasing the impact of shadowing due to obstructions in the path between the ground terminal and the satellite. Diversity is also exploited for soft handovers. In-orbit network diversity provides redundancy by using satellite constellations with closely spaced satellites.

9.2.2 Reliability

The reliability of a network is a measure of the reliability of the different components of the network and their interconnections. It is generally measured as the *mean time between failures* (MTBF) or *mean time to repair* (MTTR). For a reliable network, MTBF should be as high as possible and MTTR should be as low as possible. In other words, the fraction of time during which the network is down or unable to support a connection should be as low as possible and once a connection is established the probability of it being terminated due to either failure of a network component or inadequate performance should be as low as possible.

The reliability of the network may be measured by its ability to transmit data with an acceptable error rate, which can be specified for each component link of the network. The bit error test (BERT) or the block bit error rate test (BLERT) may be used to determine the reliability of different component parts of the network and hence the reliability of the total network. In the case of BERT, a known pattern of bits is transmitted and the number of single bit errors determined in the received pattern. The ratio of the number of bit errors to the total number of bits received is BERT. BLERT is the ratio of the number of block errors detected to the total number of blocks received. BLERT is more appropriate than the BERT as the link level protocols used have the ability to correct bit errors and it is more important to know how often blocks need to be retransmitted.

9.2.3 Security

Networks are the backbone of all communications (audio, video and data) that take place between government agencies, large enterprises, small and medium business houses and individuals. The objective of *network security* is to monitor and prevent unauthorized access, eavesdropping, misuse, modification or denial of use of a network and its resources to authorized users. A network security system typically relies on layers of protection that consists of multiple components, including network monitoring and security software in addition to hardware and appliances. All components work together to increase the overall security of the computer network. The terms 'network security' and 'information security' are often used interchangeably. Network security is generally taken as providing protection at the boundaries of an organization by keeping out intruders or hackers with the help of network security tools such as firewalls, virtual private networks (VPNs), data encryption, biometric devices and so on. Information security on the other hand explicitly focuses on protecting data resources from malware attack or simple mistakes by people within an organization by use of data loss prevention techniques.

Network security is jeopardized by a range of threats including viruses, Trojan horse programs, vandals, attacks, data interception and social engineering.

Viruses are computer programs designed to replicate themselves and infect computers when triggered by an event. A virus might corrupt or even delete data on affected computers. They often spread through attachments in email messages or instant messaging messages. They also spread through downloads on the internet and can be hidden in illicit software or other files that are downloaded.

A *Trojan horse* is a program in which malicious or harmful code is contained inside seemingly harmless useful program or data in such a way that it can get to do its chosen form of damage such as destroying the file allocation table (FAT) on your hard disk. Trojan horse programs act as delivery vehicles for the destructive code. They are a non-self-replicating type of malware, which gets privileged access to the operating system. They appear to perform a desirable function but drop a malicious payload that often allows unauthorized access to the targeted computer.

Computer vandalism is a program that performs malicious functions such as extracting a user's password or other data. It can be distinguished from a virus in the sense that it doesn't attach itself to an existing executable program; instead it is an executing entity itself, which can be downloaded from internet through an email attachment, a Java applet or a browser plug-in.

Attacks of various kinds are also a threat to network security. Different types of attacks include *reconnaissance attacks*, which are information gathering activities later used to compromise networks, *access attacks*, which exploit network vulnerabilities to gain an unauthorized access to email or databases, and *denial-of-service attacks*, which prevent an authorized access.

Data interception is another network security issue and may be used for the malicious purpose of eavesdropping on communications or altering data packets being transmitted. Network security is also seriously jeopardized by maliciously obtaining network security information by posing as a technical support person.

There are a number of tools that can be used to make a network secure. *Antivirus software packages* such as Symantec endpoint solution, Kaspersky anti-virus, and Norton anti-virus counter most virus threats provided they are regularly updated and properly maintained. Network hardware and software also have certain features that ensure network security. For example, switches and routers have hardware and software features that provide protection against intrusion, secure connection, identity services and security management. A VPN allows individual users or remote offices of an organization to access their organization's network without the risk of the data being intercepted or hacked. The objective of a VPN is to provide the organization with the security features of a public network at a much lower cost. *Data encryption* ensures that data are not intercepted or read by anyone other than the authorized recipient. Identity services such as passwords, digital certificates, digital authentication keys and so on help to identify the users and control their activities on the network. Various security tools need to be layered to make the network safe and secure from different kinds of threats and vulnerabilities.

9.2.4 Throughput

Network throughput is defined as the average rate at which data is transferred through the network in a given time. It is measured in bits per second (bps) and also sometimes in data packets per second or data packets transferred in a given time slot. The throughput is typically less than the transmission speed of the individual links. In a network with multiple links, the overall throughput is equal to or less than the lowest throughput value.

The *theoretical maximum throughput* can be considered synonymous with the digital bandwidth capacity and the *effective throughput* allows the subscriber to access data, The latter is always less than the former. For example, a communication link using 9600 bps modems may yield an effective throughput of 6000 bps as there are always some extra bits in the overall bit stream used for functions such as error checking and the information bits are always less than the total number of bits being transmitted. Throughput may be measured as *transfer rate of information bits* (TRIB) by taking the ratio of the number of information bits transferred to the time required for the information bits to be transferred.

The throughput of a network is constrained by factors such as network protocols and different components comprising the network, such as switches and routers, type of cabling, whether the cabling is fibre optic or Ethernet and so on. It is further constrained by the specifications of network adapters used on client systems in the case of wireless networks.

9.2.5 *Scalability*

Scalability is the capability of a network to accommodate increase data rate requirements and number of users. It also indicates how well a network can adapt itself to new applications and replacement of old components by new components with enhanced features. In other words, it is the ability of the network to handle increased quantum of work and to increase total throughput under an increased load when resources are added. In the case of a scalable network, performance improves in direct proportion to the capacity added to the network. Modularity is an important ingredient of all such networks.

9.2.6 *Topology*

Topology is the way different components of the network are connected (*physical topology*) and the logical way data passes through the network from one component or device to the next irrespective of the physical structure of the network (*logical topology*). Common physical topologies include star, bus, mesh, tree and ring. Logical topology defines how the devices communicate across the physical topologies. Two common types of logical topologies are *shared media topology* and *token based topology*.

In a *shared media topology* various devices on the network have unrestricted access to the physical media and it suffers from the problem of collisions. Ethernet is an example of shared media logical topology. In *token based topology* the sending device uses a token to send a packet of data. The token travels on the network with the data packet. After the data packet is received by the intended machine, the token travels back to the sender. It is taken off the network by the sender and a new empty token is sent out to be used by the next device. Token based networks do not have the collision problem of shared media topology but suffer from latency instead. Token based topology is best configured in physical ring topology.

9.2.7 *Cost*

Cost is the expenditure incurred in setting up and maintaining the network.

9.3 Applications and Services

Satellite networks are used for a wide range of multimedia communication, broadcast and broadband services in addition to contributing to other application domains such as remote sensing, meteorology, astronomy, global positioning systems, military communication, private data, communication services and so on. Satellite networks carry our telephone calls and emails, relay television broadcasts and play an important role in global network infrastructure. The primary objective of satellite networks is to provide services and applications. The network provides transportation services between user terminals and user terminals provide services and applications directly to users. A typical satellite network comprises more than one of the following components: fixed Earth station, mobile Earth station, portable and handheld terminals, satellites with intersatellite links and in many cases user terminals accessing the satellite link directly through a terrestrial network. The user terminals connect to the satellite network via the user Earth station and terrestrial networks interface with the satellite networks through the gateway Earth station.

Network services are provided by the user Earth station and gateway Earth station. These are classified as teleservices and bearer services. The former are high level services including telephone, fax, video and data services. The latter services are low level services provided by networks to support teleservices. Teleservices and bearer services are user centric and network centric, respectively. The network QoS (quality of service) and user QoS have conflicting requirements. An attempt to increase user QoS by decreasing load reflects a reduced network QoS due to under utilization of resources.

9.3.1 Satellite and Network Services

The terms 'applications' and 'services' are often used interchangeably. Applications are built from services. One or more network services combine to provide an application. For example tele-education is an application based on voice, video and data services. Applications and services may be further described as *satellite services*, *network services* and *internet services*. The International Telecommunications Union (ITU) defines different satellite services, including fixed satellite services (FSSs), mobile satellite services (MSSs) and broadcast satellite services (BSSs). It also defines two main classes of services, *interactive services* and *distribution services*, that are provided by the networks to the users. Internet services mainly include the *World Wide Web* (WWW), *electronic mail* (email), *file transfer protocol* (FTP), *voice over internet protocol* (VoIP), *domain name system* (DNS) and *multicast and content distribution*.

9.3.2 Satellite Services

FSSs use radio communications between Earth stations at fixed locations on the surface of Earth and one or more satellites to support all types of telecommunication and data network services, such as telephony, video, radio, television, fax, data and internet. One or more than one satellite with intersatellite links may be used to link the Earth stations. FSSs also use feeder links between an Earth station and the satellite for broadcast purposes such as broadcasting from media and sporting events or news conferences. In addition, FSS satellites provide a wide range of services including paging networks and point-of-sale support, such as credit card transactions and inventory control.

MSSs are radio communication services between mobile Earth stations and one or more satellites. They use a constellation of satellites interconnected with terrestrial networks to provide interactive mobile-to-mobile and mobile-to-fixed voice, data and multimedia communication services. BSSs make use of transmission or retransmission from satellites intended for direct reception by users. Direct-to-home (DTH) and community antenna television (CATV) are examples of broadcasting services. In addition to the above mentioned traditional fixed, mobile and broadcast satellite services, other services intended for some dedicated applications include those for military communications and intelligence, navigation, meteorology, Earth observation and space exploration.

9.3.3 Network Services

Interactive services allow a user to interact with another user through either a conversation or messaging service in real time. They also allow user interaction with servers in computers. Telephony, video telephony and video conferencing are examples of broadband *conversational services. Messaging services* offer communication between individual users. Messaging services include the short message service (SMS) and the multimedia messaging service (MMS). While SMS allows exchange of text messages only, MMS is used to send messages with multimedia content, including text messages, pictures, videos and ring tones. *Retrieval services* are used by individuals to retrieve information stored in information centres and intended for public use.

The other category of network services is *distribution services*. In one of the subclasses of distribution services, information is distributed from a central source to an unlimited number of authorized recipients connected to the network. In this case, the user does not have any control over the broadcasted information in terms of its start time and order of presentation. Television broadcast services are a common example of this type of distribution service. The other subclass of distribution service is the one in which the user can exercise control over the start and order of presentation. The information in this case is distributed as a sequence of information entities with cyclical repetition. Video-on-demand is one such service.

9.3.4 Internet Services

Common internet services and applications include the WWW, email, FTP, unicast and multicast streaming for content distribution, VoIP and DNS.

The *WWW* is a system of internet servers that support documents formatted in hyper text markup language (HTML). This supports links to audio, video, graphics and other documents. Web browsers such as Firefox and Internet Explorer are applications that allow easy access to the WWW.

Email is a method of transmitting messages electronically from one computer user to one or more recipients over a communications network. Some email systems are confined to a single network while others have gateways to other networks thus enabling users to send email anywhere in the world. In an email system each user has a private mailbox and all email messages sent to it are put in the mailbox by the email system. The messages are stored in the mailbox until they are deleted by the recipient. All internet service providers and other online services offer an email facility and most of them support gateways, allowing users connected to one network to exchange emails with users connected to other systems.

FTP is an easy way for internet users to upload files from their computers to a website or download files from a website to their computers utilizing transmission control protocol (TCP) and internet protocol (IP) systems to carry out file transfer operations in the same manner as hypertext transfer protocol (HTTP) transfers web pages and related files, and simple mail transfer protocol transfers email. The three modes of transferring data using FTP are stream mode, block mode and compressed mode. In stream mode files are transmitted as a continuous stream from one port to another with no intervening information processing, which is the case, for example, when the two computers involved use the same operating system. In the case of block mode, the data to be transferred is divided into blocks of information. Each block of data has a header, byte count and data field. In the case of compressed mode, as the name suggests, files are compressed by encoding process.

VoIP allows telephone calls to be made over a broadband internet network by converting analogue voice signals into digital data packets and using IP for the two-way transmission. VoIP services are not only possible using a computer or a dedicated VoIP phone, but can also

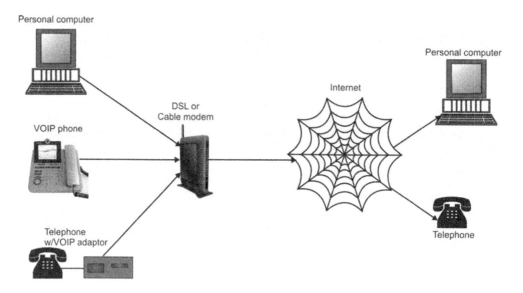

Figure 9.1 VoIP communication

be placed from a traditional land line phone using a special adapter. Figure 9.1 shows a basic schematic representation of how VoIP works. Different VoIP configurations include dedicated routers that allow the use of traditional phones to place VoIP calls, adapters (usually USB adapters) that also allow the use of traditional phones for making VoIP calls, soft phones that allow VoIP calls to be made directly from a computer equipped with a headset, microphone and sound card, and a VoIP phone that connects directly to a computer network.

Multicast streaming for content distribution is a method of transmitting information from one source to many recipients in which only one copy of the information is sent to a group address. The information reaches all the recipients who are part of the group. A client wishing to receive the multicast stream informs the multicast enabled router, which replicates the traffic to the client. Copies of the information are created only if the network topology requires it. Since only one copy is transmitted in multicast content distribution, bandwidth is conserved. Figure 9.2 illustrates the multicast streaming concept.

Multicast streaming is different from *unicast transmission*, in which the data packet is sent from a single source to a specified destination. In unicast transmission each client that connects to the server takes up additional bandwidth. The total bandwidth requirement increases in proportion to the number of clients being served. Figure 9.3 illustrates the unicast streaming concept. As is evident from the above discussion, multicast streaming is much less bandwidth intensive than unicast streaming. However, multicast streaming on the internet is generally not practical as all sections of the internet are not multicast enabled. Multicast streaming could be very effective in saving bandwidth in corporate environments where all routers can be multicast enabled. Unicast streaming continues to be the most predominant form of transmission on local area networks (LANs) and the internet. The unicast transfer mode is well supported by LANs and IP networks, and users are familiar with standard unicast applications such as HTTP, simple mail transfer protocol (SMTP) and FTP that employ TCP.

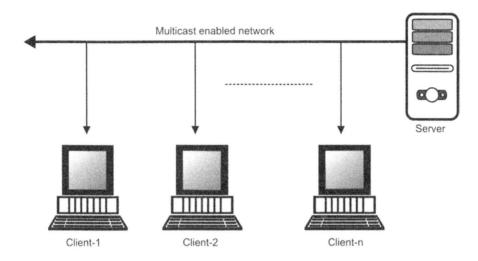

Figure 9.2 Multicast streaming

DNS is an example of application layer service and is used by many internet applications. It performs the function of translating internet and host domain names, which are alphabetical, into IP addresses. DNS automatically converts names typed in a web browser address bar to the IP addresses of the web servers hosting those sites. DNS implements a distributed database to store the domain name and IP address information for all public hosts on the internet. In addition, DNS also includes support for caching requests and redundancy.

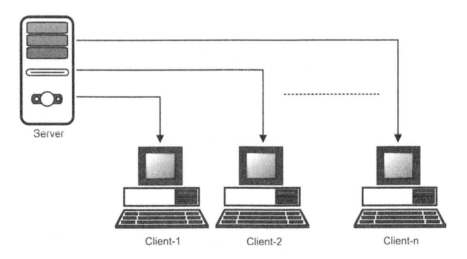

Figure 9.3 Unicast streaming

9.4 Network Topologies

The topology of a communication network is the schematic description that defines the network geometry, showing how different nodes are connected to each other and the paths the signals follow from node to node. While the way in which different nodes are connected is more precisely known as physical topology, the nature of the signal flow paths represents logical topology. Common physical topologies include bus, star, ring, mesh and tree topologies. Token based topology and shared media topology are the common logical topologies.

9.4.1 Bus Topology

In the case of bus topology all devices to be connected to the network are connected to a central cable that acts as the backbone of the network with the help of interface connectors. This is the simplest of all network topologies, in which all work stations communicate with the other devices through a common cable known as a bus. A network device intended to communicate with another device on the network sends out a broadcast message on the cable. The message is seen by all the devices on the network but is received and processed by the intended recipient only. Terminators are used at the ends of the cable to prevent bouncing of signals. The network may be extended using barrel connectors. Figure 9.4 shows the basic schematic bus topology.

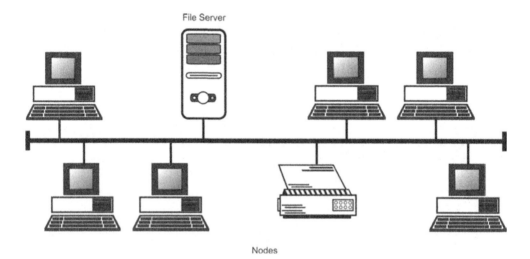

Figure 9.4 Bus topology

Bus topology is low cost for given a number of devices connected to the network, it is easy to set up and expand, it uses the least cable length and it is well suited to small networks such as LANs. The disadvantages include relatively high maintenance cost, reducing efficiency with increase in number of devices connected to the network, mandatory requirement of terminations and low security. It is also not suitable for networks with heavy traffic.

9.4.2 Star Topology

In star topology all work stations are connected to a central device with a point-to-point connection, unlike bus topology in which different devices are wired to a common cable that acts as a shared medium. The central device could be a hub, router or switch. The hub may act as a repeater or a signal booster depending on the nature of the central device. In addition to acting as a junction of the network, the hub also controls and manages the entire network. The central device can communicate with the hubs of other networks. Different devices in the star network are connected to the central device using unshielded twisted pair Ethernet cables. Each work station on the network communicates with the other workstations through the hub. Figure 9.5 shows the basic schematic diagram of a star network.

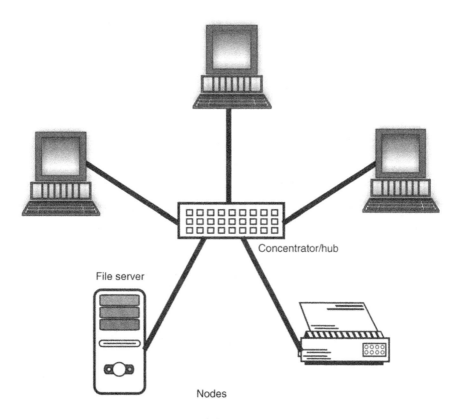

Figure 9.5 Star topology

A star network has the advantages of better performance compared to bus topology depending on the capability of the central hub and the ease with which new devices can be added or existing devices can be removed from the network without affecting the rest of the network. The network is easy to monitor and troubleshoot, and the failure of one node does not affect the rest of the network. The network, however, has some shortcomings. The whole of the network fails with failure of the central hub. The expansion capability is also governed by the capacity of the central hub.

9.4.3 Ring Topology

In ring topology all workstations are connected to one another in a closed loop, as illustrated in the basic schematic diagram of Figure 9.6. Each workstation is directly connected to one workstation on either of the two sides, as shown in the diagram. Data communication in the case of ring topology takes place with the help of a *token*. Data travels around the network in one direction. The sourcing workstation sends out the data along with token information. Every successive node examines the token to determine if the data were meant for it. If yes, data are received and an empty token is passed onto the network. If no, data along with token are passed onto next node. The process continues until the data reach their intended destination. Only those nodes equipped with token can send data. Others have to wait until an empty token reaches them.

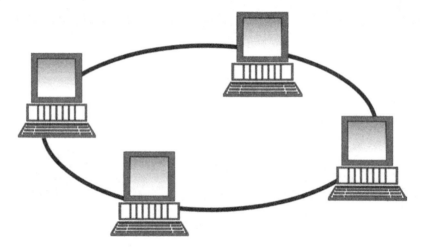

Figure 9.6 Ring topology

In ring topology data flows in one direction at high speed and the probability of collision is low as each node gets to send data only when it receives an empty token. Network performance is not affected by the addition of new devices. Each device has equal access to network resources. The topology offers better performance compared to bus topology in the case of increased load on the network. However, it is relatively slow compared to star topology as the data has to pass through all nodes between the source and destination nodes. The network is affected by a fault in any of the nodes. A failure in the cable or any device breaks the loop, bringing down the entire network. The topology is relatively more costly to implement compared to bus and star topologies.

9.4.4 Mesh Topology

In mesh topology each network node is connected to every other node. This is true mesh topology (Figure 9.7). However, mesh topology suffers from a large number of redundant connections and consequently is more expensive to implement. It is not generally used for computer networks and is preferred for wireless networks. The shortcomings of full or true

Network Topologies

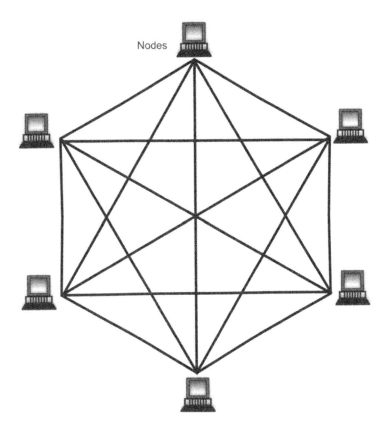

Figure 9.7 True Mesh topology

mesh topology are overcome in the case of partial mesh topology in which some of the nodes are connected, as for true mesh topology, while others are connected to one or two nodes only. In other words, different nodes are either directly or indirectly connected to every other node. This reduces redundancy and the cost of the network.

Mesh topology allows simultaneous transmission from different nodes. Also, data transmission is not affected as a result of failure of a particular node. The topology is amenable to expansion and modification without affecting existing nodes. The topology also has several disadvantages including high cost, high redundancy and is difficult to maintain and administer.

9.4.5 Tree Topology

Tree topology combines the features of star and bus topologies. In fact, tree topology is an expanded star topology in which multiple star networks are interconnected by a common bus. Figure 9.8 shows the basic schematic diagram of tree topology. It is known as tree topology for obvious reasons as the common bus represents the stem of the tree and the multiple star networks connected to the bus act like branches of the tree. Ethernet protocol is commonly used in tree topology.

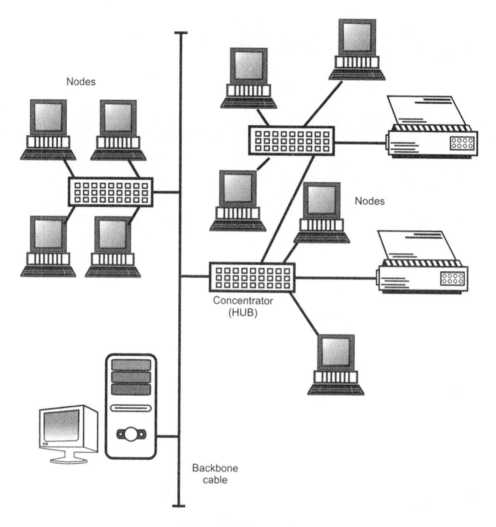

Figure 9.8 Tree topology

Tree topology allows more efficient network expansion than what is achievable with star or bus topologies individually. While star network expansion is limited by the capacity of the central hub, there is a limit to the maximum number of devices that can be connected to a bus network due to the broadcast traffic the network would generate. Tree topology overcomes these problems. It also allows easier error detection and correction, and damage to any one segment does not affect other segments of the network. The disadvantages include heavy dependence of tree topology on the central bus and the network scalability being dictated by the type of common cable.

9.4.6 *Hybrid Topology*

Hybrid topology is the result of integration of two or more of the basic network topologies discussed in the previous sections. The resultant hybrid topology combines the good and bad

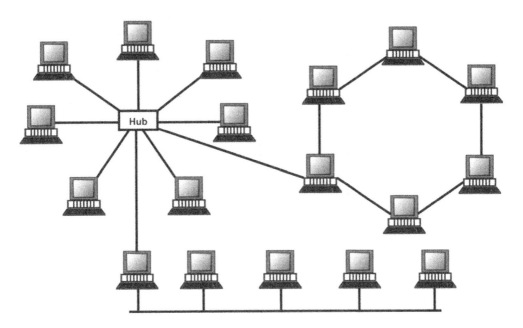

Figure 9.9 Hybrid topology

features of all constituent topologies. Different hybrid topologies may be configured depending on application requirement. For example, if the network of one office uses a star topology and that of another a bus topology and if the two networks were to be integrated in a wide area network, a star–bus hybrid topology would be selected. Another common hybrid topology is star–ring topology. Figure 9.9 shows the basic schematic diagram of a hybrid topology that is a combination of star, bus and ring topologies.

Hybrid network topologies are: scalable, as the network size can be conveniently enhanced without affecting the network architecture, reliable, as troubleshooting is relatively easier and the faulty part can be isolated without affecting the network, and flexible, as they allow design of a network customized for the requirement. Hybridization also allows the good points of the constituent topologies to be maximized and the weak points to be minimized.

9.5 Network Technologies

In the previous section we discussed the different network topologies used to connect various nodes. This section describes the technologies used to exchange information between different nodes. A common method of differentiating network technologies is on the basis of the path followed for the flow of information between the communicating devices. There are two broad categories of networking methods, namely circuit switching and packet switching, with the consequent networks known as circuit switched networks and packet switched networks. These are briefly described in the following sections.

9.5.1 Circuit Switched Networks

In a circuit switched network, a dedicated physical path is established between the communicating nodes through the network before the start of actual communication. This dedicated

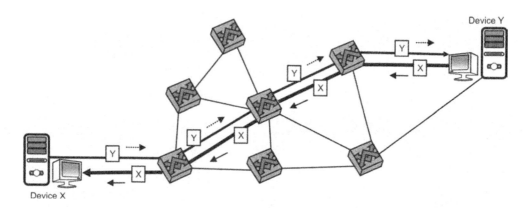

Figure 9.10 Circuit switched network depicting the dedicated path between communicating nodes

path is held for the duration of the communication session. The circuit functions as if the two nodes were physically connected by an electrical circuit. The circuit may be a fixed one that is always present or may be created on a requirement basis. It may be mentioned here that even though alternative paths may exist between the two communicating nodes or devices, the communication in the case of circuit switched networks takes place only over the identified path. The dedicated circuit cannot be used by other callers until the circuit is released and a new connection is set up. Also, the communication channel or circuit remains unavailable for use to other callers even if there is no communication actually taking place. Figure 9.10 illustrates the concept of a circuit switched network. The dedicated circuit for communication between the two devices X and Y is illustrated in the figure. As is evident from the schematic representation of the circuit switched network shown in Figure 9.10, communication between the two devices takes place over this dedicated circuit only, even though there are several alternative paths available. A telephone system such as the *public switched telephone network* (PSTN) is a classical example of a circuit switched network. The PSTN is a worldwide collection of interconnected public telephone networks primarily designed for handling voice traffic. Each time a caller makes a call, the network establishes a dedicated circuit between the calling party and the called party, and this circuit is used for the duration of the call. Next time a call is made by the same party to the same destination, the network identifies a circuit again, which may be different from the one provided earlier. Figure 9.11 depicts the communication path between devices X and Y at different times.

The circuit switched concept can also be used between two communicating nodes for the transfer of information other than voice. It should not be thought to be a technique used only for connecting analogue and digital voice circuits.

9.5.2 Packet Switched Networks

In packet switched network technology there is no dedicated communication path or circuit identified prior to the start of communication for the entire duration of the communication between the source and the destination. The data in this case is broken down into small pieces called packets and then sent over the network based on the destination address contained

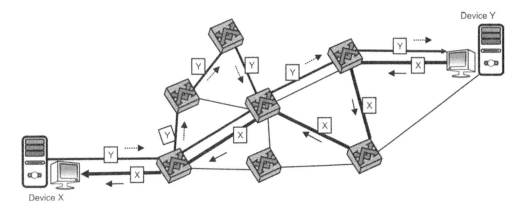

Figure 9.11 Circuit switched network with a changed dedicated path between communicating nodes

within the packet. Packets of data from a given source may take any number of different paths within the network while travelling to the destination. However, the communication circuit is dedicated for the duration of packet transmission to that packet alone and is not interrupted to transmit other packets. The data is reassembled into their original form at the receiving end. Figure 9.12 illustrates the concept of a packet switched network. Packet switching is used in transmitting data over the internet and often over LAN. The availability of multiple paths allowing the same line to be used for multiple communications simultaneously leads to improved efficiency, particularly when a large volume of traffic is to be handled. The packet switched mode is the signal transmission technology used in all internet communications.

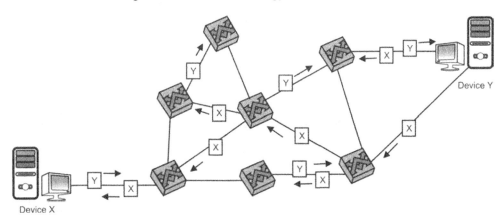

Figure 9.12 Packet switched network

9.5.3 Circuit Switched versus Packet Switched Networks

1. Circuit switched networks make use of a dedicated circuit established prior to beginning the communication between the two involved parties and this dedicated circuit is committed to the communication for its entire duration. The PSTN is an example of a circuit switching

mode of operation. On the other hand, in packet switched networks data are broken into a large number of packets with different packets following different paths to reach their eventual destination. There is a dedicated circuit only for transmission of a data packet. VoIP and internet telephony use packet switching.
2. Data transfer through a circuit switched network involves three phases: connection establishment, data transfer and connection release. This causes a call set-up delay, which increases with traffic. Delivery delay is constant irrespective of traffic.
3. In circuit switched networks data rates at the source and destination are the same; data rates in packet switched networks need not be the same.
4. Circuit switching allows a given circuit or channel to be used for only one communication at a time while packet switching allows simultaneous transmission of multiple communication signals on a single circuit using interspersed data packets.
5. The circuit switched communication mode is more costly than the packet switched communication mode. When one makes a PSTN call, which uses circuit switching, the communication circuit is basically rented and therefore the cost of making the call is directly proportional to the call duration. In a call using PSTN, the involved parties are speaking one at a time, which implies that one speaks only for half of the total call duration. With VoIP, which represents the packet switching mode of operation, multiple communication signals can be sent simultaneously over a single circuit.
6. Circuit switched networks, which use dedicated circuits for data transfer for the entire duration of the communication session, are more reliable than packet switched networks, which allow concurrent use of a given circuit by multiple communication signals. There is a good probability of loss of data packets or data packets showing up in the wrong order in packet switching. This explains the relatively lower quality of VoIP voice compared to PSTN. Packet switching is made more reliable by using TCP protocol.

9.6 Networking Protocols

A networking protocol defines the standard used for communication between different devices connected to a network, such as a LAN, intranet or the internet. Networking protocols include mechanisms for identification of network devices to make connections as well as formatting rules that specify how data is packaged into messages to be sent and received by different devices. To ensure a high performance and reliable network communication, some networking protocols also have message acknowledgement and data compression features built into them.

9.6.1 Common Networking Protocols

Hundreds of different networking protocols have been developed, each designed for specific application requirements. Each protocol has its own method of formatting data before it is sent or processing data after it is received. Data compression and error correction techniques also differ for various protocols. It is not feasible to discuss each of these hundreds of protocols, but the most commonly used ones are discussed in the following paragraphs. These include the OSI model, IP, TCP, HTTP, FTP, asynchronous transfer mode (ATM), simple mail transfer protocol and user datagram protocol (UDP).

The different protocols listed in alphabetical order include the following:

Address resolution protocol (ARP): Used with IP to map 32-bit IP addresses into MAC addresses.

Aloha and *slotted aloha*: Used for satellite and terrestrial radio transmission and are the basis for Ethernet LAN protocol.

Analog display service interface (ADSI): Allows voice and data to be heard and displayed on devices such as phones over an analogue telephone line.

Apple Talk: A proprietary networking protocol used for communication between Apple Macintosh computers and networking devices. It has been replaced by TCP/IP.

Asynchronous transfer mode (ATM): High speed networking protocol that supports voice and data communication.

Bootstrap protocol (BOOTP): Allows clients to discover certain network information such as their own IP address.

Challenge handshake authentication protocol (CHAP): An authentication protocol used over point-to-point protocol (PPP).

Common internet file system (IFS): Allows users of multiple platforms to share files with millions of other users.

Client to client protocol (CTCP): Used to obtain information such as a ping (packet internet groper), time, user information or version information.

Datagram delivery protocol (DDP): An Apple Talk protocol that ensures that data packets are sent and received properly.

Dynamic host configuration protocol (DHCP): Assigns IP addresses to computers and other devices connected to the network.

Electronic data interchange-internet integration (EDIINT): A set of protocols for conducting highly structured interorganization exchanges.

Enhanced interior gateway routing protocol (EIGRP): A proprietary CISCO routing protocol.

Ethernet: A widely used LAN protocol having a number of variants.

Full duplex handshaking protocol (FDHP): Used by duplex modems.

Frame relay (FR): Designed to replace X.25 networking protocol.

File sharing protocol (FSP): Similar to FTP and used with TCP.

File transfer protocol (FTP): A commonly used protocol for sending and receiving files between two computers.

H.323: An ITU standard for IP telephony.

High level data link protocol (HDLP): A common layer-2 protocol of the OSI model.

Hyper text transfer protocol (HTTP): A set of standards allowing users of the WWW to exchange information found on web pages.

Internet message control protocol (IMCP): An extension to IP.

Internet group management protocol (IGMP): A standard for IP multicasting involving sending data to multiple recipients.

Interior gateway protocol (IGP): A routing protocol used to route information in an autonomous system.

Internet protocol (IP): Dictates how packets of data are sent over networks.

Internet protocol security (IPSec): A set of protocols designed to support secure exchange of packets at the IP layer.

IPv4, IPv6: Versions of IP.

Internet packet exchange/sequential packet exchange (IPX/SPX): A LAN communications protocol exchanging information between network clients, applications and peripherals.

Kerberos: A network authentication protocol designed to encrypt and secure data on an insecure network.

Layer-2 tunnelling protocol (L2TP): A tunnelling protocol used with VPNs.

Network control protocol (NCP): Enables communications of protocols on PPP.

Network news transfer protocol (NNTP): Used by USENET to transfer postings between clients and servers.

Name server protocol (NSP): Used with TCP/IP and responsible for taking requests from other computers and providing name-to-number translation.

Network time protocol (NTP): Used to synchronize the time of computers and other network devices.

Open shortest path first (OSPF): A routing protocol that helps to determine the best or the shortest path to the next hop in a network.

Packet ensemble protocol (PEP): A communications protocol that allows 9600 bps modems to communicate.

Post office protocol (POP): A protocol commonly used to receive email.

Point-to-point protocol (PPP): A communications protocol that allows a dialup connection to be used to connect to other networking protocols such as TCP/IP, IPX etc.

Point-to-point tunnel protocol (PPTP): Allows companies to extend their networks through private tunnels over the internet and create a VPN.

Routing information protocol (RIP): Used by computers and other network devices such as routers to broadcast known addresses, allowing network devices to learn available routes.

Real time streaming protocol (RTSP): A standard used to control streaming data.

Secure file transfer protocol (SFTP): A version of FTP that uses encryption to enhance security.

Serial line internet protocol (SLIP): An internet protocol that allows users to gain internet access with a computer modem.

Snap: An Ethernet protocol; also refers to a feature introduced in Microsoft Windows 7 that allows users to see two windows side by side without the requirement for resizing the window.

Simple mail transfer protocol (SMTP): A communications protocol used to send email messages from one server to another over port 25.

Simple network management protocol (SNMP): Used to examine and change configuration parameters of routers, switches, repeaters, bridges and so on connected to local and wide area networks.

Simple object access protocol (SOAP): Used to exchange XML messages over a network.

Socks: Used for handling client to server communications made through a proxy server.

Secure shell (SSH): A secure protocol for remote log-ins.

Spanning tree protocol (STP): Designed to create a single path over a network thereby preventing any loop formation in the presence of multiple paths to the intended destination.

Trivial file transfer protocol (TFTP): A file transfer protocol similar to FTP used to load configuration files to routers or boot workstations without disks.

Transmission control protocol/internet protocol (TCP/IP): A combination of two protocols, TCP and IP, working together to govern all communications between computers on the internet. IP controls the mode in which packets of data are sent over networks. TCP ensures reliability of data transmission across networks connected by the internet.

User datagram protocol (UDP): an alternative protocol to TCP/IP that runs on the top of IP and known as UDP/IP.

Voice over internet protocol (VoIP): An internet protocol that allows users to make calls over the internet; also called IP telephone or internet phone.

Virtual router redundancy protocol (VRRP): Used with routers to prevent network down time.

Wireless application protocol (WAP): Defines specifications for a set of communication protocols that enable wireless devices to access the internet.

X.25: A CCITT-series standard defining layers 1, 2 and 3 of the OSI model and approved for the packet switching protocol on LANs designed for burst network traffic.

Zip: A file extension associated with software compression programs such as WinZip and PKZIP. It is a collection of software programs or files compressed into one zip file.

9.6.2 The Open Systems Interconnect (OSI) Reference Model

The OSI reference model is at the core of the OSI standard developed in 1984 by the International Organization for Standardization (ISO), an international federation of national standards organizations representing 130 countries. The OSI reference model comprises seven layers that define the different stages that data must go through when transported from one device to another in a network. Based on the nature of activities performed by different layers of the model the seven layers of the OSI reference model are divided into two sets: the *transport set* comprising layers 1 to 4 and the *application set* comprising layers 5 to 7. While the upper layers of the OSI model (layers 5, 6 and 7) manage the application level functions and represent the software that implements network services such as encryption and connection management, the protocols that operate at lower layers control end-to-end transport of data between devices and implement hardware oriented functions such as routing, addressing and flow control. In the OSI reference model when the two devices communicate with each other, the flow of information begins with the topmost layer on the sending end. The information flows down the stack of layers to the bottommost layer at the sending end. It then traverses the network connection to the bottommost layer of the receiving end, from where it moves up the stack to the topmost layer of the model. Figure 9.13 illustrates the flow of information in the OSI reference model. It may be mentioned here that when the communication is via a network of intermediate systems, only the lower three layers of the OSI protocols are used in the intermediate systems. As stated earlier, the OSI reference model has seven layers. The following sections give brief descriptions of each of these seven layers.

The seven layers of the OSI reference model, beginning with the bottommost layer, are:

1. physical layer
2. data link layer
3. network layer
4. transport layer
5. session layer
6. presentation layer
7. application layer

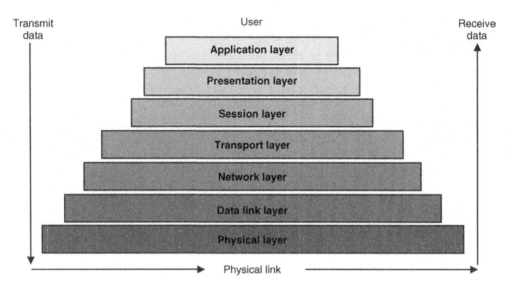

Figure 9.13 Seven layers of OSI reference model

9.6.2.1 Physical Layer

The *physical layer* (layer-1) transmits bits from one node to another and regulates the transmission of a bit stream over a physical medium. This layer defines how the cable is attached to the network adapter and what transmission technique is used to send data over the cable. The physical layer of the OSI model controls the signalling and transfer of raw bits onto the physical medium. It also describes the electrical/optical, mechanical and functional interfaces to the physical medium. The physical layer performs the following major functions. It defines the physical and electrical specifications of the interface with the physical transmission medium, protocol to establish and terminate a connection between two directly connected nodes, protocol for flow control and conversion of raw data into corresponding signals to be transmitted over the communication medium.

9.6.2.2 Data Link Layer

The *data link layer* (layer-2) provides error-free transfer of data frames from one node to another over the physical layer, allowing layers above it to assume virtually error-free transmission over the link. It performs its function by establishing and terminating the logical link between the two nodes, telling the transmitting node to back-off in the absence of any frame buffers, transmitting/receiving frames sequentially, detecting and recovering from errors occurring in the physical layer by retransmitting frames not acknowledged, creating and recognizing frame boundaries, checking received frames for integrity and providing media access management.

9.6.2.3 Network Layer

The *network layer* (layer-3) controls the inter-network communication and determines the route from the source computer to the destination computer. It performs its functions by routing frames among networks, controlling traffic by instructing a sending station to hold back frame transmission if the router's buffer were filled up and doing frame fragmentation where the router's maximum transmission unit size is less than the frame size. The network layer's two key responsibilities include providing a unique address that identifies the host and the network the host exists on (a function called *logical addressing*) and finding the best path to the destination network and routing data accordingly (a function called *routing*). Two of the most common network layer protocols include IP and inter-network packet exchange (IPX).

9.6.2.4 Transport Layer

The *transport layer* (layer-4) handles error recognition and recovery. It ensures error-free and sequential delivery of messages, and prevents losses and duplications. When necessary, it repackages long messages into small packets at the sending end and rebuilds the data packets into the original message at the receiving end. It also sends receipt acknowledgments at the receiving end. Transport layer communication falls under two categories, namely connection oriented, which requires a connection with specific agreed-upon parameters to be established before data is sent, and connectionless, which requires no connection for data to be sent. The TCP/IP suite incorporates two transport layer protocols: TCP for connection oriented communication and UDP for connectionless communication.

9.6.2.5 Session Layer

The *session layer* (layer-5) is entrusted with the task of establishing, maintaining and ultimately terminating sessions between devices. It performs the functions that allow devices to communicate over the network, performing security, name recognition, logging and so on. If a session is broken, the session layer attempts to recover it. Session layer communication falls under one of three categories: full duplex, which is simultaneous two-way communication, half duplex, which is two-way communication but not simultaneous, and simplex, which is one-way communication. The session layer often has to rely on lower layer protocols for session management as many contemporary protocol suites such as TCP/IP do not implement session layer protocols.

9.6.2.6 Presentation Layer

The *presentation layer* (layer-6) translates data from the application layer into a network format while sending information and vice versa while receiving information. Standards have been developed for the formatting of data types such as audio (MIDI, MP3 and WAV), text (RTF, ASCII and EBCDIC), video (MPEG, AVI and MOV) and images (GIF, TIF and JPG). This layer also manages security issues by providing services such as data encryption and compression. The presentation layer translates data from a format used by the application layer into a common format at the sending station and translates data from the common format to a format known to the application layer at the receiving station. In other words, it ensures that the data from the sending application can be understood by the receiving application.

9.6.2.7 Application Layer

The *application layer* (layer-7) provides the interface between the user application such as an email client or a web browser and the network. The application layer performs the functions of identifying communication partners, determining resource availability and synchronizing communication. Application layer protocols include FTP (via an FTP client), HTTP (via a web browser), POP3 and SMTP (via an email client), and Telnet. Application protocols reside at the application layer. The user interacts with the application, which in turn interacts with the application protocol.

9.6.3 Internet Protocol (IP)

IP was developed in the 1970s and is the primary network protocol used on the internet. It is generally used together with TCP on the internet and many other networks, and referred to as TCP/IP. It is a mechanism for transfer of data packets between computers by allowing computers to be connected by a variety of physical media, including modems, Ethernet cabling, fibre optics and radio and satellite links.

Each machine on the internet has a unique number with which it can be identified. This unique number is called its *IP address*. The two common versions of IP include IPv4 and IPv6. IP addresses in IPv4 have a length of four bytes (32 bits). In the newer IPv6, IP addresses are 16 bytes (128 bits) long.

IP addresses are normally expressed in decimal format as a dotted decimal number comprising four octets. Each octet can be represented by an 8-bit binary number, which means 32 bits in all for the IP address. This further implies a total of $2^{32} = 4,294,967,296$ (approximately 4.3 billion) possible unique values. Octets are used to create classes of IP addresses that can be assigned to a particular entity based on size and requirement. The octets are split into *net* and *host* sections. The net section is used to identify the network the computer belongs to and always contains the first octet. The host section is used to identify the actual computer on the network and always contains the last octet.

There are five classes of IP ranges: Class A, Class B, Class C, Class D and Class E. Each class has a range of valid IP addresses. These are shown in Table 9.1 along with the networks supported by them.

Table 9.1 Classes of IP addresses

Class	Address range	Networks supported
Class A	1.0.0.1 to 126.255.255.254	16 million hosts on each of 127 networks
Class B	128.1.0.1 to 191.255.255.254	65,000 hosts on each of 16,000 networks
Class C	192.0.1.1 to 223.255.254.254	254 hosts on each of 2 million networks
Class D	224.0.0.0 to 239.255.255.255	Reserved for multicast groups
Class E	240.0.0.0 to 254.255.255.254	Reserved for future purposes and R&D

Data on an IP network is organized into packets. Each IP packet includes a header that specifies source, destination and other information about the data other than the message data itself. The IP suite enables these data packets to be transferred between computers based on the IP addresses contained in the data packet header.

9.6.4 Transmission Control Protocol (TCP)

TCP is one of the main protocols in the IP suite and is used along with IP to send data packets between computers over the internet. While IP takes care of handling the actual delivery of the data, TCP takes care of keeping track of the individual data packets that a message is divided into for efficient routing through the internet. It enables the sending and receiving hosts to establish a connection to exchange data streams and also ensures that the packets are delivered in the same order in which they were sent. TCP is responsible for ensuring the division of data into packets at the sending end and reassembly of data packets into the original message at the receiving end.

The role of TCP/IP in data exchange between two computers can be explained with the help of the example of a web server sending an HTML file. The TCP program layer in the server divides the file into different data packets, which are then numbered. The data packets are forwarded to the IP program layer one at a time. Though all data packets have the same destination IP address, different data packets may follow different routes through the network to reach the intended destination. At the receiving end computer, TCP reassembles the individual data packets after they have all arrived and forwards them as a single file. TCP is a connection oriented protocol, that is, it establishes and maintains the connection until the data to be exchanged by application programs at the two ends has been completed. TCP works at *transport layer* (layer-4) of the OSI model.

9.6.5 Hyper Text Transfer Protocol (HTTP)

HTTP is an application layer protocol built on top of TCP. HTTP utilizes TCP port 80 by default, though other ports such as 8080 can also be used. It is the underlying protocol used by the WWW. HTTP defines how messages are formatted and transmitted. It provides a standard for communication of web browsers and servers. Web browsers (HTTP clients) and servers communicate via HTTP request and response messages. For example, when a URL is entered in the browser an HTTP command is sent to the web server directing it to fetch and transmit the requested web page. The three main HTTP message types are GET, POST and HEAD. HTML is the other standard controlling the working of the WWW. HTML covers how web pages are formatted and displayed.

Currently HTTP version 1.1, which is an improvement over HTTP version 1.0, is in widespread use. One of the shortcomings of HTTP is that it is a stateless protocol, that is, each command is executed independent of knowledge of the previous commands, which makes it difficult to implement websites that react intelligently to user input. This shortcoming is being addressed by a number of new technologies such as ActiveX, JavaScript and Cookies.

9.6.6 File Transfer Protocol (FTP)

FTP is a standard based on IP used for transferring files between computers on the internet. It is an application protocol that makes use of TCP/IP protocols like HTTP, used to transfer displayable web pages and related files, and SMTP, which transfers email. FTP is also used to refer to the process of copying files using FTP technology.

Transfer of files using FTP technology takes place as follows. As the first step an FTP client program initiates a connection to a remote computer running FTP server software. To connect to an FTP server, a username and password as set by the administrator of the server are required. Clients identify an FTP server either by its IP address or by its host name. After the client is connected to the server, copies of files can be sent and/or received singly or in groups. Publicly available files can be accessed using the user name 'Anonymous'. Though most network operating systems include simple FTP clients, many alternative third party FTP clients with enhanced performance features are available. FTP supports ASCII (plain text) and binary modes of data transfer, which can be set in the FTP client. The transferred file will not be usable by the intended recipient if it were, for instance, a binary file transferred in text mode.

9.6.7 *Simple Mail Transfer Protocol (SMTP)*

SMTP is a TCP/IP protocol used to transfer email messages between servers. It is usually used along with either of the two other protocols, namely POP3 and IMAP, which are used by messaging clients for retrieval of email messages. POP3 and IMAP allow the recipients' saved messages in their server mailbox to be downloaded periodically from the server. This helps SMTP overcome its limited ability to queue messages at the receiving end. In other words, while SMTP is typically used for sending emails, either POP3 or IMAP is used for receiving emails. SMTP usually operates over internet port 25. A large number of mail servers now support *extended simple mail transfer protocol* (ESMPT). ESMTP also allows multimedia files to be delivered as email. X.400, which is an alternative SMTP, is widely used in Europe.

SMTP is reliable and simple. In a typical SMTP transaction, a server identifies itself and announces the type of operation it is intending to perform. Once the operation has been authorized by the other server, the message is sent. If there is something wrong, such as the wrong address, the receiving server will respond with an appropriate error message.

9.6.8 *User Datagram Protocol (UDP)*

UDP was introduced in 1980 and is one of the oldest protocols in use. It is an alternative to TCP. Together with IP it is referred to as UDP/IP. UDP does not have some of the desirable features of TCP. While TCP performs the task of dividing the message into data packets at one end and reassembling them at the other end, UDP does not provide this service. Application programs that use UDP therefore must make sure that the entire message has arrived and is in the right order. Also, checksums that protect data from tampering or getting corrupted during transmission are mandatory in TCP but optional in UDP. However, like TCP, UDP also operates on the transport layer (layer-4) of the OSI reference model.

One message unit in the case of UDP network traffic is called a datagram and comprises a header section and a data section. The header section comprises four fields: source port number, destination port number, datagram size and checksum. Each field has a length of two bytes, which implies that the header information is contained in eight bytes. The size of datagrams varies depending on the operating environment but can be a maximum of 65535 bytes.

9.6.9 Asynchronous Transfer Mode (ATM)

ATM is a high-speed networking protocol that supports both voice and data communications. Faster processing and switching speeds are possible as the protocol is designed to be easily implemented by hardware rather than software. It uses asynchronous time-division multiplexing. It encodes data into small, fixed-sized cells of 53 bytes length comprising 48 bytes of data and five bytes of header information. ATM operates at the data link layer (layer-2) of the OSI model and transmits data over a physical medium such as fibre or twisted-pair cable. It differs from data link technologies such as Ethernet in the sense that it uses fixed length data packets as opposed to variable length packets in Ethernet. It does not utilize routing, and hardware devices such as ATM switches are used to establish a point-to-point connection between the end points for flow of data directly from source to destination.

The performance of ATM is often expressed in the form of *optical carrier* (OC) levels written as 'OC-xxx'. OC levels are a set of signalling rates for transmitting digital signals on optical fibres. These are designed for transmission over synchronous optical networks (SONETs) and are also applicable to ATM networks. The base rate is 51.84 Mbps (OC-1). The pre-specified bit rates are either 155.520 Mbps (OC-3) or 622.080 Mbps (OC-12). Performance levels as high as 10 Gbps (OC-192) are technically achievable. Along with SONET and several other technologies, ATM is a key component of broadband ISDN (BISDN).

9.7 Satellite Constellations

A large number of satellites have been launched individually and in groups for a variety of applications, including communications and other purposes such as remote sensing, meteorology, navigation and so on. While individual satellites in geostationary orbits provide a relatively large fixed footprint on a round the clock basis, those in lower orbits such as low Earth orbits provide a small footprint at any given time that is repeated with a certain periodicity. In both cases, individual satellites do not provide round the clock global or near global coverage. This problem is overcome by having satellite constellations. A satellite constellation is a group of satellites with their operation so synchronized as to provide coordinated ground coverage on a round the clock basis. This implies that the footprints of different satellites in the constellation sufficiently overlap to provide uninterrupted global or near global coverage on a 24 × 7 basis. There are a number of operational satellite constellations and many more are in the pipeline for a variety of applications including voice and data communication, satellite radio, messaging and navigation. Beginning with the requirements of constellation geometry, major satellite constellations are briefly described in the following paragraphs.

9.7.1 Constellation Geometry

There are a number of different constellation geometries to satisfy the intended mission requirements. Three most important orbital parameters governing satellite constellation geometry are *altitude*, *inclination* and *eccentricity*. Orbit altitude is chosen on the basis of both physical and geometric considerations, which include coverage area, time of satellite visibility and revisit periodicity, signal propagation delay, signal power and avoidance of Van Allen radiation belts. Orbit inclination is the second important parameter of a satellite constellation. The choice of orbital inclination is governed by the requirement of global coverage and the minimum angle of elevation, for example higher inclination provides more coverage to polar regions. Similarly, an inclination of around 45° allows coverage of temperate zones and populated regions of the Earth. Orbit eccentricity determines the shape of the orbit, which in turn may affect the dwell time of a satellite in the constellation over a certain specific area on the ground. For example, for a satellite in an eccentric elliptical orbit, the dwell time of the satellite can be maximized over the region of interest by adjusting the position of the apogee.

Satellite constellations are usually designed with the satellites in the constellations having similar orbits, eccentricity and inclination with the advantage that any perturbations affect each satellite in more or less the same manner. This also helps in preserving the constellation geometry and thereby minimizing station keeping and fuel usage requirements. Also, sufficient separation is maintained between adjacent satellites in the same orbital plane to avoid interference and prevent collision.

There are two major types of satellite constellation: the *polar constellation* and the *Walker constellation*. The Walker constellation has an associated notation proposed by John Walker according to which the constellation is represented by $i: t/p/f$. In this notation, i is the orbital inclination, t is the total number of satellites, p is the number of equally spaced planes and f is the relative spacing between satellites in adjacent planes. Furthermore, there is the Walker Delta constellation and the near polar Walker Star constellation. While the Galileo navigation system belongs to the former category, the Iridium satellite constellation employs near polar Walker Star geometry (Figure 9.14). Both polar and Walker constellations are designed to provide global coverage or near global coverage with the minimum number of satellites. However, each constellation has its advantages and disadvantages. The polar constellation provides global coverage including the polar region. On the other hand, the Walker constellation provides coverage only to areas below a certain latitude, which for the Globalstar constellation is $\pm 70°$. As a consequence of this the Walker constellation is capable of offering higher diversity than the polar constellation for a given number of satellites. Diversity is the average number of satellites in view simultaneously to a user on ground. Higher diversity brings with it the added benefits of higher availability, which would mean fewer dropped connections and reduced multipath fading.

9.7.2 Major Satellite Constellations

Satellite constellations are used for a variety of applications, which include voice communication (e.g. the Iridium and Globalstar constellations), satellite radio (e.g. Sirius XM Radio), broadband networking (e.g. the Teledesic and SkyBridge constellations), messaging (e.g. the Orbcomm constellation) and navigation (e.g. the global positioning system (GPS), global navigation satellite system (GLONASS), Galileo constellations). In addition to these examples of satellite constellations there are many more satellite constellations for different categories of applications. Some of the major satellite constellations are discussed in the following chapters on applications of satellites. The prominent constellations are briefly described in the following sections. An outline of those discussed in the subsequent chapters is also presented here.

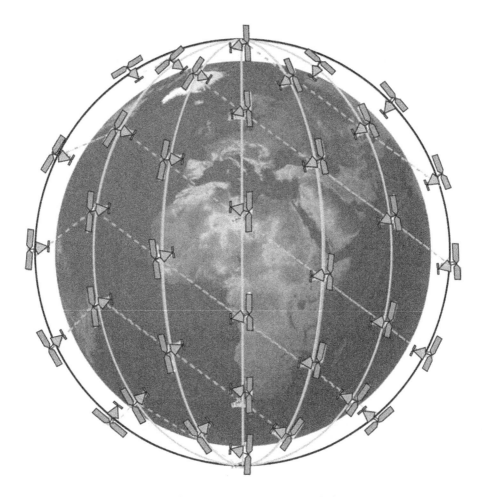

Figure 9.14 Iridium satellite constellation

9.7.2.1 Iridium Satellite Constellation

The Iridium satellite constellation is a global satellite network designed to provide voice communication, data, fax and paging services independent of the user's location in the world and of the availability of traditional telecommunication networks. The space segment of the constellation comprises 66 active satellites and 14 spare back up satellites orbiting at an altitude of 780 km in six orbital planes. The ground segment comprises gateways and a system control segment. Iridium subscriber products include phones and pagers that allow users to have access to either a compatible cellular telephone network or the Iridium network. The Iridium constellation is described in detail in section 10.10.1.3.

9.7.2.2 Globalstar Constellation

The US Globalstar satellite constellation provides global voice, data, fax and messaging services. The constellation comprises 48 satellites and an additional four in-orbit spares orbiting at an altitude of 1410 km and an inclination of 52°. The satellites are placed in eight orbital planes with six satellites in each plane (Figure 9.15). The constellation provides service on Earth between 70°N latitude and 70°S latitude covering the USA and 120 other countries.

Figure 9.15 Globalstar satellite constellation

9.7.2.3 Sirius XM Radio

Sirius XM Radio was formed by the merger of XM Satellite Radio with Sirius Satellite Radio in July 2008. The company provides two satellite radio services, Sirius Satellite Radio and XM Satellite Radio, in the USA. An affiliate company called XM Canada provides these services in Canada. Currently, the company provides music, sports, news and entertainment channels to listeners. The satellites used for the Sirius radio service are the Radiosat series comprising Radiosat-1 to Radiosat-4, to avoid confusion with Sweden's fleet of Sirius satellites. The first three satellites of the series were launched in 2000 using Proton-K Block-DM3 launch vehicles. Radiosat-4 was built as a spare satellite. The three Radiosat satellites broadcast directly to users' receivers. The satellites operate in an orbit of 23 975 km perigee and 46 983 km apogee inclined at 63.4° (tundra orbit) to provide higher angles of elevation in the northern regions of north America. Because of this highly elliptical orbit only two of the three satellites broadcast at any given time.

9.7.2.4 Teledesic Satellite Constellation

The Teledesic satellite constellation is a worldwide satellite network comprising 924 low Earth orbit satellites orbiting at an altitude of 700 km and placed in 21 orbital planes inclined at 98.2° with an adjacent plane separation of 9.5° (Figure 9.16). The constellation offers a wide range of services, including multimedia conferencing, video conferencing, voice communication, video telephony and distance learning, and provides seamless coverage to 100% of the population on Earth on a round the clock basis. The constellation has a peak load capacity of 2 000 000 simultaneous full duplex 16 kbps connections, with a service quality comparable to today's terrestrial communication systems.

Figure 9.16 Teledesic satellite constellation

9.7.2.5 SkyBridge Satellite Constellation

SkyBridge is a constellation of 80 low Earth orbit satellites divided into two symmetrical Walker sub-constellations of 40 satellites each. The satellites orbit at an altitude of 1457 km. Each satellite provides a 3000 km radius coverage divided into fixed spot beams of 350 km radius. The SkyBridge satellite constellation is designed to provide the communication infrastructure for a full range of broadband services, including interactive multimedia communication, high speed data communications and internet access.

All traffic management and routing in the SkyBridge constellation is handled on the ground, with no direct links between satellites. The gateway handles interconnections with local servers and terrestrial networks. Being a low Earth orbit satellite constellation the propagation delay is of the order of 20 milliseconds, similar to what it is in the case of landline broadband systems. This allows applications currently used for existing broadband networks to be seamlessly transmitted via SkyBridge.

9.7.2.6 Orbcomm Satellite Constellation

Orbcomm is a satellite constellation of 32 satellites located in four orbital planes A, B, C and D, with each plane having eight satellites. Planes A, B and C are inclined at 45° to the equator and the satellites in these planes orbit at an altitude of 825 km. Successive satellites in each of the planes are 45° apart. Plane D is inclined at 0° and the satellites in this orbital plane also orbit at an altitude of 825 km. There are two supplemental orbital planes, Plane F and Plane G, that contain two satellites each and have an altitude of 780 km. Planes F and G are inclined at 70° and 80°, respectively. The two satellites in each of these two planes are 180° apart. Orbcomm has a license to launch up to 48 satellites.

The Orbcomm system comprises a network control centre (NCC) to manage the operation of the overall system and three operational segments: the space segment of the satellites, the ground segment consisting of gateway Earth stations, and the control centre and subscriber segment. A fully deployed constellation is capable of providing a near real time wireless data communication service worldwide.

9.7.2.7 Global Positioning System

GPS is satellite based US global navigation system. The GPS constellation comprises 24 satellites and ground support facilities to provide three-dimensional position, velocity and timing information to users around the world on a 24 × 7 basis. The constellation is described in detail in Chapter 13.

9.7.2.8 Global Navigation Satellite System

GLONASS is a Russian satellite based navigation system. The constellation comprises 21 active satellites providing continuous global services, like GPS. The constellation is discussed in detail in Chapter 13.

9.7.2.9 Galileo Satellite Navigation System

Galileo is a global satellite navigation system being built by the European Union and the European Space Agency. It is designed to provide European countries with a fully autonomous and reliable satellite based positioning, navigation and timing capability independent of the US GPS and Russian GLONASS systems. Though the Galileo system is intended to be independent of GPS and GLONASS, systems, it is fully interoperable with them, thereby making it a fully integrated new element in global satellite navigation systems. The Galileo constellation is described in detail in Chapter 13.

9.8 Internetworking with Terrestrial Networks

The techniques of inter-networking of similar types of networks, such as inter-networking of terrestrial networks, are well established. Inter-networking of different types of networks, such as inter-networking of a satellite network with a terrestrial network, encounters multiple problems including those due to different transmission media, different data formats, different transmission speeds and different protocols. Satellite inter-networking with other types of networks such as terrestrial networks involves any of the three lower layers of the OSI reference model, namely the physical layer, the data link layer and the network layer. Repeaters, bridges, switches and routers are the commonly used interface elements. Repeaters operate at the physical layer, bridges operate at the data link layer and switches can work at any of the three lower layers, that is the physical layer, the data link layer or the network layer. Routers operate at the network layer.

9.8.1 Repeaters, Bridges, Switches and Routers

One of the issues that needs to be addressed while inter-networking a satellite network with a terrestrial network is related to use of different physical media and protocols. Inter-networking at the physical layer is at bit level. As the physical layer protocol functions are simple, inter-networking between satellite networks and terrestrial networks is relatively easy. Terrestrial networks have much higher data transmission rates compared to satellite networks, and this issue and other issues related to the use of different physical media and protocols need to be addressed by the interconnecting devices. Repeaters, bridges, switches and routers are commonly used network interconnection devices.

A *repeater* is a network device used to regenerate or replicate data signals. It relays data signals between sub-networks using different physical media and protocols. A repeater cannot do intelligent routing, which is performed by bridges and routers.

A *bridge* operates at the data link layer of the OSI model. It serves as the inter-networking unit between the satellite network and the terrestrial network. A bridge examines the incoming traffic from the satellite network and decides whether to forward it or discard it. In taking this decision the bridge may look at the source and destination addresses and even the frame size. If the frames are to be forwarded they are formatted according to the protocol of the terrestrial network. The reverse process of handling data flow from a terrestrial network to a satellite network, though similar, is more complicated. The satellite network in its interface with terrestrial networks has to deal with a large number of different types of networks and protocol translations.

Network *switches* are multiport bridges that are used to link multiple computers in a network. Network switches inspect data packets as they are received, determine the source and destination network device of each packet, and forward them to the intended device only. In this way a network switch conserves network bandwidth. Network switches operate at any of the three lower layers depending on the nature of the network.

A *router* is a network device used to connect multiple networks, either LANs or a LAN with an internet service provider's (ISP) network. A router forwards data packets along the networks using headers and routing tables to determine the best paths for forwarding them. Routers are located at the gateways where two or more networks are connected. Routers can be used to interconnect with heterogeneous terrestrial networks. In this case, all user terminals use IP. Routers operate at the network layer of the OSI model.

9.8.2 Protocol Translation, Stacking and Tunnelling

Protocol translation, stacking and tunnelling are the three commonly used techniques in inter-networking with heterogeneous networks. Protocol translation is implemented through *network address translation* (NAT) and *port translation*. NAT is the translation of an IP address used within one network to a different IP address known within another network.

Protocol stacking refers to a group of protocols that work together to allow software or hardware function. TCP/IP that uses four lower layers of the OSI model is an example.

Protocol tunnelling, also known as *port forwarding*, allows two of the same type of network to communicate with each other through other networks. The tunnelling technique can be used to carry data across an incompatible delivery network or provide a secure path through a network that is perhaps not trustworthy. One application of tunnelling is the transmission of data meant for use within a corporate network through a public network with routing nodes in the public network unaware of the fact that data belongs to the private network. Tunnelling would allow the use of the internet to convey data on behalf of private networks. There are a number of tunnelling protocols in use. Two of the better known protocols are *point-to-point tunneling protocol* (PTPP), developed by Microsoft, and *generic routing encapsulation* (GRE), developed by Cisco Systems.

9.8.3 Quality of Service

QoS is referred to as the collection of networking technologies and techniques to guarantee delivery of a certain performance level. With reference to QoS, elements of performance include network availability or uptime, latency or delay, bandwidth or throughput and error rate. In order to achieve the desired QoS, it is important that data transmission rates, error rates and other characteristics are monitored and measured, improved and to an extent guaranteed in advance. Traffic shaping techniques such as packet prioritization, application classification and queuing at congestion points can be used to improve the QoS. One such protocol that allows expedited delivery of data packets passing through a gateway host is the internet's resource reservation protocol.

There are three fundamental elements in the implementation of QoS, which are QoS identification and marking techniques to coordinate end-to-end quality of service between network elements, tools such as queuing, scheduling and traffic shaping to ensure QoS within individual network elements, and management and accounting functions to administer and control end-to-end traffic across the network.

Further Readings

Beyda, W.J. (1999) *Data Communications – From Basics to Broadband*, Prentice-Hall, New Jersey.
Sun, Z. (2005) *Satellite Networking, Principles and Protocols*, John Wiley & Sons, Chichester.
Zhang, Y. (ed.) (2003) *Internetworking and Computing over Satellite Networks*, Springer.
Sherrif, R.E. and Hu, Y.F (2001) *Mobile Satellite Communication Networks*, John Wiley & Sons, Chichester.
Richharia, M. (2001) *Mobile Satellite Communications – Principles and Trends*, Addison-Wesley Longman, Boston.
Lutz, E., Werner, M. and Jahn, A. (2000) *Satellites for Personal and Broadband Communications*, Springer, New York.
Giambene, G. (2007) *Resource Management in Satellite Networks*, Springer, New York.

Internet Sites

1. computernetworkingnotes.com
2. www.webopedia.com
3. compnetworking.about.com
4. www.howstuffworks.com
5. searchnetworking.techtarget.com
6. www.computerhope.com
7. computer.howstuffworks.com/osi.htm
8. www.iridium.com
9. www.gmat.unsw.edu.au
10. www.novatel.com/assets/Documents/Papers/GLONASSOverview.pdf

Glossary

Asynchronous transfer mode (ATM): This is a high-speed networking protocol that supports both voice and data communications
Availability: This refers to the ability of the network to respond to the requests of users intending to access the network
Bridge: A bridge serves as the internetworking unit between the satellite network and the terrestrial network. It examines the incoming traffic from the satellite network and decides on whether to forward it or discard it
Bus topology: In this network topology all devices to be connected to the network are connected to a central cable that acts as the backbone of the network with the help of interface connectors
Circuit switched network: In the case of a circuit switched network, a dedicated physical path is established between the communicating nodes through the network before the start of actual communication. This dedicated path is held for the duration of the communication session

Email: This is the short name for electronic mail and is the method of transmitting messages electronically from one computer user to one or more recipients over a communication network

Galileo constellation: This is a global satellite navigation system being built by the European Union and the European Space Agency. It is designed to provide European countries with a fully autonomous and reliable satellite based positioning, navigation and timing capability independent of the US GPS and Russian GLONASS

File transfer protocol (FTP): This is a standard based on IP used for transferring files between computers on the internet

Globalstar constellation: The US Globalstar satellite constellation comprises 48 satellites and four in-orbit spares and is designed to provide global voice, data, fax and messaging services

GLONASS constellation: GLONASS (global navigation satellite system) is a Russian satellite based navigation system. The constellation comprises 21 active satellites providing continuous global services like GPS

GPS constellation: GPS (global positioning system) comprises 24 satellites and ground support facilities to provide three-dimensional position, velocity and timing information to users around the world on a 24/7 basis

Hybrid topology: Hybrid topology is the result of integration of two or more basic network topologies. The resultant hybrid topology combines the good and bad features of all constituent topologies

Hyper text transfer protocol (HTTP): This is an application layer protocol built on the top of TCP. HTTP defines how messages are formatted and transmitted. The protocol provides a standard for communication of web browsers and servers

Internet protocol (IP): This is the primary network protocol used on the internet and is generally used with TCP

Iridium: This is a type of low Earth orbit satellite constellation with 66 active satellites and 14 in-orbit spares designed to provide voice communication, data, fax and paging services

Mesh topology: In mesh topology each network node is connected to every other node. This is true mesh topology

Network protocol: A network protocol defines the standard used for communication between different devices connected to a network such as a LAN, intranet or internet. Network protocols include mechanisms for identification of network devices to make connections as well as formatting rules that specify how data is packaged into messages to be sent and received by different devices

Open systems interconnect (OSI) reference model: The OSI model is at the core of open systems interconnect standard developed in 1984 by International Organization for Standardization (ISO). It comprises seven layers that define the different stages that data must go through when transported from one device to another in a network

Orbcomm satellite constellation: This is a satellite constellation of 32 satellites. A fully deployed constellation is capable of providing near real time wireless data communication services worldwide

Packet switched network: In packet switched network technology there is no dedicated communication path or circuit identified prior to the start of communication for the entire duration of communication between the source and destination

Quality of service: Quality of service is a collection of networking technologies and techniques to guarantee delivery of a certain performance level

Reliability: The reliability of a network is a measure of the reliability of different components of the network and their interconnections. It is generally measured as mean time between failures (MTBF) or mean time to repair (MTTR)

Repeater: This is a device used to connect two segments of the same network to extend its coverage. The primary function of a repeater is to regenerate data signals

Ring topology: In ring topology all workstations are connected to one another in a closed loop

Router: A router is the network device used to connect multiple networks, either LANs or a LAN with an internet service provider (ISP) network. A router forwards data packets along the networks using headers and routing tables to determine the best paths for forwarding them

Satellite constellation: A satellite constellation is a group of satellites with their operation synchronized to provide coordinated ground coverage on a round the clock basis. This implies that the footprints of different satellites in the constellation sufficiently overlap to provide uninterrupted global or near global coverage on a 24/7 basis

Scalability: This is the capability of a network to accommodate increased data rate requirements and number of users. It also shows how well a network can adapt itself to new applications and the replacement of old components by new components with enhanced features

Security: The objective of network security is to monitor and prevent unauthorized access, eavesdropping, misuse, modification or denial of use of a network and its resources to authorized users

Simple mail transfer protocol (SMTP): This is a TCP/IP protocol used to transfer email messages between servers

Sirius XM Radio: This is a satellite radio constellation formed by the merger of XM Satellite Radio with Sirius Satellite Radio and comprises the Radiosat series of satellites (Radiosat-1 through Radiosat-4). It provides two satellite radio services, Sirius Satellite Radio and XM Satellite Radio in the USA

SkyBridge satellite constellation: This is a constellation of 80 low Earth orbit satellites divided into two symmetrical Walker sub-constellations of 40 satellites each. The SkyBridge satellite constellation is designed to provide the communications infrastructure for a full range of broadband services, which includes interactive multimedia communication, high speed data communications and internet access

Star topology: In this topology, all work stations are connected to a central device with a point-to-point connection, unlike bus topology in which different devices are wired to a common cable that acts as a shared medium

Switch: Switches are multiport bridges that are used to link multiple computers in a network

Teledesic satellite constellation: This is a worldwide satellite network comprising 924 low Earth orbit satellites. The constellation offers a wide range of services, including multimedia conferencing, video conferencing, voice communication, video telephony and distance learning, providing seamless coverage to 100% of the population on Earth on a round the clock basis

Throughput: This is defined as the average rate at which data is transferred through the network in a given time and is measured in bits per second (bps) and also sometimes in data packets per second or data packets transferred in a given time slot

Topology: This shows the way different components of the network are connected and the logical way data passes through the network from one component or device to the next irrespective of the physical structure of the network

Transmission control protocol (TCP): This is one of the main protocols in the internet protocol suite and is used along with the IP to send packetized data between computers over the internet

Tree topology: Tree topology combines the features of star and bus topologies. In fact tree topology is an expanded star topology in which multiple star networks are interconnected by a common bus

User datagram protocol (UDP): This is one of the oldest protocols in use and is an alternative to TCP. Together with IP protocol it is referred to as UDP/IP

Voice over internet protocol (VoIP): This allows telephone calls to be made over a broadband internet network by converting analogue voice signals into digital data packets and using IP for the two-way transmission

World Wide Web (WWW): This is a system of internet servers that support documents formatted in HTML. It supports links to audio, video, graphics and other documents. Web browsers such as Firefox and Internet Explorer are applications that allow easy access to the World Wide Web

Part II

Satellite Applications

10

Communication Satellites

Since the launch of Sputnik-1 in the year 1957, over 8500 satellites have been launched to date for a variety of applications like communication, navigation, weather forecasting, Earth observation, scientific and military services. The term 'satellite' has become a household word today as the horizon of its applications has touched the life of everyone, whether it be talking to someone thousands of kilometres away within the comforts of one's own house in a matter of a few seconds, watching a variety of TV programmes or having access to the world news and weather forecast on a routine basis. Satellites are also being used as navigational aids by vehicles on land, in the air or on the sea; in remote sensing applications to unearth the hidden mineral resources which may otherwise have remained untapped, in astronomical research and in exploring the atmosphere. Because of its growing application potential, satellite technology which was originally confined to the developed countries is finding new outlets in the developing countries of the world.

Based on the intended applications, the satellites are broadly classified as communication satellites, navigation satellites, weather forecasting satellites, Earth observation satellites, scientific satellites and military satellites. In the following chapters, there will be a focus on this ever-expanding vast arena of satellite applications. The emphasis is on the underlying principles, the application potential and the contemporary status of these application areas. This chapter, in particular, focuses on communication satellites.

10.1 Introduction to Communication Satellites

Satellite telecommunication stands out as the most prominent one among other applications of satellites, both in terms of application potential and the number of satellites launched in each category. The application areas of communication satellites mainly include television broadcasting, international telephony and data communication services. Communication satellites act as repeater stations that provide either point-to-point, point-to-multipoint or multipoint interactive services.

The concept of using satellites for communication evolved back in 1945, when Arthur C. Clarke, a famous science fiction writer, described how the deployment of artificial satellites in geostationary orbit could be used for the purpose of relaying radio signals. The concept turned into reality in the year 1962, with the launch of Telstar-1, which established the first intercontinental link between USA and Europe, providing telephony as well as television services. In the past more than forty years, since the launch of Telstar-1, communication satellite technology has made progress by leaps and bounds. To date more than 3000 communication satellites have been launched, out of which more than 1000 satellites were launched in the decade 2000–2010. This is quite large when compared to the number of satellites launched in earlier decades: 150 satellites during 1960–1970, 450 satellites during 1970–1980, 650 satellites during 1980–1990 and 750 during 1990-2000. These electronic birds have tied the whole world together and made it look like a global village.

10.2 Communication-related Applications of Satellites

Telecommunication satellites provide a varied range of services mainly including television broadcasting, international telephony and data communication services, most of these services being multipurpose in nature. Traditionally, satellite applications included television broadcasting and fixed and mobile telephony services but now newer dimensions are being added to the spectrum of the satellite applications with the advent of services like the internet and multimedia. However, satellites are facing tough competition from terrestrial networks in general, with fibre optics in particular.

Satellite TV refers to the use of satellites for relaying TV programmes from a point where they originate to a large geographical area. GEO satellites in point-to-multipoint configuration are employed for satellite TV applications. There are primarily two types of satellite television distribution systems, namely the television receive-only (TVRO) and the direct broadcasting satellite (DBS) systems.

In satellite telephony, satellites provide both long distance (especially intercontinental) point-to-point or trunk telephony services as well as mobile telephony services, either to complement or to bypass the terrestrial networks. They are particularly advantageous when the distances involved are large or when the region to be covered is sparsely populated or has a difficult geographical terrain. Point-to-point satellite links are used for satellite telephony networks.

Satellites also provide data communication services including data, broadcast and multimedia services such as data collection and broadcasting, image and video transfer, voice, internet, two-way computer interactions and database inquiries. Satellites in this case provide multipoint interactive connectivity, enabling the user terminals to exchange information with the central facility as well as other user terminals. Low cost very small aperture terminals (VSATs), with each VSAT supporting a large number of user terminals, are used for implementing such a network.

Communication satellites can be GEO satellites or a constellation of LEO, MEO or HEO (highly elliptical orbit) satellites. GEO satellites maintain a key role in setting up national programmes and distributing traditional services such as television or more novel services such as access to the internet. New trends in mobile communication have led to the development of constellations of non-GEO satellites in the LEO, MEO and HEO. These constellations guarantee flexible links to users, without requiring Earth-based installations at all points on the globe. Hence, broadcasting services like TV, radio and telephony communication services mainly remain in the domain of GEO satellites while the newer services like messaging, voice, fax, data and video conferencing facilities are well suited to LEO, MEO or HEO satellite constellations.

10.2.1 Geostationary Satellite Communication Systems

The geostationary orbit has been the preferred orbit for satellite communication systems and provides most of the revenue for satellite system operators. The first geostationary communication satellite, named Early Bird (Intelsat 1) was launched by INTELSAT in 1965. Commercial satellites launched in the 1970s and 1980s were all geostationary satellites. These satellites were used for international, regional and domestic telephone and video distribution services. Some of the important geostationary satellite missions include Intelsat, Inmarsat, Telstar, Asiasat, Arabsat, Galaxy, GE, Superbird, Eutelsat, Astra, Palapa and so on. New trends in the field of satellite communication include the launch of satellites in non-geostationary orbits for some specialized applications. The most important applications of geostationary communication satellites in the current scenario include DTH satellite television broadcasting services and VSAT services.

10.2.2 Non-geostationary Satellite Communication Systems

Non-geostationary satellite communication systems are emerging to provide mobile communication services as well as other services like messaging, video, fax and data communication. Constellations of satellites orbiting in LEO or MEO orbits can provide global mobile communication services. However, the cost of building such a constellation of satellites is huge as compared to having a geostationary satellite. Therefore, these systems have not made great progress and are still in the developmental stage. IRIDIUM, Orbcomm, Globalstar and ICO systems are some of the non-geostationary satellite communication systems.

10.3 Frequency Bands

Satellite communication employs electromagnetic waves for transmission of information between Earth and space. The bands of interest for satellite communications lie above 100 MHz including the VHF, UHF, L, S, C, X, Ku and Ka bands. Frequency allocation and the coordination mechanism were explained in detail in the previous chapter.

10.4 Payloads

Transponder is the key payload of any communication satellite. A brief outline on the basic satellite communication link set-up would not be out of place. Basic elements of a satellite communication system (Figure 10.1) include the ground segment and the space segment. The ground segment comprises the transmitting and the receiving Earth stations together with their associated instruments, antennae, electronic circuits, etc. These Earth stations provide access to the space segment by transmitting and receiving information from the satellite, interconnect

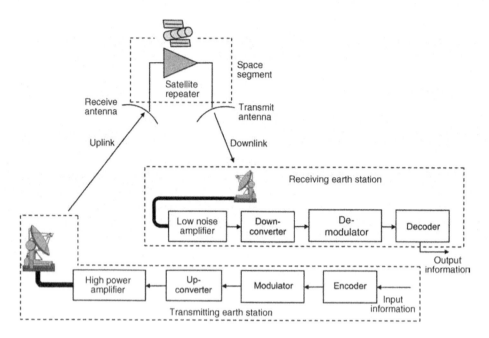

Figure 10.1 Basic elements of a satellite communication system

users with one another and with the terrestrial network. The space segment comprises one or more satellites, which act as repeater stations providing point-to-point, point-to-multipoint or multipoint interactive services.

The information to be transmitted (including voice channels for a telephone service, a composite video signal or digital data, and so on) is modulated using analogue or digital means, up-converted to the desired microwave frequency band of transmission (VHF, UHF, L, S, C, X, Ku or Ka), amplified to the required power level and then beamed up to the satellite from the transmitting Earth station (uplink). The received signals are amplified by the satellite, down-converted to a different frequency and then retransmitted towards Earth (downlink). The device on board the satellite that performs the amplification and frequency conversion is referred to as a transponder and is the main payload of any communication satellite. Satellites carry a number of these transponders, varying from 10 to as many as 100 on a high capacity satellite. The downlink signal, received either by an Earth station, a DTH receiver or a mobile receiver, is weak and is first amplified to bring it to a level where it can be processed. The signal is then down-converted, demodulated and converted back into a base band signal.

Hence, a transponder is the key element in the satellite communication network and is essentially a repeater which receives a signal transmitted from the Earth station on the uplink, amplifies the signal and retransmits it on the downlink at a different frequency from that of the received signal. This frequency conversion is done in order to avoid interference between the uplink and the downlink signals. Moreover, as the atmospheric propagation losses are less for lower frequencies and due to the limitation of available power on board the satellite, the downlink frequency is kept lower than the uplink frequency. An exception to this is the Iridium constellation of satellites which uses the same frequency both for uplink as well as for downlink.

The first generation satellites used single channel repeaters providing a single channel of transmission within the satellite. However, the satellites developed thereafter had multiple repeaters with each repeater capable of carrying several channels. The available satellite bandwidth, typically 500 MHz for the C and Ku bands and 2000 MHz for the Ka band, is divided into various frequency channels, typically 30 to 80 MHz wide, each of which is handled by a separate repeater. This repeater, known as a transponder, is responsible for handling a complete signal path through the satellite. Typical transponder bandwidths are 27 MHz, 36 MHz, 54 MHz and 72 MHz, of which 36 MHz is the most common as it is the bandwidth required to transmit one analogue video channel.

The term 'transponder equivalent (TPE)' is used to define the total transmission capacity available on satellites in terms of transponders having a bandwidth of 36 MHz. For example, a 72 MHz transponder will be equal to two TPEs.

10.4.1 Types of Transponders

Transponders may be broadly classified into two types depending upon the manner in which they process the signal:

1. Transparent or bent pipe transponders
2. Regenerative transponders

10.4.1.1 Transparent or Bent Pipe Transponders

Transparent transponders process the uplink satellite signal in such a way that only their amplitude and the frequency are altered; the modulation and the spectral shape of the signal are not affected. They are also referred to as 'bent pipe' transponders as they simply transmit the information back to Earth.

Transparent transponders comprise an input filter, low noise amplifier (LNA), down converter, input multiplexer, channel amplifiers, high power amplifiers and output de-multiplexer (Figure 10.2). The uplink section of the transponder, comprising the input filter, LNA and the down converter is common to all the channels and is shared by all the transponders. The down converter is basically a mixer which provides a fixed frequency translation corresponding to the exact frequency difference between the centre of the uplink and the downlink frequency bands. For example, the down converter for a C band transponder provides a frequency translation of 2.225 GHz as the difference between the centre of the uplink frequency band (5.925–6.425 GHz) and the downlink frequency band (3.7–4.2 GHz) in this case is 2.225 GHz.

The full bandwidth is separated into individual transponder channels by a bank of RF filters called the input multiplexer (IMUX), with each filter being tuned to pass the full bandwidth of a particular channel and reject all other channels. The output of each IMUX filter is then amplified by separate power amplifiers. The power amplifiers employed are travelling wave tube amplifiers (TWTA) for higher power levels (50 W or more) and at higher frequency bands (Ku and Ka bands) and solid state power amplifiers (SSPAs) for lower power applications. The output of all the transponder channels is then combined in an output de-multiplexer, which is composed of specially designed low-loss waveguide filters and then fed to a common transmitting antenna for down-beaming the signal on to Earth. The transponder always has redundant equipment to ensure its proper performance in the case of failure of any one of them.

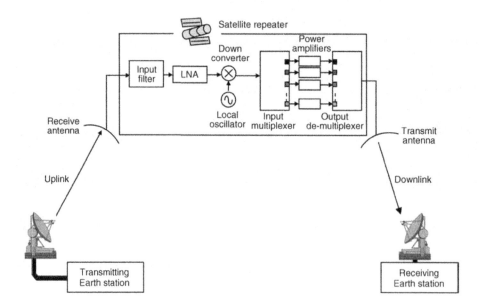

Figure 10.2   Transparent transponders

10.4.1.2 Regenerative Transponders

Regenerative transponders are those in which some onboard processing is done and the received signal is altered before retransmission. This onboard processing helps to improve the throughput and error performance by restoring the signal quality prior to retransmission to the Earth. These repeaters are also called digital processing repeaters as they use various digital techniques like narrowband channel selection and routing, demodulation, error correction, reformatting of data, etc. for processing the received signal.

Transparent transponders, although they are simplest to design and can handle all three multiple-access methods, that is FDMA, TDMA and CDMA, cannot be used to optimize the transmission link for a particular service, for reducing link noise or for improving the satellite link performance. Regenerative transponders offer the flexibility of link design for optimizing satellite performance as they actively alter the signal before retransmission to the Earth. Three types of regenerative transponders are currently being used in communication satellites. They are satellite-switched TDMA transponders employing wideband RF and IF switching, narrowband digital processing transponders with channel routing and digital beam forming and the demod-remod transponders, which demodulate the received signal and completely restore the information before retransmission.

10.4.2   Transponder Performance Parameters

The product of the transmit antenna gain and the maximum RF power per transponder defines the most important technical parameter of a communication satellite, the transmit effective isotropic radiated power (EIRP). As a general trend, EIRP values for the C band and Ku band satellites are around 40 dBW and 55 dBW respectively. EIRP defines the downlink performance of a transponder and specifies the coverage area of a satellite. The uplink performance of a satellite is defined by the parameter called the relative gain-to-noise temperature (G/T) ratio. It is the ratio of the receive antenna gain and the noise temperature of the satellite receiving system.

10.5 Satellite versus Terrestrial Networks

Satellites, initially conceived to provide support services to terrestrial communication networks, have made a great deal of progress in the last fifty years. Satellites have established themselves as a pioneering element of communication networks. However, with the advances made in the field of terrestrial communication network technology, like the advent of fibre optic technology, satellites are facing tough competition from the terrestrial networks. When compared with each other, both satellites as well as terrestrial networks have certain advantages and disadvantages with regard to each other. Some of the important ones are outlined below:

10.5.1 Advantages of Satellites Over Terrestrial Networks

Satellites offer certain advantages over terrestrial networks. Some of the advantages are as follows:

1. **Broadcast property – wide coverage area.** Satellites, by virtue of their very nature, are an ideal means of transmitting information over vast geographical areas. This broadcasting property of satellites is fully exploited in point-to-multipoint networks and multipoint interactive networks. The broadcasting property is one of the major plus points of satellites over terrestrial networks, which are not so well suited for broadcasting applications.
2. **Wide bandwidth – high transmission speeds and large transmission capacity.** Over the years, satellites have offered greater transmission bandwidths and hence more transmission capacity and speeds as compared to terrestrial networks. However, with the introduction of fibre optic cables into terrestrial cable networks, they are now capable of providing transmission capabilities comparable to those of satellites.
3. **Geographical flexibility – independence of location.** Unlike terrestrial networks, satellite networks are not restricted to any particular configuration. Within their coverage area, satellite networks offer an infinite choice of routes and hence they can reach remote locations having rudimentary or nonexistent terrestrial networks. This feature of satellite networks makes them particularly attractive to Third World countries and countries having difficult geographical terrains and unevenly distributed populations.
4. **Easy installation of ground stations.** Once the satellite has been launched, installation and maintenance of satellite Earth stations is much simpler than establishing a terrestrial infrastructure, which requires an extensive ground construction plan. This is particularly helpful in setting up temporary services. Moreover, one fault on the terrestrial communication link can put the entire link out of service, which is not the case with satellite networks.
5. **Uniform service characteristics.** Satellites provide a more or less uniform service within their coverage area, better known as a 'footprint'. This overcomes some of the problems related to the fragmentation of service that result from connecting network segments from various terrestrial telecommunication operators.

6. **Immunity to natural disaster.** Satellites are more immune to natural disaster such as floods, earthquakes, storms, etc., as compared to Earth-based terrestrial networks.
7. **Independence from terrestrial infrastructure.** Satellites can render services directly to the users, without requiring a terrestrial interface. Direct-to-home television services, mobile satellite services and certain configurations of VSAT networks are examples of such services. In general, C band satellites usually require terrestrial interfaces, whereas Ku and Ka band systems need little or no terrestrial links.
8. **Cost aspects – low cost per added site and distance insensitive costs.** Satellites do not require a complex infrastructure at the ground level; hence the cost of constructing a receiving station is quite modest – more so in case of DTH and mobile receivers. Also, the cost of satellite services is independent of the length of the transmission route, unlike the terrestrial networks where the cost of building and maintaining a communication facility is directly proportional to the distances involved.

Hence, by virtue of their broadcast nature coupled with uniform services offered within the coverage area and easy installation of ground stations, satellites remain the most flexible means for providing links between all points on the globe with a minimum of terrestrial facilities.

10.5.2 Disadvantages of Satellites with Respect to Terrestrial Networks

For certain applications, satellites are at disadvantage with respect to terrestrial networks:

1. **Transmission delay.** Transmission delays of the order of a quarter of a second are involved in transmission of signals from one Earth station to another via a geostationary satellite. It may be mentioned here that for satellite-based data communication services, the data communication protocols that require acknowledgement feedback further add to the delay. Hence, GEO satellites are not suited for certain applications like interactive media, which require small transmission delays. Large transmission delays also have an adverse impact on the quality of voice communication and data transmission at high data rates.
2. **Echo effects.** The echo effect, in which the speaker hears his or her own voice, is more predominant in satellite-based telephone networks as compared to terrestrial networks. This is due to larger transmission delays involved in the case of satellites. However, with the development of new echo suppressors, satisfactory link quality has been provided in the case of single-hop GEO satellite networks. However, for double-hop GEO networks, the problem of echo still exists.
3. **Launch cost of a satellite.** Although the cost of a satellite ground station is less than that of terrestrial networks and the cost of satellite services are independent of the distances involved, the cost of launching a satellite is huge.

To conclude, although satellites have an edge over terrestrial networks in terms of quality, connectivity and reliability of services offered, the problem they face is that the terrestrial networks already exist and transferring the service over to a satellite network becomes a complex and difficult task. Moreover, due to improvements in the terrestrial network technology, satellites are facing tough competition from terrestrial networks. In fact, satellites correspond to only a small part of communications as a whole, around 2 %. Current trends in the field of telecommunication favour space systems that complement terrestrial networks rather than maintaining their independence from them.

10.6 Satellite Telephony

Satellites provide both long distance point-to-point trunk telephony services as well as mobile telephony services, either to complement or to bypass terrestrial networks. Potential users of these services include international business travellers and people living in remote areas. Satellite telephones either allow the users to access the regular terrestrial telephone network or place the call through a satellite link. Satellite telephony networks employ point-to-point duplex satellite links enabling simultaneous communication in both the directions. Single GEO satellites or a constellation of LEO, MEO and GEO satellites are used for providing telephony services. Telephone satellite links generally employ circuit-switched systems offering a constant bit rate services, but only for the limited duration of the call. However, sometimes dedicated or preassigned bandwidth services are used, in which the communication is maintained continuously for an extended period of time, for heavy telephone trucking applications. Some of the major satellite systems offering voice services are Intelsat, Eutelsat, Inmarsat, Globalstar, Iridium, ICO, Ellipso and Odyssey systems.

Figure 10.3 shows a variety of satellite point-to-point telephone networks having either single-user or shared multiuser Earth stations. Various steps in making a call through a satellite

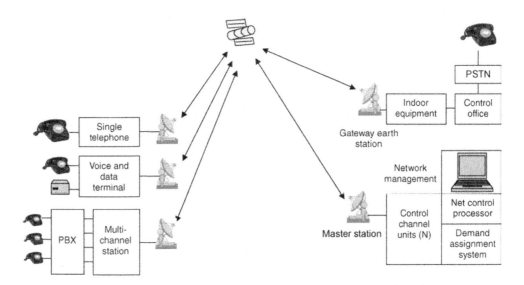

Figure 10.3 Satellite point-to-point telephone networks (PBX, private branch exchange; PSTN, public switched telephone network)

network are outlined below. This is just a conceptual explanation, the actual procedure is much more complicated:

1. The user lifts the receiver when he or she wants to make a call. This sends a request to the local Earth station, which in turn sends a service request to the master station.
2. If the master station is able to provide the satellite capacity, it sends a confirmation signal to the local Earth station, resulting in a dial tone in the telephone instrument.
3. The user then dials the destination number, which is transferred to the control station, which determines the destination Earth station and signals it that a connection needs to be established.
4. The destination Earth station then signals the called party of the incoming call by ringing that telephone instrument.
5. The satellite capacity is allocated to the connection and the telephone link is established once the called party lifts the handset.
6. Once the conversation is over, the calling party hangs up the receiver, hence indicating to the local Earth station to terminate the call.

In the case of a telephony network using satellite constellation, the call may involve connection through multiple satellites and cross-links.

10.6.1 Point-to-Point Trunk Telephone Networks

One of the traditional applications of satellites includes long distance, especially intercontinental trunk telephony services, also referred to as thin-route satellite telephony services. Thin route services are used in those regions where installation of terrestrial networks is not feasible either due to low density of population or because of difficult geographical terrain. These services are particularly useful for establishing connections between the companys headquarters and its remote offices, through gateway Earth stations.

Trunk telephony services come under the domain of fixed satellite services (FSS), mainly utilizing C and Ku bands. Generally, GEO satellites are utilized for providing these services. Intelast, Europestar, Eutelsat, PamAmSat are examples of some of the satellites used for the purpose. Although, route telephony services provide reliable and secure communication, but with the expansion of new technologies like fibre optics they are becoming less and less popular.

10.6.2 Mobile Satellite Telephony

One of the important services provided by mobile satellite services (MSS) is the interactive voice communication to mobile users. This service is referred to as mobile satellite telephony. The satellite phones target two specific markets. The first is that of international business users requiring global mobile coverage. Satellites provide them with truly global mobile services with a single mobile phone, which is impossible with terrestrial systems due to the difference in cellular mobile phone standards from region to region. The second market is the unserved regions where the basic telecommunication services are not present.

A glance at the history of mobile satellite services tells us that the first mobile service experiment began in the year 1977 using NASA's satellite ATS-6. Year 1982 saw the launch of the first civilian mobile satellite, Inmarsat. Since then, there has been a steady and gradual growth in the field of MSS until the early 1990s, after which many new MSS services were launched, mainly to provide mobile satellite telephony services. The third quarter of the decade saw the business failure of many MSS operators, but now again there is a spurt in the aggregate worldwide demand for satellite telephones. MSS satellites launched in the periods 1980–1990 and 1990–1998 were GEO satellites, categorized as Generation-I and Generation-II satellites respectively, mainly providing telephony services to relatively large mobile terminals. Third generation mobile satellites, comprising constellations of LEO, MEO, HEO and GEO satellites, provide voice and multimedia services to mobile and hand-held terminals. Moreover, these third generation mobile satellite services have entered the realm of personal communications and are also referred to as global mobile personal communication services (GMPCS).

GMPCS is a personal communication system providing transnational, regional or global two-way voice, fax, messaging, data and broadband multimedia services from a constellation of satellites accessible with small and easily transportable terminals. There are several different types of GMPCS systems: GEO systems, small LEO systems, big LEO systems, MEO systems, HEO systems and broadband GMPCS systems. Except for small LEO satellite systems, which offer only messaging services, all the other systems provide mobile satellite telephony services. Moreover, all these systems operate in the L and S bands allocated for mobile services, except for the broadband GMPCS systems which operate in the Ku band, where MSS services have been allocated a secondary status. Table 10.1 enumerates the features of these various GMPCS systems.

Table 10.1 Features of the various GMPCS systems

Types of GMPCS	Services offered	Frequency range	Terrestrial counterpart	Examples
Little LEO (data only GMPCS)	Data services like messaging in the store-and-forward mode	Below 1 GHz	Messaging services like paging and mobile data services	Orbcomm
Big LEO including LEO, HEO and MEO satellites (narrowband GMPCS)	Real time voice and data services	1–3 GHz	Cellular telephone	Iridium, Globalstar (LEO orbit), ICO constellation (MEO orbit) and Ellipso constellation (HEO orbit)
GEO (narrowband/ broadband MSS)	Both store-and-forward and real time voice, data and video services	1.5–1.6 GHz and around 2 GHz	Cellular ISDN	Inmarsat, ACeS (Asia cellular satellite), APMT (Asia-Pacific mobile telecommunications), ASC and Thuraya satellite systems
Boadband GMPCS (broadband FSS)	Real time multimedia including voice and data	Above 10 GHz	Fibre optics	Sky Bridge Teledesic constellation

10.7 Satellite Television

Satellite television is the most widely used and talked about application area of communication satellites. In fact, it accounts for about 75 % of the satellite market for communication services. Satellite television basically refers to the use of satellites for relaying TV programmes from a central broadcasting centre to a large geographical area. Satellites, by their very nature of covering a large geographical area, are perfectly suited for TV broadcasting applications. As an example, satellites like GE and Galaxy in the US, Astra and Hot Bird in Europe, INSAT in India and JCSAT (Japanese communications satellite) and Superbird in Japan are used for TV broadcasting applications. The five Hot Bird satellites provide 900 TV channels and 560 radio stations to 24 million users in Europe. Other means of television broadcasting include terrestrial TV broadcasting and cable TV services. Satellites can provide TV transmission services either directly to the users or in conjunction with the cable and terrestrial broadcasting networks. This will be explained in detail in the paragraphs to follow.

10.7.1 A Typical Satellite TV Network

Satellite television employs GEO satellites acting as point-to-multipoint repeaters receiving a certain telecast from the transmission broadcasting centre and retransmitting the same after frequency translation to the cable TV operators, home dishes, and so on, lying within the footprint of the satellite. Satellites can provide TV programmes either directly to the users (direct-to-home television) or indirectly with the help of cable networks or terrestrial broadcasting networks, where the satellite feeds the signal to a central operator who in turn transmits the programmes to the users either using cable networks or through terrestrial broadcasting. A typical satellite TV network, like any other satellite network, can be divided into two sections: the uplink section and the downlink section.

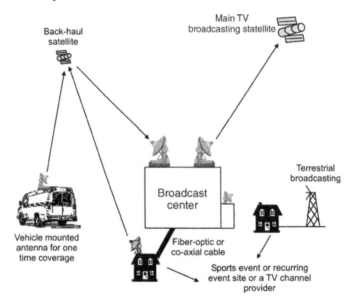

Figure 10.4 Uplink section of satellite TV networks

The uplink section (Figure 10.4) comprises three main components: the programming source, the broadcasting centre and the main broadcasting satellite. The programming source comprises TV channel networks, cable TV programmers, and so on, that provide various TV programming signals, like TV channels, sports coverage, news coverage or local recorded TV programmes, to the broadcasting centre either through terrestrial means, like using the line-of-sight microwave communication and the fibre optic cable, or using satellites referred to as back-haul satellites. As an example, for one-time events like various news events, a vehicle-mounted Earth station generally operating in the Ku band is driven to the site and then the programmes are transmitted to the main broadcast centre using a back-haul satellite on a point-to-point connectivity basis [known as satellite news gathering (SNG)]. In the case of a live telecast of certain events like sports, the signals picked up by the cameras are transmitted to the main broadcasting centre either with the help of a microwave link, a fibre optic link or a point-to-point backhaul satellite link. The broadcasting centre is the hub of the satellite TV system and it processes and beams the signal to the main broadcasting satellite. It also adds commentary or advertisements to the signals from the various programming sources. Generally, the signals are transmitted using analogue techniques in the C band or using a digital format employing various compression techniques in the Ku band. The signals are also generally encrypted before transmission to prevent unauthorized viewing.

The satellite downlink comprises the main broadcasting satellite and the TV receiving network. In fact, the main broadcasting satellite is common in both the uplink and the downlink sections. The receiver network in the case of satellite distributing programmes to the terrestrial broadcast network comprise various terrestrial broadcasting centres that receive the satellite signal and transmit them to the users in the VHF and the UHF bands using terrestrial broadcasting. The user end has directional Yagi antennas to pick up these signals. In the case of satellite distributing programmes to a cable operator, the downlink section comprises the cable–TV head ends and the cable distribution network. For DTH services, receive-only satellite dishes are mounted at the user's premises to receive the TV programmes directly from the satellite.

10.7.2 Satellite–Cable Television

As mentioned earlier, cable TV refers to the use of coaxial and fibre optic cables to connect each house through a point-to-multipoint distribution network to the head end distribution station. Cable TV, originally referred to as CATV (community antenna television) stood for a single head end serving a particular community, like various houses in a large building. The present day cable TV system is more complex and involves a larger distribution area. The head ends receive programming channels from either a local broadcasting link or through satellites. The use of satellites to carry the programming channels to the cable systems head ends is referred to as satellite–cable television (Figure 10.5). The head end in this case consists of various receive-only Earth stations with the capability of receiving telecast from two to six satellites. These Earth stations either have multiple receiving antennas or, a single dish antenna with multiple feeds, with each feed so aligned as to receive telecast from a different satellite.

The transmission from the satellite is either in the analogue format (mainly in the C band) or in the digital format (mainly in the Ku band). In analogue format of transmission, each receiver is tuned to a different transponder channel and the signals from various receivers

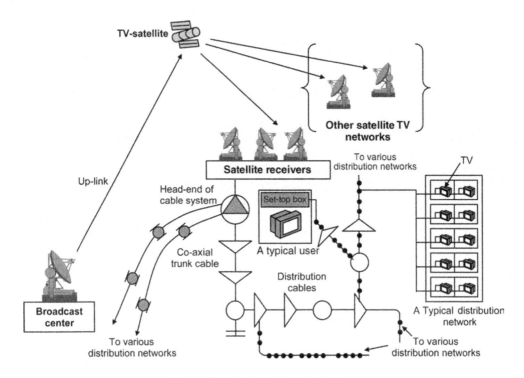

Figure 10.5 Satellite cable television

are multiplexed for transmission to the users. The channels received in the digital format can be transmitted either digitally or in the analogue form as mentioned above. This processed digital or analogue information is then transmitted over a typical cable distribution network to a large number of houses known as subscribers, who pay a monthly fee for the service. The cable operators scramble their programmes to prevent unauthorized viewing. The receiving end then consists of a set top box to descramble and retrieve the original signal. The cable TV operators also transmit the videotaped recorded programmes from other sources in addition to showing programmes received from the satellites.

10.7.3 Satellite--Local Broadcast TV Network

It is the same as the satellite–cable TV network except for the fact that here the satellite distributes programming to local terrestrial broadcasting stations instead of distributing it to the cable head end stations. The broadcasting stations use powerful antennas to transmit the received signals to various users within the line-of-sight (50–150 km) using UHF and VHF microwave bands. The users receive these TV signals using directional antennas like Yagi antennas, reflector antennas or dipole antennas. A typical satellite–local broadcast network is shown in Figure 10.6. Sometimes, a combination of both the satellite-cable TV and the satellite-local broadcast TV networks is used for distributing TV programmes to the users. As an example, one of the possible configurations is where the satellite sends the signals to the local broadcasting stations, which in turn broadcast them to the cable operators.

Satellite Television

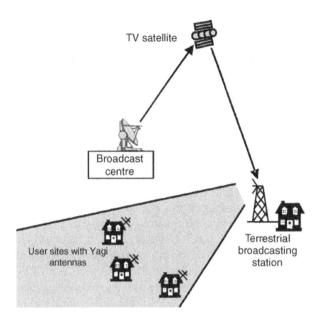

Figure 10.6 Typical satellite local-broadcast TV network

10.7.4 Direct-to-Home Satellite Television

Direct-to-home (DTH) satellite television refers to the direct reception of satellite TV programmes by the end users from the satellite through their own receiving antennas (Figure 10.7).

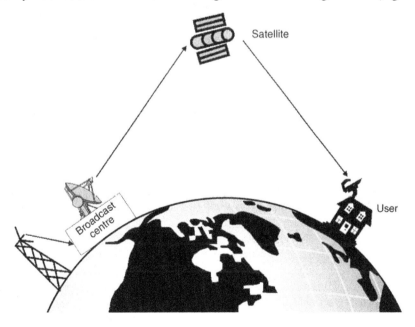

Figure 10.7 Direct-to-home satellite television

DTH services can be broadly classified into two types, namely the television receive-only (TVRO) and the direct broadcasting satellite (DBS) services, depending upon the frequency band utilized and the size of the receiving antennas. TVRO systems operate in the C band whereas the DBS systems operate in the Ku band. In the present context, when DTH systems are mentioned, more often than not it refers to the DBS systems only.

10.7.4.1 Television Receive-Only (TVRO) Services

TVRO systems employ large dishes (6 to 18 feet across) placed in the user's premises for the reception of analogue signals from the satellite operating in the C band. The antenna size is larger in this case as compared to DBS systems as the wavelength at C band frequencies is larger than at the Ku band frequencies. In addition, international and domestic regulations limit C band power because of the possibilities of RF interference between the satellite and the microwave links operating in the C band. Generally, each C band transponder provides one analogue TV channel, and hence a satellite with 16 such transponders will be able to support only 16 TV channels. Hence, for complete channel coverage, the TVRO receiver antenna must have a steerable dish. These systems are made user friendly by using microprocessor control, allowing the viewer to select the desired channel with a remote control unit. The antenna then moves automatically using electronic control methods to point to the desired satellite. TVROs are based on open standard equipment and provide the largest variety of TV programmes, including cable TV programmes, foreign stations, free programming channels and live unedited feeds between broadcasting stations like news, sports, etc.

A look into the development of DTH services suggests that TVROs were the original form of DTH satellite TV reception. The concept of using dishes at the user's premises to view satellite television directly started in early 1980s, when people in the USA started putting dishes in their backyards for direct satellite reception. TVROs reached their peak around the year 1994 and have slowly given way to direct broadcasting satellite (DBS) services. However, TVRO systems still exist and are being updated to receive digitally scrambled programming channels from Ku band satellites.

10.7.4.2 Direct Broadcasting Satellite (DBS) Services

The DBS service is a relatively recent development in the world of television distribution. The first DBS service, Sky Television, was launched in the year 1989. The DBS service uses high powered Ku band satellites that send digitally compressed television and audio signals to relatively small (45–60 cm across) fixed satellite dishes. DBS satellites transmit signals to Earth in the BSS segment of the Ku band (between 12.2 and 12.7 GHz), making use of MPEG-2 (Moving Picture Experts Group) digital compression techniques. The channel capacity per transponder is five to twelve channels depending upon the data rate and the compression parameters, and hence they can provide about 200 channels from one satellite. Hence, the dishes for DBS services need not be steerable. Figure 10.8 (a) shows the pictorial representation of a typical DBS receiver set-up. The receiver [Figure 10.8 (b)] basically consists of a descrambler that descrambles the digital signals received by the antenna and a converter module that converts the digitally compressed bit stream into analogue TV channels. Then

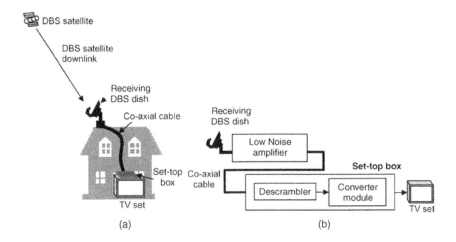

Figure 10.8 (a) DBS receiver set-up (b) Block diagram of a DBS receiver

depending upon the channel the user has chosen, that particular channel is split out and sent to the TV screen. Hence one cannot have two television sets viewing different programmes from the same receiver, as in the case of cable TV. They also provide a fully interactive TV guide and automatically feed the billing information to the local computer of the service provider.

DBS systems are completely closed systems that employ some form of encryption techniques, hence enabling only conditional access by authorized users. This also implies that there are no free channels available on DBS systems. Some of the DBS service providers include DirecTV, Echostar, PrimeStar of the USA, TataSky and DishTV of India and Star Choice of Canada. The FCC has allocated eight orbital slots at 61.5°, 101°, 110°, 148°, 157°, 166° and 175° west longitude GEO locations for DBS systems. TVRO systems have better picture quality than DBS or digital cable systems, which tend to use larger amounts of digital signal compression. However, DBS systems are easy to install and are cheaper as compared to TVRO systems.

10.7.4.3 Newer Satellite TV Services

Digital DBS TV offers users with a lot of services like HDTV (high definition television), which is a high resolution digital TV service, interactive programme watching in which the user can interact with the programme and create his or her own programme, do interactive shopping by tuning in to the shopping channel and choosing what to buy and order it, personal video recording in which the user can record the programme and play it later. Other services offered include video-on-demand, in which the viewer can view at any moment the programme of his choice, near video-on-demand, in which the viewer can view the programme of his choice at a latter scheduled time, pay TV in which the viewer is charged according to the programmes he views. Another important service offered is a high speed Internet connection through the satellite TV link.

10.7.5 Digital Video Broadcasting (DVB)

DVB is a family of technology standards defining digital broadcasting and designed to facilitate broadcasting of audio, images and multimedia to allow a large degree of user interaction. These standards are designed to use existing satellite, cable and terrestrial infrastructures for broadcasting.

10.7.5.1 Evolution

The development of DVB standards and the subsequent introduction of services is coordinated by a project on DVB called the DVB Project. Before the 1990s, when all television broadcasts were analogue in nature, the use of digital systems was considered not feasible due to complexity of the digital signal processing they required. With significant advances in digital signal processing techniques and integrated circuit technology, television broadcasting exploiting digital techniques soon became a reality. It started with the formation of a consortium of various organizations named the *Electronics Launching Group* (ELG), which discussed ways and means of moving forward in this direction. A memorandum of understanding was signed in 1993 and the group was renamed the DVB Project. The project was committed to the task of developing technologies and standards for digital broadcasting using existing satellite, cable and terrestrial means, and also with the early introduction of these services. The consortium currently has more than 270 members spread over 80 countries. DVB today is a synonym for digital television and data broadcasting around the world and DVB services have already been introduced in North and South America, Europe, Asia, Africa and Australia.

10.7.5.2 Salient Features

In contrast to the television broadcasting systems of the 1970s to the 1990s, which can be considered as closed systems, DVB is an open system. Open systems such as DVB allow subscribers to choose different content providers, while closed systems are content provider specific. Closed systems are optimized for television alone, but open systems facilitate the integration of televisions and PCs, interactive viewing, banking, private network broadcasting and so on. Features such as restricting access to subscribers, thereby minimizing loss of revenue due to unauthorized viewing, availability of high quality television in public and private transport, the possibility of expanding services regardless of geographical location etc are some of the other highlights of DVB.

10.7.5.3 Data Compression and Conditional Access

Two key elements of DVB are *data compression* and *conditional access*. Compression of audio and video signals allows transmission of digital signals using existing satellite, cable and terrestrial infrastructures. One such commonly employed data compression standard is MPEG-2, which is one of the series of MPEG standards for compression of audio and video signals. The current digital television formats are standard definition television (SDTV) and HDTV. While SDTV gives DVD-like audio and video quality, HDTV is far superior, providing cinema-like quality. Conditional access provides secure access and prevents external piracy. There are a number of conditional access systems for content providers to choose from depending on their requirements. The conditional access system basically is a security module that scrambles and encrypts data before transmission. The security module is either embedded in the receiver or is available in the form of a detachable printed circuit card. The receiver also has a smart card containing subscribers' access information. The module contains the unscrambling algorithms. Once the authorization code is verified, the conditional access module unscrambles the data stream for the receiver to process the same and output it for viewing. Conditional access not only prevents piracy, but detachable cards also allow subscribers to use DVB services anywhere this technology is supported.

10.7.5.4 Different Forms of DVB

The major forms of DVB are briefly described as follows:

1. *DVB-C* (DVB-Cable) is the standard for delivery of video services via cable networks. Modulation schemes used by this standard include different variants of quadrature amplitude modulation (QAM) such as 16-QAM, 32-QAM, 64-QAM, 128-QAM and 256-QAM.
2. *DVB-H* (DVB-Handheld) is used to provide services to handheld devices such as mobile phones etc. This standard was published in November 2004.
3. *DVB-S* (DVB-Satellite) is the standard for delivery of television/video via satellite. The modulation schemes used in DVB-S (SHF) are QPSK, 8PSK or 16-QAM.
4. *DVB-SH* (DVB-Satellite Handheld) is used for delivery of DVB services to handheld devices via satellite.
5. *DVB-S2* (DVB-Satellite Second Generation) is the standard for the second generation of DVB services via satellite. The modulation schemes used for DVB-S2 are QPSK, 8PSK, 16APSK and 32APSK.
6. *DVB-T* (DVB-Terrestrial) is the standard for broadcasting DVB services using terrestrial means. DVB-T (VHF/UHF) uses 16-QAM or 64-QAM (or QPSK) in combination with coded OFDM (orthogonal frequency division multiplexing).
7. *DVB-T2* is an enhanced version of DVB-T standard.
8. *DVB-RCS* (DVB-Return Channel Satellite) provides satellite DVB services with a return channel to facilitate interactivity.
9. *DVB-RCS2*: Second generation DVB-RCS standard.

10.7.6 DVB-S and DVB-S2 Standards

DVB-S was the first digital television satellite standard released by the DVB Project in 1994, providing digital satellite television services to over 100 million subscribers across the world. The world's first digital satellite television services, launched in South Africa and Thailand towards the end of 1994, employed the DVB-S standard. DVB-S2 is an upgraded version of the DVB-S standard. It is the second generation digital satellite transmission system developed by the DVB Project. DVB-S2 is gradually replacing DVB-S because HDTV services are offered by the new standard. Advanced techniques used for channel coding, modulation and error correction have made it possible for many new services to be commercially viable. The launch of HDTV services has been possible due to the availability of the state-of-the-art video compression technology. Some of the key technical features of DVB-S2 are as follows:

1. Modulation schemes used in DVB-S2 include any of the four schemes, namely QPSK, 8PSK, 16APSK and 32APSK. The first two schemes (QPSK and 8PSK) are mainly used for broadcast applications where satellite transponders are driven close to saturation. For professional applications such as news gathering, interactive services etc. requiring a higher carrier-to-noise (C/N) ratio, 16APSK or 32APSK are used.
2. DVB-S2 offers optional backwards compatibility so that DVB-S compliant receivers are not rendered obsolete while the DVB-S2 compliant receivers have the benefit of additional services offered by the new standard.
3. DVB-S2 achieves a high degree of performance in the presence of high levels of noise and interference by using a powerful forward error correction scheme.
4. DVB-S2 uses adaptive coding and modulation, which allow change in transmission parameters on a frame-to-frame basis depending on the delivery path conditions of individual users. This feature is particularly useful in professional applications and in providing interactive services.

Table 10.2 gives a comparison of DVB-S and DVB-S2 satellite television standards for satellite EIRP of 51 dBW and 53.7 dBW.

A large number of major satellite television broadcasters in Europe and the USA are using DVB-S2 in conjunction with MPEG-4 advanced video coding for delivery of HDTV services. DVB-S2 is also used by direct-to-home (DTH) operators in Asia, the Middle East and Africa. DVB-S and DVB-S2 together are offering DVB services to more than 250 million subscribers across the world. Some of the key broadcasters' names in Europe and the USA include DirecTV in the USA, Sky in Italy, Premiere in Germany and BSkyB in UK and Ireland.

Table 10.2 Comparison of DVB-S and DVB-S2 performance

Standard	DVB-S	DVB-S	DVB-S2	DVB-S2
Satellite EIRP (dBW)	51	53.7	51	53.7
Modulation and coding	QPSK 2/3	QPSK 7/8	QPSK 3/4	8PSK 2/3
C/N (in 27.5 MHz) (dB)	5.1	7.8	5.1	7.8
Symbol rate (Mbaud)	27.5 ($\alpha = 0.35$)	27.5 ($\alpha = 0.35$)	30.9 ($\alpha = 0.20$)	29.7 ($\alpha = 0.25$)
Useful bit rate (Mbps)	33.8	44.4	46 (gain = 36%)	58.8 (gain = 32%)
No. of SDTV programmes	7 MPEG-2 15 AVC	10 MPEG-2 20 AVC	10 MPEG-2 21 AVC	13 MPEG-2 26 AVC
No. of HDTV programmes	1-2 MPEG-2 3-4 AVC	2 MPEG-2 5 AVC	2 MPEG-2 5 AVC	3 MPEG-2 6 AVC

10.7.7 DVB-RCS and DVB-RCS2 Standards

DVB-RCS is one of the digital television broadcast standards in the DVB family of standards. Though originally intended to be a broadcast technology, its unique features have interested many users belonging to the wide spectrum of broadband communications. Most of the terrestrial broadband standards are based on the one-to-many concept and therefore involve only a one-way transmission. DVB-RCS, on the other hand, uses a return or uplink channel to enable two-way transmission. The DVB-RCS standard was formulated by the DVB Project in 1999 to provide interactive communication using satellites.

DVB-RCS defines the air interface specification for a two-way satellite broadband scheme and uses a VSAT terminal to provide to the user an ADSL (asymmetric digital subscriber line) type of link that enables two-way communication without the need for a terrestrial network of cables. Interactive broadcasting has been done by having cable connectivity. While this may be economically feasible in urban and built-up areas, providing cable connectivity in remote areas turns out to be far too expensive with very little probability of recovering installation costs for a very long time. DVB-RCS today is an established satellite communication standard created in an open environment with a highly efficient bandwidth management.

To be able to make use of DVB-RCS, the hardware required is a combination of a suitable satellite dish antenna and a satellite interactive terminal (SIT) or satellite modem. The whole process of accessing desired data in the DVB-RCS standard is as follows. The interested user receives the multimedia stream via the satellite downlink. The user then sends the service request signal through their SIT and the return or uplink channel to the satellite, from where it is routed to the service provider. The service provider then responds to the request and the routing from service provider to the satellite is via a standard uplink station. From the satellite, the information is routed to the user's SIT via the satellite's downlink channel.

DVB-RCS is the only multivendor VSAT standard enabling users to keep the choice of terminal vendor open after initial procurement. DVB-RCS has been accepted worldwide for a variety of applications, some of the major ones being in cellular backhaul, voice over IP services, corporate networking, telemedicine, remote monitoring like SCADA and so on.

DVB-RCS2 is the second generation DVB-RCS standard and is more efficient and flexible than DVB-RCS. The new version, published in 2012, adds support for mobility and meshed networks. Table 10.3 gives a comparison of the DVB-RCS and DVB-RCS2 standards.

10.7.8 DVB-T and DVB-T2 Standards

The DVB-T standard, first published in March 1997, is the most widely used digital television broadcast standard worldwide for delivering high quality video. DVB-T allows transmission of several television broadcast channels and audio channels on a single transmission link. The modulation scheme used is *orthogonal frequency division multiplexing* (OFDM), which gives it the capability to recover signal strength against selective fading from multi-path effects. OFDM uses a large number of carriers modulated with low rate data. The orthogonal nature of the signals makes them immune to any mutual interference. OFDM also allows the network to implement what is termed a *single frequency network*. A single frequency network allows a number of transmitters to operate on the same frequency without causing

Table 10.3 Comparison of the DVB-RCS and DVB-RCS2 standards

S.No.	Feature	DVB-RCS	DVB-RCS2
1	Harmonized IP-level quality of service	None	Yes
2	Harmonized management and control	None	Yes (optional)
3	Multiple virtual network support	None	Yes
4	Security	Single security solution	Multiple security solutions
5	Return link access scheme	TDMA, continuous carrier	TDMA, continuous carrier, random access
6	Modulation techniques used	QPSK	*Linear*: BPSK, QPSK, 8PSK and 16QAM *Constant envelope*: CPM
7	Channel coding	RS/convolutional, 8-state PCCC turbo code	16-state PCCC turbo code (linear modulation), SCCC (CPM)
8	Burst spread spectrum	Burst repetition	Direct sequence
9	Return link adaptivity	Limited support	Inherent in air interface
10	Bandwidth efficiency	Not applicable	Improvement of 30 % over DVB-RCS

any interference. The standard also allows variation of various transmission parameters. This feature can be used by network operators to find the right balance between DVB-T transmission capacity and robustness. These include three modulation options, namely QPSK, 16-QAM and 64-QAM, five different forward error correction (FEC) rates of 1/2, 2/3, 3/4, 5/6 and 7/8, four guard interval options of 1/4, 1/8, 1/16 and 1/32, two carrier options of 2k or 8k, three channel bandwidth options of 6, 7 or 8 MHz and two video refresh rates of 50 Hz or 60 Hz.

Another unique feature of the DVB-T standard is its hierarchical modulation facility, which allows transmission of two completely different data streams on a single DVB signal. A high priority data stream is embedded within a low priority data stream. The operators can use this feature to target two different types of receivers with different services.

DVB-T2 is the second generation DVB-T standard. It offers backwards compatibility so that DVB-T standard compliant receivers are not rendered obsolete. In addition, it provides additional features and services. DVB-T2, like DVB-T, uses OFDM. Table 10.4 gives a comparison of DVB-T and DVB-T2 performance specifications.

10.7.9 DVB-H and DVB-SH Standards

The DVB-H standard provides video and television services for cellular phones and handsets. DVB-H is derived from the well established and widely accepted DVB-T standard. As

Table 10.4 Comparison of the DVB-T and DVB-T2 standards

S.No.	Parameter	DVB-T	DVB-T2
1	Number of carriers	2k, 8k	1k, 2k, 4k, 8k, 16k and 32k
2	Modulation schemes	QPSK, 16-QAM, 64-QAM	QPSK, 16-QAM, 64-QAM, 256-QAM
3	Error correction	Convolutional + reed solomon 1/2, 2/3, 3/4, 5/6 and 7/8	LPDC + BCH 1/2, 3/5, 2/3, 3/4, 4/5, 5/6
4	Guard interval	1/4, 1/8, 1/16, 1/32	1/4, 19/128, 1/8, 19/256, 1/16, 1/32, 1/128
5	Scattered pilots	8 % OF TOTAL	1 %, 2 %, 4 %, 8 % of total
6	Continual pilots	2.6 % of total	0.35 % of total

compared to conventional terrestrial television services, conditions for handheld devices are considerably different:

1. The receiver antennas need to be very small and often integrated in the handset unlike the directional antennas used in television sets.
2. Handheld devices experience high levels of interference and also encounter signal variations due to multipath effects.
3. Handheld devices are often on the move whereas domestic television sets are static.
4. Handheld devices run on batteries whereas domestic television sets operate on AC mains. Battery life is a major concern in handheld devices whereas current consumption is not an issue in domestic television sets.

DVB-H, like DVB-T, uses OFDM. It supports different types of modulation formats, such as QPSK, 16-QAM and 64-QAM, within the OFDM signal. The choice of modulation format is usually a trade-off between data rate and the required signal strength for error-free reception. While QPSK offers better reception under low signal and high noise conditions but at a lower data rate, 64-QAM offers a higher data rate but also requires a higher signal level. The time slicing feature of the DVB-H standard allows the power consumption of a mobile set to be reduced by 90 %, which significantly minimizes the drain on the battery and gives a much longer battery life between charges.

The DVB-SH standard delivers audio, video and data services to handheld devices using frequencies within the S-band from either satellite or terrestrial networks. It complements the DVB-H standard, which delivers mobile video from terrestrial networks in UHF television bands. DVB-SH is targeted for both satellite and terrestrial delivery. While satellite delivery achieves coverage of large areas, terrestrial coverage can be used to fill gaps in built-up areas in cities where tall buildings may shield the satellite signal. Two possible architectures, namely SH-A and SH-B, are used in DVB-SH. SH-A uses OFDM on both satellite and terrestrial links. SH-B uses TDM on the satellite link and OFDM on the terrestrial link. The choice of architecture is governed by satellite characteristics and regulatory considerations.

10.8 Satellite Radio

A satellite providing high fidelity audio broadcast services to the broadcast radio stations is referred to as a satellite radio and is a major revolution in the field of radio systems. Sound quality is excellent in this case due to a wide audio bandwidth of 5–15 kHz and low noise provided over the satellite link. Satellite radio like the satellite TV employ GEO satellites and the network arrangement for the satellite radio is more or less identical to that used for TV broadcasting. Using point-to-multipoint connectivity, the audio signals from various music channels, news and sports centres are transmitted by the satellite to a conventional AM or FM radio station. The signal is then de-multiplexed and the local commercials and other information is added here in the same way as in a TV network and then sent to the users using terrestrial broadcasting topology. The satellite can also transmit the signal directly to the user's radio sets. Some of the major providers of satellite radio services include Sirius and XM Radio of the USA.

10.9 Satellite Data Communication Services

The role of communication satellites is expanding from the traditional telephony and TV broadcast services to newer horizons like secured user oriented data communication services. Data communication via satellites refers to the use of satellites as a communication channel to transmit data between two computers or data processing facilities located at different places. Data communication services are provided either by GEO satellites or by a constellation of LEO, MEO or HEO satellites. Some of these satellites are part of the global mobile personal communication system (GMPCS). GEO satellites provide broadcast, multicast and point-to-point unidirectional or bidirectional data services through special networks called VSAT networks. GMPCS satellites provide data services like messaging services, pager services, facsimile services, and so on. Satellites also provide low data rate mobile data communication services allowing the transmissions of alarm and distress messages and message transmission between the mobile terminals. Terrestrial networks can also provide data broadcast services, but they do it by stringing point-to-point links together. Satellites, being inherently broadcast in nature, offer significant advantages as compared to terrestrial networks.

Mostly when data communication services are discussed, reference is made to unidirectional or bidirectional services provided by GEO satellites through VSAT networks. Hence, the discussion here will mainly be about the data communication services using VSATs. Various data services offered by other satellite constellations are explained in brief towards the end.

10.9.1 Satellite Data Broadcasting

Satellite data broadcasting refers to the use of satellites in point-to-multipoint or multipoint interactive configurations for the transmission of information in digital form. Large multi-national companies or international organizations having offices in remote areas make use of satellite broadcasting services for data collection and broadcasting, image and voice transfer, two-way computer interactions and database inquiries between these remote stations and the main head centre.

Point-to-multipoint broadcast services refer to unidirectional data transmission from a single uplink to a large number of remote receiving points within the coverage area of the satellite. A multipoint interactive network is similar to the point-to-multipoint network except for the fact that the remote terminals in this case also have the transmitting capability. Hence, these networks are bidirectional in nature. The broadcast half transmits the bulk information to all the remote points. These points in turn transmit their individual requests back to the main broadcasting station. In general, the amount of data transmitted from the central station to the remote terminals (outbound direction) far exceeds the data transmitted from the remote terminals to the central station (inbound direction). Hence these networks are asymmetrical in form, having higher data rates in the outbound direction as compared to the inbound direction. Interactive data communication is the foundation of most corporate and government networks. Figures 10.9 (a) and (b) show the configurations of typical point-to-multipoint and multipoint interactive networks respectively.

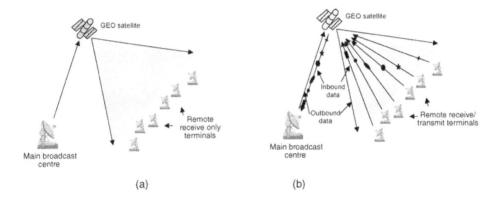

Figure 10.9 (a) Typical point-to-multipoint network (b) Typical multipoint interactive network

10.9.2 VSATs (Very Small Aperture Terminals)

VSATs, as mentioned above, stand for very small aperture terminals and are used for providing one-way or two-way data broadcasting services, point-to-point voice services and one-way video broadcasting services. VSAT networks are ideal for centralized networks with a central host and a number of geographically dispersed terminals. Typical examples are small and medium businesses with a central office, banking institutions with branches all over the country, reservation and airline ticketing systems, etc. VSATs offer various advantages, like wide geographical area coverage, high reliability, low cost, independence from terrestrial communication infrastructure, flexible network configurations, etc. However, VSATs suffer from a major problem of delay between transmission and reception of data (around 250 ms) due to the use of GEO satellites.

10.9.2.1 VSAT Network

The ground segment of a typical VSAT network consists of a high performance hub Earth station and a large number of low performance terminals, referred to as VSATs. The space segment comprises of GEO satellites acting as communication links between the hub station and the VSAT terminals. A typical VSAT network is shown in Figure 10.10. VSAT networks using non-GEO satellites are still in their conceptual stage. It may be mentioned here that VSATs employ a high performance central station so that the various remote stations can be simpler and smaller in design, thus enabling the VSAT networks to be extremely economical and flexible.

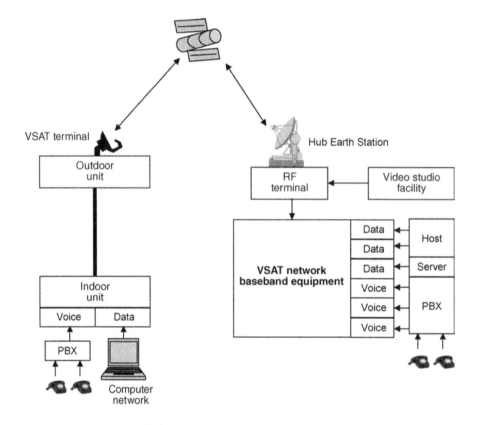

Figure 10.10 A typical VSAT network

The hub station is usually a large, high performance Earth station comprising an outdoor antenna (with a diameter of between 6 to 9 metres) for transmission, RF terminals for providing a wideband uplink of one digital carrier per network, base band equipment comprising modems, multiplexers and encoders, a control centre for managing the network and various kinds of interfacing equipment to support a wide variety of terrestrial links. These terrestrial links connect the hub station to the head office or to the data processing centre, from where the data has to be broadcasted. In the case of bidirectional networks, the outdoor antenna is also configured for reception of signals and the RF equipment comprises several narrowband downlink channels for reception from various remote VSAT terminals. VSAT terminals are smaller and simpler in design as compared to the hub centre and comprise an outdoor antenna (0.5 to 2.4 m in diameter), an RF terminal comprising an LNB (low noise block) for reception and base band equipment. They also comprise an up-converter and power amplifier for uplinking in the case of bidirectional networks. VSAT networks employ either C band or Ku band frequencies for transmission and reception. Ku band VSAT networks have smaller antenna diameters as compared to C band networks.

It may be mentioned here that most VSAT systems operate in the Ku band with the antenna diameter of the Earth stations being as small as 1 to 2 m. The Earth stations are connected in star network topology. The next decade is expected to see the growth of VSAT networks operating in the Ka band. These VSAT networks may operate in direct-to-home configuration for internet and multimedia applications.

Data transmission through VSATs, as mentioned earlier, is generally asymmetrical in nature because the amount of outbound data to be transmitted far exceeds the inbound data. Generally, VSAT networks can transmit at a rate of 64–1024 kbps (64 kbps per remote terminal) in the outbound direction and 64–256 kbps (1.2 to 16 kbps per remote terminal) in the inbound direction. Hence, VSAT networks generally support data, video and voice services in the outbound direction and only data and voice services in the inbound direction. However, some VSAT networks offer compressed digital video services in the inbound direction also.

10.9.2.2 VSAT Network Topologies

VSAT networks come in various topologies, but the most commonly used topologies are star topology for both unidirectional and bidirectional networks and mesh topology for bidirectional networks.

Unidirectional star networks (Figure 10.11) are those in which the information is transmitted only in one direction from the hub station to the remote terminals. There is no information transfer from the remote station to the hub station or to other remote stations. The Broadcast

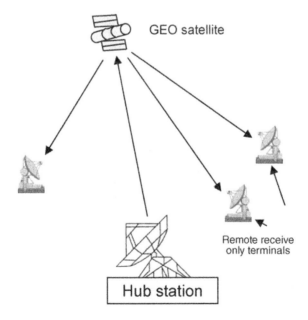

Figure 10.11 Unidirectional star networks

satellite service (BSS), makes use of this topology. The introduction of digital technology allows the service provider and the user much greater flexibility in the operation of a broadcast network. Different subscribers can access different portions of the downlink transmission meant for them by using proprietary software. This process is referred to as narrowcasting. Bidirectional star networks allow the transmission of information in both the directions, but in this case the information cannot be transmitted directly from one VSAT terminal to another but is routed through the hub station. Figure 10.12 shows part of such a network. It can be seen from the figure that the information from station A to station B (shown by regular line) has to first go to the central hub station and from there it is routed to station B. The same holds for transmission from station B to station A (shown by the dotted line). In the case of mesh VSAT networks, the remote terminals can transmit data directly to each other without passing though the hub (Figure 10.13). These networks are particularly appropriate for large corporations where local facilities need to be in contact with facilities in other regions.

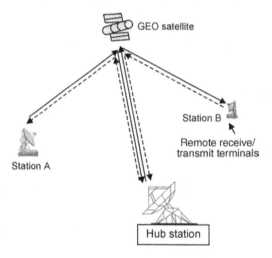

Figure 10.12 Bidirectional star networks

Mesh topology is also more effective if the network is to be mainly used for telephony or video-teleconferencing applications. In certain networks, the hub is owned by a service provider and is shared among large number of users. These networks are referred to as shared hub networks. Each user in these networks is allocated a particular time slot. There are certain networks referred to as mini-hub networks in which each user has a mini-hub, which is smaller than the conventional hub. Thus the user has control over his own communication link. Overall management of the complete network is provided by the service provider who has a super-hub.

Another topology used by the VSAT networks is wherein the high capacity downlink stream is not complemented by an uplink capability from the user terminal. The user transmits via the uplink by employing some other communication channel (such as a telephone line). The VSAT terminal in this case does not require transmit capability which significantly reduces its size and complexity.

A VSAT can either use dedicated bandwidth services or dynamic bandwidth allocation services. Some networks that provide continuous data transfer for critical real time processes employ dedicated bandwidth services, referred to as PAMA (permanently assigned multiple

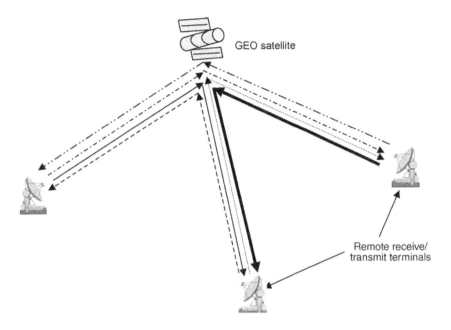

Figure 10.13 Mesh networks

access). Most networks employ dynamic bandwidth allocation services, also referred to as demand assigned multiple access (DAMA), using packet-switching techniques in which the data is broken down into small packets and then transmitted in the form of these packets. Moreover, VSAT networks generally employ a TDM/TDMA scheme for transmission of data. Hence, in most VSAT networks, the outbound data is sent nearly continuously in the form of data packets using the TDM (time division multiplexing) scheme. Each packet contains the source and the destination address and is transmitted through the common outbound link. At the receiving end, each VSAT terminal identifies its packet using the destination address. The inbound data is transmitted from various remote stations using TDMA (time division multiple access), hence allowing many (10–1000) VSATs to share the same communication link. Each VSAT terminal transmits data only for a small time interval in either a preassigned inbound channel slot or in any inbound channel slot, depending on the manufacturer. The main inbound transmission modes are ALOHA, slotted ALOHA, fixed assignment and dynamic assignment. Other schemes used for implementing the VSAT networks include SCPC and CDMA. These schemes have been briefly explained in detail in Chapters 5 and 6.

Non-GEO satellite systems also provide data services like messaging services, pager services, internet services, data services in the store-and-forward mode and some real time data services etc. The non-GEO systems include the little LEO, big LEO and MEO systems. Little LEO satellite systems offer two-way messaging services (including e-mail and paging) in the store-and-forward mode, limited internet, facsimile services and remote data services mainly for emergency situations. Big LEO and MEO systems offer global Internet, fax, real time data services and even broadband multimedia services. Moreover, VSAT networks using LEO satellites will be operational in the near future.

10.10 Important Missions

Various satellite missions are broadly classified into three categories namely, international, regional and domestic systems, depending on the scope of these missions. As the name suggests, international systems provide global coverage, regional systems provide services to a particular region, continent or to a group of countries and national systems provide coverage to a particular country that owns the satellite. In this section some of these major systems are described in detail. For information on other systems reference can be made to the compendium provided at www.wiley.com/go/maini.

10.10.1 International Satellite Systems

The first and most demonstrable need for commercial satellites is to provide international communication services. Initially all international satellites were GEO satellites but now certain non-GEO satellite constellations that provide global coverage have come to the market. Some of the international satellite missions include Intelsat, PamAmSat, Orion, Intersputnik, Inmarsat, etc., in the category of GEO systems and the Iridium and Globalstar constellations in the non-GEO category. Intelsat, Inmarsat and the Iridium satellite systems are described in detail in the following paragraphs.

10.10.1.1 Intelsat Satellite System

Intelsat Limited is the world's largest commercial satellite communications service provider. Originally, it was formed as the International Telecommunication Satellite Organization (INTELSAT) in 1964 to own and manage a constellation of GEO satellites that could provide international communication services mainly including video, voice and data services to the telecom, broadcast, government and other communications markets. It was an intergovernmental consortium initially having 11 members. In 2001, it became a private company and acquired PamAmSat in 2006. Today, it is the world's largest provider of fixed satellite services, operating a fleet of more than 50 satellites.

To date, more than 80 Intelsat satellites have been launched with each satellite offering a significant upgrade in terms of capability and the quality of services offered over its predecessor. As an example, Intelsat 1 and 2 employed a single isotropic antenna; Intelsat 3 had a de-spinning directional antenna so as to maintain an intense beam on the surface of the Earth. Further innovations were made in Intelsat 4 satellites to shape the beam so that it does not cover the ocean areas. Transponder capacity has also increased with each generation; Intelsat 1 had one C band transponder whereas Intelsat 4 had 12 C band transponders. Intelsat 5 used 4 Ku band transponders in addition to the 21 C band transponders. Intelsat 10, the latest series of Intelsat satellites, has 45 C band and 16 Ku band transponders. Moreover, the services offered have increased many fold in the last four decades: Intelsat 1 had the capability of handling 240 telephone calls or a single TV channel, Intelsat VIII can handle more than 120 000 telephone calls or 500 TV channels. In February 2007, Intelsat changed the names of 16 of its satellites formerly known under the Intelsat Americas and PamAmSat series to Galaxy and Intelsat series respectively. As of October 2013, it operated 28 satellites and supports more than 30 DTH platforms world-wide. Table 10.5 enumerates salient features of the various satellites owned by Intelsat Limited.

Table 10.5 Intelsat satellites

Satellite	Transmission Capability	Stabilization	Location
Intelsat 1 (comprising of one satellite Intelsat 1 1)	1 transponder (240 circuits or one TV channel)	Spin	Intelsat 1 1 (332°E) over AOR
Intelsat 2 (comprising of four satellites Intelsat 2 1, 2 2, 2 3 and 2 4)	2 VHF transponders each (240 two-way telephone circuits or one two-way TV channel)	Spin	Intelsat 2 1, 2 2 and 2 4 over the POR, 2 3 over the AOR
Intelsat 3 (comprising of 8 satellites Intelsat 3 1, 3 2, 3 3, 3 4, 3 5, 3 6, 3 7, 3 8)	1500 voice or 4 TV channels	Spin	Intelsat 3 2, 3 6, 3 7 over the AOR, 3 3 over the IOR, 3 4 over the POR. Intelsat 3 1, 3 5 and 3 8 were launch failures
Intelsat 4 (comprising of 8 satellites Intelsat 4 1, 4 2, 4 3, 4 4, 4 5, 4 6, 4 7 and 4 8)	12 C band transponders each (4000 voice circuits or 2 TV channels each)	Spin	Intelsat 4 1 initially over IOR then moved to AOR, Intelsat 4 2, 4 3 and 4 7 over AOR, Intelsat 4 4 and 4 8 over POR and Intelsat 4 5 over IOR
Intelsat 4A (comprising of 6 satellites Intelsat 4A 1, 4A 2, 4A 3, 4A 4, 4A 5 and 4A 6)	20 C band transponders each (7250 voice or 2 TV channels each)	Spin	Intelsat 4A 1, 4A 2 and 4A 4 over the AOR, Intelsat 4A 3 over the IOR
Intelsat 5 (comprising of 9 satellites Intelsat 501, 502, 503, 504, 505, 506, 507, 508 and 509)	21 C band and 4 Ku band transponders each (12000 voice + 2 TV channels)	3-axis	Intelsat 501 first over AOR then POR, 502, 506 over AOR, 503 over AOR then POR, 504, 505, 507, 508 over IOR
Intelsat 5A (comprising of 6 satellites Intelsat 510, 511, 512, 513, 514 and 515)	26 C band and 6 Ku band transponders each (15000 voice + 2 TV channels)	3-axis	Intelsat 510 over POR, Intelsat 511, 515 over IOR and Intelsat 512 and 513 over AOR
Intelsat 6 (comprising of 5 satellites Intelsat 601, 602, 603, 604, 605 and 606)	38 C band and 10 Ku band transponders each (120000 two-way telephone calls + three television channels each)	Spin	Intelsat 602 (178°E) and 605 (174°E) over POR, 604 (60°E) and 601 (47.5°E) over IOR, 603 (340°E) over AOR
Intelsat 7 (comprising of 6 satellites Intelsat 701, 702, 703, 704, 705 and 709)	26 C band and 10 Ku band transponders each (18000 telephone calls and 3 color TV broadcasts simultaneously or up to 90000 telephone circuits using digital circuit multiplication equipment (DCME))	3-axis	Intelsat 701 (180°E) over POR, 702 (55°E), 703 (57°E), 704 (66°E) and 706 (52°E) over IOR, 705 (310°E) over AOR, 709 (85°E) over APR

(continued)

Table 10.5 (*Continued*)

Satellite	Transmission Capability	Stabilization	Location
Intelsat 7A (comprising of 3 satellites 706, 707 and 708)	26 C band transponders and 14 Ku band transponders each (22500 telephone calls and 3 color TV broadcasts simultaneously or up to 112500 telephone circuits using DCME)	3-axis	Intelsat 706 (50°E) over IOR, 707 (307°E) and 708 (310°E) over AOR
Intelsat 8 (comprising of 4 satellites Intelsat 801, 802, 803 and 804)	38 C band transponders and 6 Ku band transponders each (22000 telephone calls and 3 color TV broadcasts simultaneously or up to 112500 telephone circuits using DCME)	3-axis	Intelsat 801 (328.5°E), 803 (310°E) over AOR, 802 (33°E) over IOR and 804 (64°E) over IOR
Intelsat 8A (comprising of 2 satellites Intelsat 805 and 806)	28 C band transponders and 3 Ku band transponders each	3-axis	Intelsat 805 (304.5°E), 806 (319.5°E) over AOR
Intelsat 9 (comprising of 7 satellites Intelsat 901, 902, 903, 904, 905, 906 and 907)	44 C band transponders and 12 Ku band transponders each	3-axis	Intelsat 901 (342°E), 903 (325.5°E), 905 (335.5°E) and 907 (332.5°E) over AOR, Intelsat 902 (62°E), 904 (60°E) and 906 (64°E) over IOR
Intelsat 10 (comprising of 1 operational satellite Intelsat 10-02)	45 C band and 16 Ku band transponders	3-axis	Intelsat 10-02 (359°E) over AOR
Intelsat 1R (former PAS 1R)	36 C band and 36 Ku band transponders	3-axis	315°E over AOR
Intelsat 2 (former PAS 2)	20 C band and 20 Ku band transponders	3-axis	169°E over POR
Intelsat 3R (former PAS 3R)	20 C band and 20 Ku band transponders	3-axis	317°E over AOR
Intelsat 4 (former PAS 4)	20 C band and 30 Ku band transponders	3-axis	72°E over APR
Intelsat 7 (former PAS 7)	14 C band and 30 Ku band transponders	3-axis	68.5°E over IOR
Intelsat 8 (former PAS 8)	24 C band and 24 Ku band transponders	3-axis	166°E over POR
Intelsat 9 (former PAS 9)	24 C band and 24 Ku band transponders	3-axis	302°E over AOR
Intelsat 10 (former PAS 10)	24 C band and 24 Ku band transponders	3-axis	68.5°E over IOR
Insalsat 11 (former PAS 11)	16 C band and 18 Ku band transponders	3-axis	317°E over AOR

Table 10.5 (*Continued*)

Satellite	Transmission Capability	Stabilization	Location
Intelsat 12 (former PAS 12)	30 Ku band transponders	3-axis	45°E over IOR
Galaxy 3C	24 C band and 53 Ku band transponders	3-axis	95°W over AOR
Galaxy 11	24 C band and 40 Ku band transponders	3-axis	33°W over IOR
Galaxy 14	20-24 C band transponders	3-axis	125°W over AOR
Galaxy 15	20-24 C band and one L band transponders	3-axis	133°W over AOR
Galaxy 16	24 C band and 24 Ku band transponders	3-axis	99°W over AOR
Galaxy 17	24 C band and 24 Ku band transponders	3-axis	91°W over AOR
Galaxy 18	24 C band and 24 Ku band transponders	3-axis	123°W over AOR
Galaxy 19	24 C band and 28 Ku band transponders	3-axis	97°W over AOR
Galaxy 23	24 C band, 32 Ku band and 2 Ka band transponders	3-axis	121°W over AOR
Galaxy 25	24 C band and 28 Ku band transponders	3-axis	93°W over AOR
Galaxy 26	24 C band and 28 Ku band transponders	3-axis	93°W over AOR
Galaxy 27	24 C band and 24 Ku band transponders	3-axis	129°W over AOR
Galaxy 28	22 C band, 36 Ku band and 24 Ka band transponders	3-axis	89°W over AOR
Horizons 1	24 C band and 24 Ku band transponders	3-axis	127°W over AOR
Horizons 2	20 Ku band transponders	3-axis	74°W over AOR
Intelsat 14	40 C and 22 Ku band transponders, IRIS	3-axis	315°EL
Intelsat 15	22 Ku band transponders	3-axis	85°EL
Intelsat 16	24 Ku band transponders	3-axis	58°W
Intelsat 17	28 C and 46 Ku band transponders	3-axis	66°E
Intelsat 18	24 C and 12 Ku band transponders	3-axis	180°E
Intelsat 19	24 C and 34 Ku band transponders	3-axis	166°E
Intelsat 20	24 C, 54 Ku and 1 Ka band transponders	3-axis	68.5°E
Intelsat 21	24 C and 36 Ku band transponders	3-axis	302°E
Intelsat 22	24 C, 18Ku and 18UHF transponders	3-axis	72°E

(*continued*)

Table 10.5 (*Continued*)

Satellite	Transmission Capability	Stabilization	Location
Intelsat 23	24 C and 15 Ku band transponders	3-axis	53°W
Intelsat 24	9 Ku band transponders	3-axis	31°E
Intelsat 25	22 Ku and 38 C band transponders	3-axis	31.5°W
Intelsat 26	12 C and 28 Ku band transponders	3-axis	50°E
Intelsat 27 (launch failure)	20 C, 20 Ku and 20 UHF transponders	3-axis	304.5°E (planned)
Intelsat 28 (New Dawn)	28 C and 24 Ku band transponders	3-axis	32.8°E

These satellites basically serve four regions including the Atlantic Ocean Region (AOR) – covering North America, Central America, South America, India, Africa and western portions of Europe; the Indian Ocean Region (IOR) – covering Eastern Europe, Africa, India, South East Asia, Japan and Western Australia; the Asia Pacific Region (APR) – covering Eastern Europe, the former USSR and all the regions from India to Japan and Australia; and the Pacific Ocean Region (POR) – covering Southeast Asia to Australia, the Pacific and the western regions of America and Canada. These coverage regions overlap with each other providing truly global services covering almost every country.

10.10.1.2 Inmarsat Satellite System

INMARSAT, an acronym for the International Maritime Satellite Organization, is an international organization, currently having 85 member countries that control satellite systems in order to provide global mobile communication services. It was established in the year 1979 to serve the maritime industry by providing satellite communication services for ship management, distress and safety applications, but now the horizon of its applications has expanded from providing maritime services to providing land, mobile and aeronautical communication services. INMARSAT operates a global satellite system that is used by independent service providers to offer a range of voice and multimedia communication services for customers on the move and in remote locations. They serve customers from diverse markets including merchant shipping, fisheries, airlines and corporate jets, land transport, oil and gas sector, news media and businessmen whose executives travel beyond the reach of conventional terrestrial communication boundaries. Currently, more than 125 000 Inmarsat mobile terminals are in use.

INMARSAT began its operation in the year 1982 by leasing capacity from the MARISAT, MARECS and the INTELSAT satellites. This formed the first generation of INMARSAT satellites. The first generation of INMARSAT satellites was phased out in the year 1991. The second generation was comprised of four satellites (INMARSAT 2F1, 2F2, 2F3 and 2F4). The third generation of INMARSAT satellites comprises five satellites (INMARSAT 3F1, 3F2, 3F3, 3F4 and 3F5) and the fourth generation comprises of three satellites (INMARSAT 4F1, 4F2 and 4F3). It is being planned to launch three satellites (INMARSAT 5F1, 5F2 and 5F3) in the INMARSAT fifth generation of satellites. INMARSAT has made an agreement with the

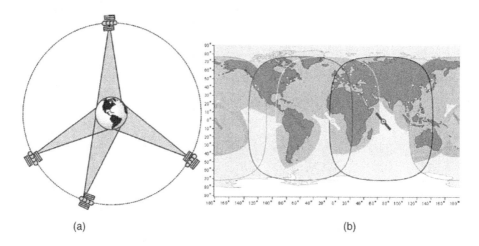

Figure 10.14 (a) Space segment of the Inmarsat satellite system. (b) World coverage provided by Inmarsat satellites [Reproduced by permission of © INMARSAT (satellite coverage areas might vary and INMARSAT cannot guarantee satellite network access near or at the edge of the coverage footprint)]

European Space Agency (ESA) and developed the Alphasat (INMARSAT 4A F4) satellite, which complements the fourth generation INMARSAT satellites.

The Inmarsat satellite system comprises:

1. **Space segment:** It consists of a constellation of four prime GEO satellites strategically placed at one of the four ocean regions to provide a global coverage [Figure 10.14 (a)]. Two satellites are placed over the Atlantic Ocean Regions East and West (AORE 15.5° W and AORW 55.5° W respectively), one over the Indian Ocean Region (IOR 64.5° E) and one over the Pacific Ocean Region (POR 180° E). There is an overlap between the footprints of each of the satellites. Hence, in many parts of the world the services are provided by two Inmarsat satellites [Figure 10.14 (b)]. Each of these four satellites is backed up by spare operational satellites so that the services are not blocked due to failure of the operational satellites.
2. **Ground segment:** It comprises a large number of fixed Earth stations (gateways) and mobile Earth stations (MES). Gateways, referred to as land Earth stations (LES) or coastal Earth stations (CES) by the maritime community and as ground Earth stations (GES) by the aeronautical community, serve as interfaces to the terrestrial public switched networks. The ground segment also comprises an Inmarsat network control centre (NCC) and three satellite control centers (SCC). The NCC located in the UK, monitors and controls the complete network of LES, MES and the satellites. SCC are responsible for the physical management of Inmarsat satellites.
3. **Subscriber units:** These include the satphones, facsimile, telex and data terminals.

Figure 10.15 shows a typical communication network using Inmarsat satellites. The calls can be made between two mobile users and between mobile users and terrestrial phones. Standard Inmarsat satphones and telex terminals are available to make and receive calls using the Inmarsat satellite network. All the calls to the terrestrial phones are routed via the satellite

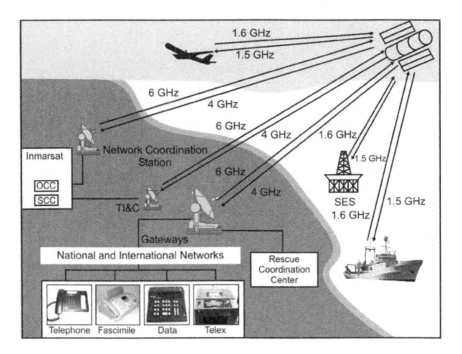

Figure 10.15 Typical communication network using Inmarsat satellites

to the gateways from where they are sent directly to the terrestrial public-switched networks. Satellite gateway links, known as feeder links, employ a 6 GHz band in the uplink direction and a 4 GHz band in the downlink direction. The mobile links use 1.6 GHz /1.5 GHz in the uplink/downlink directions. Inmarsat satellites provide various services, namely Inmarsat-A, B, C, D, D+, E, M, mini-M, GAN, R-BGAN, Aero, BGAN, M2M communications, global voice services (ISat Phone Pro, Isat Pro-link, FleetPhone), MPDS (mobile packet data services), XpressLink, FleetPhone, BGAN M2M, Isat Data Pro and other, with each service targeted towards a particular niche market. Table 10.6 enumerates salient features of these services. It may be mentioned here that INMARSAT has withdrawn the Inmarsat-A, E and R-BGAN services.

10.10.1.3 Iridium Satellite Constellation

The Iridium network is a global mobile communication system designed to offer voice communication services to pocket-sized telephones and data, fax and paging services to portable terminals, independent of the user's location in the world and of the availability of traditional telecommunications networks. Iridium is expected to provide a cellular-like service in areas where a terrestrial cellular service is unavailable or where the public switched telephone network (PSTN) is not well developed. It has revolutionized communication services for business professionals, travellers, residents of rural or undeveloped areas, disaster relief teams and others to whom it provides global communication services using a single mobile.

Table 10.6 Services offered by the Inmarsat satellites

Service	Communication Capability	Applications
INMARSAT A	Analogue telephony, data and compressed video through desktop PC type terminals	Land and maritime commercial, social and safety related applications
INMARSAT B	Digital telephony, fax, data and full video through briefcase type terminals	Land and maritime commercial, social and safety related applications with better services than INMARSAT A
INMARSAT C	Low bit rate store and forward data communication through briefcase sized terminals	International e-mail services, database access and global telex services
INMARSAT D & D+	Low bit rate two way data communication using personal CD player sized terminals	Data broadcast e.g. financial data, vehicle tracking, personal messaging
INMARSAT M	World's first personal, portable mobile satellite system providing digital telephony and data services using briefcase sized terminals	Remote and rural fixed communications, mobile communications e.g. business travellers, police and emergency services
INMARSAT mini-M	INMARSAT's most popular service employing the smallest, lightest and the cheapest terminals to provide digital voice and data services	Used by journalists, workers, business people, emergency services, rural telephony
INMARSAT Aero C	Same services as INMARSAT C to aircraft	Messaging services to corporate aircraft
INMARSAT Aero-L	Real time duplex digital telephone, fax and data services to aircraft	Real-time flight and passenger related communication e.g. engine monitoring
INMARSAT Aero-H	Telephone, data and fax services to passenger air cabs and cockpits	Medium bit rate real time voice, data and fax communication
INMARSAT Aero-I	Telephone, data and fax services to short and medium-haul aircraft	Passenger voice telephone, facsimile, cockpit voice and data, air traffic control, secure voice access to major air traffic control centers
INMARSAT E	Provides global maritime distress alerting services	Distress and safety-GMDSS compliant
Fleet	Voice, data, ISDN	Ocean-going and coastal vessels
Swift 64	Voice, fax, ISDN and MPDS	Private, business and commercial aircraft
R-BGAN	"Always-on" IP-data service	Land-mobile market
BGAN	internet and intranet solutions, video on demand, video-conferencing, fax, e-mail, telephone and high-speed LAN access	Land-mobile market
FleetBroadband (FB)	Offers similar services to BGAN	Maritime service

(*continued*)

Table 10.6 (*Continued*)

Service	Communication Capability	Applications
SwiftBroadband (SB)	Global voice and high speed data simultaneously at speeds up to 432 kbps per channel	Aeronautical service
BGAN HDR (High Data Rate)	High data rate streaming	Broadcasting, media organizations and global governments
XpressLink	High speed broadband for a fixed monthly fee	Maritime service
FleetPhone	Satellite phone service or use when beyond the range of land-based networks	Maritime industry
IsatPhone Pro	Low cost handheld satellite phone	Global services
IsatPhone Link	Low cost global satellite phone service	Rural and remote areas outside cellular coverage
M2M (machine to machine service) includes BGAN M2M, Isat Data Pro and Isat M2M	Two-way data connectivity for messaging, tracking and monitoring of fixed and mobile assets	Global services
BGAN M2M	Global two-way IP data service for long term M2M management of fixed assets	Global services
IsatData Pro	Global two-way SMS service for M2M communication	Global services
Isat M2M	Store and forward low data rate global messaging services to and from remote areas	Global services

The Iridium network can support 172 000 simultaneous users, providing each of them with 2.4 kbps fully duplex channels. The Iridium satellite system comprises the following three principal segments:

1. Space segment
2. Ground segment
3. Iridium subscriber products.

The space segment comprises 66 active communication satellites and 14 spare back-up satellites revolving around Earth, in six LEO orbital planes having 11 satellites each, at an altitude of 780 km. The system was originally supposed to have 77 active satellites (hence the name Iridium, as the element iridium has an atomic number of 77). Each satellite is cross-linked with four other satellites, providing flexibility and independence from the terrestrial networks.

The ground segment comprises gateways and a system control segment. Gateways are large fixed Earth stations connected to fixed and mobile terrestrial networks providing an interface between the terrestrial network and the satellites. Terrestrial users access the Iridium satellites through these gateways. They also verify the user account, record call duration and user

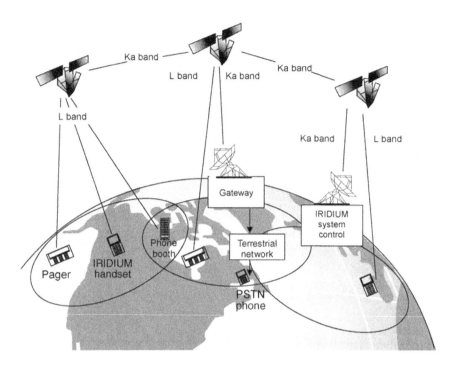

Figure 10.16 Typical communication set-up of the Iridium satellite constellation

location for billing purposes. The control segment comprises a ground station that performs TT&C functions, data routing and frequency planning operations.

Iridium subscriber products include phones and pagers that allow users to have access to either the compatible cellular telephone network or the Iridium network.

Figure 10.16 shows a typical communication set-up of the Iridium satellite constellation. As mentioned earlier, the Iridium network allows the users to have global mobile services using a single subscription number. To accomplish this, each user is associated with a gateway called the 'home gateway,' which maintains a record of its profile and location and looks after its services. Each user also has links with two satellites at a time. As the satellite moves out of the range of the user, the link from the user to the satellite is handed off from the satellite, leaving the user's area to the one entering the area.

The call from the user is picked up by the satellite and it authenticates the subscriber's account through the nearest land-based gateway. If the user is out of the home gateway region, the gateway will recognize that it is a visiting subscriber and sends information to its home gateway through intersatellite links (ISLs) to view the subscriber's profile and takes permission for making the call. The user-to-satellite service uplink and downlink operate in the same frequency ranges of 1.616 to 1.6265 GHz in the L band. In order to ensure that there is no interference between the uplink and the downlink signals, uplinking and downlinking are not done simultaneously. In order to make efficient use of the limited spectrum, the Iridium system employs a combination of FDMA and TDMA signal multiplexing. The satellite–gateway links, called the feeder links, employ 27.5–30 GHz in the Ka band for uplink signals and 18.8–20.2 GHz in the Ka band for downlink signals. If the destination phone is part of the PSTN, the call is routed from the original satellite to the nearest gateway for transmission through the terrestrial network. If the call is for another Iridium user, it is routed to the neighbouring satellite and so on through ISLs until it reaches the called user's home gateway, which determines the user's location. Depending upon the location of the user, whether he or she is within the home gateway region or in the visiting gateway region, the call is routed to the satellite directly above the user through the home gateway or the visiting gateway. A point to be noted here is that the call set-up information travels through the gateways and the voice information can completely travel over the Iridium ISLs. ISL links employ frequencies of 22.55–23.55 GHz (Ka band) as these frequencies are heavily absorbed by water and hence are not useful for establishing Earth–satellite links. In order to allow easy integration with terrestrial systems and to benefit from the advances made in the field of terrestrial mobile telephony, the architecture of the Iridium system follows the well-established terrestrial standard, the GSM (global systems for mobile communications).

The Iridium system once failed financially, mainly due to insufficient demand for the service, the bulkiness and cost of the hand-held devices as compared to cellular mobile phones and the rise of cellular GSM roaming agreements during Iridium's decade-long construction period. The Earth stations were shut down, however, the Iridium satellites were retained in orbit. Their services were re-established in 2001 by the newly founded Iridium Satellite LLC, partly owned by Boeing and other investors. Iridium NEXT, which are second generation Iridium satellites, are scheduled to be launched from 2015.

Another mobile satellite system similar to Iridium is the Globalstar satellite system of the USA. It provides global voice, data, fax and messaging services through a constellation of 48 satellites. These satellites orbit at an altitude of 1410 km and are inclined at an angle of 52°.

10.10.2 Regional Satellite Systems

One of the drawbacks of international satellite systems is that they are not optimized to the needs of the individual countries. The first step to meet the focused needs of the countries was to have a regional system that provided services to countries on a regional basis rather than on a global basis. Regional satellite missions were established with the aim of strengthening the communication resources of the countries belonging to the same geographical area. Some of the regional satellite systems include Eutelsat, Arabsat, AsiaSat, Measat, ACeS (Asia cellular satellite), Thuraya, etc. EUTELSAT operates a fleet of satellites that provide communication services to Europe, the Middle East, Africa and large parts of the Asian and American continents. Arabsat satellites provide satellite communication services to the Middle East, Africa and large parts of Europe. The Asia Satellite Telecommunications Company Limited (AsiaSat) and Measat systems are Asia's regional satellite operators, providing satellite services to the Asia Pacific region.

ACeS is another satellite-based regional communication system providing services to Asia. It provides fully digital video, voice and data services throughout Asia. The Thuraya system provides mobile communication services to the Middle East, North and Central Africa, Europe, Central Asia and the Indian subcontinent. In the following paragraphs, the Eutelsat system is described in detail.

10.10.2.1 EUTELSAT (European Telecommunication Satellite Organization)

The EUTELSAT organization was formed in the year 1977 to commission the design and construction of satellites and to manage the operation of regional satellite communication services in Europe. The first communication satellite to be launched by EUTELSAT was the orbital test satellite (OTS) in the year 1978, which carried out link tests with small Earth stations with the help of a powerful antenna on board the satellite. Then came the ECS-1 satellite in the year 1983 to provide communication services to post office and telecommunication administration and to broadcast TV programmes. The ECS satellite programme was renamed the Eutelsat satellite programme. Eutelsat satellites provide television, telephony and data transmission services on a regional basis. The more advanced satellites in this series also provided specific services like business communication services and mobile communication services. In addition to the Eutelsat satellites, other series of satellites, namely the Hot Bird, Eurobird and Atlantic Bird series, were launched to expand the horizon of the services offered and the coverage area of the satellites of the EUTELSAT organization. In 2012, EUTELSAT renamed all the satellite series under the brand name of Eutelsat. Today, the Eutelsat satellite system provides services that include television and radio broadcasting, professional video broadcasting, networking, Internet services, mobile communication services and broadband services for local communities, businesses and individuals.

Table 10.7 enumerates the salient features of the Eutelsat series of satellites.

10.10.3 National Satellite Systems

National satellite systems, also referred to as domestic satellite systems, provide services to a particular country. National satellite systems were originally established by developed countries like the USA, USSR and Canada to serve their country's population according to their specific needs. Today, in addition to these developed countries, some developing nations like India, China, Japan, and so on, also have their own national satellite systems. Some of the domestic satellite systems include Galaxy, Satcom, EchoStar and Telestar of the USA, Brasilsat of Brazil, INSAT of India, Optus of Australia and Sinosat of China. The INSAT system is briefly described below.

10.10.3.1 INSAT (Indian National Satellite)

Owned by the Indian Department of Space, named the Indian Space Research Organization (ISRO), INSAT is one of the largest domestic communication satellite networks in the world, providing services in the areas of telecommunications, television broadcasting, mobile satellite services and meteorology including disaster warning. INSAT is a joint venture of the Department of Space (DOS), Department of Telecommunications (DOT), Indian Meteorological Department (IMD), All India Radio (AIR) and Doordarshan. Making a modest beginning with the launch of INSAT-1A in 1982, the INSAT satellite programme has come a long way today. INSAT-1A belonged to the INSAT-1 series, further comprising INSAT-1B, 1C and 1D satellites. The INSAT-1 series was followed by INSAT-2 and INSAT-3 series of satellites. They were superceded by the INSAT-4 series of satellites. In addition to the INSAT series of satellites, GSAT satellites also provide communication services. Table 10.8 lists the salient features of the INSAT series of satellites.

Table 10.7 Eutelsat series of satellites

Satellite	Transmission capability	Stabilization	Location
Eutelsat 12 West A	24 Ku band transponders	3-axis	12.5°W
Eutelsat 8 West A	26 Ku band transponders	3-axis	8°W
Eutelsat 8 West C	28 Ku and 4 Ka band transponders	3-axis	7.8°W
Eutelsat 7 West A	56 Ku band transponders	3-axis	8°W
Eutelsat 5 West A	35 Ku and 10 C band transponders	3-axis	5°W
Eutelsat 3A	24 Ku band transponders	3-axis	3.3°E
Eutelsat 3C	64 Ku band transponders	3-axis	3°E
Eutelsat 3D	53 Ku and 3 Ka band transponders	3-axis	3.1°E
Eutelsat 4A	28 Ku band transponders	3-axis	4°E
Eutelsat 4B	20 Ku band transponders	3-axis	4°E
Eutelsat 7A	38 Ku and 2 Ka band transponders	3-axis	7°E
Eutelsat 9A	38 Ku band transponders	3-axis	9°E
Eutelsat Ka-Sat 9A	82 Ka band spotbeams transponders	3-axis	9°E
Eutelsat 10A	46 Ku and 10 C band transponders and S band payload	3-axis	10°E
Eutelsat Hotbird 13B	64 Ku band transponders	3-axis	13°E
Eutelsat Hotbird 13C	64 Ku band transponders	3-axis	13°E
Eutelsat Hotbird 13D	64 Ku band transponders	3-axis	13°E
Eutelsat 16A	53 Ku and 3 Ka band transponders	3-axis	16°E
Eutelsat 16B	20 Ku band transponders	3-axis	15.8°E
Eutelsat 16C	18 Ku band transponders	3-axis	16°E
Eutelsat 21A	24 Ka band transponders	3-axis	21.5°E
Eutelsat 21B	40 Ku band transponders	3-axis	21.5°E
Eutelsat 25B	3 Ku and 14 Ka band transponders	3-axis	25.5°E
Eutelsat 25C	24 Ku band transponders	3-axis	25.5°E
Eutelsat 28A	24 Ku band transponders	3-axis	28.5°E
Eutelsat 28B	32 Ku band transponders	3-axis	28.5°E
Eutelsat 33A	20 Ku band transponders	Spin	33°E
Eutelsat 36A	31 Ku band transponders	3-axis	36°E
Eutelsat 36B	70 Ku band transponders	3-axis	35.9°E
Eutelsat 48A	20 Ku band transponders	3-axis	48.2°E
Eutelsat 48C	24 Ku band transponders	3-axis	48°E
Eutelsat 70B	48 Ku band transponders	3-axis	70.5°E
Eutelsat 172A	22 C band and 26 Ku band transponders	3-axis	172°E
SESAT2	12 Ku band transponders leased to Eutelsat	3-axis	53°E
TELSTAR 12	38 Ku band transponders	3-axis	15°W

Due to the emergence of private satellite operators, the boundaries between the international, national and regional systems are diminishing very fast. As time progresses, the dividing lines classifying these systems will become less and less clear.

10.11 Future Trends

Satellites have been used for communication applications since the launch of the first communication satellite SCORE in 1958. The technology and applications of communication

Table 10.8 INSAT series of satellites

Satellite	Transponders	Position	Stabilization
INSAT 1A	12 C band and 2 S band transponders and VHRR (very high resolution radiometer) meteorological payload	74°E	3-axis
INSAT 1B	12 C band and 2 S band transponders and VHRR meteorological payload	74°E	3-axis
INSAT 1C	12 C band and 2 S band transponders and VHRR meteorological payload	93.5°E	3-axis
INSAT 1D	12 C band and 2 S band transponders	83°E	3-axis
INSAT 2A	12 C band, 6 extended C band and 2 S band transponders, 1 data relay transponder, 1 search and rescue transponder and VHRR meteorological payload	74°E	3-axis
INSAT 2B	12 C band, 6 extended C band and 2 S band transponders, 1 data relay transponder, 1 search and rescue transponder and VHRR meteorological payload	93.5°E	3-axis
INSAT 2C	12 C band, 6 extended C band, 3 Ku band, 2 S band BSS and 1 S band MSS transponders	93.5°E	3-axis
INSAT 2DT	25 C band and 1 S band BSS transponders	55°E	3-axis
INSAT 2E	12 C band, 5 extended C band transponders, meteorological payloads VHRR and CCD camera. 11 of the C band transponders have been leased to the INTELSAT organization	83°E	3-axis
INSAT 3B	12 extended C band, 3 Ku band and 1 S band MSS transponders and 1 Ku band beacon	83°E	3-axis
INSAT 3C	24 C band, 6 extended C band, 2 S band BSS transponders and a MSS transponder operating in S band for uplink and C band for downlink	74°E	3-axis
INSAT 3A	12 C band, 6 extended C band, 1 S band, 6 Ku band transponders, satellite aided search and rescue (SAS&R) transponder, meteorological payloads of VHRR, CCD camera and 1 data relay (DR) transponder	93.5°E	3-axis
INSAT 3E	24 C band and 12 extended C band transponders	55°E	3-axis
INSAT 4A	12 C band and 12 Ku band transponders	83°E	3-axis
INSAT 4B	12 C band and 12 Ku band transponders	93.5°E	3-axis
INSAT 4CR	12 Ku band transponders	74°E	3-axis
GSat 4 (failure)	Multi-beam Ka band transponders	3-axis	82°E
GSat 5P (failure)	36 G/H band transponders	3-axis	55°E
GSat 8 (INSAT 4G)	24 Ku band transponders	3-axis	55°E
GSat 12	12 extended C band transponders	3-axis	83°E
GSat 10	12C, 6 extended C and 12 Ku band transponders	3-axis	83°E
GSat 7 (INSAT 4F)	11 transponders (UHF, S, C and Ku band)	3-axis	74°E

satellites have grown manyfold in the last five decades or so. As a comparison the Syncom 2 satellite launched in the year 1963 had a launch mass of 68 kg, lasted three years and had a transmit amplifier power of a few watts. The Eutelsat W7 satellite launched in the year 2009 has a launch mass of 5000 kg, a design life of 15 years and has 70 Ku band transponders and a transmitter power of 12 kW catering to both global area and spot area applications. Moreover, satellites which were used for providing point-to-point trunk telephony and television services 30 years ago are now being used for mobile communications, real time on-demand data, multimedia and internet services and sound and video broadcasting applications.

The future trend in the field of communication satellites is towards launching more satellite constellations in low altitude orbits, designing complex satellite platforms with more on-board power, increased support to personal communication services (PCS) users, use of higher frequency bands and shift from RF spectrum to light quantum spectrum. Key technology areas in this field include development of large-scale multi-beam antennas to allow intensive reuse of frequencies, USAT terminals to replace VSAT terminals, development of signal processing algorithms to perform intelligent functions on-board the satellite including signal regeneration, to overcome the signal fading problem due to rain and to allow the use of smaller antennas. Flexible cross-link communication between satellites will be developed to allow better distribution of traffic between the satellites.

10.11.1 Development of Satellite Constellations in LEO Orbits

The future will see the trend of replacing a large single satellite in the geostationary orbit with a large number of co-located and interlinked smaller satellites in LEO orbits as the geostationary satellites cannot support very high data rate services due to an increase in the free space propagation loss with distance. Each satellite will perform its limited function and optical intersatellite links will be used to interface high-speed links between small satellites with a result that as a whole the satellite constellation will work as a single large satellite. Small terminals with antenna diameters of few centimetres will establish Gbits/s links between small satellites and the tracking requirements are minimized as beams with larger beam divergence can be used due to short distances. The services that will be provided by these satellites include broadcasting services to portable hand-held devices, two-way mobile broadband services for the land/aeronautical and maritime sectors and IPTV services.

10.11.2 Development of Personal Communication Services (PCS)

The advent of PCS has given rise to the need to have lower data rate systems supporting many users. The deployment philosophy in this case is to employ tens or hundreds of satellites in LEO orbits rather than employing three to five satellites in the geostationary orbits. The trend will be to have more on-board processing on the satellite and the satellite will support more services including routing, flow control, packet error detection/correction, and so on. In other words, the trend is to increase the application of higher layer protocols on the satellite. In addition to all this, satellite-satellite cross-linking will increase so as to make the satellite based PCS system feasible and realizable. Also, with the advances made in technology, the size of the repeaters on board the satellites will reduce significantly while their capability will increase manyfold.

10.11.3 Use of Higher Frequency Bands

Satellites will employ higher frequency bands like the Ka band to provide their intended services. More sophisticated satellite systems operating in the 21 GHz band will be realized around 2020. With use of higher frequency spectrum becoming common, extension of services provided by VSAT including GSM, IP trunking and broadband services will be possible. Development of robust access technologies will aid in this process. One of the possible developments that can take place is the construction of a common platform based on the internet protocol (IP) including the integration of telephone systems and television broadcasting systems into the internet.

10.11.4 Development of Light Quantum Communication Techniques

Further innovation will be facilitated by the advent of conversion from communication based on electromagnetic waves to that based on light quantum communications. This will offer very secure communication providing high capacity and confidentiality in information communication. The outstanding feature of quantum cryptography communication is that in case some quantum signals are stolen during transmission, the state of quanta changes instantaneously making the data meaningless while at the same time it is recorded at the detector that the data has been stolen. Quantum communications technology utilizes properties such as quantum entanglement and quantum superposition and is quite different from the RF communication technologies being used to date.

10.11.5 Development of Broadband Services to Mobile Users

There will be advancement in the services offered by communication satellites to providing broadband services to mobile users located on aircraft, space planes, boats and vehicles such as high speed trains, buses and cars. Some of the technologies that will be utilized to make these services possible include spatial diversity (such as two receive antenna configuration), time diversity (channel interleaving/ spreading techniques) and upper layer FEC (forward error correction). FEC at upper layers is under study to protect transmitted packets and avoid retransmissions due to limited access or absence of a return channel for a potentially high number of users and to look into the challenges posed by mobile services.

10.11.6 Development of Hybrid Satellite/Terrestrial Networks

Future satellite communications cannot be oblivious to the evolution of terrestrial broadcast and broadband communication systems. Cooperative communication techniques are also being considered for hybrid satellite/terrestrial networks with the aim of extending the satellite coverage and of supporting terrestrial networks unable to provide their services because of lack of coverage, network overloads, terrain constraints or emergency situations. In this perspective, development of new satellite systems tends to align with that of terrestrial communications leading to a full integration between the two networks which enables higher data rates and high quality of service anywhere anytime.

10.11.7 Advanced Concepts

In the paragraphs above, we mentioned the applications of communication satellites that will become a reality in a decade or so. We now shift focus to some interesting concepts which are fiction right now but may become realizable in next two decades or so. Satellites can be used for communicating with submarines to transmit coded information and for establishing an inter-planetary television link and so on. We present in brief these concepts so as to give the readers a glimpse of the potential of this majestic piece of equipment.

Satellite-to-Submarine Communication: Satellites can be used to communicate with many submarines that are submerged in sea water at depths of 100 m or so. This would eliminate the need to have submarines come to the surface to establish communication thus reducing their vulnerability.

The concept is highlighted in Figure 10.17. Satellites in geostationary orbit are used and transmit a large number of narrow beams to create random spots on the ocean, with each beam transmitting encrypted data. Large numbers of spots are generated so as to create empty positions and not give away the location of submarines. Blue-green wavelength laser is used for maximal penetration in sea water.

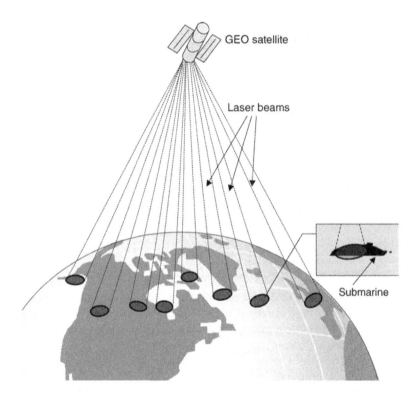

Figure 10.17 Satellite-to-submarine communication

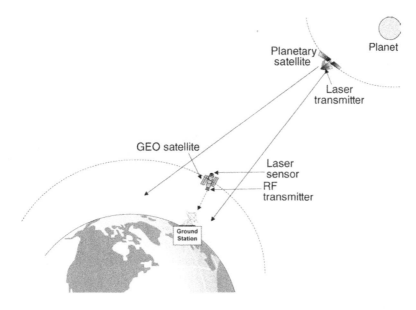

Figure 10.18 Interplanetary TV link

Interplanetary TV Link: The concept of the interplanetary TV link is shown in Figure 10.18. The set up makes use of a satellite orbiting around a planet with which the link has to be established and a satellite orbiting in geostationary orbit around the Earth. The planetary satellite makes use of a low power laser to transmit the signals. The Earth orbiting satellite will have a sensor to receive the optical signal and process it and convert it into the microwave signal. The signal is converted from the optical spectrum to the microwave spectrum as the optical signals do not penetrate clouds and are highly attenuated by rain. The conversion therefore allows establishing a non-interruptive link. This link would allow monitoring the events happening on different planets on a real-time basis.

Further Readings

Calcutt, D. and Tetley, L. (1994) *Satellite Communications: Principles and Application*, Edward Arnold, a member of the Hodder Headline Group, London.
Chartant, R.M. (2004) *Satellite Communications for the Nonspecialist*, SPIE, Washington.
Elbert, B.R. (1997) *The Satellite Communication Applications Handbook*, Artech House, Boston, Massachusetts.
Elbert, B.R. (1999) *Introduction to Satellite Communication*, Altech House, Boston, Massachusetts.
Gatland, K. (1990) *Illustrated Encyclopedia of Space Technology*, Crown, New York.
Gedney, R.T., Schertler, R. and Gargione, F. (2000) *Advanced Communication Technology Satellite: Insider's Account of the Emergence of Interactive Broadband Services in Space*, SciTech, New Jersey.

Gedney, R.T., Schertler, R. and Gargione, F. (2000) *The Advanced Communication Technology*, SciTech, New Jersey.
Geoffrey, L.E. (1992) *Communication Services via Satellite*, Butterworth Heinemann, Oxford.
Iida, T., Peltomn, J.N. and Ashford, E. (2003) *Satellite Communications in the 21st Century*, American Institute of Aeronautics and Astronautics, Inc., Virginia.
Kadish, J.E. (2000) *Satellite Communications Fundamentals*, Artech House, Boston, Massachusetts.
Levitan, B. and Harte, L. (2003) *Introduction to Satellite Systems: Technology Basics, Market Growth, Systems, and Services*, Excerpted from *Wireless Communications Basics*, Althos Publishing.
Lutz, E., Werner, M. and Jahn, A. (2000) *Satellite Systems for Personal and Broadband Communications*, Springer, New York.
Maral, G. (2003) *VSAT Networks*, John Wiley and Sons, Ltd, Chichester.
Maral, G. and Bousquet, M. (2002) *Satellite Communication Systems: Systems, Techniques and Technology*, John Wiley & Sons, Ltd, Chichester.
Miller, M.J., Vucetic, B. and Berry, L. (1993) *Satellite Communications: Mobile and Fixed Services*, Kluwer Academic Publishers, Massachusetts.
Perez, P. (1998) *Wireless Communications Design Handbook: Space Interference*, Academic Press, London.
Richharia, M. (1999) *Satellite Communication Systems*, Macmillan Press Ltd.
Richharia, M. (2001) *Mobile Satellite Communications: Principles and Trends*, Addison Wesley.
Sherrif, R.E. and Hu, Y.F. (2001) *Mobile Satellite Communication Networks*, John Wiley & Sons, Ltd, Chichester.
Verger, F., Sourbes-Verger, I., Ghirardi, R., Pasco, X., Lyle, S. and Reilly, P. (2003) *The Cambridge Encyclopedia of Space*, Cambridge University Press.

Internet Sites

1. www.alcatel.com
2. www.arabsat.com
3. www.asiasat.com
4. www.astrium-space.com
5. www.boeing.com
6. http://electronics.howstuffworks.com/satellite-tv.htm/printable
7. http://en.wikipedia.org/wiki/Satellite_television
8. http://www.ias.ac.in/meetings/annmeet/69am_talks/ksdgupta/img3.html
9. http://www.ddinews.com/DTH
10. http://www.angelfire.com/electronic/vikram/tech/vsattut.html
11. www.esa.int
12. www.eutelsat.org
13. www.globalstar.com
14. www.hughes.com
15. www.inmarsat.org
16. www.intelsat.com
17. www.eads.net

18. www.isro.org
19. www.ssloral.com
20. www.skyrocket.de

Glossary

Broadcast Satellite Services (BSS): This refers to the satellite services that can be received at many unspecified locations by relatively simple receive-only Earth stations

Communication satellites: Communication satellites act as repeater stations that relay radio signals providing either point-to-point, point-to-multipoint or multipoint interactive services. The application areas of communication satellites mainly include television broadcast, international telephony and data communication services

Digital video broadcasting (DVB): DVB is a family of technology standards defining digital broadcasting and designed to facilitate broadcasting of audio, images and multimedia facilitating a large degree of user interaction. These standards are designed to use existing satellite, cable and terrestrial infrastructures for broadcasting

Digital Video Broadcast – Handheld Standard (DVB-H): The DVB-H standard provides video and television services for cellular phones and handsets

Digital Video Broadcast with Return Channel via Satellite Standard (DVB-RCS): DVB-RCS is one of the digital television broadcast standards in the DVB family of standards providing interactive broadcasting

Digital Video Broadcast-Satellite services to Handheld Standard: The DVB-SH standard delivers audio, video and data services to handheld devices using frequencies within the S-band from either satellite or terrestrial networks. It complements the DVB-H standard that delivers mobile video from terrestrial networks in UHF TV bands

Digital Video Broadcast-Terrestrial Standard: The DVB-T standard is the most widely used digital television broadcast standard worldwide for delivering high quality video. DVB-T allows transmission of several television broadcasts and audio channels on a single transmission

Direct Broadcasting Services (DBS): DBS uses specially high powered Ku band satellites that send digitally compressed television and audio signals to relatively small (45 to 60 cm across) fixed satellite dishes

Direct-to-home satellite television: Direct-to-home (DTH) satellite television refers to the direct reception of satellite TV programmes by the end users from the satellite through their own receiving antennas

DVB-RCS2 Standard: Second generation DVB-RCS standard

DVB-Satellite Standard: DVB-S was the first digital television satellite standard released by the DVB Project in 1994 and provides digital satellite television services

DVB-S2 Standard: Second generation DVB-S standard

DVB-T2 Standard: Second generation DVB-T standard

EUTELSAT (European Telecommunication Satellite Organization): The EUTELSAT Organization is responsible for commissioning the design and construction of satellites and for managing the operation of regional satellite communication services to Europe

Fixed Satellite Services (FSS): This refers to the two-way communication between Earth stations at fixed locations via a satellite. It supports the majority of commercial applications including satellite telephony, satellite television and data transmission services. The FSS primarily uses two frequency bands: C band (6/4 GHz) and Ku band (14/11 GHz, 14/12 GHz)

Frequency bands: Satellite communication employs electromagnetic waves for transmission of information between Earth and space. The bands of interest for satellite communications lie above 100 MHz including the VHF, UHF, L, S, C, X, Ku and Ka bands

Gateways: Gateways are large fixed Earth stations connected to fixed and mobile terrestrial networks providing an interface between the terrestrial network and the satellites

GMPCS (Global Mobile Personal Communication System): This is a personal communication system providing transnational, regional or global two-way voice, fax, messaging, data and broadband multimedia services from a constellation of satellites accessible with small and easily transportable terminals

Inmarsat Satellite System: Inmarsat, an acronym for the International Maritime Satellite Organization, is an international organization currently having 85 member countries that control satellite systems in order to provide global mobile communication services

INSAT (Indian National Satellites): Owned by the Indian Department of Space, called the Indian Space Research Organisation (ISRO), INSAT is one of the largest domestic communication satellite systems in the world, providing services in the area of telecommunications, television broadcasting, mobile satellite services and meteorology including disaster warnings to India

Iridium satellite constellation: The Iridium network, comprising 66 active satellites, is a global mobile communication system designed to offer voice communication services to pocket-sized telephones and data, fax and paging services to portable terminals, independent of the user's location in the world and of the availability of traditional telecommunications networks

Intelsat satellite system: The International Telecommunication Satellite Organisation (INTELSAT) is the world's largest international communications service provider and provides services including video, voice and data services to the telecom, broadcast, government and other communications market

International satellite systems: The international satellite systems provide global communication services. Some of the international satellite missions include Intelsat, PamAmSat, Orion, Intersputnik, Inmarsat, etc

International Telecommunication Union (ITU): The ITU is a specialized institution formed in the year 1865 for ensuring the proper allocation of frequency bands as well as the orbital positions of the satellites in the GEO

Mobile Satellite Services (MSS): This refers to the reception by receivers that are in motion, like ships, cars, lorries, etc. Increasingly, MSS networks are providing relay communication services to portable handheld terminals. L band (2/1 GHz) and S band (4/2.5 GHz) are mainly employed for MSS

Mobile satellite telephony: This refers to the interactive voice communication to mobile users and is one of the important services provided by Mobile Satellite Services (MSS)

Multipoint interactive network: Such networks are bidirectional data broadcasting networks

Orthogonal frequency division multiplex (OFDM): This is a type of modulation format that gives immunity against selective fading from multipath signals

Point-to-multipoint broadcast services: These refer to unidirectional data broadcast networks where the transmission occurs from a single uplink to a large number of remote receiving points within the coverage area of the satellite

Regenerative transponders: Regenerative transponders are those in which some onboard processing is done and the received signal is altered before retransmission

Relative gain-to-noise temperature (G/T) ratio: The relative gain-to-noise temperature (G/T) ratio is the ratio of the receive antenna gain and the noise temperature of the satellite receiving system. It defines the uplink performance of any satellite

Satellite–cable television: Satellite–cable television refers to the use of satellites to carry the programming channels to the cable system head-ends for further distribution to end users via co-axial and fibre optic cables

Satellite data broadcasting: Satellite data broadcasting refers to the use of satellites in the point-to-multipoint or multipoint interactive configuration for the transmission of information in digital form

Satellite radio: Satellites provide high fidelity audio broadcast services to the broadcast radio stations, referred to as satellite radio

Satellite telephony: Satellites provide long distance (especially intercontinental) point-to-point or trunk telephony services as well as mobile telephony services, either to complement or to bypass the terrestrial networks

Satellite TV: Satellite TV refers to the use of satellites for relaying TV programmes from a point where they originate to a large geographical area. GEO satellites in point-to-multipoint configuration are employed for satellite TV applications

Television receive-only (TVRO) services: TVRO systems refer to the satellite television services operating in the C band and requiring large dishes (6 to 18 feet across) placed at the user's premises for TV reception

Transmit effective isotropic radiated power (EIRP): This is the most important technical parameter of a communication satellite and is defined as the product of the transmit antenna gain and the maximum RF power per transponder

Transparent or bent pipe transponders: Transparent transponders process the uplink satellite signal in such a way that only the amplitude and the frequency are altered; the modulation and the spectral shape of the signal are not affected

Transponder: A transponder is the key element in the satellite communication network and is essentially a repeater which receives the signal transmitted from Earth on the uplink, amplifies the signal and retransmits it on the downlink at a different frequency from that of the received signal

Transponder equivalent (TPE): This defines the total transmission capacity available on satellites in terms of transponders having a bandwidth of 36 MHz

VSATs (Very small aperture terminals): VSATs are used for providing one-way or two-way data broadcasting services, point-to-point voice services and one-way video broadcasting services

11
Remote Sensing Satellites

Remote sensing is a technology used for obtaining information about the characteristics of an object through the analysis of data acquired from it at a distance. Satellites play an important role in remote sensing. Some of the important and better known applications of satellites in the area of remote sensing include providing information about the features of the Earth's surface, such as coverage, mapping, classification of land cover features such as vegetation, soil, water, forests, etc. In this chapter, various topics related to remote sensing satellites will be covered, including their principle of operation, payloads on board these satellites and their use to acquire images, processing and analysis of these images using various digital imaging techniques and finally interpreting these images for studying different features of Earth for varied applications. The chapter gives a descriptive view of the above-mentioned topics with relevant illustrations wherever needed. Some of the major remote sensing satellite systems used for the purpose will also be introduced towards the end of the chapter.

11.1 Remote Sensing – An Overview

Remote sensing is defined as the science of identifying, measuring and analysing the characteristics of objects of interest without actually being in contact with them. It is done by sensing and recording the energy reflected or emitted by these objects and then processing and interpreting that information. Remote sensing makes use of the fact that every object has a unique characteristic reflection and emission spectra that can be utilized to identify that object. Sometimes, the gravitational and magnetic fields are also employed for remote sensing applications. One of the potential advantages that this technology offers is that through it various observations can be made, measurements taken and images of phenomena produced that are beyond the limits of normal perception. Remote sensing is widely used by biologists, geologists, geographers, agriculturists, foresters and engineers to generate information on objects on Earth's land surface, oceans and atmosphere. Applications include monitoring natural and agricultural resources, assessing crop inventory and yield, locating forest fires and assessing the damage caused, mapping and monitoring of vegetation, air and water quality, and so on.

Satellite Technology: Principles and Applications, Third Edition. Anil K. Maini and Varsha Agrawal.
© 2014 John Wiley & Sons, Ltd. Published 2014 by John Wiley & Sons, Ltd. Companion Website: www.wiley.com/go/maini3

A brief look into the history of remote sensing suggests that this technology dates back to the early 19th century when photographs of Earth were taken from the ground. The idea of aerial remote sensing emerged in the early 1840s when the pictures were taken from cameras mounted on balloons, pigeons, and so on. During both World Wars, cameras mounted on airplanes and rockets were used to provide aerial views of fairly large surface areas that were invaluable for military reconnaissance. Until the introduction of satellites for remote sensing, aerial remote sensing was the only way of gathering information about Earth. The idea of using satellites for remote sensing applications evolved after observing the images taken by the TIROS weather forecasting satellite. These images showed details of the Earth's surface where the clouds were not present. The first remote sensing satellite to be launched was Earth Resources Technology Satellite (ERTS-A) in the year 1972. Today satellites have become the main platform for remote sensing applications as they offer significant advantages over other platforms, which include radio-controlled aeroplanes and balloon kits for low altitude applications as well as ladder trucks for ground-based applications. In the paragraphs to follow, a brief comparison will be given of aerial and satellite remote sensing.

11.1.1 Aerial Remote Sensing

In aerial remote sensing, as mentioned before, sensors are mounted on aircraft, balloons, rockets and helicopters. Cameras mounted on aircraft have been used to monitor land use practices, locate forest fires and produce detailed and accurate maps of remote and inaccessible locations of the planet. Weather balloons and rockets are used for obtaining direct measurements of the properties of the upper atmosphere. Aerial systems are less expensive and more accessible options as compared to the satellite systems and are mainly used for one-time operations. The advantage of aerial remote sensing systems as compared to the satellite-based systems is that they have a higher spatial resolution of around 20 cm or less. However, they have a smaller coverage area and higher cost per unit area of ground coverage as compared to the satellite-based remote sensing systems. They are generally carried out as one-time operations whereas satellites offer the possibility of continuous monitoring of Earth.

11.1.2 Satellite Remote Sensing

Satellites are the main remote sensing platforms used today and are sometimes referred to as 'eyes in the sky'. Use of satellites for remote sensing applications has brought a revolution in this field as they can provide information on a continuous basis of vast areas on the Earth's surface day and night. Constellations of satellites continuously monitor Earth and provide even minute details of the Earth's surface. This ever-expanding range of geographical and geophysical data help people to understand their planet better, monitor various parameters more minutely and hence manage and solve the problems related to Earth more efficiently. Satellites have become the main platform for carrying out remote sensing activities as they offer a number of advantages over other platforms. Some of these advantages include:

1. Continuous acquisition of data
2. Frequent and regular re-visit capabilities resulting in up-to-date information

3. Broad coverage area
4. Good spectral resolution
5. Semi-automated/computerized processing and analysis
6. Ability to manipulate/enhance data for better image interpretation
7. Accurate data mapping

A point worthy of a mention here is that the aerial and satellite-based remote sensing systems use those regions of the electromagnetic spectrum that are not blocked by the atmosphere. These regions are referred to as 'atmospheric transmission windows'. In this chapter satellite remote sensing will be discussed; hereafter, whenever remote sensing is mentioned, it will refer to satellite remote sensing unless otherwise stated.

11.2 Classification of Satellite Remote Sensing Systems

The principles covered in this section apply equally well to ground-based and aerial platforms but here they will be described in conjunction with satellites. Remote sensing systems can be classified on the basis of (a) the source of radiation and (b) the spectral regions used for data acquisition.

As mentioned before, satellite remote sensing is the science of acquiring information about the Earth's surface by sensing and recording the energy reflected or emitted by the Earth's surface with the help of sensors on board the satellite. Based on the source of radiation, they can be classified as:

1. Passive remote sensing systems
2. Active remote sensing systems

Passive remote sensing systems either detect the solar radiation reflected by the objects on the surface of the Earth or detect the thermal or microwave radiation emitted by them. Active remote sensing systems make use of active artificial sources of radiation generally mounted on the remote sensing platform. These sources illuminate the objects on the ground and the energy reflected or scattered by these objects is utilized here. Examples of active remote sensing systems include microwave and laser-based systems. Depending on the spectral regions used for data acquisition, they can be classified as:

1. Optical remote sensing systems (including visible, near IR and shortwave IR systems)
2. Thermal infrared remote sensing systems
3. Microwave remote sensing systems

11.2.1 Optical Remote Sensing Systems

Optical remote sensing systems mostly make use of visible (0.3–0.7 μm), near IR (0.72–1.30 μm) and shortwave IR (1.3–3.0 μm) wavelength bands to form images of the Earth's surface. The images are formed by detecting the solar radiation reflected by objects

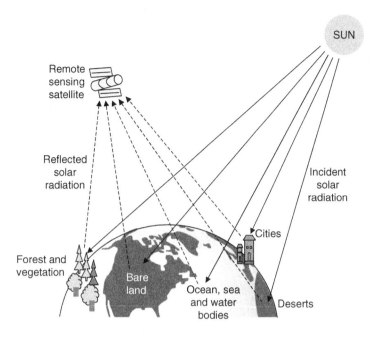

Figure 11.1 Optical remote sensing

on the ground (Figure 11.1) and resemble the photographs taken by a camera. However, some laser-based optical remote sensing systems are also being employed in which the laser beam is emitted from the active sources mounted on the remote sensing platform. The target properties are analysed by studying the reflectance and scattering characteristics of the objects to the laser radiation. Optical remote sensing systems employing solar energy come under the category of passive remote sensing systems and the laser-based remote sensing systems belong to the category of active remote sensing systems. Passive optical remote sensing systems work only during the day as they rely on sensing reflected sunlight. This phenomenon makes them weather dependent because during cloudy days the sunlight is not able to reach Earth.

Solar energy based optical remote sensing systems work on the principle that different materials reflect and absorb differently at different wavelengths in the optical band; hence the objects on the ground can be differentiated by their spectral reflectance signatures (Figure 11.2) in the remotely sensed images. As an example, vegetation has a very strong reflectance in the green and the near IR band and it has strong absorption in the red and the blue spectral bands. Moreover, each species of vegetation has a characteristic different spectral reflectance curve. Vegetation under stress from disease or drought also has a different spectral reflectance curve as compared to healthy vegetation. Hence, the vegetation studies are carried out in the visible and the near IR bands. Minerals, rocks and soil have high reflectance in the whole optical band, with the reflectance increasing at higher wavelengths. Their studies are carried out in the 1.3 to 3.0 µm band. Clear water has no reflectance in the IR region but has very high reflectance in the blue band and hence appears blue in colour. Water with vegetation has very

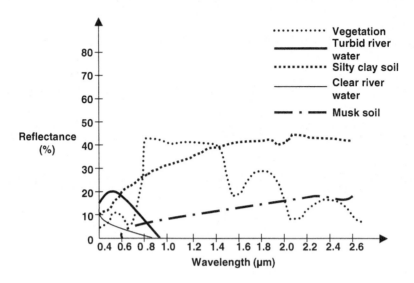

Figure 11.2 Spectral reflectance signatures of different objects

high absorption in the blue band and high reflectance in the green band and hence it appears greenish in colour. Muddy water has a yellow-brownish colour due to the strong reflectance in that band. Hence, water studies are made in the visible band. Table 11.1 enumerates the optical bands employed for various applications.

Table 11.1 Optical bands employed for various applications

0.45–0.52 µm	Sensitive to sedimentation, deciduous/coniferous forest colour discrimination, soil vegetation differentiation
0.52–0.59 µm	Green reflectance by heavy vegetation, vegetation vigour, rock–soil discrimination, turbidity and bathymetry in shallow water
0.62–0.68 µm	Sensitive to chlorophyll absorption, plant species discrimination, differentiation of soil and geological boundary
0.77–0.86 µm	Sensitive to the green biomass and moisture in the vegetation, land and water studies, geomorphic studies

In the optical band, panchromatic or black and white images can also be taken, where different shades of grey indicate different levels of reflectivity. The most reflective surfaces are light or nearly white in colour while the least reflective surfaces are represented as black.

11.2.2 Thermal Infrared Remote Sensing Systems

Thermal infrared remote sensing systems employ the mid wave IR (3–5 µm) and the long wave IR (8–14 µm) wavelength bands. The imagery here is derived from the thermal radiation emitted by the Earth's surface and objects. As different portions of the Earth's surface are at different temperatures, thermal images therefore provide information on the temperature of the ground and water surfaces and the objects on them (Figure 11.3). As the thermal infrared remote sensing systems detect the thermal radiation emitted from the Earth's surface, they come under the category of passive remote sensing systems. The 10 μm band is commonly employed for thermal remote sensing applications as most of the objects on the surface of the Earth have temperatures around 300 K and the spectral radiance for a temperature of 300 K peaks at a wavelength of 10 μm. Another commonly used thermal band is 3.8 μm for detecting forest fires and other hot objects having temperatures between 500 K and 1000 K.

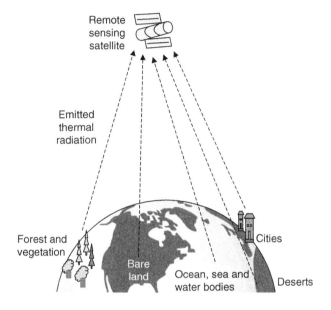

Figure 11.3 Thermal remote sensing

Colder surfaces appear darker in the raw IR thermal images, but the general remote sensing concept for IR images is to invert the relationship between brightness and the temperature so that the colder objects appear brighter as compared to the hotter ones. Hence, clouds that are colder than the Earth's surface appear in darker shades against the light background of the Earth in raw thermal images, but appear in lighter shades against the darker background in the processed thermal images. Also, low lying clouds appear darker as compared to clouds at high altitude in the processed image as the temperature normally decreases with height. Thermal systems work both during the day and night as they do not use solar radiation, but they suffer from the disadvantage that they are weather-dependent systems. Other than remote sensing satellites, weather forecasting satellites also make extensive use of the thermal IR bands.

11.2.3 Microwave Remote Sensing Systems

Microwave remote sensing systems generally operate in the 1 cm to 1 m wavelength band. Microwave radiation can penetrate through clouds, haze and dust, making microwave remote sensing a weather independent technique. This feature makes microwave remote sensing systems quite attractive as compared to optical and thermal systems, which are weather dependent. Microwave remote sensing systems work both during the day as well as at night as they are independent of the solar illumination conditions. Another advantage that a microwave remote sensing system offers is that it provides unique information on sea wind and wave direction that cannot be provided by visible and infrared remote sensing systems. However, the need for sophisticated data analysis and poorer resolution due to the use of longer wavelength bands are the disadvantages of microwave remote sensing systems.

Shorter microwave wavelength bands are utilized for the analyses of hidden mineral resources as they penetrate through the Earth's surface and the vegetation, whereas longer wavelength bands are utilized for determining the roughness of the various features on the Earth's surface. Microwave remote sensing systems can be both passive as well as active. Passive microwave remote sensing systems work on a concept similar to that of thermal remote sensing systems and detect the microwave radiation emitted from the objects. The characteristics of the objects are then formed on the basis of the received microwave power as the received power is related to their characteristics, such as temperature, moisture content and physical characteristics.

Active microwave remote sensing systems provide their own source of microwave radiation to illuminate the target object (Figure 11.4). Images of Earth are formed by measuring the microwave energy scattered by the objects on the surface of the Earth. The brightness of every point on the surface of the Earth is determined by the intensity of the microwave energy scattered back to the radar receiver on the satellite from them. The intensity of this backscatter is dependent on certain physical properties of the surface such as slope, roughness and the dielectric constant of the surface materials (dielectric constant depends strongly on the moisture content), on the geometric factors such as surface roughness, orientation of the

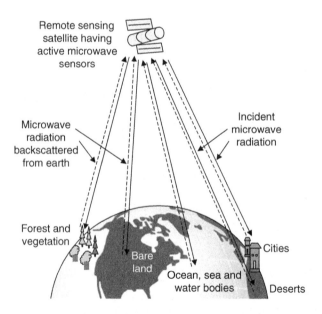

Figure 11.4 Active microwave remote sensing

objects relative to the radar beam direction and the types of land cover (soil, vegetation or man-made objects). The backscatter radiation is also dependent on the incident microwave beam parameters like frequency, polarization and angle of incidence.

Examples of passive microwave systems include altimeters and radiometers. Real aperture and synthetic aperture radar are active microwave remote sensing systems.

11.3 Remote Sensing Satellite Orbits

Remote sensing satellites have sun-synchronous subrecurrent orbits at altitudes of 700–900 km, allowing them to observe the same area periodically with a periodicity of two to three weeks. They cover a particular area on the surface of the Earth at the same local time, thus observing it under the same illumination conditions. This is an important factor for monitoring changes in the images taken at different dates or for combining the images together, as they need not be corrected for different illumination conditions. Generally the atmosphere is clear in the mornings; hence to take clear pictures while achieving sufficient solar illumination conditions, remote sensing satellites make observations of a particular place during morning (around 10 a.m. local time) when sufficient sunlight is available. As an example, the SPOT satellite has a sun-synchronous orbit with an altitude of 820 km and an inclination of 98.7°. The satellite crosses the equator at 10:30 a.m. local solar time.

Satellite orbits are matched to the capabilities and objectives of the sensor(s) they carry. Remote sensing satellites generally provide information either at the regional level or at the local area level. Regional level remote sensing satellite systems have a resolution of 10 m to 100 m and are used for cartography and terrestrial resources surveying applications, whereas local area level remote sensing satellite systems offer higher resolution and are used for precision agricultural applications like monitoring the type, health, moisture status and maturity of crops, and so on, for coastal management applications like monitoring photo planktons, pollution level determination, bathymetry changes, and so on.

As the satellite revolves around the Earth, the on board sensors it see a certain portion of the Earth's surface. The area imaged on the surface is referred to as the swath. The swath width for space-borne sensors generally varies between tens of kilometres to hundreds of kilometres. The satellite's orbit and the rotation of Earth work together to allow the satellite to have complete coverage of the Earth's surface.

11.4 Remote Sensing Satellite Payloads

11.4.1 Classification of Sensors

The main payloads on board a remote sensing satellite system are sensors that measure the electromagnetic radiation emanating or reflected from a geometrically defined field on the surface of the Earth. Sensor systems on board a remote sensing satellite can be broadly classified as:

1. Passive sensors
2. Active sensors

A passive system generally consists of an array of sensors or detectors that record the amount of electromagnetic radiation reflected and/or emitted from the Earth's surface. An active system, on the other hand, emits electromagnetic radiation and measures the intensity of the return signal. Both passive and active sensors can be further classified as:

1. Scanning sensors
2. Non-scanning sensors

This mode of classification is based on whether the entire field to be imaged is explored in one take as in the case of non-scanning sensors or is scanned sequentially with the complete image being a superposition of the individual images, as in the case of scanning sensors. Figure 11.5 enumerates the various types of sensors on board remote sensing satellites.

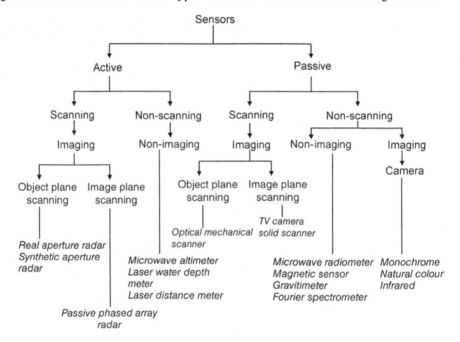

Figure 11.5 Various types of sensors on board remote sensing satellites

Scanning sensors have a narrow field of view and they scan a small area at any particular time. These sensors sweep over the terrain to build up and produce a two-dimensional image of the surface. Hence they take measurements in the instantaneous field-of-view (IFOV) as they move across the scan lines. The succession of scan lines is obtained due to the motion of the satellite along its orbit. It may be mentioned here that the surfaces are scanned sequentially due to the combination of the satellite movement as well as that of the scanner itself (Figure 11.6). The scanning sensors can be classified as image plane scanning sensors and object plane scanning sensors depending upon where the rays are converged by the lens in the optical system.

A non-scanning sensor views the entire field in one go. While the sensor's overall field-of-view corresponds to the continuous movement of the instantaneous field-of-view in the case of scanning sensors, for non-scanning sensors the overall field-of-view coincides with its instantaneous field-of-view. Figure 11.7 shows the conceptual diagram of a non-scanning satellite remote sensing system.

Remote Sensing Satellite Payloads

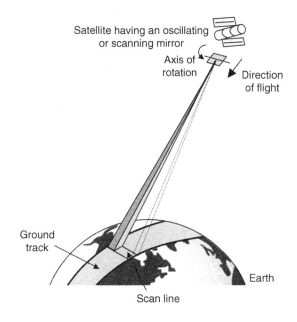

Figure 11.6 Scanning satellite remote sensing system

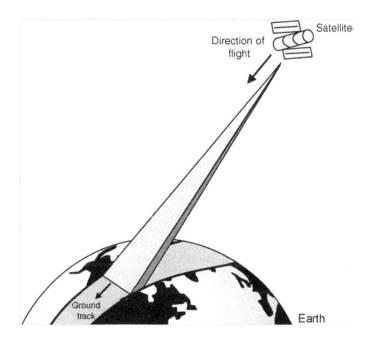

Figure 11.7 Non-scanning satellite remote sensing system

The sensors can be further classified as imaging or non-imaging sensors. Imaging sensors build up images of Earth using substances like silver in the film or by driving an image producing device like a TV or a computer monitor, whereas non-imaging sensors do not build images of Earth in any literal sense as they measure the radiation received from all points in the sensed target area and integrate it.

Before the working principle of various types of sensors is discussed, various sensor parameters will be described to help to give a better understanding of the working fundamentals of sensors.

11.4.2 Sensor Parameters

Sensor parameters briefly described in the following paragraphs include:

1. Instantaneous field-of-view (IFOV)
2. Overall field-of-view
3. S/N ratio
4. Linearity
5. Wavelength band
6. Swath width
7. Dwell time
8. Resolution

1. Instantaneous field-of-view (IFOV). This is defined as the solid angle from which the electromagnetic radiation measured by the sensor at a given point of time emanates.
2. Overall field-of-view. This corresponds to the total size of the geographical area selected for observation. In the case of non-scanning sensors, the instantaneous and the total field-of-view are equal and coincide with one another, whereas for scanning sensors, the overall field-of-view is a whole number multiple of the instantaneous field-of-view.
3. S/N ratio. This defines the minimum power level required by the sensor to identify an object in the presence of noise.
4. Linearity. Linearity refers to the sensor's response to the varying levels of radiation intensity. The linearity is generally specified in terms of the slope of the sensor's response curve and is referred to as 'gamma'. A gamma of one corresponds to a sensor with a linear response to radiation. A gamma that is less than one corresponds to a sensor that compresses the dark end of the range, while a gamma greater than one compresses the bright end. Sensors based on solid state circuitry like CCDs are linear over a wide range as compared to other sensors like vidicon cameras.
5. Wavelength band. Sensors employ wavelength bands for remote sensing applications: the optical band, the thermal band and the microwave band.
6. Swath width. The swath width of the sensor is the area on the surface of the Earth imaged by it.
7. Dwell time. The sensor's dwell time is defined as the discrete amount of time required by it to generate a strong enough signal to be detected by the detector against the noise.
8. Resolution. Resolution is defined as the ability of the entire remote sensing system (including the lens, antenna, display, exposure, processing, etc.) to render a sharply defined

image. Resolution of any remote sensing system is specified in terms of spectral resolution, radiometric resolution, spatial resolution and temporal resolution. These are briefly described as follows:

(a) Spectral resolution. This is determined by the bandwidth of the electromagnetic radiation used during the process. The narrower the bandwidth used, the higher is the spectral resolution achieved. On the basis of the spectral resolution, the systems may be classified as panchromatic, multispectral and hyperspectral systems. Panchromatic systems use a single wavelength band with a large bandwidth, multispectral systems use several narrow bandwidth bands having different wavelengths and hyperspectral systems take measurements in hundreds of very narrow bandwidth bands. Hyperspectral systems are the ones that map the finest spectral characteristics of Earth.
(b) Radiometric resolution. Radiometric resolution refers to the smallest change in intensity level that can be detected by the sensing system. It is determined by the number of discrete quantization levels into which the signal is digitized. The larger the number of bits used for quantization, the better is the radiometric resolution of the system.
(c) Spatial resolution. Spatial resolution is defined as the minimum distance the two point features on the ground should have in order to be distinguished as separate objects. In other words, it refers to the size of the smallest object on the Earth's surface that can be resolved by the sensor. Spatial resolution depends upon the instantaneous field-of-view of the sensor and its distance from Earth. In terms of spatial resolution, the satellite imaging systems can be classified as: low resolution systems (1 km or more), medium resolution systems (100 m to 1 km), high resolution systems (5 m to 100 m) and very high resolution systems (5 m or less). It should be mentioned here that higher resolution systems generally have smaller coverage areas.
(d) Temporal resolution. This is related to the repetitive coverage of the ground by the remote sensing system. It is specified as the number of days in which the satellite revisits a particular place again. Absolute temporal resolution of the satellite is equal to the time taken by the satellite to complete one orbital cycle. (The orbital cycle is the whole number of orbital revolutions that a satellite must describe in order to be flying once again over the same point on the Earth's surface in the same direction.) However, because of some degree of overlap in the imaging swaths of adjacent orbits for most satellites and the increase in this overlap with increasing latitude, some areas of Earth tend to be re-imaged more frequently. Hence the temporal resolution depends on a variety of factors, including the satellite/sensor capabilities, the swath overlap and latitude.

11.5 Passive Sensors

Passive sensors, as described in earlier paragraphs, record the amount of electromagnetic radiation reflected and/or emitted from the Earth's surface. They do not emit any electromagnetic radiation and hence are referred to as passive sensors. Passive sensors can be further classified as scanning and non-scanning sensors.

11.5.1 Passive Scanning Sensors

The multispectral scanner (MSS) is the most commonly used passive scanning sensor. It operates in a number of different ways and can be categorized into three basic types depending upon the mechanism used to view each pixel. These include optical mechanical scanners, push broom scanners and central perspective scanners.

11.5.1.1 Optical Mechanical Scanner

This is a multispectral radiometer (a radiometer is a device that measures the intensity of the radiation emanating from the Earth's surface) where the scanning is done in a series of lines oriented perpendicular to the direction of the motion of the satellite using a rotating or an oscillating mirror (Figure 11.8). They are also referred to as across-track scanners. As the platform moves forward over the Earth, successive scans build up a two-dimensional image of the Earth's surface. Hence optical mechanical scanners record two-dimensional imagery using a combination of the motion of the satellite and a rotating or oscillating mirror scanning perpendicular to the direction in which the satellite is moving.

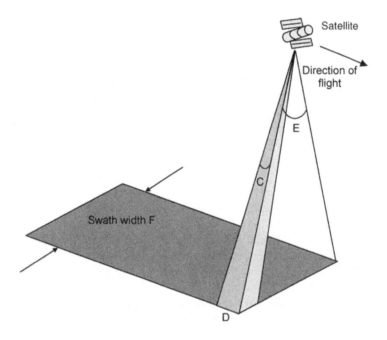

Figure 11.8 Optical mechanical scanner

These scanners comprise the following subsystems:

1. An optical system for collecting the radiation from the ground. It comprises reflective telescope systems such as Newton and Cassegrain telescopes.
2. A spectrographic system comprising a dichroic mirror, grating prisms or filters for separating the incoming radiation into various wavelength bands.

3. A scanning system comprising a rotating or an oscillating mirror for scanning the ground.
4. A detector system comprising various kinds of photodetectors for converting the optical radiation into an electrical signal.
5. A reference system for the calibration of electrical signals generated by the detectors. It comprises light sources or thermal sources with a constant intensity or temperature.

After passing through the optical system, the incoming reflected or emitted radiation is separated into various wavelength bands with the help of grating prisms or filters. Each of the separated bands is then fed to a bank of internal detectors, with each detector sensitive to a specific wavelength band. These detectors detect and convert the energy for each spectral band in the form of an electrical signal. This electrical signal is then converted to digital data and recorded for subsequent computer processing.

Referring to Figure 11.8, the IFOV (C) of the sensor and the altitude of the platform determine the ground resolution of the cell (D), and thus the spatial resolution. The swath width (F) is determined by the sweep of the mirror (specified as the angular field of view E) and the altitude of the satellite. Remote sensing satellites orbiting at a height of approximately 1000 km, for instance, have a sweep angle of the order of 10° to 20° to cover a broad region. One of the problems with this technique is that as the distance between the sensor and the target increases towards the edges of the swath, the cells also become larger and introduce geometric distortions in the image. Moreover, the dwell time possible for these sensors is small as they employ a rotating mirror, and this affects the spatial, spectral and radiometric resolution of the sensor.

The multispectral scanner (MSS), the thematic mapper (TM) and the enhanced thematic mapper (ETM) of Landsat satellites and the advanced very high resolution radiometer (AVHRR) of the NOAA satellites are examples of optical mechanical scanners.

11.5.1.2 Push Broom Scanners

A push broom scanner (also referred to as a linear array sensor or along-track scanner) is a scanner without any mechanical scanning mirror but with a linear array of semiconductor elements located at the focal plane of the lens system, which enables it to record one line of an image simultaneously (Figure 11.9). It has an optical lens through which a line image is detected simultaneously perpendicular to the direction of motion of the satellite. Each individual detector measures the energy for a single ground resolution cell (D). Thus the distance of the satellite from the Earth's surface and the IFOV of the detectors determines the spatial resolution of the system. A separate linear array is required to measure each spectral band or channel. For every scan line, the energy detected by individual detectors of the linear array is sampled electronically and then digitally recorded. Charged coupled devices (CCDs) are the most commonly used detector elements for linear array sensors. HRV (high resolution vidicon) sensor of SPOT satellites, MESSR (multispectral electronic self-scanning radiometer) of MOS-1 satellite (Marine observation satellite) are examples of linear CCD sensors. MESSR of MOS-1 satellite has 2048 elements with an interval of 14 mm.

As compared to push broom scanners, optical mechanical scanners offer narrower view angles and small band-to-band registration error. However, push broom scanners have a longer dwell time and this feature allows more energy to be detected, which improves the radiometric resolution. The increased dwell time also facilitates a smaller IFOV and narrower

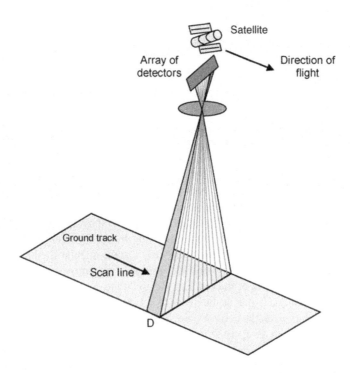

Figure 11.9 Push broom scanner

bandwidth for each detector. Thus, a finer spatial and spectral resolution can be achieved in case of push broom scanners without impacting the radiometric resolution. Moreover, they are cheaper, lighter and more reliable as they do not have any moving part and also require less power. However, calibration of detectors is very crucial in the case of push broom sensors to avoid vertical stripping in the images.

11.5.1.3 Central Perspective Scanners

These scanners employ either the electromechanical or linear array technology to form image lines, but images in each line form a perspective at the centre of the image rather than at the centre of each line. In this case, during image formation, the sensing device does not actually move relative to the object being sensed. Thus all the pixels are viewed from the same central position in a manner similar to a photographic camera. This results in geometric distortions in the image similar to those that occur in photographic data. In satellite-derived imagery, however, radial displacement effects are barely noticeable because of the much smaller field-of-view relative to the orbital altitude. The early frame sensors used in vidicon cameras such as the return beam vidicon in Landsat-1, 2 and 3 satellites operated from a central perspective.

It may be mentioned here that some scanners have the capability of inclining their viewing axes to either side of the nadir, which is referred to as oblique viewing. In the case of oblique viewing, the median line corresponding to the centre of the field (referred to as the instrumentation track) is offset from the satellite ground track. This concept is further explained in Figure 11.10.

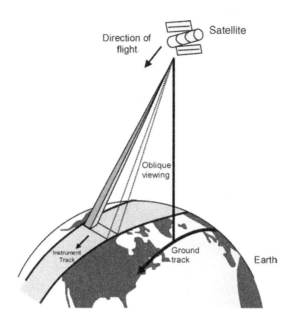

Figure 11.10 Oblique viewing

11.5.2 Passive Non-scanning Sensors

These sensors are further subdivided into imaging and non-imaging sensors depending upon the methodology by which the image is acquired.

Passive non-scanning non-imaging sensors include microwave radiometers, magnetic sensors, gravimeters, Fourier spectrometers, etc. A microwave radiometer is a passive device that records the natural microwave emission from Earth. It can be used to measure the total water content of the atmosphere within its field-of-view.

Passive non-scanning imaging sensors include still multispectral and panchromatic cameras and television cameras. Camera systems employ passive optical sensors, a lens system comprising of a number of lenses to form an image at the focal plane and a processing unit. The ground coverage of the image taken by them depends on several factors, including the focal length of the lens system, the altitude of the satellite and the format and size of the film. Some of the well-known examples of space cameras are the metric camera on board the Space Shuttle by ESA (European space agency), the large format camera (LFC) also on board the Space Shuttle by NASA (National aeronautics and space administration) and the KFA 1000 on board Cosmos by Russia. Figure 11.11 shows the LFC camera. It comprises a film magazine, a device for fixing the camera level and for forward motion compensation, a lens and filter assembly and a thermal exposure system.

Multispectral cameras take images in the visible and the reflective IR bands on separate films and are mainly used for photointerpretation of land surface covers. Panchromatic cameras have a wide field-of-view and hence are used for reconnaissance surveys, surveillance of electrical transmission lines and supplementary photography with thermal imagery.

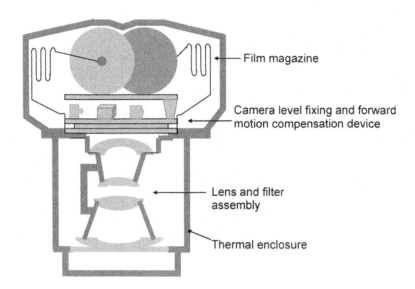

Figure 11.11 LFC camera

11.6 Active Sensors

Active sensor systems comprise both a transmitter as well as a receiver. The transmitter emits electromagnetic radiation of a particular wavelength band, depending upon the intended application. The receiver senses the same electromagnetic radiation reflected or scattered by the ground. Similar to passive sensors, active sensors can also be further categorized into two types, namely the scanning and non-scanning sensors.

11.6.1 Active Non-scanning Sensors

Active non-scanning sensor systems include microwave altimeters, microwave scatterometers, laser distance meters and laser water depth meters. Microwave altimeters or radar altimeters are used to measure the distance between the satellite and the ground surface by measuring the time delay between the transmission of the microwave pulses and the reception of the signals scattered back from the Earth's surface. Some applications of microwave altimetry are in ocean dynamics of the sea current, geoid and sea ice surveys. Microwave scatterometers are used to measure wind speed and direction over the ocean surface by sending out microwave pulses along several directions and recording the magnitude of the signals backscattered from the ocean surface. The magnitude of the backscattered signal is related to the ocean surface roughness, which in turn is dependent on the surface wind condition. Some of the satellites having scatterometers as their payloads include Sesat, ERS-1 and 2 and QuickSat. Laser distance meters are devices having the same principle of operation as that of microwave altimeters except that they send laser pulses in the visible or the IR region instead of the microwave pulses.

11.6.2 Active Scanning Sensors

The most common active scanning sensor used is the synthetic aperture radar (SAR). In synthetic aperture radar imaging, microwave pulses are transmitted by an antenna towards

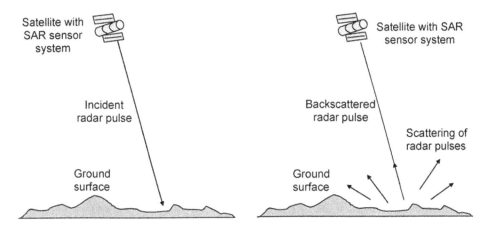

Figure 11.12 Principle of operation of synthetic aperture radar

the Earth's surface and the energy scattered back to the sensor is measured (Figure 11.12). SAR makes use of the radar principle to form an image by utilizing the time delay of the backscattered signals.

Generally, oblique viewing or side-looking viewing is used in the case of SAR and is often restricted to one side of the satellite trajectory, either left or right (Figure 11.13). The microwave

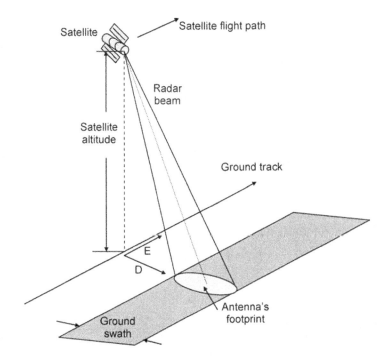

Figure 11.13 Oblique viewing for synthetic aperture radar

beam is transmitted obliquely at right angles to the direction of flight and hence the swath (C) covered by the satellite is offset from the nadir. The SAR produces a two-dimensional image of the object. The range (D) refers to the across-track dimension perpendicular to the flight direction, while the azimuth (E) refers to the along-track dimension parallel to the flight direction. The across-track resolution or the ground swath resolution depends upon the length of the microwave pulse and the angle of incidence and hence the slant range. Two distinct targets on the surface will be resolved in the range dimension if their separation is greater than half the pulse length. The azimuth or along-track resolution is determined by the angular width of the radiated microwave beam and the slant range distance. The narrower the beam, the better is the ground resolution. As the beam width is inversely proportional to the size of the antenna, the longer the antenna the narrower is the beam. It is not feasible for a spacecraft to carry a very long antenna, which is required for high resolution imaging of the Earth's surface. To overcome this limitation, the SAR capitalizes on the motion of the satellite to emulate a large antenna (1–5 km) from the small antenna (2–10 m) it actually carries on board. Some of the satellites having SAR sensors include Radarsat, ERS and Sesat.

SAR can make use of vertical (V) or horizontal (H) polarization for emission and reception. Four types of images can thus be obtained, two with parallel polarization (HH and VV) and two with cross-polarization (HV and VH). It is also possible to use the phase difference between two radar echoes returned by the same terrain and received along two trajectories very close together for reconstructing terrain relief. This technique is referred to as interferometry.

11.7 Types of Images

The data acquired by satellite sensors is processed, digitized and then transmitted to ground stations to construct an image of the Earth's surface. Depending upon the kind of processing used, the satellite images can be classified into two types, namely primary and secondary images. Secondary images can be further subcategorized into various types. In this section, various types of satellite images will be discussed in detail.

11.7.1 Primary Images

The raw images taken from the satellite are referred to as primary images. These raw images are seldom utilized directly for remote sensing applications but are corrected, processed and restored in order to remove geometric distortion, blurring and degradation by other factors and to extract useful information from them.

11.7.2 Secondary Images

As mentioned above, the primary images are processed so as to enhance their features for better and precise interpretation. These processed images are referred to as secondary images. Secondary images are further classified as monogenic images and polygenic images, depending upon whether one or more primary images have been used to produce the secondary image.

11.7.2.1 Monogenic Secondary Images

Monogenic image, also referred to as panchromatic image, is produced from a single primary image by applying changes to it like enlargement, reduction, error correction, contrast adjustments, etc. Some of the techniques used for producing monogenic images are changing the pixel values in the primary image with either the difference in values between the adjacent pixels for gradient images or by the pixel occuring most commonly for smoothened images. Monogenic images are mostly displayed as a grey scale image or sometimes as a single colour image.

In the case of optical images, the brightness of a particular pixel in the image is proportional to the solar radiation reflected by the target. For the thermal and microwave images, the brightness of the pixel in the image is proportional to the thermal characteristics of the object and the microwave radiation scattered or emitted by the object respectively. Figures 11.14 (a) and (b) respectively show an image and its corresponding monogenic secondary image in which the grey scale adjustment has been made in such a way that the darkest grey is seen as black and the lightest grey is seen as white.

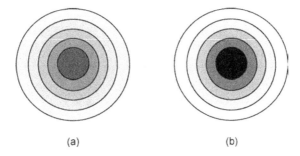

Figure 11.14 (a) Primary image and (b) corresponding monogenic secondary image

11.7.2.2 Polygenic Secondary Images

Polygenic secondary images are composite images formed by combining two or three primary images in order to make the extraction of information from the image easier and more meaningful. The images used can be either optical, thermal or microwave images. Polygenic images are further classified as multispectral images or multitemporal images depending upon whether the primary images used for generating the polygenic image are acquired simultaneously or at different times.

In multispectral images, three images taken in different spectral bands are each assigned a separate primary colour. Depending upon the assignment of the primary colours to these images, they can be further classified as natural colour composite images, false colour composite images and true colour composite images.

In a true colour composite image, the spectral bands correspond to the three primary colours and are assigned a display colour the same as their own colour. The R colour of the display is assigned to the red band, the G colour to the green band and the B colour to the blue band, resulting in images similar to that seen by a human eye. Hence, a true colour composite image always uses the red, green and blue spectral bands and assigns the same display colour as the spectral band. For example, the image in Figure 11.15 (a) is a true colour composite image of San Francisco Bay taken by the Landsat satellite. This image is made by assigning R, G and B colours of the display to the red, green and blue wavelength bands (bands 3, 2 and 1) respectively of the Landsat thematic mapper. Hence, the colours of the image resemble what is observed by the human eye. The salt pond appears green or orange-red due to the colour of the sediments they contain. Urban areas in the image, Palo Alto and San Jose, appear grey and the vegetation is either dark green (trees) or light brown (dry grass).

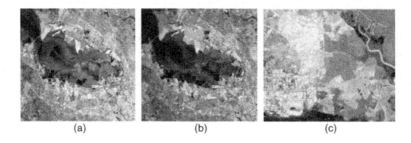

Figure 11.15 (a) True colour composite and (b) False colour composite images of San Francisco Bay taken by the Landsat satellite (Reproduced by permission of © NPA Ltd., www.npagroup.com). (c) Natural colour composite image taken by the SPOT satellite (Reproduced by permission of SPOT Image – © CNES). The images shown in Figure 11.15 are the grey scale versions of the original images and therefore various colours have been reproduced in corresponding grey shades. **Original images are available on the companion website at www.wiley.com/go/maini**

If the spectral bands in the image do not correspond to the three primary colours, then the resulting image is called a false colour composite image. Hence, the colour of an object in the displayed image has no resemblance to its actual colour. It may be mentioned here that microwave radar images can also be used here. There are many schemes for producing false colour composite images with each one best suited for a particular application. Figure 11.15 (b) shows a false colour composite image corresponding to the true colour composite image shown in Figure 11.15 (a). Here the red colour is assigned to the near IR band, the green colour to the red band and the blue colour to the green band. This combination is best suited for the detection of various types of vegetation. Here the vegetation appears in different shades of red (as it has a very strong reflectance in the near IR band) depending upon its type and condition. This image also provides a better discrimination between the greener residential areas and the other urban areas.

In case of sensors not having one or more of the three visible bands, the optical images lack these visual bands. In this case, the spectral bands may be combined in such a way that the appearance of the displayed image resembles a visible colour photograph. Such an image is termed the natural colour composite image. As an example, the SPOT HRV multispectral sensor does not have a blue band. The three bands correspond to green, red and near IR bands. These bands are combined in various proportions to produce different images. These images are then assigned the red, green and blue colours to produce natural colour images as shown in Figure 11.15 (c).

Multitemporal images, as mentioned earlier, are secondary images produced by combining two or more primary images taken at different times. Each primary image taken at a particular time is assigned one colour. Then, these various primary images are combined together. The resultant image is referred to as a multitemporal image. Multitemporal images are particularly helpful in detecting land cover changes over a period of time. Figure 11.16 shows a multitemporal image formed from two images of the River Oder in Frankfurt taken by the SAR sensor on the ERS satellite on 21 July 2001 and 6 August 2001. The picture highlights the flooded areas on 6 August 2001 as compared to those on 21 July 2001.

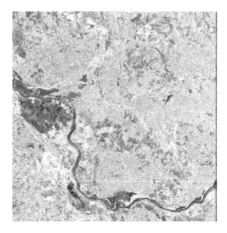

Figure 11.16 Multitemporal image (Reproduced by permission of © ESA 1997. Processed by Eurimage ESA ESRIN Earth Watching Team). The image shown in Figure 10.16 is the greyscale version of the original image in colour. **The original image is available on the companion website at www.wiley.com/go/maini**

11.8 Image Classification

The processed satellite images are classified using various techniques in order to categorize the pixels in the digital image into one of the several land cover classes. The categorized data is then used to produce thematic maps of the land cover present in the image. Normally, multispectral data are used to perform the classification and the spectral pattern present within the data for each pixel is used as the numerical basis for categorization. The objective of image classification is to identify and portray, as a unique grey level (or colour), the features occurring in an image in terms of the object or type of land cover they actually represent on the ground. There are two main classification methods, namely:

1. Supervised Classification
2. Unsupervised Classification

With supervised classification, the land cover types of interest (referred to as training sites or information classes) in the image are identified. The image processing software system is then used to develop a statistical characterization of the reflectance for each information class.

Once a statistical characterization has been achieved for each information class, the image is then classified by examining the reflectance for each pixel and making a decision about which of the signatures it resembles the most.

Unsupervised classification is a method that examines a large number of unknown pixels and divides them into a number of classes on the basis of natural groupings present in the image values. Unlike supervised classification, unsupervised classification does not require analyst-specified training data. The basic principle here is that data values within a given class should be close together in the measurement space (that is have similar grey levels), whereas for different classes these values should be comparatively well separated (that is have very different grey levels). Unsupervised classification is becoming increasingly popular with agencies involved in long term GIS (geographic information system) database maintenance. Unsupervised classification is useful for exploring what cover types can be detected using the available imagery. However, the analyst has no control over the nature of the classes. The final classes will be relatively homogeneous but may not correspond to any useful land cover classes.

11.9   Image Interpretation

Extraction of useful information from the images is referred to as image interpretation. Interpretation of optical and thermal images is more or less similar. However, interpretation of microwave images is quite different. In this section various techniques used to interpret different types of images are described.

11.9.1   Interpreting Optical and Thermal Remote Sensing Images

These images mainly provide four types of information:

1. Radiometric information
2. Spectral information
3. Textural information
4. Geometric and contextual information

Radiometric information corresponds to the brightness, intensity and tone of the images. Panchromatic optical images are generally interpreted to provide radiometric information. Multispectral or colour composite images are the main sources of spectral information. The interpretation of these images requires understanding of the spectral reflectance signatures of the objects of interest. Different bands of multispectral images may be combined to accentuate a particular object of interest. Textural information, provided by high resolution imagery, is an important aid in visual image interpretation. The texture of the image may be used to classify various kinds of vegetation cover or forest cover. Although all of them appear to be green in colour, yet they will have different textures. Geometric and contextual information is provided by very high resolution images and makes the interpretation of the image quite straightforward. Extraction of this information, however, requires prior information about the area (like the shape, size, pattern, etc.) in the image.

11.9.2 Interpreting Microwave Remote Sensing Images

Interpretation of microwave images is quite different from that of optical and thermal images. Images from active microwave remote sensing systems images suffer from a lot of noise, referred to as speckle noise, and may require special filtering before they can be used for interpretation and analysis. Single microwave images are usually displayed as greyscale images where the intensity of each pixel represents the proportion of the microwave radiation backscattered from that area on the ground in the case of active microwave systems and the microwave radiation emitted from that area in the case of passive microwave systems. The pixel intensity values are often converted to a physical quantity called the backscattering coefficient, measured in decibel (dB) units, with values ranging from +5 dB for very bright objects to −40 dB for very dark surfaces. The higher the value of the backscattering coefficient, the rougher is the surface being imaged. Flat surfaces such as paved roads, runways or calm water normally appear as dark areas in a radar image since most of the incident radar pulses are specularly reflected away. Trees and other vegetations are usually moderately rough on the wavelength scale. Hence, they appear as moderately bright features in the image. Ships on the sea, high rise buildings and regular metallic objects such as cargo containers, built-up areas and many man-made features, and so on, appear as very bright objects in the image. The brightness of areas covered by bare soil may vary from very dark to very bright depending on their roughness and moisture content. Typically, rough soil appears bright in the image. For similar soil roughness, the surface with a higher moisture content will appear brighter.

Multitemporal microwave images are used for detecting land cover changes over the period of image acquisition. The areas where no change in land cover occurs will appear in grey while areas with land cover changes will appear as colourful patches in the image.

11.9.3 GIS in Remote Sensing

The geographic information system (GIS) is a computer-based information system used to digitally represent and analyse the geographic features present on the Earth's surface. Figure 11.17 shows the block diagram of a typical GIS system. The GIS is used to integrate

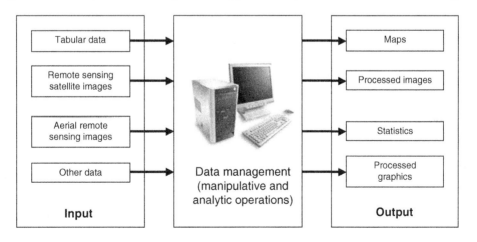

Figure 11.17 Block diagram of a typical GIS system

the remote sensing data with the geographic data, as it will help to give a better understanding and interpretation of remote sensing images. It also assists in the automated interpretation, detecting the changes occurring in an area and in map revision processes. For example, it is not enough to detect land cover change in an area, as the final goal is to analyse the cause of the change or to evaluate the impact of the change. Hence, the remote sensing data should be overlaid on maps such as those of transportation facilities and land use zoning in order to extract this information. In addition, the classification of remote sensing imagery will become more accurate if the auxiliary data contained in the maps are combined with the image data.

The history of the GIS dates back to the late 1950s, but the first GIS software came in the late 1970s from the laboratory of the Environmental Systems Research Institute (ESRI), Canada. Evolution of the GIS has transformed and revolutionized the ways in which planners, engineers, managers, and so on, conduct database management and analysis.

The GIS performs the following three main functions:

1. To store and manage geographic information comprehensively and effectively.
2. To display geographic information depending on the purpose of use.
3. To execute query, analysis and evaluation of geographic information effectively.

The GIS uses the remote sensing data either as classified data or as image data. Land cover maps or vegetation maps classified from remote sensing data can be overlaid on to other geographic data, which enables analysis for environmental monitoring and its change. Remote sensing data can be classified or analysed with other geographic data to obtain a higher accuracy of classification. Using the information available in maps like the ground height and slope gradient information, it becomes easier to extract relevant information from the remote sensing images.

11.10 Applications of Remote Sensing Satellites

Data from remote sensing satellites is used to provide timely and detailed information about the Earth's surface, especially in relation to the management of renewable and non-renewable resources. Some of the major application areas for which satellite remote sensing is of great use are the assessment and monitoring of vegetation types and their status, soil surveys, mineral exploration, map making and revision, production of thematic maps, planning and monitoring of water resources, urban planning, agricultural property management planning, crop yield assessment, natural disaster assessment, and so on. Some of the applications are described in detail in this section.

11.10.1 Land Cover Classification

Land cover mapping and classification corresponds to identifying the physical condition of the Earth's surface and then dividing the surface area into various classes, like forest, grassland, snow, water bodies, and so on, depending upon its physical condition. Land cover classification helps in the identification of the location of natural resources. Figure 11.18 (a) is a digital satellite image showing the land cover map of Onslow Bay in North Carolina taken by Landsat's thematic mapper (TM) in February 1996. Figure 11.18 (b) is the land cover classification map derived from the satellite image shown in Figure 11.18(a) using a variety of techniques and tools, dividing the area into 15 land cover classes.

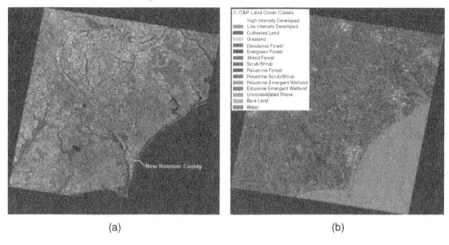

(a)   (b)

Figure 11.18 (a) Digital satellite image showing the land cover map of Onslow Bay in North Carolina taken by Landsat's thematic mapper (TM) in February 1996 and (b) the land cover classification map derived from the satellite image in Figure 11.18(a) (Courtesy: National Oceanic and Atmospheric Administration (NOAA) Coastal Services Center, www.csc.noaa.gov, USA). These images are greyscale versions of original colour images. **Original images are available on the companion website at www.wiley.com/go/maini**

11.10.2 Land Cover Change Detection

Land cover change refers to the seasonal or permanent changes in the land cover types. Seasonal changes may be due to agricultural changes or the changes in forest cover and the permanent changes may be due to land use changes like deforestation or new built towns, and so on. Detection of permanent land cover changes is necessary for updating land cover maps and for management of natural resources. Satellites detect these permanent land cover changes by comparing an old image and an updated image, with both these images taken during the same season to eliminate the effects of seasonal change.

Figure 11.19 shows three photographs of Kuwait taken by the Landsat satellite. Figures 11.19 (a), (b) and (c) show Kuwait City before, during and after the Gulf War respectively. The red part of the images show the vegetation and the bright areas are the deserts. Clear, deep water looks almost black, but shallow or silty water looks lighter. The Landsat image during the war shows that the city is obscured by smoke plume from burning oil wells. There were around 600 oil wells that were set on fire during the war. The third image was acquired after the fires had been extinguished. It shows that the landscape had been severely affected by the war. The dark grey patched areas in the third image are due to the formation of a layer of hardened 'tarcete' formed by the mixing of the sand and gravel on the land's surface with oil and soot. Black pools within the dark grey tarcrete are the oil wells that were formed after the war. It was detected from the satellite images that some 300 oil wells were formed. Satellite images provided a cost effective method of identifying these pools and other changes to quantify the amount of damage due to the war, in order to launch an appropriate clean-up programme.

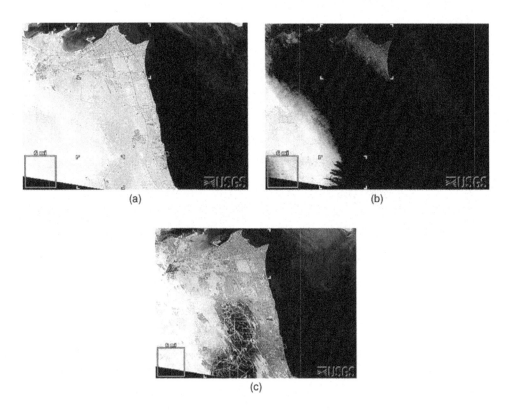

Figure 11.19 Image of Kuwait City taken by Landsat satellite (a) before the Gulf War in August 1990, (b) during the Gulf War in February 1991 and (c) after the Gulf War in November 1991 (Data available from US Geological Survey). These images are greyscale versions of original colour images. **Original images are available on the companion website at www.wiley.com/go/maini**

11.10.3 Water Quality Monitoring and Management

Satellite imagery helps in locating, monitoring and managing water resources over large areas. Water resources are mapped in the optical and the microwave bands. Water pollution can be determined by observing the colour of water bodies in the images obtained from the satellite. Clear water is bluish in colour, water with vegetation appears to be greenish-yellow while turbid water appears to be reddish-brown. Structural geographical interpretation of the imagery also aids in determining the underground resources. The changing state of many of the world's water bodies is monitored accurately over long periods of time using satellite imagery. Figure 11.20 shows a false colour composite image taken from the IKONOS satellite, displaying the water clarity of the lakes in Eagan, Minnesota. Scientists measured the water quality by observing the ratio of blue to red light in the satellite data. Water quality was found to be high when the amount of blue light reflected off the lakes was high and that of red light was low. Lakes loaded with algae and sediments, on the other hand, reflect less blue light and more red light. Using images like this, scientists created a comprehensive water quality map for the water bodies in the region.

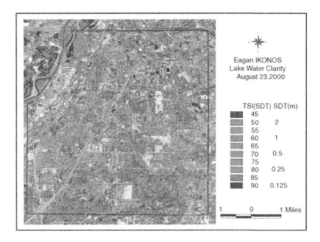

Figure 11.20 False colour composite image taken by the IKONOS satallite displaying the water clarity of the lakes in Eagan, Minnesota (Reproduced by permission of the University of Minnesota, Remote Sensing and Geospatial Analysis Laboratory). The image is the greyscale version of the original colour image. **Original image is available on the companion website at www.wiley.com/go/maini**

11.10.4 Flood Monitoring

Satellite images provide a cost effective and potentially rapid means to monitor and map the devastating effects of floods. Figures 11.21 (a) and (b) show the false colour composite images of the Pareechu River in Tibet behind a natural dam forming an artificial lake, taken by the advanced space-borne thermal emission and reflection radiometer (ASTER) on NASA's Terra satellite on 1 September 2004 and 15 July 2004 respectively. From the two images it is evident that the water levels were visibly larger on 1 September 2004 than they were on 15 July 2004. The lake posed a threat to communities downstream in northern India, which would have been flooded if the dam had burst.

Figure 11.21 Flood monitoring using remote sensing satellites (Courtesy: NASA). These images are greyscale versions of original colour images. **Original images are available on the companion website at www.wiley.com/go/maini**

11.10.5 Urban Monitoring and Development

Satellite images are an important tool for monitoring as well as planning urban development activities. Time difference images can be used to monitor changes due to various forms of natural disasters, military conflict or urban city development. Figures 11.22 (a) and (b) show Manhattan before and after the 11 September 2001 attacks on the World Trade Center. These images have a resolution of 1 m and were taken by the IKONOS satellite. Remote sensing data along with the GIS is used for preparing precise digital basemaps of the area, for formulating proposals and for acting as a monitoring tool during the development phase. They are also used for updating these basemaps from time to time.

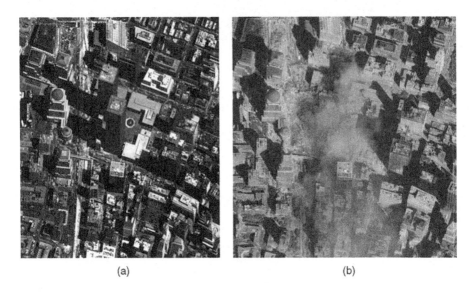

Figure 11.22 Images of Manhattan (a) before and (b) after the 11 September 2001 attacks, taken by the IKONOS satellite (Satellite imagery courtesy of GeoEye). These images are greyscale versions of original colour images. **Original images are available on the companion website at www.wiley.com/go/maini**

11.10.6 Measurement of Sea Surface Temperature

The surface temperature of the sea is an important index for ocean observation, as it provides significant information regarding the behaviour of water, including ocean currents, formation of fisheries, inflow and diffusion of water from rivers and factories. Satellites provide very accurate information on the sea surface temperatures. Temperature measurement by remote sensing satellites is based on the principle that all objects emit electromagnetic radiation of different wavelengths corresponding to their temperature and emissivity. The sea surface temperature measurements are done in the thermal infrared bands. Figure 11.23 shows the sea surface temperature map derived from the thermal IR image taken by the GMS-5 satellite.

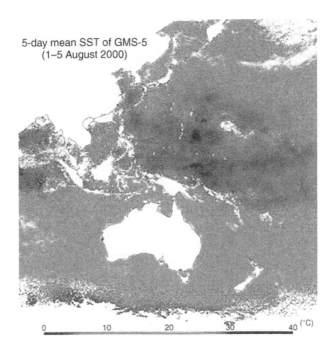

Figure 11.23 Sea surface temperature map derived from the thermal IR image taken by the GMS-5 satellite (Reproduced by permission of © Japan Meteorological Agency). The image is the greyscale version of the original colour image. **Original image is available on the companion website at www.wiley.com/go/maini**

11.10.7 Deforestation

Remote sensing satellites help in detecting, identifying and quantifying the forest cover areas. This data is used by scientists to observe and assess the decline in forest cover over a period of several years. The images in Figure 11.24 show a portion of the state of Rondônia, Brazil, in which tropical deforestation has occurred. Figures 11.24 (a) and (b) are the images taken by the multispectral scanners of the Landsat-2 and -5 satellites in the years 1975 and 1986 respectively. Figure 11.24 (c) shows the image taken by the thematic mapper of the Landsat-4 satellite in the year 1992. It is evident from the images that the forest cover has reduced drastically.

11.10.8 Global Monitoring

Remote sensing satellites can be used for global monitoring of various factors like vegetation, ozone layer distribution, gravitational fields, glacial ice movement and so on. Figure 11.25 shows the vegetation distribution map of the world formed by processing and calibrating 400 images from the NOAA remote sensing satellite's AVHRR sensor. This image provides an unbiased means to analyse and monitor the effects of droughts and long term changes from possible regional and global climate changes. Figure 11.26 shows the global ozone distribution taken by the global ozone monitoring experiment (GOME) sensor on the ERS-2 satellite. Measurement of ozone distribution can be put to various applications, like informing people

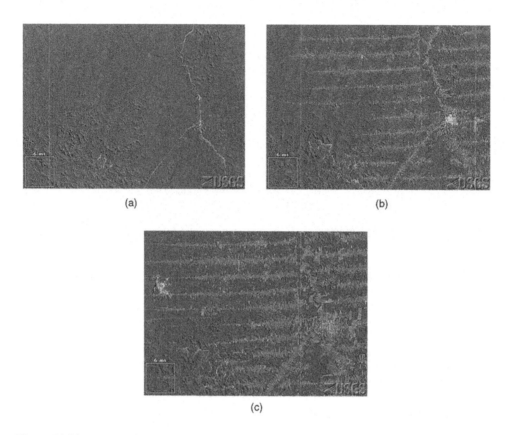

Figure 11.24 Images taken by the multispectral scanners of (a) Landsat-2 satellite in the year 1975 and (b) Landsat-5 satellite in the year 1986. (c) Image taken by the thematic mapper of the Landsat-4 satellite in the year 1992 (Data available from US Geological Survey). These images are greyscale versions of original colour images. **Original images are available on the companion website at www.wiley.com/go/maini**

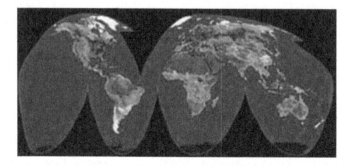

Figure 11.25 Vegetation distribution map of the world (Reproduced by permission of © NOAA/NPA). The image is the greyscale version of the original colour image. **Original image is available on the companion website at www.wiley.com/go/maini**

of the fact that depletion of the ozone layer poses serious health risks and taking measures to prevent depletion of the ozone layer. The figure shows that the ozone levels are decreasing with time. Satellites also help us to measure the variation in the gravitational field precisely, which in turn helps to give a better understanding of the geological structure of the sea floor. Gravitational measurements are made using active microwave sensors.

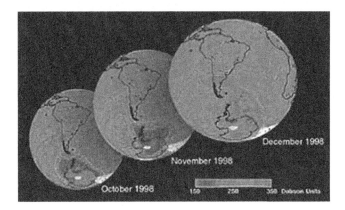

Figure 11.26 Global ozone distribution (Reproduced by permission of © DLR/NPA). The image is the greyscale version of the original colour image. **Original image is available on the companion website at www.wiley.com/go/maini**

11.10.9 Predicting Disasters

Remote sensing satellites give an early warning of the various natural disasters like earthquakes, volcanic eruptions, hurricanes, storms, and so on, thus enabling the evasive measures to be taken in time and preventing loss of life and property. Geomatics, a conglomerate of measuring, mapping, geodesy, satellite positioning, photogrammetry, computer systems and computer graphics, remote sensing, geographic information systems (GIS) and environmental visualization is a modern technology, that plays a vital role in mitigation of natural disasters.

11.10.9.1 Predicting Earthquakes

Remote sensing satellites help in predicting the time of the earthquake by sensing some precursory signals that the earthquake faults produce. These signals include changes in the tilt of the ground, magnetic anomalies, swarms of micro-earthquakes, surface temperature changes and a variety of electrical field changes prior to the occurrence of earthquakes. As an example, the French micro-satellite Demeter detects electromagnetic emissions from Earth that can be used for earthquake prediction. The NOAA/AVHRR series of satellites take thermal images and can be used to predict the occurrence of earthquakes. It is observed that the surface temperature of the region where the earthquake is to happen increases by 2–3 °C prior to the earthquake (7–24 days before) and fades out within a day or two after the earthquake has occurred. As an example, Figure 11.27 shows the infrared data images taken by the moderate resolution imaging spectroradiometer (MODIS) on board NASA's Terra satellite of the region surrounding Gujarat, India. These images show a 'thermal anomaly' appearing on 21 January 2001 [Figure 11.27 (b)] prior to the earthquake on 26 January 2001. The anomaly disappears shortly after the earthquake [Figure 11.27 (c)]. The anomaly area appears yellow-orange. The boxed star in the images indicates the earthquake's epicentre. The region of thermal anomaly is southeast of the Bhuj region, near to the earthquake's epicentre. Multitemporal radar images of Earth can also be used to predict the occurrence of earthquakes by detecting the changes in the ground movement.

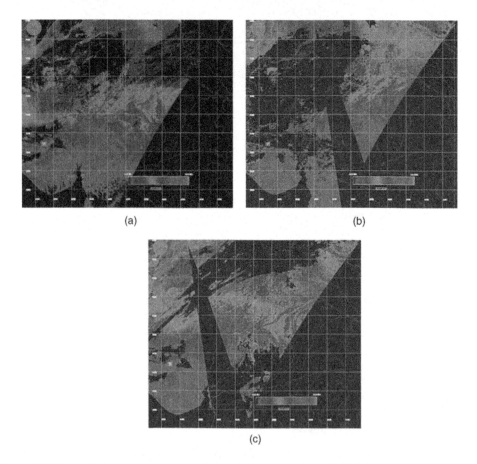

Figure 11.27 Predicting earthquakes using infrared data images from NASA's Terra satellite (Courtesy: NASA). The images are greyscale versions of original colour images. **Original images are available on the companion website at www.wiley.com/go/maini**

In addition to predicting the time of occurrence of an earthquake, remote sensing satellites are also used for earthquake disaster management. High resolution satellites having a resolution better than 1 m have revolutionized the concept of damage assessments, which were earlier carried out by sending a team of specialists into the field in order to visually assess the damage. This helps to speed up the process of damage assessment and could provide invaluable information for the rescue or assessment team when they are en route to a disaster site in a cost effective and unbiased fashion. This information also helps in planning the evacuation routes and designing centres for emergency operations. Figure 11.28 shows the use of remote sensing images for mapping damage due to an earthquake with Figure 11.28 (a) showing badly damaged buildings (red mark) and Figure 11.28 (b) showing less damaged buildings (yellow mark). It may be mentioned here that global positioning satellites (GPS) are also used for predicting earthquake occurrences by detecting precise measurements of the fault lines.

(a) (b)

Figure 11.28 Use of images from remote sensing satellites for mapping building damage due to earthquake (Image source: GIS Development Private Limited). The images are the greyscale versions of the original colour images. **Original images are available on the companion website at www.wiley.com/go/maini**

11.10.9.2 Volcanic Eruptions

Satellites are an accurate, cost effective and efficient means of predicting volcanic eruptions as compared to ground-based methods. Remote sensing satellites and global positioning satellites are used for predicting the occurrence of volcanic eruptions. Remote sensing satellites use thermal, optical and microwave bands for predicting volcanic eruptions.

Before a volcano erupts, it usually has increased thermal activity, which appears as elevated surface temperature (hot spots) around the volcano's crater. Early detection of hot spots and their monitoring is a key factor in predicting possible volcanic eruptions. Hence by taking infrared images, a medium term warning can be made about an eruption that may be several days to some weeks away. The method is particularly useful for monitoring remote volcanoes. Active microwave techniques can be used to detect how the mountains inflate as hot rock is injected beneath them. As an example, the TOMS (total ozone mapping spectrometer) satellite produced an image of the Pinatubo volcanic cloud (emitted during the eruption in 1991) over a nine day period, showing the regional dispersal of the sulfur dioxide plume.

Precision surveying with GPS satellites can detect bulging of volcanoes months to years before an eruption, but is not useful for short term eruption forecasting. Two or more satellite images can be combined to make digital elevation models, which also help in predicting volcanic eruptions. Satellite images taken after the volcanic eruption help in the management of human resources living near the volcanic area by predicting the amount and direction of the flow of lava.

11.10.10 Other Applications

Other important applications of remote sensing satellites include the computation of digital elevation models (DEM) using two satellite images viewing the same area of Earth's surface from different orbits. Yet another application of remote sensing satellites is in the making of topographic maps that are used for viewing the Earth's surface in three dimensions by drawing landforms using contour lines of equal elevation. Remote sensing satellites are also used for making snow maps and for providing data on snow cover, sea ice content, river flow and so on.

11.11 Major Remote Sensing Missions

In this section three major remote sensing satellite missions will be described, namely the Landsat, SPOT and Radarsat satellite systems. The Landsat and SPOT systems operate in the optical and the thermal bands while Radarsat is a microwave remote sensing satellite system.

11.11.1 Landsat Satellite System

Landsat is a remote sensing satellite programme of the USA, launched with the objective of observing Earth on a global basis. It is the longest running enterprise for the acquisition of imagery of Earth from space. The Landsat programme comprises a series of optical/thermal remote sensing satellites for land observation purposes. Landsat imagery is used for global change research and applications in agriculture, water resources, urban growth, geology, forestry, regional planning, education and national security. Scientists use Landsat satellites to gather images of the land surface and surrounding coastal regions for global change research, regional environmental change studies and other civil and commercial purposes.

Eight Landsat satellites have been launched to date. All these satellites are three-axis stabilized orbiting in near polar sun-synchronous orbits. The first generation of Landsat satellites [Figure 11.29 (a)] comprised three satellites, Landsat-1, -2 and -3, also referred to as Earth resource technology satellites, ERTS-1, -2 and -3 respectively. Landsat-1 was launched on 23 July 1972, with a design life of one year but it remained in operation for six years until January 1978. Landsat-1 carried on board two Earth viewing sensors – a return beam vidicon (RBV) and a multispectral scanner (MSS). Landsat-2 and -3 satellites launched in 1975 and 1978 respectively, had similar configurations.

The second generation of Landsat satellites [Figure 11.29 (b)] comprised of Landsat-4 and -5 satellites, launched in 1982 and 1984 respectively. Landsat-4 satellite carried a MSS and a thematic mapper (TM). Landsat-5 satellite was a duplicate of Landsat-4 satellite. Landsat-6 satellite was launched in October 1993 but failed to reach the final orbit. It had an enhanced thematic mapper (ETM) payload. Landsat-7 was launched in the year 1999 to cover up for the loss of Landsat-6 satellite. Landsat-7 satellite [Figure 11.29 (c)] comprised an advanced ETM payload referred to as the enhanced thematic mapper plus (ETM+). Currently, Landsat-7 and Landsat-8 satellites are operational. Landsat-8 satellite was launched on 11 February 2013. It sends 400 images every day, which are used for various applications. Table 11.2 enumerates the salient features of the Landsat satellites.

11.11.1.1 Payloads on Landsat Satellites

1. Return beam vidicon (RBV). Landsat-1, -2 and -3 satellites had the RBV payload. RBV is a passive optical sensor comprising an optical camera system. The sensor comprises

Major Remote Sensing Missions 559

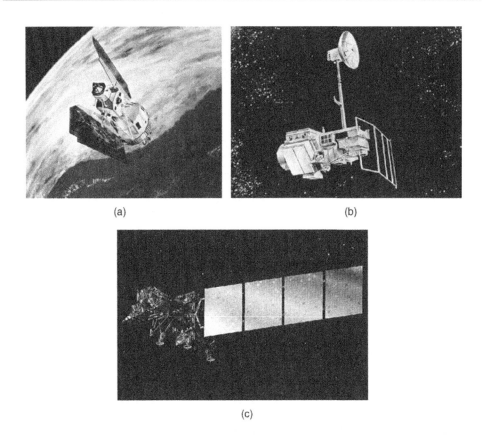

Figure 11.29 (a) First generation Landsat satellites, (b) second generation Landsat satellites and (c) Landsat-7 satellite (Courtesy: NASA)

Table 11.2 Salient features of Landsat satellites

Satellites	Orbit	Altitude (km)	Orbital period (min)	Inclination (degrees)	Temporal resolution (days)	Equatorial crossing (a.m)	Sensors
Landsat-1	Sun-synchronous	917	103	99.1	18	9:30	RBV, MSS
Landsat-2	Sun-synchronous	917	103	99.1	18	9:30	RBV, MSS
Landsat-3	Sun-synchronous	917	103	99.1	18	9:30	RBV, MSS
Landsat-4	Sun-synchronous	705	99	98.2	16	9:30	MSS, TM
Landsat-5	Sun-synchronous	705	99	98.2	16	9:30	MSS, TM
Landsat-6	Sun-synchronous	705	99	98.2	16	10:00	ETM
Landsat-7	Sun-synchronous	705	99	98.2	16	10:00	ETM+
Landsat-8	Sun-synchronous	705	99	98.2	16	10:00	OLI, TIRS

three independent cameras operating simultaneously in three different spectral bands from blue-green (0.47–0.575 µm) through yellow-red (0.58–0.68 µm) to near IR (0.69–0.83 µm) to sense the reflected solar energy from the ground. Each camera contained an optical lens, a 5.08 cm RBV, a thermoelectric cooler, deflection and focus coils, a mechanical shutter, erase lamps and sensor electronics (Figure 11.30). The cameras were similar except for the spectral filters contained in the lens assemblies that provided separate spectral viewing regions. The RBV of Landsat-1 satellite had a resolution of 80 m and that of Landsat-2 and -3 satellites had a resolution of 40 m.

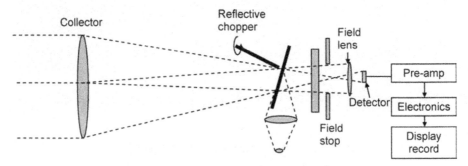

Figure 11.30 Return beam vidicon (RBV)

2. Multispectral scanner (MSS). Landsat-1 to -5 satellites had the MSS payload. The resolution of the MSS sensor was approximately 80 m with radiometric coverage in four spectral bands of 0.5 to 0.6 μm (green), 0.6 to 0.7 μm (red), 0.7 to 0.8 μm (near IR) and 0.8 to 1.1 μm (near IR) wavelengths. Only the MSS sensor on Landsat-3 satellite had a fifth band in the thermal IR. MSS is a push broom kind of sensor comprising of 24-element fibre optic array which scans from west to east across the Earth's surface, while the orbital motion of the spacecraft provides a natural north-to-south scanning motion. Then, a separate binary number array for each spectral band is generated. Each number corresponds to the amount of energy reflected into that band from a specific ground location. In the ground processing system, the binary number arrays are either directly interpreted by image classification software or reconstructed into images.

3. Thematic mapper (TM). Landsat-4 and -5 satellites had this payload. TM sensors primarily detect reflected radiation from the Earth's surface in the visible and near IR wavelengths like the MSS, but the TM sensor provides more radiometric information than the MSS sensor. The wavelength range for the TM sensor is from 0.45 to 0.53 μm (blue band 1), 0.52 to 0.60 μm (green band 2), 0.63 to 0.69 μm (red band 3), 0.76 to 0.90 μm (near IR band 4), 1.55 to 1.75 μm (shortwave IR band 5) through 2.08 to 2.35 μm (shortwave IR band 7) to 10.40 to 12.50 μm (thermal IR band 6) portion of the electromagnetic spectrum. Sixteen detectors for the visible and mid IR wavelength bands in the TM sensor provide 16 scan lines on each active scan. Four detectors for the thermal IR band provide four scan lines on each active scan. The TM sensor has a spatial resolution of 30 m for the visible, near IR and mid IR wavelengths and a spatial resolution of 120 m for the thermal IR band.

4. Enhanced thematic mapper (ETM). This instrument was carried on the Landsat-6 satellite which failed to reach its orbit. ETM operated in seven spectral channels similar to the TM (six with a ground resolution of 30 metres and one, thermal IR, with a ground resolution of 120 metres). It also had a panchromatic channel providing a ground resolution of 15 metres. ETM is an optical mechanical scanner where the mirror assembly scans in the west-to-east and east-to-west directions, whereas the satellite revolves in the north–south direction, hence providing two-dimensional coverage.

5. Enhanced thematic mapper plus (ETM+). This instrument was carried on board the Landsat-7 satellite. The ETM+ instrument is an eight-band multispectral scanning radiometer capable of providing high resolution image information of the Earth's surface. Its spectral bands are similar to those of the TM, except that the thermal IR band (band 6) has an improved resolution of 60 m (versus 120 m in the TM). There is also an additional panchromatic band operating at 0.5 to 0.9 µm with a 15 m resolution.
6. Operational land imager (OLI). OLI uses a push broom sensor instead of the whisk broom sensors that were used in earlier Landsat satellites. It has over 7000 detectors per spectral band, which gives it significantly enhanced sensitivity, fewer moving parts and improved land surface information.
7. Thermal infrared sensor (TIRS). The TIRS is used for thermal imaging. Its focal plane array uses gallium arsenide quantum well infrared photo detector arrays for detection of infrared radiation. Like OLI, it also employs a push broom sensor design and has a 185 km of cross-track field-of-view.

11.11.2 SPOT Satellite System

SPOT (satellite pour l'observation de la terre) is a high resolution, optical imaging Earth observation satellite system run by Spot Image company of France. SPOT satellites provide Earth observation images for diverse applications such as agriculture, cartography, cadastral mapping, environmental studies, urban planning, telecommunications, surveillance, forestry, land use/land cover mapping, natural hazard assessments, flood risk management, oil and gas exploration, geology and civil engineering. The SPOT program was initiated in the 1970s by Centre national d'études spatiales (CNES), a French space company and was developed in association with Belgium and Swedish space companies.

Since the launch of SPOT's first satellite SPOT-1 in 1986, the SPOT system has constantly provided improved quality of Earth observation images. Each of SPOT-1, -2 and -3 [Figure 11.31 (a)], launched in the years 1986, 1990 and 1993 respectively carried two identical

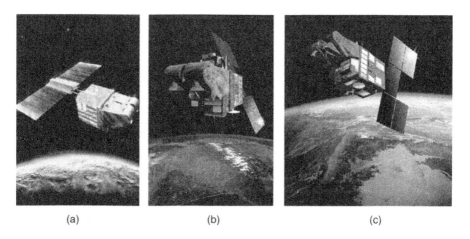

(a) (b) (c)

Figure 11.31 (a) SPOT-1, -2 and -3 satellites (Reproduced by permission of © CNES), (b) SPOT-4 satellite (Reproduced by permission of © CNES/ill. D. DUCROS, 1998) and (c) SPOT-5 satellite (Reproduced by permission of © CNES/ill. D. DUCROS, 2002)

HRV (high resolution visible) imaging instruments and two tape-recorders for imaging data. They had a design life of three years and are out of service now. Currently, two of the SPOT satellites, SPOT-5 and SPOT-6, launched in the years 2002 and 2012 respectively, are operational. SPOT-4 satellite [Figure 11.31(b)], functional till July 2013, carried two high resolution visible infrared (HRVIR) imaging instruments and a vegetation instrument. SPOT-5 satellite [Figure 11.31 (c)] has two high resolution spectroscopic (HRS) instruments and a vegetation instrument. SPOT satellites move in a circular, sun-synchronous orbit at an altitude of 832 km. SPOT-6 is the latest operational satellite in the family of SPOT satellites. SPOT-6 was launched on 9 September 2012 on board PSLV-C21. SPOT-7 is scheduled for launch in 2014. Both these satellites are designed to provide continuity of high-resolution, wide-swath Earth imaging data up to 2024. Table 11.3 enumerates the salient features of these satellites.

Table 11.3 Salient features of SPOT satellites

Satellites	Orbit	Altitude (km)	Orbital period (min)	Inclination (degrees)	Temporal resolution (days)	Equatorial crossing (a.m)	Sensors
SPOT-1	Sun-synchronous	832	101	98.7	26	10:30	2 HRV
SPOT-2	Sun-synchronous	832	101	98.7	26	10:30	2 HRV
SPOT-3	Sun-synchronous	832	101	98.7	26	10:30	2 HRV
SPOT-4	Sun-synchronous	832	101	98.7	26	10:30	2 HRVIR, vegetation instrument
SPOT-5	Sun-synchronous	832	101	98.7	26	10:30	2 HRS, vegetation instrument
SPOT-6	Sun-synchronous (In quadratic phase with Pleiades satellites)	695	98.79	98.2	26	10:00	2NAOMI

11.11.2.1 Payloads on Board SPOT Satellites

1. High resolution visible (HRV) instrument. SPOT-1, -2 and -3 satellites carried the HRV push broom linear array sensor (Figure 11.32). HRV operates in two modes, namely the panchromatic mode and the multiband mode. In the panchromatic mode, the operational wavelength band is quite broad, from 0.51 to 0.73 µm, having a resolution of 10 m. The multiband mode operates in three narrow spectral bands of 0.50 to 0.59 µm (XS1 band green), 0.61 to 0.68 µm (XS2 band red) and 0.79 to 0.89µm (XS3 band near IR), with a resolution of 20 m per pixel. Data acquired in the two modes can also be combined to form multispectral images. These sensors also have the capability of oblique viewing (with a viewing angle of 27° relative to the vertical) on either side of the satellite nadir and hence offering more flexibility in observation, enabling the acquisition of stereoscopic images.
2. High resolution visible infrared (HRVIR) instrument. The SPOT-4 satellite carried this instrument. The instrument has a resolution of 20 m and operates in four spectral bands of 0.50 to 0.59 µm (B1 band green), 0.61 to 0.68 µm (B2 band red), 0.78 to 0.89 µm (B3 band

Figure 11.32 HRV instrument (Reproduced by permission of Spot Image - © CNES)

near IR) and 1.58 to 1.75 µm (B4 band shortwave IR). In addition to these bands, there is a monospectral band (M band) operating in the same spectral region as the B2 band but having a resolution of 10 m.

3. High resolution stereoscopic (HRS) instrument. The HRS payload flown on SPOT-5 satellite is dedicated to taking simultaneous stereo pair images (Figure 11.33). It operates in the same multispectral bands (B1, B2, B3 and B4) of the HRVIR instrument on the SPOT-4 satellite but has a resolution of 10 m in the B1, B2 and B3 bands and 20 m in the B4 band. It also has a panchromatic mode of operation in the spectral band of 0.48 to 0.71 µm having a resolution of 2.5 to 5 m.

Figure 11.33 High resolution stereoscopic (HRS) instrument (Reproduced by permission of Spot Image - © CNES)

4. Vegetation instrument. The vegetation instruments were flown on board SPOT-4 and SPOT-5 satellites with the instrument on SPOT-4 satellite referred to as Vegetation-1 and the instrument on the SPOT-5 satellite referred to as Vegetation-2. These instruments are four channel instruments with three channels having the same spectral band as the B2, B3 and B4 bands of the HRVIR instrument and the fourth channel referred to as the B0 channel operating in the 0.43 to 0.47 µm band for oceanographic applications and atmospheric corrections.
5. New AstroSat optical modular instrument (NAOMI): The NAOMI sensor on board SPOT-6 satellite is a high resolution optical imager and operates in five spectral bands of 0.45 to 0.745 µm (Band PAN VIS), 0.45 to 0.52 µm (Band 1 Blue), 0.53 to 0.59 µm (Band 2 Green), 0.625 to 0.695 µm (Band 3 Red) and 0.76 to 0.89 µm (Band 4 NIR).

11.11.3 Radarsat Satellite System

Radarsat is a Canadian remote sensing satellite system with two operational satellites namely Radarsat-1 and Radarsat-2. Both the satellites carry on-board SAR sensors and orbit in sun-synchronous orbits with an altitude of 798 km and inclination of 98.6°. Radarsat-1 (Figure 11.34) was the first satellite launched in this system. It was launched on 4 November 1995 with the aim of studying the polar regions, to aid in maritime navigation, natural resource identification, management of agricultural and water resources, and monitoring of environmental changes. Radarsat-2, the second satellite of the Radarsat series was launched on 14 December 2007. It is used for a variety of applications including sea ice mapping and ship routing, iceberg detection, agricultural crop monitoring, marine surveillance for ship and pollution detection, terrestrial defence surveillance and target identification, geological mapping, land use mapping, wetlands mapping and topographic mapping. The Radarsat constellation mission is a follow-on project to Radarsat-2. It will comprise three satellites and is proposed to be functional by 2018.

Figure 11.34 Radarsat-1 satellite (Reproduced by permission of © Canadian Space Agency, 2006)

11.11.3.1 Radarsat Satellite Payloads

Both of the Radarsat satellites have SAR sensors operating in the on board C band. The SAR sensor on Radarsat-1 satellite has the unique capability to acquire data in any one of the possible seven imaging modes. Each mode varies with respect to swath width, resolution, incidence angle and number of looks. Because different applications require different imaging modes, the satellite gives users tremendous flexibility in choosing the type of SAR data most suitable for their application. It operates in the C-band at a frequency of 5.3 GHz with HH polarization. The ground resolution varies from 8 m to 100 m and the swath width varies from 50 km to 500 km for different imaging modes.

The Radarsat-2 SAR payload ensures continuity of all existing Radarsat-1 modes, and offers an extensive range of additional features ranging from improvement in resolution to full flexibility in the selection of polarization options to the ability to select all beam modes in both left and right looking modes. The different polarization modes offered are HH, HV, VV and VH. The ground resolution varies from 3 m to 100 m and the swath width is selectable from 20 km to 500 km. Other salient features include high downlink power, secure data and telemetry, solid-state recorders, on-board GPS receiver and the use of a high-precision attitude control system. The enhanced capabilities are provided by a significant improvement in instrument design, employing a state-of-the-art phased array antenna composed of an array of hundreds of miniature transmit-receive modules. The antenna is capable of being steered electronically over the full range of the swath and can switch between different operating modes virtually instantaneously.

11.11.4 Indian Remote Sensing Satellite System

The Indian remote sensing (IRS) satellite system comprises one of the largest constellations of Earth observation satellites in use for a wide variety of remote sensing and Earth observation applications at both national and international level. These satellites are designed, launched and maintained by the Indian Space Research Organization (ISRO). Some of the application areas these remote sensing satellites are designed for include crop area assessment and production estimation of major crops, flood risk zone mapping and flood damage assessment, snow melt run-off estimates for planning use of water resources, drought monitoring and assessment, urban planning, forest survey, mineral prospecting, coastal studies, land use and land cover mapping, and so on. Different payloads on the remote sensing spacecraft provide day and night high resolution imaging in several spectral bands.

The different remote sensing satellite missions of the IRS satellite system include IRS-1A, IRS-1B, IRS-P1, IRS-P2, IRS-1C, IRS-P3, IRS-1D, IRS-P4 (also called Oceansat-1), Oceansat-2, Oceansat-3, Technology Experiment Satellite (TES), IRS-P5 (also called Cartosat-1), Cartosat-2, Cartosat-2A, Cartosat-2B, Cartosat-3, IRS-P6 (also called Resourcesat-1), Resoucesat-2, Resourcesat-3, IMS-1, Risat-1, Saral and Megha-Tropiques. IRS-P1 was a launch failure. Of the remaining satellites, some of them have successfully completed their intended missions and others are still in service. Those that have completed their missions include IRS-1A, IRS-1B, IRS-P2, IRS-1C, IRS-P3, IRS-1D, IRS-P4 and TES. Some of the more recent remote sensing satellites are briefly described in the following paragraphs. The descriptions are mainly confined to spacecraft launch details and payloads on board the spacecraft.

Resourcesat-1, also known as IRS-P6, was launched on board indigenously developed polar satellite launch vehicle PSLV-C5 on 17 October 2003. It was placed in a 817 km high polar sun synchronous orbit. The payload comprises three cameras, including a high resolution LISS-4 (linear imaging self scanner) camera operating in three spectral bands in the visible and near infrared region with 5.8 m spatial resolution and steerable up to +26° across track to obtain stereoscopic images with a revisit time of 5 days, a medium resolution LISS-3 camera operating in three spectral bands in the visible and near infrared region and one in the shortwave infrared (SWIR) band with 23.5 m spatial resolution and 142 km swath with a revisit time period of 24 days and an advance wide field sensor (AWiS) operating in three spectral bands in the visible and near infrared region and one band in the shortwave infrared with 56 m spatial resolution. Two such cameras (AWiS-A and AWiS-B) achieve a swath of 730 km. The satellite was intended to continue to provide the remote sensing services earlier provided by IRS-1C and IRS-1D with enhanced resolution and data quality.

Resourcesat-2 is a follow-up satellite to Resourcesat-1 and is intended to provide remote sensing data services to global users provided by Resourcesat-1. Resourcesat-2 cameras have enhanced performance as compared to that on board Resourcesat-1. For example, LISS-3 has improved radiometric accuracy from 7 to 10 bits, LISS-4 has enhanced multispectral swath from 23 km to 70 km and improved radiometric accuracy from 7 bits to 10 bits, and AWiS has improved radiometric accuracy from 10 to 12 bits. *Resourcesat-3* is a follow-up satellite to Resourcersat-2 and is scheduled to be launched in 2015. The LISS-3 camera on board the satellite is LISS-3-WS, which is an advanced version of LISS-3 having wider swath equal to that of the AWiS sensor. LISS-3_WS therefore has the spatial resolution of LISS-3 and swath width of AWiS. It also carries an atmospheric correction sensor.

Cartosat-1 was launched on 5 May 2005 on board PSLV-C6 into a 617 km polar sun synchronous orbit. Two panchromatic cameras, namely PAN (fore) and PAN (aft), with 2.5 m resolution and a swath coverage of 30 km constitute the payload. The data from Cartosat-1 is used for the preparation of cartographic maps. A solid state recorder on board the satellite provides global data storage of areas not visible to the ground station. *Cartosat-2* (Figure 11.35) is a follow-up satellite to Cartosat-1 and was launched on 10 January 2007. It is an advanced remote sensing satellite capable of providing scene-specific spot imagery. The payload is a single panchromatic camera capable of providing better than 1 m spatial resolution with

Figure 11.35 Cartosat-2 remote sensing satellite (Courtesy: ISRO)

a swath of 9.6 km. *Cartosat-3* is a further advanced version of Cartosat-2. The camera on board the satellite will have a resolution of 30 cm and a swath of 6 km. It is scheduled to be launched during 2014 and will be used for cartography, weather mapping and some strategic applications.

Oceansat-1 was launched on 26 May 1999. The payload of Oceansat-1 includes an *ocean colour monitor* operating in 402–422, 433–453, 480–500, 500–520, 545–565, 660–689, 745–785 and 845–885 nm wavelength bands with a spatial resolution of 360 m and a *multi-frequency scanning microwave radiometer*. The ocean colour monitor has a swath of 1420 km. It is intended to study physical and biological aspects of oceanography. *Oceansat-2* was launched on 23 September 2009. In addition to providing service continuity to the users of the ocean colour monitor of Oceansat-1, it has potential applications in other areas too. The payload of Oceansat-2 includes an *ocean colour monitor*, a *Ku-band pencil beam scatterometer* and a *radio occultation sounder for atmosphere*. Oceansat-3 is the latest in the series of Oceansat remote sensing satellites and is scheduled for launch during 2014. It is mainly intended for ocean biology and sea state applications. Oceansat-3's planned payload includes a *12-channel ocean colour monitor*, a *thermal infrared sensor*, a *scatterometer* and a *passive microwave radiometer*.

Risat-1 (Radar Satellite-1) is a state-of-the-art microwave remote sensing satellite with imaging capability during both day and night and under all weather conditions. It was launched on 26 April 2012 into a circular polar sun synchronous orbit of 536 km altitude. Microwave synthetic aperture radar operating in C-band is the payload. Active microwave remote sensing provides cloud penetration and day–night imaging capability. The all-weather and day/night imaging capability of Risat-1 is put to use in applications in agriculture, forestry, soil moisture, geology, sea ice, coastal monitoring, object identification and flood monitoring. Figure 11.36 shows a photograph of Risat-1. *Risat-2* was built by Israel Aerospace Industries and was launched before Risat-1 on 20 April 2009. It was India's first radar imaging satellite. The payload is an X-band synthetic aperture radar used to monitor Indian borders to check counter-insurgency and counter-terrorism.

Figure 11.36 Risat-1 remote sensing satellite (Courtesy: ISRO)

Problem 11.1

In Figure 11.37, the path of the satellite carrying a camera with its lens at C is shown by the arrow. The camera is at a height H above the ground and has a focal length f. Determine the scale factor of the image.

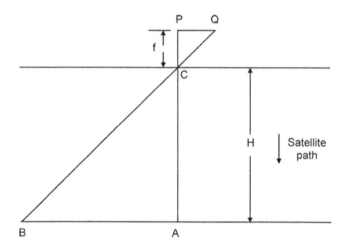

Figure 11.37 Figure for Problem 11.1

Solution: In Figure 11.37, AB is the line on the ground and PQ is its image on the film of the camera. The scale factor is defined as the ratio (PQ/AB), assuming that the picture is taken vertically. Since $\triangle ABC$ and $\triangle PQC$ are similar triangles,

$$\frac{PQ}{AB} = \frac{f}{H}$$

Hence, the scale factor is determined by the ratio of the focal length of the sensor system (f) and its height above the Earth's surface (H).

Problem 11.2

In Problem 11.1, it is given that the satellite is orbiting at a height of 1000 km and the sensor focal length is 15 cm. Determine the scale factor of the image taken from the satellite.

Solution:

$$\text{Scale factor} = \frac{f}{H}$$

where

f = focal length of the sensor system
H = height of the sensor system above the ground, which is the same as the satellite altitude
Therefore,

$$\text{Scale factor} = 15\,\text{cm}/1000\,\text{km}$$
$$= 15 \times 10^{-2}/1000 \times 10^3 = 15/10^8$$
$$= 1.5 \times 10^{-7}$$

Problem 11.3
Determine the smallest actual length on the Earth's surface whose image can be measured by the photograph taken from the satellite system of Problem 11.2, if the measurement of the film can be done up to 0.1 μm.

Solution: Let the smallest actual length measured by the system be L. Then

$$\frac{\text{Smallest image length}}{\text{Smallest actual length}} = \frac{f}{H}$$

Therefore,

$$1 \times 10^{-7}/L = 1.5 \times 10^{-7}$$
$$L = (10^{-7}/1.5 \times 10^{-7}) = 1/1.5$$
$$= 0.667\,\text{m}$$

The smallest length that can be measured by any system is also referred to as the resolution of the system.

Problem 11.4
A spacecraft is orbiting Earth in a sun-synchronous orbit at an altitude of h km from the Earth's surface. The satellite can see only a portion of the Earth's surface, referred to as the horizon 'cap'. Determine the size of this horizon 'cap' observed by the satellite. Also calculate the same for the Landsat-2 satellite having an altitude of 916 km. Assume the radius of Earth to be 6000 km.

Solution: Refer to Figure 11.38. In the figure,

S = position of the satellite
C = centre of the Earth
H = point on the horizon circle seen by the satellite
P = subsatellite point on Earth (intersection of the Earth's surface with the line joining the Earth's centre to the satellite)
Q = centre of the horizon circle
ρ = angular separation of the horizon seen by the satellite from the Earth's centre
λ = angle subtended at the Earth's center by the radius of the horizon circle

The circle formed by the boundary of the horizon cap is called the horizon circle. The size of the horizon cap is specified by the size of the angular radius λ of the horizon circle

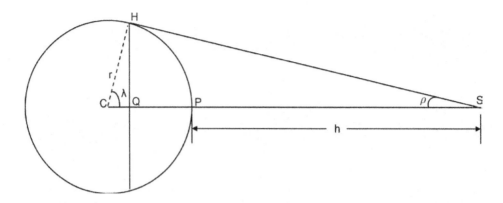

Figure 11.38 Figure for Problem 11.4

seen by the satellite. △CHS is a right-angled triangle at H. Also, SP = h, CH = CP = r, the radius of the Earth. Therefore,

$$\sin \rho = \cos \lambda = \frac{r}{r+h}$$

Landsat-2 satellite has an altitude of 916 km and the radius of the Earth is given as 6000 km.

Therefore, the angular radius λ of the horizon circle seen by the satellite is

$$\lambda = \cos^{-1}(6000/6916) = 29°$$

Problem 11.5

Refer to Figure 11.39 (a). The satellite has to image a rectangle formed by the equator, 10° parallel latitude and 60° and 90° west meridians of longitude, shown by the thick black lines. The satellite sensors will image the rectangle as shown in Figure 11.39 (b). Give reasons why this distortion occurs in the image and the measures that can be taken to correct it.

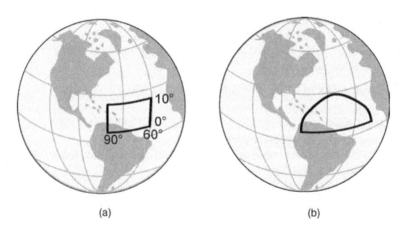

Figure 11.39 Figures for Problem 11.5

Solution: In observing Earth from space using satellite sensors, distortions are introduced into the image because of the spherical shape of Earth. Refer to Figure 11.40. It is supposed that a point R on the surface of the Earth is to be imaged by the satellite. The satellite sensors can measure the angle at which point R on Earth is observed but they cannot measure their distance from point R. So, all the points are intercepted as if they are lying on the same plane, the plane of the horizon circle. Hence, R is imaged by the satellite sensor as if it was at R', in the plane of the horizon circle. All the images taken by the satellite will therefore be distorted as it takes images by intercepting as if everything is lying on the horizon plane rather than on the surface of the Earth.

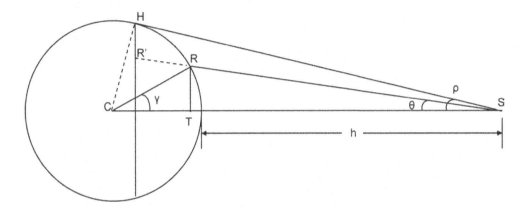

Figure 11.40 Figure for Problem 11.5

The image is corrected using softwares so that the information relayed to Earth is free of distortion. This is done by expressing the relationship between the angle of observation of R (θ), the angular deviation of R from the line joining the Earth's centre to the satellite (γ) and the angle of observation of the horizon (ρ).
As $\triangle$ RTS is right angled at T,

$$\tan\theta = RT/TS = \frac{RT}{CS - CT}$$

$$= \frac{r\sin\gamma}{(r+h) - (r\cos\gamma)}$$

As already discussed in Problem 11.4, $\sin\rho = r/(r+h)$. Therefore,

$$\tan\theta = \frac{\sin\rho\sin\gamma}{(1 - \sin\rho\cos\gamma)}$$

Problem 11.6
A satellite system uses a scanning sensor, with the scanning done in the direction orthogonal to the flight path. The sensor employs mirrors and lenses that rotate around an axis parallel to the flight path. Although the scanning system rotates at a constant rate, the images formed by it are distorted. Figure 11.41 (a) shows the actual pattern and

Figure 11.41 (b) shows the distorted image produced by the satellite. Give reasons as to why this distortion occurs and give a solution for correcting the distortion.

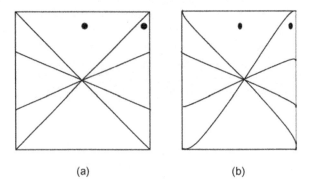

Figure 11.41 Figures for Problem 11.6

Solution: Although the scanning system rotates at a constant rate, the rate at which the scanning beam moves along the ground depends on the angle it makes with the vertical, as shown in Figure 11.42. In order to produce an undistorted picture, the actual recording of the images must be done at the Earth scan rate rather than at the satellite rotation rate. The distortion shown in the figure can be corrected by making the Earth scan rate constant, i.e. making dx/dt constant instead of making the rotation rate $d\theta/dt$ constant. If the satellite is orbiting Earth at an altitude h, then

$$\tan\theta = \frac{x}{h}$$

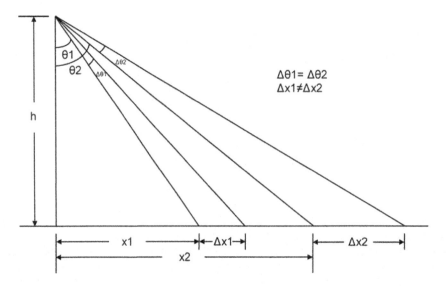

Figure 11.42 Figure for Problem 11.6

Differentiating with respect to time t gives

$$\frac{dx}{dt} = h \sec\theta \left(\frac{d\theta}{dt}\right)$$

Therefore, the solution lies in making dx/dt constant, i.e. making $h \sec\theta \, (d\theta/dt)$ constant rather than making $(d\theta/dt)$ constant.

11.12 Future Trends

Since the launch of the first remote sensing satellite, Landsat-1, in the early 1970s, remarkable growth has been made in the field of remote sensing satellites both in terms of technological developments and potential applications. The aim of the remote sensing satellite missions of today is to provide highly reliable and accurate data and launch satellites with long life times having a high level of redundancy and sensor stability in order to cater to the needs of critical remote sensing applications. These applications mostly require long-term observation with precise and accurate information. To minimize the risk of mission failure, a maximum redundancy approach is being pursued. In a nutshell, the focus is to launch satellites with longer lifetimes having on board them sophisticated sensors with maximum redundancy possible and keeping the satellite mass to the minimum possible.

Technological advances have led to development of new sensors, improvement in the resolution of the sensors, increase in observation area and reduction in access time, that is time taken between the request of an image by the user and its delivery. Future trends are to further improve each of these parameters to have more accurate and precise remote sensing data. Future missions will make use of new measurement technologies such as cloud radars, lidars and polarimetric sensors that will provide new insights into the key parameters of atmospheric temperature and moisture, soil moisture and ocean salinity. Recent developments in the field of lidar technology and laser terrain mapping systems will drastically reduce the time and efforts needed to prepare digital elevation models. Improvements in sensor technology especially in the resolution of the sensors have led to the development of hyper-spectral and ultra-spectral systems. These systems image the scene over a large number of discrete and contiguous spectral bands resulting in images over the complete reflectance spectrum. Several new gravity field missions aimed at more precise determination of the marine geoid will also be launched in the future. These missions will also focus on disaster management and studies of key Earth system processes the water cycle, carbon cycle, cryosphere, the role of clouds and aerosols in global climate change and sea level rise.

Other than improvements in the sensor technology, great advances have been made in the image compression and image analysis techniques. Image compression techniques have made it possible to transfer voluminous image data. The latest image compression techniques are image pyramids, fractal and wavelet compression. The image analysis techniques that will be used extensively in the future include image fusion, interoferometry and decision support systems and so on. Image fusion refers to merging data from a large number of sensors in hyper-spectral and ultra-spectral systems to improve system performance, to generate sharpened images, improve geometric corrections, provide stereo-viewing capabilities for stereo-photogrammetry, to enhance certain features not visible in single data images, detect changes using multi-temporal data, replace defective data and substitute missing information from one image with information from another image. Fusion of image data is done at three processing levels namely at the pixel level, feature level and decision level. Radar interferometry is a rapidly developing field in which two or more images of the same location are processed together to derive the digital elevation model. Decision support system refers to interactive, flexible and adaptable computer based information system that aids in storing and processing the image data and aids in the decision-making process.

Further Readings

Allan, T.D. (1983) *Satellite Microwave Remote Sensing* John Wiley & Sons, Inc., New York.
Berman, A.E. (1999) *Exploring the Universe through Satellite Imagery*, Tri-Space, Inc.
Bromberg, J.L. (1999) *NASA and the Space Industry*, John Hopkins University Press, Baltimore, Maryland.
Clareton, A.M. (1991) *Satellite Remote Sensing in Climatology*, CRC Press, Florida.
Conway, E.D. (1997) *Introduction to Satellite Image Interpretation*, John Hopkins University Press, Baltimore, Maryland.
Denegre, J. (1994) *Thematic Mapping from Satellite Imagery: Guide Book*, Pergamon Press, Oxford.
Gatland, K. (1990) *Illustrated Encyclopedia of Space Technology*, Crown, New York.
Gurney, R.J., Foster, J.L. and Parkinson, C.L. (1993) *Atlas of Satellite Observations Related to Global Change*, Cambridge University Press.
Hiroyuki, F. (2001) *Sensor Systems and Next Generation Satellites IV*, SPIE – The International Society for Optical Engineering, Bellingham, Washington.
Lillesand, T.M., Kiefer, R.W. and Chipman, J.W. (2004) *Remote Sensing and Image Interpretation*, John Wiley & Sons, Inc., New York.
Sanchez, J. and Canton, M.P. (1999) *Space Image Processing*, CRC Press, Boca Raton, Florida.
Fernand Verger, Isabelle Sourbes-Verger, Raymond Ghirardi, Xavier Pasco, Stephen Lyle, Paul Reilly *The Cambridge Encyclopedia of Space* Cambridge University Press 2003

Internet Sites

1. www.gisdevelopment.net
2. www.astronautix.com
3. http://www.crisp.nus.edu.sg/~research/tutorial/spacebrn.htm
4. http://www.crisp.nus.edu.sg/~research/tutorial/image.htm
5. http://www.crisp.nus.edu.sg/~research/tutorial/optical.htm
6. http://www.crisp.nus.edu.sg/~research/tutorial/opt_int.htm
7. http://www.crisp.nus.edu.sg/~research/tutorial/infrared.htm
8. http://www.crisp.nus.edu.sg/~research/tutorial/mw.htm
9. http://www.crisp.nus.edu.sg/~research/tutorial/sar_int.htm
10. http://www.crisp.nus.edu.sg/~research/tutorial/process.htm
11. http://www.gisdevelopment.net/tutorials/tuman008.htm
12. http://rst.gsfc.nasa.gov/Front/tofc.html

13. www.isro.org
14. www.nasda.go.jp
15. www.noaa.gov
16. www.orbiimage.com
17. www.spot4.cnes.fr
18. www.spotimage.fr
19. www.spaceimaging.com
20. www.skyrocket.de

Glossary

Active remote sensing: Active remote sensing involves active artificial sources of radiation generally mounted on the remote sensing platform that are used for illuminating the objects. The energy reflected or scattered by the objects is recorded in this case

Aerial remote sensing: Aerial remote sensing uses platforms like aircraft, balloons, rockets, helicopters, etc., for remote sensing

Central perspective scanners: The central perspective scanners utilizes either electromechanical or linear array technology to form image lines, but images in each line form a perspective at the centre of the image rather than at the centre of each line

False colour composite image: If the spectral bands in the image do not correspond to the three primary colours, the resulting image is called a false colour composite image. Hence, the colour of an object in the displayed image has no resemblance to its actual colour

Geographic Information System (GIS): The Geographic Information System (GIS) is a computer-based information system used to represent digitally and analyse the geographic features present on Earth's surface and the events taking place on it

Indian remote sensing satellite (IRS) system: This is one of the largest constellation of Earth observation satellites and is used for a wide variety of remote sensing and Earth observation applications at both national and international level. These satellites are designed, launched and maintained by the Indian Space Research Organization (ISRO).

Instantaneous field-of-view (IFOV): This is defined as the solid angle from which the electromagnetic radiation measured by the sensor at a given point of time emanates

Landsat satellite system: The Landsat satellite system is the USA's remote sensing satellite programme launched with the aim of observing Earth on a global basis. It comprises seven optical/thermal remote sensing satellites for land observation purposes

Microwave radiometer: The microwave radiometer is a passive device that records the natural microwave emission from Earth

Microwave remote sensing systems: Microwave remote sensing systems utilize the microwave band, generally from 1 cm to 1 m for remote sensing applications

Monogenic images (panchromatic images): Monogenic images are produced from a single primary image by applying some changes to it, like enlargement, reduction, error correction, contrast adjustments, in order to extract maximum information from the primary image

Multispectral images: In multispectral images, the final image is produced from three images taken in different spectral bands and by assigning a separate primary colour to each image

Multitemporal images: Multitemporal images are secondary images produced by combining two or more primary images taken at different times

Natural colour composite image: Natural colour composite images are those images in which the spectral bands are combined in such a way that the appearance of the displayed image resembles a visible colour photograph

Non-scanning systems: Non-scanning systems explore the entire field in one take

Optical mechanical scanner: An optical mechanical scanner is a multispectral radiometer where the scanning is done in a series of lines oriented perpendicular to the direction of the motion of the satellite using a rotating or an oscillating mirror

Optical remote sensing systems: Optical remote sensing systems mainly makes use of visible (0.3–0.7 μm), near-IR (0.72–1.30 μm) and shortwave-IR (1.30–3.00 μm) bands to form images of the Earth's surface. Some optical remote sensing systems also use laser radars, laser distance meters, etc.

Passive remote sensing: Passive remote sensing refers to the detection of reflected or emitted radiations from natural sources like the sun, etc., or the detection of thermal radiation or the microwave radiation emitted by objects

Polygenic secondary images: These are composite images formed by combining two or three primary images in order to make the extraction of information from the image easier and more meaningful

Push broom scanners: A push broom scanner is a scanner without any mechanical scanning mirror but with a linear array of solid semiconductor elements located at the focal plane of the lens system, which enables it to record one line of an image at one time

Radarsat satellite system: Radarsat is a Canadian remote sensing satellite system

Radiometric resolution: Radiometric resolution refers to the smallest change in the intensity level that can be detected by the remote sensing system

Resolution: Resolution is defined as the ability of the entire remote sensing system (including the lens, antennas, display, exposure, processing, etc.) to render a sharply defined image

Spatial resolution: Spatial resolution is defined as the minimum distance the two point features on the ground should have in order to be distinguished as separate objects

Spectral resolution: This is determined by the bandwidths of the electromagnetic radiation of the channels. The narrower the bandwidth used, the higher is the spectral resolution achieved

SPOT satellites: SPOT is the French satellite programme, with Belgium and Sweden as minority partners. The system is designed by the French Space Agency (CNES) and is operated by its subsidiary, Spot Image

Synthetic aperture radar: Synthetic aperture radar (SAR) uses a technique of synthesizing a very large array antenna over a finite period of time by using a series of returns from a relatively much smaller physical antenna that is moving with respect to the target

Temporal resolution: Temporal resolution is specified as the number of days in which the satellite revisits a particular place again

Thermal remote sensing systems: Thermal remote sensing systems sense the thermal radiations emitted by objects in the mid-IR (3–5 μm) and the long IR (8–14 μm) bands

True colour composite image: In the true colour composite image, the three primary colours are assigned to the same coloured spectral bands, resulting in images similar to that seen by a human eye

Wind scatterometer: The wind scatterometer is used to measure wind speed and direction over the ocean surface by sending out microwave pulses along several directions and recording the magnitude of the signals backscattered from the ocean surface

12

Weather Satellites

Use of satellites for weather forecasting and prediction of related phenomena has become indispensable. Information from weather satellites are used for short term weather forecasts as well as for reliable prediction of the movements of tropical cyclones, allowing rerouting of ships and a preventive action in zones through which hurricanes pass. Meteorological information is also of considerable importance for conducting of military operations such as reconnaissance missions. Due to the inherent advantages of monitoring from space, coupled with developments in the sensor technology, satellites have brought about a revolution in the field of weather forecasting. The end result is that there is a reliable forecast of weather and other related activities on a routine basis. In this chapter, a closer look will be taken at various aspects related to evolution, operation and use of weather satellites. Some of the major weather satellite missions are covered towards the end of the chapter. Like previous chapters, this chapter also contains a large number of illustrative photographs.

12.1 Weather Forecasting – An Overview

Weather forecasting, as people call it, is both a science as well as an art. It is about predicting the weather, which can be both long term as well as short term. Generally, short term predictions are based on current observations whereas long term predictions are made after understanding the weather patterns, on the basis of observations made over a period of several years. Weather watching began as early as the 17th century, when scientists used barometers to measure pressure. Weather forecasting as a science matured in the early 1900s when meteorological kites carrying instruments to measure the temperature, pressure and the relative humidity were flown. After that came the era of meteorological aircraft and balloons carrying instruments for weather forecasting.

The year 1959 marked a significant beginning in the field of satellite weather forecasting, when for the first time a meteorological instrument was carried on board a satellite, Vanguard-2, which was launched on 17 February 1959. The satellite was developed by National Aeronautics and Space Administration (NASA) of USA. Unfortunately, the images taken by the instrument could not be used as the satellite was destroyed while on mission.

The first meteorological instrument that was successfully used on board a satellite was the Suomi radiometer, which was flown on NASA's Explorer-7 satellite, launched on 13 October 1959. All these satellites were not meteorological satellites; they just carried one meteorological instrument.

The first satellite completely dedicated to weather forecasting was also developed by NASA. The satellite was named TIROS-1 (television and infrared observation satellite) and was launched on 1 April 1960. It carried two vidicon cameras, one having low resolution and the other with a higher resolution. Both these cameras were adaptations of standard television cameras. Though the satellite was operational for only 78 days, it demonstrated the utility of using satellites for weather forecasting applications. The first picture was transmitted by the TIROS-1 satellite on 1 April 1960. It showed the cloud layers covering the Earth [Figure 12.1 (a)]. The first useful transmitted weather pictures were that of the Gulf of St Lawrence [Figure 12.1 (b)]. The images, taken during 1–3 April 1960, showed the changing state of the pack ice over the Gulf of St Lawrence and the St Lawrence River.

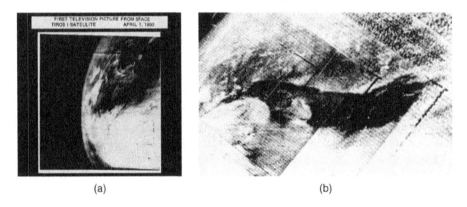

Figure 12.1 (a) First picture transmitted by the TIROS-1 satellite (Courtesy: NESDIS/National Climatic Data Center/NOAA) and (b) first useful weather image transmitted by the TIROS-1 satellite [Courtesy: US Department of Commerce, National Oceanic and Atmospheric Administration (NOAA)]

Since then there has been no looking back and nine additional TIROS satellites were launched in the next five years. TIROS-10, the last satellite of the TIROS series was launched in the year 1965. The first eight satellites in the series, TIROS-1 to TIROS-8, were launched into prograde inclined low Earth orbits. TIROS-9 and TIROS-10 satellites were launched into polar sun-synchronous orbits. The TIROS programme marked the first space-borne programme that demonstrated the feasibility and capability of observing the weather patterns from space. Alongside the TIROS programme, came the Nimbus satellite programme. Nimbus satellites orbited in polar sun-synchronous orbits. Nimbus-1, the first satellite in the Nimbus series, was launched on 28 August 1964. It was also the first sun-synchronous weather forecasting satellite. In total, seven Nimbus satellites were launched under the Nimbus satellite programme. The observational capability of the TIROS as well as the Nimbus satellites improved with time as the newer satellites launched carried better payloads than their predecessors.

The world's first polar weather satellite system, referred to as the TIROS operational system (TOS), became a reality in the year 1966 with the launch of ESSA-1 and ESSA-2 satellites, on

3 February 1966 and 28 February 1966 respectively. The system comprised of a pair of ESSA satellites in sun-synchronous polar orbits. A total of nine ESSA satellites were launched in a span of three years between 1966 and 1969. It was succeeded by the improved TIROS (ITOS) satellite system, referred to as the second generation of polar weather satellite systems. The first satellite in this series was ITOS-1, launched on 23 January 1970. Five other satellites were launched in this series, namely NOAA-1 to NOAA-5, with the last one, NOAA-5, launched in the year 1976. They had sensors with better resolution as compared to the earlier satellites and provided improved infrared and visible observations of cloud cover. They also provided solar proton and global heat data on a daily basis.

The TIROS-N (new generation TIROS) series of satellites marked the third generation of weather satellites. The TIROS-N system provided global meteorological and environmental data for the experimental World Weather Watch (WWW) programme. The first satellite in this series, TIROS-N, was launched in the year 1978. It was followed by three more satellites, NOAA-6, NOAA-B and NOAA-7 launched in the years 1979, 1980 and 1981 respectively. All these satellites carried a radiometer, a sounding system and a solar proton monitor. The fourth generation of weather satellites, the advanced TIROS-N (ATN) series, became operational with the launch of NOAA-8 in the year 1983. To date, 12 satellites have been launched in this series, namely NOAA-8, 9, 10, 11, 12, 13, 14, 15, 16, 17, 18 and 19. Currently, six ATN satellites namely NOAA-13, 15, 16, 17, 18 and 19 are operational. They are discussed in detail later in the section on international weather satellite systems.

The Russians came up with their polar sun-synchronous meteorological system called Meteor, nine years after the USA launched its first weather satellite. The first satellite in the Meteor series, Meteor-1, was launched in the year 1969. To date, the Meteor-1, 2, 3 and 3M series of satellites have been launched. The Meteor-2 series comprised 21 satellites launched over a period of 18 years from 1975 to 1993. The Meteor-3 series comprised seven satellites. The first satellite in this series, Meteor-3-1(a), was launched in the year 1984 and the last one, Meteor-3-6, in the year 1994. Meteor-3M-1 satellite, launched on 10 December 2001, was the only satellite to be launched in the Meteor-3M series. Meteor M is the new generation Russian meteorological satellite series. The first satellite of the series, Meteor-M1, was launched in 2009 and three more satellites in the series are proposed to be launched in the next two years.

The first geostationary satellite carrying meteorological payloads was application technology satellite 1 (ATS-1), launched by NASA in the year 1966. The first geostationary meteorological satellites were the synchronous meteorological satellites (SMS-1 and SMS-2), launched in the years 1974 and 1975 respectively. The SMS satellites were superceded by the geostationary operational environmental satellites (GOES) system of satellites. Both the SMS and the GOES satellite systems were developed by NASA. GOES-1 (A), launched on 16 October 1975, was the first satellite of the GOES satellite system. The GOES satellite system is a two-satellite constellation that views nearly 60% of the Earth's surface. Fifteen more GOES satellites have been launched since then. Currently, four GOES satellites, GOES-12 (M), GOES-13 (N), GOES-14 (O) and GOES-15(P) are in use. GOES-12 (M), GOES-13(N) and GOES-15(P) are the operational satellites while GOES-14 (O) is an in-orbit spare satellite. The GOES satellite system is discussed in detail in the section on international weather satellite systems.

Today, many countries of the world other than the USA and Russia have their own weather forecasting satellite systems to monitor the weather conditions around the globe. Japan, Europe, China and India have launched their own weather forecasting satellite systems, namely the

GMS, Meteosat, Feng Yun and INSAT satellite systems respectively. GMS, Meteosat and INSAT satellite systems employ geostationary satellites whereas the Feng Yun system has satellites orbiting both in LEO polar orbits and geostationary orbits.

12.2 Weather Forecasting Satellite Fundamentals

Weather forecasting satellites are referred to as the third eye of meteorologists, as the images provided by these satellites are one of the most useful sources of data for them. Satellites measure the conditions of the atmosphere using onboard instruments. The data is then transmitted to the collecting centres where it is processed and analysed for varied applications. Weather satellites offer some potential advantages over the conventional methods as they can cover the whole world, whereas the conventional weather networks cover only about 20 % of the globe. Satellites are essential in predicting the weather of any place irrespective of its location. They are indispensable in forecasting the weather of inaccessible regions of the world, like oceans, where other forms of conventional data are sparse. As a matter of fact, forecasters can predict an impending weather phenomenon using satellites 24 to 48 hours in advance. These forecasts are accurate in more than 90 cases out of 100.

Satellites offer high temporal resolution (15 minutes to 1 hour between images) as compared to other forecasting techniques. However, their spatial resolution is less, of the order of 1 to 10 km. Moreover, satellites are not forecasting devices. They merely observe the atmosphere from above. This implies that the data collected by satellites needs to be further processed, so that it can be converted into something meaningful. Satellites have poor vertical resolution as it is difficult to assign features to particular levels in the atmosphere and low-level features are often hidden.

12.3 Images from Weather Forecasting Satellites

Weather forecasting satellites take images mainly in the visible, the IR and the microwave bands. Each of these bands provides information about different features of the atmosphere, clouds and weather patterns. The information revealed by the images in these bands, when combined together, helps in better understanding of the weather phenomena. The images in the visible band are formed by measuring the solar radiation reflected by Earth and the clouds. The IR and the microwave radiation emitted by the clouds and Earth are used for taking IR and microwave images respectively. Some images are formed by measuring the scattering properties of the clouds and Earth when microwave or laser radiation is incident on them. This is referred to as active probing of the atmosphere. In this section, various types of images will be discussed in detail.

12.3.1 Visible Images

Satellites measure the reflected or scattered sunlight in the wavelength region of 0.28 to 3.0 µm. The most commonly used band here is the visible band (0.4 to 0.9 µm). Visible images represent the amount of sunlight being reflected back into space by clouds or the Earth's surface in the visible band. These images are mainly used in the identification of clouds. Mostly, weather satellites detect the amount of radiation without breaking it down to individual colours. So these images are effectively black and white. The intensity of the image depends on the reflectivity (referred to as albedo) of the underlying surface or clouds

Different shades of grey indicate different levels of reflectivity. The most reflective surfaces appear in white tones while the least reflective surfaces appear in shades of dark grey or black. In general, clouds have a higher reflectivity as compared to the Earth's surface and hence they appear as bright (white) against the darker background of the Earth's surface. Visible images give information on the shape, size, texture, depth and movement of the clouds. Brighter clouds have larger optical depth, higher water or ice content and smaller average cloud droplet size than darker looking clouds. Visible band is also used for pollution and haze detection, snow and ice monitoring and storm identification. Almost all satellites have instruments operating in the visible band. Examples include the GOES (the GOES imager has one channel in the visible band of 0.52 to 0.72 µm and the GOES sounder also operates in the visible band), Meteosat [the SEVIRI (spinning enhanced visible and infrared imager) on the MSG-2] and the ATN [AVHRR (advanced very high resolution radiometer) has one channel in the 0.58 to 0.68 µm band] satellites.

Other than the visible band, weather satellites also measure the reflected solar light in the near-IR, shortwave-IR and UV bands. The near-IR band provides useful information for water, vegetation and agricultural crops. The shortwave IR band is used for identification of fog at night and for discrimination between water clouds and snow or ice clouds during daytime. Measurements of the amount and vertical distribution of the atmospheric ozone are carried out in the UV band.

Figure 12.2 shows a visible image taken by the GOES satellite. The continental outlines have been added to the image. The bright portions of the image indicate the presence of clouds. It can be inferred from the image that about half of the image is covered by clouds. The other half, which is not covered by clouds, is the Earth's surface. From the visible images, it can also be identified as to whether it is a land or water area.

Visible images are very frequently used for weather forecasting. Sometimes they provide information that may not appear in IR images. Two objects having the same temperatures

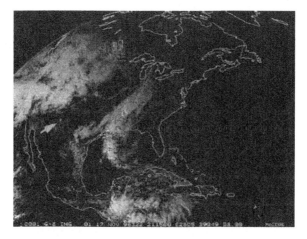

Figure 12.2 Visible image taken by the GOES satellite (Reproduced by permission of John Nielson-Gammon, Texas A&M University)

can be discriminated using a visible image but not from an IR image. For instance, if the temperature of fog is the same as that of land, then they will appear similar on the IR image, but will appear different on the visible image as they have different albedo. However, one of the main limitations of using visible images is that they are available only during the daytime. It is also difficult to distinguish between low, middle and high level clouds in a visible satellite image, since they can all have a similar albedo. Similarly, it is difficult to distinguish between clouds and ground covered with snow. Thin clouds do not appear on visible images and hence they cannot be detected using these images. Infrared satellite images are more useful for such applications. Satellites also carry instruments to do active probing in the visible band. Active probing is discussed later in the section.

12.3.2 IR Images

Another common type of satellite imagery depicts the radiation emitted by the clouds and the Earth's surface in the IR band (10 to 12 μm). IR images provide information on the temperature of the underlying Earth's surface or cloud cover. This information is used in providing temperature forecasts, in locating areas of frost and freezes and in determining the distribution of sea surface temperatures offshore. Since the temperature normally decreases with height, IR radiation with the lowest intensity is emitted from clouds farthest from the Earth's surface. The Earth's surface emits IR radiation with the highest intensity. Hence, in IR images clouds appear dark as compared to Earth. Moreover, high lying clouds are darker than low lying clouds. High clouds indicate a strong convective storm activity and hence IR images can be used to predict storms.

One of the potential advantages of IR imagery is that it is available 24 hours a day, as the temperatures can be measured regardless of whether it is day or night. However, IR images generally cannot distinguish between two objects having the same temperature. For instance, using IR images it may not be possible to distinguish between thin and thick clouds present at the same altitude, especially if they are present at higher altitudes. Also, IR images have poorer resolution than visible images. This is so because the emitted IR radiation is weaker in intensity than the visible radiation. Therefore, the payload on the satellite has to sense radiation from a broader area so as to be able to detect it. Examples of satellite sensors operating in the IR band include the GOES imager and AVHRR sensor on the ATN satellites.

IR images can be grey-scale images or can be colour-enhanced images with different colours for features having different temperatures. Grey scale images are black and white images where darker shades correspond to lower temperatures. The normal convention is to reverse the appearance of these images so as to make them consistent with the visible images. Hence in the reversed images, lighter shades will correspond to lower temperatures, so the clouds appear as white against the darker background of the Earth's surface. Figures 12.3 (a) and (b) show the IR images taken by the GOES satellite, both in the raw and the normal conventional formats respectively. The portions that appear as bright in the raw image are the warmest areas. They are the Mexican deserts and the oceans. Dark patches in the raw image correspond to the clouds. In the conventional IR image shown in Figure 12.3(b), the pattern is reversed. Here, the Mexican deserts and the oceans appear dark whereas the clouds appear bright. In coloured IR images, features having the same temperature are assigned a particular colour. This is done in order to extract more information from the images. These images are discussed in detail in Section 12.6 on image processing and enhancement.

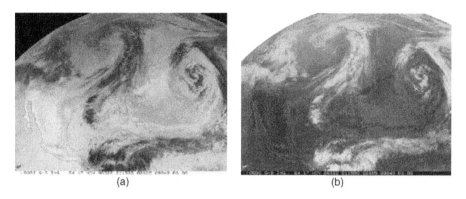

Figure 12.3 IR image taken by the GOES satellite in (a) the raw format and (b) the normal conventional format (Reproduced by permission of © John Nielson-Gammon, Texas A&M University)

Satellites also carry instruments to do active probing in the IR band. Active probing is discussed latter in the section.

12.3.3 Water Vapour Images

The visible and the IR images discussed so far are passed unobstructed through the Earth's atmosphere. These images tell little about the atmosphere as for these wavelengths the atmosphere is transparent. Satellite images are also constructed using IR wavelengths that are absorbed by one or more gases in the atmosphere, like water vapour, carbon dioxide, and so on. Radiation around the wavelength band of 6.5 μm is absorbed as well as emitted by water vapour. The water vapour channel on weather forecasting satellites works around this wavelength. It detects water vapour in the air, primarily from a height of 10 000 feet to 40 000 feet up from the Earth's surface. The level of brightness of the image taken in this band indicates the amount of moisture present in the atmosphere. The radiation emitted from the bottom of the water vapour layer is absorbed by the water vapour present above it. Very little radiation is emitted from the top of the layer, because there is very little water vapour there. Most of the radiation comes from the middle of the water vapour layer. The areas with the most water vapour in the middle atmosphere show up as white or in light shades of grey. The drier areas appear in darker shades of grey and the driest areas are black. Generally, cold areas have a lot of water vapour and hence they appear in shades of white or light grey, whereas the warm areas have little water vapour in the upper atmosphere and are in shades of dark grey or black.

Measurement of water vapour movement is used to calculate upper air winds. This helps in forecasting the location of thunderstorm outbreaks and the potential areas for heavy rains. However, water vapour images show upper level moisture only. For determination of low level moisture and surface humidity, ground-based measurements are done. Moreover, water vapour imagery is useful only in those areas where there are no clouds. Examples of instruments operating in the water vapour band are the VISSR (visible and infrared spin scan radiometer) sensor on the GOES satellite. Channels 9 and 10 of this sensor operate in the 7.3 and 6.7 μm bands respectively.

Figure 12.4 shows the water vapour image taken by the GOES satellite. The continental map has been overlaid on the image to identify the locations. The image is in the normal conventional format, so the areas with lots of water vapour (cold areas) appear light and those with little water vapour in the upper atmosphere (warm areas) appear dark. In the image, high clouds are seen as bright areas. The image also shows several light and dark streaks over Mexico. These streaks correspond to bands of water vapour being carried from the subtropics toward the southeastern United States.

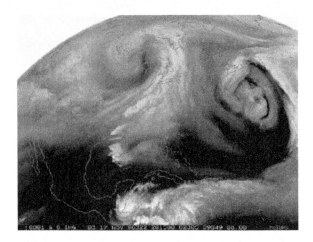

Figure 12.4 Water vapour image taken by the GOES satellite (Reproduced by permission of © John Nielson-Gammon, Texas A&M University)

12.3.4 Microwave Images

Weather satellites also utilize the microwave band, mostly within the wavelength region from 0.1 to 10 cm. They use both passive as well as active techniques for making measurements in the microwave band. Passive techniques measure the amount of microwave radiation emitted from the Earth's surface and clouds. Active microwave probing involves the use of sensors that emit microwave radiation towards the Earth and then record the scattered or the reflected radiation from the clouds, the atmosphere and the Earth's surface. In the following paragraphs, passive techniques will be discussed. Active probing is discussed latter in the section.

As already mentioned in the chapter on remote sensing satellites, the amount of microwave radiation emitted by an object is related to its temperature. Hence, measurements in the microwave band help in determining the temperature of the clouds and Earth's surface. Microwaves can penetrate water vapour and clouds, enabling the sensors to take measurements even under complete cloud cover. Microwave sensors work both during the day as well as during night. Microwave images can be either displayed as grey scale images or as colour enhanced images.

Measurements in the microwave band also help in determination of quantities such as snow cover, precipitation and thunderstorms, the temperature of lower and upper atmosphere over a period of years and the amount of rainfall. Rain droplets interact strongly with microwave radiation and microwave radiometers can detect the size and volume of the droplets. Therefore, microwave images are a more direct method for determining precipitation than visible or IR images. Microwave emission from water vapour is used for measuring the atmospheric humidity and that from oxygen for determining the bulk temperature of the troposphere and the lower stratosphere and in estimating the wind speed.

The main drawback of microwave images is that they have poor spatial and temporal resolution as compared to the visible and IR images. Moreover, interpretation of microwave images is more difficult, especially if the image is that of the land surface. Examples of sensors operating in the microwave band include AMSU-A (advanced microwave sounding unit A) and AMSU-B (advanced microwave sounding unit B) sounders on ATN satellites. Figure 12.5 is the microwave image taken by the TRMM satellite, showing the tropical storm Blanca near Mexico. The wind speed, direction and the amount of rainfall were predicted using such images of the storm.

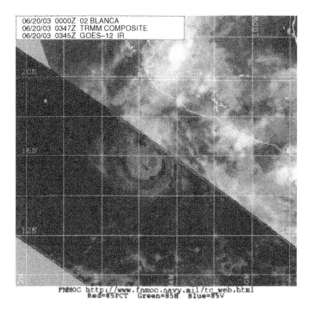

Figure 12.5 Microwave image taken by TRMM satellite [Courtesy: US Department of Commerce, National Oceanic and Atmospheric Administration (NOAA)]. The image is the greyscale version of the original colour image. **Original image is available on the companion website at www.wiley.com/go/maini**

12.3.5 Images Formed by Active Probing

Until now, the images that have been discussed correspond to the radiation reflected or emitted by clouds and the Earth's surface. Satellites also carry instruments that actively probe the atmosphere. Generally, active probing is done in the microwave wavelength band. Some satellites also do active probing in the visible and the IR bands using lasers. Active microwave probing involves the use of active microwave sensors. Radar is the most commonly used active microwave sensor. It emits microwave pulses towards the ground and measures the echo backscattered by the clouds, rain particles and the Earth's surface.

Medium to longer microwave wavelengths in frequency bands near 3, 5, 10 and 15 GHz are primarily utilized for observation of rainfall. Examples of such systems include synthetic aperture radar (SAR) systems operating near 5 GHz and 10 GHz bands. Examples of such a radar is the precipitation radar (PR) on the TRMM (Tropical Rainfall Measuring Mission) satellite used for measuring the rain profile. It emits microwave pulses at 13.796 GHz and 13.802 GHz towards the surface of the Earth. The echo backscattered by the rain is used to calculate the rain profile as the strength of the echo is proportional to the square of the volume of falling water. Millimetre wavelengths around 35 GHz and 94 GHz are used to determine the cloud characteristics and in studying the role of the effect of clouds on climate. Reflection from the land surface determines the land topography.

Measurements in the microwave region are also used to determine the sea state and the direction of surface winds. A rough ocean surface returns a stronger signal towards the satellite as the waves reflect more of the radar energy as compared to a smooth ocean surface. Examples of sensors used for this purpose include the SeaWinds scatterometer on the QuikSCAT satellite, which measures the ocean near-surface wind speed and direction.

Laser-based active weather probing systems work in the visible or the IR region. They use active sensors called lidars (laser detection and ranging) or laser altimeters for performing different measurements. The working principle of lidar is the same as that of radar except that they emit laser pulses in the visible or the IR region rather than microwaves pulses. Laser-based measurements are used to determine the distance to clouds and aerosol layers, to detect meteoric and volcanic debris in the stratosphere and to predict the formation of clouds. They are used to generate high resolution vertical profiles of atmospheric temperature and pressure, for profiling winds, measuring concentration of trace species, etc. Laser altimeters are used to determine the Earth's surface features, differential absorption lidar (DIAL) for determining temperature, moisture profiles and species concentration and Doppler lidar for wind speed and direction.

Laser-based systems offer better spatial resolution as compared to their microwave counterparts. However, laser probing is a relatively new concept in weather forecasting. Very few such systems have been used on satellites due to high costs and limited availability of such high power laser sources.

12.4 Weather Forecasting Satellite Orbits

Weather forecasting satellites are placed into either of the two types of orbits, namely the polar sun-synchronous low Earth orbit and the geostationary orbit. Polar sun-synchronous weather forecasting satellites revolve around the Earth in near polar low Earth orbits, visiting a particular place at a fixed time so as to observe that place under similar sunlight conditions. These orbits are similar to those discussed earlier in the case of remote sensing satellites. Polar weather forecasting satellites, due to their low altitudes, have better spatial resolution as compared to the geostationary satellites. Hence they help in a detailed observation of the weather features like the cloud formation, wind direction, and so on. However, these satellites have a poorer temporal resolution, visiting a particular location only one to four times a day. Hence, only a few weather satellite systems have satellites in these orbits. Some of the weather forecasting satellites in

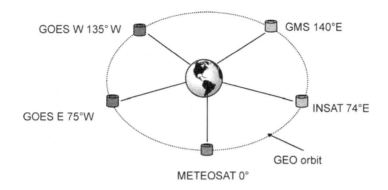

Figure 12.6 Weather forecasting satellites in geostationary orbit

near polar orbits include Feng Yun satellites of China and Advanced TIROS satellites of the USA.

Most weather forecasting satellites employ a geostationary orbit as these satellites offer better temporal resolution as compared to that provided by the polar weather forecasting satellites. Geostationary weather forecasting satellites are the basis of the weather forecasts we see on television. As discussed in earlier chapters, satellites in a geostationary orbit revolve around Earth in equatorial circular orbits having an altitude of around 36 000 km. These satellites have an orbital period of 24 hours, which is also the rotation rate of the Earth. Hence, these satellites appear to be fixed over a single spot. Weather forecasting satellites in geostationary orbits have fairly coarse resolution when compared to those in the polar orbits. However, this resolution is sufficient for most of the weather forecasting applications. A single weather satellite in the geostationary orbit covers around 40 % of the Earth's surface.

The geostationary weather forecasting satellite system of the USA is named the GOES (geostationary operational environment satellite) system. It operates two satellites simultaneously, one over the west coast, located at 135° W (referred to as GOES West), and the other over the east coast located at 75° W (referred to as GOES East). In addition, other countries like India, Europe and Japan have their own geostationary weather forecasting satellite systems. India has the INSAT series of satellites (74° E), Japan operates the GMS series (140° E) and Europe operates the Meteosat series (0° E). Together, these five satellites located at approximately 70° longitude intervals form a global network of geostationary weather forecasting satellites (Figure 12.6). This global network provides an almost complete coverage of Earth except for the polar region.

12.5 Weather Forecasting Satellite Payloads

Weather forecasting satellites carry instruments that scan Earth to form images. These instruments usually have a small telescope or an antenna, a scanning mechanism, a detector assembly that detects the incoming radiation and a signal processing unit that converts the output of the detectors into the required digital format. The processed output is then transmitted to receiving stations on the ground. The most commonly used instrument on a weather forecasting satellite is the radiometer.

12.5.1 Radiometer

A radiometer is an instrument that makes quantitative measurements of the amount of electromagnetic radiation incident on it from a given area within a specified wavelength band. The most commonly used bands, as mentioned earlier, are the visible, thermal and IR bands.

The radiometer comprises an optical system, a scanning system, an electronic system and a calibration system. The optical system consists of an assembly of lenses and is used for viewing radiation from a small field and focusing that radiation on to the detectors. The scanning system comprising oscillating or rotating mirrors is used for performing the scanning operation. The typical swath width for weather forecasting satellites extends to around 1500 km on either side of the orbit. The incoming radiation is separated into a number of optical beams having different wavelengths using optical filters, and each beam is focused on to a separate detector of the detector array assembly. The electronic system comprises an array of different detectors and a signal processing unit. The detector array located at the focal plane of the optical system is used for sensing the incoming radiation and converting it into an electrical voltage. This electrical voltage is fed to the signal processing unit, where it undergoes amplification, filtering, and so on. It is then converted into the desired digital format for transmission to the control centres on Earth. The radiometers also comprise a calibration system which views on board sources of known temperatures for calibration purposes.

Radiometers can operate in one of two modes, namely the imaging mode and the sounding mode. Radiometers operating in the imaging mode are referred to as imagers and those operating in the sounding mode as sounders. Imagers measure and map sea-surface temperatures, cloud-top temperatures and land-surface temperatures. In this mode, the satellite sensors scan across segments of the Earth's surface and atmosphere, collecting radiance data to produce the satellite image. As an example, the imager on board second generation GOES satellites (Figure 12.7) is a five channel (one visible and four IR channels) imaging radiometer designed to sense reflected solar energy from sampled areas of Earth. The imager comprises

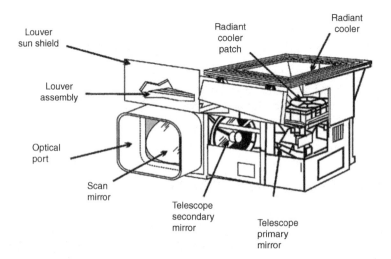

Figure 12.7 Imager onboard second generation GOES satellites [Courtesy: US Department of Commerce, National Oceanic and Atmospheric Administration (NOAA)]

the electronics, power supply and sensor modules. The sensor module contains a telescope, scan assembly and detectors. The electronics module performs command, control and signal processing functions and provides redundant circuitry. The power supply module contains the converters, fuses and power control for interfacing with the spacecraft's electrical power subsystem.

Sounder is a special kind of radiometer, which measures changes in the atmospheric temperature due to change in water vapour content of the atmosphere with height. In this mode, the sensors mainly make vertical soundings of the atmosphere by detecting the thermal radiation emitted from various levels of the atmosphere over a particular point. These upwelling radiance fluxes depend upon the absorption and emission properties of the atmosphere at different heights for a given wavelength band, which in turn depends on the atmospheric temperature and moisture. These radiance fluxes are processed using complex computer algorithms to produce a vertical temperature profile of the atmosphere. As an example, the sounder on board second generation GOES satellites (Figure 12.8) is a 19 channel radiometer covering the spectral range from the visible band up to 15 µm in the longwave IR band. It comprises a sensor module, a detector assembly, an electronics unit and a power supply unit. The incoming radiation is separated into various wavelength bands by passing it through a set of filters. Each of these beams is then passed through separate detectors. It has four sets of detectors operating in visible, shortwave IR, mediumwave IR and longwave IR bands. The outputs of the detectors are fed to the electronics unit, which processes them and produces the desired digital output.

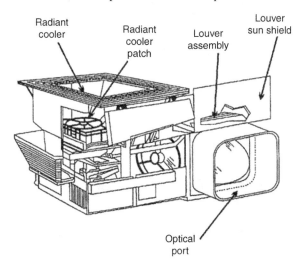

Figure 12.8 Sounder onboard second generation GOES satellites [Courtesy: US Department of Commerce, National Oceanic and Atmospheric Administration (NOAA)]

12.5.2 Active Payloads

Satellites also contain active payloads that emit their own radiation and measure the backscattered portion of this emitted radiation. Two such instruments carried by satellites include the radar and the lidar. Three types of radar are most commonly used on weather satellites. These are altimeters, scatterometers and synthetic aperture radar (SAR). All of them work on the same principle of sending out a pulse of microwave radiation and measuring the return signal as a function of time. They also measure the intensity and the frequency of the return pulse. By knowing the time taken by the beam to return, the distribution of reflecting particles (mostly water droplets and ice crystals) in the atmosphere is determined. The amplitude of the return signal gives information on the kind of particles present in the atmosphere. As an example, larger pieces of ice reflect strongly; hence a strong return signal indicates their presence in the atmosphere. Change in the frequency of the return signal gives information on the wind speed and direction.

12.5.2.1 Altimeter

An altimeter sends a very narrow pulse of microwave radiation with a duration of a few nanoseconds vertically towards Earth (Figure 12.9). The time taken by the reflected signal to reach the satellite determines the distance of the satellite from Earth with an accuracy of the order of few centimetres. This helps in calculating the surface roughness of the land surface, strength of ocean currents, wave heights, wind speeds and other motion over the oceans.

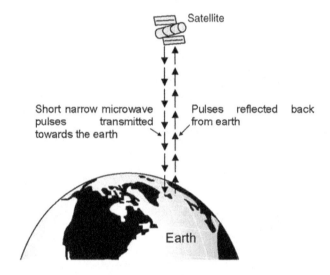

Figure 12.9 Principle of operation of an altimeter

12.5.2.2 Scatterometer

A scatterometer is a microwave radar sensor used to measure the reflection or scattering produced while scanning the surface of the Earth using microwave radiation. It emits a fan-shaped microwave pulse having a duration of the order of a few milliseconds and measures the frequency and the intensity profile of the scattered pulse (Figure 12.10). A rough ocean surface returns a stronger signal because the ocean waves reflect more of the radar energy back towards the scatterometer whereas a smooth ocean surface returns a weaker signal because less energy is reflected back in this case. This helps in determining the direction and size of the

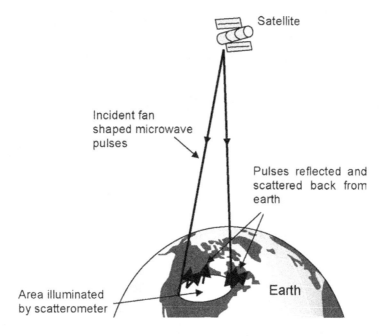

Figure 12.10 Principle of operation of a scatterometer

ocean waves and hence in estimating the wind speed and direction. As an example, SeaWinds scatterometer on board the QuikSCAT satellite is a microwave radar operating at 13.4 GHz and is designed specifically to measure ocean near-surface wind speed and direction.

12.5.2.3 Synthetic Aperture Radar (SAR)

Synthetic aperture radar (SAR) is the most commonly used radar on weather forecasting satellites. SAR is a special type of radar that uses the motion of the spacecraft to emulate a large antenna from a physically small antenna. It works on the same principle as that of a conventional radar. It also sends microwave pulses and measures the intensity, time delay and frequency of the return pulse. The intensity of the return pulse is dependent on the scattering properties of the area being viewed. This in turn depends on the characteristics of the reflecting surface (surface roughness, and so on), the dielectric constant of the surface, the frequency and the angle of incidence of the radar signal. As all other parameters are known, the characteristics of the Earth's surface or the clouds can be determined. Time and frequency information are used for determining the distribution and the motion of the atmospheric particles.

12.5.2.4 Lidar

Lidar has the same principle of operation as that of a radar, except that it sends laser pulses rather than microwave pulses. Lidar sends a beam of laser light through the atmosphere. The particles present in the path of the beam scatter it. A portion of the scattered beam returns to the receiver. The time delay involved between the transmission and reception of the beam as well as the amplitude and the frequency of the return beam is measured by the lidar receiver. An advantage of using laser pulses is that they offer better resolution than their microwave counterparts. Hence lidar can detect even small particles, such as very thin layers of haze. This helps in predicting the regions where clouds will form, even before they are actually formed. Lidar measurements are used to determine the distance to clouds and aerosol layers, to detect meteoric and volcanic debris in the stratosphere and to predict the formation of clouds. The first laser-based system on board a satellite was LITE (laser-in-space technology experiment) used on the Space Shuttle. Both lidar and radar based systems also make use of Doppler-effect-based measurements to determine the intensity of storms by measuring the velocity of wind circulation.

12.6 Image Processing and Analysis

The sensors on board the weather satellites convert the incoming radiation into electrical signals. These electrical signals are further converted into a digital stream of data and then transmitted back to Earth. Typically they are transmitted using UHF band (around 400 MHz) or S band (1600 to 2100 MHz). The information corresponding to the observations forms the major portion of the data transmitted by the satellite to Earth. Other information transferred includes telemetry data for the ground control of the satellite and so on.

At the receiving control centres, the relevant data containing the information is separated from the other data. The data is then processed using various techniques to extract the maximum information from it. For instance, atmospheric sounding measurements are converted into temperature profiles by using the fact that the observed signal is proportional to the fourth-power of temperature. The visible data is converted in terms of reflectivity using the basic relation that the brightness of the image is linearly proportional to the object reflectivity. Sometimes the data collected by two or more satellites is processed together. The information collected by the satellites can also be combined with ground-based observations and information from other platforms, which act as input to the weather forecasting centres.

12.6.1 Image Enhancement Techniques

Images formed on the basis of data sent from the weather forecasting satellites are subjected to various enhancement techniques to make the interpretation of information easier. Longitude and latitude lines are superimposed on the images for better identification of the location of various places on these images. Other techniques employed are colour coding of images. For thermal-IR and passive microwave images, different colours are assigned to various portions of the image, depending on the temperature. All the features having the same temperature are given the same colour in the image. In the case of visible and active microwave images, features having the same reflection characteristics are assigned a single colour. These images are referred to as false colour composite images. These images help in identifying various weather phenomena more precisely. As an example, these images are very helpful in the prediction of hurricanes and extra tropical depressions. Figure 12.11 (a) shows a greyscale image taken in the IR band and its corresponding colour image shown in Figure 12.11 (b).

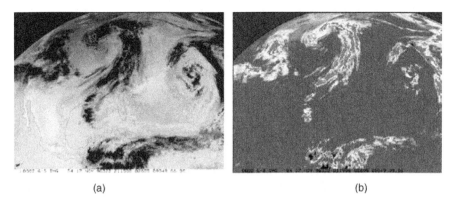

Figure 12.11 (a) Greyscale image taken in the IR band and (b) colour image of the greyscale image shown in part (a) (Reproduced by permission of © John Nielson-Gammon, Texas A&M University). The image shown in Figure 12.11 (b) is the grey scale print image of the original colour image. **The original image is available on the companion website at www.wiley.com/go/maini**

In the enhanced image, hot areas are red, cooler areas are blue and really cold areas are shaded white to black. As can be seen from the figures, the enhancement scheme makes the variations in the intensity of infrared radiation much more prominent. Very cold temperatures, for example, often indicate the tops of tall thunderstorms and the enhanced image makes these cold temperature regions stand out.

Other enhancement processes used to highlight features that are of particular interest include concentrating on a narrow range of temperatures to isolate developments at a certain level in the atmosphere. Another commonly used technique is to string a series of images together to show the movement of clouds and air which helps in easier and better understanding of various weather phenomenon. Moving three-dimensional weather images are also produced. These moving images help in observing weather patterns over an extended period of time.

12.7 Weather Forecasting Satellite Applications

Satellites play a major role in weather forecasting. All the daily weather forecast bulletins which we hear every day are broadcasted on the basis of data sent by weather forecasting satellites. Satellites have helped in predicting the paths of tropical cyclones far more reliably than any other weather forecasting tool. They also help in predicting the frost, rainfall, drought and fog and so on that is of immense help to farmers. Various combinations of satellite images are used to identify clouds and determine their approximate height and thickness. The cloud and water vapour patterns are used to identify cyclones, frontal systems, outflow boundaries, upper level troughs and jet streams. As a matter of fact, not even a single tropical cyclone has gone unnoticed since the use of satellites for weather forecasting. Moreover, these satellites provide early frost warnings, which can save millions of dollars a day for citrus growers. They also play an important role in forest management and fire control. In this section some of the major applications of weather forecasting satellites will be discussed.

12.7.1 Measurement of Cloud Parameters

Satellite imagery enables meteorologists to observe clouds at all levels of the atmosphere, both over land and the oceans. Generally, both visible and IR images are used together for the identification of clouds. Visible images give information on thickness, texture, shape and pattern of the clouds. Information on cloud height is extracted using IR images. False colour IR images are used for a detailed analysis of clouds. Information from visible and IR images can be combined to identify the types of clouds and the weather patterns associated with them. This helps in the prediction of rainfall, thunderstorms and hurricanes. Moreover, information on the movement of clouds is a valuable input in predicting the wind speed and direction.

Figures 12.12 (a) and (b) show a visible image and IR image respectively, taken from the GOES satellite. The map of the area is overlaid on the image to help in locating the places. The clouds marked as A and D in the images appear to be fairly bright in the visible image and are barely seen in the IR image. This indicates that these clouds are low lying warm clouds of medium thickness. The clouds marked as B are very bright in the visible image but they are not seen in the IR image. These clouds again are low lying warm clouds. But, they are thicker than those clouds marked as A and D as they appear to be much brighter in the visible image. Clouds marked as C appear bright both in the visible and IR images and hence they are high-lying thick cold clouds. This information on the types of clouds is further used for estimating rainfall, thunderstorms and hurricanes.

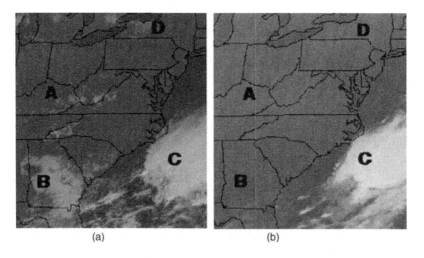

(a)   (b)

Figure 12.12 (a) Visible image taken by GOES satellite used for determining cloud parameters and (b) IR image taken by GOES satellite used for determining cloud parameters (Courtesy: National Oceanic and Atmospheric Administration (NOAA)/Maryland Space Grant Consortium)

12.7.2 Rainfall

Imagery from space is also used to estimate rainfall during thunderstorms and hurricanes. This information forms the basis of flood warnings issued by meteorologists. Satellite images of the clouds are processed and analysed to predict the location and amount of rainfall. As mentioned before, it is possible to determine the cloud thickness and height using visible and IR images respectively. Both these images are combined to predict the amount of rainfall, as it depends both on the thickness and height of clouds. Thick and high clouds result in more rain. Moreover, clouds in their early stage of development produce more rain. Therefore, regular observations from GEO satellites, which can track their development, are used for rainfall prediction. Measurements in the microwave band help in determining the intensity of rain as scattering depends on the number of droplets in a unit volume and their size distribution. As an example, during Hurricane Diana, using images from a GOES satellite, it was calculated that there would be nearly 20 inches of rainfall over North Carolina in a two-day period. The actual recorded rainfall was 18 inches.

Figure 12.13 shows the distribution of rain intensity in different regions of Hurricane Charley on the basis of images taken by the TRMM satellite on 10 August 2004.

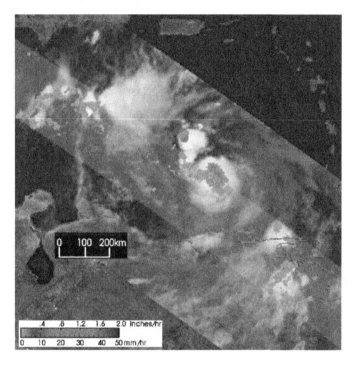

Figure 12.13 Derivation of rainfall rates on the basis of images taken by the TRMM satellite (Courtesy: NASA). The image is the greyscale version of the original colour image. **Original image is available on the companion website at www.wiley.com/go/maini**

12.7.3 Wind Speed and Direction

Determination of wind speed and direction is essential to provide an accurate picture of the current state of the atmosphere. Wind information can be determined by tracking cloud displacements in successive IR and visible images taken from geostationary weather forecasting satellites. However, these measurements can only be taken when the cloud cover is present. To overcome this, successive water vapour channel images are used to track the movement of wind fields. However, both of these methods are not accurate. A more accurate method is to make simultaneous measurements of both the temperature profile as well as the position of the cloud tops. The VISSR atmospheric sounder (VAS) instrument on the GOES satellite is used to perform such measurements.

12.7.4 Ground-level Temperature Measurements

Satellite data cannot produce detailed information about the temperature profile of the lowest few hundred metres of the atmosphere, but it can provide some physically important observations. Infrared radiometers can make widespread observations of maximum and minimum temperatures. High resolution IR satellite imagery is used to produce heat maps of Earth. However, where standard ground-based measurements are available, satellite measurements are generally not used. They are used at those places where ground-based measurements are not feasible. However, in some conditions satellite measurements of the ground-level temperature are more accurate than the ground-based measurements. For example, when it is exceptionally cold and the radiative contribution of the atmosphere is minimal, satellite observations can provide considerably more information than ground-based measurements.

12.7.5 Air Pollution and Haze

Air pollution and haze are recognizable in visible imagery by their grey appearance. Satellite images have shown that the pollution level is low in the morning and increases as the day passes by. Satellite data is also used to infer the effect of air pollution on weather. Using satellite data, it has been found that haze bands may act as boundaries along which thunderstorm activities can develop. Satellite measurements have indicated that the increase in air pollution leads to an increase in the amount of rainfall.

12.7.6 Fog

Fog is detected using visible satellite imagery. Fog appears as a flat textured object with sharp edges in these images. The level of brightness of the image is a measure of the thickness of the fog. Satellite images also provide information on the clearance of fog during the day.

12.7.7 Oceanography

Weather forecasting satellites are a useful tool for oceanography applications. Satellite images are used to map locations of different ocean currents and to measure ocean surface temperatures accurately. Polar orbiting satellites compute around 20 000 to 40 000 global ocean temperature measurements daily. This information on the ocean surface temperature is utilized by meteorologists to observe ocean circulation, to locate major ocean currents and to monitor its effect on climate and weather changes. Moreover, observation of these temperatures before and after the occurrence of hurricanes helps to show the way in which these hurricanes pick up energy from oceans. This helps to predict their behaviour and to improve forecasts of their motion. Satellite observations have shown that hurricanes result in cooling of the ocean surface. The stronger the hurricane, the more cooling effect it has on the temperature of the ocean surface.

Satellite IR imagery is used to detect ocean thermal fronts in the surface layer of the oceans. Satellites also measure the surface roughness of the oceans using microwave measurements, which helps in determining wind speed and direction. As an example, satellite imagery provides timely information about the occurrence of El Niño. El Niño is a cyclic weather phenomenon that results in widespread warming of water off the west coast of South America. It results in reversal of weather patterns and has dramatic effects throughout the Pacific region. Figure 12.14 is a false colour enhanced image that shows the sea surface temperatures in the eastern Pacific region. The red areas indicate increased surface water temperatures, which is an indication of the presence of the El Niño effect.

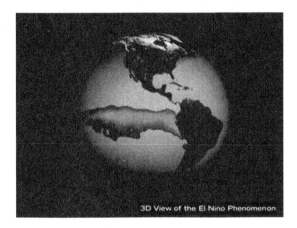

Figure 12.14 False colour enhanced image showing the sea surface temperatures in the eastern Pacific region highlighting the El Niño effect (Courtesy: NASA). The image is the greyscale version of the original colour image. **Original image is available on the companion website at www.wiley.com/go/maini**

12.7.8 Severe Storm Support

One of the most important applications of weather forecasting satellites is in the prediction of hurricanes, tropical storms, cyclones and so on. Satellites are crucial to detecting and tracking intense storms through their various stages of development. This allows meteorologists to issue advanced warnings before the storms actually hit. These advanced warnings have saved lives of millions of people.

The development of these storms is analysed by studying the cloud patterns and by determining how they change with time. Repeated images provide information on the rate of growth or decay of the storm. Hurricanes are predicted and monitored by observing their centre core, which is a low pressure area with little winds, clear skies and no rainfall. It is also referred to as the 'eye' of the hurricane. The shape, spiralling and intensity of this core give information on the development stage of the hurricane. Hurricanes also have a circular high speed wind pattern around the eye. By observing the wind speed, the intensity of the hurricane can be predicted. The direction of motion of the hurricane is known by observing the movement of the eye of the hurricane. Typhoons, cyclones and thunderstorms are also analysed on a similar basis. Figure 12.15 shows the image of the hurricane Katrina taken by the GOES satellite on

Figure 12.15 Image of hurricane Katrina taken by GOES satellite [Courtesy: US Department of Commerce, National Oceanic and Atmospheric Administration (NOAA)]. The image is the greyscale version of the original colour image. **Original image is available on the companion website at www.wiley.com/go/maini**

29 August 2005. By observing such repeated images, meteorologists were able to determine the strength and movement of the hurricane.

12.7.9 Fisheries

Commercial fishery operations have also benefited from data supplied by weather satellites. Information on ocean currents and sea temperatures help in finding the location of tuna or salmon fishes. It also assists in tracking the movement of fish eggs and larvae. Satellite data can be used to study hypoxia, a condition of severe lack of oxygen at deep sea levels that can completely block the growth and development of sea life.

12.7.10 Snow and Ice Studies

Weather satellites are used to observe snow cover on land surfaces and to monitor ice on lakes, rivers and other water bodies. These data help meteorologists to estimate the climate of the place and to plan irrigation and flood control methodologies. Snow cover estimates are especially helpful in mountain regions where a large part of the water supply comes from melting of snow. It is also used to issue winter storm warnings. Satellite ice monitoring provides useful information to the shipping industry. Information on the progression of freezing seasonal temperatures allows farmers to take timely measures to protect their crops.

Both visible and IR images are used in the identification of ice. Both appear in light shades of grey in visible imagery. Fresh snow resembles cloud cover in these images. They are distinguished by examining a series of images. Clouds are in motion while the snow appears to be fixed. Moreover, overlaying the map on the images helps to distinguish snow and ice from clouds. Figure 12.16 shows an image of Lake Tahoe, USA, taken by the MODIS sensor on the Terra satellite in March 2000. Snow covered areas are shown in white colour. MODIS is a high resolution instrument having the capability to discriminate between snow and clouds.

Figure 12.16 Image of Lake Tahoe, USA taken by MODIS sensor on Terra satellite (Courtesy: NASA). The image is the greyscale version of the original colour image. **Original image is available on the companion website at www.wiley.com/go/maini**

12.8 Major Weather Forecasting Satellite Missions

The weather forecasting satellite family comprises a core structure of five geostationary satellites complemented by a host of polar orbiting satellites. The geostationary satellite system comprises GOES East and GOES West satellites from the USA, INSAT satellites from India, Meteosat satellites from Europe and GMS satellites from Japan. The polar orbiting satellites include the Feng Yun satellites of China, POES (polar operational environmetal satellite) satellites of the USA comprising the Defense Meteorological Satellite Program (DMSP) and the NOAA polar operational environmental satellite (NPOES) Program, and Meteor satellites of Russia. In this section, there will be a discussion of the GOES satellites of the USA and Meteosat satellites of Europe in the category of geostationary weather satellites and the ATN NOAA satellites currently operational under the POES satellite program of the USA in the polar orbiting satellites category.

12.8.1 GOES Satellite System

GOES (geostationary operational environmental satellite) is a weather forecasting satellite system designed by NASA for the National Oceanic and Atmospheric Administration (NOAA) of the USA. GOES satellites form the backbone of the US meteorological department for weather monitoring and forecasting. They provide the meteorological department with frequent, small scale imaging of the Earth's surface and cloud cover and have been used extensively by them for weather monitoring and forecasting for over 20 years. GOES satellites orbit around Earth in geostationary orbits. The GOES program maintains two satellites operating in conjunction to provide observational coverage of 60 % of Earth. One of the GOES satellites is positioned at 75° W longitude (GOES East) and the other is positioned at 135° W longitude (GOES West). Each satellite views almost a third of the Earth's surface: GOES East monitors North and South America and most of the Atlantic ocean, while GOES West looks down at North America and the Pacific ocean basin. The two operate together to send a full-face picture of Earth every 30 minutes, day and night.

The first GOES satellite, GOES-1 (A), was launched back in the year 1975. Since then 14 GOES satellites have been launched, with GOES-14, launched in the year 2009, being the latest one. The first generation of the GOES satellite system consisted of seven satellites from GOES-1 (launched in 1975) to GOES-7 (launched in 1992). Due to their design, these satellites were capable of viewing the Earth for only 10 % of the time. The second generation satellites became operational with the launch of GOES-8 (launched in 1994) and offers numerous technological improvements over the first series. The second generation of GOES satellite system comprises five satellites namely GOES-8 (Figure 12.17), 9, 10, 11 and 12 satellites. They provide near-continuous observation of Earth, allowing more frequent imaging (as often as every 15 minutes). This increase in temporal resolution coupled with improvements in the spatial and radiometric resolution of the sensors provides timely information and improved data quality for forecasting meteorological conditions. Three satellites, GOES-13 (N), GOES-14 (O) and GOES-15(P), of the third-generation GOES satellite series have been launched. The last satellite of the GOES series, GOES-15(P) was launched on 4th March 2010.

Figure 12.17 GOES-8 satellite (Courtesy: NASA)

Currently, one second generation and three third-generation GOES satellites, GOES-12 (M), GOES-13 (N), GOES-14 (O) and GOES-15(P), are operational. GOES-13 and GOES-15, positioned at 60° W and 75° W respectively, are the operational satellites. GOES-14 satellite is currently in storage at 90°W. Figure 12.18 shows the coverage areas of the GOES-13 (N) and GOES-15 (P) satellites, also referred to as GOES East and GOES West satellites respectively.

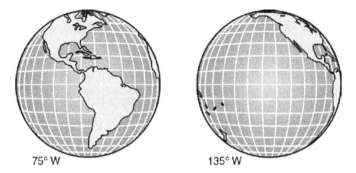

Figure 12.18 Coverage areas of the GOES East and GOES West satellites [Courtesy: US Department of Commerce, National Oceanic and Atmospheric Administration (NOAA)]

Future plans include launch of GOES-R and GOES-S satellites in 2015 and 2017, respectively. The satellites will provide a more accurate location of severe storms and other weather phenomena, resulting in more precise weather forecasts. Table 12.1 enumerates the salient features of the GOES satellites.

Table 12.1 Salient features of GOES satellites

Satellite	Launch date	Position	Stabilization	Payloads
GOES-1	16 October 1975	Directly over the equator	Spin	Visible and infrared spin scan radiometer (VISSR), WEFAX, space environment monitor (SEM) and data collection system (DCS)
GOES-2	6 June 1977	60° W	Spin	Same as GOES-1
GOES-3	16 June 1978	Directly over the equator	Spin	Same as GOES-1
GOES-4	9 September 1980	135° W	Spin	VISSR atmospheric sounder (VAS), SEM, WEFAX and DCS
GOES-5	22 May 1981	75° W	Spin	Same as GOES-4
GOES-6	28 April 1983	136° W	Spin	Same as GOES-4
GOES-7	26 February 1987	75° W	Spin	Same as GOES-4
GOES 8	13 April 1994	75° W	Three-axis	Imager, sounder, WEFAX, SEM and SARSAT
GOES-9	23 May 1995	135° W	Three-axis	Same as GOES-8
GOES-10	25 April 1997	60° W	Three-axis	Same as GOES-8
GOES-11	3 May 2000	135° W	Three-axis	Same as GOES-8
GOES-12	23 July 2001	75° W	Three-axis	Same as GOES-8
GOES-13	25 May 2006	75° W	Three-axis	Imager, sounder, solar X-ray imager, SEM
GOES-14	27 June 2009	135° W	Three-axis	Same as GOES-13
GOES-15	04 March 2010	135° W	Three-axis	Same as GOES-14

12.8.1.1 Payloads on GOES Satellites

First generation: The first generation GOES satellites carry the visible and infrared spin scan radiometer (VISSR), weather facsimile transponders (WEFAX), space environment monitor (SEM), VISSR atmospheric sounder (VAS) and data collection system (DCS).

1. Visible and infrared spin scan radiometer (VISSR). The instrument carried on-board GOES-1, 2 and 3 satellites provided high-quality day/night cloud cover data and made radiance temperature measurements of the Earth/atmosphere system. The VISSR instrument consists of a scanning system, a telescope and infrared and visible sensors.
2. Space environment monitor (SEM). SEM measured the proton, electron and solar X-ray fluxes and magnetic fields.
3. Data Collection System (DCS). DCS relayed processed data from central weather facilities to small APT-equipped regional stations and collected and re-transmitted data from remotely located Earth-based platforms.
4. VISSR Atmospheric Sounder (VAS). GOES-4, 5, 6 and 7 satellites were equipped with an improved VISSR incorporating a VISSR atmospheric sounder (VAS). VAS measured vertical temperature versus altitude cross-sections of the atmosphere. From these cross-sections the altitudes and temperatures of clouds were determined and a three-dimensional picture of their distribution was drawn for more accurate weather prediction.
5. Weather facsimile transponders (WEFAX). They are used to transmit low-resolution imagery sectors as well as conventional weather maps to users with low-cost reception equipment. Images of the GOES satellites and the images received from the polar orbiting satellites are processed in the ground stations and then radioed back up to the GOES satellite for broadcast in graphical form as weather fascimile or WEFAX. WEFAX images are received by ground stations on land as well as on ships.

Second generation: The second generation of GOES satellites has identical payloads, comprising of an imager, sounder, space environment monitor (SEM), weather facsimile transponders (WEFAX) and search and rescue transponders (SARSAT, or search and rescue satellite-aided tracking):

1. Imager. The GOES imager is a multichannel instrument designed to sense radiant and solar-reflected energy from sampled areas of Earth. The imager operates in one visible band of 0.52–0.72 µm, and four IR bands of 3.78–4.03 µm, 6.47–7.02 µm, 10.2–11.2 µm and 11.5–12.5 µm with a resolution of 4 km in the 0.52–0.72 µm, 3.78–4.03 µm and 10.2–11.2 µm bands and of 8 km in the 6.47–7.02 µm and 11.5–12.5 µm bands. The imager of GOES-12 satellite has a 12.9–13.7 µm band, instead of the 11.5–12.5 µm band. The resolution of the 6.47–7.02 µm band for GOES-12 satellite is 4 km instead of 8 km.

    The 0.52–0.72 µm band is used for cloud, pollution, haze and storm detection. The 3.78–4.03 µm band is used for identification of fog at night, discriminating water clouds and snow or ice clouds during daytime, detecting fires and volcanoes and for night-time determination of sea surface temperatures. The 6.47–7.02 µm band is used for estimating regions of mid-level moisture content and for tracking mid-level atmospheric motion. The 10.2–11.2 µm band is used for identifying cloud-drift winds, severe storms and heavy rainfall and the 11.5–12.5 μm band is used for identification of low-level moisture, determination of sea surface temperature and detection of air-borne dust and volcanic ash. For details on the imager construction read the Section 12.5 on payloads.

2. Sounder. The GOES sounder is a 19-channel discrete-filter radiometer, covering the spectral range from the visible channel wavelengths to around 15 μm in the long-IR band. It has one channel in the visible band, six channels in the shortwave IR band, five channels in the mediumwave IR band and seven channels in the longwave IR band. All 19 channels have a spatial resolution of 8 km and 13-bit radiometric resolution. The sounder provides data to determine the atmospheric temperature and moisture profiles, surface and cloud-top temperatures and pressures, and ozone distribution. All this information is extracted from the data using various mathematical analytic techniques. It operates both independently and simultaneously with the imager. The construction details are given in Section 12.5 on payloads.

3. Space environment monitor (SEM). The space environment monitor (SEM) studies the activities of the sun and monitors its effect on the near-Earth environment. It basically measures the condition of the Earth's magnetic field, the solar activity and radiation around the spacecraft and transmits this data to a central processing facility. The SEM is a suite of several instruments including the energetic particle sensor (EPS), X-ray sensor (XRS), solar X-ray imager (SXI), extreme ultraviolet sensor (EUV) and a magnetometer. The EPS includes the energetic proton, electron and alpha detector (EPEAD) and the magnetic electron detector (MAGED). The EPS detects electron and proton radiation trapped by the Earth's magnetic field and the direct solar protons, alpha particles and cosmic rays. The magnetometer measures the three components of the Earth's magnetic field and monitors the variations caused by ionospheric and magnetospheric current flows. The XRS and SXI instruments monitor X-ray activities of the sun and the EUV sensor measures solar ultraviolet radiation.

4. Weather facsimile transponders. They are used to transmit low resolution imagery sectors as well as conventional weather maps to users with low-cost reception equipment. Images of the GOES satellites and the images received from the polar orbiting satellites are processed in the ground stations and then radioed back up to the GOES satellite for broadcast in graphical form as a 'weather facsimile', or WEFAX. WEFAX images are received by ground stations on land as well as on ships.

5. Search and rescue transponders (SARSAT). GOES satellites also carry search and rescue transponders which can relay distress signals at all times. GOES satellites cannot locate these distress signals. Only the low altitude polar orbiting satellites can compute their location. The two satellites work together to create a search and rescue system, allowing a message to be intercepted and relayed by a GOES satellite, even though the polar satellite may be outside the control centre's 'line-of-sight.'

Third generation: Three satellites of the third generation GOES series [GOES 13 (N), GOES 14 (O) and GOES 15(P)] have been launched. These satellites have on board them imager, sounder, solar X-ray imager and space environment monitor (SEM) payloads. Each third generation satellite has one downlink and five uplink channels in the S-band, eight downlink channels in the L-band and one downlink and two uplink channels in the UHF band.

1. Imager. The imager of the third generation GOES satellites is a multi-channel instrument designed to sense radiant and reflected solar energy from sampled areas of the Earth. It operates in five spectral bands namely channel 1 (0.52–0.71 µm), channel 2 (3.73–4.07 µm), channel 3 (13.0–13.7 µm), channel 4 (10.2–11.2 µm) and channel 5 (5.8–7.3 µm). The multi-element spectral channels simultaneously sweep east-west and west-east directions along a north-to-south path by means of a two-axis mirror scan system. The instrument can produce full-Earth disc images, sector images that contain the edges of the Earth and various sizes of area scans completely enclosed within the Earth scene. The instrument is used for cloud cover detection, determination of water vapour and sea surface temperatures, wind determination and detection of fires and smoke.
2. Sounder. The sounder used on-board third generation GOES satellites is a 19-channel discrete-filter radiometer covering the spectral range from the visible channel wavelengths to far IR band up to 15 microns. It is designed to provide data from which atmospheric temperature and moisture profiles, surface and cloud-top temperatures and ozone distribution is deduced through mathematical computation and analysis. It uses a flexible scan system similar to that of the imager and operates in the independent mode as well as simultaneously with the imager. The sounder's multi-element detector array assemblies simultaneously sample four separate fields or atmospheric columns.
3. Solar X-ray imager (SXI). The Solar X-Ray imager (SXI) is essentially a small telescope that is used to monitor the solar conditions and activities. Every minute the SXI captures an image of the sun's atmosphere in X-ray band, providing space weather forecasters with the necessary information in order to determine when to issue forecasts and alerts of conditions that may harm space and ground systems.
4. Space environment monitor (SEM). SEM consists of three instrument groups namely an energetic particle sensor (EPS) package, two magnetometer sensors and a solar X-ray sensor (XRS) and an extreme ultraviolet sensor (EUV). SEM provides real-time data to the Space Environment Center (SEC) in Colorado, USA. SEC receives, monitors and interprets a wide variety of solar terrestrial data and issues reports, alerts, warnings and forecasts for special events such as solar flares and geomagnetic storms. EPS accurately measures the number of particles over a broad energy range, including protons, electrons and alpha particles, and are the basis for operational alerts and warnings of hazardous conditions. It comprises magnetosphere electron detector (MAGED), energetic proton, electron, and alpha detector (EPEAD), magnetosphere proton detector (MAGPD) and high energy proton and alpha detector (HEPAD).

The magnetometer sensors can operate independently and simultaneously to measure the magnitude and direction of the Earth's geomagnetic field, detect variations in the magnetic field near the spacecraft, provide alerts of solar wind shocks or sudden impulses that impact the magnetosphere and assess the level of geomagnetic activity. The second magnetometer sensor serves as a backup in case the first magnetometer sensor fails and provides for better calibration of the magnetometer data channel. XRS is an X-ray telescope that observes and measures solar X-ray emissions in two ranges - one from 0.05 to 0.3 nm and the second from 0.1 to 0.8 nm. The five-channel EUV telescope is new on the third generation GOES satellites. It measures solar extreme ultraviolet energy in five wavelength bands from 10 nm to 126 nm. The EUV sensor provides a direct measure of the solar energy that heats the upper atmosphere and creates the ionosphere.

12.8.2 Meteosat Satellite System

Meteosat satellite network is a European weather forecasting satellite system, currently operated by EUMETSAT (European Organisation for Meteorological Satellites). Meteosat satellites aid the forecasters in swift recognition and prediction of various weather phenomena such as thunderstorms, fog, rain, depressions, wind storms and so on. Meteosat satellites provide improved weather forecasts to Europe, the Middle East and Africa. They also play a vital role in contributing to the global network of weather satellites that continuously monitor the globe.

Meteosat system of satellites became operational in the year 1977, with the launch of Meteosat-1 satellite. The system was maintained and operated by the ESA (European Space Agency). Two generations of Meteosat satellites have been launched to date. The first generation of Meteosat satellites (Figure 12.19) comprise seven satellites, namely Meteosat-1, 2, 3, 4, 5, 6 and 7. All the first generation satellites were developed by ESA. However, the maintenance of these satellites was given to the EUMETSAT in the year 1995. The second generation Meteosat satellites (MSG, or Meteosat Second Generation) are an enhanced follow-on to the first generation satellites. They are jointly developed by the ESA and EUMETSAT. Three satellites have been launched in this series, MSG-1 (Meteosat 8) (Figure 12.20), MSG-2 (Meteosat 9) and MSG-3 (Meteosat 10). Two more satellites in the series are being planned to be launched in the near future.

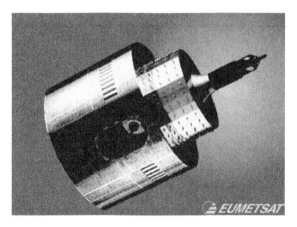

Figure 12.19 First generation of Meteosat satellites (Reproduced by permission of Copyright 2005 © EUMETSAT)

Meteosat satellites are spin-stabilized satellites. They are placed in geostationary orbits at 0° longitude. The first generation Meteosat satellites contained a three-band imaging radiometer operating in the visible, IR and water vapour bands. The second generation Meteosat satellites carry the spinning enhanced visible and infrared imager (SEVIRI) radiometer and the geostationary Earth radiation budget (GERB) payloads. They offer significant advantages over the first generation Meteosat satellites. They provide images every 15 minutes in 12 visible and IR channels as compared to 30 min images in three channels on the first generation Meteosat satellites. Their spatial resolution is also twice as compared to that of the first generation satellites. Table 12.2 lists the salient features of these satellites.

Figure 12.20 MSG-2-1 satellite (Reproduced by permission of © EADS SPACE)

Table 12.2 Salient features of Meteosat satellites

Satellites	Launch	Orbit	Payloads
First generation			
Meteosat-1	23 November 1977	GEO	Three-band imaging radiometer
Meteosat-2	19 June 1981	GEO	Three-band imaging radiometer
Meteosat-3	15 June 1988	GEO	Three-band imaging radiometer
Meteosat-4	6 March 1989	GEO	Three-band imaging radiometer
Meteosat-5	3 March 1991	GEO	Three-band imaging radiometer
Meteosat-6	20 November 1993	GEO	Three-band imaging radiometer
Meteosat-7	2 September 1997	GEO	Three-band imaging radiometer
Second generation			
MSG-1 (Meteosat-8)	28 August 2002	GEO	SEVIRI, GERB
MSG-2 (Meteosat-9)	21 December 2005	GEO	SEVIRI, GERB
MSG-3 (Meteosat-10)	5 July 2012	GEO	SEVIRI, GERB

12.8.2.1 Payloads Onboard Meteosat Satellites

First Generation. An imaging radiometer on board the first generation Meteosat satellites operates in three spectral bands, namely the visible band (0.5–0.9 µm), IR band (10.5–12.5 µm) and water vapour band (5.7–7.1 µm). Resolution for the visible band is 2.5 km, while that for the other two bands is 5 km.

Second Generation. The second generation Meteosat satellites had the SEVIRI and the GERB payloads. SEVIRI is an advanced imaging radiometer operating in 12 channels of four visible/near-IR bands and eight thermal-IR bands. The instrument comprises an optical assembly, a scan assembly, a calibration unit, a detection electronic assembly and a cooler assembly. The optical assembly consists of a telescope that focuses the incoming radiation on to the detectors. The scan assembly provides continuous bidirectional scanning of Earth.

The detector assembly comprises 12 sets of detector arrays at the focal plane of the optical assembly. Silicon detectors are used for sensing visible radiation, InGaAs detectors for near-IR radiation and HgCdTe detectors for thermal-IR radiation. The detector output is converted into an electrical signal by means of a preamplifier and a main detection unit. The electrical signal is sampled, digitized and transmitted back to the control centres on Earth. It provides images every 15 minutes. Figure 12.21 shows the anatomy of the instrument.

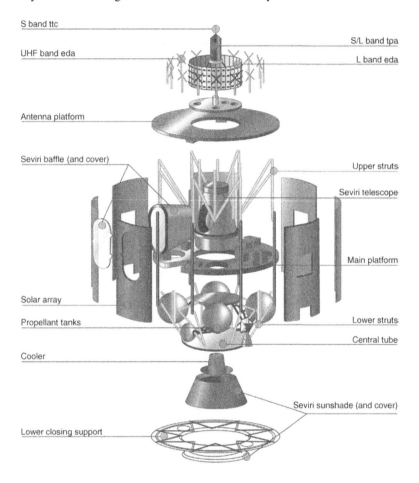

Figure 12.21 SEVIRI payload onboard second generation Meteosat satellites (Reproduced by permission of Copyright 2005 © EUMETSAT)

GERB instrument (Figure 12.22) monitors the Earth's radiation budget at the top of the atmosphere. It is a scanning radiometer operating in two broadband channels, one covering the solar spectrum (0.3–4.0 µm) and the other covering the mid-IR and long-IR spectrum (4.0–30 µm). It has an accuracy level of 1% for IR channels and 0.5 % for solar channels. Measurements in these bands allow calculations of the shortwave and longwave radiation, which is essential for understanding Earth's climate. It comprises a telescope assembly, scanning mechanism, linear detector array, quartz filters, a calibration unit and a signal processing unit.

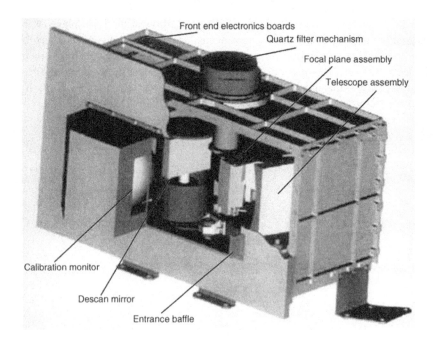

Figure 12.22 GERB payload onboard second generation Meteosat satellites (Reproduced by permission of © CCLRC)

12.8.3 Advanced TIROS-N (ATN) NOAA Satellites

The ATN NOAA series of satellites mark the fourth generation of polar weather forecasting satellites in the Polar Operational Environmental Satellite (POES) program of USA. The first satellite in this series was NOAA-8, launched on 23 March 1983. A total of 12 satellites have been launched in this series since then, namely NOAA-8 (E), 9 (F), 10 (G), 11 (H), 12 (D), 13 (I), 14 (J), 15 (K), 16 (L), 17 (M) (Figure 12.23), 18 (N) and 19 (N′). Currently, six ATN satellites, namely NOAA-13, 15, 16, 17, 18 and 19 are operational.

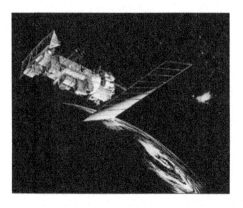

Figure 12.23 NOAA-17 satellite (Courtesy: NOAA and NASA)

These satellites have polar, sun-synchronous low Earth orbits. Two satellites operate simultaneously at all times. One satellite is placed at an altitude of 833 km in a morning orbit (crossing the equator at 7:30 a.m. local time) and the other at 870 km in an afternoon orbit (crossing the equator at 1:40 p.m. local time). These satellites observe the polar areas and send more than 16 000 measurements of atmospheric temperature and humidity, surface temperature, cloud cover, water–ice–moisture boundaries, space proton and electron fluxes on a daily basis. They also have the capability of receiving, processing and retransmitting data from search and rescue beacon transmitters, free-floating balloons, buoys and globally distributed remote automatic observation stations. Table 12.3 enumerates the salient features of the ATN NOAA satellites.

Payloads on board ATN NOAA satellites include the advanced very high resolution radiometer (AVHRR), AVHRR/2, AVHRR/3, advanced microwave sounding unit A (AMSU-A), advanced microwave sounding unit B (AMSU-B), HIRS/2, HIRS/3, space environment monitor (SEM), high resolution picture transmission (HRPT), automatic picture transmission (APT), direct sounder broadcast (DSB), Earth radiation budget experiment (ERBE), solar backscatter ultraviolet radiometer (SBUV/2), TIROS operational vertical sounder (TOVS), microwave sounder unit (MSU), stratospheric sounding unit (SSU), search and rescue

Table 12.3 Salient features of Advanced TIROS-N NOAA satellites

Satellite	Launch date	Orbit	Payloads
NOAA-8	23 March 1983	785 × 800 km × 99°	AVHRR, TOVS, SEM, SARSAT, MSU, SSU, HIRS/2, HRPT, DCS, APT, DSB
NOAA-9	12 December 1984	833 × 855 km × 99°	AVHRR/2, HIRS/2, MSU, DCS, SARSAT, ERBE, SBUV/2
NOAA-10	17 September 1986	795 × 816 km × 99°	AVHRR, HIRS/2, SSU, MSU, DCS, SARSAT, ERBE, SEM
NOAA-11	24 September 1988	838 × 854 km × 99°	AVHRR/2, HIRS/2, SSU, MSU, DCS, SARSAT, SBUV/2
NOAA-12	14 May 1991	804 km altitude, 98.7°	AVHRR, HIRS, MSU, SEM
NOAA-13	9 August 1993	845 × 861 km × 99°	AVHRR/2, HIRS/2, SSU, MSU, SARSAT, SBUV/2, SEM
NOAA-14	30 December 1994	845 × 861 km × 99°	AVHRR/2, HIRS/2, MSU, DCS, SARSAT
NOAA-15	13 May 1998	847 × 861 km × 99°	AMSU/A, AMSU/B, AVHRR/3, HIRS/3, OCI, SARSAT, APT, HRPT, DSB, SEM, SBUV/2, ARGOS, DCS
NOAA-16	21 September 2000	807 × 824 km × 99°	AMSU/A, AMSU/B, AVHRR/3, HIRS/3, SBUV/2, OCI, SARSAT, DCS, ARGOS, APT, HRPT, DSB, SEM
NOAA-17	4 June 2002	853 × 867 km × 99°	AMSU/A, AMSU/B, AVHRR/3, HIRS/3, SBUV/2, OCI, SARSAT, DCS, ARGOS, APT, HRPT, DSB, SEM
NOAA-18	20 May 2005	853 × 872 km × 98.9°	AVHRR/3, HIRS/3, AMSU/A, MHS, SBUV/2, SEM/2, DCS/2, SARR, SARP, DDR
NOAA-19	6 Feb 2009	870 km altitude, 98.73°	AMSU/A, MHS, AVHRR/3, HIRS/4, SBUV/2, SEM, A-DCS, SARSAT, SARR, SARP

transponders (SARSAT) processor data collection system (DCS/2), microwave humidity sounder (MHS), search and rescue repeater and sounder (SARR and SARP) and digital data recoder (DDR).

12.8.3.1 Important Payloads Onboard ATN NOAA Satellites

1. Advanced very high resolution radiometer (AVHRR). AVHRR is an imager used for determining cloud cover and surface temperature of Earth, clouds and water bodies. The first model of the AVHRR operated on five channels (0.58–0.68 μm, 0.725–1.10 μm, 3.44–3.93 μm, 10.3–11.3 μm and 11.5–12.5 μm). It was carried on board NOAA-8 and NOAA-10 satellites. The second version of AVHRR, referred to as AVHRR/2, is a five channel instrument operating at 0.58–0.68 μm, 0.725–1.10 μm, 3.55–3.93 μm, 10.3--11.3 μm and 11.5–12.4 μm bands. It was carried on board the NOAA-9, 11, 13 and 14 satellites. The latest version of the instrument is AVHRR/3, operating on six channels (0.58–0.68 μm, 0.72–1.00 μm, 1.58–1.64 μm, 3.55–3.93 μm, 10.3–11.3 μm and 11.5–12.5 μm bands). NOAA-15, 16, 17, 18 and 19 satellites carried the AVHRR/3 payload.

   AVHRR/3 is a cross-track scanning radiometer used for detailed analysis of hydroscopic, oceanographic and meteorological parameters. It comprises five modules, namely the optical module, scanning assembly, detector module, radiant cooler module and the processing electronics unit. The optical assembly consists of a reflective Cassegrain- type telescope and various optical filters for splitting the incoming radiation into six optical bands. Scanning is achieved by using an assembly of rotating mirrors. The detector assembly comprises of different detectors for sensing different wavelengths. It has silicon detectors for the visible band, InGaAs detectors for the near-IR band, InSb detectors for the mid-IR band and the HgCdTe detector for the longwave-IR bands. The output of the detector assembly is amplified, sampled and digitized in the processing unit.

2. Advanced microwave sounding unit A (AMSU/A). AMSU/A is a multichannel microwave radiometer that is used for measuring global atmospheric temperature profiles and atmospheric water in all its forms. It is a cross-track, line-scanned instrument designed to measure radiance in 15 discrete frequency channels. The operating frequency bands are 23.800 GHz, 31.400 GHz, 50.300 GHz, 52.800 GHz, 53.596 GHz ±115 MHz, 54.400 GHz, 54.940 GHz, 55.500 GHz, 57.290 344 GHz (f_0), $f_0 \pm 217$ MHz, $f_0 \pm 322.2$ MHz ± 48 MHz, $f_0 \pm 322.2$ MHz ± 22 MHz, $f_0 \pm 322.2$ MHz ± 10 MHz, $f_0 \pm 322.2$ MHz ± 4.5 MHz and 89.000 GHz. It was carried on NOAA-15, 16, 17, 18 and 19 satellites. It comprises four major subsystems, namely the antenna/drive/calibration, receiver, signal processor and structural/thermal subsystems. It is configured as a combination of two units, namely AMSU/A1 and AMSU/A2. AMSU/A1 module provides a complete and accurate vertical temperature profile of the atmosphere from the Earth's surface to a height of approximately 45 km. AMSU/A2 module is used to study atmospheric water in all its forms, with the exception of small ice particles.

3. Advanced microwave sounding unit B (AMSU/B). AMSU/B is a five channel cross-track line-scanned microwave radiometer used to receive and measure radiation from a number of different layers of the atmosphere in order to obtain global data on humidity profiles. It works in conjunction with the AMSU/A instrument to provide a 20 channel microwave radiometer. It was carried on board the NOAA-15, 16 and 17 satellites.

   The AMSU/B instrument consists of the following subsystems, namely the parabolic reflector antenna, the quasi-optical front assembly and a receiver assembly. The parabolic reflector scans the Earth for the incoming radiation. The incoming radiation is separated into various frequency bands in the optical front assembly. The receiver assembly amplifies, samples and digitizes the signals. The digitized receiver output is sent to control centres on Earth.

4. High resolution infrared sounder (HIRS). HIRS is a discrete stepping line scan radiometer used mainly for calculating the vertical temperature profile from the Earth's surface to a height of around 40 km. The first HIRS instrument was developed and flown in 1975 on the Nimbus-6 satellite. The improved version of the instrument, HIRS/2, was flown on the TIROS-N, NOAA-8, 9, 10, 11, 12, 13 and 14 satellites. Additional improvements and operational changes resulted in the design of HIRS/3. NOAA-15, 16, 17 and 18 satellites carried the HIRS/3 payload. The HIRS/4 design is a modification of the HIRS/3 design, and is carried onboard the NOAA-19 satellite.

  HIRS/3 works in 20 spectral bands including one visible channel (0.69 μm), seven shortwave-IR channels (3.7–4.6 μm) and 12 longwave-IR channels (6.5–15 μm). It calculates the vertical temperature profile of the atmosphere. The front end optical assembly comprising a telescope and a rotating filter wheel separates the incoming radiation into 20 channels. The receiver assembly comprises a detector assembly, processing electronics and command and telemetry units. The detector assembly uses a silicon photodiode for the visible energy, an indium antimonide detector to sense shortwave-IR energy and a mercury cadmium telluride detector for longwave-IR energy. The detector output is a weak signal that is amplified using low noise amplifiers. The amplified signal is then sampled, multiplexed and digitized and sent to the ground station. HIRS/4 is a 20 channel scanning radiometric sounder. It operates in one visible channel (0.69 μm), seven shortwave-IR channels (3.7–4.6 μm) and 12 longwave-IR channels (6.7–15 μm).

5. Space energy monitor (SEM). The SEM is a multichannel spectrometer that senses flux of charged particles over a broad range of energies. This helps in understanding the solar–terrestrial environment. It was carried on NOAA-8, 10, 12, 13, 15, 16, 17 and 19 satellites. SEM/2 is an improved version of the SEM, which will be carried on the new satellites to be launched. It measures the flux over a broader range of energies. SEM/2 consists of two detectors, namely the total energy detector (TED) and the medium energy proton and electron detector (MEPED).

6. Solar backscatter ultraviolet radiometer (SBUV). This is a nadir-pointing, non-spatial, spectrally scanning, ultraviolet radiometer comprising of a sensor module and an electronics module. The instrument measures solar irradiance and Earth radiance (backscattered solar energy) in the near ultraviolet spectrum.

7. Search and rescue satellite-aided tracking system (COPAS-SARSAT). This transmits the location of emergency beacons from ships, aircraft and people in distress around the world to the ground stations.

8. Microwave humidity sounder (MHS). This is a five-channel microwave instrument intended primarily to measure profiles of atmospheric humidity. Additionally, it measures the liquid water content of clouds and provides qualitative estimates of the precipitation rate.

9. Search and rescue repeater (SARR). SARR is used for receiving and re-broadcasting the 406 MHz signals to a ground station where they can be detected and located by measuring their Doppler shift.

10. Search and rescue processor (SARP). This provides stored data interleaved with the real-time data sent through the downlink transmitter.

12.9 Future of Weather Forecasting Satellite Systems

Future weather forecasting satellites will carry advanced payloads including multispectral imagers, sounders and scatterometers with better resolution. Hyperspectral measurements from newly developed interferometers will be possible in the near future. These instruments will have more than thousand channels over a wide spectral range. Also, the satellite data download rates are expected to exceed several terabytes per day. Therefore, the information content will vastly exceed that of the current measuring devices. Emerging new technologies including the use of rapidly developing visualization tools will be employed. All of these technological advancements will help in unlocking the still unresolved mysteries towards improving our understanding and prediction of atmospheric circulation systems such as tropical cyclones. In addition, there will be an increase in integrated use of satellite data and conventional meteorological observations for synoptic analysis and conventional forecast to extract critical weather information.

Further Readings

Bromberg, J.L. (1999) *NASA and the space industry*, John Hopkins University Press, Baltimore, Maryland.

Burroughs, W.J. (1991) *Watching the World's Weather*, Cambridge University Press.

Carr, M. (1999) *International Marine's Weather Predicting Simplified: How to Read Weather Charts and Satellite Images*, International Marine/Ragged Mountain Press.

Ellingson, R.G. (1997) *Satellite Data Applications: Weather and Climate (Satellite Data Appplications)*, Elsevier Science, Oxford.

Gatland, K. (1990) *Illustrated Encyclopedia of Space Technology*, Crown, New York.

Gurney, R.J., Foster, J.L. and Parkinson, C.L. (1993) *Atlas of Satellite Observations Related to Global Change*, Cambridge University Press.

Hiroyuki, F. (2001) *Sensor Systems and Next Generation Satellites IV*, SPIE – The International Society for Optical Engineering, Bellingham, Washington.

Hodgson, M. (1999) *Basic Essentials: Weather Forecasting*, The Globe Pequot Press.

Sanchez, J. and Canton, M.P. (1999) *Space Image Processing*, CRC Press, Boca Raton, Florida.

Santurette, P. and Georgiev, C. (2005) *Weather Analysis and Forecasting: Applying Satellite Water Vapour Imagery and Potential Vorticity Analysis*, Academic Press.

Vasquez, T. (2002) *Weather Forecasting Handbook*, Weather Graphics Technologies, Texas.

Verger, F., Sourbes-Verger, I., Ghirardi, R., Pasco, X., Lyle, S. and Reilly, P. (2003) *The Cambridge Encyclopedia of Space*, Cambridge University Press.

Internet Sites

1. www.astonautix.com
2. http://en.wikipedia.org/wiki/Weather_satellite
3. http://www.met.tamu.edu/class/ATM0203/tut/sat/satmain.html
4. www.skyrocket.de
5. http://www.science.edu.sg/ssc/detailed.jsp?artid=3673&type=4&root=140&parent=140 &cat=239

Glossary

Altimeter: An altimeter is a type of radar that sends a very narrow pulse of microwave radiation of duration of a few nanoseconds vertically towards Earth. The time taken by the reflected signal to reach the satellite helps in determining the distance of the satellite from Earth with an accuracy of a few centimetres

False colour composite images: False colour composite images are colour enhanced IR images where all the features having the same temperature or same reflectivity are assigned a particular colour

Geostationary orbit: This is a circular equatorial orbit with an altitude of appoximately 36 000 km. Satellites in this orbit remain stationary with respect to Earth

GOES: The GOES (geostationary operational environmental satellite) system is a geostationary weather satellite system of the USA

Imager: An imager is an instrument that measures and maps sea-surface temperatures, cloud-top temperatures and land temperatures

IR images: Infrared images measure radiation emitted by the atmosphere and Earth in the IR band. They provide information on the temperature of the underlying surface or the cloud

Lidar: Lidar is an active sensor that emits laser pulses and measures the time of return of the scattered beam

Microwave image: Microwave images are taken in the wavelength region of 0.1 to 10 cm

Polar sun-synchronous orbit: Polar sun-synchronous orbits are near polar low Earth orbits in which the satellite visits a particular place at a fixed time in order to observe that place under similar solar conditions

Radar: Radar is an active microwave instrument that works on the principle of sending out a pulse of microwave radiation and measuring the return signal as a function of time

Radiometer: The Radiometer is an instrument that makes quantitative measurements of the amount of electromagnetic radiation incident on a given area within a specified wavelength band

Synthetic aperture radar (SAR): SAR is a special type of radar that uses the motion of the spacecraft to emulate a large antenna from a physically small antenna

Scatterometer: A scatterometer is a microwave radar sensor used to measure the reflection or scattering effect produced while scanning the surface of the Earth. They emit a fan-shaped radar pulse of the duration of the order of a few milliseconds and measure the frequency and intensity profile of the scattered pulse

Sounder: The sounder is a special kind of radiometer that measures changes of atmospheric temperature with height, and also changes in the water vapour content of the air at various levels

Visible images: Visible images are formed by measuring the reflected or scattered sunlight in the visible wavelength band (0.4–0.9 μm). The intensity of the image depends on the reflectivity (referred to as albedo) of the underlying surface or clouds

Water vapour image: Water vapour images are constructed using the IR wavelength around 6.5 μm that is absorbed by water vapour in the atmosphere. These images detect invisible water vapour in the air, primarily from around 10 000 feet up to 40 000 feet. Hence, the level of brightness of the image taken in this band indicates the presence or absence of moisture

13

Navigation Satellites

Navigation is the art of determining the position of a platform or an object at any specified time. Satellite-based navigation systems represent a breakthrough in this field that has revolutionized the very concept and application potential of navigation. These systems have grown from a relatively humble beginning as a support technology to that of a critical player used in a vast array of economic, scientific, civilian and military applications. Two main satellite-based navigation systems in operation today are the Global Positioning System (GPS) of the USA and the Global Navigation Satellite System (GLONASS) of Russia. The GPS navigation system employs a constellation of 24 satellites and ground support facilities to provide the three-dimensional position, velocity and timing information to all the users worldwide 24 hours a day. The GLONASS system comprises 21 active satellites and provides continuous global services like the GPS. These navigation systems are used in various domains such as surveying and navigation, vehicle tracking, automatic machine guidance and control, geographical surveying and mapping and so on. The chapter gives a brief outline on the development of satellite-based navigation systems and a descriptive view of the fundamentals underlying the operation of the GPS and GLONASS navigation systems and the future trends in satellite based navigation systems.

13.1 Development of Satellite Navigation Systems

Various navigation methods have been used over the ages including marking of trails using stones and twigs, making maps, making use of celestial bodies (sun, moon and stars), monumental landmarks and using instruments like the magnetic compass, sextant, and so on. These traditional methods were superceded by ground-based radio navigation techniques in the early 20th century. These ground-based systems were widely used during World War II. These systems, however, could only provide accurate positioning services in small coverage areas. Accuracy reduced with an increase in the coverage area. Satellite-based navigation systems were developed to provide accurate as well as global navigation services simultaneously. These systems emerged on the scene in the early 1960s. They provided an accurate universal reference system that extends everywhere over land as well as sea and in near space regardless of weather

---

Satellite Technology: Principles and Applications, Third Edition. Anil K. Maini and Varsha Agrawal.
© 2014 John Wiley & Sons, Ltd. Published 2014 by John Wiley & Sons, Ltd. Companion Website: www.wiley.com/go/maini3

conditions. These systems were originally developed for military operations, but their use for civilian applications soon became commonplace.

Initially, the systems developed were based on the 'Doppler effect'. Later 'trileration'-based systems came into the picture. In this section, the various development stages of both these systems will be discussed, with more emphasis on the 'trileration'-based systems, as the contemporary satellite navigation systems use this technique. Moreover, the focus of this chapter is also on the 'trileration'-based systems.

13.1.1 Doppler Effect based Satellite Navigation Systems

The first satellite navigation system was the Transit system developed by the US Navy and John Hopkins University of the USA back in the early 1960s. The first satellite in the system, Transit I, was launched on 13 April 1960. It was also the first satellite to be launched for navigation applications. The system was available for military use in the year 1964 and to civilians three years later in 1967. The system employed six satellites (three active satellites and three in-orbit spares) in circular polar LEOs at altitudes of approximately 1000 km. The last Transit satellite was launched in the year 1988. The main limitations of the system were that it provided only two-dimensional services and was available to users for only brief time periods due to low satellite altitudes. Moreover, high speed receivers were not able to use the system. The system was terminated in the year 1996. The Transit system was followed by the Nova navigation system, which was an improved system having better accuracy.

Russians launched their first navigation satellite, Kosmos-158, in the year 1967, seven years after the first American navigation satellite launch. The satellite formed a part of the Tsyklon system. It provided services similar to that provided by the Transit system. It was operational until 1978. The system was superceded by the Parus and the Tsikada systems. Parus is a military system comprising satellites in six orbital planes spaced at 30° longitude intervals, thus having an angular coverage of 180°. Ninety-eight satellites have been launched in the Parus system with the last satellite Parus 98 launched on 21 July 2009. The system is operational and is mainly used for data relay and store-dump communication applications. Tsikada is a civilian system covering the rest of the 180°. It comprised satellites in four orbital planes at 45° intervals. Twenty Tsikada satellites have been launched with the last satellite launched on 21 January 1995.

All the systems discussed above are based on the 'Doppler effect'. The satellite transmits microwave signals containing information on its path and timing. The pattern of the Doppler shift of this signal transmitted by the satellite is measured as it passes over the receiver (Figure 13.1). The Doppler pattern coupled with the information on the satellite orbit and timing establishes the location of the receiver station precisely. One satellite signal is sufficient for determining the receiver locations. However, the systems mainly transmit two frequencies as it improves the accuracy of the system. As an example, the Transit system transmitted signals at 150 MHz and 400 MHz frequencies. The positioning accuracy was around 500 m for single frequency users and 25 m for dual frequency users.

13.1.2 Trilateration-based Satellite Navigation Systems

Doppler-based navigation systems have given way to systems based on the principle of 'trileration', as they offer global coverage and have better accuracy as compared to the

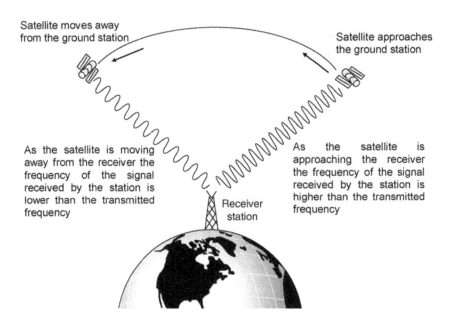

Figure 13.1 Principle of operation of Doppler effect based satellite navigation systems

Doppler-based systems. In this case, the user receiver's position is determined by calculating its distance from three (or four) satellites whose orbital and the timing parameters are known. The receiver is at the intersection of the invisible spheres, with the radius of each sphere equal to the distance between a particular satellite and the receiver, with the centre being the position of that satellite (Figure 13.2). Two such systems are in operation today, namely the Global Positioning System (GPS) of the USA and the Global Navigation Satellite System (GLONASS) of Russia. Another trilateration-based navigation system is the European system named Galileo. It is currently in the development phase. The first test satellite of the constellation was launched on

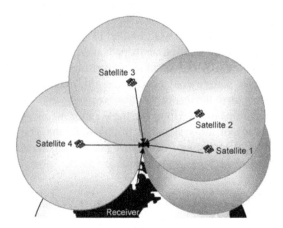

Figure 13.2 Principle of operation of trileration-based satellite navigation systems

28 December 2005, while the second test satellite was launched on 26 April 2008. The third test satellite will be launched in the near future. These satellites will characterize the critical technologies of the system.

13.1.2.1 Development of the Global Positioning System (GPS)

The first effort in this area began in the year 1972, with the launch of Timation satellites. These satellites provided time and frequency transfer services. Three satellites were launched in this series. The third satellite acted as a technology demonstrator for the Global Positioning System (GPS) Program, also known as the Navigation Satellite Timing and Ranging (NAVSTAR) Program. The GPS was the first operational navigation system that provided continuous positioning and timing information anywhere in the world. It was developed by the US Department of Defense (DoD) for use in military operations. It is now a dual-use system, used for both military as well as civilian applications. The GPS receivers calculate their location on the basis of ranging, timing and position information transmitted by GPS satellites [the GPS satellites transmit information at two frequencies, 1575.42 MHz (L1) and 1227.6 MHz (L2)].

The first GPS satellite was launched on 22 February 1978. It marked the beginning of first generation GPS satellites, referred to as Block-I satellites (Figure 13.3). Eleven satellites were launched in this block and were mainly used for experimental purposes. These satellites were out of service by the year 1995. The second generation of GPS satellites (Figure 13.4) comprised Block-II and Block-IIA satellites. Block-IIA satellites were advanced versions of Block-II satellites. A total of 28 Block-II and Block-IIA satellites (nine satellites in the Block-II series and 19 satellites in the Block-IIA series) were launched over the span of eight years, from 1989 to 1997. The GPS system was declared fully functional on 17 July 1995, ensuring the availability of at least 24 operational, non-experimental GPS satellites.

Currently, third generation GPS satellites, referred to as Block-IIR (Figure 13.5) satellites, are being launched. The first satellite in this series was launched in the year 1997. Twelve

Figure 13.3 Block-I GPS satellite (Courtesy: NASA)

Figure 13.4 Block-II and -IIA GPS satellites (Courtesy: NASA)

Figure 13.5 Block-IIR GPS satellites (Courtesy: Lockheed Martin Corporation)

Block-IIR satellites had been launched by August 2013. One of the potential advantages of Block-IIR satellites over the Block-II and -IIA satellites is that they have reprogrammable satellite processors enabling upgradation of satellites while in orbit. These satellites can calculate their own positions using intersatellite ranging techniques. Moreover, they have more stable and accurate clocks on board as compared to the Block-II and Block-IIA satellites. Block-IIR satellites have three Rubidium atomic clocks (having an accuracy of 1 second in 300 000 years), whereas Block-II and Block-IIA satellites have two Caesium atomic clocks (having an accuracy of 1 second in 160 000 years) and two Rubidium atomic clocks (having an accuracy of 1 second in 300 000 years). Eight of the planned Block-IIR satellites have been improved further and are renamed Block-IIR-M satellites. These satellites will carry a new military code on both the frequencies (L1 and L2) and a new civilian code on the L2 frequency. The dual codes will provide increased resistance to jamming and the new civilian code will provide better accuracy to civilian users by increasing capability to compensate for atmospheric delays. Eight Block-IIR-M satellites have been launched by August 2013.

Block-IIR-M satellites will be followed by Block-IIF satellites (Figure 13.6). The first Block-II F satellite was launched in May 2010. Since August 2013 there have been four operational Block-II F satellites in GPS constellation. These satellites will have a third carrier signal, L5, at 1176.45 MHz. They will also have a larger design life, fast processors with more memory and a new civilian code. The GPS-III phase of satellites are in the planning stage. These satellites will employ spot beams, enabling the system to have better position accuracy (less than a meter). They will be positioned in three orbital planes having non-recurring orbits. As of August 2013, Block III satellites are in production and deployment phase. As many as 32 Block III satellites with a design life of 15 years have been planned. First Block-III satellite (Block IIIA-1) is scheduled for launch in 2014.

Figure 13.6 Block-IIF GPS satellites (Reproduced by permission of © Aerospace Corporation)

13.1.2.2 Development of the GLONASS Satellite System

GLONASS is a Russian satellite navigation system managed by the Russian space forces and operated by the Coordination Scientific Information Centre (KNIT) of the Ministry of Defence, Russia. The system is a counterpart to the GPS system of the USA. Moreover, both systems have the same principle of operation in data transmission and positioning methodology.

The first GLONASS satellite was launched on 12 October 1982, four years after the launch of the first GPS satellite. Two GLONASS satellites were launched into MEO orbits at an altitude of 19 100 km to characterize the gravitational fields at these orbit heights. The launch marked the beginning of the experimental phase or the pre-operational phase (Block-I) of the GLONASS satellite system. Eighteen satellites were launched in this phase between the years 1982 and 1985. Block-I experimental satellites were followed by the operational Block-IIa, -IIb and -IIv satellites. Six Block-IIa satellites were launched in the span of two years between

1985 and 1986. They had a longer operational life and more accurate and stable clocks than the Block-I satellites. Of 12 Block-IIb satellites launched, six were lost during the launch phase. Block-IIV satellites followed the block-IIb satellites. A total of 25 Block-IIV satellites were launched between the years 1988 and 2000. Block-IIa, -IIb and -IIV satellites are referred to as the first generation GLONASS satellites.

First generation GLONASS satellites (Figure 13.7) were launched in two phases, namely Phase-I and Phase-II. It was planned that during Phase-I, 10 to 12 satellites would be launched by the year 1989-1990. Phase-II marked the deployment of the complete 21 satellite system by the year 1991. However, Phase-I satellites were launched only by the year 1991 and the deployment of a full constellation of 21 satellites was completed by the end of the year 1995. The system was officially declared operational on 23 September 1993. Thereafter, the number of operational satellites decreased due to the short lifetime of operational satellites and because no new satellites was being launched. There were only 10 operational satellites in August 2000.

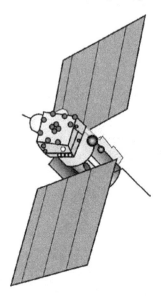

Figure 13.7 First generation GLONASS satellites

In 2001, the Russian government took steps to revive and enhance the GLONASS system to a constellation of 24 satellites within a decade. The development work for the second generation GLONASS satellites, also referred to as GLONASS-M (Uragan-M) (Figure 13.8) satellites, started in the 1990s. The first satellite of the GLONASS-M series was launched in the year 2001. Forty-one satellites in the GLONASS-M series have been launched by July 2013. Another eight satellites are planned to be launched by 2015. GLONASS-K satellites (Figure 13.9) are the third generation GLONASS satellites. The first satellite of the GLONASS-K series was launched on 26 February 2011. Second and third generation GLONASS satellites have improved lifetimes over first generation GLONASS satellites (GLONASS-M has a design lifetime of seven years and GLONASS-K of 10 to 12 years). GLONASS-K satellites will offer an additional L-band navigational signal.

After this brief description on the development of the GPS and the GLONASS satellite systems, both systems are discussed at length in the sections to follow.

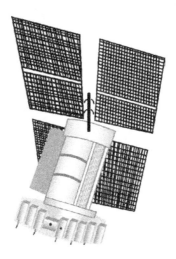

Figure 13.8 GLONASS-M satellites

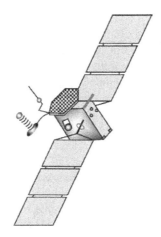

Figure 13.9 GLONASS-K satellites

13.2 Global Positioning System (GPS)

The GPS comprises of three segments, namely the space segment, control segment and user segment. All the three segments work in an integrated manner to ensure proper functioning of the system. In this section, we discuss these three segments in detail.

13.2.1 Space Segment

The space segment comprises of a 28 satellite constellation out of which 24 satellites are active satellites and the remaining four satellites are used as in-orbit spares. The satellites are placed in six orbital planes, with four satellites in each plane. The satellites orbit in circular medium

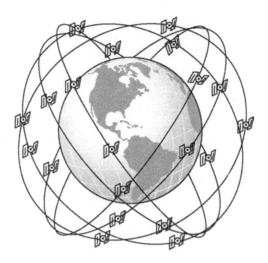

Figure 13.10 Space segment of GPS

Earth orbits (MEO) at an altitude of 20 200 km, inclined at 55° to the equator (Figure 13.10). The orbital period of each satellite is around 12 hours (11 hours, 58 mins). The MEO orbit was chosen as a compromise between the LEO and GEO orbits. If the satellites are placed in LEO orbits, then a large number of satellites would be needed to obtain adequate coverage. Placing them in GEO orbits would reduce the required number of satellites, but will not provide good polar coverage. The present constellation makes it possible for four to ten satellites to be visible to all receivers anywhere in the world and hence ensure worldwide coverage.

All GPS satellites are equipped with atomic clocks having a very high accuracy of the order of a few nanoseconds (3 ns in a second). These satellites transmit signals, synchronized with each other on two microwave frequencies of 1575.42 MHz (L1) and 1227.60 MHz (L2). These signals provide navigation and timing information to all users worldwide. The satellites also carry nuclear blast detectors as a secondary mission, replacing the 'Vela' nuclear blast surveillance satellites. The satellites are powered by solar energy. They have back-up batteries on board to keep them running in the event of a solar eclipse. The satellites are kept in the correct path with the help of small rocket boosters, a process known as 'station keeping'.

13.2.2 Control Segment

The control segment of the GPS system comprises a worldwide network of five monitor stations, four ground antenna stations and a master control station. The monitor stations are located at Hawaii and Kwajalein in the Pacific Ocean, Diego Garcia in the Indian Ocean, Ascension Island in the Pacific Ocean and Colorado Springs, Colorado. There is a master control station (MCS) at Schriever Air Force Base in Colorado that controls the overall GPS network. The ground antenna stations are located at Diego Garcia in the Indian Ocean, Kwajalein in the Pacific Ocean, Ascension Island in the Pacific Ocean and at Cape Canaveral, USA. Figure 13.11 shows the locations of the stations of the control segment.

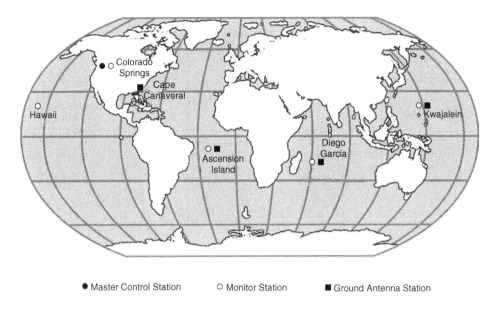

● Master Control Station   ○ Monitor Station   ■ Ground Antenna Station

Figure 13.11 Control segment of GPS

Each of the monitor stations is provided with high fidelity GPS receivers and a Caesium oscillator to continuously track all GPS satellites in view. Data from these stations is sent to the MCS which computes precise and updated information on satellite orbits and clock status every 15 minutes. This tracking information is uploaded to GPS satellites through ground antenna stations once or twice per day for each satellite using S band signals. This helps to maintain the accuracy and proper functioning of the whole system. The ground antenna stations are also used to transmit commands to satellites and to receive satellite telemetry data. Figure 13.12 describes the functioning of the control segment.

13.2.3 User Segment

The user segment includes all military and civil GPS receivers intended to provide position, velocity and time information. These receivers are either hand-held receivers or installed on aircraft, ships, tanks, submarines, cars and trucks. The basic function of these receivers is to detect, decode and process the GPS satellite signals. Some of the receivers have maps of the area stored in their memory. This makes the whole GPS system more user-friendly as it helps the receiver to navigate its way out. Most receivers trace the path of the user as they move. Certain advanced receivers also tell the user the distance they have travelled, their speed and time of travel. They also tell the estimated time of arrival at the current speed when fed with destination coordinates. Moreover, there is no limit to the number of users using the system simultaneously. Today many companies make GPS receivers, including Garmin, Trimble, Eagle, Lorance and Magellan. Figure 13.13 shows the photograph of a commonly used GPS receiver.

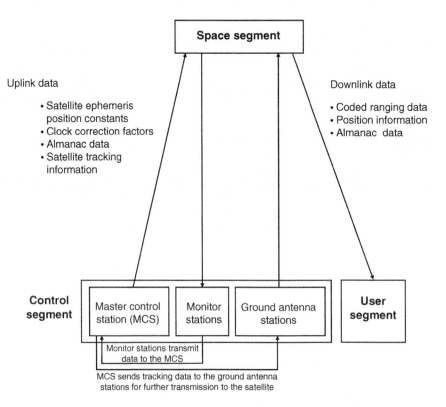

Figure 13.12 Operation of the control segment of GPS system

Figure 13.13 GPS receiver (Reproduced by permission of © Randy Bynum/www.nr6ca.org)

GPS receivers comprise three functional blocks:

1. Radio frequency front end. The front end comprises one or more antennas to receive the GPS signal, filters and amplifiers to discriminate the wanted signal from noise and a down-converter to remove the carrier signal. Simple receivers process one GPS signal at a time using multiplexing techniques. Sophisticated receivers comprise multiple channels for processing the signals from various satellites simultaneously.
2. Digital signal processing block. It correlates the signals from satellites with signals stored in the receiver to identify the specific GPS satellite and to calculate pseudoranges.
3. Computing unit. This unit determines position, velocity and other data. The display format is also handled by the computing unit.

Figure 13.14 explains the functionality of the GPS system. GPS satellites transmit coded information that is used for range calculations. Range and satellite position information from three or four satellites is used to calculate the position of the receiver. A detailed description of the working principle is discussed in the sections to follow.

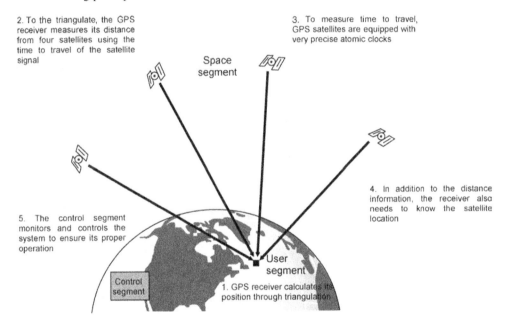

Figure 13.14 Operation of the GPS navigation system

13.3 Working Principle of the GPS

13.3.1 Principle of Operation

The basic principle of operation of the GPS is that the location of any point can be determined if its distance is known from four objects or points with known positions. Theoretically, if the distance of a point is known from one object, then it lies anywhere on a sphere with the

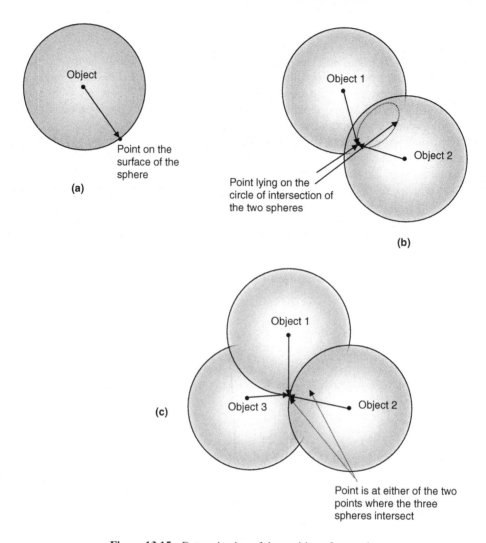

Figure 13.15 Determination of the position of any point

object as the centre having a radius equal to the distance between the point and the object [Figure 13.15 (a)]. If the distance of the point is known from two objects, then it lies on the circle formed by the intersection of two such spheres [Figure 13.15 (b)]. The distance from the third object helps in knowing that the point is located at any of the two positions where the three spheres intersect [Figure 13.15 (c)]. The information from the fourth object reveals the exact position where it is located, that is at the point where the four spheres intersect.

In the GPS, the position of any receiver is determined by calculating its distance from four satellites. This distance is referred to as the 'Pseudorange'. (The details of calculating the pseudorange are covered later in the chapter.) The information from three satellites is sufficient for calculating the longitude and the latitude positions; however, information from the fourth satellite is necessary for altitude calculations. Hence, if the receiver is located on

Earth, then its position can be determined on the basis of information of its distance from three satellites. For airborne receivers the distance from the fourth satellite is also needed. In any case, GPS receivers calculate their position on the basis of information received from four satellites, as this helps to improve accuracy and provide precise altitude information. The GPS is also a source of accurate time, time interval and frequency information anywhere in the world with unprecedented precision.

The GPS uses a system of coordinates called WGS-84, which stands for World Geodetic System 1984. It produces maps having a common reference frame for latitude and longitude lines. The system uses time reference from the US Naval Observatory in Washington DC in order to synchronize all timing elements of the system.

13.3.2 GPS Signal Structure

The GPS signal contains three different types of information, namely the pseudorandom code, ephemeris data and almanac data. The pseudorandom code (PRN code) is an ID (identity) code that identifies which satellite is transmitting information and is used for 'pseudorange' calculations. Each satellite transmits a unique PRN code. Ephemeris data contains information about health of the satellite, current date and time. Almanac data tells the GPS receiver where each satellite should be at any time during the day. It also contains information on clock corrections and atmospheric data parameters. All this information is transmitted at two microwave carrier frequencies, referred to as L1 (1575.42 MHz) and L2 (1227.60 MHz). It should be mentioned here that all satellites transmit on the same carrier frequencies, however different codes are transmitted by each satellite. This enables GPS receivers to identify which satellite is transmitting the signal. The signals are transmitted using the code division multiple access (CDMA) technique.

Pseudorandom codes (PRN codes) are long digital codes generated using special algorithms, such that they do not repeat within the time interval range of interest. GPS satellites transmit two types of codes, namely the coarse acquisition (C/A code) and the precision code (P code). C/A code is an unencrypted civilian code while the P code is an encrypted military code. During military operations, the P code is further encrypted, known as the Y code, to make it more secure. This feature is referred to as 'antispoofing'. Presently, the C/A code is transmitted at the L1 carrier frequency and the P code is transmitted at both L1 and L2 carrier frequencies. In other words, the L1 signal is modulated by both the C/A code and the P code and the L2 signal by the P code only. The codes are transmitted using the BPSK (binary phase shift keying) digital modulation technique, where the carrier phase changes by 180° when the code changes from 1 to 0 or 0 to 1.

The C/A code comprises 1023 bits at a bit rate of 1.023 Mbps. The code thus repeats itself in every millisecond. The C/A code is available to all users. GPS receivers using this code are a part of standard positioning system (SPS). The P code is a stream of 2.35×10^{14} bits having a modulation rate of 10.23 Mbps. The code repeats itself after 266 days. The code is divided into 38 codes, each 7 days long. Out of the 38 codes, 32 codes are assigned to various satellites and the rest of the six codes are reserved for other uses. Hence, each satellite transmits a unique one-week code. The code is initiated every Saturday/Sunday midnight crossing. Precise positioning systems (PPS), used for military applications, use this code. SPS and PPS services are discussed later in the chapter.

Other than these codes, the satellite signals also contain a navigation message comprising the ephemeris and almanac data. This provides coordinate information of GPS satellites as a function of time, satellite health status, satellite clock correction, satellite almanac and atmospheric data. The navigation message is transmitted at a bit rate of 50 kbps using BPSK technique. It comprises 25 frames of 1500 bits each (a total of 37 500 bits). Figure 13.16 shows the structure of the GPS satellite signal.

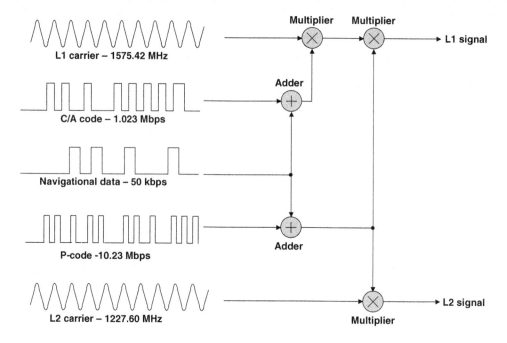

Figure 13.16 GPS satellite signal structure

13.3.3 Pseudorange Measurements

As mentioned before, the fundamental concept behind the GPS is to make use of simultaneous distance measurements from three (or four) satellites to compute the position of any receiver. The GPS receiver calculates its distance from the GPS satellites by timing the journey of the signal from the satellite to the receiver, that is measuring the time interval between transmission of the signal from the satellite and its reception by the receiver. As mentioned in the previous section, each GPS satellite transmits a unique long digital pattern called the pseudorandom code (PRN code). The receiver also runs the same code in synchronization with the satellite. When the satellite signal reaches the receiver, it lags behind the receiver's pattern depending upon the distance between the satellite and the receiver (Figure 13.17). This time delay is calculated by comparing and matching the satellite code sequence received by the receiver with that stored in the receiver, using correlation techniques. Delay in the arrival of the signal

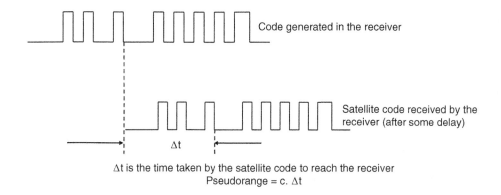

Figure 13.17 Pseudorange measurements

is equal to its travel time. The pseudorange is calculated by multiplying this time with the velocity of the electromagnetic signal. (Velocity of electromagnetic wave is same as that of velocity of light.)

For such calculations to be effective, both the receiver and the satellites need to have accurate atomic clocks. However, placement of accurate clocks on every receiver is not feasible, as these clocks are very expensive. Receivers are equipped with inexpensive normal clocks which they reset with the help of satellite clocks. This is done by making corrections on the basis that four spheres will not intersect at one point if the measurements are not correct. As the distances are measured from the same receiver, they are proportionally incorrect. The receiver performs the necessary corrections to make the four spheres intersect and resets its clock constantly based on these corrections. This makes receiver clocks as accurate as atomic clocks on the satellites. In this way, the measurement of the distance of the receiver from the fourth satellite helps in correcting for receiver clock errors and hence in increasing the accuracy of the system. In fact, a GPS receiver calculates pseudoranges from all satellites visible to it.

Pseudorange measurements can also be done using carrier phase techniques. Range in this case is the sum of the total number of full carrier cycles plus fractional cycle between the receiver and the satellite, multiplied by the carrier wavelength (Figure 13.18). Carrier phase measurements are more accurate than measurements done using PRN codes and are used for high accuracy applications. However, they can be only used for differential GPS positioning as they require a second receiver also (differential GPS is discussed later in the chapter). The pseudorange can be defined as

$$\text{Pseudorange} = c\Delta t \qquad (13.1)$$

where Δt is the time taken by the satellite code to reach the receiver.

13.3.4 Determination of the Receiver Location

After calculating pseudoranges from four satellites, the receiver determines the position/time solution of four ranging equations for generation of its position and time information.

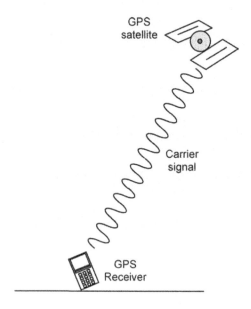

Figure 13.18 Carrier-phase measurements

$$(x_1 - U_x)^2 + (y_1 - U_y)^2 + (z_1 - U_z)^2 = (PR_1 \pm EC)^2 \qquad (13.2)$$

$$(x_2 - U_x)^2 + (y_2 - U_y)^2 + (z_2 - U_z)^2 = (PR_2 \pm EC)^2 \qquad (13.3)$$

$$(x_3 - U_x)^2 + (y_3 - U_y)^2 + (z_3 - U_z)^2 = (PR_3 \pm EC)^2 \qquad (13.4)$$

$$(x_4 - U_x)^2 + (y_4 - U_y)^2 + (z_4 - U_z)^2 = (PR_4 \pm EC)^2 \qquad (13.5)$$

where,

x_n, y_n, z_n = x, y and z coordinates of the nth satellite
U_x, U_y, U_z = x, y and z coordinates of the user receiver
PR_n = pseudorange of the user receiver from the nth satellite
EC = error correction

The x, y and z coordinates of satellites are calculated from the altitude, latitude and the longitude information of the satellite on the basis of complex three-dimensional Pythagoras equations. All these calculations along with position determination calculations are carried out in the GPS receiver using special algorithms.

After discussing the GPS system and its operation, readers are in a position to understand the various services offered by the GPS system. In the next section, the positioning modes and services offered by the GPS will be discussed.

Problem 13.1

Compute the range in accuracies of the GPS system using (a) C/A code and (b) P code.

Solution: (a) The modulation rate of the C/A code is 1.023 Mbps. Therefore, the duration of one bit $= 1/1.023 \times 10^6 \cong 1\ \mu s$. As, distance = velocity × time and the velocity of the electromagnetic wave $= 3 \times 10^8$ m/s. Therefore, the distance inaccuracy $= 3 \times 10^8 \times 1 \times 10^{-6} = 300$ m.

(b) The modulation rate of the P code is 10.23 Mbps. Therefore, the duration of one bit $= 1/10.23 \times 10^6 \cong 0.1\ \mu s$ and the distance inaccuracy $= 3 \times 10^8 \times 0.1 \times 10^{-6} = 30$ m. Therefore, the distance inaccuracy in the case of GPS system using the P code is 10 times less than the GPS system using the C/A code.

Problem 13.2

The code pattern generated by a transmitter is given in Figure 13.19. The same pattern is also generated in the receiver. The pattern of the transmitter is received by the receiver after some time delay. The receiver pattern and delayed received pattern are shown in the figure. Calculate the distance between the transmitter and the receiver, if the bit rate is 1 Mbps.

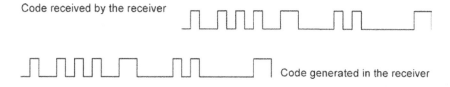

Figure 13.19   Figure for Problem 13.2

Solution: Let us assume that the bit pattern is received by the receiver after a time delay of (Δt). As can be seen from Figure 13.19, the bit pattern received by the receiver is shifted wrt the pattern stored in the receiver by a time equal to the time period of 18 bits. Therefore, $\Delta t = 18 \times$ one bit interval. As the bit rate is 1 Mbps, the bit period is $1/1 \times 10^6 = 1\ \mu s$. Therefore the time taken by the pattern to reach the receiver is 18 μs and the distance between the receiver and the transmitter $= 3 \times 10^8 \times 18 \times 10^{-6} = 5400$ m. This is just a simple illustration of how the pseudorange calculations are done. However, in actual practice, cross-correlation techniques are used to calculate the time delay (Δt).

13.4   GPS Positioning Services and Positioning Modes

13.4.1   GPS Positioning Services

There are two levels of GPS positioning and timing services, namely the precision positioning service (PPS) and the standard positioning service (SPS). The PPS, as the name suggests, is the most precise and autonomous service and is accessible by authorized users only. SPS is less accurate than PPS and is available to all users worldwide, authorized or unauthorized.

13.4.1.1 Standard Positioning System (SPS)

SPS is a positioning and timing service available to all GPS users worldwide, on a continuous basis without any charge. It is provided on L1 frequency using the C/A code. It has horizontal position accuracy of the order of 100 to 300 m, vertical accuracy within 140 m and timing accuracy better than 340 ns. SPS was previously intentionally degraded to protect US national security interests using a scheme called 'selective availability'. Selective availability (SA) is a random error introduced into the ephemeris data to reduce the precision of the GPS receivers. However, the scheme was turned off on 1 May 2000. With discontinuation of SA, SPS autonomous positioning accuracy is presently at a level comparable to that of PPS.

13.4.1.2 Precision Positioning System (PPS)

PPS is a highly accurate military positioning, velocity and timing service which is available only to authorized users worldwide. It is denied to unauthorized users by use of cryptography. PPS service was mainly designed for US military services and is also available to certain authorized US federal and allied government users. It uses the P code for positioning and timing calculations. The expected positioning accuracy is 16 m for the horizontal component and 23 m for the vertical component at 95 % probability level.

13.4.2 GPS Positioning Modes

Positioning with GPS can be performed in either of the following two ways:

1. Point positioning
2. Relative positioning

13.4.2.1 Point Positioning

Point positioning employs one GPS receiver to do the measurements. Here the receiver calculates its position by determining its pseudoranges from three (or four) satellites using the codes transmitted by the satellite (Figure 13.20). It is used for low accuracy applications like the recreation applications and for low accuracy navigation.

13.4.2.2 Relative Positioning

GPS relative positioning, also referred to as differential positioning, employs two GPS receivers simultaneously for tracking the same satellites. They are used for high accuracy applications such as surveying, precision landing systems for aircraft, measuring movement of the Earth's crust, mapping, GIS and precise navigation. Special receivers known as differential GPS (DGPS) receivers are required for using this service. Pseudorange in this case can be measured using either PRN codes (for medium accuracy applications) or by performing carrier phase measurements (for high accuracy applications).

Differential GPS systems employ a receiver at a known position (known as the 'base receiver') to determine the inaccuracy of the GPS system. The basic idea is to gauge GPS

GPS Positioning Services and Positioning Modes   633

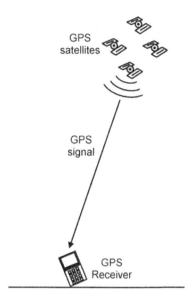

Figure 13.20 Point positioning

inaccuracy at the base receiver and then to make corrections accordingly. The corrections are based on the difference between the true location of the base receiver and the location determined by the GPS system. This correction signal is then broadcasted to all DGPS equipped receivers in the area either through tower-based or satellite-based systems. Typical accuracy of DGPS systems is of the order of 1 to 10 m. Figure 13.21 shows the conceptual working of the DGPS system.

DGPS services are provided free of cost by government agencies or at an annual fee by commercial providers. An example of DGPS service in operation is the network of land-based broadcast towers near major navigable bodies of water established by the United States Coast Guard (USCG). The system is free of cost but it has a limited number of base stations; hence the coverage area is very limited. Several commercial companies have established DGPS systems. The base stations are located at areas of interest. Corrections from these stations are transmitted to the users via communication satellites (not via GPS satellites). These DGPS systems offer a better coverage area but they charge an annual fee for their services.

Another free differential GPS service is the Wide Area Augmentation System (WAAS) developed by the FAA (Federal Aviation Administration) for aviation users. It is a regional augmentation DGPS employing a network of 25 ground reference stations that cover the USA, Canada and Mexico. Each reference station is linked to a master station, which puts together a correction message and broadcasts it via a satellite. WAAS capable receivers will have accuracies of the order of 3 to 5 m horizontally and 3 to 7 m in altitude.

DGPS systems can either be real time differential systems or post-process differential systems depending upon whether the position information is determined instantaneously

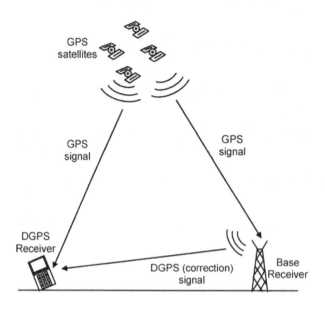

Figure 13.21 Relative positioning

or at a later time. In real time systems, the position of the receiver is determined instantaneously whereas in the post-processing differential systems, the position information is computed later. Differential GPS systems are further classified as static GPS surveying, fast static GPS surveying, stop-and-go GPS surveying, RTK (real time kinematic) GPS and real time differential GPS. The explanation of these systems is beyond the scope of the book.

13.5 GPS Error Sources

GPS measurements are affected by several types of random errors and biases. These errors may be due to inaccuracies in the receiver, errors in orbital positions of the satellites, receiver and satellite clock errors, errors during signal propagation and multiple path errors. In addition to these errors, the accuracy of the computed GPS position is also affected by geometric locations of GPS satellites as seen by the receiver. The accuracy of civilian systems was degraded intentionally, by employing 'selective availability'. In this section, the main sources of errors are discussed and ways of treating them are introduced.

The main sources of error are as follows:

1. Signal propagation errors.   Signal propagation errors include delays in the GPS signal as it passes through the ionospheric and tropospheric layers of the atmosphere. The ionosphere acts like a dispersive medium that bends the GPS radio signals and changes their speed as they pass through various ionospheric layers to reach the receiver. A change in speed causes significant range error, whereas error due to bending is more or less negligible. The

ionosphere speeds up propagation of the carrier phase while slows down the propagation of the PRN code. Moreover, this delay is frequency dependent: the lower the frequency, the greater is the delay. Hence, delay for the L2 signal is greater than that for the L1 signal. The ionospheric error can be corrected using the differential GPS or by combining P code measurements for both the L1 and L2 carriers. The troposphere acts as a non-dispersive medium for GPS signals and delays the GPS carrier and the codes identically. The delay is computed using various mathematical models.

2. Multipath reflections. Multipath reflections are also a major source of GPS errors. Multipath error occurs when the GPS signal arrives at the receiver through different paths (Figure 13.22). Reflections of the GPS signal from objects such as tall buildings, large rock surfaces or from the ground surface near the receiver provide multiple signals slightly shifted in time, which results in errors. These errors can be reduced by using receivers having ring antennas that attenuate the reflected signals or by selecting an observation site with no reflecting objects in the vicinity of the receiver antenna, etc.

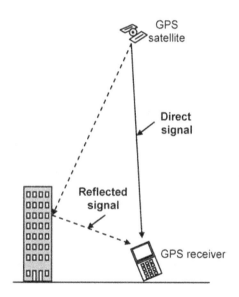

Figure 13.22 Multipath reflections

3. Clock errors. As mentioned before, the receiver clock is not as accurate as the atomic clock on board GPS satellites. Although correction algorithms are employed to reduce the errors, there are still some timing errors present. Inaccuracies in satellite clocks also lead to range errors. However, as satellites use accurate atomic clocks, errors due to their inaccuracies are not appreciable.

4. Ephemeris errors. These errors are due to inaccuracies in the satellite's reported location. Generally, ephemeris error is of the order of 2 to 5 m. Ephemeris errors can be reduced by using differential GPS techniques.

5. Number of satellites visible. The errors decrease as the number of satellites visible to the receiver increases. Buildings, terrain, rocks and electronic interference block signal reception, causing position errors. GPS units typically do not work indoors, underwater or underground.
6. Satellites geometry. The error also depends on satellite geometry, that is on the geometry of locations of GPS satellites as seen by the receiver. If satellites are located at wide angles relative to each other, then the errors are less [Figure 13.23 (a)]. However, if they are located in tight grouping, then the errors increase [Figure 13.23 (b)]. In fact, if satellites are clustered near each other, then one metre of error in the measuring distance may result in tens or hundreds of metres of errors in position. For satellites scattered in the sky, position error is of the order of a few metres for every metre of error in measuring distances. This effect of the geometry of satellites on position error is called 'geometric dilution of precision' (GDOP).

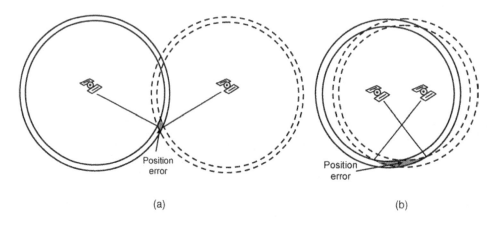

Figure 13.23 Errors caused by satellite geometry

7. The GPS signal is weak and hence can be jammed by a low power transmitter. Multiple separated antennas assist in rejecting the signals coming from the ground. Increasing the power of satellites will reduce the jamming probability. The plan is to use focused spot beams in GPS-III satellites, to ensure a much higher signal power.
8. Selective availability. Selective availability (SA) was introduced by the US Department of Defense to reduce the accuracy of GPS services for unauthorized users. SA introduces two types of errors, namely the delta error and the epsilon error. The delta error results from dithering of the clock signal and epsilon error is an additional slow varying orbital error. With SA turned on, nominal horizontal and vertical errors were 100 m and 156 m respectively, at 95 % probability level. SA was discontinued on 1 May 2000, resulting in much improved autonomous GPS accuracy (horizontal accuracy of the order of 22 m and vertical accuracy of the order of 33 m at 95 % probability level). Table 13.1 lists the typical values of errors caused due to these sources for the GPS and the DGPS systems.

Table 13.1 Typical values of various errors for the GPS and DGPS systems

Error source	Typical range error for the GPS (m)	Typical range error for the DGPS (m)
Selective availability	10–25	–
Ionosphere delay	7–10	–
Troposphere delay	1–2	–
Satellite clock error	1	–
Satellite ephemeris error	1	–
Multipath error	0.5–2	0.5–2
Typical horizontal dilution of precision	1.5	1.5
Range error	15–25	2

Problem 13.3
Calculate the range inaccuracy for a GPS system, where the synchronization between the receiver clock and the satellite clock is off by 100 ps.

Solution: The range inaccuracy due to non-synchronization between the satellite and the receiver clocks is given by

$$\Delta R = \Delta T \times c$$

where

ΔR = range inaccuracy
ΔT = Non-synchronization between the satellite and the receiver clocks
c = speed of light

Therefore

$$\Delta T = 100 \text{ ps} = 10^{-10} \text{s}$$

$$\Delta R = 10^{-10} \times 3 \times 10^8 = 0.03 \text{ m}.$$

The range inaccuracy is 0.03 m.

13.6 GLONASS Satellite System

GLONASS is a satellite navigation system developed by Russia. Like the GPS, it works on the principle of 'trileration'. Most of the concepts explained earlier *vis-à-vis* the GPS system are also applicable to the GLONASS system. Hence the system will be discussed in brief with more emphasis on features that are specific to the GLONASS system. Table 13.2 shows the comparison between the GPS and the GLONASS navigation systems.

Table 13.2 Comparison between the GPS and the GLONASS systems

Features	GPS	GLONASS
Number of satellites (operational till August 2013)	32	29
Number of orbital planes	6	3
Orbital inclination	55°	64.8°
Orbit altitude	20 180 km	19 100 km
Period of revolution	11 h 58 min 00s	11 h 15 min 44s
Geodetic datum	WGS-84	PZ-90
Geodetic time reference	UTC (United States Naval Observatory)	UTC (Russia)
Signalling	CDMA	FDMA
L1 carrier frequency	1575.42 MHz	1602–1615.5 MHz
L2 carrier frequency	1227.60 MHz	1246–1256.5 MHz
Chip rate	1.023 Mbps for C/A code 10.23 Mbps for P code	511 kbps for C/A code 5.11 Mbps for P code
Number of code elements	1023 for C/A code 2.35×10^{14} for P code	511 for C/A code 5.11×10^{6} for P code

13.6.1 GLONASS Segments

The GLONASS system also comprises of three segments, namely the space segment, the control segment and the user segment.

13.6.1.1 Space Segment

The nominal constellation of the GLONASS system consists of 21 operational satellites plus three spares in circular medium Earth orbits at a nominal altitude of 19 100 km (Figure 13.24). The satellites are arranged in three orbital planes inclined at an angle of 64.8°. Each plane comprises eight satellites displaced at 45° with respect to each other. The orbital period of each of these satellites is 11 hours, 15 minutes and 44 seconds. Each satellite is identified by its slot number, which defines the orbital plane and its location within the plane. The first orbital plane has slot numbers 1 to 8, the second orbital plane has slot numbers 9 to 16 and the third orbital plane has slot numbers 17 to 24. Each of the GLONASS satellites carries three caesium atomic clocks. The system time is derived from the UTC (Russian time). The GLONASS system provides better coverage as compared to the GPS system at higher latitude sites as GLONASS satellites are placed at a higher inclination than the GPS satellites. Each of the satellites transmits range, timing and positioning information. The GLONASS system uses the Earth parameter system 1990 (PZ-90) to determine the position of its satellites.

13.6.1.2 Control Segment

The control segment comprises a small number of control tracking stations (CTS), a system control centre (SCC) and various quantum optical tracking stations (QOTS). CTS are

GLONASS Satellite System

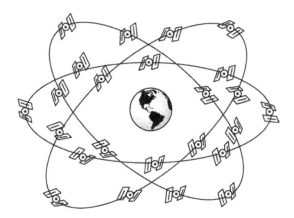

Figure 13.24 Space segment of the GLONASS satellite system

monitoring ground stations located at various places in the former Soviet Union. SCC is the main monitoring station located at Moscow. QOTS are laser ranging systems used for periodically calibrating the measurements done at CTS.

CTS track the GLONASS satellites in view, calculate the satellite ranges and receive satellite navigation messages. This data is fed to the SCC where clock corrections, satellite status messages and navigation messages are generated. These data are sent to CTS for further transmission to the satellites. The measurements made by CTS are periodically calibrated with measurements at QOTS sites.

13.6.1.3 User Segment

The user segment comprises military and civilian GLONASS receivers for providing positioning and timing information. These receivers, like the GPS receivers, can be hand-held or platform-mounted and work on similar lines as the GPS receivers.

13.6.2 GLONASS Signal Structure

GLONASS satellites transmit signals carrying three types of information, namely the pseudorandom code, almanac data and ephemeris data (Figure 13.25). Like the GPS system, the GLONASS system also carries two types of pseudorandom codes, namely the C/A code and the P code. They also transmit the C/A code on the L1 carrier and the P code on both the L1 and L2 carriers. The navigation message is transmitted on both carriers. Each satellite transmits the same pseudorandom code, but the carrier frequencies are different for each satellite. Their carrier frequencies are in the range of 1602-1615.5 MHz for the L1 band and 1246-1256.5 MHz for the L2 band depending upon the channel number. The GLONASS system uses the frequency division multiple access (FDMA) technique for transmitting the signals.

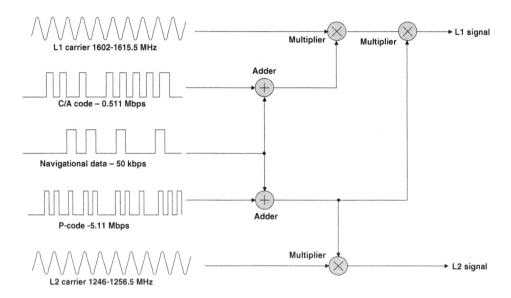

Figure 13.25 GLONASS signal structure

The nominal carrier frequencies for the L1 band are given by

$$f_{k1} = f_{01} + k\Delta f_1 \qquad (13.6)$$

where

$f_{01} = 1602$ MHz
$\Delta f_1 = 562.5$ kHz

k (frequency number used by the GLONASS) = 1 to 24

Similarly, for the L2 band, the carrier frequencies are given by

$$f_{k2} = f_{02} + k\Delta f_2 \qquad (13.7)$$

where

$f_{02} = 1246$ MHz
$\Delta f_2 = 437.5$ KHz

k (frequency number used by the GLONASS) = 1 to 24

Earlier, each satellite had a separate carrier frequency. However, the higher frequencies interfered with the reserved radio astronomy bands. Therefore, the frequency pattern was changed in the year 1998. A pair of satellites was assigned the same L1 and L2 frequencies. Hence the value of k now varies from 1 to 12. Satellites carrying the same frequency were placed in antipodal positions, that is on opposite sides of the Earth so that the user can not see them simultaneously. Future plans are to shift the L1 and L2 bands to 1598.0625–1604.25 MHz and 1242.9375–1247.75 MHz respectively to avoid interference with radio astronomers and operators of low Earth orbiting satellites. As mentioned before, GLONASS codes are the same for all the satellites. The chipping rate for the C/A code and the P code are 0.511 Mbps and 5.11 Mbps respectively. The GLONASS navigation message is a 50 kbps data stream which provides information on satellite ephemeris and channel allocation. They are modulated onto the carrier using BPSK techniques.

The ranging measurements, position calculation methodology and positioning modes are the same as that of the GPS system. The GLONASS system also offers standard positioning services (SPS) and precise positioning services (PPS), similar to that offered by the GPS system. However, the GLONASS system does not employ the 'selective availability' feature of intentional degradation of the civilian code.

13.7 GPS-GLONASS Integration

Integration of the GPS and the GLONASS systems has improved positioning accuracy as well as system reliability, as the integrated space segment has a larger number of satellites. It has also increased the coverage area of the system. Increase in the total number of satellites is particularly useful for urban areas as satellite visibility is poor in these areas due to tall buildings. Figure 13.26 shows that the availability of satellites increases with integration. Figure 13.26 (a) shows satellite availability with one system operational and Figure 13.26 (b) shows the increase in the number of available satellites after integration.

The integration process is complex and faces two main problems. Firstly, the GPS and the GLONASS systems use different coordinate frames (the GPS uses the WGS 84 system and the GLONASS uses the Earth parameter system 1990 named PZ-90). Secondly, the systems use different reference times. Hence, the receiver in this case needs to take measurements from a minimum of five satellites. The working principle of receivers remains the same, as

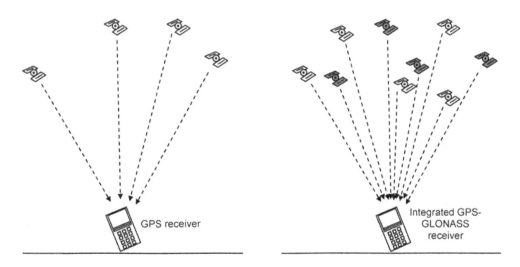

Figure 13.26 Availability of satellites increases with integration of the GPS and the GLONASS systems

discussed earlier. The accuracy level of the GPS-GLONASS integrated system is of the order of 7 m. Accuracies of the order of 50 cm can be achieved if integrated differential GPS-GLONASS or GPS-differential GLONASS receivers are used. If the differential GPS and the differential GLONASS receivers are integrated, then accuracy levels of the order of 35 cm can be achieved.

Some companies have made receivers that compute position using data from both GPS and GLONASS satellites. One such company is Ashtech, which has designed the GG24 GPS-GLONASS receiver (Figure 13.27).

Figure 13.27 GPS-GLONASS receiver (Reproduced by permission of © Thales Navigation Inc.)

13.8 EGNOS Satellite Navigation System

The European Geostationary Navigation Overlay Service (EGNOS) is the first ever pan European satellite based navigation system and it is a joint venture of three organizations: the European Space Agency (ESA), the European Commission and Eurocontrol. It is a precursor to Galileo, which is a full global satellite navigation system. EGNOS system was developed by a consortium of companies led by Alcatel Space (now Thales Alenia Space) of France, contracted by the ESA. The European Commission looked after international cooperation and coordination work, and ensured that aspects of all modes of transport were fed into the design and implementation of the system. Eurocontrol, which is a European organization for the safety of air navigation, defined the needs of civil aviation and the mission requirements of the system. The Alcatel led consortium successfully developed the system in April 2009 and it was made operational for non-safety critical applications in October 2009. Its usage in safety critical applications such as aircraft landing began in March 2011. EGNOS operations are currently managed by the European Commission through a contracted operator in France called the European Satellite Service Provider (ESSP). ESSP was created by a group of companies including AENA (Spain), DSNA (France), ENAV (Italy), DFS (Germany), NATS (UK), Skyguide (Switzerland) and NAV-EP (Portugal), who signed bilateral agreements with the ESA.

The EGNOS system is used to augment the Global Positioning System (GPS) system of USA. It assists in improving its accuracy of position measurements from about 5 metres to better than 2 metres to make it suitable for safety critical applications such as those of flying aircraft or navigating ships through narrow channels. It can also be made use of by

Figure 13.28 Safety critical navigation service offered by EGNOS

trains and other modes of transport (Figure 13.28). It consists of three geostationary satellites and a network of ground stations. It transmits a signal containing information on the accuracy and reliability of positioning signals sent out by the GPS and thereby informs users of the errors in position measurements and also warns them of any disruption in the satellite signal within 6 seconds. EGNOS positioning data is available across Europe to all users equipped with EGNOS enabled GPS receivers. Other similar navigation systems include the US Wide Area Augmentation System (WAAS) and the Multi-functional Satellite Augmentation System (MSAS) of Japan. However, these systems are exclusively designed for air navigation.

The different elements of the EGNOS system are (a) ranging and integrity monitoring stations (RIMS) to pick the GPS signals, (b) master control centres to process all the data delivered to them by RIMS and (c) uplink stations that send the signal to the three geostationary

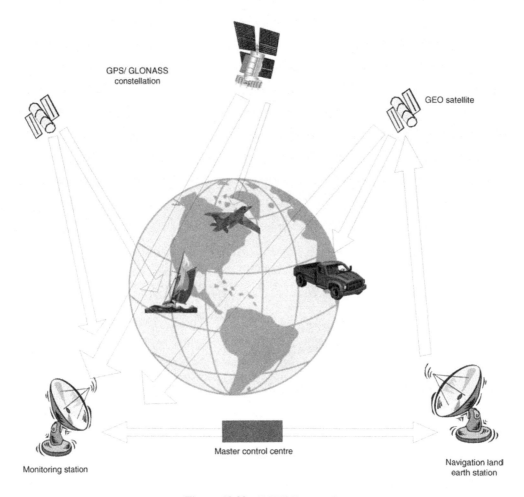

Figure 13.29 EGNOS network

satellites. The geostationary satellites in turn relay the signal back to the users on the ground. The EGNOS complete network comprises 34 RIMS, four master control centres and six uplink stations. The three geostationary satellites used are Inmarsat 3-F2 covering Atlantic Ocean Region – East, Inmarsat 4-F2 covering Europe, the Middle East and Africa, and Inmarsat 3-F1 covering the Indian Ocean. Figure 13.29 shows a pictorial representation of the EGNOS system network.

The EGNOS system provides three basic services: (a) an open service, (b) the EGNOS data access service and (c) a safety critical service. The open service provides positioning data that is available to all users across Europe equipped with EGNOS-enabled GPS receivers. The data access service offers the enhanced performance required for professional use and available on a controlled access basis such as through the internet or mobile phones. The safety critical service, for civil aviation applications, offers enhanced and guaranteed performance.

13.9 Galileo Satellite Navigation Systems

Currently, most users of satellite navigation systems around the world mainly depend on the US GPS system and the Russian GLONASS. The Indian Regional Navigational Satellite System (IRNSS) and Chinese Compass are the other satellite navigation systems in use. Galileo is a global satellite navigation system being built by the European Union and the European Space Agency. It is intended to provide European countries with a fully autonomous and reliable satellite based positioning, navigation and timing capability independent of the US GPS and Russian GLONASS. It may be mentioned here that though Galileo system is intended to be independent of the GPS and GLONASS systems it is fully interoperable with GPS and GLONASS, thereby making it a fully integrated new element in global satellite navigation systems. This implies that users with GPS or GLONASS compliant receivers will be able to make use of high precision and reliable positioning services, particularly in high rise cities. A defining feature of the Galileo system is its civilian control, unlike GPS and GLONASS, which are under the control of the military. Also, because there is a greater inclination of the satellites to the equatorial plane, the Galileo navigation system provides much better coverage for European countries having higher latitudes, which are not well covered by GPS.

13.9.1 Three-Phase Development Programme

Galileo navigation system development has been structured into the following three phases. The *first phase* was the experimental phase. The main objective of the first phase was to experiment with and verify the critical technologies needed for the Galileo system to operate in the medium earth orbit (MEO) environment. Two experimental satellites, namely GIOVE-A (Galileo In-Orbit Validation Element-A) and GIOVE-B, were launched on 28 December 2005 and 27 April 2008, respectively. The primary aim of GIOVE-A was to claim the radio frequencies set aside by the International Telecommunications Union (ITU) for the Galileo system. GIOVE-A carried two environment monitors to collect vital data about the Galileo intermediate circular orbit environment. The data was to be subsequently used to design the full constellation. Another satellite, GIOVE-A2, was launched to extend the mission life of its predecessor. GIOVE-B was the first satellite that transmitted Galileo signals. The ground test facilities used to analyze Galileo signals included the GIOVE-B Control Centre in Italy, the Galileo Processing Centre at ESA's European Space Research and Technology Centre in the Netherlands, the ESA Ground Station in Belgium and the Rutherford Appleton Laboratory in the UK.

The *second phase* is the in-orbit validation (IOV) phase. The main objective of the IOV phase is validation of system design using a scaled down constellation of only four satellites, which along with a limited number of ground stations is the minimum number needed to provide positioning and timing data. The four satellites were launched in pairs. The first pair of satellites was launched on 21 October 2011 and the second pair was launched on 12 October 2012. Each of the four satellites broadcast precise time signals, ephemeris and other data. Figure 13.30 shows the third IOV satellite, which was launched on 12 October 2012.

The *third phase* will achieve full operational capability with 30 operational satellites including the four in-orbit validation satellites. Of the 30 satellites, 27 will meet the full functional requirement and three are spare satellites. These satellites are placed in three orbital planes spaced 120° apart, with 10 satellites (nine operational satellites and one spare satellite) in

Figure 13.30 IOV satellite of the Galileo system (Courtesy: NASA)

each of the three orbital planes. All these satellites have medium earth circular orbits with altitude of 23,222 km and orbital inclination 56°. This phase will also lead the establishment of the remaining ground infrastructure. The third phase will be accomplished through an intermediate milestone with 18 satellites, comprising four IOV satellites and 14 full operational capability satellites. Services with reduced performance from this intermediate milestone will be available from mid 2014 onwards. The complete system with 27 operational satellites and three spare satellites is expected to be realized by 2019.

13.9.2 Services

The Galileo satellite navigation system is designed to provide different categories of services to cater to the needs of a wide range of users. Different categories of service provided by Galileo system include the following:

(a) An open service (OS) that will provide positioning and synchronization information free of charge. This service is mainly intended for high volume satellite navigation applications.
(b) A service for safety critical applications provided by Galileo's open service signals and/or in cooperation with other satellite navigation systems.
(c) A commercial service with improved performance and value addition to the open service for professional or commercial use.
(d) A public regulated service that uses strong and encrypted signals, and is restricted to government authorized users for sensitive applications.
(e) Assistance to search and rescue service of the Cospas-Sarsat system.

The different categories of services offered by the Galileo system will cover a wide range from location based services to individual level to emergency, security and humanitarian services, from scientific research in meteorology, geology and geodesy to space science, from civil aviation management to management of pedestrian traffic, from agriculture to the fisheries industry, and from wireless telecommunications to finance and banking.

13.10 Indian Regional Navigational Satellite System (IRNSS)

The IRNSS is an independent regional satellite navigation system intended to provide uninterrupted navigation services including a standard positioning service for civilian use and an encrypted service for authorized users. The system is currently under development by the Indian Space Research Organization (ISRO). The system is designed to provide real time position, navigation and time services on 24 × 7 basis to users on a variety of platforms. It is designed to offer position accuracy of better than 10 m over India and the region extending up to 1500 km around India.

Full constellation of IRNSS comprises seven satellites. Three of these satellites will be placed in geostationary equatorial orbit with locations at 34°E, 83°E and 131.5°E. Another two satellites will be placed in geosynchronous orbit at an equatorial inclination of 29° and equator crossings at 55°E and 111.5°E. The remaining two satellites are planned as spares. The first of the seven satellites, called IRNSS-1A, was launched onboard PSLV-C22 on 1 July 2013. The full constellation is expected to be in place by 2014. The ground segment of the IRNSS constellation consists of a master control centre (MCC), ranging and integrity monitoring stations (RIMS) and TT&C stations. Figure 13.31 illustrates the IRNSS architecture.

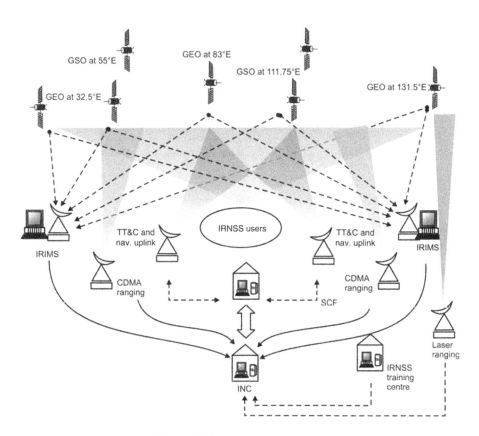

Figure 13.31 IRNSS architecture

13.11 Compass Satellite Navigation System

Compass is the second generation of the BeiDou satellite navigation system of China, also known as BeiDou-2. The first generation system was called the BeiDou satellite navigation experimental system, also known as BeiDou-1. The BeiDou system constellation comprises four satellites: three working satellites and one back-up satellite. Unlike the GPS, GLONASS and Galileo systems, which employ satellites in medium earth orbits, BeiDou-1 satellites are placed in geostationary orbits. As a result of this, the system requires a relatively smaller constellation of satellites for a given coverage, though the coverage will be limited to areas having satellite visibility. The three functional satellites of the BeiDou-1 system were launched between 2000 and 2003, with first satellite (BeiDou-1A) launched on 31 October, 2000, the second (BeiDou-1B) on 21 December 2000 and the third (BeiDou-1C) on 21 May 2003. The successful launch of BeiDou-1C established the BeiDou-1 satellite navigation system. The BeiDou-1 system has been providing navigational services since 2000 to users in China and neighbouring regions. The fourth satellite for back-up purpose (BeiDou-1D) was launched on 2 February 2007.

The Compass satellite navigation system is intended to be a global navigation system and is currently under development. It is expected to provide global navigation services similar to the US GPS, the Russian GLONASS and the European Galileo systems by 2020. The full constellation will comprise 35 satellites. Five of these satellites will be placed in a geostationary equatorial orbit for backward compatibility with the BeiDou-1 system. Of the remaining 30 satellites, 27 will be placed in medium earth orbits and three in inclined geosynchronous orbits. The first satellite of the Compass (BeiDou-2) global navigation system, called Compass-M1, was successfully launched on 14 April 2007 with the objective of validating the frequencies chosen for the BeiDou-2 constellation. The satellite was launched into a near circular medium earth orbit of 21530 km altitude and an orbital inclination of $55.26°$. The next four satellites of the constellation were launched between April 2009 and December 2011. Another five satellites were launched during 2012. The remaining satellites are proposed to be launched during the period 2013–2020.

The Compass navigation system is designed to provide two types of services. One of these is an open service to civilian users with a location tracking accuracy of 10 m, speed measurement accuracy of better than 0.2 m/s and a synchronized clock with an accuracy of 10 ns. The second service is a licensed service to government and military users offering a location accuracy of 10 cm.

13.12 Hybrid Navigation Systems

Hybrid navigation or positioning systems overcome the limitations of global satellite navigation systems in respect of sub-optimal performance in dense urban areas. The positioning accuracy offered by these global systems is very high in open areas, but is reduced heavily indoors and in areas between tall buildings. Hybrid positioning systems make use of a combination of various positioning technologies including those offered by satellite based global navigation systems, cellular tower signals, wireless internet signals, Bluetooth sensors and so on. Signals from cell towers are not adversely affected by the presence of tall buildings nor are they affected by adverse weather conditions. Hybrid systems combine the advantages of satellite navigation systems which provide precise position information and other positioning technologies such as Wi-Fi, which is more effective indoors and in adverse weather conditions.

A large number of companies are providing location based products and services based on Wi-Fi technologies, which when combined with systems like GPS provide hybrid positioning services. It may be mentioned here that Wi-Fi positioning uses wireless access points for determining position and is based on measuring the intensity of the received signal. The accuracy depends on the number of positions that have been entered into the database. Some of the well known service providers include Navizon, Skyhook Wireless, Google and Combain Mobile. Navizon, initially known as Mexens Technology, provides location based services that enable the determination of the geographical location of a mobile device. It makes use of location information from cell phone towers and Wi-Fi access points. Figure 13.32 illustrates the Navizon concept. Skyhook Wireless, previously known as Quarterscope, also provides services to determine geographical location using Wi-Fi positioning technologies. This system is capable of determining the position of a mobile device to an accuracy of better than 20 m. When integrated with GPS this system provides hybrid positioning services capable of accuracy greater than 20 m indoors and in dense urban areas with 100 % reliability. Google Maps from Google provides many map-based services, including the Google Maps website, Google Ride Finder, Google Transit etc. Combain Mobile also offers mobile positioning via cell towers and Wi-Fi technologies.

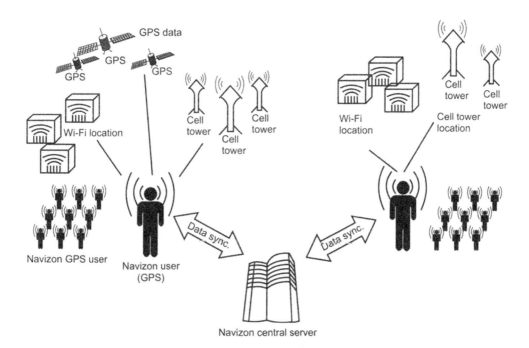

Figure 13.32 Navizon concept

13.13 Applications of Satellite Navigation Systems

Satellite navigation represents one of the dual use space technologies that has found extensive applications both in military and civilian fields. These systems have been in use over the past two and a half decades and have replaced the conventional navigation methods in most cases. Some of the main military application areas include weapon guidance, navigation, tracking, and so on. Civilian applications include construction and surveying, seismic surveying, airborne mapping, vehicle navigation, automotive, marine, military and aviation surveying. They are also used in endeavours like aerial refuelling, rendezvous operations, geodetic surveying and various search and rescue operations. In this section, these applications will be briefly discussed.

13.13.1 Military Applications

Satellite navigation systems have proved to be a valuable aid for military forces. Military forces around the world use these systems for diverse applications including navigation, targeting, rescue, disaster relief, guidance and facility management, both during wartime as well as peacetime. GPS and GLONASS receivers are used by soldiers and also have been incorporated on aircraft, ground vehicles, ships and spacecraft. Some of the main applications are briefly described in the following paragraphs.

1. Navigation. Navigation systems are invaluable for soldiers to navigate their way in unfamiliar enemy territory (Figure 13.33). They are replacing the conventional magnetic compass used by soldiers for navigation. They can also be used by special forces and crack teams to reach and destroy vital enemy installations. As an example, GPS receivers were used extensively by the US soldiers during Operation Desert Storm and Operation Iraqi Freedom. The soldiers using this system were able to move to different places in the desert terrain even during sandstorms or at night. More than 9000 such receivers were used during the mission.

Figure 13.33 Soldiers using GPS receivers (Reproduced by permission of © Aerospace Corporation)

Figure 13.34 GPS-based inertial guidance system (Reproduced by permission of © Aerospace Corporation)

2. Tracking. The services of navigation satellites are also utilized to track potential targets before they are declared hostile to be engaged by various weapon platforms. The tracking data is fed as input to modern weapon systems such as missiles and smart bombs, and so on.
3. Bomb and missile guidance. The GPS and GLONASS systems are used to guide bomb and missiles to targets and position artillery for precise fire even in adverse weather conditions. Cruise missiles commonly used by the USA use multichannel GPS receivers to determine accurately their location constantly while in flight. The multiple launched rocket system (MLRS) vehicle uses GPS-based inertial guidance to position itself and aim the launch box at a target in a very short time (Figure 13.34). GPS system was also used extensively by US military during the Balkans bombing campaign in 1999, the Afghanistan campaign in 2001–2002 and in Iraq in 2003.
4. Rescue operations. Satellite navigation systems prove invaluable to the military for determining the location of causality during operations and in navigating rescue teams to the site.
5. Map updation. These systems augment the collection of precise data necessary for quick and accurate map updation.

13.13.2 Civilian Applications

Initially developed for military applications, satellite navigation systems soon became commonplace for civilian applications as well. In fact, civil applications outnumber military uses in terms of range of applications, number of users and total market value. Satellite navigation systems are finding newer and newer commercial applications due to decreasing cost, size and introduction of new features. Civilian applications include marine and aviation navigation, precision timekeeping, surveying, fleet management, mapping, construction & surveying, aircraft approach assistance, geographic information system (GIS), vehicle tracking, natural resource and wildlife management, disaster management and precision agriculture and so on.

1. For mapping and construction. Mapping, construction and surveying companies use satellite navigation systems extensively as they can provide real time submetre and centimetre

Figure 13.35 Use of GPS satellites for mapping and construction (Reproduced by permission of © Leica Geosystems)

level positioning accuracy in a cost-effective manner. They are mainly used in road construction, Earth moving and fleet management applications. For these applications, receivers along with wireless communication links and computer systems are installed on board the Earth moving machines (Figure 13.35). The required surface information is fed to this machine. With the help of real time position information, an operator obtains information as to whether the work is in accordance with the design plan or not. As an example, the tunnel under the English Channel was constructed with the help of the GPS system. The tunnel was constructed from both ends. The GPS receivers were used outside the tunnel to check their positions along the way and to make sure that they met exactly in the centre. Satellite navigation systems are also used for telecom power placement, laying of pipelines, flood plane mapping, oil, gas and mineral exploration and in glacier monitoring.
2. Saving lives and property.   Many police, fire and emergency medical service units employ GPS receivers to determine which available police car, fire truck or ambulance is nearest to the emergency site, enabling a quick response in these critical situations. GPS-equipped aircraft monitor the location of forest fires exactly, enabling the fire supervisors to send firefighters to the required spot on time.
3. Vehicle tracking and navigation.   Vehicle tracking is one of the fastest growing satellite navigation applications today. Many fleet vehicles, public transportation systems, delivery trucks and courier services use GPS and GLONASS receivers to monitor their locations at all times. These systems combined with digital maps are being used for vehicle navigation applications. These digital maps contain information like street names and directions, business listings, airports and other important landmarks. Such units provide useful information about the car's position and the best travel routes to a given destination by linking itself to a built-in digital map (Figure 13.36).

One of the emerging uses of the GPS and GLONASS systems is for air traffic control (ATC). Here these systems are used for navigating and tracking the aircraft while in flight (Figure 13.37). This helps in efficient routing (and hence in saving fuel) and in closer spacing of plane routes in the air. They are also used in maritime navigation applications for vehicle tracking and traffic management, and so on.

Figure 13.36 Satellite navigation systems for cars (Courtesy: Paul Vlaar)

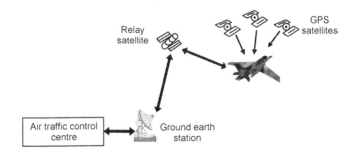

Figure 13.37 Use of navigation satellites in air traffic control

4. *Environmental monitoring.* GPS-equipped balloons monitor holes in the ozone layer across the globe. Buoys tracking major oil spills transmit data using the GPS to guide the clean-up operations. GPS systems are also used in wildlife management and insect infestation. They are also used for determination of forest boundaries.
5. *Monitoring structural deformations.* Navigation systems are used for measuring deformations on the Earth's crust. This helps in the prediction of earthquakes and volcanic eruptions. Geophysicists have been exploiting the GPS since the mid-1980s to measure continental drift and the movement of the Earth's surface in geologically active regions. They are also used for monitoring the deformation of dams, bridges and TV towers.
6. *Archaeology.* Archaeologists, biologists and explorers are using the satellite navigation systems to locate ancient ruins, migrating animal herds and endangered species.
7. *Utility industry.* Navigation systems are of tremendous help to the utility industry companies like electric, gas, water companies, and so on. Up-to-date maps provided by the navigation systems help these companies to plan, build and maintain their assets.
8. *Precision farming.* Farming systems employ navigation receivers to provide precise guidance for field operations and in the collection of map data on tillage, planting, weeds, insect and disease infestations, cultivation and irrigation.
9. *Precise timing information.* GPS satellites provide precise and accurate timing information within 100 ns of the universal time coordinated (UTC) atomic clock. The receiver required for this application is different as well as more expensive than the standard GPS

receivers. It is used in applications such as telecommunications and scientific research. It is also used for precise transfer of time between the world's timing centres and helps in tracking deep space vehicles. It is a source of precise time for various military and intelligence operations.

13.14 Future of Satellite Navigation Systems

Satellite based navigation systems are being further modernized so as to provide more accurate and reliable services. The modernization process includes development of newer satellite navigation systems, launch of new more powerful satellites, use of new codes, enhancement of ground system, and so on. In fact satellite based systems will be integrated with other navigation systems so as to increase their application potential.

GPS system is being modernized so as to provide more accurate, reliable and integrated services to the users. The first efforts in modernization began with the discontinuation of selective availability feature, so as to improve the accuracy of the civilian receivers. In continuation of this step, Block IIRM satellites carry a new civilian code on the L2 frequency. This helps in further improving the accuracy by compensating for atmospheric delays and ensures more navigation security. Moreover, these satellites carry a new military code (M-code) on both the L1 and L2 frequencies. This provides increased resistance to jamming. These satellites also have more accurate clock systems.

Block-IIF satellites have a third carrier signal, L5, at 1176.45 MHz. They also have longer design life, fast processors with more memory and a new civil signal. Third phase of GPS satellite system (GPS-III) are in the planning stage. These satellites will employ spot beams. Use of spot beams results in increased signal power, enabling the system to be more reliable and accurate, with the system accuracy approaching a metre.

Plans for GLONASS modernization in the next decade include shifting from FDMA system to CDMA system like GPS.

In addition to the continuing modernization of GPS and GLONASS systems, several new satellite constellations are expected to take shape over the next decade or so. Two types of navigation systems are being planned to be launched. The first type of system is an independent system comprising satellites in MEO orbits which can operate independently of GPS or GLONASS systems. These include the European Galileo system and the Chinese Compass system. The second type of system will comprise smaller sets of satellites in MEO or GEO orbits and will augment the GPS or GLONASS system.

The first Galileo satellite was launched on 28 December 2005 and the second one on 26 April 2008. Four Galileo IOV satellites have been launched. Once this In-Orbit Validation (IOV) phase has been completed, the remaining operational satellites will be placed in orbit so as to reach the full operational capability. The fully operational Galileo system will comprise 30 satellites (27 operational and three active spares), positioned in three circular Medium Earth Orbit (MEO) planes at an altitude of 23 222 km above the Earth and with each orbital plane inclined at 56 degrees to the equatorial plane. The system will be operational in the near future.

China is launching its own navigation system, referred to as Compass. Four satellites have been launched in the system. The Compass navigation system will comprise 30 MEO satellites and five GEO satellites. The Japanese Quasi Zenith Satellite System (QZSS) is being designed to augment the GPS satellite navigation system and also to operate independently.

QZSS satellites will occupy inclined, eccentric MEO orbits chosen specifically for optimal high-elevation visibility for users in Japan. India is planning a navigational system, IRNSS.

Navigation systems from different countries will be compatible and interoperable with the GPS system, creating a truly robust, world-wide, multi-component global navigation satellite system (GNSS). This will result in improved navigation services and the users will be able to get position information with the same receiver from any of the satellites of these systems resulting in improved accuracy, better reception and altogether new applications. The recent signing of a cooperative agreement between the United States and the European Union will expand the GPS system, laying the foundation for a compatible and interoperable GNSS. Hybrid positioning systems making use of combination of various positioning technologies (satellite based, cellular tower signals, wireless internet signals, Bluetooth sensors etc.) will provide truly global and optimized services.

All these developments will expand the horizon of the applications of satellite navigation systems to newer dimensions. In fact, the future of satellite navigation systems is as unlimited as one's imagination.

Further Readings

El-Rabbany, A. (2002) *Introduction to GPS: The Global Positioning System*, Artech House, Boston, Massachusetts.

Gatland, K. (1990) *Illustrated Encyclopedia of Space Technology*, Crown, New York.

Kaplan, E.D. (1996) *Understanding GPS: Principles and Applications,* Artech House, Boston, Massachusetts.

Larijani, L.C. (1998) *GPS for Everyone: How the Global Positioning System Can Work for You,* American Interface Corporation.

Leick, A. (2003) *GPS Satellite Surveying,* John Wiley & Sons, Inc., New York.

McNamara, J. (2004) *GPS for Dummies,* For Dummies.

Prasad, R. and Ruggieri, M. (2005) *Applied Satellite Navigation Using GPS, GALILEO and Augmentation Systems,* Mobile Communications Series, artech House, Boston, Massachusetts.

Rycroft, M.J. (2003) *Satellite Navigation Systems: Policy, Commercial and Technical Interaction,* Springer.

Vacca, J. (1999) *Satellite Encription,* Academic Press, California.

Verger, F., Sourbes-Verger, I., Ghirardi, R., Pasco, X., Lyle, S. and Reilly, P. (2003) *The Cambridge Encyclopedia of Space,* Cambridge University Press.

Internet Sites

1. www.aero.org
2. www.astronautix.com
3. http://electronics.howstuffworks.com/gps.htm/printable
4. http://www.aero.org/education/primers/gps/GPS-Primer.pdf
5. http://www.trimble.com/gps/
6. http://www.lowrance.com/Tutorials/GPS/gps_tutorial_01.asp
7. http://www.gisdevelopment.net/tutorials/tuman004.htm

8. http://www.glonass-center.ru/frame_e.html
9. www.gorp.away.com
10. www.skyrocket.de

Glossary

Almanac data: Almanac data tell the GPS receiver where each satellite should be at any time during the day

Coarse acquisition code (C/A code): The C/A code is an unencrypted civilian code comprising 1023 bits and having a bit rate of 1.023 Mbps

Compass satellite navigation system: Compass is the second generation of the BeiDou satellite navigation system of China, also known as BeiDou-2. The Compass satellite navigation system is intended to be a global navigation system and is currently under development.

Control segment: The control segment comprises a network of monitor stations for the purpose of controlling the satellite navigation system

Differential GPS: Differential GPS systems employ a receiver at a known position and then transmit corrections based on the measurements for this receiver to other receivers in the area

Ephemeris data: Ephemeris data contain information about the health of the satellite, current date and time

European Geostationary Navigation Overlay Service (EGNOS): EGNOS is the first pan European satellite based navigation system and is a joint venture of three organizations namely the European Space Agency (ESA), the European Commission and Eurocontrol.

Galileo satellite navigation system: Galileo is a global satellite navigation system being built by the European Union and the European Space Agency under the Galileo programme. It is intended to provide to the European countries a fully autonomous and reliable satellite based positioning, navigation and timing capability independent of the US GPS and Russian GLONASS.

Global Navigation System (GLONASS): The GLONASS is a Russian navigation system comprising 21 active satellites and provides similar continuous global positioning information as the GPS

Global Positioning System (GPS): The GPS is an American satellite-based navigation system that employs a constellation of 24 satellites to provide three-dimensional position, velocity and timing information to all users worldwide 24 hours a day

Hybrid navigation: A hybrid navigation or positioning system makes use of a combination of various positioning technologies, including those offered by satellite based global navigation systems such as GPS, cellular tower signals, wireless internet signals, Bluetooth sensors and so on. It overcomes the limitations of global satellite navigation systems such as GPS in respect of sub-optimal performance in dense urban areas.

Indian Regional Navigational Satellite System (IRNSS): The IRNSS is an independent regional satellite navigation system intended to provide uninterrupted navigation services including the standard positioning service for civilian use and an encrypted service for authorized users.

Navigation: Navigation is the art of determining the position of a platform or an object at any specified time

Point positioning systems: In a point positioning system, the GPS receiver calculates its location using the satellite ranging information, without the help of any other receiver or equipment

Precision positioning system (PPS): The PPS is a highly accurate military positioning, velocity and timing GPS service that is available to only authorized users worldwide

Precision code (P code): The P code is the encrypted military code comprising a stream of 2.35×10^{14} bits at a modulation rate of 10.23 Mbps

Pseudorandom code (PRN code): PRN codes are long unique digital patterns transmitted by each GPS satellite

Pseudorange: The distance between the receiver and the satellite used to determine the position of the receiver

Space segment: The space segment consists of a constellation of navigation satellites that send navigation signals to the users

Standard positioning system (SPS): The SPS is a positioning and timing service available to all GPS users worldwide, on a continuous basis without any charge

Transit system: Transit was the first satellite-based navigation system

User segment: The user segment includes all military and civilian receivers used to provide position, velocity and time information

14

Scientific Satellites

Scientific satellites provide space-based platforms to carry out fundamental research about the world we live in, our near and far space. Prior to the development of satellite-based scientific missions, our access to the universe was mainly from ground-based observations. Use of satellites for scientific research has removed constraints like attenuation and blocking of radiation by Earth's atmosphere, gravitational effects on measurements and difficulty in making *in situ* or closed studies imposed by Earth-based observations. Moreover, satellite-based scientific research is global by nature and helps in understanding the various phenomena at a global level. There are two approaches to scientific research in space. Firstly, scientific instruments are carried on board satellites whose primary mission is not scientific in nature. For instance, these instruments have been put on board the remote sensing and weather forecasting satellites. Secondly, a large number of dedicated satellites have been launched for the purpose. They have the advantage that all their parameters are optimized keeping the scientific mission in mind.

This chapter focuses on scientific applications of satellites covering in detail the contributions made by these satellites to Earth sciences, solar physics, astronomy and astrophysics. Major scientific satellite missions launched for each of these applications are listed. Payloads carried by these satellites are also discussed. Like previous chapters on satellite applications, this chapter also contains a large number of illustrative photographs.

14.1 Satellite-based versus Ground-based Scientific Techniques

Satellites have added a new dimension to scientific research as they have enabled scientists to study the entire Earth and its atmosphere and have revealed the truly violent nature of our vast universe. Ground-based observations are severely limited by Earth's atmosphere as it absorbs a large part of the electromagnetic spectrum, including lower frequency radio waves, extreme frequency UV radiation, X-rays and gamma rays. Hence, study of distant planets, stars and galaxies that are based on the electromagnetic radiation emitted by these celestial bodies had been restricted to a very narrow band of the electromagnetic spectrum. With the advent of satellites for scientific missions, the whole electromagnetic spectrum is available for making observations. In fact, the clarity, finesse and depth with which the universe is known today is due to the use of satellites.

In addition, Earth-based studies are also severely hampered by bad weather conditions, pollution and background heat radiation emitted by the Earth. Moreover, satellites have enabled great progress to be made in the field of material and life sciences as they provide platforms to carry out research under microgravity conditions, enabling the development of newer crystals, better understanding of the various life phenomena and so on.

On the other hand, the ground-based measurements have the advantage of low cost and relative simplicity plus the ability in many cases to obtain data continuously. Hence, newer ground-based techniques are being developed even today and in many situations they complement the data collected by the satellites.

14.2 Payloads on Board Scientific Satellites

Scientific satellites carry a variety of payloads depending upon their intended mission. The main application areas of these satellites include space geodesy (study of Earth), study of Earth's atmosphere, the solar system and the universe. As the application spectrum of scientific satellites is very large, therefore the range of payloads carried by them is innumerable. In this section, we describe in brief the types of payloads carried by satellites intended for various categories of scientific applications. Their detailed description is beyond the scope of the book. A start will be made with payloads on board satellites studying Earth geodesy, followed by satellites used for studying Earth's upper atmosphere and lastly, the astronomical satellites.

14.2.1 Payloads for Studying Earth's Geodesy

Satellites used for space geodesy (study of Earth) applications are GPS satellites or satellites that carry synthetic aperture radar (SAR), radar altimeters, laser altimeters, accelerometers and corner reflectors, and so on. Radar altimeters are used to measure the distance between the satellite and the ground surface by measuring the time delay between the transmission of a microwave pulse and its reception after scattering back from the Earth's surface. Laser altimeters work on the same principle as radar altimeters, but they use lasers instead of using radar. Radar and laser altimeters are carried on board satellites employing space altimetry techniques for geodynamic studies. Some of the recently launched satellites having radar altimeters are GeoSat (geodetic satellite) Follow-on, Jason and EnviSat (enviromental satellite) satellites. Accelerometers are mechanical or electromechanical devices used for measuring gravity and are used on board satellites for space gradiometery applications. One of the important satellite missions carrying accelerometers is the European Space Agency's GOCE project (gravity field and steady state ocean circulation explorer). Corner reflectors, as the name suggests, are reflectors or mirrors that reflect radiation back in the original direction from where it came. Satellites carrying corner reflectors make use of laser ranging for geometrical geodesy applications. Examples include the French Starlette and Stella satellites, the American Lageos-1 and 2, EGS (experimental geodetic satellite), Etalon-1 and 2 and GFZ (Geo Forschungs Zentrum) satellites.

14.2.2 Payloads for Earth Environment Studies

Earth environment studies include studies of the ionosphere, magnetosphere and upper atmosphere. In the following paragraphs, payloads used for these applications will be discussed.

14.2.2.1 Ionospheric and Magnetospheric Studies

Satellites for studying the ionosphere have payloads like ionospheric sounders, spectrometers, spectrographs, photometers, imagers, charged particle detectors, plasma detectors, radar, telescopes, and so on. Ionospheric sounders comprise a transmitter–receiver pair that is used to measure the effective altitude of the ionospheric layers by measuring the time delay between transmission and reception of radio signal. Radio signals are generally stepped or swept in the frequency domain in order to obtain the response of the ionosphere to the whole frequency spectrum. Satellites employing ionospheric sounders include the Alouette-1 and 2 and ISIS-1 (international satellites for ionospheric studies) and 2 of Canada.

Charged particle detectors are used to measure composition and concentration of charged particles in the ionosphere. Charged particle detectors commonly used include mass and energy spectrometers and time-of-flight spectrometers. Mass spectrometers are devices that apply magnetic force on charged particles to measure mass and relative concentration of atoms and molecules. Solar Observatory (SOHO), Cassini orbiter and Orbiting Geophysical Observatory (OGO) spacecraft have experiments based on mass spectroscopy. Time-of-flight spectrometers measure the time taken by charged sample molecules to travel a known distance through a calibrated electric field. Photometers are instruments used for measuring visible light. In the current context, they measure light produced by chemical reactions in the upper atmosphere, (mainly for auroral imaging). OSO, TERRIERS satellites carried photometers on board them.

Satellites for studying the magnetosphere carry instruments similar to those carried by satellites studying the ionosphere. In addition, they have magnetometers for measuring the strength of the magnetic fields. Magnetometers also provide information on polar auroras.

14.2.2.2 Study of the Upper Atmosphere – Measuring the Ozone Profile

Satellites measure the ozone profile and ozone levels over the entire globe on a near-daily basis by using instruments that either scan the IR radiation emitted by ozone [LIMS (limb infrared monitor of the stratosphere) instrument on Nimbus-7] or compare incident solar radiation to the radiation backscattered from the atmosphere [SBUV instrument on Nimbus-7, NOAA-9, 11, 13 and 16, TOMS instrument on Nimbus-7, Meteor 3-05, ADEOS-1 (advanced Earth observing satellite) and 2 and the TOMS Earth probe] or by measuring the decrease in solar intensity caused by ozone absorption as solar rays pass through the atmosphere to the spacecraft during sunrise and sunset [SAGE (stratospheric aerosol and gas experiment) on AEM-2 (Application Explorer satellite), Meteor and ERBS (Earth radiation budget satellite) satellites]. Figure 14.1 shows a photograph of the SAGE instrument.

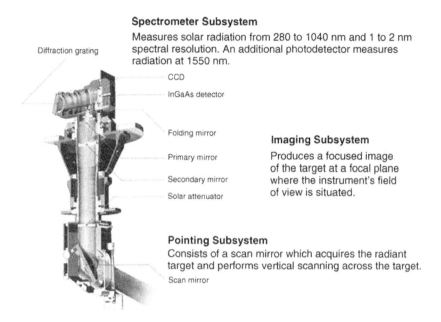

Figure 14.1 SAGE instrument (Courtesy: NASA Ames Research Center)

14.2.2.3 Study of the Upper Atmosphere – Earth's Radiation Budget

Earth's radiation budget (ERB) measurement instruments basically measure and compare visible and UV solar radiation with the radiation reflected from Earth and the thermal IR radiation of the atmosphere.

14.2.3 Payloads for Astronomical Studies

Astronomical satellites study various celestial bodies in the universe by detection and analysis of electromagnetic radiation and photons. Studies are carried out in all bands including the optical, IR, UV, radio, X-ray, gamma ray and cosmic ray bands. These studies are carried out either by space observatories or by space probes. Space observatories are satellites that orbit around the Earth and mainly comprise a telescope for making astronomical observations. Space probes are missions launched into space to orbit a particular celestial body other than Earth (including other planets of the solar system, comets, moon, asteroids, etc.) or to study various planets as fly-by missions while passing through the solar system.

Basic configuration of space observatories includes a very large telescope to gather radiation, scientific instruments to convert the gathered radiation into electrical signals and a processing unit to convert these electrical signals into the desired digital format. Telescopes are devices used in astronomy to see distant planets, galaxies, stars, and so on. Generally when telescopes are mentioned, more often than not it is optical telescopes that are referred to. However, telescopes operating in the infrared band, UV band and even the radio band are also in use. Radiation collected by the telescope is focused on to scientific instruments which convert the radiation into electrical signals. These signals are then digitized for transmission on to Earth.

Optical telescopes use lenses and mirrors to magnify the light coming from deep in space to make the objects look bigger and closer. Optical telescopes are classified into three types based on their design, namely the reflecting, refracting and catadioptric types. In a refracting telescope, light is collected by a two-element objective lens and brought to the focal plane. A reflecting telescope uses a concave mirror for this purpose. Catadioptric telescopes (also referred to as mirror-lens telescopes) employ a combination of both mirrors and lenses. All telescopes use an eyepiece (located behind the focal plane) to magnify the image formed by the primary optical system. Hubble space telescope is a major space observatory operating in the optical region. It has a reflecting compound telescope (Figure 14.2) with two mirrors based on the Ritchey–Chretien design. Other optical telescopes in space include the advanced camera for surveys (ACS), large angle and spectrographic cornograph (LASCO) for the SOHO satellite, microvariability and oscillation of stars (MOST) and so on. The James Webb space telescope (JWST) will replace the Hubble space telescope around 2018.

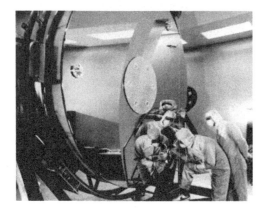

Figure 14.2 Reflecting compound telescope used in Hubble space telescope (Courtesy: NASA)

IR telescopes are similar to optical telescopes in design. It should be mentioned here that IR telescopes and their detector instruments have to be cooled to a temperature of a few Kelvin by immersing them in liquid nitrogen or helium in order to remove the background noise. Some of the observatories using IR telescopes include infrared space observatory (ISO), infrared astronomical satellite (IRAS), Spitzer space telescope, formerly known as the space infrared telescope facility (SIRTF), and the wide-field infrared explorer (WISE). The IRAS contained a 0.6 m Ritchey–Chretien telescope (Figure 14.3) cooled by helium to a temperature of near 10 K.

X-ray telescopes use mirrors that are nearly parallel to the telescope's line-sight so that X-rays enter at a grazing angle to these mirrors. The most commonly used materials for making these mirrors are gold and nickel. Several designs are based on the fact that X-rays can be focused by first reflecting them off a parabolic mirror followed by a hyperbolic mirror. Some of the well known designs are Kirkpatrick–Baez design and a couple of designs by Wolter. Skylab space station, orbiting solar observatory (OSO), high energy astrophysics observatory series (HEAO-1, 2 and 3), Chandra X-ray observatory, X-ray multi mirror satellite observatory (XMM–Newton), Smart-1 spacecraft and Nuclear Spectroscopic Telescope Array (NUSTAR) are examples of some of the space missions that make use of X-ray telescopes.

Figure 14.3 IRAS (Courtesy: NASA)

Gamma ray telescopes, unlike optical, IR and X-ray telescopes, do not work on the principle of the reflection of light as the gamma rays cannot be captured and reflected by mirrors. They are basically two-level instruments working on the principle of Compton scattering. Compton scattering occurs when a photon strikes an electron and transfers some of its energy to the electron. In the top level of the instrument, a gamma ray photon scatters off an electron in a scintillator. This photon travels down into the second level of the scintillator material which completely absorbs the photon. The interaction points at the two layers and energy deposited at each layer are determined by phototubes. Using this information, the angle of incidence of the cosmic photon can be determined. NASA's compton gamma ray observatory uses such a telescope named the compton telescope (COMPTEL). Figure 14.4 shows the

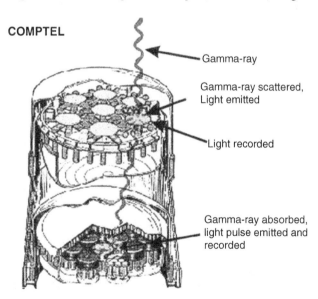

Figure 14.4 COMPTEL telescope (Courtesy: NASA)

conceptual diagram of the COMPTEL telescope. Other satellite missions employing gamma ray telescopes include the high energy transient explorer (HETE-2) mission, the international gamma ray astrophysics laboratory (INTEGRAL), Ulysses, the low energy gamma ray imager (LEGRI), the Fermi gamma-ray space telescope (GST), formerly known as the gamma-ray large area space telescope (GLAST), etc. Swift is a multi-wavelength observatory dedicated to the study of gamma-ray bursts.

Radio telescopes are instruments usually shaped like large antennas for collecting radio waves from celestial objects (such as pulsars and active galaxies). Incoming radiation is reflected from a parabolic dish antenna to a dipole situated at the focus of the dish antenna from which the signals are fed to a radio receiver. The output of the radio receiver provides information on frequency, power and timing of the emissions from various objects. They have been used in tracking the space probe 'deep space network'.

There are a variety of scientific instruments placed at the focal plane of telescopes to convert the radiation gathered and focused by these telescopes into electrical signals. These include electronic imagers, spectrographs, spectrometers, gamma ray detectors, X-ray instruments and sensitive radio receivers to name a few. Electronic imagers are electronic analogues of photographic films. Charged coupled devices (CCD) are the most commonly used imagers in the optical and UV regions of the spectrum. They have replaced photographic films for almost all space astronomical applications.

Spectrometers are optical instruments that measure various properties of light. Spectrometers generally operate in the near IR and UV bands. Near-IR spectrometers are used to map various planets and their satellites, looking for different minerals across their surface. They also study the cloud structure and gas composition of the atmosphere of these celestial bodies. UV and extreme UV spectrometers study the material composition of celestial bodies, structure and evolution of their upper atmosphere and physical properties of their clouds. Spectrometers are generally classified as spectroscopes and spectrographs. Spectrographs have superceded spectroscopes for scientific applications. Spectrographs are special astronomical instruments that, instead of taking pictures of an object, split up light into different colours of the spectrum for detection by various detectors. This is very useful in determining different properties of celestial bodies, including their chemical composition, temperature, radial velocity, rotational velocity and magnetic fields.

Hubble space telescope carries five instruments, which basically are cameras, spectrometers and spectrographs. Each instrument uses either a CCD or a photographic film to capture light. The light detected by the CCDs is digitized for storage on onboard computers for relaying back to Earth. IRAS, an infrared space observatory, has an array of 62 detectors used for an all-sky survey to detect infrared flux in bands centred at 12, 25, 60 and 100 µm. In addition to the array, a low resolution spectrometer and 60 and 100 µm chopped photometric channels are also present.

Gamma ray detectors can be broadly divided into two broad classes based on their principle of operation. The first class works on the same principle as the spectrometers and photometers used in optical astronomy and the second class is that of imaging devices. The detectors in the first class are further classified into scintillators and solid state detectors depending upon whether they transform gamma rays into optical or electronic signals for the purpose of recording. Gamma rays produce charged particles in the scintillator crystals that interact with these crystals and emit photons. These lower energy photons are subsequently collected by photomultiplier tubes. Solid state detectors are semiconductor devices (Ge, CdZnTe) that work on the same principle as that of scintillators, except that here electron–hole pairs are created rather than electron–ion pairs. Inorganic scintillators (made of inorganic materials like NaI or CsI) are the most commonly used for space applications. Some of the space missions employing inorganic scintillators are Compton gamma ray observatory (CRGO), High energy astrophysics observatory series (HEAO-1 and 2) and Rossi X-ray timing explorer (RXTE) mission. The imaging type of detectors work on the principle of Compton scattering (photoelectric ionization of the material by gamma rays) to calculate the direction of arrival of the incident photon or use a device that allows the image to be reconstructed.

X-ray instruments used in astronomy have to detect a weak source against a fairly strong background. Hence, source detection is done on photon-to-photon basis. For the kind of energies X-ray space detectors receive, photoelectric absorption is the main process on which these detectors work. Photoelectric absorption is the process where X-rays transfer their energies to electrons. Ionization detectors collect and count these electrons. Other detectors measure the heat released by these excited electrons when they go back to their original states. Various types of electron detectors used are proportional counters, microchannel plates, semiconductor detectors, scintillators, negative electron affinity detectors (NEAD) and single photon calorimeters.

Probes launched into space to study other planets or stars carry instruments similar to space observatories or satellites observing Earth's atmosphere, depending upon whether their intended mission is to take images of the celestial bodies or to study their atmospheres. Voyager satellites, launched to study the surface of Saturn and Jupiter and their atmospheres, carried payloads like the UV spectrometer, IR interoferometer spectrometer, magnetometer, charged particle analyser and cosmic ray system. Pioneer missions also launched to study Jupiter and Saturn carried instruments like non-imaging telescope for meteorite/asteroid detection, magnetometer for studying the magnetic fields, cosmic ray telescope, Geiger tube telescope, radiation detector, plasma analyser and photometer.

14.3 Applications of Scientific Satellites – Study of Earth

After presenting a brief overview of payloads carried by scientific satellites, attention will now be focused on the application potential of these satellites. Space constitutes an exceptional laboratory for most of the scientific missions. Study of Earth, its atmosphere, solar system and the universe are main application areas. In the present section and in sections to follow, some of these applications will be discussed in detail. The current section focuses on Earth science applications. Also discussed in brief will be the various scientific missions launched for carrying out experiments related to these fields.

Study of Earth includes investigations into Earth's gravitational field, its shape and structure, determination of sea levels and study of continental topography. The science dealing with all these measurements is referred to as geodesy and, when these measurements are carried out from space, they form what is known as space geodesy. Other than space geodesy, satellites also study the tectonics, internal geodynamics of Earth and terrestrial magnetic fields on Earth.

14.3.1 Space Geodesy

Geodesy is defined as the science of measurement of Earth. Space geodesy studies the shape of Earth from space, its internal structure, its rotational motion and geographical variations in its gravitational field. Two main techniques employed for space geodesy are geometric geodesy and dynamic geodesy.

Geometric geodesy determines the shape of the Earth, location of objects on its surface and the distribution of Earth's gravitational field by measuring the distances and angles between a large numbers of points on the surface of the Earth. Figure 14.5 (a) shows the concept of geometric geodesy measurements. Satellites are used to link these points (referred to as tracking stations) together to determine directions of these stations with respect to one another. This is done by determining the direction of tracking station–satellite vectors [Figure 14.5 (b)]. In fact, a network of the polyhedron of directions is obtained. The scale of the network is determined by measuring the length of at least one of the sides or by determining the exact positon of tracking stations. Both ground-based techniques as well as satellites can be used to determine the scale of the network.

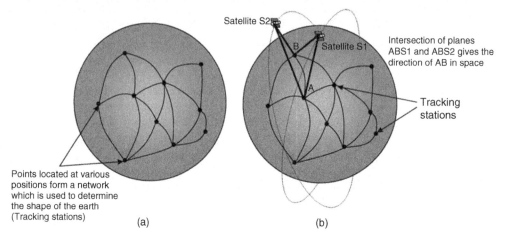

Figure 14.5 Use of satellites for performing geometric geodesy measurements

In the case of determining the position of tracking stations using satellites, use is made of the Doppler effect system or the GPS. Laser ranging techniques are used in the case of distance measurements using satellites. Specialized satellites launched for the purpose include the French satellites, Starlette and Stella (Figure 14.6), launched in the years 1975 and 1993 respectively, the American satellite Lageos-1 (1976) and 2 (1996), EGS (1989), Etalon-1 (1989) and 2 (1989) and the GFZ (1995) satellites. These are spherical satellites covered by corner reflectors to return a part of the laser beam back to its emitting station on Earth. By measuring the time of return of the signal, the distance between the tracking station (emitting station) and satellite is calculated. Several such measurements, when correlated, determine the distance between the tracking stations.

Dynamic geodesy aims to study variations in the Earth's gravitational field on the surface of the Earth by conducting a detailed analysis of satellite orbits. Earth's interior structure leads to complicated satellite orbits. By observing these orbits and with the knowledge of Earth's gravitational field distribution, the value of the gravitational field can be calculated along the orbit. Using this information, gravitational equipotential surfaces are constructed. The equipotential surfaces corresponding to the mean sea level, referred to as 'geoid', is then determined (Figure 14.7).

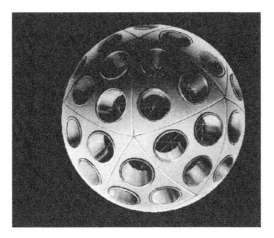

Figure 14.6 Stella satellite (Courtesy: Committee on Earth Observation (CEO))

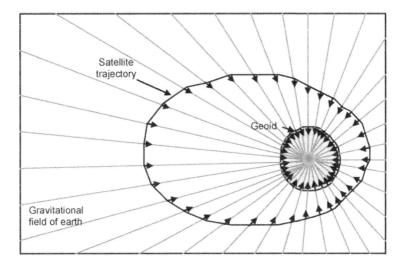

Figure 14.7 Equipotential surface

There are several global geoid models based on these two methods. The first such model was constructed way back in the year 1965 by USA. The best-known models today are the joint gravity model (JGM) of NASA and Texas University designed in the year 1996 and the GRIM-5-S1 model developed in 1999 by groups in France and Germany. The GRIM-5-S1 model is derived from satellite orbit perturbation analysis of 21 satellites.

Newer methods are being developed for more accurate gravity field measurements. These include space altimetry, space gradiometry and direct measurements of the gravitational potential using two satellites. Space altimetry is used to map the average surface of oceans using a radar altimeter on a satellite. The radar altimeter sends high repetition rate pulses towards the ocean and the time taken by these pulses for the return trip yields the distance between the satellite and the ocean surface. As the orbit of the satellite is known at all times, the distance between the satellite and the centre of the Earth can be determined at any particular instant of time. The distance between the centre of the Earth and the average surface of oceans is determined by vector subtraction of the distance between the satellite and the Earth's centre and the distance between the satellite and the ocean surface. The surface of the geoid is then determined by modelling the topography of the ocean surface. Using ground-based techniques it was not possible to include the ocean surface in geodetic networks, but satellites have made this possible. Figure 14.8 shows the use of satellite altimetry for determining the geoid height. Dynamic sea surface topography refers to the average difference between the actual surface of the Earth and the geoid. It is caused by a steady state ocean current field in the ocean. Some of the satellites used for the purpose include NASA's Skylab-4 (1974), GEOS-3 (geostationary scientific satellite) (1975), Seasat (1978), GeoSat (1985), ERS-1 (1991), ERS-2 (1995), TOPEX-Poseidon (an ocean topography experiment) (1992), GeoSat Follow-on (1998), Jason (2001) and EnviSat (2002) satellites.

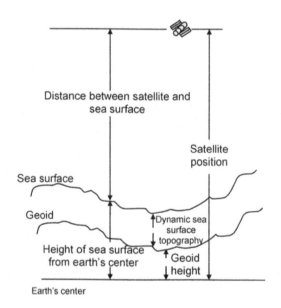

Figure 14.8 Use of satellite altimetry for determining geoid height

Space gradiometery is used for mapping fine variations in the Earth's gravitational potential by measuring the gradient (or derivative) of the gravitational field (in all three directions) on a single satellite (Figure 14.9). Satellites for this application are placed in LEO orbits at altitudes of 200 to 300 km and have ultrasensitive accelerometers for determining the gravity at various points. The European Space Agency's GOCE Project works on the principle of gradiometery and measures the gravitational field to an accuracy level of 1 mGal (milligallon) and the local level of geoid up to an accuracy of 1 cm. The project is used for geodynamics, tectonics, oceanography and glaciology.

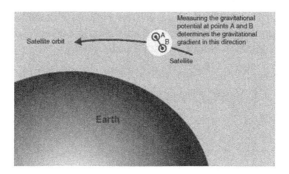

Figure 14.9 Space gradiometry

Variations in the radial velocities of two identical satellites orbiting very close to each other on the same orbit (Figure 14.10) can be used for gravity field measurements. The satellites are in the LEO having an altitude of approximately 200 km. The variations in velocity are proportional to relative variations of the gravitational potential at the satellite altitudes. One example of such a mission is the US–German GRACE (gravity recovery and climate experiment) project launched in March 2002. It has placed two satellites in the same LEO at a distance of 220 km from each other. The satellites communicate with each other via microwave links to perform these studies.

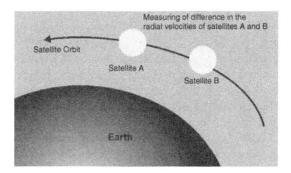

Figure 14.10 Variation in velocities of two satellites in the same orbit being used to measure the Earth's gravitational field

14.3.2 Tectonics and Internal Geodynamics

Scientific satellites are also used for tectonics and internal geodynamic measurements. In tectonics, slight movement of tectonic plates, fault systems and landslides are detected by calculating the precise positions of a network of beacons on the surface of the Earth using precise distance measurements of these beacons from the satellites.

In geodynamics, precise knowledge of Earth's combined gravitational and centrifugal force field is used to study slow and deep motions of the planet including rising land masses, variations in the sea level, subduction of oceanic plates and convection cells in the mantle.

This throws light on the history of continental movements. Fluctuations in the Earth's axis of rotation and angular velocity are measured by observing displacements of Earth stations through distance measurements from Earth stations to satellites.

14.3.3 Terrestrial Magnetic Fields

Several satellites have been launched to study the magnetic field of Earth and its variations. MagSat (magnetic field satellite), launched by NASA in the year 1979, was the first satellite to produce a complete instantaneous survey of Earth's magnetic field. Danish Oersted satellite launched in the year 1999 studies the terrestrial magnetism. CHAMP (Challenging minisatellite payload) and SAC-C (Sat.de aplicaciones cientificas) satellites, both launched in the year 2000, carried magnetometers and accelerometers for mapping the Earth's gravitational field and its variations.

14.4 Observation of the Earth's Environment

Phenomena on the surface of the Earth and its atmosphere are closely linked with each other. Earth's environment explorations include investigations into the dynamics and physicochemistry of the stratosphere, mesosphere and ionosphere, assessing the depletion of the ozone layer, studying the effect of solar radiation on the atmosphere, establishing the radiation budget of Earth at the top of the atmosphere, studying the cloud cover, atmospheric circulation and wind activity. In the past, observations related to Earth and its atmosphere were done separately, but today they are done together with the aim of producing a more global model of the Earth as a system.

Many specialized satellites have been launched in the last 35 years for better understanding of the atmosphere. They carry a number of detectors for diverse experimental measurements. In addition to data from these satellites, data collected by multipurpose satellites carrying one odd scientific instrument, Earth observation and weather forecasting satellites and ground instruments are also used. Large scale programmes aiming to study the Earth system as a whole make models based on these data. Some of these programmes are the Earth science enterprise (ESE) of NASA and Cornerstones of ESA. ESE [initially named the Mission to planet Earth (MTPE) and then ESE in 1998] aims to study how the Earth's system of air, land, water and life interact with each other. Satellites launched under this programme include mainly the Terra, Aqua and Aura satellites to study the Earth's environment. Data from Landsat-7 satellite also provides additional input to the data from these satellites. MTPE programme initiated in 1991 generated data about areas of environmental concern by launching satellites like UARS, shuttle-based space radar laboratories, TOPEX/POSIEDON, Sesat satellites and the TOMS spectrometer instrument flown on several satellites. Cornerstones programme of ESA has launched two satellites, namely SOHO and Cluster to study the activities of the sun and their effect on the Earth's environment.

SOHO and Cluster satellites also form a part of the ISTP (intersolar terrestrial physics) programme. ISTP was an international mission developed in the year 1977 to have a global and comprehensive understanding of the Earth–sun interaction and to further explore the Earth's atmosphere. Members of the programme included NASA, ESA, ISAS (Institute of space and astronautical science), IKI (Russian Space Research Institute) and more than 100 universities and research centres in 16 countries. Satellites launched under this programme include the Geotail, Wind, Polar, Equator-S, SOHO and Cluster satellites.

In the following paragraphs major areas of study carried out by scientific satellites *vis-à-vis* Earth's environment are discussed. The areas are broadly covered under the headings of study of the ionosphere and magnetosphere, study of the upper atmosphere and study of the interaction between Earth and its environment.

14.4.1 Study of the Earth's Ionosphere and Magnetosphere

Radio sounding techniques, used in the early 20th century to probe the Earth's atmosphere had established that Earth has an ionized atmosphere. However, the electromagnetic waves are blocked at around 300 km in the atmosphere. Hence, the only way to study the Earth's atmosphere above 300 km is through satellites. As a matter of fact, the first satellites to be launched, namely the Sputnik and Explorer satellites, studied radiation belts in the magnetosphere (called Van Allen belts). Satellites launched thereafter have studied the composition of the magnetosphere, ionospheric plasma and plasma waves, polar auroras, interaction with solar and cosmic radiation, etc. The magnetosphere, ionosphere and upper atmosphere are studied in order to understand the large scale flow of plasma and energy transfers at all heights of the atmosphere throughout the year. Activities in the ionosphere and magnetosphere are interrelated with each other; hence most scientific missions make observations on both the magnetosphere and the ionosphere.

14.4.1.1 Study of the Ionosphere

Ionosphere is the layer of the atmosphere between 50 and 500 km from the surface of the Earth that is strongly ionized by UV and X-rays of solar radiation. Satellites have studied the composition of the ionospheric plasma (plasma is the mix of positively and negatively charged ions, electrons and various gases) and plasma waves, polar auroras, interactions with solar winds and the effect of ionosphere on propagation of electromagnetic waves.

Ionospheric composition. Satellites have provided information on the electron and ion distribution, their temporal and spatial variations, their irregularities and resonances, the influence of incoming charged particles, cosmic and solar noise, polar cap absorption, solar wind penetration and ion species in the Earth's atmosphere. Figure 14.11 shows the spatial profile of main ions present in the ionosphere established using the data collected by the satellites. Data have also shown that the electron density undergoes a strong daily variation according to the position of the sun. Ionosphere is divided into various layers depending upon the electron density. A detailed description of these layers is beyond the scope of the book.

Polar aurora. Satellites have also helped to provide a global view of the polar aurora phenomenon and have helped to perform *in situ* measurements. The Earth's magnetic field prevents the solar wind from entering the Earth's atmosphere. However, some electrons in the solar wind are able to diffuse into the magnetic tail and are able to descend down to altitudes of around 100 to 300 km from the surface of the Earth. These electrons collide with oxygen and nitrogen atoms or molecules present at these altitudes in the atmosphere, raising them to excited states. When these atoms and molecules come to their normal states, they emit light

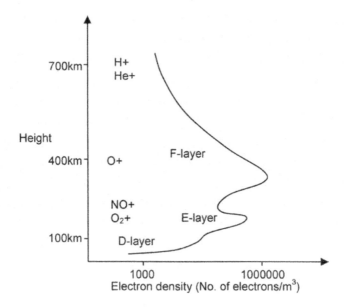

Figure 14.11 Spatial profile of the main ions present in the ionosphere

rays of a well-defined characteristic wavelength. This emission of light rays is named the polar aurora. Polar auroras are also caused by solar flares.

Colours of these auroras depend on the energies of the precipitating electrons as the energy levels decide the depth of their penetration into the atmosphere. The typical colours of the auroras are green (557.7 nm) and red (630 nm) from atomic oxygen (O) and blue (391.4 and 427.8 nm) from molecular nitrogen (N_2). Figure 14.12 shows a photograph of an aurora named aurora australis taken by an astronaut aboard Space Shuttle Discovery (STS-39) in the year 1991. Some of the important observations of polar auroras were made from the IMAGE

Figure 14.12 Photograph of aurora australis taken by an astronaut aboard Space Shuttle Discovery (STS-39) in 1991 (Courtesy: NASA)

(Imager for magnetopause-to-aurora global exploration) satellite, which detected both electron and proton aurora, and the Polar satellite, which observes X-rays from aurora.

Moreover, the auroral phenomenon mainly occurs at high latitudes having discontinuities in the magnetic field lines. The regions where it occurs frequently form an oval, slightly off-centre with respect to the Earth's magnetic pole, referred to as the auroral oval. Figure 14.13 shows the image of the auroral oval taken by the Dynamics Explorer satellite. Some of the major satellite missions launched to study the ionosphere include AEROS, AEROS B, POLAR Dynamics explorer and TIMED satellites launched by the United States as well as Alouette-1 and -2 and two ISIS satellites launched by Canada.

Two dynamic ionosphere cubesat experiment (DICE) satellites have been launched to map geomagnetic storm enhanced density (SED) plasma bulge and plasma formations in the ionosphere. The space shuttles launched by the United States also make observations on the ionosphere. ICON (Ionospheric connection) satellite is being planned to be launched in 2017 to further study the ionosphere.

Figure 14.13 Image of auroral oval taken by Dynamics Explorer satellite (Courtesy: NASA)

14.4.1.2 Study of the Magnetosphere

Magnetosphere is the region of the atmosphere that extends from the ionosphere to about 40 000 miles. It is the region where the Earth's magnetic field is enclosed. Several scientific missions have been launched since the 1960s for detailed studies of the magnetosphere. They have helped scientists understand the interaction between solar wind and Earth's magnetic field and the distribution of magnetic field lines and charged particles in the magnetosphere. In fact, all the present knowledge about the Earth's magnetosphere has been acquired by satellites. Satellites studying the magnetosphere generally have elliptical inclined orbits with a high apogee, with the exception of those satellites used for observing polar clefts and the lower magnetosphere. The inclination of these satellites is in the range of 61°–65° for satellites launched from Russia and in the range of 28°–34° for satellites launched from USA and Japan.

The different satellite missions have broadly studied the following main features of the magnetosphere.

Structure of the magnetosphere. Experiments carried out in space by satellites launched in the 1960s and 1970s had revealed the structure of the magnetosphere (Figure 14.14). They proved the theory that Earth is a magnetized planet and is surrounded by a geomagnetic field.

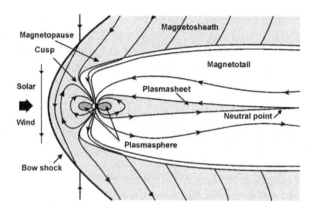

Figure 14.14 Structure of the Earth's magenetosphere (Courtesy: NASA)

This field interacts with the solar winds, comprising electrons and protons travelling away from the sun at 300–1000 km/s. When this wind comes in contact with the Earth's magnetic field, electric current flows. This current prevents the charged particles in the solar wind from entering the atmosphere and also prevents the geomagnetic field from spreading into the interplanetary space. This region where the magnetic field is confined is referred to as the magnetosphere. It is compressed on the dawn side (sunward side) and stretched away in the opposite direction towards the dusk side (night side). The sunward side of the magnetosphere is only 6 to 10 times the radius of Earth and the night side is around 200 times the Earth's radius (magnetotail).

Solar winds create an electric field of the order of several tens of mV/m, directed from the dawn side to the dusk side of the magnetosphere. The electric potential difference created by the field is around 60 kV to 150 kV between the two sides. Moreover, since charged particles of the solar wind are not able to cross the Earth's magnetic field lines, a shock wave is created at the magnetosphere boundary on the sun side. Some of the satellite missions that have made these observations include NASA's Interplanetary monitoring programme (IMP), the joint NASA–ESA International Sun–Earth explorer (ISEE) programme, the Soviet Prognoz satellite series and ISTP's (International solar terrestrial physics) Geotail satellite. The Geotail (Figure 14.15) satellite launched on 24 July 1992 studied the structure and dynamics of the tail region of the Earth's magnetosphere. It orbited at altitudes between 8 and 210 times the radius of the Earth to study the boundary region of the magnetosphere.

Charged particles in the magnetosphere. The Earth's magnetosphere is populated by energetic charged particles including high energy electrons and protons. These particles seldom penetrate the atmosphere as they are trapped by the Earth's magnetic field. They have complex orbits with a spiralling motion along the field line, a bouncing motion to and fro from north to south and back along the field line between two 'mirror' points and a drift in the longitude caused by nonuniformity in the magnetic field. These discoveries were made by numerous satellites launched between 1960 and 1980. Some of these satellite missions include Equator-S, IMAGE, IMEX (Inner magnetosphere explorer), THEMIS (Time history of events and macroscale interactions during substorms), MagSat, Oersted, IMP-8 and IMP-J satellites.

Determining the characteristics of these particles, including their distribution in space and movement in time, is very helpful in understanding the damage these particles cause to

Figure 14.15 Geotail satellite (Reproduced by permission of the Japan Aerospace Exploration Agency (JAXA))

electronic devices in space and their effect on the health of astronauts. Moreover, studying the origin of these particles gives us detailed information about how the continuously varying solar parameters control the Earth's space environment.

Thermal plasma in the magnetosphere. Outer regions of the magnetosphere are covered by a thermal component of the plasma (referred to as thermal plasma), which is a continuation of the ionospheric plasma. The region where this plasma exists with density greater than 50 electrons/cm^3 is called the 'plasmasphere'. The temperature of plasma corresponds to energies between fractions of an electron volt to several electron volts. It is composed of equal numbers of electrons and ions. The ions are mostly of hydrogen, and some of helium and oxygen. Thermal plasma moves along certain flow lines under the influence of the electric field and magnetic fields of the magnetosphere and Earth's rotation.

The GEOS (geostationary scientific satellite), ISEE, IMAGE and Viking satellites have made *in situ* measurements of the thermal plasma. These measurements have improved understanding of relationships that exist between the electric field, ionospheric conductivity and movement of ionospheric and magnetospheric plasma. The plasma has an important role in determining the electric charge present on Earth orbiting satellites. Figure 14.16 shows the false colour image of electrified plasma inside the Earth's magnetic field. The image is taken in the UV band by the extreme ultraviolet imager (EUV) on the IMAGE satellite. The sphere in the centre of the image is the Earth. The motion of plasma is traced using such images taken at different times. This in turn helps in forming global views of Earth's magnetic field and magnetic storms.

Magnetospheric waves. The Earth's magnetosphere is covered by electromagnetic waves of several types and frequencies, originating from Earth's environment, that of other planets and the solar wind. These signals are emitted over a wide frequency range, starting from tens of mHz to several MHz depending upon their origin. They are neither emitted in a continuous fashion nor are they observable in all regions of space or at the same time. These emissions are represented by the frequency–time graphs, where time is along the x axis and frequency

Figure 14.16 Electrified plasma (Courtesy: NASA). The image is the greyscale version of the original colour image. **The original image is available on the companion website at www.wiley.com/ go/maini**

along the y axis. Satellites carry special antennae to measure these low intensity waves in space. Some of the satellite missions, which have measured magnetospheric wave parameters, include GEOS-2, Ulysees probe, MagSat, Oersted, Cluster-II (Figure 14.17), the radiation belt storm probes (RBSP) mission and the time history of events and macroscale interactions during substorms (THEMIS) satellites.

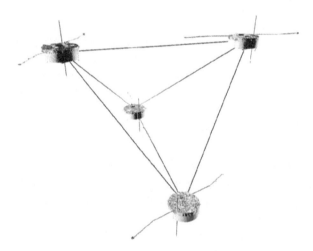

Figure 14.17 Cluster-II satellites (Copyright: ESA)

Future satellite missions include the magnetospheric multiscale mission (MMM) scheduled for launch in 2014. It is a four satellite solar–terrestrial probe designed to study the magnetic reconnection, charged particle acceleration and turbulence in the boundary regions of the Earth's magnetosphere. Another mission planned is the MAGCaT (magnetospheric constellation and tomography) mission comprising 16 satellites orbiting in the same plane and using radio frequency to probe the atmosphere.

14.4.2 Study of the Earth's Upper Atmosphere (Aeronomy)

The upper atmosphere includes the upper mesosphere, thermosphere and lower ionosphere up to an altitude of 600 km. It is characterized by a sharp increase in temperature due to UV absorption and stratification of constituent neutral gases. Several satellite missions have been launched to carry out diverse experiments for studying the upper atmosphere. These missions have increased knowledge of the mechanisms and processes taking place in the upper atmosphere, which in turn has improved forecasting of 'space weather'.

Measuring properties of the Earth's upper atmosphere. Satellites have helped measure the density, temperature, pressure and chemical composition of the upper atmosphere. Some of the important satellites for the purpose include the MAPS (measurement of air pollution from satellites) programme, UARS (upper atmosphere research satellite), Atmospheric explorer satellites, Air density explorer satellites, Terra and A-train constellation of satellites (comprising of four active satellites Aqua, CloudSat, CALIPSO and Aura satellites).

As an example, the MAPS instrument produced the first global measurements of atmospheric carbon monoxide in 1981 when it was flown aboard the Space Shuttle Columbia (STS-2). Figure 14.18 shows the measurements taken by the MAPS instrument in October 1984 when it was flown on the Space Shuttle Challenger (STS-41G). The image showed large concentrations of atmospheric carbon monoxide, caused by biomass burning in South America and southern Africa. Similar measurements from other satellites have enabled scientists to have detailed knowledge of properties of the upper atmosphere.

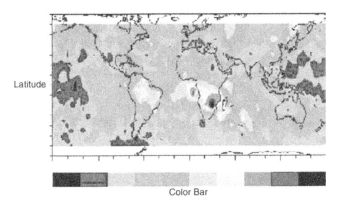

Figure 14.18 Carbon monoxide measurement by MAPS (Courtesy: Measurement of Air Pollution from Satellites (MAPS), NASA Langley Research Center, Hampton, Virginia VA23681)

Study of the influence of solar radiation on Earth's upper atmosphere. Earth's upper atmosphere is affected by the whole spectrum of solar radiation, unlike the lower atmosphere which is not affected by UV and X-ray radiations as they do not penetrate to the lower atmosphere. Moreover, heating effects of the sun influence the density of the atmosphere and its altitude distribution. When the atmosphere is heated, it expands and the density at high altitude increases, which exerts a further drag on satellites.

Satellites have also contributed significantly to understand the influence of solar activity on Earth's upper atmosphere. Some of the satellites that have contributed in this field include

Figure 14.19 UARS satellite (Courtesy: NASA)

the UARS (Upper atmosphere research satellite) (Figure 14.19), TIMED (Thermosphere ionosphere mesosphere energetics and dynamics), AMPTE (Active magnetospheric particle tracer explorers), SMM (Solar maximum mission), ERBS (Earth radiation budget satellite), SORCE satellite (Solar radiation and climate experiment) etc. Figure 14.20 shows images taken by the SABER (Sounding of the atmosphere using broadband emission radiometers) instrument on the TIMED satellite before a solar storm on 10 April 2002 and during the storm on 18 April 2002. The images show the levels of nitric oxide (an important cooling agent in the upper atmosphere) at 110 km altitude changing from dramatically low levels before the storm to high levels during the storm. The image taken on 18 April 2002 shows the effects of nitric oxide being transported from polar auroral regions towards the equator by upper atmospheric winds. The movement of nitric oxide can be used to track upper atmospheric wind patterns. These data show how the upper atmosphere's temperature structure and wind patterns change during solar storms. Such images from satellites help scientists to understand the sun–Earth connections better.

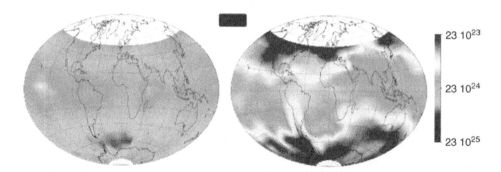

Figure 14.20 Images taken by the SABER instrument on the TIMED satellite before a solar storm on 10 April 2002 and during the storm on 18 April 2002 (Reprinted from John Hopkins APL Technical Digest by permission (TIMED Science: First Light, Vol. 24, number 2)

14.4.3 Study of the Interaction between Earth and its Environment

Satellites measure the profile of the ozone layer, Earth's radiation budget and so on to help scientists have a better understanding of the Earth's environment and its interaction with Earth.

Ozone measurements. One of the major contributions of satellites to Earth's environment studies is in the detection of the ozone hole. The first satellite-based ozone measurements were done by the Echo-I satellite back in the year 1960. Since then, satellites have measured ozone over the entire globe every day in all types of weather conditions, even over the remotest areas. They are capable of measuring the total ozone levels, ozone profiles and elements of atmospheric chemistry. In fact, satellites have played a major role in revealing that the average temperatures are increasing while the ozone layer is depleting. The Nimbus-7 satellite confirmed the results of the earlier ground-based experiments that there was a hole in the ozone layer. Some of the major satellite missions that have contributed significantly to the ozone layer observations include the Nimbus-7, ADEOS, ERBS, SPOT-3 and 4, UARS and Aqua satellites. Figure 14.21 shows the size of the ozone hole measured by the TOMS instrument during the period 1980 to 2005. It can be easily inferred from the figure that the size of the ozone hole has increased over the years.

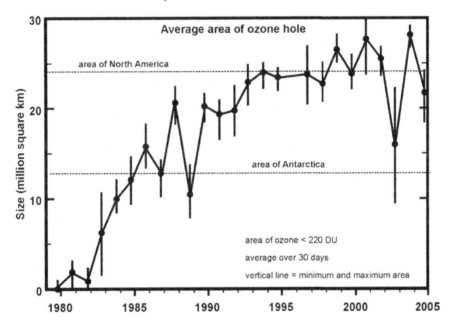

Figure 14.21 Size of the ozone hole measured by the TOMS instrument from 1980 to 2005 (Courtesy: NASA)

Earth's radiation budget. The Earth's radiation budget represents the balance between the incoming energy from the sun and the outgoing thermal (longwave IR) and reflected (shortwave IR) energy from Earth. It indicates the health of the global climate. The absorbed shortwave IR radiation (incident minus reflected) fuels the Earth's climate and biosphere systems. The longwave IR radiation represents the exhaust heat emitted to space. It can be used to estimate the insulating effect of the atmosphere (the greenhouse effect). It is also a useful indicator of the cloud cover and activity.

Several satellites with experiments to measure the Earth's radiation budget have been launched. Satellites launched for measuring the Earth's radiation budget follow LEO orbits, with the exception of the MSG satellite, which has a geostationary orbit. The MSG satellite carried onboard the ERBE (Earth radiation budget experiment) instrument and provided continuous observations in comparison to intermittent observations done by LEO satellites. Consequently, the ERBE instrument has helped scientists worldwide to have a better understanding of how clouds and aerosols as well as some chemical compounds in the atmosphere (greenhouse gases) affect the Earth's daily and long term weather. In addition, the ERBE data have helped scientists better understand how the amount of energy emitted by the Earth varies from day to night. These diurnal changes are also very important aspects of daily weather and climate. Figure 14.22 shows the total solar energy reflected back to space measured by the CERES (Cloud and Earth radiant energy sensor) instrument. White pixels in the image represent high reflection, green pixels represent intermediate reflection and blue ones show low reflection. It was inferred from the image that the presence of aerosols, particularly over the oceans, increases the amount of energy reflected back into space.

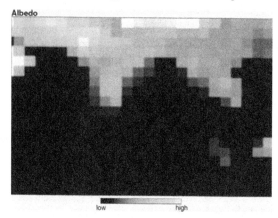

Figure 14.22 Total solar energy reflected back to space measured by the CERES instrument (Courtesy: NASA). The image is the greyscale version of the original colour image. **The original image is available on the companion website at www.wiley.com/go/maini**

Earth's surface and interface with the atmosphere. Satellites also study the interface parameters between the Earth's surface and its atmosphere. The data related to atmospheric physics and land surface studies collected by scientific satellites, remote sensing satellites and ground-based techniques are combined together to model the working of Earth as a system. Some of the satellites used for the purpose include multipurpose satellites like Terra, EnviSat, EO1, EOS (Earth observing system) PM Aqua and ALOS (advanced land observing satellite).

14.5 Astronomical Observations

Astronomy is the science involving observation and explanation of events occurring beyond the Earth and its atmosphere. These observations are done to study the position of astronomical objects in the universe (astrometry), study the physics of the universe including the physical properties of astronomical objects (astrophysics), study the origin of the universe and its evolution (cosmology) and so on. Astronomical studies are carried out by detection and analysis of electromagnetic radiation and are subclassified into various types depending upon the wavelength band in which the observations are made. Optical astronomy deals with the optical band, infrared astronomy with the IR band, radio astronomy with waves in the millimetre and decametre bands and high energy astronomy with X-rays, gamma rays, extreme UV rays, neutrino and cosmic rays. Space-based techniques have increased the pace of studies in these fields manyfold as they have made it possible to make observations in the whole wavelength spectrum. Gamma rays, X-rays, UV and IR waves are either partially or wholly blocked out by the atmosphere. Hence, ground-based astronomical observations are carried out only in the optical and radio frequency bands. In other words, space-based techniques have made it possible to carry out studies in the gamma-ray, X-ray, UV and IR bands, that are either partially or wholly blocked out by the atmosphere.

In the initial stages of the space era, satellites for making astronomical observations mostly had LEO orbits with inclinations depending upon the location of the launch base. Today, they are launched in all possible orbits, including elliptical orbits having a large apogee distance, sun-synchronous orbits and so on. These satellites, referred to as space observatories, carry sophisticated instruments for observation of distant planets, galaxies and outer space objects. Examples include COS-B (Cosmic ray satellite), EXOSAT (European X-ray observatory satellite), Astron-1, IUE (International ultraviolet explorer), ISO (Infrared space observatory), Granat, Prognoz-9 satellites, IRAS (Infrared astronomical satellite), COBE (Cosmic background explorer), HST (Hubble space telescope), Herschel space observatory, HETE-2 (High energy transient explorer), INTEGRAL (International gamma ray astrophysics laboratory), AGILE, XMM-Newton and so on. Several space probes orbiting around other planets, sun, asteroids and comets have also been launched for carrying out close-up studies of these astronomical bodies. Some probes have been launched in interplanetary orbits as fly-by missions to study various planets during their flight. This section talks about missions launched for making solar observations, while the next two sections discuss missions to study the solar system and the celestial bodies outside the solar system.

14.5.1 *Observation of the Sun*

Solar observations form the most important component of astronomical studies as sun is the nearest star to Earth and has a very strong influence on the Earth's environment. Moreover, a study of solar radiation and magnetism is an essential factor in understanding the structure of the terrestrial atmosphere. In fact, the sun was the first celestial body to be studied in the space era. Satellites enabled the scientists to make continuous observations of various solar phenomena, including a simultaneous long term observation of solar radiation over a range of different wavelengths, together with measurements of the magnetic field and the Doppler–Fizeau effect. Moreover, space-based solar observations have better resolution than those obtained using ground techniques. Earth-based observations have a maximum angular resolution of 1 arc sec whereas satellite-based instruments can have an angular resolution up to 0.1 arc sec.

Instruments used for making solar observations essentially consist of a photon collector of variable angular resolution and a light analyser to separate the wavelength regions. Observational quality depends on the pointing accuracy and the stability of the platforms on which the instruments are placed. Solar observations are carried out mainly in the visible, UV and X-ray wavelength bands with only a very few observations in the IR band. A systematic study of the sun was initiated with the launch of the OSO (Orbiting solar observatory) series of satellites. Eight OSO satellites, namely OSO-1 to OSO-8, were launched in a span of 13 years between 1962 and 1975. Since then a large number of space missions have been launched for the purpose. Some of the important missions include the Apollo telescope mount (ATM) of Skylab (1973–1974), the Solar maximum mission (SMM) (1980), the Wind satellite (1994), SOHO (1995), ACE (Advanced composition explorer) (1997), TRACE (Transition region and coronal explorer) (1998), HST (1990), Ulysses (1990). RHESSI (Reuven Ramaty high energy solar spectro scopic imager) (2002), STEREO (Solar terrestrial relations observatory) (2006), Hinode (2006), CORONAS Photon (Complex orbital observations of near Earth activity of the sun) (2009), SDO (Solar dynamics observatory) (2010) and IRIS (Interface region imaging spectrograph) (2013). Figures 14.23 and 14.24 show photographs of the SOHO solar observatory and the ACE satellite.

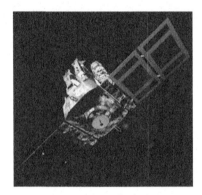

Figure 14.23 SOHO solar observatory (Courtesy: SOHO (ESA and NASA))

Figure 14.24 ACE satellite (Courtesy: NASA)

It may be mentioned here that some of the major satellite missions launched for solar studies, orbit around the sun rather than orbiting around the Earth. They are placed at Lanrange point L1, a point on the Earth–sun line where the gravitational pull from both bodies is equal. It is a very useful position for making continuous observations of various solar phenomena. Satellites placed at the L1 point include the Wind, SOHO, ACE, Triana and Genesis satellites. Figure 14.25 shows the orbit of the SOHO satellite. These missions carried out studies in the areas of solar physics, monitoring solar activity and the effect of solar radiation on the Earth's environment.

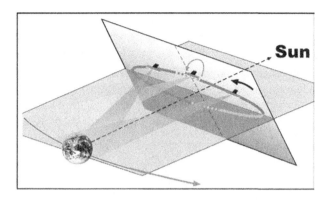

Figure 14.25 Orbit of the SOHO satellite (Courtesy: SOHO/CELIAS MTOF (ESA and NASA))

14.5.1.1 Solar Physics

The study of solar physics mainly includes the study of the dynamics and structure of the sun's interior and properties of the solar corona. All solar activities and variabilities are driven by the sun's internal magnetic field and by fluid motions that shear and twist that field. The sun's interior comprises a core, a radiative zone and a convective zone. Above this surface is the atmosphere, which comprises the photosphere, chromosphere and outer corona. Figure 14.26 shows a composite image of the sun formed by combining images taken by all the instruments on board the SOHO mission. The interior of the image, taken by the Michelson Doppler imager (MDI) of the satellite illustrates rivers of plasma underneath the solar surface. The surface was imaged with the Extreme ultraviolet imaging telescope (EIT) at 304 Å. Both the images were superimposed on a Large angle spectroscopic coronograph (LASCO) C2 image, which blocks the sun so that the corona is visible.

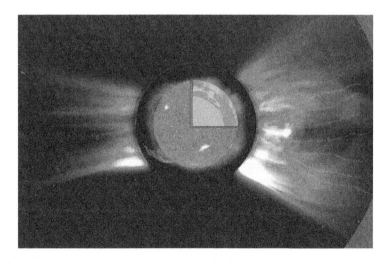

Figure 14.26 Composite image of the sun taken by the SOHO satellite (Courtesy: SOHO (ESA and NASA))

14.5.1.2 Solar Activity

Satellites have also helped in making observations of various features of the sun, such as sunspots, solar prominences and solar flares. Sunspots are dark, cool areas on the photosphere that always appear in pairs and through which intense magnetic fields break through the surface. Field lines leave through one sunspot and re-enter through another one. The magnetic field is caused by movements of gases in the sun's interior. Figure 14.27 shows the movement of various sunspots on an active region of the sun taken by the TRACE satellite.

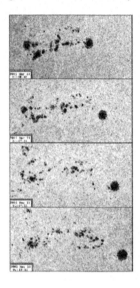

Figure 14.27 Movement of sunspots taken by the TRACE satellite (Courtesy: Transition Region and Coronal Explorer, TRACE, is a mission of the Stanford–Lockheed Institute for Space Research (a joint programme of the Lockheed–Martin Advanced Technology Center's Solar and Astrophysics Laboratory and Stanford's Solar Observatories Group) and part of the NASA Small Explorer Program)

Solar activity follows an 11 year cycle called the solar cycle, with periods of varying levels of solar activity from maximum to minimum. NASA launched the Solar maximum mission (SMM) (Figure 14.28) on 14 February 1980 to study the sun during the period of maximum solar activity. The SMM enabled scientists to examine in great detail the solar flares, which are considered to be the most violent aspect of solar activity. Sunspot activity also occurs as part of this 11 year cycle. Figure 14.29 shows the sun's 11 year solar cycle as reflected by the number of sunspots recorded to date and the projected (dotted line) number of sunspots.

Solar prominences are arches of gases that rise occasionally from the chromosphere and orient themselves along the magnetic lines from sunspot pairs. Prominences generally last two to three months and can extend up to 50 000 km or more above the sun's surface. Upon reaching this height above the surface, they can erupt for a time period of a few minutes to a few hours and send large amounts of material into space at speeds of around 1000 km/s. These eruptions are called coronal mass ejections (CME). Figure 14.30 shows images of two coronal mass ejections taken by the SMM satellite. The image is taken by blocking light from the sun using a black disc, creating an artificial eclipse in order to observe the dim light from the CME. Each row shows the evolution of CME with time.

Figure 14.28 Solar Maximum Mission (Courtesy: NASA)

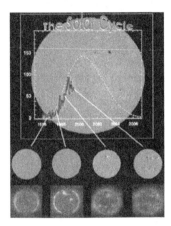

Figure 14.29 Sun's 11-year solar cycle (Courtesy: SOHO (ESA and NASA))

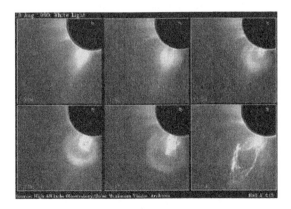

Figure 14.30 Coronal mass ejection (CME) (Courtesy: NASA)

Sometimes in complex sunspot groups, abrupt violent explosions occur from the sun due to sudden magnetic field changes. These are called solar flares. They are accompanied by the release of gases, electrons, visible light, UV light and X-rays. Figure 14.31 shows a photograph of the solar flare taken by the TRACE satellite on 22 November 1998. Several such images have helped to give an understanding of the activities of the sun in a more comprehensive way.

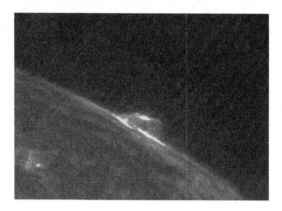

Figure 14.31 Image of solar flare taken by the TRACE satellite (With kind permission of Springer Science and Business Media (Journal: Solar Physics, Year: 2001, Vol.: 200, Issue: 1/2, Editors: Svestka, Engvold, Harvey, Authors: C.J. Schrijver and A.M. Title)

14.5.1.3 Effect of Solar Phenomena on the Earth's Atmosphere

Satellites study the processes that control the transfer of energy and momentum from solar wind, solar flares and CME to the magnetosphere and further into the near-Earth space environment. Solar winds, as mentioned before, cause electric currents to flow in the magnetosphere and ionosphere. Solar wind is constantly varying, making the system of currents highly dynamic. Detailed mapping of these currents and tracing their relations to processes in the solar wind, magnetosphere and ionosphere are the key factors in understanding the space weather.

CME and solar flares both affect the Earth's environment. On collision with Earth's atmosphere, CME can produce a geomagnetic storm above the magnetosphere. These storms can cause electrical power outages and damage to communication satellites. Solar flares, on the other hand, directly affect the ionosphere and radio communications on Earth and also release energetic particles into space. Several satellite missions have been launched to study these effects. Some of these missions are SOHO, Cluster, STEREO, and CORONOS Photon satellites.

14.6 Missions for Studying Planets of the Solar System

Several space probes have been launched to study the planets in the solar system. These probes are launched either to orbit around a particular planet (orbiters), land on their surface (landers) or orbit in interplanetary orbits to study various planets by moving closely through them (fly-by missions). Orbiters orbit around the planet in a manner similar to how satellites orbit around Earth. Landers, on the other hand, as the name suggests, are made to land on the planet in order to take a closer look at its surface and take samples of the soil to study them in detail. Planets are also observed from space observatories orbiting Earth. A brief description of some of the major missions is presented in the following paragraphs. Table 14.1 lists some of the important planetary missions launched for studying the various planets of the solar system and their objectives and major findings.

Table 14.1 Important planetary missions

Spacecraft	Country/Year	Mission objectives
Spacecraft for studying Mercury		
Mariner-10	USA/ 1994	First probe to study the planet Mercury.
MESSENGER	USA/ 2004	To study the surface composition, geologic history, core and mantle, magnetic field and atmosphere of Mercury.
Spacecraft for studying Venus		
Mariner-2	USA/ 1962	First successful Venus probe
Venera-4 (Atmospheric probe)	Soviet Union/ 1967	The first probe to enter another planet's atmosphere and return direct measurements. It showed that atmosphere of Venus contains 95% CO_2. Together with Mariner-5 probe, it showed that surface pressure of Venus was between 75 and 100 atmospheres
Venera-7 (Lander)	Soviet Union/ 1970	First successful landing on Venus. It relayed that the surface temperatures of Venus are of the order of 450–500°C
Venera-8 (Lander)	Soviet Union/ 1972	Measured the pressure and temperature profiles of Venus. It also studied the cloud layer and analyzed the chemical composition of the crust of the planet.
Venera-9 (Orbiter and Lander)	Soviet Union/ 1975	First artificial satellite to orbit Venus. It returned information about the planet's clouds, ionosphere, magnetosphere, as well as performing bistatic radar measurements of its surface. The lander took first pictures of the surface and analyzed the crust. It also took measurements of clouds on Venus
Venera-10 (Orbiter and lander)	Soviet Union/ 1975	Same studies as that conducted by Venera-9
Pioneer Venus Orbiter	USA/ 1978	It carried 17 instruments to study Venus's atmosphere, clouds, solar winds etc.
Pioneer Venus Multiprobe	USA/ 1978	The Pioneer Venus multiprobe carried one large and three small atmospheric probes to carry out extensive study of the planet.
Venera-11 (Fly-by and Lander)	Soviet Union/ 1978	Lander discovered a large proportion of chlorine and sulphur in the Venetian clouds.
Venera-12 (Fly-by and Lander)	Soviet Union/ 1978	Lander discovered a large proportion of chlorine and sulphur in the Venetian clouds.

(*continued*)

Table 14.1 (*Continued*)

Spacecraft	Country/Year	Mission objectives
Venera-13 (Fly-by and Lander)	Soviet Union/ 1981	Studied the soil samples of the planet and results showed that rocks similar to potassium-rich basalt rock were present on the planet.
Venera-14 (Fly-by and Lander)	Soviet Union/ 1981	Studied the soil samples of the planet and results showed that rocks similar to potassium-rich basalt rock were present on the planet.
Venera-15 (Orbiter)	Soviet Union/ 1983	It analyzed and mapped the upper atmosphere. Venera- 15 and -16 provided the first detailed understanding of the surface geology of Venus, including the discovery of unusual massive shield.
Venera-16 (Orbiter)	Soviet Union/ 1983	Venera-15 and -16 provided the first detailed understanding of the surface geology of Venus, including the discovery of unusual massive shield.
Magellan probe (Orbiter)	USA/ 1989	Magellan probe created the first high resolution mapping images of the planet.
MESSENGER (flyby)	USA/ 2004	Collect scientific data on flybys
Venus Express	European Space Agency/ 2005	It provided the temperature map of the southern hemisphere of the planet. It also made observations about the atmosphere of the planet.
Akatsuki (Orbiter)	Japan/ 2010	Failed
Spacecraft for studying Mars		
Mariner-4	USA /1964	It returned 22 close-up photos of the planet showing a cratered surface. The thin atmosphere was confirmed to be composed of CO_2 having 5–10 mbar pressure. A small intrinsic magnetic field was also detected.
Mars-2	USSR/1971	Mars-2 had both lander and the orbiter. The lander crashed-landed because its breaking rockets failed and hence was not able to transmit any data but it created the first human artifact on Mars. Orbiters of Mars-2 and -3 measured that the magnetic field of the planet was of the order of 30 nanotesla.
Mars-3	USSR/1971	Mars-3 also had both the lander and the orbiter. Its lander made the first successful landing on Mars. It failed after relaying 20 seconds of video data to the orbiter. The Mars-3 orbiter made measurements of the surface temperature and atmospheric composition and measured the magnetic field of the planet
Mariner-9	USA/1971	This was the first US spacecraft to enter an orbit around a planet other than Earth. It photographed features of the Martian surface. It also obtained images to help scientists choose suitable landing sites for the Viking probes

(*continued*)

Table 14.1 (*Continued*)

Spacecraft	Country/Year	Mission objectives
Mars-5	USSR/1973	It acquired imaging data for the Mars-6 and -7 missions.
Mars-6	USSR/1973	Mars-6 entered into orbit and launched its lander. The lander returned atmospheric descent data, but failed on its way down.
Mars-7	USSR/ 1973	Mars-7 failed to go into orbit about Mars and the lander missed the planet. Mars-7 found that there was a small amount of water vapour in the Martian atmosphere and that an inert gas was also present in the atmosphere
Viking-1 and 2	USA/ 1975 (both missions)	Viking-1 and -2 consisted of an orbiter and lander. Landers of both the missions had experiments to search for Martian micro-organism. The results of these experiments are still being debated. The landers provided detailed colored panoramic views of the Martian terrain. They also monitored the Martian weather. The orbiters mapped the planet's surface, acquiring over 52,000 images.
Phobos-1	USSR/ 1988	To study the moon of Mars named Phobos. Failed.
Phobos-2	USSR/ 1988	To study the moon of Mars named Phobos. Failed.
Mars observer	USA/ 1992	To study the geoscience and climate of the planet. Failed.
Mars global surveyor	USA/ 1996	Mars global surveyor was designed to orbit Mars over a two-year period and collect data on the surface morphology, topography, composition, gravity, atmospheric dynamics and magnetic field.
Mars 1996	Russia/ 1996	Consisted of an orbiter, two landers, and two soil penetrators. Failed.
Mars pathfinder	USA/ 1996	Comprised of a lander and surface rover. Mars pathfinder returned 2.6 billion bits of information, including more than 16,000 images from the lander and 550 images from the rover, as well as more than 15 chemical analyses of rocks and extensive data on winds and other weather factors.
Nozomi	Japan/ 1998	This is the first Japanese spacecraft to reach another planet.
2001 Mars odyssey	USA/ 2001	The 2001 Mars odyssey orbiter was launched to orbit Mars for three years, with an objective of conducting a detailed mineralogical analysis of the planet's surface from the orbit and measuring the radiation environment. The mission has as its primary science goals to gather data to help determine whether the environment on Mars was ever conducive to life, to characterize the climate and geology of Mars, and to study potential radiation hazards to possible future manned missions.

(*continued*)

Table 14.1 (*Continued*)

Spacecraft	Country/Year	Mission objectives
Mars express	European space agency/ 2003	Comprised of an orbiter and a lander. The lander was lost.
Mars exploration rover (MER)	USA/ 2003	Twin Rover vehicles to explore two sites on Mars, searching for signs of water and to explore Martian surface and geology.
Mars reconnaissance orbiter (MRO)	USA/2005	To conduct reconnaissance and exploration of the planet from the orbit
Phoenix (lander)	USA/2007	To study the geologic history of water and to evaluate potential habitability in the ice-still boundary
Dawn (flyby)	USA/2007	To take images of the planet on flyby
Fobos-Grunt (lander/sample return)	Russia/2011	To study Phobos, one of the moons of Mars. Failed
Yinghuo-1 (orbiter/sample return)	China/2011	To study the planet's surface, atmosphere, ionosphere and magnetic field (sample return spacecraft to visit Mars' moon Phobos and return sample)
MSL Curiosity (rover)	USA/2011	Robotic space probe, successfully landed Curiosity rover on Mars. To investigate Mars's habitability, study its climate and geology and collect data for manned mission to Mars
Spacecraft for studying outer planets of the solar system		
Pioneer-10	USA/1972	Pioneer-10 flew by Jupiter on 1 December 1973. It returned over 500 images of Jupiter and its moons. Pioneer-10's greatest achievement was the data collected on Jupiter's magnetic field, trapped charged particles and solar wind interactions
Pioneer-11	USA/1973	Pioneer-11 flew by Jupiter on 1 December 1974. It took better pictures than Pioneer-10, and measured Jupiter's intense charged-particle and magnetic-field environment. As it flew by Jupiter it was given a gravity assist which swung it onto course for Saturn. On 1 September 1979, Pioneer-11 flew past the outer edge of Saturn's A ring. It studied the magnetosphere of Saturn and its magnetic field and also studied its various moons. It has now left the solar system
Voyager-1	USA/1977	Voyager-1 flew by Jupiter in the year 1979 and by Saturn in the year 1980. It took high-resolution images of the rings, magnetic field, radiation environment and moons of the Jupiter system. It also detected the complex structures in Saturn's rings and studied its atmosphere and that of its moon, Titan.

(*continued*)

Table 14.1 (*Continued*)

Spacecraft	Country/Year	Mission objectives
Voyager-2	USA/1977	Voyager-2 flew by Jupiter in the year 1979, by Saturn in the year 1981, by Uranus in the year 1986 and by Neptune in the year 1989. It discovered few rings around Jupiter, studied the Great Red Spot on the planet and the volcanic activities in one of its moon. It measured the temperature and density profiles of the atmosphere of Saturn. It studied the atmosphere of Uranus, its ring structure and discovered ten moons of the planet.
Ulysses	USA/1990	The Ulysses spacecraft is an international project to study the poles of the Sun and interstellar space above and below the poles.
Galileo	USA/1989	Galileo was designed to study Jupiter's atmosphere, satellites and the surrounding magnetosphere for 2 years.
Hubble space telescope	USA/ 1990	The Hubble space telescope has taken photographs of Jupiter and other planets. In July 1994, it photographed the collision of the comet Shoemaker-Levy 9 with Jupiter.
Cassini/ Huygens	USA and Europe/ 1997	The aim of the mission was to study whole Saturn system – the planet, its atmosphere (rings and magnetosphere) and some of its moons. The Cassini mission consisted of the NASA-provided Saturn orbiter coupled with ESA's Huygens probe, which was dropped into the atmosphere of the biggest moon of Saturn named Titan.
New Horizons	USA/ 2006	The aim of the mission is to study planet Pluto and its moons
Juno (Orbiter)	USA/ 2011	To study the planet's composition, gravity field, magnetic field and polar magnetosphere

14.6.1 Mercury

Mariner-10 probe was launched on 3 November 1994 into an elliptical solar orbit crossing Venus in order to study the planet Mercury. It made use of the gravitational field of Venus to reach Mercury. It carried five principle scientific experiments and was in service for 17 months. It provided information on the atmospheric pressure, surface temperature, magnetic field and surface structure of the planet, which has been heavily cratered by meteorites. It also provided information on the cloud circulation on Venus. Figure 14.32 (a) shows the orbit of Marnier-10 and Figure 14.32 (b) shows the photograph taken by Mariner-10 of the surface of planet Mercury. The photograph shows that the surface of the planet is covered by faults. Another probe named Mercury Surface, Space Environment, Geochemistry and Ranging (MESSENGER) was launched on 3 August 2004 to study the surface composition, geologic history, core and mantle, magnetic field and tenuous atmosphere of Mercury. It

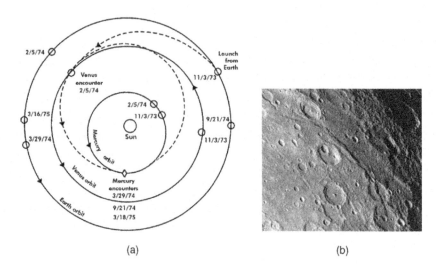

Figure 14.32 (a) Orbit of Mariner-10 probe (Courtesy: NASA). (b) Photograph of Mercury taken by Mariner-10 probe (Courtesy NASA/JPL-Caltech)

entered into Mercury's orbit in March 2011 after performing fly-bys of Earth in February 2005, two fly-bys of Venus in October 2006 and October 2007, and three fly-bys of Mercury in January 2008, October 2008 and September 2009. It has yielded significant data, including characterization of Mercury's magnetic field and discovery of water ice at the planet's north pole. Future mission to be launched for studying Mercury is Bepi colombo, a joint mission of Japan and ESA scheduled to be launched in August 2013. It includes two satellites namely Mercury planetary orbiter (MPO) and Mercury magnetospheric orbiter (MMO).

14.6.2 Venus

The first successful spacecraft to visit Venus was Mariner-2 in 1962. More than 20 spacecraft have been launched to date for probing the planet. The main missions include NASA's Pioneer Venus series (Pioneer Venus Multiprobe and Pioneer Venus Orbiter) and the Magellan probe (Figure 14.33), the Soviet Union's Venera series (16 Venera satellites, Venera-1 to Venera-16, were launched in a span of 22 years from 1961 to 1983) and European space agency's Venus Express spacecraft launched in the year 2005. The first successful spacecraft to land on the planet was Venera-7 and the first artificial satellite of Venus was Venera-9.

All these missions have helped scientists to study the planet's surface and its atmosphere. Most of Venus' surface consists of gently rolling plains covered by lava flows, with two large highland areas deformed by geological activity. Figure 14.34 shows images taken by the Magellan probe of the planet. Figure 14.34 (a) shows the global view of Venus made from a mosaic of radar images from the Magellan spacecraft and Figure 14.34 (b) shows the image of a volcano named the Sif Mons volcano on the planet. Recent missions have indicated that Venus is still volcanically active but only in a few hot spots.

Data from the Magellan probe indicates that Venus' crust is stronger and thicker than had previously been assumed. In addition, data from various missions have confirmed that the planet has an almost neglible magnetic field. The atmosphere mostly comprises CO_2 and

Figure 14.33 Magellan probe (Courtesy: NASA)

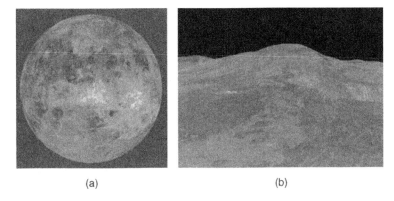

Figure 14.34 (a) Global view of Venus taken by Magellan probe (Courtesy: NASA). (b) Image of Sif Mons volcano on Venus taken by Magellan probe (Courtesy: NASA/JPL-Caltech)

small amounts of N_2. The Venera-4 probe for the first time showed that CO_2 accounts for around 95 % of the atmosphere. The pressure of Venus' atmosphere at the surface is around 90 atmospheres. Venera-4 and -5 missions provided data on the atmospheric pressure of the planet. The atmosphere has several layers of clouds many kilometres thick composed of sulfuric acid. Venera-8 provided information on the cloud layers of the planet. The dense atmosphere raises the surface temperature of the planet. Venera-7 confrimed that the surface temperature of Venus was of the order of 450–500 °C.

Dense clouds prevented scientists from uncovering the geological nature of the surface. Developments in radar telescopes and radar imaging systems orbiting the planet have made it possible to see through the clouds to observe the surface below. Figure 14.35 shows two different perspectives of Venus. Figure 14.35 (a) shows the image acquired by the Mariner-10 spacecraft in February 1974. The image shows thick cloud coverage that prevents an optical observation of the planet's surface. The surface of Venus remained a mystery until the year 1979 when the Pioneer Venus-1 mission used radar to map the planet's surface. The Magellan spacecraft launched in August 1990 mapped the planet's surface in great detail. The

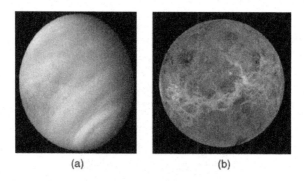

Figure 14.35 Images of Venus (a) taken by Mariner-10, (b) taken by Magellan probe (Courtesy: NASA)

image [Figure 14.35 (b)] shows the planet's features taken from radar on the Magellan probe. The Venus Express probe launched in 2005 by European space agency (ESA) has mapped the temperature profile of the southern hemisphere of the planet. Japan launched a spacecraft named Akatsuki (formerly known as Venus Climate Orbiter) to explore the planet in May 2010, but it failed to enter the orbit around Venus. It is being planned to attempt orbital insertion when the probe returns to Venus in 2015. Future missions include NASA's Venus in-situ explorer (VISE), to be launched in the near future, and Russian Venera-D spacecraft, to be launched in 2016. In addition, the Bepicolombo mission to study solar corona will do fly-bys of Venus.

14.6.3 Mars

The first satellite mission successfully launched for studying the planet Mars was the Mariner-4 probe that passed over the surface of Mars on 14 July 1965 at an altitude of just under 10 000 km. The Mariner-4 mission measured the magnetic dipole movement of the planet to be less than three ten-thousandths that of Earth. This result indicated that Mars does not have a metallic core. The probe also studied the structure of the Martian atmosphere and sent the first pictures of the Martian surface. Since then, several missions comprising orbiters and landers have been launched. Figure 14.36 (a) shows an image of Mars taken by the Viking-1 orbiter and Figure 14.36 (b) shows the first photograph of the surface of Mars taken by the Viking-1 lander. It should be mentioned here that the largest number of missions have been launched to study Mars as compared to any other planet in the solar system. All these satellite missions have helped scientists to study in detail the Martian surface and its environment.

Images taken from orbiters and landers have enabled scientists to divide the surface of Mars into three major regions: southern highlands, northern plains and polar regions. Satellite images from Mariner-3, -4, -6 and -7 showed that the surface of Mars was covered by craters. These regions are the southern highlands. Figure 14.37 shows an image of one of the craters on the southern highlands taken by the Mars Global Surveyor. The northern or low lying regions are covered with lava flows, small cinder cones, dunes, wind streaks and major channels and basins similar to dry 'river valleys'. The Mariner-9 spacecraft for the first time imaged these features of the planet. Figure 14.38 shows an image of a large network of canyons (called

Missions for Studying Planets of the Solar System

Figure 14.36 (a) Image of Mars taken by Viking-1 orbiter (Courtesy: NASA). (b) Photograph of the surface of Mars taken by Viking-1 lander (Courtesy: NASA-JPL)

Figure 14.37 Image of one of the craters on Mars (Courtesy: NASA JPL-Caltech)

Valles Marineris) in the northern plains taken by the Viking-1 orbiter spacecraft. The polar regions are covered with polar ice caps made mostly of frozen carbon dioxide (dry ice).

The rover of the Pathfinder mission (Figure 14.39) has provided detailed information about the surface composition of the planet. An X-ray spectrometer on board the rover performed 15 separate chemical analyses of the Martian soil and identified large amounts of silicate minerals, suggesting that Mars had a similar geological history to that of Earth. Satellite

Figure 14.38   Image of Valles Marineris on Mars (Courtesy: NASA/JPL-Caltech)

Figure 14.39   Rover of the Pathfinder mission (Courtesy: NASA/JPL-Caltech)

missions have been launched to study if any form of life exists on Mars. The Mariner-6 and -7 missions provided information on the atmospheric composition of the planet. Mariner-7 found that there was a small amount of water vapour in the atmosphere. Many probes have seen geological features that look like river valleys and water erosion, which indicates that water might have once been present on the planet. Satellites have also provided information on Martian weather. Figure 14.40 shows an image of the Martian clouds taken by the Mars pathfinder. The Mars odyssey mission detected hydrogen on the surface of Mars, which is thought to be contained in water ice. Mars express spacecraft detected methane in the Martian atmosphere, and the Mars exploration rovers detected that liquid water existed at some time in the past on the planet. Phoenix is a robotic spacecraft on a space exploration mission to Mars. It landed on Mars on 25 May 2008. It provided in-depth information on the landscape, weather, climate cycles and surface chemistry of the planet. The Mars science laboratory (MSL) is a robotic space probe mission to Mars launched by NASA in November 2011. It successfully landed Curiosity, a Mars rover, in Gale Crater in August 2012. It has provided information on the biological, geochemical and planetary processes and the surface radiation pattern of the planet. Some of the future missions to study the planet include Phobos-Grunt and Maven.

Figure 14.40 Image of Martian clouds taken by Mars pathfinder (Courtesy: NASA/JPL-Caltech)

14.6.4 Outer Planets

In this section, the missions launched for studying the planets beyond Mars will be discussed, that is Jupiter, Saturn, Uranus and Neptune; Jupiter, Saturn, Uranus and Neptune together are called the 'Jovian planets'. In addition, missions launched to study Pluto will also be covered. Pluto was one of the planets of the solar system but in the year 2006, it was declared a dwarf planet.

14.6.4.1 Jupiter

The planets Jupiter and Saturn were explored as part of two American space programmes – Pioneer and Voyager (Figure 14.41). Each of these missions had two spacecraft, Pioneer-10 and -11 in the Pioneer programme and Voyager-1 and -2 in the Voyager programme. The main objectives of the Pioneer mission were to explore the interplanetary medium beyond Mars, to examine the asteroid belt and to explore Jupiter with Pioneer-10 spacecraft and Saturn with Pioneer-11 spacecraft. The Voyager spacecraft were launched to visit Jupiter and Saturn in order to examine their magnetosphere and moons, in particular Titan (Saturn's largest moon).

Figure 14.41 Voyager spacecraft (Courtesy: NASA/JPL)

Other missions launched for studying Jupiter include Ulysses and Galileo. The Pioneer and Voyager missions were fly-by missions; Galileo [Figure 14.42 (a)] was an orbiter mission and was inserted into orbit around Jupiter in December 1995. It also launched a probe called the Galileo probe into Jupiter's atmosphere [Figure 14.42 (b)]. The planet is also regularly monitoried by the Hubble space telescope. Future mission to be launched to study Jupiter include the joint NASA/ESA's Europe Jupiter System Mission (EJSM).

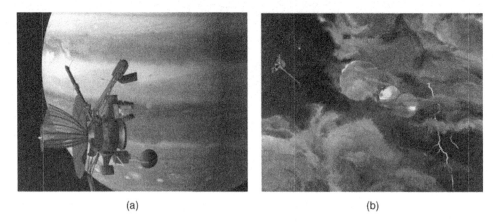

(a) (b)

Figure 14.42 (a) Galileo Orbiter Mission (Courtesy: NASA). (b) Galileo probe (Courtesy: NASA)

Juno was launched in August 2011 to study the planet's composition, gravitational and magnetic fields and polar magnetosphere. It will arrive at Jupiter in 2016. EJSM will comprise of NASA-led Jupiter Europe Orbiter and ESA-led Jupiter Ganymede Orbiter and will be launched in 2020.

Knowledge of the features of Jupiter is mostly accquired by indirect means. The only probe launched to study the planet, Galileo's atmospheric probe, went to only about 150 km below the cloud layers. The planet appears to be covered by coloured bands. Figure 14.43 shows an image of the planet taken by the Hubble space telescope. The Voyager mission has provided

Figure 14.43 Image of Jupiter taken by HST (Courtesy: NASA/JPL-Caltech)

detailed information on the boundaries between these bands. The planet is perpetually covered with a layer of clouds. Galileo's probe has made observations on the cloud layers of the planet. It has provided information that bands of clouds at different latitudes flow in opposing directions due to the prevailing winds. The interactions of these conflicting circulation patterns cause storms and turbulence having wind speeds of up to 600 km/h. The best-known feature of Jupiter, the Great Red Spot (GRS), is a violent storm having a size of about three times Earth's diameter. Figure 14.44 shows the image of the Great Red Spot taken by the Voyager-1 spacecraft on 25 February 1979.

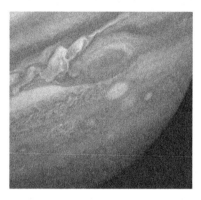

Figure 14.44 Image of Great Red Spot taken by Voyager-1 (Courtesy: NASA/JPL-Caltech)

The first close-up pictures of the atmosphere of Jupiter were provided by the Voyager missions. Jupiter has a very large and powerful magnetosphere. Pioneer probes confirmed that the planet's magnetic field is 10 times stronger than the Earth's magnetic field. Studies on the planet's magnetosphere were also carried out by the Ulysees solar probe. Figure 14.45 shows the image of Jupiter's magnetosphere, which is conceived on the basis of a large number of observations made by these probes. Jupiter has rings like those of Saturn. They were totally unexpected and were only discovered by the Voyager-1 spacecraft. The Galileo spacecraft

Figure 14.45 Image of Jupiter's magnetosphere (Courtesy: NASA/JPL/John Hopkins)

also took images of these rings. Jupiter has at least 63 moons. The Galileo mission performed several fly-bys of Jupiter's moons. Figure 14.46 shows the image of Jupiter and its four moons, photographed by Voyager-1 spacecraft. (They are not to scale, but are in their correct relative positions.)

Figure 14.46   Image of Jupiter and its four moons taken by Voyager-1 spacecraft (Courtesy: NASA)

It is worth mentioning here that in July 1994, Comet Shoemaker Levy-9 collided with Jupiter. The Galileo spacecraft made this observation. Figure 14.47 shows one such image showing the impact of the collision.

Figure 14.47   Jupiter's image by the Galileo spacecraft after its collision with comet Shoemaker Levy-9 (Courtesy: NASA)

14.6.4.2 Saturn

As mentioned earlier, Saturn was explored as part of two American space programmes – Pioneer and Voyager. Pioneer-11 spacecraft and both the Voyager spacecraft, Voyager-1 and -2, studied the planet's atmosphere, its ring structure and its moons. After that, the Cassini/Huygens probe has been launched to study the planet in further detail and also to study the largest moon of Saturn, called Titan. Cassini/Huygens (Figure 14.48) is a joint NASA/ESA probe and consists of the Cassini orbiter, which reached Saturn in July 2004, and

Figure 14.48 Cassini/Huygens probe (Courtesy: NASA/JPL-Caltech)

a small probe named Huygens, which landed on the surface of Titan on 14 January 2005. It made several findings, including evidence of liquid water reservoirs that erupt in geysers on Saturn's moon Enceladus, a planetary ring, first proof of hydrocarbon lakes near Titan's north pole and four new moons of Saturn.

Figure 14.49 shows the image of Saturn taken by Voyager-2 spacecraft in July 1981. The clouds are present low in the atmosphere. Saturn is surrounded by planetary rings extending from 6630 km to 120 700 km above Saturn's equator. The Voyager spacecraft discovered the rings to have an intricate structure of thousands of thin gaps and ringlets. Until 1980, the structure of the rings of Saturn was explained exclusively on the basis of the action of gravitational forces. The Voyager spacecraft found dark radial features in the B ring, called spokes, which made scientists believe that the ring structure is connected to electromagnetic interactions.

Figure 14.49 Image of Saturn taken by the Voyager-2 spacecraft (Courtesy: NASA/JPL-Caltech)

Saturn has 31 officially recognized moons. The most famous of them is Titan. It was studied by the Voyager and Cassini spacecraft. Cassini made 30 fly-by operations across Titan. It also launched a probe named Huygens to study Titan's atmosphere and map its surface. Figure 14.50

Figure 14.50 Image of Titan taken by the Cassini spacecraft (Courtesy: NASA/JPL/Space Science Institute)

shows an image of Titan taken by the Cassini spacecraft during its fly-by operation of the moon. The image taken in the UV band shows Titan as a softly glowing sphere. Future missions include NASA/ESA's Titan Saturn system mission (TSSM) for exploring the planet and its moons Titan and Enceladus.

14.6.4.3 Uranus

NASA's Voyager-2 is the only spacecraft to have studied Uranus. The spacecraft made its closest approach to Uranus on January 1986 before continuing its journey to Neptune. However, observations made by the Hubble space telescope and other such instruments have helped scientists to study the planet. Uranus is known to have extreme seasonal variations. Figure 14.51 shows the image of Uranus taken by the Voyager-2 spacecraft in January 1986.

Figure 14.51 Image of Uranus taken by the Voyager-2 spacecraft (Courtesy: NASA/JPL-Caltech)

14.6.4.4 Neptune

Neptune has been visited by only one spacecraft, Voyager-2, which flew by the planet on 25 August 1989. Neptune is a dynamic planet and has several large dark spots similar to those of Jupiter, caused by hurricane-like storms. The largest spot, known as the Great Dark Spot, is about the size of Earth and is similar to the Great Red Spot on Jupiter. Figure 14.52 shows an image of Neptune taken by the Voyager-2 spacecraft in August 1989. The image shows the Great Dark Spot in the centre. The Hubble space telescope image taken in 1994 found that the Great Dark Spot is missing. These dramatic changes in the weather system are not completely understood but they reveal the dynamic nature of the planet's atmosphere. NASA has proposed launching the Neptune Orbiter with probes to explore the planet in 2016.

Figure 14.52 Image of Neptune taken by the Voyager-2 spacecraft (Courtesy: NASA/JPL-Caltech)

14.6.4.5 Pluto

Little is known about Pluto because of its great distance from Earth and also because no exploratory spacecraft has yet visited the dwarf planet. Originally the Voyager-1 probe was planned to visit Pluto, but was redirected for a close fly-by of Saturn's moon Titan. In 2006, NASA launched a mission 'New Horizons' to study Pluto and its moons. It is expected to reach Pluto in the year 2015.

14.6.5 Moon

The first spacecraft to reach the moon was the unmanned Soviet probe, Luna-2 which crashed on its surface in September 1959. The first probe to land on the surface of the moon and transmit pictures was Luna-9. The probe, launched by the Soviet Union, landed on the moon in February 1966. The first artificial satellite of the moon was Luna-10 launched by the Soviet Union in March 1966.

The moon became the first celestial body to be visited by humans on 20 June 1969 when astronauts from Apollo-11 mission (Figure 14.53) landed there. In the next three years, six missions went sent to the moon under the Apollo programme. They carried a total of 12 humans. Samples from the surface of the moon have been brought back to Earth from these six Apollo missions as well as from the three Luna missions. The Clementine spacecraft was sent to the moon in the year 1994. It was a joint US defence department/NASA spacecraft and

Figure 14.53 Apollo-11 mission (Courtesy: NASA)

it sent the first near global topographic map of moon and the first multispectral images of its surface. Another mission was the Lunar Prospector launched by NASA in 1998. It indicated the presence of excess hydrogen at the lunar poles. Lunar Reconnaissance Orbiter and Lunar Crater Observation and Sensing Satellite were launched in June 2009. Two missions named GRAIL (Gravity recovery and interior laboratory) and LADEE (Lunar atmosphere and dust environment explorer) were launched by NASA in 2011 and 2013, respectively. GRAIL's mission is to do high quality gravitational field mapping of the moon to determine its internal structure. LADEE will orbit the moon's equator and study the lunar exosphere and dust in the moon's vicinity. The USA plans to launch a manned mission to moon again in 2020. The European spacecraft Smart-1 was launched on 27 September 2003 to carry out an extensive survey of the moon. It was in lunar orbit from November 2004 to September 2006. It surveyed the lunar environment and sent close-up images of its surface.

China has the Chang'e programme for lunar exploration. The first spacecraft under this programme, Chang'e-1, was launched in October 2007 to look for the isotope helium-3 for use as an energy source on Earth. China launched the Chang'e-2 lunar orbiter in October 2010. Japan launched a lunar orbiter fitted with a high resolution camera and two small satellites named SELENE in September 2007. India launched its lunar spacecraft, Chandrayaan-1, in October 2008 with objectives to create a three-dimensional atlas of the moon and do chemical and meteorological mapping of its surface. Chandrayaan-2 is planned to be launched by 2014.

The missions have shed light on the composition of the surface of the moon, its atmosphere and other properties. The moon is covered with tens of thousands of craters having diameters of at least one kilometre. Figure 14.54 shows an image of the moon's heavily cratered far side taken by the astronauts of the Apollo-11 mission in 1969. The Apollo and the Luna missions returned 382 kilograms of rock and soil from the surface. The sample studies showed three types of rock: regolith (fine-grained debris formed by micrometeorite bombardment), maria (dark, relatively lightly cratered) and terrae (relatively bright, heavily cratered highlands). The dark patches seen from the Earth are due to maria.

The recently launched missions have confirmed that there are traces of water on the moon. In July 2008, small amounts of water were found in the interior of the volcanic pearls from

Figure 14.54 Image of the moon's heavily cratered far side taken by the astronauts of the Apollo-11 mission in 1969 (Courtesy: NASA)

the moon bought back to the Earth by the astronauts of the Apollo 15 mission in 1971. The Chandrayaan-1 mission founded evidence of water on the surface of the moon. These observations were confirmed by NASA's Lunar Crater Observation and Sensing Satellite.

The moon is in synchronous rotation with the Earth; hence one side of the moon (the 'near side') is permanently turned towards the Earth. The other side (the 'far side'), mostly cannot be seen from the Earth. Four nuclear-powered seismic stations were installed during the Apollo project to collect seismic data about the interior of the moon. It was found that there is only residual tectonic activity due to cooling and tidal forcing and other moonquakes have been caused by meteor impacts and artificial means.

14.6.6 Asteroids

Asteroids are rocky and metallic objects that orbit the sun but are too small to be considered as planets. Most asteroids are contained within a belt that exists between the orbits of Mars and Jupiter. Before the year 1991, the only information obtained on asteroids was through Earth-based observations. The first asteroid to be photographed in close-up by a spacecraft was 951 Gaspra in 1991 imaged by the Galileo probe en route to Jupiter (Figure 14.55). The spacecraft imaged another asteroid 243 Ida in 1993.

The first dedicated probe launched to study asteroids was NEAR (Near Earth asteroid rendezvous) Shoemaker, which photographed 253 Mathilde in 1997, before entering into orbit around 433 Eros, finally landing on its surface in 2001. Other asteroids briefly visited by spacecraft en route to other destinations include 9969 Braille (by Deep Space-1 in 1999) and 5535 Annefrank (by Stardust in 2002). The Hayabusa mission was launched by Japan to return a sample of material from a small near Earth asteroid 25143 Itokawa on 9 May 2003. It landed on the asteroid in November 2005 and collected samples in the form of tiny grains of asteroid material. The samples were returned to Earth in a recovery capsule in June 2010. The European Rosetta probe studied 2867 Steins asteroid in 2008 and the Lutetia asteroid in 2010. NASA launched a robotic spacecraft, named Dawn, on 27 September 2007 to study the asteroid

Figure 14.55 Image of 951 Gaspra asteroid taken by the Galileo probe (Courtesy: NASA)

Vesta and the dwarf planet Ceres. It explored Vesta from July 2011 to September 2012 and will orbit Ceres in 2015. China's Chang'e-2 flew within 2 miles of asteroid 4179 Toutatis as an extended exploration mission. Future explorations include JAXA's Hayabusa-2 space probe planned to be launched in 2015 and NASA's OSIRIS-REx sample return mission to be launched in 2016.

14.6.7 Comets

Comets are small, fragile, irregularly shaped bodies composed of a mixture of nonvolatile grains and frozen gases. Images of comets have been taken by many satellites including the HST, ROSAT (Roentgen satellite), Deep space mission (DS-1), Giotto, Vega-1 and -2, CONTOUR (Comet nucleus tour), etc. The European Giotto and the Russian Vega-1 and -2 imaged Halley's comet. The HST imaged the Shoemaker Levy-9 comet's collision with Jupiter in July 1994. It also imaged the Hyakutake comet. Figure 14.56 shows the image of the Hyakutake comet taken by the HST on 25 March 1996 when the comet passed at a

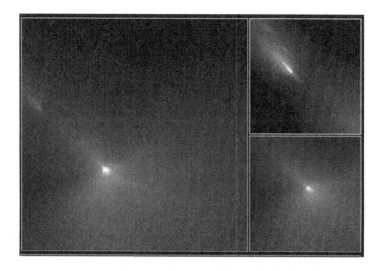

Figure 14.56 Image of the Hyakutake comet taken by the HST (Courtesy: NASA/JPL-Caltech)

distance of 9.3 million km from Earth. The image provides an exceptionally clear view of the near-nucleus region of the comet. The Hyakutake comet was also imaged by the NEAR spacecraft in March 1996. The Stardust spacecraft collected particles from the coma of comet Wild 2 in January 2004 and returned samples to Earth in a capsule in January 2006. In 2005, Deep Impact probe blasted a crater on comet Tempel 1 to study its interior. Later, it was renamed EPOXI and made a fly-by of Comet Hartley 2 in November 2010. In 2014, Rosetta probe will orbit comet Churymov-Gerasimenko and place a small lander on its surface.

14.7 Missions Beyond the Solar System

Space missions help scientists to determine the position and movement of heavenly bodies (like the position of stars, other galaxies, and so on), study the structure of radio galaxies, and supernova remnants, determine the amount of water and oxygen molecules in dense interstellar clouds, observe the birth of stars and galaxies and the early stages in their evolution and so on. These observations are mainly done by space observatories. NASA has launched four space observatories under the 'Great Observatories' programme. These include the Hubble space telescope (HST), Compton gamma ray observatory (CGRO), Chandra X-ray observatory and the Spitzer space telescope (SST). Fermi gamma-ray space telescope launched in the year 2008 in a follow-on to the CGRO. Other observatories include the Infrared astronomy satellite (IRAS), Infrared space observatory (ISO), Solar observatory (SOHO), High energy astronomy observations (HEAO) and High precision parallax collecting satellite (HIPPARCOS). In the following paragraphs a few important missions for space astronomical applications are briefly outlined, touching upon some of the important observations made by them.

The Hubble space telescope (HST) (Figure 14.57) launched in the year 1990 by NASA in collaboration with ESA represents the most important and prestigious space astronomical mission. It is a 2.4 m, f/24 telescope with Ritchey Chretian design having an effective focal length of 57.6 m, orbiting in a LEO orbit at an altitude of 610–620 km. It makes observations

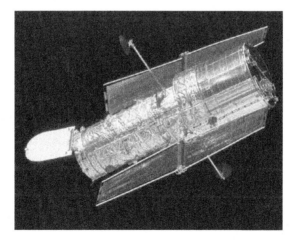

Figure 14.57 Hubble space telescope (HST) (Courtesy: NASA/STscI)

in the visible, near UV and near IR wavelength bands. Observations from HST have helped in partially confirming the theory that most galaxies have a black hole in their nucleus. The current model of the accelerating universe has taken inputs from the images provided by the HST. HST observations have confirmed that there are also planets revolving around other stars. It has imaged large portions of the universe and strengthened the belief of scientists that the universe is uniform over large scales. Figure 14.58 shows an image taken by the HST of the Eagle Nebula. The image taken in the visible wavelength band showed the detailed structure of many light-year long pillars. A surprising finding made by the Hubble space telescope was that the pillars are covered with a large number of small bumps and protrusions. These objects were not seen in the previous observations of the nebula made by ground-based measurements. The HST will be replaced by the James Webb Space Telescope (JWST) around 2018. Other observatories operating in the visible band include Astrosat, COROT, Kepler mission and MOST.

Figure 14.58 Image of Eagle Nebula taken by the HST (Courtesy: NASA)

The Compton gamma ray observatory (CGRO) launched in the year 1991 was operational until 2000. Observations in the gamma-ray region are relevant to violent processes that occur in stellar and galactic evolution, revealed as gamma bursts. One of the important accomplishments of the observatory was the discovery of terrestrial gamma-ray sources that came from thunderclouds. Fermi gamma-ray space telescope, launched in the year 2008 is a follow-up to CGRO and is used to perform gamma-ray astronomy observations from LEO orbit. Other observatories operating in the gamma-ray band include Astrorivelatore gamma ad immagini leggero (AGILE), High energy transient explorer (HETE-2), International gamma ray astrophysics laboratory (INTEGRAL), Low energy gamma-ray imager (LEGRI), Swift gamma-ray burst explorer and Fermi gamma ray space telescope (FGST).

The Chandra X-ray observatory (Figure 14.59), launched in the year 1999, makes observations in the X-ray band. X-ray astronomy deals with the stellar coronas, supernova remnants, active galactic nuclei, quasars and accretion phenomena under black holes. It gave much information on the formation of galaxies, supernova remnants and far away stars. Figure 14.60 shows the X-ray image of Cassiopeia A, remnant of a star that exploded 320 years ago. The image shows an expanding shell of hot gas produced by the explosion. Some of the other

Figure 14.59 Chandra X-ray Observatory (Courtesy: NASA)

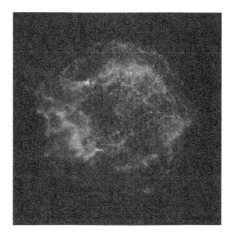

Figure 14.60 X-ray image of Cassiopeia-A taken by Chandra X-ray observatory (Courtesy: NASA)

observatories operating in the X-ray band include A broadband imaging X-ray all-sky survey (ABRIXAS), Advanced satellite for cosmology and astrophysics (ASCA), AGILE, Array of low energy X-ray imaging sensors (ALEXIS), Astrosat, BeppoSAX, HETE-2, International gamma-ray astrophysics laboratory (INTEGRAL), XMM-Newton and Nuclear spectroscopic telescope array (NUSTAR).

The Spitzer space telescope (SST) is an infrared space observatory launched in the year 2003. Another observatory operating in the infrared band is the Infrared astronomy satellite (IRAS). It provided a good insight into the birth of stars and galaxies and their early stages of evolution. It surveyed the whole sky and found 250 000 infrared sources in the universe. It was followed by the Infrared space observatory (ISO) launched by the ESA. It had better resolution than the IRAS and made measurements between 2.5 mm and 200 mm. The satellite HIPPARCOS (High precision parallax collecting satellite), launched in the year 1989, established the position coordinates and components of the proper motion of 120 000 stars. Other infrared space observatories include

IRAS, Infrared space observatory (ISO), Solar observatory (SOHO), AKARI, Hershel space observatory, Wide-field infrared explorer (WIRE), Wide-field infrared survey explorer (WISE) and the HIPPARCOS satellite.

Some of the important missions taking measurements in the UV region include the International ultraviolet explorer (IUE), Extreme ultraviolet explorer (EUVE), Far ultraviolet spectroscopic explorer (FUSE), High energy transient explorer (HETE-2), Galaxy evolution explorer (GALEX), Astro-2, Astrosat, Cosmic hot interstellar spectrometer (CHIPS) and Korea advanced institute of science and technology satellite 4 (Kaistsat-4). EUVE is NASA's explorer class satellite mission launched in the year 1992. It was operational for nine years until 2001. It made observations in the wavelength range from 70 to 760 Å. The EUVE mission was divided into two phases. The first phase (six months) was dedicated to an all-sky survey using imaging instruments. The second phase was dedicated to pointed observations using mainly spectroscopic instruments. The FUSE was NASA's mission launched in the year 1999 into a 768 km circular orbit inclined at 25°. It was launched to determine the abundance of deuterium in a wide range of the galactic, to study the Milky Way disc and to explore the nature and distribution of the hot intergalactic medium (IGM). Space observatories planned to be launched are the Darwin mission, X-ray evolving universe spectroscopy mission (XEUS), Tel Aviv university ultraviolet explorer (TAUVEX), SIM lite astrometric observatory, James Webb space telescope, constellation-X and Laser interferometer space antenna (LISA). Constellation-X and LISA are a part of NASA's Beyond Einstein programme. This programme is designed to explore the limits of Einstein's theory of general relativity and will include two space observatories (constellation-X and LISA) and a number of observation probes.

14.8 Other Fields of Investigation

Scientific satellites also carry out research in the fields of microgravity, cosmic rays and fundamental physics. Several satellite experiments have been launched to carry out these studies. These are briefly discussed in the following paragraphs.

14.8.1 Microgravity Experiments

Microgravity, as the name suggests, is a condition where the effects of gravity are either nonexistent or present on a very small scale. Ideally, it is a state of 'weightlessness' created by balancing the gravitational force with an equivalent acceleration force. In practice, however, an exact equilibrium state is difficult to achieve and a very small gravitational force is always present. Hence, the term 'microgravity' rather than weightlessness is more common. Gravity influences most of the physical processes on Earth, including convection, sedimentation, hydrostatic pressure, buoyancy, and so on. Under microgravity conditions, these processes are significantly altered or even removed. The sedimentation process affects crystal growth, convection currents affect flames and human bodies are affected by buoyancy and so on. Thus, by creating a microgravity environment, various phenomena related to these processes can be studied in a better way.

Microgravity experiments can be conducted using drop towers or tubes through various heights, the KC-135 aircraft, sounding rockets, several space shuttles and the ISS (International space station). However, the microgravity conditions in tubes and aircraft exist for only a few seconds. Space shuttles and the ISS are the main platforms for carrying out long term experimentation under microgravity conditions.

Space shuttles, which are reusable launch vehicles designed for carrying and bringing back astronauts and experimental setups, act as temporary research platforms in low Earth orbits and can provide up to 17 days of high quality micro gravity conditions. They can accommodate a wide range of experimental apparatus and provide a laboratory environment in which scientists can conduct relatively short term investigations. The USA has three space shuttles, Atlantis, Discovery and Endeavor. These space shuttles had a reusable laboratory named Spacelab, developed by the ESA to carry out microgravity experiments. It was used on 25 shuttle missions between 1983 and 1997. It was decommissioned in 1998 but was again commissioned in 1999 and was carried on space shuttles launched in the years 2000, 2001 and 2008. Other space stations include Skylab, Salyut-5, Salyut-6, Mir space station, and so on.

ISS is a permanent facility placed in LEO orbit that can maintain microgravity conditions for years. ISS enables scientists to conduct their experiments in microgravity conditions over a period of several months without having to return the entire laboratory to Earth each time an experiment is completed. Microgravity experiments on space shuttles and ISS are mainly carried out in the fields of life sciences and material sciences. Other fields of investigation include combustion studies and fluid physics. Figure 14.61 shows the photograph a space shuttle during flight and Figure 14.62 shows the photograph of ISS.

Figure 14.61 Space Shuttle during flight (Courtesy: NASA)

14.8.2 Life Sciences

Life science studies include understanding the effects of microgravity and cosmic rays on the lives of human beings and to compare biological processes on Earth (in the presence of gravity) with those occurring in space (in microgravity conditions). Scientists in the space environment are able to study the adaptation of life to the space environment and gain new knowledge about basic life processes. Moreover, they are able to study life from the simplest, one-celled forms, such as bacteria, to the larger, more complex life forms such as animals and humans.

Figure 14.62 ISS (Courtesy: NASA)

14.8.2.1 Human Physiology

Various aspects of human physiology, like musculoskeletal, metabolic, pulmonary, human behaviour and performance, have been studied under microgravity conditions. Space shuttle STS-78 conducted experiments in these fields. The scientists in the Russian Mir station studied their heart and lung behaviour and the digestion process during their long stays on the station. These studies are done for a wide range of potential usage including ensuring astronaut health, improving health care on Earth and the production of new and more potent medicines. Various changes occur in the human body when in space, like weakening of bones, shifting of fluids towards the upper body and disruption of body rhythms, and so on. Studying these changes help the scientists to understand the various human processes better and also help the astronauts to stay on a space station.

14.8.2.2 Biological Processes

One of the important missions included study of plants and animals in the absence of gravity. As an example, the STS-78 mission carried three space biology experiments to study the growth of pine saplings, development of fish embryos and bone changes in laboratory rats. Scientists on the Russian Mir station grew wheat seeds to discover the effects of microgravity on their growth. They also performed experiments on egg development in Mir's incubator in 1996. The results were compared with the same egg development phases on Earth. Some space shuttles also studied the ecological life support systems similar to that on Earth. This is done by growing plants to make oxygen and remove carbon dioxide, raising animals and creating an environment to duplicate the ecological system on Earth.

14.8.3 Material Sciences

Microgravity experiments provide scientists with an opportunity to study how materials behave outside the influence of Earth's gravity. Studies in material science include growing various alloys, crystals, proteins and viruses to better understand their structure and to produce materials that cannot be produced on Earth. These studies are used in the field of drug research, electronics and semiconductors and fluid physics. Material science research has been conducted on a large number of Spacelab missions including Spacelab-1, -2, -3, -D1, -D2, -J, IML-1 (International microgravity lab), IML-2, USMP-1, USMP-2, USML-1 and USML-2.

14.8.3.1 Growing Crystals, Alloys, and so on

Gravity alters the way atoms come together to form crystals. Near-perfect crystals can be formed in microgravity conditions. Such crystals yield better semiconductors for faster computers or more efficient drugs to combat diseases. Alloys of metals that do not combine on Earth can also be formed under microgravity conditions. Figure 14.63 shows one material science experiment, the drop dynamics module (DPM), carried out on board the Spacelab-3 in 1985. It studied the behaviour of liquid drops in microgravity. The experiment has also been carried out on several subsequent flights.

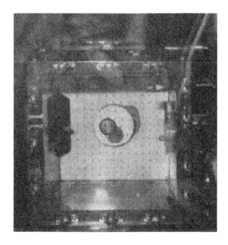

Figure 14.63   Material science experiment on board Spacelab-3 (Courtesy: NASA)

14.8.3.2 Protein Growth in Space

Through experiments in space, it is found that larger, higher quality protein crystals can be created in microgravity conditions. NASA has established a protein crystal growth programme to explore the formation and growth of these crystals in space. More than 40 protein growth payloads have been carried on the space shuttles and there is a protein growth research payload on the ISS. The crystals grown in space are returned to Earth and three-dimensional models of these crystals are created using X-ray mapping.

14.8.4   Cosmic Ray and Fundamental Physics Research

14.8.4.1 Cosmic Ray Research

Earth is constantly subjected to cosmic ray radiation coming from an unknown source in the universe. Since the dawn of the space age, the main focus of cosmic ray research has been directed towards astrophysical investigations of where these cosmic rays originate, how they propagate in space and what role they play in the dynamics of the galaxy. Satellite-based experiments allow the measurements to be done before these rays are slowed down and broken up by the atmosphere. The first satellite missions that made observations in this field were the Soviet Luna-1 and -2 spacecraft carrying instruments to measure the total electric charge of arriving ions. Other satellites carrying cosmic ray experiments include the Explorer-VII, IMP-8 and ACE satellites launched in the years 1959, 1973 and 1997 respectively.

14.8.4.2 Fundamental Physics

Satellites are also used in the field of fundamental physics to prove Einstein's general theory of relativity. LAGEOS (Laser geodynamics satellite) satellites have been used for this purpose. They are covered by reflectors and are used for laser ranging purposes. Scientists charted the path of LAGEOS-1 and -2 satellites over a period of 11 years, using laser range finding technique with a precision of a few millimetres. The satellite orbits dragged out of position by about 2 m each year, which was in accordance with Einstein's general theory of relativity. NASA launched the Gravity Probe B in April 2004 carrying four gyroscopes to study Einstein's theory with even higher accuracy.

14.9 Future Trends

Satellites have been used for scientific applications since the 1950s. In fact, the first satellite to be launched, Sputnik-1, was also a scientific satellite. It provided information on the density and temperature of the upper atmosphere. Since then remarkable progress has been made in the field of scientific satellites both in terms of technological development and application potential.

Technological advances have led to the development of new sensors, improvement in the resolution of the sensors, increase in information output and enhancement in the efficiency of the information delivery mechanism. Future trends are to further improve each of these parameters so as to have more accurate and precise data to help understand our universe better. The efficiency of the information delivery mechanism has increased by making use of new advanced image compression techniques which include image pyramids, fractal and wavelet compression.

In addition to improvements in sensor technology and information processing and compression techniques, the focus of the scientists is to reduce the size of individual missions by splitting up the payload components to allow smaller, dedicated and more focused satellites to be flown. Future trends include the launch of cluster of micro, nano and pico satellites to replace one large satellite. These satellites make use of technologies like application specific integrated micro-instruments (ASIM), micro electro-mechanical systems (MEMS), and so on. These missions will ensure cost-effectiveness and will also reduce the complexity of the satellite sub-systems as well as the satellite development time.

Many manned missions are being planned to explore the different planets and other celestial bodies of the solar system. NASA is building the next fleet of vehicles to service the international space station and to launch manned missions to the moon, Mars and beyond. Many other countries including Russia, China and India are also planning manned missions to the moon.

Further Readings

Bromberg, J.L. (1999) *NASA and the Space Industry*, John Hopkins University Press, Baltimore, Maryland.

Davies, J.K. (1988) *Satellite Astronomy: The Principles and Practice of Astronomy from Space*, Ellis Horwood Library of Space Science and Space Technology, Halsted Press.

Evans, B. and Harland, D.M. (2003) *NASA's Voyager Missions: Exploring the Outer Solar System and Beyond*, Springer.

Fazio, G. (1988) *Infrared Astronomical Satellite and the Space Infrared Telescope Facility*, Taylor & Francis, London.

Gatland, K. (1990) *Illustrated Encyclopedia of Space Technology*, Crown, New York.

Kramer, H.J. (2001) *Observation of the Earth and Its Environment: Survey of Missions and Sensors* (hardcover), Springer.

National Research Council (US) (1999) National Research Task Group on Sample Return from Solar System Bodies, *Evaluating the Biological Potential in Samples Returned from Planetary Satellites and Small Solar System Bodies: Framework for Decision Making*, National Academies Press.

Verger, F., Sourbes-Verger, I., Ghirardi, R., Pasco, X., Lyle, S. and Reilly, P. (2003) *The Cambridge Encyclopedia of Space*, Cambridge University Press.

Voit, M. (2000) *Hubble Space Telescope: New Views of the Universe*, Harry N. Abrams.

Internet Sites

1. http://science.howstuffworks.com/hubble.htm
2. http://en.wikipedia.org/wiki/Geodesy
3. http://www.jqjacobs.net/astro/geodesy.html
4. http://www.unistuttgart.de/gi/education/analytic_orbit/InABkSat.pdf
5. http://www-spof.gsfc.nasa.gov/Education/Intro.html
6. http://www.windows.ucar.edu/cgi-bin/tour_def/earth/Magnetosphere/overview.html
7. http://asd-www.larc.nasa.gov/erbe/ASDerbe.html
8. http://eosweb.larc.nasa.gov/EDDOCS/whatis.html
9. http://science.howstuffworks.com/sun.htm/printable
10. http://www.solarviews.com/eng/toc.htm
11. http://science.howstuffworks.com/mars.htm/printable
12. http://www.windows.ucar.edu/tour/link=/space_missions/space_missions.html
13. www.eos-am.gfsc.nasa.gov
14. www.eospso.gfsc.nasa.gov
15. www.jpl.nasa.gov
16. www.msl.jpl.nasa.gov
17. www.solarviews.com
18. www.skyrocket.de

Glossary

Aeronomy: This is the science dealing with physics and chemistry of the upper atmosphere

Astronomy: This is the science of the universe, which deals with studies of various celestial bodies of the universe

Charged particle detectors: They are used to measure the composition and number of charged particles in the ionosphere. Charged particle detectors commonly used include mass and energy spectrometers and time-of-flight spectrometers

Dynamical geodesy: This is the study of the variations in the Earth's gravitational field on the surface of the Earth by conducting detailed analysis of the satellite orbits

Earth radiation budget: The radiation budget represents the balance between incoming energy from the sun and the outgoing thermal (longwave IR) and reflected (shortwave) energy from Earth

Geodesy: Geodesy is defined as the science of measurement of the shape of the Earth

Geometrical geodesy: This determines the shape of the Earth by measuring the distances and angles between a large numbers of points on the surface of the Earth

Ionosphere: The ionosphere is the layer of the atmosphere between 50–500 km from the surface of the Earth that is strongly ionized by UV and X-rays of the solar radiation

Ionospheric sounder: This comprises a transmitter–receiver pair that is used to measure the effective altitude of an ionospheric layer by measuring the time delay between the transmission and reception of a radio signal

Magnetometer: An instrument used for measuring the strength of magnetic fields

Magnetosphere: The magnetosphere is that region of the atmosphere that extends from the ionosphere to about 40 000 miles where the Earth's magnetic field is enclosed

Mass spectrometer: A device that applies magnetic force on charged particles to measure mass and relative concentration of atoms and molecules

Microgravity: Ideally this is a state of 'weightlessness' created by balancing the gravitational force with an equivalent acceleration force. It is a condition where the effects of gravity are either non-existent or present on a very small scale. In practice, however, an exact equilibrium state is difficult to achieve and a very small gravity force does always remain

Polar aurora: This is a luminous phenomenon observed in the atmosphere from 100 to 300 km around the polar region. It is caused by excitation of atoms or molecules in the atmosphere to higher energy levels, emitting light during the process of falling back to normal states

Solar flares: Solar flares are abrupt and violent explosions that take place in complex sunspot groups

Solar physics: This is the study of the dynamics and structure of the sun's interior and the properties of the solar corona

Space geodesy: Space geodesy studies the shape of the Earth from space, its internal structure, its rotational motion and the geographical variations in its gravitational field

Space gradiometery: Space gradiometery is used for mapping fine variations in the Earth's gravitational potential by measuring the gradient of the gravitational field (in all three directions) on a single satellite

Sunspots: Sunspots are dark, cool areas on the photosphere, which always appear in pairs through which intense magnetic fields break through the sun's surface

15

Military Satellites

Military systems of today rely heavily on the use of satellites both during war as well as peacetime. Military satellites provide a wide range of services including communication services, gathering intelligence data, weather forecasting, early warning, providing navigation information and timing data and so on. Military satellites have been launched in large numbers by many developed countries of the world, but more so by the USA and Russia.

In the last five chapters, mainly civilian applications of satellites have been discussed. In this chapter deliberation will be given to various facets of military satellites related to their development and application potential. The chapter begins with an overview of military satellites, followed by a description of various types of military satellites.

15.1 Military Satellites – An Overview

Military satellites are considered as 'Force Multipliers' as they form the backbone of most of the modern military operations. They facilitate rapid collection, transmission and dissemination of information, which is a major requisite in modern-day military systems. Space-based systems offer features like global coverage, high readiness, non-intrusive forward presence, rapid responsiveness and inherent flexibility. These features enable them to provide real-time or near real-time support for military operations in peacetime, crisis and throughout the entire spectrum of the conflict. They are also very useful during the planning phase of military operations as they provide information on enemy order of the battle, precise geographical references and threat locations.

The application sphere of military satellites extends from providing communication services to gathering intelligence imagery data, from weather forecasting to early warning applications, from providing navigation information to providing timing data. They have become an integral component of military planning of various developed countries, more so of USA and Russia. As a matter of fact, the USA has the maximum number of military satellites in space, even more than the rest of the world put together. The USA used the services of military satellites extensively during its military campaign in Iraq in 2003, against Afghanistan in 2001 and

Yugoslavia in 1999. In the following paragraphs, the different applications of military satellites will be discussed.

15.1.1 Applications of Military Satellites

1. Military communication satellites. These satellites link communication centres to the front line operators.
2. Reconnaissance satellites. Reconnaissance satellites, also known as spy satellites, provide intelligence information on the military activities of foreign countries. There are basically four types of reconnaissance satellites:
   (a) Image intelligence or IMINT satellites
   (b) Signal intelligence or ferret or SIGINT satellites
   (c) Early warning satellites
   (d) Nuclear explosion detection satellites
3. Military weather forecasting satellites. They provide weather information, which is very useful in planning military operations.
4. Military navigation satellites. Navigation systems pinpoint the exact location of soldiers, military aircraft, military vehicles, and so on. They are also used to guide a new generation of missiles to their targets.
5. Space weapons. They are weapons that travel through space to strike their intended target.

15.2 Military Communication Satellites

Satellite communication has been a vital part of the military systems of the USA and Russia. These satellites provide reliable, continuous, interoperable, mobile, secure and robust communication services between the various military units and between these units and the command centres. They help streamline military command and control and ensure information superiority in the battlefield. Services provided by these satellites include:

1. Reliable networks

    (a) Secured network of voice and broadband data services for command and control
    (b) Secured telephony backbone services for remote locations and wide area networking for data applications

2. Field services

    (a) Voice, data, broadband and video services between military forces in the deployment areas and headquarters

3. Terrestrial back-up

    (a) Back-up communication services for disaster areas where the existing infrastructure is damaged
    (b) Back-up technical coordination links for critical locations

4. Air traffic control

(a) Secure and reliable communication among control towers as well as relaying information between pilots and towers

5. Video conferencing and tele-medicine network

    (a) Secure broadband communication between field medical crews and major hospitals
    (b) Full support of file transfers (X-ray, medical files) and video conferencing equipment for virtual meetings

6. Border control and custom network

    (a) Secure global communication services for surveillance operation inside and outside the country
    (b) Full support of captured surveillance video images

Military communication satellite systems serve a large number of users, ranging from those who have medium to high rate data needs using large stationary ground terminals to those requiring low to medium data rate services using small, mobile terminals and to those users who require extremely secure communication services. Each of these user groups has different requirements and is characterized by their own satellite and Earth terminal designs. Depending upon the intended user group, military communication satellite systems can be further subcategorized as:

1. Wideband satellite systems
2. Tactical satellite systems
3. Protected satellite systems

Wideband satellite systems provide point-to-point or networked moderate to high data rate communication services at distances varying from in-theater to inter-continental distances. Typical data rates for these systems are greater than 64 kbps. Users of the wideband segment primarily have fixed and mobile transportable land-based terminals with a few terminals on large ships and aircraft.

Tactical satellite systems are used for communication with small mobile land-borne, airborne and ship-borne tactical terminals. Such systems offer low to moderate data rate services at distances ranging from in-theater to trans-oceanic. Tactical satellites employ high power transmitters as they communicate with small terminals.

Protected satellite systems provide communication services to mobile users on ships, aircraft and land vehicles. These systems require an extremely protected link against physical, nuclear and electronic threats. They generally offer low to moderate data rate services.

15.3 Development of Military Communication Satellite Systems

Since the development of the first military communication satellite system in the late 1960s, satellite technology has made unprecedented progress in this field. Military satellites of today are far more advanced in terms of transmission capability, robustness, anti-jamming capability and so on as compared to their predecessors. Earlier only the USA and Russia had these systems but now many other countries including Israel, France, UK, and so on, have developed their own military satellite communication systems. In this section, the evolution process of military communication satellites will be discussed.

The first military communication satellite systems were developed by the USA in the 1960s. The systems developed initially were experimental in nature and demonstrated the feasibility of employing satellites for military communication applications. They also provided the basic experience required for the development of sophisticated systems meeting all the stringent military requirements, whether it is the anti-jamming feature or the reliability and the maintainability aspect and so on. The experimental systems included SCORE (Signal communication by orbiting relay equipment) Courier (Figure 15.1), Advent, LES (Lincoln experimental satellites) and West Ford satellites.

Figure 15.1   Courier satellite (Courtesy: US Army)

The first operational military satellite communication system was developed by the USA in the late 1960s. It was named the Initial defense communications satellite program (IDCSP). A total of 28 satellites were launched under the program, in a period of three years from 1966 to 1968. Each satellite had a single repeater with a capacity of around 10 voice circuits or a 1 Mbps data communication rate. The system was used during the Vietnam war in 1967 to transmit data from Vietnam to Hawaii through one satellite and on to Washington DC through another. The complete system was declared operational in the year 1968 and its name changed to the Initial defense satellite communication system (IDSCS). IDSCS was a wideband system used for strategic communication applications between fixed and transportable ground stations and large ship-borne equipment, all having large antennae.

In the 1970s and 1980s, only USA and Russia had military communication satellites. However, today many other developed countries of the world like the UK, France, Italy, Israel, China, and so on, have such systems. In the following paragraphs, the military satellites developed by various nations are discussed.

15.3.1 American Systems

MILSATCOM (Military satellite communications) architecture was proposed by USA in the year 1976 to guide the development of the military communication satellite systems in the country. Three types of military systems were proposed to be developed under this architecture, namely the wideband systems, mobile and tactical systems (or narrowband systems) and protected systems (or nuclear capable systems). The wideband systems developed were the Defense satellite communication systems (DSCS-II and III) and the Global broadcast service (GBS) payload on the UHF follow-on (UFO) satellite.

Systems developed under the category of the mobile and tactical segment include the Fleet satellite communication system, the LEASAT (Leased satellite) program and the UFO program. Satellites developed under the category of protected systems include the MILSTAR (Military strategic and tactical relay satellite) system, Air force satellite communication (AFSATCOM) and the Extremely high frequency (EHF) payloads. Figure 15.2 shows the satellites in the three types of systems.

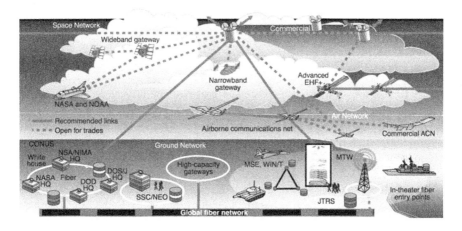

Figure 15.2 MILSATCOM architecture (Courtesy: Aerospace Corporation)

15.3.1.1 Wideband Systems

IDCSP satellites mentioned above represented phase I of the Defence satellite communication system (DSCS) program. Phase II of the program, named DSCS-II, began with the launch of six satellites launched in pairs, with the first pair launched in the year 1971. These satellites suffered some major technical problems and hence failed to operate after one or two years of their launch. Certain modifications were made in the next launches in order to remove these problems. By the year 1989, a total of 16 satellites were launched. The DSCS-II constellation comprised at least four active and two spare satellites. DSCS-II satellites offered increased capability over DSCS-I satellites and also had longer lifetimes.

The DSCS programme was initially developed to provide long distance communication services between major military locations. However, by the 1990s, DSCS satellites served a large number of small, transportable and ship-borne terminals. DSCS-III (Figure 15.3) satellites were developed to operate in this diverse environment. They had increased communication capacity, particularly for mobile terminal users, and improved survivability. The first DSCS-III satellite was launched in the year 1982. A total of 14 satellites have been launched to date.

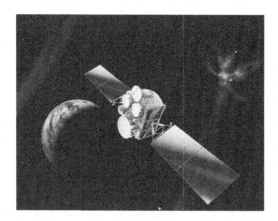

Figure 15.3 DSCS-III Satellite (Courtesy: Lockheed Martin Corporation)

The DSCS-III satellite system has a constellation of five operational satellites providing the required coverage.

The Global broadcast service (GBS) is another part of MILSATCOM's wideband architecture. It provides a high data rate intelligence, imagery, map, video and data communication services to tactical forces using small portable terminals. GBS broadcasts via communication payloads on two UFO Ka-band augmented satellites, four commercial satellites (Ku-band) and wideband global satcom (WGS) satellites (Ka- and X-band).

The Wideband Global Satcom program, initially referred to as the Wideband gapfiller satellite program (Figure 15.4) supplements the military X-band communications capability currently provided by the DSCS satellite system and the military Ka band capability of the GBS. In addition, the programme supports the warfighter with newer and far greater capabilities. Six WGS satellites have been launched between 2007 and 2013. This programme will be succeeded by the Advanced wideband system, which is in the planning stages.

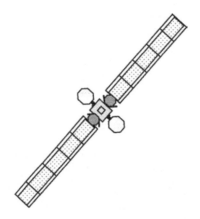

Figure 15.4 Wideband gapfiller satellite

15.3.1.2 Mobile and Tactical Systems

Developmental testing of tactical communication satellite systems began in the late 1960s with the launch of the Lincoln experiment satellites LES-5 and -6 and the tactical communications satellite named TACSAT-I. All the three satellites operated in the UHF and SHF frequency bands. These satellites tested the feasibility of supporting small, mobile antenna users. The Fleet satellite communication (FLTSATCOM) system was the USA's first operational military satellite system for tactical users. A total of five FLTSATCOM satellites were launched in a span of three years, between 1978 and 1981.

The FLTSATCOM system was followed by the Leasat satellite system (Figure 15.5). The first operational Leasat satellite was launched in the year 1984. Leasat satellites operated in the UHF band. Five satellites were launched in the Leasat system in a span of six years. These satellites primarily served the US navy, air force, ground forces and mobile users. Leasat satellites have been replaced by UFO satellites. Four blocks of UFO satellites were launched in a span of ten years from 1993 to 2003, namely the Block-I, II, III and IV satellites. Block-I and III comprised three satellites each, while Block-II and IV had four and one satellite respectively. UFO satellites support the global communications network of the US navy and a variety of other US fixed and mobile military terminals. They are compatible with ground- and sea-based terminals already in service. Figure 15.6 shows the photograph of the Block-IV UFO satellite (UFO-II).

Figure 15.5 Leasat satellite [Courtesy: NASA Johnson Space Center (NASA-JSC)]

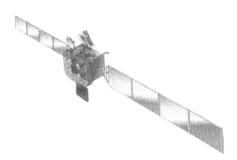

Figure 15.6 Block-IV UFO (UFO-11) satellite (Courtesy: Aerospace Corporation)

15.3.1.3 Protected Satellite Systems

Protected satellite systems serve the nuclear capable forces. These satellites provide global coverage and have maximum survival capability. Satellites developed under the category of protected systems include the Milstar System, AFSATCOM programme and the EHF payloads. The Milstar system was designed to provide increased robustness and flexibility to the users. The Milstar system includes two Block-I and four Block-II satellites. The Block-I satellites were launched in the years 1994 and 1995. The first Block-II satellite was lost during launch. The second one was launched in the year 2001 and the third and fourth satellites were launched in the years 2002 and 2003 respectively.

Protected satellite systems of the future include the Advanced extremely high frequency satellite system and the Enhanced polar satellite system. The AEHF satellite system (Figure 15.7), also referred to as the Milstar-3 system, will be fully operational by the year 2010. It will have 12 times the total throughput as compared to the Milstar-II system in some scenarios. Single-user data rates will increase to 8 Mbps. The system will also provide a large increase in the number of spot beams, which will improve user accessibility The Advanced Wideband satellite (AWS) system is the next generation communication system, providing highly secure and high capacity survivable communication services to US warfighters. Three AEHF satellites (AEHF-1, -2 and -3) were launched between 2010 and 2013. The Enhanced polar satellite system will have two satellites in highly inclined Molniya orbits to provide communication services to the polar regions.

Figure 15.7 Advanced Extremely High Frequency (AEHF) System (Courtesy: Lockheed Martin Corporation)

15.3.2 Russian Systems

Military communication satellites developed by Russia include the Parus, Potok (Geizer), Raduga (Gran), Raduga-1 (Globus), Raduga-1M, Strela-1, Strela-1M, Strela-2, Strela-2M, Strela-3 and Strela-3M series. The Parus satellite system was the first military communication satellite system of Russia and is currently operational. It was developed to provide location information for the Parus navigation system. Parus communication satellites also provide store-and-dump communication services and relay data for ocean surveillance satellites. A total of 96 Parus satellites have been launched in a span of 31 years between 1974 and 2005. The last satellite of the series, Parus-96, was launched in January 2005.

The first satellite of the Raduga system was launched in the year 1976. A total of 34 Raduga satellites have been launched since then, with the last satellite having been launched in the year 1999. Raduga-1 satellites are improved versions of Raduga satellites. Eight Raduga-1 satellites have been launched between the years 1989 and 2009. Raduga-1M satellites are further improved versions of Raduga-1 satellites. Two satellites have been launched in the Raduga-1M series in the years 2007 and 2010.

The Potok series, code-named Geizer, were military relay satellites designed to handle communications between the ground stations and the electro-optical reconnaissance satellite, Yantar. The first Potok satellite was launched in the year 1982. Ten Potok satellites have been launched to date, with the last satellite, Potok-10, launched in the year 2000.

The Strela series of satellites are Russian tactical communication satellites. The Strela communication satellite system comprised a constellation of medium orbit store-dump satellites that provided survivable communications for Soviet military and intelligence forces. Under the Strela-1 series, 21 experimental satellites were launched in a span of one year, between 1964 and 1965. Strela-1 satellites were followed by Strela-1M satellites. Around 370 Strela-1M satellites were launched between the years 1970 and 1992.

Five satellites were launched in the Strela-2 series, with the first launch taking place in the year 1965 and the last launch occurring in 1968. They were followed by the Strela-2M satellite series. The Strela-2M series comprised 52 satellites. The first Strela-2M satellite was launched in the year 1970 and the last satellite of the series was launched in the year 1994. All these satellites represent the first generation of strategic store-dump military communication satellites of Russia.

Strela-3 satellite system represent the second generation of Russian strategic store-dump military communication satellites. The operational constellation comprised 12 spacecraft in two orbital planes, spaced 90° apart. The first satellite in the series was deployed in 1985 and the system was accepted into military service in 1990. In the Strela-3 series, 143 satellites were launched. The last satellite of the Strela-3 series was launched in the year 2012.

15.3.3 Satellites Launched by other Countries

Many other countries including the UK, Italy, Israel, China and France have launched their own military communication satellites. The UK operates the Skynet series of satellites. It has launched 14 satellites during the time period from 1969 to 2012. Italy has launched two communication satellites named SICRAL-1 and SICRAL-1b in the years 2000 and 2009 respectively. It is planning to launch two satellites, namely Athena-Fidus and SICRAL-2, in the near future. SICRAL-2 is a joint French and Italian defence project. Israel has launched five satellites, AMOS-1 (Affordable modular optimized satellite), -2 (Figure 15.8) -3, -4 and -5 in the GEO orbit between 1996 and 2013. France operates the Telecom-1 and -2 series of military communication satellites. The Telecom-1 series comprises three satellites, namely Telecom-1A, -1B and -1C, launched between 1984 and 1988. Telecom-2 series is an advanced version of Telecom-1 series and comprise four satellites namely Telecom-2A, -2B, -2C and -2D, launched during the period 1991 to 1996. France has also launched Syracuse-3A and -3B satellites in the years 2005 and 2006 respectively. It also plans to launch Syracuse-3C satellite in the near future.

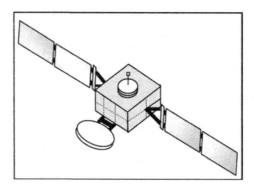

Figure 15.8 AMOS-2 satellite

China has launched several series of military communication satellites, including the DFH-1 (Dong Fang Hong), DFH-2, DFH-2A, DFH-3 (Figure 15.9), FH-1 (Feng Huo), FH-2, Spacenet-1, -2, -3, -3R series, ST-1, ST-2 and ZX-7 (Zhongxing) series.

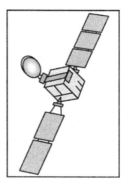

Figure 15.9 DFH-3 satellite

15.4 Frequency Spectrum Utilized by Military Communication Satellite Systems

As mentioned in the chapter on satellite communication applications, the bands of interest for satellite communications lie above 100 MHz, including the VHF, UHF, L, S, C, X, Ku, Ka and Q bands. Out of these bands, the main bands of interest for military satellite systems are the X, K, Ka and Q bands. It must be emphasized here that the military communication needs are fundamentally distinct from those of commercial communications. Military spectrum requirements are based on the need for high volume communications with continuous uninterrupted service during wartime. Table 15.1 lists the various bands used by both commercial and military satellite systems.

Use of high frequencies (K, Ka and Q bands) helps military satellites achieve a high degree of survivability during both electronic warfare and physical attack. It also offers advantages like

Table 15.1 Frequency bands used by commercial and military satellite systems

Segment	Band	Bandwidth used	User	Satellites
UHF	200–400 MHz	160 KHz	Military	FLTSAT, LEASAT
	L (1.5–1.6 GHz)	47 MHz	Commercial	Marisat, Inmarsat
SHF	C (6/4 GHz)	200 MHz	Commercial	Intelsat, DOMSATs, Anik E
	X (8/7 GHz)	500 MHz	Military	DSCS, Skynet and Nato
	Ku (14/12 GHz)	500 MHz	Commercial	Intelsat, DOMSATs, Anik E
	Ka (30/20 GHz)	2500 MHz	Commercial	JCS
	Ka (30/20 GHz)	1000 MHz	Military	DSCS-IV
EHF	Q (44/20 GHz)	3500 MHz	Military	Milstar
	V (64/59 GHz)	5000 MHz	Military	Crosslinks

reliable communication services in the nuclear environment, minimal susceptibility to enemy jamming and eavesdropping, and the ability to achieve smaller secure beams with modest-sized antennas. The military communication satellites of the USA operate in three main operational frequency segments, namely the UHF, SHF and EHF segments. The frequency band of interest in the UHF segment is the 200–400 MHz band. The X band (8/7 GHz) and Ka band (30/20 GHz) in the SHF segment and the Q band (44/20 GHz) in the EHF segment are also used extensively for these applications. Mobile and tactical military communication satellite systems operate in the UHF band (200–400 MHz). Wideband satellite systems operate in the X band (8/7 GHz) and the Ka band (30/20 GHz). Protected satellite systems operate in the EHF spectrum (44/20 GHz).

Russian military communication satellites mainly include the Raduga and the Strela series. The Raduga satellites operate in the C band (6.2/3.875 GHz).

15.5 Dual-use Military Communication Satellite Systems

Communication satellites intended for military applications are quite different from their civilian counterparts. They have better protection against jamming, better flexibility to rapidly extend services to new regions of the globe and to reallocate system capability as needed. Moreover, they employ better encryption techniques, enhanced TTC&M (Tracking, telemetry command and monitoring) security, hardening against radiation and so on. They use special frequencies for transmitting the signals. Because of these unique design features, they cost as much as three times as compared to their equivalent civilian counterpart satellites. Due to the high costs of military communication satellites, commercial satellite systems have also been used for non-strategic and non-tactical military applications.

Since the mid-1990s, many commercial civilian communication satellites are being used for military services of non-tactical nature. These satellites are used for providing radio and television services to the armed forces, telephone or other services that allow the overseas forces to talk to their relatives and many other services which do not require any special security protection. Keeping in mind their possible military usage, commercial communication satellite systems have adapted to this situation in terms of capacity availability, flexibility of geographical coverage and various types of security and encryption requirements. Digital video broadcast (DVB) services have also been developed to meet the military requirements.

15.6 Reconnaisance Satellites

Reconnaissance satellites, also known as spy satellites, provide intelligence information on the military activities of foreign countries. They can also detect missile launches or nuclear explosions in space. These satellites can catch and record radio and radar transmissions while passing over any country. Reconnaissance satellites can be further subcategorized into the following four types, depending upon their applications:

1. Image intelligence (IMINT) or photosurveillance satellites
2. Signal intelligence (SIGINT) or ferret satellites
3. Early warning satellites
4. Nuclear explosion detection satellites

IMINT and SIGINT satellites are collectively referred to as surveillance satellites.

15.6.1 Image Intelligence or IMINT Satellites

Image intelligence satellites provide detailed high resolution images and maps of geographical areas, military installations and activities, troop positions and other places of military interest. These satellites constitute the largest category of military satellites. They are generally placed in low, near-polar orbits at altitudes of 500–3000 km as they take high resolution close-up images. The resolution of images provided by these satellites is of the order of a few centimetres. Due to large atmospheric drag at these altitudes, image intelligence satellites generally have small lifetimes of the order of a few weeks. These satellites were widely used by the USA during operation Desert Storm in 1992. They provided warning of the Iraqi invasion of Kuwait nearly a week before it occurred, including both the timing and the magnitude of the assault. It should be mentioned here that some high resolution non-military Earth observation satellites have also been used for military applications. These include the ORBIMAGE-4 (Orbital imaging corporation) and the QuickBird series of satellites.

IMINT satellites can be classified as close-look IMINT satellites and area survey IMINT satellites, depending upon their mode of operation. Close-look IMINT satellites provide high resolution photographs that are returned to Earth via a re-entry capsule, whereas area survey IMINT satellites provide lower resolution photographs that are transmitted to Earth via radio. Recently launched IMINT satellites have the capability to take both close-look images as well as area images.

IMINT satellites can also be classified into the following three types depending upon their wavelength band of operation:

1. PHOTOINT or optical imaging satellites
2. Electro-optical imaging satellites
3. Radar imaging satellites

15.6.1.1 PHOTOINT or Optical Imaging Satellites

These satellites have visible light sensors that detect missile launches and take images of enemy weapons on the Earth's surface. These satellites can either be film-based or television-based. Film-based systems employ a film for recording the images and were the first type of systems to be used by reconnaissance satellites. They are no longer in use now. The system comprised two parts: the camera and the recovery capsule. In this case, after the pictures were taken, the film would spool-up in the return capsule. The capsule was released from the orbiting satellite once it had taken all the pictures. The capsule was then recovered in the Earth's atmosphere by an aircraft. The whole process of film retrieval took around one to three days. The image was then processed and analysed. Due to this time lag, they were used for strategic planning rather than in tactical combat situations. Moreover, these images could not be taken in cloudy conditions or in darkness and are susceptible to camouflaging. Figure 15.10 shows the operation of film-based PHOTOINT satellites.

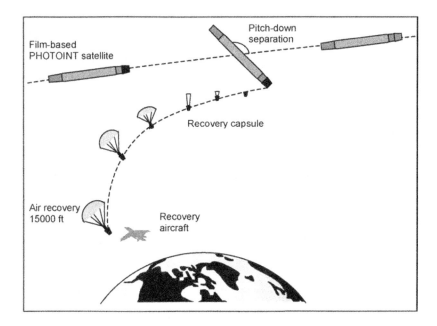

Figure 15.10 Operation of the film-based PHOTOINT satellites

Another type of PHOTOINT satellite system includes the television-based systems that take pictures in the conventional manner. After the images are taken, the film is scanned for electronic retransmission back to Earth. Due to the complexities involved, the system was phased out rather quickly. Some of the famous PHOTOINT satellites include the USA's KH-1 (KeyHole), KH 2, KH-3, KH-4, KH-4A, KH-4B, KH-5, KH-6, KH-7, KH-8 and KH-9 satellites and Russian Araks, Orlets, Yantar and Zenit series.

15.6.1.2 Electro-optical Imaging Satellites

Electro-optical imaging satellites provide full-spectrum photographic images in the visible and the IR bands. They use CCD cameras to take images. The CCD camera assigns different digital number values to represent varying light levels in the image. Digital enhancement techniques are used to further sharpen the images and remove the background noise. The digital information is then transmitted to the ground station via electronic communication links and the image is then 'reassembled' by the ground station computer. Electro-optical imaging satellites are able to image heat sources during the night but not objects having normal temperatures. Moreover, they do not work in cloudy conditions and are only slightly less susceptible to camouflaging as compared to the PHOTOINT satellites. The USA's KH-11 and KH-12 satellites and Russian Yantar-4KS1 and 4KS2 satellites are examples of electro-optical imaging satellites.

15.6.1.3 Radar Imaging Satellites

Both the PHOTOINT and the electro-optical imaging satellites were unable to take images under cloudy conditions. Radar imaging satellites overcome this problem. However, their resolution is poor when compared with the PHOTOINT and the electro-optical imaging satellites. Moreover, they suffer from the problem of 'backscatter noise' and are susceptible to active jamming.

These satellites mostly employ synthetic aperture radar (SAR) to take images in the microwave band. Here, microwave pulses are transmitted towards the Earth's surface by SAR. These pulses penetrate the cloud cover and hit the various objects on the Earth's surface. Taking into consideration the time taken by the reflected pulses to reach the satellite and the signal strength of the return beam, images are created. Different digital numbers are assigned to various light levels and then this information is transmitted electronically to Earth in the same manner as that for the electro-optical satellites. Other radar-based technologies employed by these satellites include the Doppler radar technology and the GMTI (Ground moving target indication) radar. Doppler radar technology is used to spot the movement of ships and aircraft and GMTI radar is useful for detecting ground movement of vehicles. Some of the radar imaging satellites include the USA's Lacrosse, Quill and Indigo satellites and Russian Almaz series of satellites.

After briefly introducing the types of IMINT satellites, the development of these satellites will be discussed in the following section.

15.6.1.4 Development of IMINT Satellites

The first IMINT satellites were launched by the USA followed by the erstwhile Soviet Union. IMINT satellites launched initially belonged to the PHOTOINT category. The first PHOTOINT satellite systems were the Discoverer and the Satellite and missile observation system (SAMOS) of the USA. Discoverer satellites circled the Earth in polar orbits. They used photographic films that were returned to Earth through a re-entry capsule. The first Discoverer satellite was launched in the year 1959. There were 38 public launches of the satellites under the program. Discoverer-14 satellite, launched on 18 August 1960, was the first satellite to successfully return film from orbit. This satellite marked the beginning of the age of satellite reconnaissance. The Discoverer programme officially ended in the year 1962 with the launch of Discoverer-38 satellite. However, the programme continued under the secret code name Corona until the year 1972, carrying out a total of 148 launches. Corona's major accomplishment was to provide photographs of missile launch complexes of the Soviet Union. It also identified the Plesetsk missile test range of Soviet Union and provided information on the types of missiles being developed, tested and deployed by Soviet Union. The SAMOS programme launched heavier payloads to collect photographic and electromagnetic reconnaissance data, which was transmitted electronically back to Earth.

The National reconnaissance office (NRO) was formed in the year 1961 to design, build, operate and manage the US reconnaissance satellites. Even today, it manages and operates all the reconnaissance satellites launched by the USA. The Corona programme lasted for 13 years and comprised four satellite generations named KH-1, KH-2, KH-3 and KH-4. The KH-4 family of satellites was further classified as KH-4, KH-4A and KH-4B. The KH (Key Hole) designation is used to refer to all photographic American reconnaissance satellites. The KH-1 satellites are sometimes referred to as the USA's first 'spy' satellites. The satellites launched initially had a resolution of the order of 10 m and a lifetime of around a week, which was later improved to 3 m and 19 days respectively in the KH-4B series.

The SAMOS and the Corona programmes were the first generation of IMINT satellites that returned high resolution images to Earth using re-entry capsules. Other first generation satellites included the Argon and the Lanyard series of satellites. Argon was the code name given to the KH-5 satellites, designed for large scale map-making. Lanyard satellites, or the KH-6 satellites, were used for gathering important intelligence information. Twelve satellites were launched in the KH-5 series and three satellites were launched in the KH-6 series. The KH-6 series was followed by KH-7, KH-8 and KH-9 series. All the satellites from KH-1 to KH-9 were film-based 'close-look PHOTOINT' satellites that returned high resolution images to Earth using small re-entry capsules and were part of the KeyHole (KH) series of satellites. They orbited in low Earth orbits at an altitude of around 200 km. Around 150 satellites were launched in the KH-1 to KH-9 series during the period 1960 to 1972.

The use of PHOTOINT satellites employing return capsules was discontinued in the early 1980s. Satellites that took wide-area images were advanced versions of IMINT satellites and transmitted images back to Earth via an electronic telemetry link. These satellites were referred to as the 'electro-optical' imaging satellites. The first electro-optical imaging satellite series was KH-11 series, code named Crystal/Kennan. The first satellite under this series was launched in December 1976. 16 satellites have been launched under the series in a span of 36 years from 1976 to 2013. KH-11 satellites orbited in higher orbits compared to their predecessors. They had the capability to take visible, near-IR and thermal-IR images.

The Russian IMINT satellites include the Zenit series, Yantar series, Orlets-1 and -2 and Araks series. The Zenit series comprised Zenit-2, -2M, -4, -4M, -4MK, -4MKM, -4MKT, -4MT, -6 and -8 series of satellites. The Zenit-2 series was the first to be launched, with 21 Zenit-2 satellites launched in a span of 29 years from 1961 to 1990. These satellites were film-based low resolution photo-intelligence satellites. Satellites in other Zenit series were high resolution film-based satellites.

The Yantar series comprised the Yantar-1K, -2K, -4K1, -4K2, -4K2M, -4KS1 and -4KS1M series of satellites. Yantar-1K, -2K, -4K1, -4K2 and -4K2M series of satellites comprised film-based photo-intelligence satellites whereas Yantar-4KS1 and -4KS1M were electro-optical imaging satellites. Orlets-1 and -2 were also film-based reconnaissance satellites. Eight satellites in the Orlets-1 series and two in the Orlets-2 series have been launched. Araks is the most recent reconnaissance satellite series having a resolution of 2–10 m. Two satellites have been launched in the series.

The development of radar-based reconnaissance satellites started in the 1970s. Quill was the first radar based reconnaissance satellite. It was launched by the USA in the year 1964. It was followed by the Indigo satellite launched in the year 1976. The most important radar-based intelligence satellite project was an American project named Lacrosse, whose first satellite was launched in the year 1988. It is an active radar imaging satellite system using synthetic aperture radar for observing tactical and strategic military targets. It also uses GMTI radar. The Lacrosse constellation comprises two operational satellites orbiting in low Earth orbits at an altitude of around 650 km. Five Lacrosse satellites have been launched, with the last one launched in April 2005. The USA launched the NROL 21 satellite in 2006, and the Topaz 1 and 2 satellites in 2010 and 2012, respectively. The erstwhile Soviet Union launched its first radar imaging satellite series known as the Almaz series in the late 1980s. Three satellites were launched in the series in a span of five years from 1986 to 1991. It also launched the Kondor-1 satellite in 2013.

Countries like the UK and Japan have launched their own IMINT satellites. Japan has launched four electro-optical imaging satellites named IGS (Intelligence gathering satellite)-Optical-1, -2, -3V and -5V in the years 2003, 2006, 2007 and 2009 respectively. It has also launched four radar reconnaissance satellites IGS-Radar-1, -2, -3 and -4 in the years 2003, 2007, 2011 and 2013 respectively. The UK has also launched its first reconnaissance satellite, TopSat-1 (Topographic satellite), in the year 2005. TopSat-1 is a photo-imaging satellite. Israel has Ofeq-3, -4, -5, -6, -7 and -9 optical imaging satellites. These satellites orbit in unusual retrograde orbits.

Helios is an European optical reconnaissance satellite system funded by France, Italy and Spain. It comprises the Helios-1 and -2 series each having two satellites, namely Helios-1A (1995), -1B (1999), -2A (2004) and -2B (2009) respectively. China has launched several optical reconnaissance satellites since the 1970s. It has launched the FSW-0 (Fanhui Shi Weixing), FSW-1, FSW-2, FSW-3 and FSW-4 series of satellites. In addition, it has also launched high resolution military imaging satellites named ZY-2A, ZY-2B, ZY-2C, Yaogan 2, 4, 5, 7, 8, 9, 11, 12, 14, 15, 16 and 17 satellites.

Israel has launched its radar-based reconnaissance satellite named TECHSAR. Other technology demonstrator satellites named Ofeq-1 and -2 were launched in 1988 and 1990 respectively. Ofeq-3, -5, -7 and 9 satellites are spy satellites of Israel. Germany has launched radar reconnaissance satellites SAR-Lupe -1, -2, -3, -4 and -5 with resolution less than 1 m. China has launched radar imaging satellites Yaogan-1, -3, -6, -10 and -13.

15.7 SIGINT Satellites

Signal intelligence or SIGINT satellites detect transmissions from broadcast communication and non-communication systems such as radar, radio and other electronic systems. These satellites intercept and decrypt government, military and diplomatic communications transmitted by radio, intercept ESM signals, receive telemetry signals during ballistic missile tests and relay radio messages from CIA agents in foreign countries. These satellites are essentially super-sophisticated radio receivers that can capture radio and microwave transmissions emitted from any country and send them to sophisticated ground stations equipped with supercomputers for analysis. SIGINT is considered to be the most sensitive and important form of intelligence. These satellites provided one of the first warnings of the possibility of an Iraqi invasion of Kuwait. SIGINT satellites, however, are not capable of intercepting landline communications.

SIGINT satellites need to intercept radio communications over a very large frequency range, typically from 100 MHz to 25 GHz. It is difficult to cover this wide frequency range in one satellite, hence different types of SIGINT satellites operating in different parts of the radio frequency spectrum are operated simultaneously. The USA employs Rhyolite, Chalet, Vortex and Aquacade satellites, all operating in different parts of the radio frequency spectrum. Intercepted radio data are transmitted to Earth on a 24 GHz downlink using a narrow-beam antenna.

The main missions carried out by these satellites are outlined below:

1. Interception and decryption of governmental, military and diplomatic communications transmitted by radio
2. Interception of ESM (electronic support measure) signals that characterize the operating modes of the higher command organizations, installations of air defence, missile forces and also the combat readiness of foreign armed forces
3. Reception of telemetry signals during ballistic missile tests
4. Relay of radio messages from CIA agents in foreign countries

SIGINT satellites can be further categorized as communication intelligence (COMINT) or electronic intelligence (ELINT) satellites depending upon their intended function.

COMINT or communication intelligence satellites perform covert interception of foreign communications in order to determine the content of these messages. As most of these messages are encrypted, they use various computer-processing techniques to decrypt the messages. The information collected is used to obtain sensitive data concerning individuals, government, trade and international organizations. COMINT satellites of today collect economic intelligence information and information about scientific and technical developments, narcotics trafficking, money laundering, terrorism and organized crime.

ELINT or electronic intelligence satellites are used for the analysis of non-communication electronic transmissions. This includes telemetry from missile tests (TELINT) or radar transmitters (RADINT). The most common ELINT satellites are designed to receive radio and radar emanations of ships at sea, mobile air defence radar, fixed strategic early warning radar and other vital military components for the purpose of identification, location and signal analysis.

15.7.1 Development of SIGINT Satellites

The USA and Russia have the largest number of SIGINT satellites. However, some other countries like France and China have also developed their own SIGINT satellite systems.

15.7.1.1 USA Satellites

The first SIGINT satellites were launched by the USA in the early 1960s. These satellites orbited in LEO orbits. The limited and intermittent operation of these satellites suggested that for continuous monitoring and interception of communication channels, these satellites need to be placed at higher altitudes. In addition, satellites orbiting at higher attitudes are able to carry out both COMINT and ELINT operations.

The USA developed SIGINT satellites called 'Jumpseat' in the 1970s to be placed in the Molniya orbit. The basic task of these satellites was to intercept radio communications transmitted by communications satellites of the erstwhile Soviet Union orbiting in Molniya orbit. From 1971 to 1987, seven Jumpseat satellites were launched. Another series of satellites launched in the Molniya orbit was the Trumpet series. Three satellites were launched in the series in a span of three years from 1994 to 1997. The USA launched another series of satellites named Spook Bird series, beginning in 1968, for radio interception from satellites of erstwhile Soviet Union orbiting in the geosynchronous orbit. Spook Bird satellites were launched in quasi-stationary orbits having an inclination of 3° to 10°, apogee distances of 39 000 km to 42 000 km and perigee distances of 30 000 km to 33 000 km. Spook Bird satellite move in a complex elliptical trajectory, enabling them to view broad regions. After two experimental launches, production models of these satellites, named Rhyolite, were launched. Rhyolite constellation consisted of four operational satellites intercepting signals in the lower frequency UHF and VHF bands. They carried out a wide variety of missions in intercepting microwave communication transmissions and missile telemetry data from erstwhile Soviet Union and China. Four Rhyolite satellites were launched between 1970 and 1978. These satellites were later renamed Aquacade.

Another SIGINT satellite series developed by the USA during the 1970s was named Chalet. The first satellite of the series was launched in 1978 to intercept conversations carried on UHF radio links. The name of Chalet satellites was changed to Vortex in 1981. Six satellites were launched between 1978 and 1989. Vortex satellites (Figure 15.11) were modernized versions of Chalet satellites with better onboard equipment for the purpose of expanding the range of interceptable radio frequencies in the direction of the centimetric band. Mercury satellites (Advanced Vortex satellites), successor to the Chalet/Vortex satellites, are used to pinpoint radar locations. These satellites are in the GEO orbits as opposed to quasi-stationary orbits of Chalet/Vortex satellites. Three satellites were launched in this series in the years 1994, 1996 and 1998.

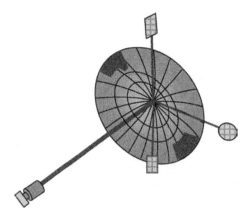

Figure 15.11 Vortex satellite

Magnum/Orion satellites were deployed at the end of the 1980s to replace Rhyolite series of satellites as they reached the end of their operating lifetimes. Targets for these satellites include telemetry, VHF radio, cellular mobile phones, paging signals and mobile data links. Two Magnum satellites were launched, one in the year 1985 and the other in the year 1989. The Magnum series of satellites were replaced by the Mentor satellites. Six Mentor satellites were launched in a span of 17 years between 1995 and 2012. The USA had six to eight operational SIGINT satellites during the 1980s and 1990s. The frequency of launches of SIGINT satellites dropped at the beginning of the 1990s.

The first space-based ELINT system of the USA was named GRAB (Galactic radiation and background). A total of five GRAB satellites were launched between 1960 and 1962. A primary ELINT programme named 'White Cloud' is a satellite constellation that is the US Navy's principal means of over-the-horizon reconnaissance and target designation for its weapons systems.

15.7.1.2 Russian Satellites

The first SIGINT satellite launched by the erstwhile Soviet Union was an ELINT satellite named Cosmos 189, launched in the year 1967. Until now, more than 200 SIGINT satellites have been launched under the Tselina satellite system (Figure 15.12). It basically comprised the low-sensitivity Tselina-O satellites and the high-sensitivity Tselina-D satellites. Tesleina-O and -D satellites represent the first generation of Russian ELINT satellites. Tselina-2, Tselina-OK and Tselina-R represent the second generation of Tselina satellite system. The Tselina satellites detected and located the source of radio transmissions as well as determined the type, characteristics and performance modes of their targets.

Figure 15.12 Tselina satellite

15.7.1.3 Other Countries

France has launched several SIGINT satellites. The first SIGINT satellite of France was the Cerise satellite launched in the year 1995. It was a technology demonstrator satellite. It was followed by the Clementine satellite launched in the year 1999 and the Essaim series of satellites. Essaim is a system of four microsatellites that analyses the electromagnetic environment of the Earth's surface. All four Essaim satellites were launched in the year 2004. China has also launched SIGINT satellites, named the JSSW (Ji Shu Shiyan Weixing) series of satellites, comprising six satellites launched between 1973 and 1976.

15.8 Early Warning Satellites

Early warning satellites constitute a significant part of military systems. They provide timely information on the launch of missiles, military aircraft and nuclear explosions by the enemy to military commanders on the ground. This information enables them to ensure treaty compliance as well as provide an early warning of missile attack for appropriate action. Space-based infrared satellite systems are also being developed, which could track ballistic missiles throughout their trajectory and provide the earliest possible trajectory estimates to the command centre. In other words, these satellites would provide the earliest information of the start of a major missile attack and will be used to track long term patterns of space programmes of foreign countries.

Early warning (EW) satellites constitute an important part of the missile defence program of USA, which aims to intercept and destroy missiles by shooting them down before they hit the target. EW satellites detect the launch of the missile, track the initial trajectory of the missile and relay this information to a missile defence command centre on the ground.

The USA and Russia have developed extensive early warning satellite systems. In the following paragraphs, these systems will be discussed briefly.

15.8.1 Major Early Warning Satellite Programmes

The first early warning satellite system developed by the USA was the MIDAS system. It employed 24 satellites in low Earth orbits for detecting the launch of inter-continental ballistic missiles (ICBM) by Russia. However, the MIDAS programme was not very successful. The attention then shifted to launching early warning satellites in GEO orbits, as only four GEO satellites would be required for global coverage. The first geostationary early warning satellite system was the Defence support program (DSP) of USA. DSP satellites detected the launch of intercontinental and submarine launched ballistic missiles, using IR and optical sensors. They also provided information on nuclear explosions. Over 19 DSP satellites (Figure 15.13) have been launched during 1970 to 1984. During the Persian Gulf War, DSP satellites provided effective warning of the launch of Scud missiles by Iraq.

Figure 15.13 DSP satellite (Courtesy: US Air Force)

The Space-based infrared system (SBIRS) is intended to be the next-generation missile warning and tracking system. It will replace the DSP satellite system. The system (Figure 15.14) comprises a constellation of 24 satellites orbiting in three types of orbits, namely the GEO, HEO and LEO orbits. The constellation will have four satellites in the GEO orbit, two satellites in the HEO orbit and 18 satellites in the LEO orbit. The GEO and

Early Warning Satellites

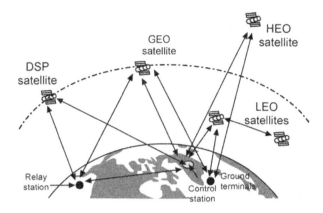

Figure 15.14 SBIRS architecture

HEO satellites constitute the SBIRS–high component (Figure 15.15) and the LEO satellites form the SBIRS–low component. Two SBIRS-GEO satellites have been launched in 2011 and 2013. Two satellites SBIRS HEO-1 and -2 have been launched 2006 and 2008 in HEO orbit. The SBIRS–low component has been renamed Space tracking and surveillance system (STSS). SBIRS system is a part of the National missile defence (NMD) programme of the USA. SBIRS–high satellites are three-axis stabilized satellites and their sensors monitor the ground continuously, thereby providing much more accurate data. They will replace the DSP satellites. STSS will track missiles as they fly above the horizon, offering much more accurate information on their trajectories. Such information is necessary for an effective anti-ballistic missile defence.

Figure 15.15 SBIRS-High component (Courtesy: Lockheed Martin Corporation)

The first early warning satellite launched by the erstwhile Soviet Union was a test satellite launched in the year 1972. The first operational early warning satellite was launched five years later in 1977. These satellites, named Prognoz, orbited in Molniya orbits. This orbit enabled the satellite sensors to view the missiles against the cold background of space rather than the warm background of Earth. However, nine satellites were required to make the constellation fully operational. The Soviet Union government was unable to maintain the system and in the 1990s only half of the constellation was working. In the mid-1980s Soviet Union launched geostationary early warning satellites named Oko satellites, but they were not very successful. France has launched its early warning satellite programme named SPIRALE (Système Préparatoire Infrarouge pour Alerte, or Preparatory System for IR Early Warning), comprising two satellites, SPIRALE-A and -B, to detect ballistic missiles in their boost phase.

15.9 Nuclear Explosion Satellites

Vela satellites were developed by the USA to detect nuclear explosions on Earth and in space in order to monitor worldwide compliance with the 1963 nuclear test ban treaty. A total of 12 Vela satellites were launched during the period 1963 to 1970. In the 1970s, the nuclear explosion detection mission was taken over by the DSP system, and in the late 1980s, by the GPS system. The programme is now referred to as the Integrated operational nuclear detection system (IONDS).

Two experimental satellites, namely the Array of low energy X-ray imaging sensors (ALEXIS) satellite and the Fast on-orbit recording of transient events (FORTÉ) satellite, were launched by the USA in the years 1993 and 1997 respectively. The ALEXIS satellite sensors provide near real-time information on transient, ultra-soft X-rays. In addition, they also offer unique astrophysical monitoring capabilities. The FORTÉ satellite features an electromagnetic pulse sensor. The sensor provides wideband radio frequency signal detection. The FORTÉ satellite integrates with related technology to help discriminate between natural (such as lightning) and man-made signals.

15.10 Military Weather Forecasting Satellites

Weather forecasting satellites provide high quality weather information to the operational commanders in the battlefield. This helps in effective deployment of weapon systems, protection of Department of Defence (DoD) resources and for exploits deep in enemy territory. The weather forecasting satellites provided useful information to the American forces during the Persian Gulf War.

The Defence meteorological satellite program (DMSP), originally known as the Defense system applications program (DSAP), is the USA's military weather satellite programme to monitor the meteorological, oceanographic and solar-geophysical environment of Earth in order to support DoD operations. It provides visible and IR cloud cover imagery and other meteorological, oceanographic, land surface and space environmental data. The first DMSP satellite was launched in the year 1966. Since then 12 series of DMSP satellites, namely DMSP-1A, -2A, -3A, -3B, -4A, -4B, -5A, -5B, -5C, -5D1, -5D2 and -5D3 (Figure 15.16), have been launched. All satellites launched have had tactical (direct readout) as well as strategic (stored data) capacity. The satellites orbit in near-polar sun-synchronous orbits. The DMSP constellation comprises a constellation of two active satellites. In December 1972, DMSP data were declassified and made available to the civil/scientific community.

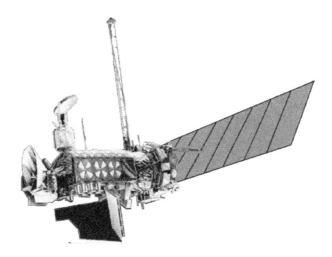

Figure 15.16 DMSP-5D3 satellite (Courtesy: NASA)

15.11 Military Navigation Satellites

Satellite navigation systems have proved to be a valuable aid for military forces. Military forces around the world use these systems for diverse applications including navigation, targeting, rescue, disaster relief, guidance and facility management both during wartime as well as peacetime. The main satellite navigation systems operational today are the GPS system of the USA and the GLONASS system of Russia. The GPS and GLONASS receivers are used by soldiers and also have been incorporated on aircraft, ground vehicles, ships and spacecraft. In addition, the Galileo navigation satellite system of Europe, the Beidou navigation satellite system of China and the IRNSS system of India are recent satellite based navigation systems. The Galileo navigation system will comprise 30 satellites in three planes in medium earth orbit. Four Galileo IOV satellites have so far been launched. The Beidou satellite navigation system comprises two separate satellite constellations: a limited test system and a truly global navigation system, also referred to as Compass. Compass will comprise 35 satellites and will be fully operational by 2020. The basic Beidou satellite constellation requires three satellites. BD-1, BD-2, BD-2I and BD-2M series of navigation satellites have been launched by China to date. Four satellites have been launched in BD-1 series, six satellites in BD-2 series, two-satellites in BD-2I series and five in BD-2M series. IRNSS is being developed by India. It will comprise a constellation of seven satellites and a ground control segment. The first satellite of the constellation IRNSS-1A was launched in July 2013 and the complete constellation will be operational by 2014–15.

Military applications of navigation satellites have been discussed in detail in Chapter 13.

15.12 Space Weapons

Space weapons are categorized as weapons that travel through space to strike their intended targets. The intended target may be located on the ground, in the air or in space. Space weapons include anti-satellite weapons that can target the space systems of the adversary from a ground based, aerial or space borne weapon system and also space based weapon systems that attack targets on the ground or intercept missiles travelling through space. Space weapons have been the subject of intense discussion and debate among scientists, technologists, Defence strategists and policy makers for more than 50 years. It began during pre-cold war days, when it was triggered by the possibility of bombardment of satellites carrying nuclear weapons. The second time was during the period that followed the end of the cold war and this time it involved the possibility of space-based defence against nuclear missiles. This period witnessed the Strategic Defence Initiative (SDI) programme of the United States. Today it is again an area of focused research and development activity for developed and some developing countries to offer defence against ballistic missiles, safeguard space assets and project force. In the following sections we describe the different types of space weapons in terms of the technologies involved, international status, capabilities, limitations and deployment issues. Some prominent systems which are briefly discussed in terms of their features and facilities are also discussed in detail towards the end.

15.12.1 Classification of Space Weapons

Space weapons may be classified on the basis of physical location of the weapon and intended target as follows. Each of the three above mentioned categories includes both kinetic as well as directed energy weapons.

1. Space-to-Space weapons
2. Earth-to-Space weapons
3. Space-to-Earth weapons

15.12.1.1 Space-to-Space Weapons

The idea of using space platforms for military purposes has its origin in the cold war era and was the brain child of the USA and the erstwhile Soviet Union. The Almaz programme of the then Soviet Union and the MOL programme of the USA exemplify the idea of use of manned space platforms for carrying out military missions. The Almaz programme of the Soviet Union comprised a series of military space stations called Orbital piloted stations (OPS). These space stations were launched under the cover of the Salyut programme as the Soviet authorities didn't want to disclose the existence of the top-secret Almaz programme. As a consequence, the Almaz orbital piloted stations OPS-1, OPS-2 and OPS-3 were named Salyut-2, Salyut-3 and Salyut-5 respectively. Figure 15.17 shows the Almaz manned space station. OPS-1 (Salyut-2) was launched on 3 April 1973 from Baikonur, but days after the launch, an accident left the spacecraft disabled and depressurized. OPS-2 (Salyut-3) was launched on 25 June 1974. OPS-2 was also deorbited in January 1975. OPS-3 (Salyut-5) was launched on 22 June 1976. The space station was visited by two crews during 1976–1977. OPS-3 finally burned up in the Earth's atmosphere on 8 August 1977. The next Almaz space station, OPS-4. that promised a number of upgrades never became a reality. This space station was to be the first space station to be launched with synthetic aperture radar (SAR) and a manned reusable return vehicle and there was a plan to replace the Shchit-1 defence gun with Shchit-2 space-to-space cannon. The space station has remained grounded with the result that OPS-3 remained the last manned space station under the Almaz programme.

Each of the Almaz space stations was equipped with a reconnaissance payload that comprised a colossal telescope called Agat-1, an optical sight that permitted the crew to come to a standstill over a facility and infrared and topographic cameras. The telescope was

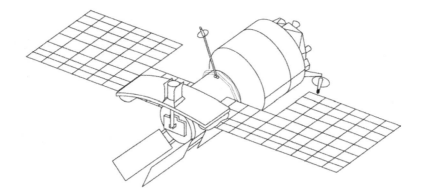

Figure 15.17 Almaz manned space station

approximately one metre in diameter and had a focal length of 6.4 metres. The reconnaissance payload was used to take images of military installations such as airfields, missile complexes with a resolution better than 50 cm. The data from the reconnaissance payload could also be relayed to the ground via a radio link. It appears that the camera films were developed on board. These were then scanned and transmitted to ground via the link.

In addition to the reconnaissance payload described above, Almaz space stations were also reported to have been equipped with a 23 mm Nudelman-Rikhter (NR-23) rapid-fire self-lubricating cannon capable of firing 950 rounds per minute. However, the entire station had to be reoriented towards the threat in order to aim the gun. It is reported that OPS-2 carried out a successful test firing on a target satellite.

The Manned orbital laboratory (MOL) was proposed by the United States Air Force and was initially intended to test the military worthiness of humans in orbit. Figure 15.18 shows the MOL. The programme was planned as a successor to the cancelled X-20 Dyna-Soar project. It was thought having a man in loop would facilitate in-orbit repair, target selection and ability to shoot through cloud cover. The space station was configured around a modified Gemini-B spacecraft that could be attached to a laboratory vehicle. The space station was planned to be launched on board the Titan IIIC rocket. The space station was equipped with optical telescope and gyro stabilized cameras to be operated by astronauts to gather photo intelligence on Soviet military assets. The programme was launched in December 1963. One mock-up mission was launched on 3 November 1966. The proposed missions under the MOL programme included MOL-1 (1 December 1970), MOL-2 (1 June 1971), MOL-3 (1 February 1972), MOL-4 (1 November 1972), MOL-5 (1 August 1973), MOL-6 (1 May 1974) and MOL-7 (1 February 1975). MOL-1 and MOL 2 were proposed as unmanned missions while MOL-3, MOL-4, MOL-5, MOL-6 and MOL-7 were proposed as manned missions. The mission was cancelled in June 1969 due to budget constraints and the escalating war in Vietnam. Another reason for premature closure of the programme was the feeling that the features and facilities of unmanned spy satellites that followed thereafter met or exceeded the capabilities of manned MOL missions.

15.12.1.2 Earth-to-Space Weapons

Earth-to-space weapons are anti-satellite weapons that are designed to incapacitate or destroy satellites intended for strategic military applications. These satellites are mainly in low Earth

Figure 15.18 Manned Orbital Laboratory (Courtesy: NASA)

orbits. Countries like the United States, Russia and China are believed to have developed and successfully field tested either kinetic energy or directed energy weapon systems or both for anti-satellite applications. These weapon systems are both land-based as well as mounted on aerial platforms. These countries in the past have used these weapon systems to destroy their own satellites that have malfunctioned while in orbit and were rendered useless. Some of these experiments are briefly mentioned in the following paragraphs.

One such test was conducted by the United States on 13 September 1985, when an anti-satellite missile ASM-135 was used to destroy US satellite P78-1. P78-1, also known as Solwind, was launched on 24 February 1979 and was of the type of Orbiting solar observatory (OSO) with a solar oriented sail. The payload comprised of a gamma ray spectrometer, a high latitude particle spectrometer, a white light spectrograph, an ultraviolet spectrometer, an aerosol monitor and an X-ray monitor. The satellite was the backbone of coronal research for more than six years. The satellite was brought down on 13 September 1985 using ASM-135 missile. ASM-135 (Figure 15.19) is an air-launched anti-satellite multi-stage missile that was first produced in 1984 and had a kinetic energy hit-to-kill warhead. On 13 September 1985, ASM-135 was fired from an F-15A aircraft about 200 miles west of Vandenberg Air Force base and destroyed the Solwind satellite flying at an altitude of 345 miles.

Another test of same type was carried out on 21 February 2008 when the US spy satellite USA-193 was brought down using the RIM-161 standard missile 3 (RIM-161 SM-3). USA-193, also called NRO-21, was a US spy satellite launched on 14 December 2006 aboard Delta-II rocket. The satellite malfunctioned shortly after deployment and was brought down intentionally on 21 February 2008. The satellite was shot down using RIM-161 SM-3 missile (Figure 15.20) fired from a US warship near Hawaii. The exact purpose of the satellite was kept as a closely guarded secret, but it is believed that the satellite carried high resolution

Space Weapons 743

Figure 15.19 ASM-135 anti-satellite missile (Attribution: Lorax)

Figure 15.20 RIM-161 SM-3 missile (Courtesy: US Navy)

radar to generate images for National Reconnaissance Office. RIM-161 is basically a ship borne anti-ballistic missile that evolved from the well proven SM-2 Block-IV design. Like the Block-IV missile, it uses the same booster and a dual thrust rocket motor for the first and second stages. It also uses the same steering control section and guidance mechanism. The missile is equipped with a hit-to-kill kinetic energy warhead.

Russia too has been experimenting with the use of land based and aerially delivered anti-satellite weapons of both kinetic energy and directed energy types. The erstwhile Soviet Union tested ground based lasers from the 1970s onwards for anti-satellite applications. A number of US spy satellites were reportedly blinded temporarily during the 1970s and 1980s. The Terra-3 programme is an example. The Terra-3 complex was a laser testing centre built in the 1970s on the Sary Shagan anti-ballistic missile testing range in Kazakhstan. The complex was equipped with high power/energy carbon dioxide and ruby lasers for anti-ballistic and anti-satellite applications. However, the laser energy from these sources was not sufficient for any anti-ballistic applications. Initial use therefore was limited to anti-satellite applications primarily to blind sensors. One such experiment was executed on 10 October 1984 when a low energy laser beam was directed at US space shuttle Challenger (OV-99) causing some of the on-board equipment to malfunction and also causing discomfort to crew members. The Soviet Union also researched directed energy weapons under the Fon project from 1976 onwards. They also started development of air-launched anti-satellite weapons in early 1980s. Modified MiG-31 Foxhounds were used as the launch platform.

China has also successfully tested the anti-satellite missile, named SC-19, with a kinetic kill warhead. SC-19 has been reported to be based on a modified DF-21 ballistic missile or its commercial derivative KT-2. The ASAT missile is guided by an infrared imaging seeker. The test demonstrated use of a ground platform launched kinetic kill anti-satellite missile to destroy a near Earth orbit satellite. The satellite was a defunct Chinese weather satellite FY-1C of Feng yun series and the test was carried out on 11 January 2007 when the satellite in its polar orbit at 865 km was destroyed by a kinetic kill vehicle travelling with a speed of 8 km/s in the opposite direction. The missile was launched from a Transporter-erector-launcher (TEL) vehicle. Figure 15.21 shows the orbital planes of the space debris of the satellite one month after its disintegration by the Chinese ASAT missile.

The Strategic Defensive Initiative (SDI) programme of the United States, nicknamed 'Star Wars', proposed by the then US President Ronald Reagan on 23 March 1983 and having the objective of developing a defensive system to offer protection against enemy inter-continental ballistic missiles (ICBM) has also given a major boost to the ASAT programmes of the United States and Russia. The SDI programme is discussed in detail in section 15.13.

While the ASAT projects were adapted for anti-ballistic missile applications; the reverse was also true. It may be noted that interception of a satellite with a static orbit is a much easier proposition than intercepting a warhead on a ballistic trajectory. This is mainly due to the low level of uncertainty encountered in the case of satellite orbits and also due to the availability of relatively much longer tracking and manouevring times in an anti-satellite intercept.

15.12.1.3 Space-to-Earth Weapons

In the category of space-to-earth weapons, concepts of orbital weaponry and orbital bombardment have been designed by both the United States and the Soviet Union during the cold war era. The fractional orbital bombardment weapon system deployed by the Soviet Union during 1968–1983 is one such system. In this system, a nuclear warhead could be placed in a low

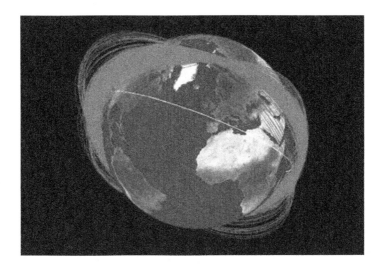

Figure 15.21 Orbital planes of the space debris of FY-1C weather satellite (Courtesy: NASA)

Earth orbit and then later at the time of strike deorbited to hit any location on the surface of the Earth. Presently, there are no known operative orbital weapons. This has been largely due to the coming into existence of several international treaties prohibiting deployment of weapons of mass destruction in space. Fractional orbital bombardment system was also phased out in 1983. However, other weapons like kinetic bombardment weapons do exist as they don't violate these treaties.

The Space Based Laser (SBL) programme of the United States is a technology demonstration programme with the objective of establishing the capability of shooting down a ballistic missile in its boost phase with a space based high power laser. SBL is aimed at providing global boost-phase intercept of ballistic missiles. Under the programme, it is proposed to put an experimental high power laser system into space and follow it up with the experiment of shooting down a missile. The outcome of this experiment, known as the Integrated Flight Experiment (IFX), is likely to determine the efficacy of SBL to protect the United States and its allies from ballistic missile threat as a part of layered defence.

Another space based laser programme aimed at converting solar energy to laser light in space is the collaborative effort of the Japan aerospace exploration agency (JAXA) and Osaka University. This space generated laser light could then be transmitted to the Earth to generate electricity or to power a massive 'death ray'. It is estimated to put this novel laser system into space by 2030. Figure 15.22 shows the concept.

15.13 Strategic Defence Initiative

The Strategic Defence Initiative (SDI) was the brain child of the then US President Ronald Reagan. The programme was unveiled by him on 23 March 1983 through which he proposed to use ground-based and space-based systems to offer protection to the United States and its allies from strategic nuclear warhead equipped ballistic missiles. The programme was nick-named the 'Star Wars' programme after the popular 1977 film by George Lucas. The SDI programme as envisaged by the US President was studied in detail by the Strategic Defence

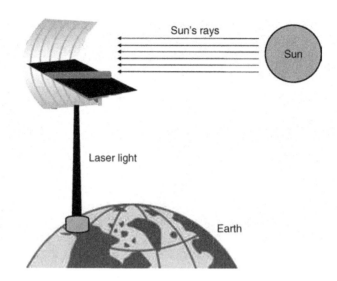

Figure 15.22 Collaborative Space based laser concept of JAXA and Osaka University

Initiative Organization (SDIO) set up in 1984 within the United States Department of Defence. Defence strategists and scientists described the programme as highly ambitious and felt that its implementation was not feasible with the then existing technology. Subsequently in 1993, the programme was renamed as the Ballistic Missile Defence Organization (BMDO) by the then US President Bill Clinton. The programme was modified with the emphasis shifting from national missile defence to theatre missile defence and its scope reduced from global coverage to regional coverage. Though the programme was never fully realized as envisaged; the research work carried out and the technologies developed under the programme have led to development of some of the contemporary anti-ballistic missile systems. The SDI programme witnessed the initiation and development of many technologies and products, some successful and some not-so-successful and some unsuccessful, which included ground-based programmes, directed energy weapon (DEW) programmes, space programmes, sensor programmes and countermeasures programmes. In the following paragraphs, the major technologies and systems initiated under the SDI programme are briefly discussed.

15.13.1 Ground-based Programmes

Prominent ground-based programmes included the Extended Range Interceptor (ERINT), Homing Overlay Experiment (HOE) and Exoatmospheric Re-entry Vehicle Interception System (ERIS). Each one of these is briefly described in the following paragraphs.

15.13.1.1 Extended Range Interceptor (ERINT)

The ERINT programme was an extension of the Flexible lightweight agile guided experiment (FLAGE) involving the development of a small, agile radar homing hit-to-kill vehicle. FLAGE was tested successfully by targeting a MGM-52 Lance missile in flight. The test was conducted at the White Missile Range in 1987. ERINT was the follow-on to the FLAGE experiment. ERINT used a new solid propellant rocket motor, which allowed the missile to fly faster and higher than FLAGE. ERINT also had an upgraded design including addition of aerodynamic manoeuvring fins and attitude control motors, which increased the range. With the new guidance technology, the missile was designed to be used primarily against manoeuvring tactical missiles and secondly against air-breathing aircraft and cruise missiles. The first flight test of ERINT (Figure 15.23) was conducted at the White Sands Missile Range in June 1992 followed by another successful test in August 1992. These two preliminary tests did not attempt to hit target missiles. Preliminary testing with three direct hits simulating theatre missile defence was concluded in November 1993. This was followed by another test in June 1994 where it was used to destroy a drone to establish the accuracy of its guidance system. ERINT was subsequently selected as the new missile for the Patriot advanced capability-3

Figure 15.23 Flight test of ERINT (Photo Courtesy of U.S. Army)

system (PAC-3) mainly because of its increased range, accuracy and lethality, all in a smaller package.

15.13.1.2 Homing Overlay Experiment (HOE)

The HOE of the US army was the first to demonstrate the concept of exoatmospheric hit-to-kill to intercept and destroy ballistic missiles. The US army started a technology demonstration programme in the mid 1970s to validate the emerging technologies designed to have non-nuclear hit-to-kill intercepts of Soviet ballistic missiles in space. Planning began in 1976 and the contract for development of interceptor was awarded to Lockheed Martin in August 1978. The interceptor of the HOE programme consisted of Minuteman-I launch stages carrying the homing and kill vehicle. The Kinetic Kill Vehicle (KKV) was equipped with an infrared seeker, guidance electronics and a propulsion system. The infrared seeker allowed the interceptor to guide itself into the path of an incoming ballistic missile warhead and collide with it.

Four flight tests were carried out in February 1983, May 1983, December 1983 and June 1984. Each of the four tests involved launching a target from Vandenberg Air Force Base in California and an HOE interceptor from the Kwajalein Missile Range in the Republic of the Marshall Islands in the Pacific. Figure 15.24 shows the HOE test vehicle. The first three tests did not achieve a successful intercept, with the targeted vehicle. In the fourth test in

Figure 15.24 Homing Overlay Experiment (HOE) test vehicle

June 1984, the kinetic kill vehicle interceptor did find the Minuteman intercontinental ballistic missile re-entry vehicle in space and guided itself to an intercept and finally destroyed the target through collision. Both target and interceptor had sensors, which along with ground-based radars and airborne optical sensors produced data to show that the target was destroyed by the collision of the interceptor and not by an explosive charge after a near miss. The data also produced evidence that the interceptor guided itself to the target with the help of its infrared homing sensor. Using an explosive charge to destroy the target in the event of a near miss was a part of the deception programme, which was reportedly discontinued before the third flight test. In the first two tests; it could not alter the result as the interceptor missed the target by large distances.

The intercept vehicle had a fixed fragment net intended to increase the lethal radius of the interceptor. It consisted of 36 aluminium ribs with stainless steel fragments that increased the interceptor size to achieve greater probability of target hit. The structure of the ribs was kept folded in flight and was deployed shortly before intercept. Once deployed, this umbrella-like web had a spread of about 4 m diameter.

15.13.1.3 Exoatmospheric Re-entry Vehicle Interception System (ERIS)

Development of ERIS began in 1985. The ERIS programme was an extension of the HOE programme and was built on the technologies tested during the HOE programme. ERIS was made up of the second and third stages of Minuteman ICBM and had a kill vehicle equipped with a long wave infrared scanning seeker. The sensor and guidance technology of the ERIS KKV (kinetic kill vehicle) was based on the experience learned from the HOE tests. The ERIS KKV with its inflatable octagonal kill enhancer was significantly smaller and lighter than the HOE KKV.

The first test of the ERIS KKV was conducted on 28 January 1991. The intercept vehicle successfully detected and intercepted a mock ICBM warhead launched from Vandenberg Air Force Base. It was the first time that an SDI experiment attempted an interception in a countermeasures environment by discriminating against decoys. The target re-entry vehicle deployed two balloon decoys on either side. The KKV was pre-programmed to hit the centre target that was the warhead.

The second and final test was conducted on 13 May 1992, when the intercept vehicle was targeted against a Minuteman-I ICBM. Though the test was a partial failure and the kill vehicle did not achieve a direct intercept, nevertheless the test met the primary targeted objectives of collection of radiometric data on the target and decoys, acquisition and resolution of threat and demonstration of target handover. Two of the originally planned four tests were cancelled. Due to the change in the global situation after the end of the cold war, the SDI programme was reoriented in the early 1990s and the ERIS programme was not developed into an operational system. The experiences of the ERIS programme were used to advantage in the successful development and deployment of the next generation of exoatmospheric kill vehicles.

15.13.2 *Directed Energy Weapon Programmes*

The prominent directed energy weapon programmes included a nuclear explosion powered X-ray laser cluster aimed at targeting multiple warheads simultaneously, a chemical laser for use as anti-ballistic missile and anti-satellite weapon, a particle beam accelerator and a hyper-velocity rail gun. Each of these programmes is briefly described in the following paragraphs.

15.13.2.1 Nuclear Explosion Powered X-Ray Laser

The programme involved development of a nuclear explosion powered cluster of X-ray lasers that would be deployed using a series of submarine launched missiles or satellites. This curtain of nuclear energy powered X-ray lasers was intended to be used to shoot down many incoming warheads simultaneously. The first test, known as the Cabra event, was performed in March 1983 and was a failure. The failure of the first test was one of the primary reasons for opposition to the programme from critics who argued that X-ray lasers would not offer any significant advantage as an option for ballistic missile defence. However the programme offered many spin-off benefits. The knowledge gained from the programme led to the development of X-ray lasers for biological imaging, 3D holograms of living organisms and advanced materials research.

15.13.2.2 Chemical Lasers

Under this programme, SDIO (Strategic Defence Initiative Organisation) funded the development of a Deuterium Fluoride (DF) laser system called Mid Infrared Advanced Chemical Laser (MIRACL). The MIRACL system (Figure 15.25) was first tested in 1985 in a simulated set up at the White Sands Missile Range. The test set up simulated the conditions the booster was likely to be in during the boost phase of its launch. The laser was subsequently tested on drones simulating cruise missiles with some success. The laser was also tested on an US Air Force satellite to demonstrate its capability as anti-satellite weapon, though with mixed results. The technologies developed during the MIRACL programme were subsequently used

Figure 15.25 MIRACL system (Courtesy: US Army)

to develop the Tactical High Energy Laser (THEL) system, which is in use against artillery shells. Airborne Laser (ABL) and Advanced Tactical Laser (ATL) are the other key chemical laser systems that have been successfully developed and tested after the closure of SDI. Both ABL and ATL are Chemical Oxy-iodine Laser (COIL) systems configured on aerial platforms. These are described in section 15.14.5 on 'Important laser sources'.

15.13.2.3 Particle Beam Accelerator

This is a programme aimed at establishing the operation of particle beam accelerators in space called BEAR (Beam Experiment Aboard Rocket) using a sounding rocket to carry a neutral particle beam accelerator into space. The experiment conducted in July 1989 successfully established that a particle beam would propagate in space as predicted. A spin-off of the technology was its use for management of nuclear waste by reducing the half life of nuclear waste using transmutation technology driven by an accelerator.

15.13.2.4 Hypervelocity Rail Gun

The SDI hypervelocity rail gun experiment was named the Compact High Energy Capacitor Module Advanced Technology Experiment (CHECMATE). A hypervelocity rail gun is similar to a particle accelerator in the sense that it converts electrical potential energy into kinetic energy that is imparted to the projectile. It differs from conventional mass accelerators as here no gases are used. It differs from conventional electromagnetic accelerators in the sense that in the case of rail gun, the magnetic field trails behind the projectile at all times. A conductive pellet, which constitutes the projectile in this case, is attracted down the rails by the magnetic forces produced as a result of gigantic current impulse of the order of hundreds of thousands of amperes flowing through the rail thereby generating muzzle velocities greater than 35 km per second.

Hypervelocity rail guns were considered as an attractive alternative to the space-based defence system because of their projected capability to quickly shoot at multiple targets. There are however many technological challenges. Early prototypes were essentially single-use weapons due to rapid erosion of rail surfaces as a result of very high values of current and voltage. Another challenge is the survivability of projectile, which experiences an acceleration force of greater than 100 000 g. Any on-board guidance system would also need to withstand same level of acceleration force.

15.13.3 Space Programmes

Space based programmes under the SDI saw the development of space based interceptors. One such activity was a non-nuclear system of satellite based miniature missiles called Brilliant Pebbles. These mini missiles used high velocity kinetic energy warheads. The system was designed to operate in conjunction with the Brilliant Eyes sensor system to detect and destroy the target missiles. The Brilliant Pebbles system was designed and developed by Lawrence Livermore National Laboratory during the period 1988–1994.

15.13.4 Sensor Programmes

Prominent activities under the SDI's sensor programme included the Boost Surveillance and Tracking System (BSTS), Space Surveillance and Tracking System (SSTS) and Brilliant Eyes. BSTS was designed to assist detection of missiles during the boost phase. SSTS was originally designed to track ballistic missiles during the mid-course phase. The Brilliant Eyes system was a derivative of SSTS and was designed to operate in conjunction with the Brilliant Pebbles system. Yet another programme that was used to test several sensor related technologies was the Delta 183 programme. The programme was so named as per the designation of the launch vehicle. The Delta 183 programme was initially conceived as a collaborative effort between the erstwhile Soviet Union and the United States. The Soviet Union subsequently withdrew from the programme and the United States proceeded without Soviet participation. The programme was reconfigured to carry several sensor payloads, which included an ensemble of imagers and photo sensors covering visible and ultraviolet bands, long wave infrared imager, laser detection and ranging device and a UV intensified CCD video camera. Figure 15.26 shows the exploded view of the Delta Star spacecraft. The long wave infrared imager was adapted from the guidance and control section of a Maverick missile. Different sensor payloads on board Delta Star were used to observe several missile launches. A great deal of data was generated on the performance of sensors. In some of these experiments, sensor performance was evaluated in the presence of countermeasures. The countermeasures scenario was created by the release of liquid propellant during launch of the missile.

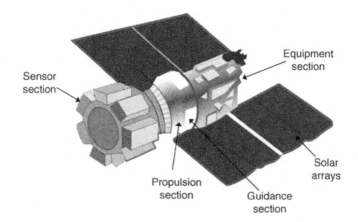

Figure 15.26 Delta Star spacecraft

15.14 Directed Energy Laser Weapons

Kinetic energy weapons transport mass to target in order to cause the destructive effect. Kinetic energy weapons, unguided or guided, have their respective advantages and disadvantages. However they have a common drawback, which is inherent in the mode of their travel from source to target and the mechanism of transfer of energy to the target. Both types transfer the energy to the target through a physical object such as a projectile, which must travel a certain distance through the medium from source to target. One would like the time taken by the projectile to travel from the launch source to the target to be as short as possible. However, practical considerations put a limit on the maximum possible projectile velocity and hence the minimum achievable travel time. Efforts are on to increase the projectile velocity by developing a device called a rail gun that employs plasma driven by a magnetic field to accelerate the projectile to velocities exceeding 40 km/s.

Use of high energy laser weapons overcomes all the limitations of conventional kinetic energy weapons besides offering many new advantages. Belonging to the category of directed energy weapons, these high energy laser weapons once deployed on a mass scale will render obsolete many weapon systems hitherto considered unbeatable.

15.14.1 Advantages

The main advantages of laser-based directed energy weapons include speed-of-light delivery, multiple target engagement and rapid re-targeting capability, deep magazine, low incremental cost per shot, no effect of gravity and immunity to electromagnetic interference.

- **Speed-of-light delivery.** Laser weapons engage targets at the speed of light with essentially no time of flight required as compared to projectile weapons. This feature makes them highly effective.
- **Multiple target engagements and rapid re-targeting.** As laser weapons are constantly powered by recharging their chemical or electrical energy stores, they can engage multiple targets very quickly. Shifting from one target to another involves only re-pointing and re-focusing of the beam directing optical system.
- **Deep magazines.** The total number of shots a laser can fire is only limited either by the amount of chemical fuel (in case of chemical lasers) or electrical power (in case of solid-state lasers).
- **Low incremental cost per shot.** Projectile weapon systems, guided missile systems in particular, expend a lot of expensive hardware (that is, rocket motors, guidance systems, avionics, seekers, airframes and so on) every time they fire. In the case of laser weapons, the cost of each laser firing is essentially the cost of the chemical fuel or the electrical power consumed and tends to be quite low.
- **No influence of gravity.** Laser pointing is practically without any inertia and a light bullet has no mass and hence no mid-course correction is needed.
- **Immunity to electromagnetic interference.** Generation and transfer of lethal laser power to the target is purely in the optical spectrum and hence is immune to any electromagnetic interference and jamming.

15.14.2 Limitations

Limitations of laser-based directed energy weapons include atmospheric attenuation and turbulence, requirement of finite dwell time, line-of-sight dependence, large size and weight, high power consumption and minimal effects on hardened structures.

- **Atmospheric attenuation and turbulence.** The effectiveness of the laser weapon is highly affected by atmospheric conditions due to attenuation (absorption and scattering by airborne particles and gas molecules) and turbulence that deforms the laser beam wave front and increases the laser beam spot size at the target.
- **Finite dwell time.** Unlike projectile weapons that instantly destroy the target upon impact, laser weapons require a minimum dwell time of the order of 3 to 5 seconds to deposit sufficient energy for target destruction.
- **Line-of-sight dependence.** Laser weapons require direct line-of-sight to engage a target. Their effectiveness is reduced or neutralized by the presence of any object or structure in front of the target that cannot be burned through.
- **Large size and weight.** Laser-based directed energy weapons are very bulky in size and hence the size and weight of these systems poses a big limitation to their usage.
- **High power consumption.** One of the major practical problems of laser-based directed energy weapons is their high power consumption. Existing laser sources are inefficient and most of the input power is wasted as heat. As a result, these lasers would need a massive power source and a complex thermal management system. The problem of the requirement of huge electrical power would be lessened if the laser weapon were installed at a static location near an electrical power plant or if it were powered by nuclear energy. Gas Dynamic Laser (GDL), chemical lasers like the Chemical Oxy-iodine Laser (COIL) and Deuterium Fluoride (DF) lasers do not suffer from this problem. In the case of GDL, the input energy required for the laser action comes from ignition of mixture of fuel and oxidizer. In the case of chemical lasers, energy is released in a chemical reaction. The chemical reaction is between hydrogen peroxide and iodine in the case of the COIL and between atomic fluorine and deuterium in the case of the DF laser. Thermal management however remains an issue in most cases.
- **Minimal effects on hardened structures.** Laser weapons will produce minimal or no effect on hardened structures, that is bunkered buildings and armoured vehicles. In these cases, they will be effective only in disabling vulnerable components used on these structures such as antennas, sensors and external fuel stores.

15.14.3 Directed Energy Laser Weapon Components

The directed energy laser weapon is an integration of many complex systems and the magnitude of complexity is proportional to the output laser power generated by the system and the deployment scenario. Figure 15.27 shows the important components of a laser weapon system. The high energy laser weapon system essentially comprises two major subsystems, namely the *Laser Source* and the *Beam Control System*. Each of the two systems comprises a number of subsystems. The overall system performance, the irradiance on target and the time to kill a target are affected by the performance of each of these subsystems.

The laser beam from the high energy laser system while travelling through the atmosphere is affected by it in different ways such as attenuation due to absorption and scattering and defocus due to blooming. *Beam transmission/propagation* describes the effects on the beam after it leaves the output aperture of the laser system and travels through the battlefield environment to the target. The optical stability of the platform and beam interactions with the particles in the atmosphere (both molecules and aerosol particles) primarily determine laser beam quality at the target. Beam quality is a measure of how effective the laser weapon system is in producing

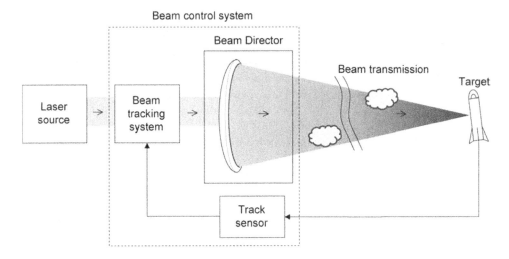

Figure 15.27 Components of a directed energy laser weapon system

the desired value of power/energy density at the target. Lethality defines the total energy and/or fluence level required to defeat specific targets. Laser energy must couple efficiently to the target material so as to exceed a certain failure threshold value. Laser output power and beam quality, atmospheric effects and laser target coupling efficiency are the key factors for determining whether the laser system has sufficient fluence to neutralize or incapacitate a specific target.

15.14.4 Important Design Parametres

Parameters that largely govern the design of a directed energy laser weapon system include operational wavelength, beam quality, telescope aperture, transmission losses and power scalability.

- **Operational wavelength.** Operational wavelength is the most important design parameter as its choice has a bearing on almost all of the other design parameters influencing the overall efficacy of the weapon system. For the same laser output power and transmitting telescope aperture, a directed energy laser system with 1 micron source such as COIL and solid-state lasers would have an operational range that is approximately ten times that of a similar system with a 10 micron laser source like a carbon dioxide laser for a given power density requirement at the target. It may be mentioned here that the amount of laser energy absorbed by the target material, called coupling efficiency, increases for shorter wavelengths.
- **Beam quality.** Beam quality is essentially a measure of how tightly the laser beam can be focused to form a small and intense spot of light on a distant target. The ideal value of beam quality (B) is 1 and it signifies that the laser spot size at the target is limited only by the laws of diffraction. For a real laser beam, the value of B is greater than one, and hence the focused spot size is larger than the diffraction limited spot size. One of the most challenging tasks is maintenance of good laser beam quality as the output power level is scaled up.

- **Telescope aperture.** This determines the focusing ability of the laser system. Increasing the aperture size of the telescope will produce tighter focusing and hence increased laser power densities at the target
- **Transmission characteristics.** The transmission characteristics of the atmosphere in relation to different wavelengths is another very important criterion. As the laser beam has to propagate over long atmospheric paths, it is important that the laser wavelength should have minimum transmission losses. The atmosphere exhibits transmission windows in 0.4–1.7 microns (Visible – NIR), 3–5 microns (MWIR) and 8–14 microns (FWIR) bands shown in Figure 15.28. It is therefore important that the operational wavelength of the laser falls within one of these transmission windows.

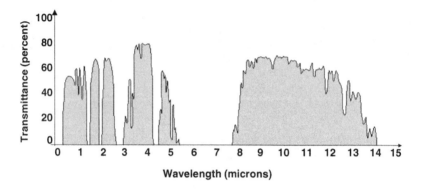

Figure 15.28 Transmission characteristics of atmosphere

- **Power scalability.** This is yet another important issue, especially when it comes to designing directed energy laser weapons for long-range strategic applications. The basic physics and technology of the laser system design and architecture should be such as to allow power scaling to the megawatt level typically needed for most of long range strategic programmes.

15.14.5 Important Laser Sources

Not all laser sources meet the requirements of power scale-up. The elite categories of laser sources that qualify for directed energy laser weapon system applications include Gas Dynamic CO_2 Laser, Chemical Lasers such as HF/DF laser and Chemical Oxy-Iodine Laser (COIL), solid state lasers and fibre lasers.

15.14.5.1 Gas Dynamic CO_2 Laser

The gas dynamic CO_2 laser is a molecular laser with carbon dioxide gas (CO_2) as the lasing medium. Nitrogen is added to enhance the population inversion in the lasing medium. The laser system essentially comprises a combustion chamber, a nozzle bank, a resonant cavity and a diffuser. In addition, an aerodynamic window is used as an interface between the cavity and the atmosphere. In the combustion chamber, the CO_2 laser gas mixture is initially generated at very high temperature and pressure by combustion of fuel and oxidizer. Toluene as fuel and air as oxidizer is commonly used. In this region, due to the elevated temperatures, the population of both the lower and the upper laser levels increase. However, the population of the lower laser level remains higher than that of the upper laser level. This gas mixture is then expanded adiabatically to very low pressures of the order of 20 to 30 torr through a bank of expansion nozzles. Due to this sudden reduction in pressure, the populations of both the levels tend to relax. However, the upper laser level relaxes slowly as the upper laser level lifetime is much longer as compared to the lifetime of the lower laser level. Due to this, the population of the upper laser level remains higher than the population of the lower laser level over a fairly extended region in the laser cavity. The gas pressure is then recovered in the diffuser from where the gases are directly exhausted into the atmosphere. Figure 15.29 shows the schematic diagram of a gas dynamic CO_2 laser.

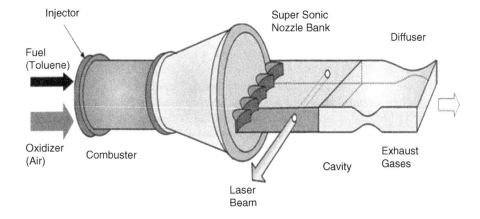

Figure 15.29 Schematic diagram of gas dynamic CO_2 laser

The gas dynamic laser is a proven technology and is safe and non-toxic to work with. Cavity pressure is relatively higher as compared to other chemical lasers, which makes it easier to exhaust the used gases to the atmosphere. However, a larger wavelength of 10.6 microns necessitates a larger telescope aperture. Also, it is affected more severely as compared to 1 micron lasers in humid conditions. An Airborne laser laboratory (ALL) configured on a modified NKC-135A aircraft employing gas dynamic CO_2 laser is the United States Air Force's test platform for directed energy laser weapon research. The aircraft also carries several low power lasers for the purpose of alignment and diagnostics. The programme is aimed at a demonstration of high energy laser weapon effectiveness in the interception and destruction of ballistic missiles. The laser source and the pointing/tracking systems of ALL are housed in the forward fuselage. Figure 15.30 shows a photograph of the system.

15.14.5.2 Chemical Lasers

Two common chemical lasers worthy of being used as directed energy laser weapons include the Hydrogen Fluoride/Deuterium Fluoride (HF/ DF) laser and the Chemical Oxy-iodine Laser (COIL). In the case of the HF/DF laser, energy liberated from an exothermic chemical reaction

Figure 15.30 Airborne laser lab (ALL) (Courtesy: US Air Force)

is used for producing population inversion. Figure 15.31 shows the basic components of a combustion driven DF laser. Fluorine atoms are produced through combustion of NF_3 and molecular hydrogen through a chemical reaction process. The atomic fluorine is supersonically expanded through a nozzle assembly producing a temperature of 300–500 K and a pressure of 5–10 torr in the gain region. Molecular deuterium is then injected into this supersonic expansion assembly through a large number of very small nozzles to enable good mixing and efficient reaction with atomic fluorine to produce vibrationally excited DF molecules for lasing. The laser beam is extracted using a resonant cavity. A diffuser assembly is used to recover the pressure for suitable exhaust of the lasing gases.

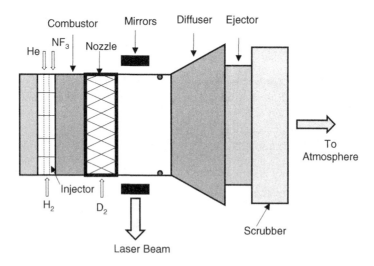

Figure 15.31 Schematic diagram of DF laser system

HF laser is realized by replacing deuterium with hydrogen. Figure 15.32 shows the block diagram of the HF laser system. DF laser output propagates well through atmosphere due to the good transmission characteristics at its operational wavelength of 3.6 microns. On the other hand, due to poor atmospheric transmission at 2.7 microns, HF laser output suffers

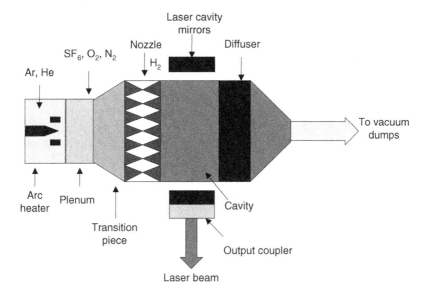

Figure 15.32 Schematic diagram of HF laser system

heavy attenuation. As a result, HF lasers are suitable only for space-based directed energy laser systems. The HF/DF lasers offer higher specific energies and relatively smaller sizes as compared to gas dynamic CO_2 lasers. The HF/DF laser however involves highly toxic and explosive gases and therefore needs very complex logistics. MIRACL (USA) with output power up to 2.2 MW is a well-known DF laser.

15.14.5.3 Chemical Oxy-Iodine Laser (COIL)

COIL due to its shorter operational wavelength of 1.3 microns is becoming increasingly popular for directed energy laser weapon applications. Figure 15.33 shows the schematic diagram of a COIL laser. The pump source is singlet oxygen produced by the chemical reaction of chlorine

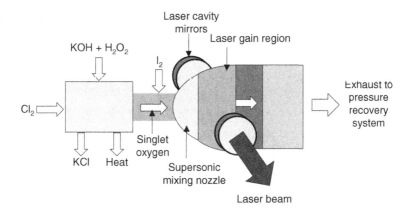

Figure 15.33 Schematic diagram of COIL laser system

gas with liquid basic hydrogen peroxide. The singlet oxygen is then routed to a nozzle bank where it is mixed with molecular iodine. The singlet oxygen transfers its energy to iodine molecules through collisional energy transfer and generates excited iodine atoms in the cavity region from where the laser is extracted.

COIL output at 1.3 microns has good transmission through the atmosphere, requires a smaller aperture for the beam director for a given laser spot size on the target and offers good target coupling efficiency. Other features include high specific energy and absence of toxicity hazards. However, due to low cavity pressure of approximately two to three torr, there is the requirement of a complex pressure recovery system for ground based applications. The pressure recovery requirement is not that stringent for the space borne COIL system due to relatively lower outside pressure.

The Air Borne Laser (ABL) configured on a modified Boeing 747 aircraft called YAL-1A using COIL with an output power of 1.2 MW is one such system designed to destroy missiles during boost phase. Figure 15.34(a) shows the cutaway of the airborne laser and Figure 15.34(b) shows the ABL system in action to destroy missile. The technologies developed under the

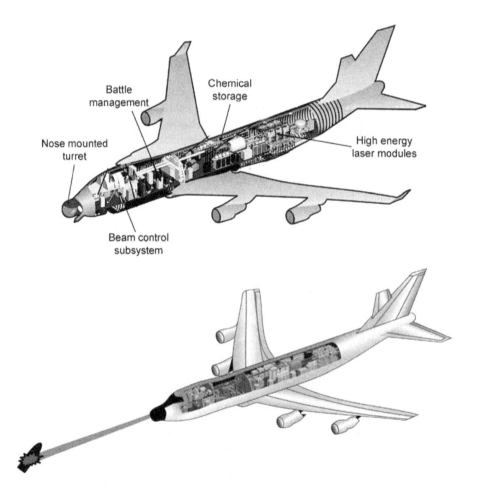

Figure 15.34 Air Borne Laser (ABL)

ABL programme could lead to development of directed energy laser systems for targeting space assets. The Advanced Tactical Taser (ATL) configured on a C-130H Hercules aircraft is another example of COIL laser (Figure 15.35). The laser was successfully test fired in May 2008. The ATL system is envisioned to offer the mobility of a small aircraft, high-resolution imagery for target identification and the ability to localize damage to a small area of less than a foot in diameter from a range of 5 to 10 km. The ATL system has the capability to disable communication lines and radio and TV broadcast antennas, neutralize satellite and radar dishes, break electrical power lines and transformers, incapacitate individual vehicles and so on.

Figure 15.35 Advanced Tactical Laser (ATL) (Courtesy: US Air Force)

15.14.5.4 Solid State Lasers

Solid-state lasers are electrically driven devices. Pumping of the gain medium for producing population inversion is achieved by semiconductor laser diode bar arrays. The all solid-state configuration offers unmatched advantages in terms of compactness, robustness, reliability and logistic simplicity. However, with the present status of solid-state laser technology, solid state lasers cannot match the output power levels of the order of megawatts possible with chemical lasers. The thrust of the present technology is to realize solid-state laser sources with an output power level of hundreds of kilowatts. These sources will be utilized for tactical battlefield operations that do not demand a megawatt class power level.

The technology of electrically driven solid-state lasers is well established with the advantages of high electrical-to-optical conversion efficiency, robustness and compact sizes. However, power scaling of solid-state lasers is limited by thermo-mechanical distortions caused by waste heat deposited in the gain medium by optical pumping. In the case of chemical lasers, the waste heat is removed and ejected out with the gas mixture at a high flow rate thereby allowing power scale up to very high levels with good laser beam quality. Chemical lasers are presently the most favored choice for long range applications because of their proven scalability to high power levels with good beam quality.

The requirement of a compact and mobile laser weapon system is driving the technology development of power scalable solid-state lasers. The recent development in heat capacity disk laser technologies and the demonstration of a heat capacity disk laser system with an output power of 67 kW, developed by the Lawrence Livermore National Laboratory, USA has generated interest internationally in solid-state laser systems as futuristic high power laser weapon systems. In addition to being robust, compact and free from the safety hazards usually associated with chemical lasers, solid state lasers offer all of the other advantages associated with shorter wavelength operation.

15.14.5.5 Fibre Lasers

The technology of fibre lasers is the most advanced among all the solid-state laser technologies available today. The basic configuration of a fibre laser as shown in Figure 15.36 comprises a gain medium in the form of a long optical fibre of suitable material doped with lasing ions. For high power operation, ytterbium doped glass fibre is typically used. The entire fibre length is pumped with a large number of single emitter fibre coupled laser diode arrays. The laser resonator cavity is formed by embedding Bragg Grating reflectors at the two ends of the fibre.

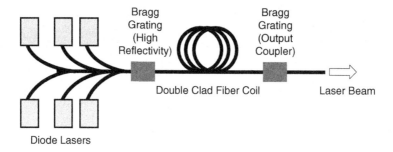

Figure 15.36 Schematic diagram of fibre laser

Due to the small aperture of the fibre of the order of few microns, the output laser beam is emitted with diffraction limited beam quality resulting in output laser intensity that is nearly two orders of magnitude higher as compared to that produced by conventional solid-state lasers with the same output power. Also, since the resonator is formed within the fibre, there is no need for free space optics thereby making the fibre laser extremely robust and reliable as compared to other lasers.

The technology of a fibre laser is extremely complex. Single mode fibre lasers with output power of 400 to 600 watts are commercially available. At IPG-Photonics, USA, the operation of a 3 kW single mode fibre laser has been established. The inherent unmatched advantages associated with fibre lasers have established it as a leading candidate for futuristic high power laser systems. Power scalability to a level of 100 kW is being targeted by coherent laser beam combination of multiple fibre laser beams.

15.14.6 Beam Control Technology

The objective of the beam control subsystem of a directed energy laser weapon system is to acquire the intended target and point and focus the laser energy precisely at the designated point on the target for a dwell time sufficient to cause the desired damage to the target. A beam pointing system with a pointing accuracy of a few micro radians is an essential requirement of a directed energy laser weapon system so as to be able to engage fast moving and manoeuvring aerial targets such as rockets, artillery shells, mortars, battlefield missiles, and aircraft and so on. The critical requirement is to aim and maintain the laser beam on the vulnerable spot on the target until a kill has been achieved. The beam control system comprises a beam transport system, beam directing telescope, target acquisition and tracking system and adaptive optical system

The *beam transport system* transfers laser radiation from the exit of the laser source where it is generated to the gimbal mounted beam directing telescope. The beam is coupled to the telescope system through a number of gimbal follower mirrors that ensure proper alignment of the laser beam axis with the telescope axis irrespective of its orientation. The *beam directing*

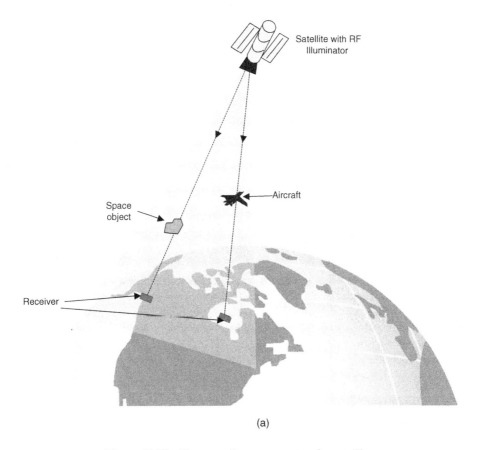

(a)

Figure 15.37 New surveillance concepts using satellites

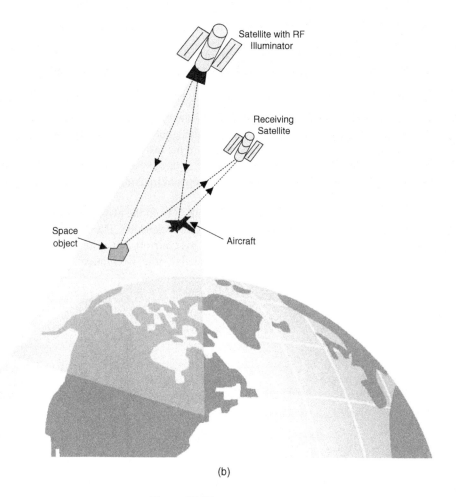

(b)

Figure 15.37 (*continued*)

telescope focuses the laser beam precisely onto the target. A bore-sighted laser range finder in closed loop operation keeps the laser beam focused on the target in the entire operating range. The telescope aperture size controls the laser spot size and hence the lethal range of the weapon system. The *target acquisition and tracking system* comprises a target acquisition video camera which is either bore sighted or shared with the telescope system. The camera acquires the target and tracks it by controlling the movement of the gimbal platform. Another important component of the beam control system is the *adaptive optical system* that senses the atmospheric aberrations and corrects them in real time.

15.15 Advanced Concepts

This section covers some of the advanced concepts that may become a reality in the next decade or so. These include new surveillance concepts using satellites, long reach non-lethal laser dazzler and long reach laser target designator.

Advanced Concepts

15.15.1 New Surveillance Concepts Using Satellites

This concept makes use of commercial or military satellites orbiting in LEO or MEO orbits as RF illumination sources. Detection is done using antennas and receivers placed on the ground [Figure 15.37(a)]. This will help to detect small and low observability airborne or near surface objects. However, in this case the target is required to be in a straight line between the transmitter and the receiver. This limitation can be removed by employing a separate receiving satellite or a UAV to collect the radar scatter to detect the airborne threats as shown in Figure 15.37(b).

15.15.2 Long Reach Non-lethal Laser Dazzler

Laser dazzlers are non-lethal weapons that are used for temporary or flash blinding enemy troops. They are line-of-sight weapons and generally have an operational range varying from 50 m to few km. Figure 15.38 shows an idea that can be used for making long reach non-lethal

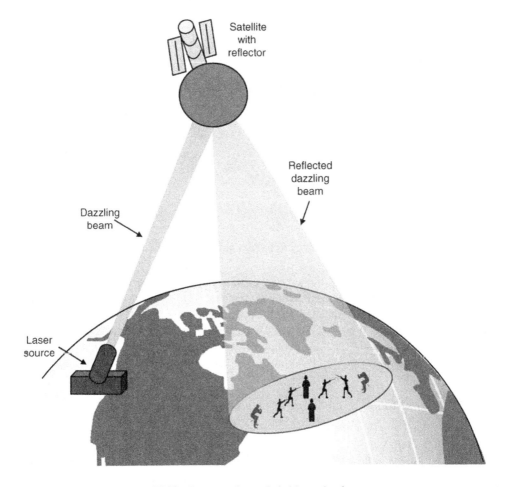

Figure 15.38 Long reach non-lethal laser dazzler concept

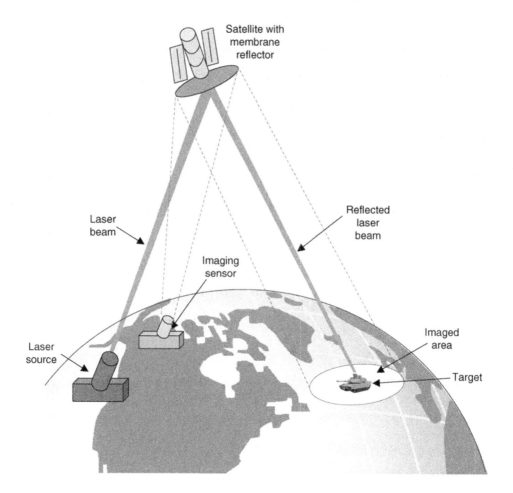

Figure 15.39 Long reach laser target designator concept

laser dazzlers. The concept highlighted here enables one to increase the range of the device many times and removes the line-of-sight requirement.

It makes use of a reflective sphere placed on the satellite in the LEO or MEO orbit. A ground based laser of sufficient power and required divergence is used to illuminate the sphere. The direction of the laser can be changed to illuminate a different portion of the sphere in order to irradiate the ground at the desired location.

15.15.3 Long Reach Laser Target Designator

In a laser guided munition deployment scenario, a laser target designator on the ground or onboard an aircraft is used to illuminate a target. This radiation is scattered from the target and the laser guided munition homes on to the target by sensing the scattered radiation. The typical distance between the designator and the target is 5 to 15 km and line-of-sight is required between the target and the designator.

A long reach laser target designator can be designed to overcome the line-of-sight and range limitation by employing the concept shown in Figure 15.39. It employs a sensor and a laser source on the ground and an optically flat mirror in a satellite orbiting in LEO or MEO orbit. This will allow the use of laser guided munitions even in the deepest denied areas. The mirror is used for both sensing as well as laser designation functions. The sensor determines the approximate location of the desired target. The area is imaged with the ground sensor by making use of the mirror onboard the satellite. The desired area is selected by using the attitude control subsystem of the satellite. When the target is detected, the laser beam from a ground-based laser designator is reflected off the same mirror and is aimed at the target. Another variation could be that the imaging sensor is placed on the satellite itself.

Further Readings

Burkett, D.L. (1989) *The U.S. Anti Satellite Program: A Case Study in Decision Making*, National Defense University, National War.

Gatland, K. (1990) *Illustrated Encyclopedia of Space Technology*, Crown, New York.

Long, F.A., Hafner D. and Boutwell, J. (1986) *Weapons in Space*, W.W. Norton & Company, New York.

Vacca, J. (1999) *Satellite Encription*, Academic Press, California.

Verger, V., Sourbes-Verger, I., Ghirardi, R., Pasco, X., Lyle, S. and Reilly, P. (2003) *The Cambridge Encyclopedia of Space*, Cambridge University Press.

Internet Sites

1. www.aero.org
2. www.skyrocket.de
3. www.armyspace.army.mil
4. www.astronautix.com
5. http://science.howstuffworks.com/question529.htm
6. http://www.aero.org/publications/crosslink/winter2002/01.html
7. http://www.aero.org/publications/crosslink/winter2002/08.html
8. www.energia.ru

Glossary

COMINT (communication intelligence) satellites: These satellites perform covert interception of foreign communications in order to determine the content of these messages. As most of these messages are encrypted, they use various computer-processing techniques to decrypt the messages

DSCS satellites: DSCS stands for Defense satellite communication systems. Launched by the USA, satellites in this series are intended for providing wideband military communication services

DMSP satellites: DMSP stands for Defence meteorological satellite program. It is an American military weather forecasting satellite programme

Early warning satellites: Early warning satellites provide timely information on the launch of missiles, military aircraft and nuclear explosions to military commanders on the ground

Electro-optical satellites: Electro-optical satellites provide full-spectrum photographic images in the visible and the IR bands

ELINT (electronic intelligence) satellites: ELINT satellites are used for the analysis of non-communication electronic transmissions. This includes telemetry from missile tests (TELINT) or radar transmitters (RADINT)

IMINT (Image Intelligence) satellites: IMINT satellites provide detailed high resolution images and maps of geographical areas, military installations and activities, troop positions and other places of military interest

Milstar satellites: Milstar satellites are American military communication satellites belonging to the category of protected satellite systems

Protected satellite systems: Protected satellite systems provide communication services to mobile users on ships, aircraft and land vehicles

PHOTOINT or optical imaging satellites: These satellites have visible light sensors that detect missile launches and take images of enemy weapons on the ground

Reconnaissance satellites: Reconnaissance satellites, also known as spy satellites, provide intelligence information on the military activities of foreign countries

SIGINT (Signal intelligence) satellites: These satellites detect transmissions from broadcast communication systems such as radar, radio and other electronic systems. They can also intercept and track mobile phone conversations, radio signals and microwave transmissions

Tactical satellite systems: Tactical satellite systems are used for communication with small mobile land based, airborne and shipborne tactical terminals

Vela satellites: Vela satellites are American satellites of the 1960s intended for detection of a nuclear explosion

Wideband satellite systems: These systems provide point-to-point or networked moderate to high data rate communication services at distances varying from in theatre to intercontinental distances

16

Emerging Trends

16.1 Introduction

In the concluding chapter of this book we will discuss an assortment of topics representing some of the emerging trends in the field of satellite technology and its applications. The assortment of topics included in this chapter could not have been justifiably discussed as separate chapters from the viewpoint of scope of the present book. The chapter begins with a discussion on some non-conventional technologies such as *space tethers* and their possible use in executing propellant-less satellite orbit manoeuvres, generation of electric power for satellites and as *space elevators*, *robotic drones* for satellite stabilization and *aerostats* for surveillance, reconnaissance and communication relay applications. This is followed up by a detailed discussion on millimetre wave satellites, emerging concepts in space station technology and satellite services.

16.2 Space Tethers

One of the most important issues related to the development, launch and station keeping of satellite and other spacecraft missions is the economics of these operations. Scientists and technologists have been working hard to find new ways of reducing the cost to orbit and also the cost of maintaining satellites in the desired orbit during the life cycle of the mission. Efforts have also been made to develop alternative technologies to fuel rocket engines if conceived applications such as space tourism were to become a reality in the not too distant future. As a simple illustration, considering the fact that using the best available launch technology today launching 1 kg into a geostationary orbit costs about US$ 20 000 and launching 1 kg into a low Earth orbit costs around US$ 2500, one can comprehend how important it is to reduce the costs. As another illustration, if we consider that a space station such as International Space Station (ISS) needs ten tonnes of propellant per year in orbit, this would add up to billions of US dollars for the operation and maintenance of the space station during its lifetime. Space tethers drastically lower the cost to orbit by significantly reducing the differential velocity that the launch rocket would need to impart to the payload to put it in the desired orbit. Space tethers can also be used to generate power or thrust, as described in the subsequent paragraphs, to carry out orbital manoeuvres.

A space tether is a long and strong cable usually made up of thin strands of high strength conducting wires or fibres. The desired tether material properties depend on the intended applications. Some common desirable properties include high tensile strength, low density and high electrical conductivity. Space tethers are prone to damage by collision with space debris or micro meteoroids. They are also adversely affected by exposure to ultraviolet radiation and atomic oxygen. A protective coating may therefore be desirable. Some of the common materials proposed for building space tethers include Kevlar, ultra high molecular weight polyethylene, carbon nano tubes, M5 fibre and diamond. Use of space tethers for applications such as spacecraft propulsion, which can significantly reduce the cost to orbit, satellite stabilization, making use of thrust or power generated by space tether systems, and maintenance of spacecraft formations is being extensively researched and experimented worldwide.

16.2.1 Space Tethers – Different Types

There are three major categories of employment of space tethers. These include *momentum exchange tethers*, *electrodynamic tethers* and *tethers for formation flying*. While momentum exchange tethers are mainly employed for orbital manoeuvring, electrodynamic tethers can be used for generation of electrical power or a mechanical thrust. Tethers used in the case of formation flying, as the name suggests, are used for maintaining a fixed distance between different spacecraft in formation flying. The three types are briefly described in the following paragraphs.

16.2.1.1 Momentum Exchange Tether

Momentum exchange tethers are used to couple two objects in space in such a way that one can transfer momentum or energy to the other. The deployment of the tether takes advantage of the gravity gradient force that exists due to the differential gravitational force between the two ends of the tether. It is this differential gravitational force, called the gravity gradient force, that keeps the tether taut and the two objects tied to the two ends of the tether pulled apart. Once the tether is deployed, if there are no other forces acting on the tether, it will attain a vertically aligned equilibrium orientation. One of the main applications of a momentum exchange tether is to adjust the orbit of a spacecraft. The tether spins and the act of spinning makes the objects tied to the ends of the tether experience a continuous acceleration with the magnitude of acceleration depending on the rotation rate of the spinning system. If either of the two end objects are released during rotation, momentum exchange will occur. The transfer of momentum from the spinning system to the released object causes the spinning system to lose orbital energy and thus lose altitude. This loss of energy can be replenished by using electrodynamic thrusting without consumption of any fuel, as described in the following paragraphs. The resultant tether system is a hybrid combination of both the momentum exchange tether and the electrodynamic tether called the *momentum exchange/electrodynamic-reboost* system, which combines the best of both types of tethers to create an effective upper stage launch station. In the hybrid tether system, the momentum exchange property is used to catch and release spacecraft, and the electrodynamic property is used to create the thrust needed to restore its orbital energy after a payload transfer.

There are various mechanisms by which momentum or energy can be exchanged between the two objects. In one of the methods, called *gravity gradient stabilization*, tether is used for attitude control. The tether in this case has a small mass tied to one of its ends and the satellite whose attitude needs to be controlled is tied to the other end. Gravity gradient forces stretch the tether between the two masses. The top and bottom portions of a long object are also pulled by different forces, which help the bottom of the object to be stretched out. The result is that the satellite attitude is stabilized with its long dimension pointing towards the radius vector of the planet it is orbiting. It may be mentioned here that although the tether is stationary in the orbital reference frame, it is rotating once per orbit in the inertial reference. Figure 16.1 illustrates the concept of gravity gradient stabilization.

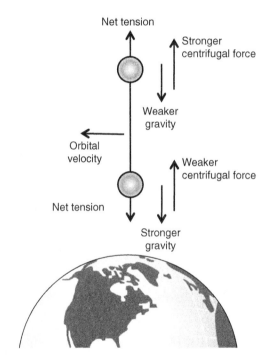

Figure 16.1 Gravity gradient stabilization

Another type of momentum exchange tether is the rotating momentum exchange tether, called a *bolo*. In this case the tether acts like a momentum energy bank and transfer of momentum and energy from the tether to and from the spacecraft with very little loss can be used for orbital manoeuvring. The mechanism can be used to either increase or decrease the altitude of the satellite orbit. The momentum exchange phenomenon could also be used to catch a payload coming in its plane of rotation from any direction at any speed less than the maximum tip speed. It could use the same momentum exchange phenomenon to launch the payload in some other direction at a different speed.

Yet another type of momentum exchange tether is a long rotating bolo called a *rotovator*. One common application of a rotovator tether is to lift or put payloads from/onto a planet or moon. Such a tether orbits in a relatively lower orbit around a planet or moon and is used from a space platform. Present day material strengths allow building such momentum exchange tethers for Mars, Mercury and most moons, including the Earth's moon.

16.2.1.2 Electrodynamic Tethers

An electrodynamic tether is essentially a long conducting cable extended from a spacecraft. The cable is kept stretched and oriented along the vertical direction by the gravity gradient force (Figure 16.2). The electrodynamic tether generates thrust through Lorentz force interaction with the planetary magnetic field. This thrust can be used to execute orbital manoeuvres without the need to carry large quantities of propellant into orbit.

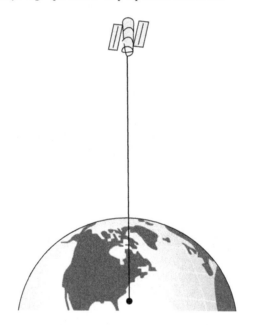

Figure 16.2 Electrodynamic tether

An electrodynamic tether operates as follows. The tether is made up of a conducting material. The motion of the conducting tether across the planetary magnetic field induces a voltage along the length of the tether and this induced voltage could be of the order of several hundred volts per kilometre of tether length. The induced voltage drives an electric current through the conducting tether. The current carrying tether cutting across the magnetic field produces a Lorentz force, which opposes the motion of the tether and also the spacecraft attached to it. The Lorentz force can be given by the expression ($J \times B$, where J is the current density and B is the magnetic flux density. This electrodynamic drag force has the effect of decreasing the orbit and the host spacecraft. Electrodynamic tethers can also be used to boost the spacecraft orbit in a similar manner. In this case, a source of current added to the tether system is used to drive a current through the tether in a direction opposite to motion induced EMF. The thrust generated in the process propels the spacecraft.

16.2.1.3 Tethers for Formation Flying

These are non-conducting tethers used to maintain a fixed distance between multiple spacecraft (Figure 16.3). Spacecraft formation flying is a widely researched subject and space interferometry is one of the key areas of research interest for scientists worldwide. Use of multiple relatively smaller apertures on multiple spacecraft in a precise formation to synthesize a much larger aperture using the concept of coherent interferometry is one such well established example of formation flying. A large synthesized aperture achieves higher resolution, which would otherwise be possible only with a prohibitively large single aperture. As an illustration, the resolution achievable with a 1 km wide single aperture could be realized with a few 2 or 3 m apertures. Different possible architectures that can be employed to realize this include the *structurally connected interferometer* (SCI), as planned for NASA's space telescope to be built jointly with Northrop-Grumman for the *Space Interferometry Mission* but subsequently cancelled in 2010, the *separated spacecraft interferometer* (SSI), as proposed for NASA's *Terrestrial Planet Finder* to construct a system of telescopes to detect extra solar terrestrial planets but subsequently cancelled in 2011 after several postponements, and the *tethered formation flight interferometer*. While the SCI architecture allows very limited baseline changes, SSI architecture requires prohibitively large quantities of propellant. Precise spacecraft formation flying has the advantages of both SCI and SSI architectures. It could be used to synthesize large apertures to build high resolution astronomical telescopes. The use

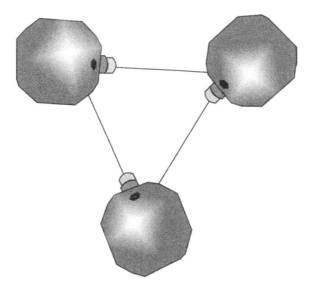

Figure 16.3 Tethers for formation flying

of tethers in spacecraft formation flying allows the mitigating need for huge quantities of propellant that would otherwise be required to maintain the precision formation flying. NASA's SPECS (Sub-millimeter Probe of the Evolution of Cosmic Structure) mission designed to carry out studies to find answers to questions related to formation of universe is an example of tethered spacecraft flying. The mission is designed to study formation of universe, evolution of galaxies, and cosmic history of energy release and so on.

16.2.2 Applications

The following paragraphs briefly describe some common applications of space tethers. The applications are categorized as those related to momentum exchange, electrodynamic and hybrid tether systems.

16.2.2.1 Momentum Exchange Tether Applications

1. One of the main applications of momentum exchange tethers is their use to launch payloads into a higher orbit such as from a low Earth orbit to a geosynchronous transfer orbit and then to a geostationary orbit without the use of expensive rockets. As outlined earlier, the momentum exchange tether loses some energy in the process, which can be compensated for by using a hybrid tether system. The hybrid system makes use of electrodynamic reboost properties to re-energize the momentum exchange tether system, enabling its use multiple times.
2. Another application of a momentum exchange tether is a possible method for transferring payloads, for example to get back small payloads from the space stations such as the ISS without the use of rockets. This operation is carried out as follows. The payload is tied down inside a mini capsule that can be ejected downwards through a robotic airlock. The capsule deploys a tether approximately 30 km long that orientates the capsule for re-entry. The tether is eventually cut and burned up in the atmosphere. The technology was demonstrated successfully in 1993 in the NASA sponsored Small Expendable Deployment System (SEDS-1).

16.2.2.2 Electrodynamic Tether Applications

1. One of the primary applications of electrodynamic tethers is in de-orbiting a satellite at the end of its useful life. When the subject satellite is to be de-orbited, a long wire antenna, which is a part of the satellite, is unfurled. A current is made to flow through the wire and the interaction of the current carrying wire with the Earth's magnetic field generates the Lorentz drag force. This drag force reduces the orbit velocity, as a result of which the satellite de-orbits to a lower orbit. It eventually enters the Earth's atmosphere and is burnt up.
2. Another important application of electrodynamic tethers is in the stabilization of the orbit of the ISS. It may be mentioned here that the cost of maintaining the ISS in the desired orbit over its planned life time is estimated to be close to US$ 1000 million considering the total quantity of propellant that is required. Electrodynamic forces experienced by the electrodynamic tether, as described earlier, could be used to hold the ISS in orbit.

3. Electrodynamic tethers may also be used as an economical means of generating electrical power in orbit, but only for providing bursts of energy in short duration experiments. The electrodynamic tether in this case can be used to convert some of the spacecraft's orbital energy into electrical power. This is good for generating short bursts of energy as the process of generating electrical energy is accompanied by a decrease in orbit altitude.

16.2.2.3 Hybrid Tether Applications

1. The primary application of hybrid tether systems also called momentum-exchange/electrodynamic-reboost tethers is to transfer payloads from a low Earth orbit into medium Earth and geosynchronous orbits. A typical system may be able to transfer a 2500 kg payload to a geosynchronous orbit and a 5000 kg payload to a medium Earth orbit. The main advantage of using tethers is their near zero propellant usage. The small quantity of propellant needed onboard the tether system is for the purpose of making minor trajectory corrections to ensure a precise payload rendezvous. An added advantage of using hybrid tethers for payload transfer to a geosynchronous Earth orbit is the very short transfer times, which would otherwise have required huge quantities of propellant thus increasing the size and cost of the mission. Re-usability of the tether system is yet another advantage. When the hybrid tether system is at the end of its life span, the electrodynamic part of it can be used to de-spin the system and reel in the tether. The electrodynamic drag force is then used to lower the orbit, forcing it to re-enter the Earth's atmosphere.
2. Another possible application of hybrid tether systems is their use as a platform for testing momentum exchange tethers in space environment with dummy payloads before they are used for high value payloads.
3. Interplanetary transfer of goods and resources could be yet another application of hybrid tethers in the future. For example, once there is a manned presence on the moon, a lunar orbiting tether along with an Earth orbiting hybrid tether may allow convenient transportation between the two. Use of tether systems could drastically reduce the cost of future space missions, making them economically viable.

16.2.3 Space Tether Missions

The following paragraphs briefly describe some important space tether missions executed worldwide in the last couple of decades and also important space tether missions planned for the coming years.

16.2.3.1 Tethered Satellite System (TSS)

The Tethered Satellite System (TSS) is a collaborative programme by NASA and Italian Space Agency (ASI) with the objective of developing a reusable multi-disciplinary facility to conduct space experiments in Earth orbit. The system consists of a satellite, a deployment system in the Space Shuttle's payload bay, an electrically conductive tether approximately 20 km long and six scientific instruments. While development of the satellite was the primary responsibility of the ASI, NASA offered the deployment system and the tether. TSS provided the capability of deploying a satellite on a long, gravity-gradient stabilized tether from the

Space Shuttle for the purpose of carrying out scientific investigations in space physics and plasma electrodynamics.

Figure 16.4 TSS-1 deployment from Space Shuttle *Atlantis* payload bay (Courtesy: NASA)

The first TSS mission, called TSS-1, was conducted aboard Space Shuttle flight Atlantis (STS-46) from 31 July to 8 August 1992. Figure 16.4 shows TSS-1 deployment from the Space Shuttle orbiter's payload bay. The TSS-1 mission was twofold, first to demonstrate the feasibility of deploying and controlling long tethers in space and second to conduct exploratory experiments in space plasma physics to evaluate the efficacy of such a system. The tethered satellite system orbited the Earth at an altitude of 296 km, placing the tether system in the electrically charged atmosphere of the ionosphere. The operations lasted for about 24 hours, after which the tether was retrieved. The TSS-1 mission was followed up by a re-flight mission named TSS-1R aboard Space Shuttle flight Columbia (STS-75). The tether was deployed on 22 February 1996. The mission intended to deploy the tether to its full length of approximately 20.7 km, but the tether suddenly snapped and burnt prior to reaching full deployment of 20.7 km. Tether dynamics could be verified only up to 19.6 km. The two missions not only validated the concept of gravity gradient tethers but also the feasibility of the generation of electrical power from orbital energy.

16.2.3.2 Small Expendable Deployer System (SEDS)

The Small Expendable Deployer System (SEDS) was developed by NASA's Marshall Space Flight Centre (MSFC), which was primarily responsible for the development of transportation and propulsion technologies, and the Tether Application Company of San Diego. Two flight experiments, SEDS-1 and SEDS-2, were carried out in 1993 and 1994. Both SEDS-1 and SEDS-2 flew as secondary payloads on Delta-II launches of GPS satellites.

The objective of the SEDS-1 experiment was to demonstrate the use of a tether to place a payload in a de-orbit trajectory and study payload re-entry after the tether was cut. This involved a downward deployment of a 20 km long non-conducting momentum exchange tether. During the experiment, following the deployment, the tether was cut at the deployer end, forcing the payload to re-enter the atmosphere, where it was burnt. The main purpose of the SEDS-2 flight, which was also a downward deployment of a 20 km long non-conducting momentum exchange tether, was to demonstrate deployment and stabilization of the tether system along the local vertical.

16.2.3.3 OEDIPUS Tethered Sounding Rocket Missions

OEDIPUS, an acronym *for Observations of Electric-field Distribution in the Ionospheric Plasma -- a Unique Strategy*, was a joint programme by the National Research Council of Canada and NASA. The programme included participation by the Communication Research Center in Ottawa, Canada, who were the Principal Investigators, various Canadian universities and the US Air Force Phillips Laboratory. Bristol Aerospace Ltd was the primary payload contractor. The mission consisted of two sounding rocket experiments that used spinning conductive tethers. The two experimental missions, namely OEDIPUS-A and OEDIPUS-C, were launched on 30 January 1989 and 6 November 1995, respectively, using three-stage sounding rockets called Black Brant-X. OEDIPUS-A was designed as a double probe for sensitive measurement of weak electric fields in the aurora. OEDIPUS-C also had two spinning payloads. The payloads were connected by a 1174 m long conductive tether. The mission objectives included understanding of the effect of charged particles associated with aurora on satellite transmissions and studying natural and artificial waves in the ionospheric plasma.

16.2.3.4 Plasma Motor Generator (PMG)

The *Plasma Motor Generator* (PMG), a 500 m long electrodynamic tether payload of a Department of Defence satellite, was launched on 26 June 1993 as a secondary payload onboard a Delta-2 rocket. The primary objectives of the PMG mission were to test the performance of a hollow cathode assembly to provide a low impedance bidirectional electrical current and demonstrate the application of an electrodynamic tether for space propulsion and conversion of orbital energy into electrical energy. The mission successfully established the use of a tether as a generator with electron current flow down the tether and as a motor with electron current driven up the tether.

16.2.3.5 Tether Physics and Survivability (TiPS)

The *Tether Physics and Survivability* (TiPS) experimental payload was built and operated by the Naval Centre for Space Technology (NCST) of the Naval Research Laboratory (NRL). It was launched on board the Titan-4 launch vehicle on 12 May 1996 and deployed on

20 June 1996. The TiPS payload is a free flying satellite comprising two end bodies separated by a 4.0 km long non-conducting tether. This is unlike other tether systems such as those flown onboard the shuttle where one end of the tether system was connected to the massive host vehicle. The TiPS satellite payload was jettisoned by the host launch vehicle on 20 June 1996. This was followed by separation of the end bodies by the tether. The primary objective of the TiPS experiment was to study the long term dynamics and survivability of tether systems in space.

16.2.3.6 Atmospheric Tether Mission (ATM)

The *Atmospheric Tether Mission* (ATM) is an experimental mission proposed to be executed from Space Shuttle in which a tethered probe will be lowered successively into different regions of the atmosphere such as the mesosphere, thermosphere and ionosphere for the purpose of understanding the atmosphere and plasma around the Earth in these regions. The measurements are proposed to be made by a set of 11 instruments housed in an end mass or spacecraft. The end mass will be lowered from the shuttle by a 90 km long tether. The instrument payload comprises an ion drift meter, a retarding potential analyzer, an ion mass spectrometer, a Langmuir probe, a neutral wind meter, a neutral mass spectrometer, an energetic particle spectrometer, E-field double probes, an infrared spectrometer, a UV photometer and a three-axis magnetometer. The experiment is proposed to be carried out over a period of six days. The orbiter will be at 220 km altitude. For the first two days of the mission, the tether is proposed to be lowered up to 170 km altitude using a 50 km tether length. Another 20 km of tether length will be lowered for the next two days and further lowered by another 20 km to its full length of 90 km and an altitude of 130 km during fifth and sixth days.

16.2.3.7 STEP-AIRSEDS

STEP-AIRSEDS is an acronym for *Space Transfer Using Electrodynamic Propulsion – Atmospheric Ionospheric Research Small Expendable Deployer System*. This mission satellite, developed by the Michigan Technic Corporation Holland, Michigan, USA, and weighing 1000 kg, including the tether, consists of two units tethered together by an approximately 6–7 km long electrodynamic tether. The upper and lower units drive the boost and deboost operations, respectively. The mission is intended to experiment with new and innovative methods of use of conducting tethers, solar power and the Earth's magnetic field to execute a range of satellite orbit manoeuvres such as moving the satellite up, down and across planes without using any propellants. It is proposed to perform these operations on a satellite operating in the altitude range of 350–1100 km for a minimum period of one year.

16.2.3.8 Space Tether Experiment (STEX)

The Space Tether Experiment (STEX) is a science and technology experiment of the Institute of Space and Aeronautical Science (ISAS) to be flown onboard the Space Flight Unit (SFU) mission. SFU is an unmanned, multi-purpose reusable platform that can be used for performing a range of science and technology experiments and carrying out flight tests of space and industrial technologies. The satellite weighs 4000 kg and has a payload capacity of 1000 kg. The STEX payload comprises a tether, a tether deployment and retraction system, and a 40 kg sub-satellite. The sub-satellite can be deployed up to 10 km. Deployment and retrieval of the sub-satellite can be done several times. The sub-satellite is equipped with instrumentation including a vacuum gauge, plasma probes and wave receivers to study the electromagnetic environment of SFU. One of the primary objectives of STEX is to assess the tether technology for future scientific missions.

16.2.3.9 Propellant Reboost for International Space Station

It is proposed to use a short and light electromagnetic tether as an alternative to propellant driven thrusters reboost for the ISS. The proposed electrodynamic tether system is approximately 10 km long and weighs 200 kg, which is small enough to cause an insignificant shift of less than 5 m in the centre of the mass of the space station. The tether when deployed would be capable of generating thrust of 0.5–1.0 Newton for an electrical power of 5.0–10 kW. This will compensate for the average aerodynamic drag of 0.3–1.1 Newton on the space station. To produce thrust of the order of 1.0 Newton in a 10 km long tether, the tether current needs to be of the order of 10 amperes. This is made possible by having a good part of the tether length as uninsulated or bare, unlike standard tethers, which collect electrons from the ionosphere only at the ends. Also, the tether design is such that it is insensitive to variations in electron density in the ionosphere, which also allows it to operate efficiently during night time.

16.2.4 Space Elevator

The dream of space transportation was realized back in 1981 with the successful launch of Space Shuttle mission Columbia, thereby establishing the use of a reusable spacecraft to travel to space. Since then, technology has matured a great deal and more than 100 such missions have been carried out with success. What has not changed is the cost of executing such missions. Whether it is Space Shuttle mission or a non-reusable spacecraft; the cost of launch continues to be about US$ 20 000 per kilogram of mass to be transported to geostationary Earth orbit and much more if it were to be sent to a more distant point in space.

A space elevator, which is related to the fundamentals of space tethers, provides a concept for the development of a new space transportation system. The proposed concept has the potential of making the transportation of tonnes of payloads and resources to geostationary Earth orbit and beyond almost a daily affair, albeit at an incredible cost of US$ 200 to 800.

A space elevator is nothing but a very long tether anchored to the surface of Earth or an off-shore sea platform at one end. A counterweight connected to the other end or an extension of the tether further into space ensures that the centre of mass is above the geostationary orbit altitude. A climber is the other essential component of a space elevator. It is a kind of robotic platform that is made to climb up the tether. The tether is kept stretched by the downward acting gravitational pull and upward acting centrifugal force due to the Earth's rotation. It may be mentioned here that once above the geostationary level the climber would experience an upward weight due to centrifugal force overpowering the gravitational pull. The tether consists of a ribbon-like thin cable made from a carbon nano tube composite material. Figure 16.5

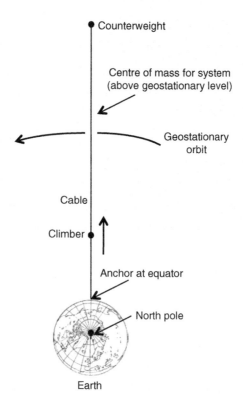

Figure 16.5 Space elevator

illustrates the operational concept of a space elevator. The diagram is self explanatory. Once the space elevator is deployed, the climber platform could be made to repeatedly ascend and descend the ribbon-like cable.

Once deployed, the tether would be ascended repeatedly by mechanical means to orbit and descended to return to the surface from orbit.

One such proposed space elevator is from LiftPort, one of the many companies engaged in the development of space elevators and their components. The proposed space elevator will be 100 000 km long, close to a quarter of the distance between the Earth and the moon. The cable of carbon nano tube composite ribbon is proposed to be anchored to an off-shore sea platform. The weight of the climber may vary from 5 to 20 tonnes. A 20 tonne climber would be capable of lifting up to 13 tonnes. The robotic lifter will use the ribbon to guide the ascent into space. A 2.4 MW free electron laser located on or near the anchoring station powers photovoltaic cells that convert light energy into electrical energy. The electrical energy in turn feeds niobium magnet DC electrical motors, which drive the lifter. Figure 16.6 is a conceptual drawing of the climber. The launch is scheduled for the year 2018. Once operational, climbers will climb up the elevator almost every day at a speed of about 190 km per hour.

Figure 16.6 Conceptual drawing of climber (Courtesy: Liftport)

16.3 Aerostat Systems

An *aerostat* may be defined as an aircraft, balloon or dirigible, deriving its lift from the buoyancy of surrounding air rather than from aerodynamic motion. The name aerostat is derived from the fact that it operates on aerostatic lift, also called buoyant force, and not aerodynamic lift, which essentially requires at least some part of the aerial body to move through the surrounding air mass. The aerostat family includes free balloons, tethered balloons, helikites and airships, though in a narrower sense aerostats are generally identified by tethered balloons. The major constituents of an aerostat system are the *aerostat*, the *tether*, the *mooring system*, the *payload*, comprising a range of *sensors* depending on the application, and the *command and control unit*. Figure 16.7 shows deployment of a tethered aerostat system.

Figure 16.7 Tethered aerostat under deployment (GNU Free Documentation License)

16.3.1 Components of an Aerostat System

An *aerostat* is an unmanned, aerodynamically shaped non-rigid or semi-rigid airship tethered to the ground by a cable and kept aloft by the buoyant force. The aerostat is made up of a large fabric envelope filled with a non-inflammable lighter-than-air gas such as helium. It is this gas fill that provides the buoyant or lifting force. The *tether*, in addition to providing the required anchoring, also contains electrical cables to supply electrical power to the aerostat and fibre optic cables for data relay between the aerostat and the ground station. The *payload* comprises a wide range of sensors and other electronic systems, depending on the intended application. Some common sensor systems generally installed on aerostat systems include surveillance sensors, observation devices, communication equipment, and military intelligence systems such as ELINT (electronic intelligence) and COMINT (communication intelligence) systems. The main functions of the *command and control centre* include controlling the aerostat platform and the onboard payload, signal processing to build up the total situational awareness picture, and initiation and control of response functions.

16.3.2 Types of Aerostat Systems

Aerostat systems are classified as free balloons, tethered or moored balloons, helikites and airships. A *free balloon* is a type of aerostat that moves with the wind and remains aloft due to its buoyancy. A capsule suspended beneath the balloon with the help of cables is used to carry observation and surveillance equipment, and communication relay equipment. It also houses mechanisms used for flight control. These are hot air balloons that get their buoyancy by heating air inside the balloon, gas balloons that are hot air balloons inflated with a gas lighter than the surrounding atmosphere, such as helium, and filled with pressure equal to or slightly more than that of the surrounding atmosphere or Rozière balloons filled with both heated and unheated lifting gases. The ECHO satellite launched in 1960 for passive relay of radio communication, PAGEOS launched in 1966 for high precision calculation of different locations on the Earth's surface and Vega-1/2 satellites of the former Soviet Union launched in 1984 releasing two balloons carrying scientific payloads to perform experiments in the atmosphere of Venus are some examples of the use of balloons in the early days of satellite launches. Rozière balloons are generally used for circumnavigation.

A *tethered balloon*, also called a *moored balloon*, is not free flying like the free balloons described in the previous paragraph. Its movement is restricted by anchoring it to a ground surface or a vehicular platform by a cable or a set of cables. A tethered balloon is a structured envelope of fabric usually filled with helium gas. Traditional tethered balloons are blimp shaped, fin stabilized and use helium gas for lift. Simple round balloons do not use any fin stabilization and rely on helium alone for lift. Hybrid balloons make use of both buoyancy and aerodynamic lift to stay aloft.

The *helikite* is one of the most commonly used aerostat designs for all weather, high altitude applications. It is a combination of a helium balloon that uses helium gas for buoyancy and a kite that exploits wind for lift. The word helikite originates from a combination of the words 'helium' and 'kite'. This kite style of aerostat has been patented by Sandy Allsopp in England. The kite-like structure provides excellent aerodynamic stabilization and the aerostat exploits both helium driven buoyancy and wind for lift.

An *airship* is an aerostat in the broader sense as defined earlier. In the narrower sense, where the word aerostat is generally associated with tethered or moored balloons, an airship may be considered as a kind of lighter-than-air aircraft that is steered with the help of thrust mechanisms such as rudders and propellers. Airships are being developed for a wide range of applications including passenger flight, flying cranes, scientific experimentation, surveillance, communication relay, and so on. A large number of companies worldwide are engaged in building airships. Some of the prominent names include Zeppelin NT, American Blimp Corporation, ABC Lightship, Raven Aerostar, Advanced Technology Group, Airship Industries and so on. Figure 16.8 is a photograph of an airship manufactured by Zeppelin NT.

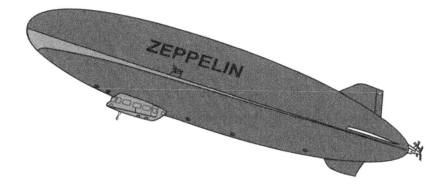

Figure 16.8   Airship (Courtesy: AngMoKio@gmx.de)

16.3.3 Applications

Aerostats were initially considered more of a spectacle than machines that could be used for important civilian and military applications. In the early days of their development their use was limited to advertising banners or event broadcasting due to their low speed and large size. With advances in technology, particularly fibre optic technology, the application spectrum of aerostats has widened to include many new areas such as early warning, communication relay and surveillance, homeland security and law enforcement and intelligence gathering. Modern aerostats are capable of flying to altitudes in excess of 5 km, have a large detection range coverage extending as far as 500 km, have payload carrying capacity exceeding 2500 kg and continuous operation of weeks before they need to be retrieved for routine maintenance.

Figure 16.9 is a photograph of a tactical aerostat from M/S Raven Aerostar. A wide variety of payloads such as communications, intelligence gathering, surveillance and reconnaissance payloads, EO/IR sensors, communication repeaters and radar sensors can be integrated on this platform to provide a total solution.

Figure 16.9 Tethered tactical aerostat, Type TIF-75K (Courtesy: Raven Aerostar)

16.4 Millimetre Wave Satellite Communication

The requirement to have larger bandwidth for audio, video and data communication services, terrestrial or satellite based, has been on the rise with the addition of new services on a continuous basis and also with the quest to enhance the quality of existing and new services. Services such as three-dimensional television, high definition television, video on-demand, virtual reality imaging and high speed data communication demand bandwidths that are a thousand times larger than what they would be in the case of traditional voice telephony and television broadcast services, notwithstanding the fact that various digital compression techniques are in use today to conserve bandwidth while providing these services. There are a host of technologies available for delivery of the required bandwidth. One of the most prominent ones used in terrestrial systems is fibre optics. However, when it comes to the cost effectiveness of fibre optic communication systems, particularly in the case of extensive geographical coverage and the need to have easy access to remote and sparsely populated areas, it is definitely not the best solution. Millimetre wave wireless communication has the potential of offering a comparable bandwidth that is currently possible with fibre optic communication technology without the economic and logistic challenges of the latter. The following paragraphs present the fundamentals of millimetre wave communication and the associated advantages and limitations. This is followed by a brief discussion of the experimental satellite missions carried out in the last three to four decades to establish the feasibility of millimetre wave satellite communication.

16.4.1 Millimetre Wave Band

The millimetre wave band extends from a wavelength of 10 mm to 1 mm, corresponding to a frequency band of 30–300 GHz. It is also referred to as the extremely high frequency (EHF) band (Figure 16.10). The terahertz region also falls within this band, which is sandwiched between the near and far infrared bands on the shorter wavelength side and the radio and microwave regions on the longer wavelength side. Although the millimetre wave band may

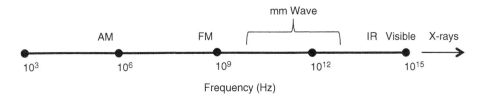

Figure 16.10 Millimetre wave band

be considered theoretically to extend from 30 to 300 GHz, in the context of wireless communication a few spectral bands around 38, 60 and 94 GHz are considered important from the viewpoint of their propagation characteristics through the atmosphere. More recently, the propagation band from 70 to 90 GHz, also known as the E-band, has been allocated for wireless communication in the public domain. Within this band there are two sub-bands around 70 GHz (71–76 GHz) and 80 GHz (81–86 GHz,) that are used for wireless communication. Other millimetre wave bands that have been opened for commercial applications include the 59–64 GHz band, commonly known as the 60 GHz band or the V-band, and the 92–95 GHz band, also referred to as the 94 GHz band or the W-band. Because of the high level of absorption by atmospheric oxygen in the 60 GHz band, this band is best suited to very short range point-to-point and point-to-multipoint communication applications. The 94 GHz band is less spectrally efficient due to an excluded band at 94–94.1 GHz. The 70 GHz and 80 GHz (E-band) bands with the availability of a 5 GHz spectrum in each of the two bands makes them an ideal choice for wireless millimetre wave communication. In fact, the spectral bandwidth offered by these two bands exceeds the total collective bandwidth offered by all allocated microwave bands.

16.4.2 Advantages

It is not just recently that the potential that the millimetre wave frequency band holds for a variety of applications, including their use in radio astronomy in the 1960s and military applications in the 1970s and now point-to-point and point-to-multipoint terrestrial and satellite communication systems, has been realized. The advantages that the millimetre waves would offer in terms of larger bandwidth, higher resolution, smaller antenna sizes with associated higher directive gain, reduced multipath interference, increased immunity to jamming by electronic countermeasures and so on were visualized back in the nineteenth century, as is evident from the millimetre wave experiments of J.C. Bose in the 1890s. While his contemporaries like Marconi were inventing all-important radio communications, Bose was busy experimenting with millimetre waves.

Millimetre wave research and development, however, remained confined to prominent universities and government funded research and development institutes for almost 50 years before millimetre wave applications began to see the light of the day first in the form of radio astronomy and then military applications. Subsequently, advances in the development of millimetre wave integrated circuits in the 1980s have opened up the commercial application domain in the form of automotive collision avoidance radar. Authorization of 70 GHz and 80 GHz bands by the Federal Communications Commission for licensed point-to-point communication in 2003 has helped the millimetre wave products reach industry.

Millimetre wave communication offers numerous advantages but also has some serious limitations relating to propagation through the atmosphere. Millimetre waves are particularly affected by rain, as described in the section 16.4.3, though efforts are in progress to devise new techniques to overcome these limitations with significant success. The advantages of millimetre wave communication as outlined earlier in this section are briefly described in the following paragraphs. Major advantages include increased bandwidth, smaller antennas for a given directive gain and narrower beam width for a given antenna aperture, immunity to jamming and interference, and finally inherent security.

16.4.2.1 Increased Bandwidth

Millimetre wave communication offers much higher bandwidth compared to microwave communication, for example the 5 GHz bandwidth available in each of the two sub-bands around 70 and 80 GHz of the E-band can be used as a single, contiguous transmission channel without the need for any channelization. This allows a throughput of 1–3 Gbps to be achieved in each of the two sub-bands with simple modulation schemes such as on–off keying or binary phase shift keying, which is even higher than the throughput possible with higher order modulation schemes at microwave frequencies. Using sophisticated higher order modulation schemes at E-band enables much higher throughput can be achieved.

16.4.2.2 Narrower Beam Width

The antenna beam width in both azimuth and elevation planes is inversely proportional to the wavelength. The half power beam width of an antenna with aperture diameter of D and operating at a wavelength of λ is given in degrees by $70\lambda/D$. This implies that the beam width decreases with increase in wavelength. This further means that a given antenna will produce a narrower beam width at millimetre wavelengths than at lower frequencies or higher wavelengths. As an illustration, using an equivalent antenna, the beam width of a 70 GHz link is four times as narrow as that of an 18 GHz link. In turn, it allows as much as 16 times the density of E-band millimetre wave links in a given area. Also, a narrower beam width means a higher directive gain, which allows operation at millimetre wavelengths to compensate for some of the propagation losses at millimetre wavelengths.

16.4.2.3 Immunity to Jamming and Interference

Narrow beam width and a highly directional radiation pattern allow multiple transmission channels to operate spatially close to each other without causing any troublesome adjacent channel interference. The use of cross-polarization techniques allows even multiple channels to be deployed along the same path.

16.4.2.4 Small Antenna Size

The gain of antenna for a given aperture area is inversely proportional to the square of the wavelength. Operation at millimetre wave frequencies, which are higher than microwave frequencies, allows use of smaller and lighter antenna structures for a given gain specification.

16.4.2.5 Communication and Information Security

Millimetre wave communication is inherently secure due to narrow beam width and also because millimetre waves are blocked by many solid structures. Any attempt to sniff millimetre wave radiation would require placing the interceptor near or in the path of electromagnetic radiation, which becomes difficult when the beam width is sufficiently narrow. Also, loss of data integrity due to interception, if any, may be used to detect the interception. In addition, data encryption techniques are available to further enhance security.

16.4.3 Propagation Considerations

Every electromagnetic wave irrespective of its frequency or wavelength has to experience less or more attenuation in signal intensity as it propagates through the atmosphere. Absorption and scattering are the primary mechanisms that cause signal attenuation. Attenuation is usually expressed in decibels loss per kilometre (dB/km) of propagation distance and therefore attenuation suffered by a communication link depends upon the length of the link.

With reference to millimetre wave propagation through the atmosphere, major parameters influencing signal attenuation are atmospheric oxygen, humidity, rain and fog. Figure 16.11 shows the curves of attenuation in dB/km as a function of frequency of operation both at sea level and at an altitude of 9150 m. The atmospheric oxygen causes peak absorption, as is evident from the curves, at 60 GHz within the millimetre wave region of interest. Peak attenuation due to atmospheric oxygen at 60 GHz is approximately 3 dB/km at 9150 m altitude. Attenuation at sea level is considerably higher. These figures are at high temperature and humidity levels. There is further attenuation of signal due to fog and clouds. Attenuation caused by fog and

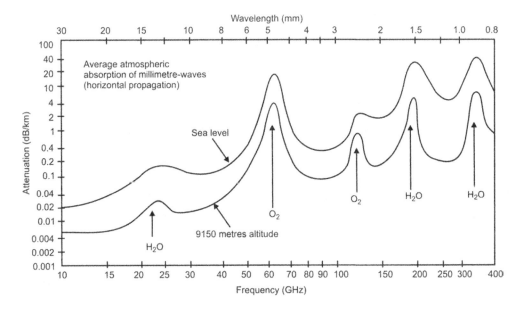

Figure 16.11 Atmospheric attenuation of millimetre waves

clouds depends upon quantity and size of liquid droplets in the air. Figure 16.12 shows the curve depicting the effect of fog corresponding to 100 m visibility conditions on millimetre wave propagation. As is shown in the curve, millimetre waves suffer attenuation in the range 0.1–1.0 dB/km.

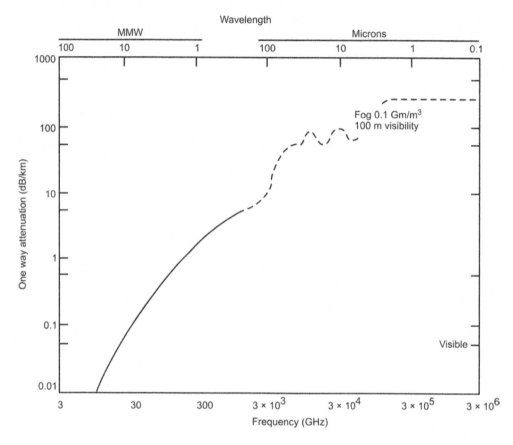

Figure 16.12 Attenuation of millimetre waves due to fog

Rain is the most predominant attenuating factor for both microwaves and millimetre waves, including atmospheric constituents like oxygen, carbon dioxide and water vapour, fog and clouds. The amount of signal loss depends upon the rate of rainfall, which is usually measured in millimetres per hour. Figure 16.13 shows the curve depicting the effect of rainfall on signal attenuation in the millimetre wave region. As is evident from the family of curves, signal attenuation in the millimetre wave region in heavy rain conditions (corresponding to 25 mm/hour) is in the range 8–15 dB/km.

16.4.4 Applications

The millimetre wave band of frequencies offers diverse applications ranging from imaging to telecommunications, and from consumer products to defence. Imaging applications mainly

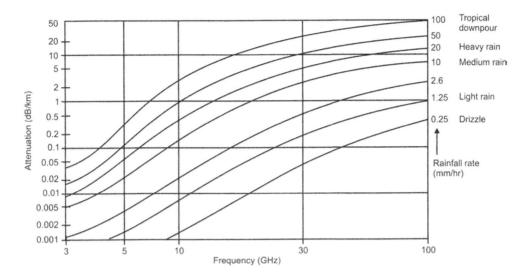

Figure 16.13 Attenuation due to rain

include check point security at airports, concerts, sports events and so on used to screen passengers and personnel, loss prevention and inventory control, through wall imaging, imaging of personnel for detection of concealed weapons and so on. Telecommunications applications mainly include campus and enterprise broadband networks, routine backhaul deployments and point-to-point communication links. The consumer market is mainly the automotive sector. Automotive radar is an established application used for adaptive cruise control and collision avoidance. There are many applications in defence and security. Some of the better known ones include millimetre wave sensors in smart munitions, satellite-to-satellite communications and radar for surveillance, perimeter protection and intrusion detection. Military intelligence is another application. Other than telecommunications, remote sensing and radio astronomy are the areas particularly important for space-based platforms, including satellites. It is not possible to discuss each and every existing and emerging application of millimetre wave bands, however some representative examples of applications relevant to satellite and space technology are presented in section 16.4.5.

16.4.5 Millimetre Wave Satellite Missions

A number of both experimental and operational space missions have been carried out since the 1970s to assess the feasibility of using millimetre wave technology and also to evaluate the performance of the technology for a variety of applications such as radio astronomy, remote sensing, imaging, meteorological observations and communication. Though the feasibility of using millimetre wave technology has been established and it has begun to be used in remote sensing, radio astronomy and meteorological applications from space-based platforms, it is not yet widely used for satellite-based telecommunications. Some representative space missions including experimental missions tried in earlier years, those carried out in recent years and also those planned for the future are briefly described in the following paragraphs.

16.4.5.1 Engineering Test Satellite-II (ETS-II)

Beginning in 1975 with the launch of ETS-I, also called KIKU-1, Japan has launched a number of engineering test satellites (ETS), with each satellite addressing the technological needs of that time. Eight satellites, ETS-I to ETS-VIII, also called KIKU-1 to KIKU-82, have been launched so far by the National Space Development Agency of Japan. ETS-II (KIKU-2) was the first geostationary satellite to carry a beacon transmitter with three coherent frequencies at 1.7 GHz, 11.5 GHz and 34.5 GHz (millimetre wave frequency) to perform propagation experiments. The signal at 34.5 GHz had 100% amplitude modulation. This mode was selected to improve rain margin. The satellite was launched in February 1977 and concluded its mission in December 1990.

16.4.5.2 Experimental Communications Satellite (ECS)

Experimental communications satellites (ECSs), also known as AYAME satellites, were experimental satellites of the National Space Development Agency (NASDA) intended to carry out communications and propagation experiments at millimetre wavelengths. Two satellites, ECS-A (also called AYAME-1) and ECS-B (also called AYAME-2), were launched on 6 February 1979 and 22 February 1980, respectively. Both satellites were lost shortly after launch during the firing of their apogee kick motors. ECS-A ceased radio transmissions 10 seconds after the apogee kick motor was fired. The last known longitude of the satellite on 13 June 1995 was 146.23°W, drifting at 33.817°E per day. ECS-B also failed in a similar manner and its last known longitude on 17 November 1988 was 146.47°W, drifting at 12.888°E per day.

16.4.5.3 FLORAD Mission

The FLORAD mission of the Italian Space Agency is a micro-satellite flower constellation of micro-satellites carrying onboard scanning millimetre wave radiometers for Earth and space observation at a regional and quasi-global scale. The research mission was carried out during 2008–2009. A flower constellation is designed using compatible orbits that allow optimization of revisiting time for the Earth regions of interest. In a flower constellation, the orbits are designed so that when a satellite leaves the petal, another satellite of the constellation takes its place. The constellation has four micro-satellites deployed in flower-like elliptical orbits and each satellite is equipped with only one payload, which is a millimetre wave radiometer. The primary objective of the mission was to study and analyze the thermal and hydrological properties of the troposphere with particular reference to water vapour profile, temperature profile, cloud liquid content, rainfall and snowfall with high spatial resolution and time repetitiveness in the terrestrial atmosphere of the Mediterranean region. In addition, the mission aimed to evaluate the performance of the flower constellation itself, passive millimetre wave sensors, advanced antenna concepts and so on. The mission investigated various configurations of millimetre wave multiband channels to find a trade-off between performance and complexity within the constraints of the micro-satellite platform.

It may be mentioned here that millimetre wave radiometers operating at 30–300 GHz score over their microwave counterparts in their reduce size and their potential in exploiting the window frequencies and different gaseous absorption bands at 60/70 GHz, 118 GHz and 183 GHz. The radiometry of the atmosphere is an established application of millimetre wave technology due to its capability to sound through clouds and detect precipitation, and it outperforms infrared sensors. The latter, however, offer higher spatial resolution but only in cloud-free areas.

16.4.5.4 Applications Technology Satellite-6 (ATS-6)

The Applications Technology Satellite-6 (ATS-6) is one of the very ambitious experimental communications satellites developed and implemented by the NASA Goddard Space Flight Centre and intended to carry out a large number of scientific and technological experiments in the geostationary space environment. The satellite carried payloads to perform a wide range of scientific experiments related to communications technology and meteorology. ATS-6, a three-axis body stabilized satellite was launched in geosynchronous orbit on 30 May 1974 from Cape Canaveral using the Titan-3C rocket. The satellite was made operational on 2 July 1974. The objectives of the scientific experiments were to study and gain a better understanding of the environment in space at the geosynchronous altitude. Figure 16.4 is a photograph of ATS-6 in orbit.

In addition to the scientific experiments, ATS-6 also supported health and education telecommunications, satellite based air traffic control and UHF television broadcasts. ATS-6 performed about 18 scientific, communications technology and meteorological experiments, most significantly millimetre wave propagation experiments, health, education and telecommunications experiments, position, location and aircraft communications experiments, radio frequency interference experiments, tracking and data relay experiments and television relay using small terminals experiments.

Figure 16.14 ATS-6 satellite in orbit (Courtesy: NASA)

With reference to the millimetre wave propagation experiment, ATS-6 provided the first direct measurements of Earth-to-space links from an orbiting geosynchronous satellite at

20 and 30 GHz. The measurements mainly focused on studies of rain attenuation effects, scintillations, depolarization, site diversity, coherence bandwidth and communications techniques. The results obtained with the direct measurements were compared with the data generated using methods of attenuation prediction with radars, rain gauges and radiometers. The experimental mission ended in July 1979. At the end of the mission, the satellite was moved 450 km out of geostationary orbit, after which it started drifting eastwards.

16.4.5.5 Communications and Broadcasting Engineering Test Satellite (COMETS)

The Communications and Broadcasting Engineering Test Satellite (COMETS), also known as KAKEHASHI, is a three-axis stabilized geostationary satellite from NASDA of Japan launched onboard the H-II launch vehicle from the Tanegashima Space Centre on 21 February 1998. The satellite had a design life of three years but its operations were terminated in August 1999. Figure 16.15 shows a photograph of the COMETS satellite. The satellite was intended to develop and evaluate futuristic communications technologies including experimenting with the use of millimetre wave frequencies.

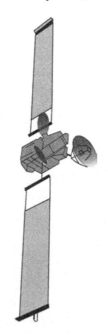

Figure 16.15 COMETS satellite

The satellite has three mission payloads. The payload related to millimetre wave band communications is an advanced mobile satellite communications system developed by the Communications Research Laboratory (CRL). The system used a millimetre wave band at 47/44 GHz and a Ka-band at 31/21 GHz for mobile satellite communications. The two other mission payloads included a 21 GHz band advanced broadcasting system developed by CRL and NASDA, and an inter-satellite communications system developed by NASDA and operating in S- and Ka-bands.

16.4.5.6 Odin Satellite Mission

Odin is an aeronomy and astronomy mini-satellite mission funded and executed by the Swedish National Space Board jointly with the space agencies of Canada, France and Finland. The spacecraft design, development and operations were carried out by the Swedish Space Corporation (SSC).

The Odin satellite was launched on 20 February 2001 into sun-synchronous polar orbit. Figure 16.16 shows the Odin satellite deployed in orbit. The satellite mission is intended for astronomy and aeronomy applications. The aeronomy related mission objectives include observation of stratospheric ozone chemistry, mesospheric ozone science and coupling of atmospheric regions. The astronomy objectives and major scientific issues relate to star formation processes, interstellar chemistry and atmospheric ozone balance. The Odin satellite is an observatory making measurements in sub-millimetre (0.5–0.6 mm) and millimetre wave bands (2.5 mm). The Odin mission completed 12 years of operation in 2013.

Figure 16.16 Odin satellite in orbit (Courtesy: Swedish National Space Board)

16.5 Space Stations

A space station may be defined as a large spacecraft orbiting Earth usually in the low Earth orbit and used as a long term base for carrying out manned operations in space. The space stations are also designed for other spacecraft used to transport people and cargo to and from the space station to dock with it from time to time. A space station is also like a huge artificial satellite and therefore deserves a brief description in a book addressing all aspects of satellite technology and applications. It is being covered as a part of the present chapter for two simple reasons. First, it could not have been covered in any of the other chapters and a separate chapter would not be justified given the scope of the text. Second, the space station continues to be an emerging and fast growing concept despite the fact that the first space station (Salyut series) was launched as early as 1971. The following sections discuss different aspects of space stations, including their importance, missions launched during the 1970s and 1980s, currently operational space stations, space stations planned to be launched in the coming years and emerging space station concepts.

16.5.1 Importance of Space Stations

Space stations are important for a variety of reasons, including the study of the effect of the microgravity environment, scientific research, space exploration and tourism. Looking at the prolonged exposure to the microgravity environment of astronauts intending to travel to other planets, it is most important to study the effect of weightlessness on the human body when the exposure is going to last from months to years.

Space stations also provide a platform to carry out research in some cutting edge technologies in an environment that cannot be matched on Earth. There are a number of phenomena that are completely different under the influence of gravity and in microgravity conditions. As an example, in crystal growth phenomenon, near perfect crystals can be grown in a microgravity environment, which can lead to better semiconductors further leading to faster computers. There are many more such examples.

Space stations by virtue of their being above Earth's atmosphere also provide an excellent platform for building manned observatories to carry out space exploration through the use of space telescopes. The merits of unmanned observatories such as the Hubble space telescope are well known.

Another application area where space stations can play an important role is in the study of the Earth's atmosphere, its land forms, vegetation, oceans and so on. Space tourism is another application of space stations as they can be used to allow tourists to stay for a brief period or an extended period of time in the not-too-distant future.

16.5.2 Space Stations of the Past

The following paragraphs briefly discuss the space stations that were launched during the 1970s and 1980s. These include Salyut, Skylab and Mir. One or more of these space stations were operational during the period 1971–1998.

16.5.2.1 Salyut

There are two broad categories of space stations, one built as one piece on Earth and subsequently launched as a single unit and the second that are modular in nature and assembled in space bit by bit. Salyut-1, the first in the Salyut series, was the first ever space station and was put into space by the former Soviet Union on 19 April 1971. Salyut-1 in fact was a combination of Soyuz and Almaz spacecraft. Almaz was originally designed for military applications and was reconfigured as a space station for the civilian role. Soyuz was used to transport crew from Earth to the space station and back. It stayed in orbit for 175 days before it was de-orbited, forcing it to make a destructive re-entry over the Pacific Ocean. The second crew launched on Soyuz-11 remained on board the space station for 23 days after the first crew launched on Soyuz-10 failed in its attempt due to malfunctioning of the docking mechanism. The second crew members unfortunately were killed during the re-entry of Soyuz-11. The Soyuz space craft was redesigned following this failure. Salyut-1 was followed up by Salyut-2, which failed to reach orbit. Salyut-3, Salyut-4 and Salyut-5 were launched subsequently. The new Soyuz spacecraft was used to ferry crew members between Earth and space stations. These space stations were manned for longer mission periods. Salyut series space stations up to Salyut-5 had only one docking port, which was used by Soyuz spacecraft. Salyut-6 launched on 29 September 1977 had a second docking port that could be used by an unmanned space station resupply craft, Progress. Salyut-7, the last in the series of Salyut space stations, was launched in 1982. It also had two docking ports. The space station remained operational till 1982 and hosted different crews for a total of 800 days. The Salyut programme was used for both civilian and military applications. Civilian applications included the study of the long term effect of the space environment on human beings and a wide range of scientific experiments related to astronomy, Earth observation and biology. Military applications included reconnaissance missions.

16.5.2.2 Skylab

Skylab-1, the first space station from the USA, was launched into orbit on 14 May 1973. The architecture of the space station was derived from the modified third stage of the Saturn-V moon rocket. It comprised an *orbital workshop* that housed working and living quarters for the crew, a *multiple docking adapter* that allowed more than one Apollo spacecraft to dock to the station, an *Apollo spacecraft* that ferried crew members between Earth and the space station, an *airlock module* that allowed access to the outside of the station and an *Apollo telescope mount* carrying telescopes. The station was damaged during the launch when a micrometeoroid shield separated from the station and in the process destroyed one of the two main solar panels and jammed the other, preventing it from fully stretching out. This resulted in very little electrical power being available on the station, which further led to a rise in inside temperature, almost threatening its shutdown. The first crewed mission, Skylab-2, carried out the first ever major in-space repair and saved the space station. In fact, three manned missions, Skylab-2, Skylab-3 and Skylab-4, were carried out between 1973 and 1974. Each of these missions was launched using the Apollo Command Service Module (CSM) onboard the Saturn-IB rocket. Also, each of these missions carried a three member crew to the space station. The space station remained in orbit till 1979, spending 2249 days in orbit including 171 manned days. Skylab was abandoned at the end of Skylab-4 (third crewed mission). It re-entered Earth's atmosphere and burned on 11 July 1979 with debris falling on parts of Western Australia. During manned missions, a number of scientific studies were carried out, which included confirmation of the existence of coronal holes in the sun and Earth observation in visible, infrared and microwave bands.

16.5.2.3 Mir

Mir was the first permanent space station owned by the former Soviet Union at the time of its launch and subsequently by Russia. The core module, also called the base block, was launched aboard the Proton rocket on 20 February 1986 from the Baikonur Cosmodrome into a low Earth orbit. It was followed up by another six modules. The space station in all consisted of a core module, seven pressurized modules and several unpressurized components. It was the first ever modular space station that was constructed in orbit between 1986 and 1996. Mir represented the third generation of the Soviet Union's space station programme following the success of the Salyut space station programme. It had a entirely new docking system with six ports that enabled the creation of a far more complex space station. The space station was capable of receiving Soyuz-TM spacecraft, unmanned cargo craft and modules carrying equipment and supplies.

Figure 16.17 Mir space station (Courtesy: NASA)

Major components of the Mir space station include *living quarters* for the crew, an *assembly compartment* housing rocket engines and fuel tanks, an *intermediate compartment* used to connect the working compartment to the rear docking port, a *transfer compartment* that allowed attachment of additional station modules, a *docking module* housing ports for Space Shuttle dockings, *progress cargo craft* carrying supplies from Earth and removing waste material from the space station, *Soyuz spacecraft* to ferry the crew, a *Kvant-1 astrophysics module* housing telescopes, a *Kvant-2 scientific and airlock module* housing equipment for scientific research, a *Kristall technological module* used for biological and material processing experiments and also containing a docking port for the US Space Shuttle, a *Spektr module* used for studying the Earth's atmosphere and monitoring its natural resources and a *Priroda remote sensing module* housing radar and spectrometers. Figure 16.17 shows a photograph of the Mir space station as observed from STS-89 flight of Space Shuttle Endeavour.

The Mir space station remained in orbit for 5519 days, including 4592 manned days, during the 10 year period it was operational. It served as a permanent research station providing a microgravity environment for research activities. The Mir space station was equipped with a whole range of scientific equipment, as outlined in the previous paragraph, to perform a wide range of scientific experiments and studies in the fields of astronomy, aeronomy, meteorology, biology and physics that need to be developed for long term sustenance in space.

The Mir space station was damaged by an onboard fire in 1994. Subsequent to this incident the Russian space agency found it hard to afford its maintenance. In February 2001, the Mir space station was de-orbited. It re-entered the atmosphere on 23 March 2001 and was burned, with its debris falling over the South Pacific Ocean in Eastern Australia, marking the end of the Mir space station.

16.5.3 Currently Operational Systems

The following paragraphs discuss currently operational space stations, including the US ISS and Tiangong-1 of China.

16.5.3.1 International Space Station

The ISS is the result of former US President Ronald Reagan's proposal in 1984 to build a permanently inhabited space station jointly with other countries. The USA, in order to meet the enormous expenditure involved in building such a massive station, joined with 14 other countries, with NASA leading the coordination efforts for the construction of the ISS with participating nations, who included Brazil, Japan, Canada and 11 members of the European Space Agency (ESA), namely the UK, France, Germany, Belgium, Italy, the Netherlands, Denmark, Norway, Spain, Sweden and Switzerland. Russia joined the consortium in 1993 after the fall of the Soviet Union. Spearheaded by NASA, the ISS mission is a joint programme of five space agencies including NASA, Roskosmos, JAXA, ESA and CSA.

The construction work on the US$ 60 billion ISS programme began in 1998 with the launch of the first module, called the *Zarya cargo module*, on 20 November 1998 aboard the Proton rocket from the Baikonur Cosmodrome in Kazakhstan. This was followed by the launch of the first US built module for the ISS, named Unity and also called Node-1, on 4 December 1998 onboard the STS-88 flight of Space Shuttle Endeavour. It mated with the in-orbit Zarya module on 6 December 1998. Unity was one of the three connecting modules, the other two being Harmony and Tranquility. Harmony, also called Node-2, was launched on 23 October 2007 onboard the STS-120 flight of *Space Shuttle Discovery* from the John F. Kennedy Space Centre. The third module, named Tranquility or Node-3, was launched onboard the STS-130 flight of Space Shuttle Endeavour on 8 February 2010 from the John F. Kennedy Space Centre. A multi-purpose laboratory module with a European Robotic Arm (ERA) is scheduled for launch in 2015 onboard the Russian Proton rocket. Between 1998 and 2013 a large number of missions were carried out to build the ISS bit by bit. Some of the prominent modules launched during this period included the *Zvezda service module* in 2000, the *Destiny laboratory module* in 2001, ESA's *Columbus laboratory* in 2008, and *multipurpose modules Leonardo* and *Robonaut* in 2011. The funding for the ISS programme is available until 2020 and the project is likely to continue until 2028. Figure 16.18 shows a recent view of the ISS.

The ISS has been in low Earth orbit maintaining an altitude between 330 and 435 km and inhabited on a continuous basis for the 13 years since the arrival of the first expedition on 2 November 2000. The space station is serviced by *Soyuz spacecraft* used to transport crew between Earth and the space station, *Progress spacecraft* to carry fuel and other supplies to the space station, *automated transfer vehicles* (ATVs) designed to supply the space station with propellant, water, air, payloads and experiments, and also reboost the space station to a higher orbit, *H-II transfer vehicles* used as a unmanned resupply spacecraft, the *Dragon spacecraft* also designed as a cargo spacecraft and *Cygnus spacecraft* designed to transport supplies.

The ISS serves as an in-space research laboratory providing a microgravity environment on a long term basis to perform a variety of scientific experiments in physics, biology, materials science, meteorology, Earth observation and astronomy. It is also an ideal platform for testing spacecraft systems and other equipment needed for planetary missions.

Figure 16.18 Recent view of the International Space Station (Courtesy: NASA)

16.5.3.2 Tiangong-1 Space Station

Tiangong-1 is the first of the series of three Chinese space stations with the ultimate goal of setting up a large modular manned space station. Tiangong-1 was launched on 29 September 2011 onboard the Chinese Long March-2F launch vehicle from the Jiuquan Satellite Launch Centre. Tiangong-1 is intended to be both a manned space laboratory and an experimental test bed to demonstrate docking and rendezvous capabilities. The space station consists of three sections including the *aft service module*, a *transition section* and the *habitable orbital module*. The service module is based on Shenzhou spacecraft and provides electrical, environmental control and propulsion subsystems. The orbital module provides living and working space for the visiting crew. The transition section is an interface between the service module and the orbital module.

Unmanned docking capabilities were established first on 3 November 2011 and again on 14 November 2011 during an unmanned mission of Shenzhou-8 flight launched on 1 November 2011 on board the modified Long March-2F launch vehicle from the Jiuquan Satellite Launch Centre. Subsequent to this, a manned docking capability was established during the Shenzhou-9 mission in 2012. Shenzhou-9 was launched on 16 June 2012 on board the Long March-2F launch vehicle, from the Jiuquan Satellite Launch Centre. The first Chinese manned spacecraft docking and rendezvous was established on 18 June 2012 during the Shenzhou-9 mission. Manned spacecraft docking capability was established again during the Shenzhou-10 mission launched on 11 June 2013. Shenzhou-10, carrying three astronauts, docked with Tiangong-1 on 13 June 2013. The crew performed a series of experiments while onboard Tiangong-1 and the spacecraft returned to Earth on 26 June 2013. The Shenzhou-10 manned spacecraft mission was the last visitor that the Tiangong-1 was scheduled to receive during its mission. Tiangong-1 had a planned mission life of two years and is planned to be de-orbited in 2014. Tiangong-1 will be followed up by the launch of Tiangong-2 and Tiangong-3, which are scheduled for launch during 2015–2016. The Tiangong-3 mission will be followed by the launch of a full scale, multi-module space station in the early 2020s.

16.5.4 Planned Space Stations

The following paragraphs briefly discuss some prominent future space station programmes, including China's *Tiangong-2*, *Tiangong-3* and the *multi-module space station*, the *Bigelow Commercial Space Station* of Bigelow Aerospace the *Almaz commercial programme* of Excalibur Almaz, the *OPSEK programme* of the Russian Space Agency and the *commercial space station* of Orbital Technologies of Russia.

16.5.4.1 Tiangong-2 and -3 and multi-module space station

Tiangong-2 was originally conceived as a back-up space laboratory to Tiangong-1. Subsequently, it was decided to have Tiangong-2 with an improved design featuring an orbital fuelling system, which will enable the space station to be refuelled by a cargo vehicle. The cargo vehicle will deliver both dry and wet cargo to the space station. This automated cargo spacecraft will be used to transport three types of cargo to the space station, including air, water and propellant for the maintenance of the station itself, food and other materials for the crew members on board the space station, and equipment for scientific experiments. The cargo vehicle may also be used to assist the space station for orbit maintenance. Tiangong-2 is also proposed to be used for testing a robotic arm, which will be used subsequently on future space stations. The spacecraft is scheduled to be launched in 2015.

Tiangong-3 is proposed to be a third generation project employing a modular space station concept. It is scheduled to be launched during 2015 following the launch of Tiangong-2 in 2013. Tiangong-3 is proposed to be visited by a number of unmanned and manned spacecraft missions. The space station is also proposed to provide continued habitation of a crew of three astronauts for as many as 40 days, a multi-docking and berthing mechanism enabling simultaneous docking of up to four spacecraft and a platform for testing regenerative life support technology. Docking and berthing mechanisms are used to connect one spacecraft to another spacecraft or space station. In a docking mechanism, one of the spacecraft uses its own propulsion to manoeuvre and connect to the other spacecraft. In the case of berthing, a robotic arm is used for the final few metres of rendezvous. The design of Tiangong-3 will also form the basis of China's multi-module space station, scheduled for launch during the early 2020s. The proposed multi-module space station will primarily comprise a *core cabin module (CCM)* analogous to the Mir core module, two *laboratory cabin modules* (LCM-1 and LCM 2) to be used to perform scientific experiments under microgravity conditions, a robotic resupply craft to be used to transport supplies and other resources to the space station, and a manned Shenzhou spacecraft to be used to ferry crew members between Earth and the space station.

16.5.4.2 Bigelow Commercial Space Station

The *Bigelow Commercial Space Station* is currently under development at a space technology start-up company in North Las Vegas, Nevada, USA. Bigelow Aerospace has already announced the development of *Commercial Space Station (CSS) Skywalker*, *Space Complex*

Alpha and *Space Complex Bravo*. CSS Skywalker comprises multiple BA330 habitat modules. Multiple modules would be inflated and connected on reaching orbit. Space Complex Alpha (Figure 16.19) is scheduled for launch during 2014. Space Complex Alpha comprises two BA330 modules and seven nations, including the UK, the Netherlands, Australia, Singapore, Japan, Sweden and the United Arab Emirates will be using the on-orbit facilities of the commercial space station. Bigelow Aerospace has also announced the launch of another commercial space station called Space Complex Bravo in 2016. Commercial operations of Space Complexes Alpha and Bravo are scheduled to begin in 2015 and 2017, respectively.

Figure 16.19 Space complex Alpha (Courtesy: NASA)

16.5.4.3 Almaz Commercial Programme

The *Almaz commercial programme* of Excalibur Almaz Inc. is intended to provide safe, reliable and competitively priced space transportation services including cargo and crew delivery and return to low Earth orbit, human space flight, microgravity experimentation and trans-lunar trajectory operations using existing flight-tested hardware components. The hardware proposed to be used consists of proven Russian space modules and space stations. The first launch is scheduled for 2015.

16.5.4.4 OPSEK (Orbital Piloted Assembly and Experimental Complex)

OPSEK (Orbital Piloted Assembly and Experimental Complex) is a third generation modular space station from Russia with scheduled launches of various modules between 2010 and 2020 from the Baikonur Cosmodrome. It will be placed in a low Earth orbit at an altitude of 370–450 km. The OPSEK space station would provide a platform to assemble components of manned interplanetary spacecraft and launch missions to the moon, Mars and other planets. The OPSEK space station is also proposed to be used for recovery of crew on such interplanetary missions before they are transported to Earth.

16.5.4.5 Commercial Space Station

Two Russian aerospace companies named Orbital Technologies and RSC Energia have teamed up for the development of an orbital space station for the commercial market. This orbital space station, called the Commercial Space Station, is proposed to be launched in a low Earth orbit with an altitude of about 350 km during 2016. This new space station, rightly called a space hotel, is proposed to have enough space to house seven passengers in four cabins with huge windows to view the turning Earth beneath. In addition, the services will also be open to private and state spaceflight exploration missions. Figure 16.20 shows a photograph of the commercial space station.

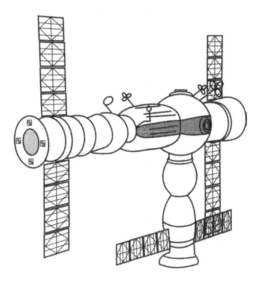

Figure 16.20 Commercial space station

16.5.5 Emerging Space Station Concepts

The process of space station design and development has been continuously evolving in the last four decades beginning with the era of Salyut- and Skylab-like monolithic space stations built on Earth and launched as one unit into space and then moving onto modular space stations such as ISS of USA built by launching individual modules and assembling them together in space. There is still a long way to go before the ultimate objective of having much larger habitable space stations and space colonies becomes a reality. While space tourism, close to being a reality, is one small but significant step in that direction, a number of technologies would need to be developed and perfected before people in large numbers could actually think of living and working in space. For example, space stations would need to have artificial gravity. For this to happen, placing the space station at Larange points where Earth's gravity and the moon's gravity counterbalance each other could be a possible solution. Similarly are many more areas related to construction of large space stations and transportation of people between Earth and space in which considerable work needs to be done. According to an estimate by a company likely to offer space travel on a commercial scale, the journey per person to a space station and a stay there for five days would cost approximately US$ 1 million. The following paragraphs briefly present some emerging space station concepts. Some of them have already been put to use in recent years and are evolving, and some are still on the drawing board.

A recent concept that has allowed the construction of large space stations is that of *modular space stations*. In the case of modular space stations, the space station is broken down into a core module and various other modules. The modules are launched individually and are then assembled in space over a period of time that could even spread to several years. Tiangong-2, Tiangong-3 and the proposed multi-module space station of China and also the US ISS are all examples of modular space stations. Advances in launch vehicle technology will allow bigger modules to be launched in future, thereby enabling construction of much bigger space stations.

The use of *space tethers*, though unconventional, is another emerging concept which in addition to performing tasks of orbit stabilization, orbit re-boost and de-orbiting would also make it very convenient to transport materials and personnel to space with ease. Space tethers are discussed in section 16.2. Space elevators, based on the space tether concept and discussed in section 16.2.4, are expected to make space transportation a daily affair in the coming years.

The concept of *inflatable space modules* will allow the launch of space modules having a bigger inner volume for a given mass thereby providing greater living space. An inflatable module will be similar to conventional cylindrical modules but with an inflatable exterior shell. This exterior shell will be made to inflate with a breathable atmosphere within once the module is in orbit. The inflated module dimension will be much larger than the dimensions of the module on Earth at the time of launch thereby providing extra living space. The TransHab concept tried by NASA for a possible manned mission to Mars is an example. TransHab design is being further developed by Bigelow Aerospace for their commercial space station programme. The concept was first tested by them in their Genesis-I module, launched on 12 July 2006. It was the first expandable space habitat technology in orbit. Genesis-I was followed by the launch of Genesis-II (Figure 16.21) on 28 June 2007. The concept of an inflatable module is proposed to be used in Space Complex Alpha of Bigelow Aerospace.

Another futuristic concept that can significantly contribute to realization of the dream of space colonies is the concept of *space docking*. The concept of a space dock originates from *dry docks* used for the construction and maintenance of ships, boats and water craft. A dry dock is a narrow vessel that allows a load to be floated in when flooded and then made to come to rest on a dry platform when drained. Space docks can be used for the construction and maintenance of space ships. A large space dock would make it possible to carry out in-orbit assembly and maintenance of large space stations. It would also allow enclosed maintenance of smaller spacecraft. In the absence of any dedicated in-orbit construction and maintenance facility, one would need to frequently lift materials from the gravity well while engaging in the construction of a large space station or make a planetary landing for the purpose of maintenance. Such an in-orbit repair facility would also be an important asset in the case of damaged spacecraft, which pose a serious risk during re-entry into the atmosphere.

The concepts briefly described in the previous paragraphs in this section are the ones that have either been realized or are close to being realized. For technologies that are not mature enough to use yet, significant progress has been made in their development towards achieving the intended objectives. In addition to these concepts, many more have been proposed and

Figure 16.21 Genesis-II space module (Courtesy: Bigelow Aerospace)

explored by science fiction writers and scientists all over the world. A large number of these concepts are proposed with the objective of achieving conditions for long term or permanent habitation of space. Overcoming weightlessness is the key requirement. Use of a *rotating wheel* as proposed by the Austro-Hungarian scientist Hermann Noordung in his 1929 book for his geostationary space station appears to be a simple solution to solving the complex problem. The centrifugal force associated with the rotating wheel sets up artificial gravity, which overcomes the problems of long term habitation in space due to weightlessness. The concept design for the building of the Cultural Centre of European Space Technologies has been derived from Noordung's habitation wheel. The concept of a spinning wheel has been extended further in the *Stanford Torus* proposal of Stanford University that examines how to build a permanent and self-sufficient orbital habitat. Some of the other well known spinning space station concepts have proposed the use of cylindrical (e.g. the *O'Neill cylinder*) or spherical (e.g. the *Bernal sphere*) structures.

Further Readings

Edwards, B.C. and Westling, E.A. (2003) *The Space Elevator: A Revolutionary Earth-to-Space Transportation System*, BC Edwards.

Ippolito, L.J. (2008) *Satellite Communications Systems Engineering*, John Wiley and Sons Ltd, Chichester.

Kao Cheng, H. and Zhaocheng, W. (2011) *Millimeter Wave Communication Systems*, Wiley-IEEE Press, New Jersey.

Mario, L., Cosmo, E. and Lorenzini, C. (1997) *Tethers in Space Handbook*, NASA Marshall Space Flight Centre.

Pelton, J.N. and Madry, S. (2013) Sergio Camacho-Lara, *Handbook of Satellite Applications*, Springer, New York.

Rappaport, T., Heath, R., Daniels, R. and Murdock, J. (2014) *Millimeter Wave Wireless Communication*, Prentice-Hall, New Jersey.

Tashirogi, T. and Yoney Ama, T. (2000) *Modern Millimeter Wave Technologies*, IOS Press, Amsterdam.

Van Pelt, M. (2009) *Space Tethers and Elevators*, Copernicus/Praxis Publishing, New York.

Internet Sites

1. www.tethers.com
2. www.nasa.gov/centers/marshall/capabilities/space_tethers.html
3. www.spacetethers.com
4. www.microwavejournal.com/.../1594-millimeter-wave-applications
5. www.scifiideas.com/sfi/technology/10-space-station-concepts
6. science.howstuffworks.com/space-station6.htm
7. www.loeacom.com/L1104-WP_Understanding%20MMWCom.pdf

Glossary

Aerostat: Aerostat is an aircraft, balloon or dirigible, deriving its lift from the buoyancy of the surrounding air rather than from aerodynamic motion. The name aerostat is derived from the fact that it operates on aerostatic lift, also called buoyant force, and not aerodynamic lift, essentially requiring at least some part of the aerial body to move through the surrounding air mass.

Airship: Airship is a lighter-than-air aircraft that is steered with the help of some thrust mechanisms such as rudders and propellers.

Applications Technology Satellite-6 (ATS-6): This is one of the very ambitious experimental communications satellites developed and implemented by the NASA Goddard Space Flight Centre and intended to carry out a large number of scientific and technological experiments in the geostationary space environment.

Atmospheric Tether Mission (ATM): An experimental mission proposed to be executed from the Space Shuttle in which a tethered probe would be lowered successively into different regions of the atmosphere such as the mesosphere, thermosphere and ionosphere for the purpose of understanding the atmosphere and plasma around the Earth in these regions.

Bigelow Commercial Space Station: A commercial space station under development at Bigelow Aerospace.

Commercial Space Station (CSS): A commercial space station of two Russian companies scheduled to be launched in 2016.

Communications and Broadcasting Engineering Test Satellite (COMETS): Also known as KAKEHASHI, COMETS is a three-axis stabilized geostationary satellite of the National Space Development Agency (NASDA) of Japan.

Electrodynamic tether: It is a long conducting cable extended from a spacecraft and kept stretched and oriented along the vertical direction by the gravity gradient force. An electrodynamic tether generates thrust through Lorentz-force interaction with the planetary magnetic field, which can be used to execute orbital manoeuvres without the need for carrying large quantities of propellant into orbit.

Engineering Test Satellite-II (ETS-II): This was the first geostationary satellite that carried a beacon transmitter with three coherent frequencies at 1.7 GHz, 11.5 GHz and 34.5 GHz (millimetre wave frequency) to perform propagation experiments.

Experimental communications satellites (ECS): Experimental satellites from NASDA intended to carry out communications and propagation experiments at millimetre wavelengths. Also known as AYAME satellites

FLORAD mission: A flower constellation of micro-satellites of the Italian Space Agency carrying onboard scanning millimetre-wave radiometers for Earth and space observation on a regional and quasi-global scale.

Helikite: One of the most commonly used aerostat designs for all-weather, high altitude applications. It is a combination of a helium balloon that uses helium gas for buoyancy and a kite that exploits wind for lift.

Hybrid tether: A momentum-exchange/electrodynamic-reboost tether used to transfer payloads from low Earth orbit into medium Earth and geosynchronous orbits.

Inflatable space station: A type of expandable space module that provides a larger inner volume for a given launch mass.

International Space Station (ISS): A permanently inhabited modular space station developed jointly with 14 countries with NASA as the nodal centre.

Millimeter wave band: Frequency band from 30 GHz to 300 GHz.

Mir: First permanent space station owned by the former Soviet Union at the time of its launch and subsequently by Russia.

Momentum exchange tether: Used to couple two objects in space in such a way that one can transfer momentum or energy to the other taking advantage of the gravity gradient force that exists due to differential gravitational forces between the two ends of the tether. It is this differential gravitational force, called the gravity gradient force, that keeps the tether taut and the two objects tied to the two ends of the tether pulled apart.

Odin: An aeronomy and astronomy mini-satellite mission funded and executed by the Swedish National Space Board jointly with the space agencies of Canada, France and Finland.

OEDIPUS: An acronym for Observations of Electric-field Distribution in the Ionospheric Plasma - a Unique Strategy, which is a joint program of the National Research Council of Canada and NASA.

Orbital Piloted Assembly and Experimental Complex (OPSEK): A third generation modular space station from Russia.

Plasma Motor Generator (PMG): An electrodynamic tether payload of a Department of Defence satellite launched on 26 June 1993 as a secondary payload onboard a Delta-2 rocket with the objective of testing the performance of hollow cathode assembly to provide a low impedance bidirectional electrical current and demonstrate the application of an electrodynamic tether for space propulsion and conversion of orbital energy into electrical energy.

Rotating wheel: A futuristic concept that allows long term habitation of space without the problems of weightlessness.

Skylab: First space station of the USA.

Salyut: A monolithic space station put into space by the former Soviet Union

Small expendable deployer system (SEDS): SEDS was developed by NASA's Marshall Space Flight Centre, which is primarily responsible for the development of transportation and propulsion technologies, and the Tether Application Company of San Diego.

Space dock: A type of space facility that allows in-orbit repair and maintenance of space stations.

Space elevator: Provides a concept for the development of a new space transportation system. The proposed concept has the potential to make transportation of tonnes of payloads and resources to geostationary Earth orbit and beyond an almost daily affair.

Space tether: A long and strong cable usually made up of thin strands of high strength conducting wires or fibres and used to generate thrust or power for satellite stabilization and maintenance of spacecraft formations.

Space Tether Experiment (STEX): STEX is a science and technology experiment of the Institute of Space and Aeronautical Science to be flown on board the Space Flight Unit (SFU) mission. SFU is an unmanned, multi-purpose reusable platform that can be used for performing a range of science and technology experiments and carrying out flight tests of space and industrial technologies.

STEP-AIRSEDS: An acronym for Space Transfer Using Electrodynamic Propulsion – Atmospheric Ionospheric Research Small Expendable Deployer System. It is a mission satellite developed by Michigan Technic Corporation, USA weighing 1000 kg and comprises of a tether. It consists of two units tethered together by an electrodynamic tether approximately 6–7 km long.

Tethered balloon: It is not free flying like a free balloon and its movement is restricted by anchoring it to a ground surface or a vehicular platform by a cable or a set of cables.

Tether Physics and Survivability (TiPS): An experimental payload built and operated by the Naval Centre for Space Technology of the Naval Research Laboratory and launched on board the Titan-4 launch vehicle on 12 May 1996.

Tethered Satellite System (TSS): A collaborative programme of NASA and the Italian Space Agency (ASI) with the objective of developing a reusable multi-disciplinary facility to conduct space experiments in Earth orbit.

Tiangong Space Station: A series of three space stations, namely Tiangong-1, Tiangong-2 and Tiangong-3. Tiangong-1 is the first of the series of three Chinese space stations with the ultimate goal of setting up a large modular manned space station. Tiangong-2 and Tiangong-3 are to follow Tiangong-1.

Index

A3E system 235–6
ABL system 760–1
Accelerometer 659
ACeS (Asia cellular satellite) 512
Access track scanner, see *optical mechanical scanner*
Active altitude control 200
Active remote sensing 526
Active sensors (*Remote sensing satellites*) 540–2
   Non-scanning sensors 540
   Scanning sensors 540–2
Active thermal control 187–8
Adaptive delta modulator 265
Adjacent channel interference 361–2
Advanced microwave sounding unit (AMSU) 585
Aerial remote sensing 525
Aerostat system 781–3
   Airship 783
   Applications 783
   Components of 782
   Free balloon 782
   Helikite 782
   Moored balloon 782
   PAGEOS 782
   Tethered balloon 782
   Types of 782
   Vega-1/2 782
Akatsuki orbiter 688
Almaz commercial programme 800
ALOHA 501
   Slotted ALOHA 501

Amplitude modulation 231–40
   Different forms 235–40
   Frequency spectrum 232
   Noise 233–4
   Power 233
Amplitude phase shift keying 268
Analogue pulse communication systems 259–61
   Pulse amplitude modulation 259–60
   Pulse position modulation 260–1
   Pulse width modulation 260
Anatomy (*Launch vehicles*) 104–6
   Liquid fuelled engine 105
   Solid fuelled engine 105
   Stages 104
Anik-A 17
Anik-B 17
Anik-C 17
Anik-D 17
Anik-E 17
Anik-F 17
Anik-G 17
Antenna gain-to-noise temperature (G/T) ratio 365
Antenna noise temperature 346–50
   Ground noise 347
   Sky noise 348–9
Antenna parameters 207–10
   Aperture 209–10
   Bandwidth 208
   Beam width 208
   EIRP 208
   Gain 207–8
   Polarization 209

Antennas 205–20
  Parameters 207–10
  Types 210–20
Antenna types 201–20
  Helical antenna 215–7
  Horn antenna 214–5
  Lens antenna 217–8
  Phased array antenna 218–20
  Reflector antennas 210–4
Apogee 47
Apogee distance 82, 87–90
Applications (*Military satellites*) 718
Applications (*Navigation satellites*) 650–4
  Civilian applications 651–4
  Military applications 650–1
Applications (*Remote sensing satellites*) 548–58
  CHAMP (Challenging mini-satellite payload) 670
  Deforestation 553
  Disaster prediction 555
  Flood monitoring 551
  Global monitoring 553–5
  GRACE (Gravity recovery and climate experiment) 669
  Land cover change detection 549–50
  Land cover classification 548–9
  Sea surface temperature measurement 552
  Urban monitoring and development 552
  Water quality management 550
Applications (Scientific satellites) 665–70
  Dynamic geodesy 666
  Geometric geodesy 666
  Internal geodynamics 669–70
  Joint gravity model 667
  Space gradiometry 668
  Space geodesy 665–9
  Tectonics 669–70
  Terrestrial magnetic fields 770
Applications (Weather forecasting satellites) 593–9
  Air pollution 596
  Cloud parameters 594
  Fisheries 598
  Fog 596
  Ground level temperature 596
  Haze 596
  Oceanography 596–7
  Prediction of hurricanes, cyclones and tropical storms 597–8
  Rainfall estimation 594–5
  Snow and ice studies 598–9
  Wind speed and direction 595–6
Arabsat-5 18, 512
ARCJET thruster 181–2
Argument of perigee 50, 81
Arthur C. Clarke 4
Ascending nodes 45
  Right ascension of 48–9
AsiaSat 512
Astronomical observations 680–6
  Apollo telescope mount (ATM) 682
  COBE (Cosmic background explorer) 681
  CORONAS (Complex orbital observations of near earth activity of the sun) 682
  Doppler-Fizeau effect 681
  EXOSAT (European X-ray observatory satellite) 681
  Extreme ultraviolet imaging telescope (EIT) 683
  HETE-2 (High energy transient explorer) 681
  Hubble space telescope 681
  INTEGRAL (International gamma ray astrophysics laboratory) 681
  IRIS (Interface region imaging spectrograph) 682
  IRAS (Infrared astronomical satellite) 681
  ISO (Infrared space observatory) 681
  LASCO (Large angle spectroscopic coronograph) 683
  Michelson Doppler imager (MDI) 683
  Observation of sun 682–6
  Orbital solar observatory (OSO) 682
  Solar activity 684–6
  Solar maximum mission (SMM) 683
  Solar physics 683
  STEREO (Solar terrestrial relations observatory) 682
  TRACE satellite 684
Asynchronous transfer mode (ATM) 459
Atlas series rockets 29, 115–7
Atmospheric tether mission (ATM) 778
Attenuation compensation techniques 341–2
  Diversity 341–2
  Power control 341
  Signal processing 341
Attitude control 199–200
Automatic tracking 412
Availability (*Networks*) 434
Azimuth angle 81

Index

B8E system 238
Backscattering coefficient 547
Baikonur cosmodrome 133–4
Balanced modulator 237–38
Balanced slope detector 252–3
Bath tub curve 225
Batteries 195–8
  Lithium ion 197–8
  Nickel-Cadmium 195–6
  Nickel hydrogen 197
  Nickel metal hydride 196–7
Bernal sphere 803
Bigelow commercial space station 799–800
Bit error rate test (BERT) (*Network Reliability*) 435
Black powder propellants 108
Block bit error rate test (BLERT) (*Network Reliability*) 435
Bridges 465
Broadcast satellite service (BSS) earth stations 382
Bukit Timah satellite earth station 426
BURAN 29, 127
Bus topology 442–3

C3F system 238
Cable TV 485
Canberra deep space communication complex 422–3
Cape Canaveral satellite launch centre 98
Candy propellants 108
Cartosat-1, 2 22
Cassini/Huygens 691
Central perspective scanner 538
Centrifugal force 39
Centripetal force 39
Circuit switched networks 447–8
Civilian applications 651–4
  Archeology 653
  Construction 651
  Environment monitoring 653
  Mapping 651–52
  Monitoring structural deformations 653
  Precision farming 653
  Precise timing information 653–4
  Saving lives 652
  Vehicle tracking 652
Classification (*Launch vehicles*) 100–4
  Expendable 101
  Heavy lift launch vehicle 102
  Large launch vehicle 102
  Medium launch vehicle 102–4

  Re-usable 102
  Small launch vehicle 104
Classification (*Remote sensing satellites*) 526–31
Code division multiple access 308–15
  Comparison DS-CDMA, FH-CDMA, TH-CDMA 314–315
  Direct sequence CDMA (DS-CDMA) 309–311
  Frequency hopping CDMA (FH-CDMA) 311–3
  Time hopping CDMA (TH-CDMA) 313–34
Commercial space station 801
Communication satellite missions 502–14
  INMARSAT satellite system 506–8
  INTELSAT satellite system 502–6
  IRIDIUM satellite constellation 508–12
  National satellite systems 513–4
  Regional satellite systems 512–3
Communication satellites 473–523
  Applications 474–5
  Frequency bands 475
  Future trends 514–9
  Missions 502–14
  Payloads 475–9
  Satellite data communication services 496–501
  Satellite radio 496
  Satellite telephony 481–3
  Satellite television 484–96
Compass satellite navigation system 648
Composite propellants 108
Compton gamma ray observatory (CGRO) 707
Computer vandalism (*Network security*) 436
Conditional access (*DVB*) 490
Conical scan 415
Constituent parts (Space launch centres) 128–9
  Control centre 128–9
  Launch complex 128–9
  Technical centre 128–9
Cost (*Networks*) 437
Courier-1B 10
Cross polarization 209

Data collection system (DCS) 602
Data compression (*DVB*) 490
Data interception (*Network security*) 436
Data signals 230
Dawn (flyby) 690
De-emphasis 249–50
Delta series rockets 29, 114–5
Demand assigned multiple access 287, 290

De-polarization compensation techniques 342
Descending nodes 45
Differential absorption lidar 586
Digital modulation techniques 267–77
   Amplitude phase shift keying 268
   Frequency phase shift keying 268–9
   Phase shift keying 269–77
Digital pulse communication systems 261–5
   Adaptive delta modulator 265
   Delta modulation 264–5
   Differential PCM 264
   Pulse code modulation 262–3
Digital video broadcasting 490–5
   DVB-C 491
   DVB-H and DVB-SH 494–5
   DVB-RCS and DVB-RCS2 493
   DVB-S and DVB-S2 491–2
   DVB-T and DVB-T2 493–4
Direct broadcasting satellite (DBS) services 488–9
Directed energy laser weapons 752–64
   Advanced tactical laser (ATL) 761
   Advantages 753
   Airborne laser (ABL) 760
   Beam control 764–5
   Chemical lasers 757–9
   Chemical oxy-iodine laser (COIL) 759–61
   Components of 754–5
   Design parameters 755–6
   Fiber laser 762
   Gas dynamic CO_2 laser 756–7
   HF/DF laser 757–8
   Laser sources 756–62
   Limitations 753–4
   Solid state laser 761–2
Direct sequence CDMA (DS-CDMA) 309–10
   Sequence asynchronous DS-CDMA 311
   Sequence synchronous DS-CDMA 310–11
Disaster prediction (Remote sensing satellites) 555–7
   Predicting earthquakes 555–7
   Predicting volcanic eruptions 557
Doppler lidar 586
Double base propellants 108
Dry dock 802

Early bird 13
Early warning satellites 735–8
   MIDAS system 736
   SBIRS (Space based infrared system) 736–7
   SPIRALE 738

Earth coverage 166–70
   Effects of altitude 169
   Effects of latitude 169
Earth's observation (Scientific satellites) 670–80
   Aeronomy 677–8
   ALOS (Advanced land observing satellite) 680
   AMPTE (Active Magnetospheric particle tracer explorer) 678
   CERES (Cloud and earth radiant energy sensor) 679
   Earth's radiation budget) 679
   ERBE (Earth radiation budget experiment) 679
   ERBS (Earth radiation budget satellite) 678
   Ionosphere 671–3
   Ionospheric composition 671
   Magnetosphere 673–6
   Magnetospheric waves 675–6
   MAPS (Measurement of air pollution from satellites) 677
   Ozone measurements 679
   Polar aurora 671–3
   SABER (Sounding of the atmosphere using broadband emission radiometers) 678
   TIMED (Thermosphere ionosphere mesosphere energetics and dynamics) 678
   UARS (Upper atmosphere research satellite) 677
Earth station 378–430
   Architecture 386–7 426
   Bukit Timah satellite earth station 426
   Canberra deep space communication complex 422–3
   Design considerations 387–8
   Environmental considerations 391–2
   Goldstone deep space communication complex 423–4
   Goonhilly satellite earth station 419–21
   Hardware 398–411
   Honeysuckle Creek tracking station 424
   Indian deep space network 429–30
   INTELSAT teleport earth station 426–27
   Kaena point satellite tracking station 426
   Madley communication centre 421
   Madrid deep space communication complex 421–2
   Makarios satellite earth station 428
   Performance parameters 388–90
   Raisting earth station 428

Site considerations 391–2
SUPARCO satellite ground station 428
Terrestrial interface 409–11
Testing 392–8
Types 380–6
Earth station characteristics 380
  Access method 380
  Antenna diameter 380
  EIRP 380
  Frequency band 380
  G/T ratio 380
  Modulation type 380
  Polarization 380
Earth station hardware 398–411
  Antenna 399–401
  High power amplifier 402–4
  IF and base band equipment 408–9
  Low noise amplifier 406–7
  RF equipment 398–407
  Terrestrial interface 409
  Up converter/Down converter 404–406
Earth station testing 392–8
  Subsystem level testing 392
  System level testing 392–8
  Unit level testing 392
Eccentricity 48
Echo-1 10
Echo effects (*Satellite networks*) 480
Eclipses 150–4
  Equinoxes 151
  Solar eclipse 151
  Umbral cone 152
ECS series satellites 17
EGNOS satellite navigation system 642–4
Electrodynamic tether 772–3
  Applications 774
EIRP-Earth station 389–90
Ekran series satellites 16
Electric propulsion 181
Energetic proton, electron and alpha detector (EPEAD) 603
Electronics launching group (ELG) 490
Energetic particle sensor (EPS) 603
Energia series rockets 28, 113–4
Equinoxes 45–6
ESSA series 19
ESTRACK network 202
EUTELSAT-II 17
EUTELSAT satellite systems 514
EUTELSAT-W 17

Explorer-1 8
Explorer-2 9
Extreme ultraviolet sensor (EUV) 603

Femto satellites 26–7
Feng Yun series 20
File transfer protocol 457–8
Final mass 106
First cosmic (or orbital) velocity 61
Fixed satellite service (FSS) earth stations 381
Flood monitoring (Remote sensing satellites) 551
FM detectors 252–7
  Balanced slope 253
  Foster-Seeley 254
  Phase locked loop based 256–7
  Ratio detector 254–5
Formation flying tethers 773–4
  Separated spacecraft interferometer (SSI) 773
  Space interferometry mission 773
  SPECS mission 774
  Structurally connected interferometer (SCI) 773
  Terrestrial planet finder 773
  Tethered formation flight interferometer 773
Frame acquisition and synchronization 305–6
  Extraction of traffic bursts 305
  Frame synchronization 305–6
  Transmission of traffic bursts 305
Frequency bands 475
Frequency considerations, satellite link design 324, 326–9
  Frequency allocation and coordination 326–9
Frequency division multiple access 288–96
  Demand assigned FDMA 290
  Multiple channel per carrier (MCPC) systems 295–6
  Pre-assigned FDMA 290–2
  Single channel per carrier (SCPC) system 293–5
Frequency modulation 241–57
  De-emphasis 249–50
  Detection of FM signals 252–7
  Frequency spectrum 243–5
  Generation of FM signals 250–2
  Narrowband FM 245–6
  Noise 246
  Pre-emphasis 249–50
  Wideband FM 245–6
Frequency shift keying 268–9
Fresnel lens 218

Future trends (*Remote sensing satellites*) 573–4
Future trends-satellites 33–5
   Communication satellites 33
   Earth observation satellites 33–4
   Military satellites 35
   Navigation satellites 34
   Weather forecasting satellites 33
Future trends (*Scientific satellites*) 714

Galileo satellite navigation system 465, 645–6
   Development program 645–6
   Services 646
Gateway stations 385
Geosynchronous satellite launch vehicle (GSLV) 124
GERB instrument 606–7
German aerospace centre 141–3
Global mobile personal communication services (GMPS) 483
Global monitoring 553–5
Global navigation satellite system 464
Global positioning system (GPS) 464, 621–37
   Control segment 622–3
   GPS error sources 634–7
   GPS positioning services 631–4
   GPS signal structure 627–8
   Operational principle 625–7
   Pseudo range measurement 628–9
   Receiver location determination 629–30
   Space segment 621–2
   User segment 623–5
Globalstar constellation 461–2
GLONASS navigation system 637–41
   Control segment 638–9
   Signal structure 639–41
   Space segment 638
   User segment 639
GOES 599–604
   Payloads 602–4
   Salient features 601
Goldstone deep space communication complex 423–4
Goonhilly satellite earth station 419–21
Gorizont series satellites 16
GPS error sources 634–7
   Clock errors 635
   Ephemeris errors 635
   Multipath reflections 635
   Satellite geometry 636

Selective availability 636
Signal propagation errors 634–5
GPS positioning modes 632–4
   Point positioning 632
   Relative positioning 632–4
GPS positioning services 631–4
   Precision positioning system 632
   Standard positioning system 632
Gravity gradient (*space tethers*) 770
Guiana space centre 133–5

H3E system 236–7
Hall thruster 182
HEAO (High energy astronomy observations) 707
High definition television (HDTV) 489
High energy composite propellants 108
High resolution infrared sounder (HIRS) 611
HIPPARCOS (High precision parallax collecting satellite) 707
Honeysuckle Creek tracking station 424
Hot air balloon 7
Hubble space telescope 691, 707
Hybrid navigation system 648–9
Hybrid propellants 107
Hybrid satellite 517
Hybrid topology 446–7
Hypergolic (*Propellants*) 108
Hyper text transfer protocol 457

Image classification (*Remote sensing satellites*) 545–6
   Supervised classification 545
   Unsupervised classification 545
Image interpretation 546–8
   Contextual information 546
   Geometrical information 546
   GIS in remote sensing 547–8
   Microwave images 547
   Optical images 546
   Radiometric information 546
   Spectral information 546
   Textural information 546
   Thermal images 546
Images 580–86
   Advanced microwave sounding unit (AMSU) 585
   Differential absorption lidar (DIAL) 586
   Doppler lidar 586
   Images formed by active probing 585–6
   IR images 582–3

Laser altimeter 586
Microwave images 584–5
Precipitation radar 586
Tropical rainfall measuring mission (TRMM) 586
Visible and infrared spin scan radiometry (VISSR) 583
Visible images 580–2
Water vapour images 583–4
Inclination 49, 80–1
Indian deep space network 429–30
Indian regional navigational satellite system (IRNSS) 647
Indian remote sensing satellite system 565–7
  Cartosat-1 566–7
  Oceansat-1 567
  Resourcesat-1 565–6
  Resourcesat-2 566
  Risat-1 567
Inflatable space modules 802
Initial mass 106
Injection point 79
Injection velocity 61
INMARSAT satellite services 509–10
INSAT series 17–18, 515
Intelligent tracking 419
INTELSAT-1 13
INTELSAT-2 15
INTELSAT-3 15
INTELSAT-4 15
INTELSAT-5 15
INTELSAT-6 15
INTELSAT-7 15
INTELSAT-8 15
INTELSAT-9 15
INTELSAT-10 15
INTELSAT satellites 503–6
INTELSAT teleport earth station 426–7
Interference related problems 353–64
  Adjacent channel interference 361–2
  Cross polarization interference 361
  Interference from adjacent satellites 357–60
  Interference from terrestrial links 357
  Inter-modulation distortion 354–7
International space station (ISS) 797
Internet protocol 456
Internet services 439–41
  Email 439
  File transfer protocol (FTP) 439
  Hypertext transfer protocol (HTTP) 439

Multicast streaming for content distribution 440
Simple mail transfer protocol (SMTP) 440–41
Transmission control protocol (TCP) 439
Voice over Internet phone (VoIP) 439–40
Interplanetary TV link 519
Inter satellite link (ISL) 511
Ionospheric effects 335–8
  Fading due to multipath signals 338–9
  Faraday Effect 335–7
  Scintillation 337–8
Ion propulsion 182
IRAS (Infrared astronomy satellite) 707
Iridium satellite constellation 461
ISTRACK 202–3

J3E system 237–8
JCSAT (Japanese communication satellite) 484
Jiuquan satellite launch centre 138–9
John F. Kennedy space centre 129–32
Juno (orbiter) 691
Jupiter-C 8

Kaena point satellite tracking station 426
Kepler's laws 41–44
  Kepler's first law 41–42
  Kepler's second law 42–44
  Kepler's third law 44

Land cover classification 549–50
Land cover change detection 549–50
Landsat 558–61
  Enhanced thematic mapper 560–1
  Multispectral scanner (MSS) 560
  Operational land imager 561
  Return beam vidicon (RBV) 558–60
  Thematic mapper 560
  Thermal infrared sensor 561
Launch sequence 95–9
  Expendable launch vehicle 95
  Launch from Baikonur 98–9
  Launch from Cape Canaveral 98
  Launch from Kourou 97
  Perigee motor 99
  Re-usable launch vehicle 95
  Space shuttle launch 99
  Types 95–9

Launch vehicles 100–27
  Anatomy 104–6
  Ariane series 109
  Atlas series 115–7
  Buran 127
  Classification 100–4
  Delta series 114–5
  Energia 113–4
  Geosynchronous satellite launch vehicle (GSLV) 124
  Long March 119–23
  Polar satellite launch vehicle (PSLV) 123–4
  Principal parameters 106–8
  Proton series 109–10
  Soyuz series 110–3
  Space shuttle 124–7
  Titan series 117–9
Liquid fuel propulsion 180–1
Liquid propellants 107
LITE (Laser-in-space technology experiment) 592
Lobe switching 413–15
Logical topology (*Networks*) 437
Long March series 119–23
Look angles 154–60
  Azimuth angle 154–5
  Elevation angle 155–6

Madley communication centre 421
Madrid deep space communication complex 421–2
Magellan probe 688
Magnetic electron detector (MAGED) 603
Makarios satellite earth station 428
Mandatory tests-Earth station 393–7
  EIRP stability 396
  Receiver figure-of-merit measurement 393–4
  Spectral shape 396
  Transmit cross-polarization isolation measurement 393
Mariner planetary mission 687
Mars 2 to 7 688–9
Mars reconnaissance orbiter 690
Mars exploration rover 690
Mars express 690
Mars global surveyor 689
Mars path finder 689
Mass fraction 106
Mean time between failures (MTBF) (*Network Reliability*) 435

Mean time to repair (MTTR) (*Network Reliability*) 435
Microwave remote sensing 529–31
Mean time between outages (MTBO) (*Networks*) 434
Medium en
Medium satellites 23
Mesh topology 444–5
Meteor series 19–20
Meteosat series 20, 605–7
  Payloads 606–7
Micro satellites 24–5
Microwave humidity sounder (MHS) 611
MIDAS 10rgy proton and electron detector (MEPED) 611
Military applications (*Navigation satellites*) 650–1
  Map updation 651
  Missile guidance 651
  Navigation 650
  Rescue operations 651
  Tracking 651
Military communication satellites 718–27
  AEHF (Advanced extremely high frequency) 724
  AFSATCOM (Air force satellite communication system) 721
  American systems 720–1
  AMOS-1 725
  Applications 718–9
  Defence satellite communication system (DSCS) 721–2
  DFH (Dong Fang Hong) series 726
  Dual use systems 727
  FH (Feng Huo) series 727
  FLTSATCOM system 723
  Frequency spectrum 726–7
  LEASAT 721
  Milstar system 724
  Mobile and tactical systems 723
  Parus 728
  Potok series 728
  Protected satellite systems 724
  Raduga series 728
  Russian systems 724–5
  SICRAL series 725
  Spacenet series 727
  Strela series 728
  TACSAT 723
  Telecom series 725
  Wideband systems 721–2

Index

Military navigation satellites 739
Military satellites 717–67
 Applications 718
 Early warning satellites 735–8
 Military communication satellites 718–27
 Military navigation satellites 739
 Military weather forecasting satellites 738
 Nuclear explosion satellites 738
 Reconnaissance satellites 728–32
 SIGINT satellites 732–5
 Space weapons 739–45
 Strategic Defence Initiative 745–52
Military weather forecasting satellites 738
Millimeter wave satellite communication 784–93
 Advantages 785
 Applications 788–9
 Millimeter wave band 784–5
 Missions 789–93
 Propagation considerations 787–8
Millimeter wave satellite missions 789–93
 Application technology satellite-6 (ATS-6) 791
 Communications and broadcasting Engineering test satellite (COMETS) 792
 Engineering test satellite-II (ETS-II) 790
 Experimental communication satellite (ECS) 790
 FLORAD 790–91
 Odin satellite mission 793
Mini satellites 23–4
Missions (*Remote sensing satellites*) 558–68
 Indian remote sensing satellite system 565–7
 Landsat 558–61
 Radarsat 564–5
 SPOT 561–4
Missions (Scientific satellites) 686–714
 Asteroids 705–6
 COMETS 706–7
 CONTOUR (Comet nucleus tour) 706
 Cosmic ray research 713–4
 Fundamental research 714
 GRAIL (Gravity recovery and interior laboratory) 704
 Jupiter 697–700
 LADEE (Lunar atmosphere and dust environment explorer) 704
 Life sciences 711
 Mars 694–96
 Material sciences 712

 Mercury 691
 MESSENGER 691
 Microgravity experiments 710–1
 Moon 703–5
 NEAR (Near earth asteroid rendezvous) 705
 Neptune 703
 Solar system 686–707
 Uranus 702
 Venus 692–3
Missions (*Weather forecasting satellites*) 599–611
 GOES 599–604
 Meteosat 605–7
 NOAA 608–11
Mobile satellite service (MSS) earth stations 383–4
Mobile satellite telephony 482–3
Molniya series rockets 28
Molniya series satellites 14–15
Momentum exchange tether 770–2, 774
 Applications 774
 Bolo 771
 Gravity gradient stabilization 771
 Rotovator 772
Monopulse track 416–19
MSL curiosity (rover) 690
Multiple access techniques 286–319
 Code division multiple access 308–15
 Frequency division multiple access 288–96
 Space division multiple access 316–9
 Time division multiple access 297–306
Multiple channels per carrier (MCPC) systems 295–6
 MCPC/FDM/FM/FDMA system 295–6
 MCPC/PCM-TDM/PSK/FDMA system 296
Multiple spot beam technology 371–4
Multiplexing techniques 277–82
 Code division multiplexing 281–82
 Frequency division multiplexing 277–9
 Time division multiplexing 279–80

Nano satellites 25–6
National satellite systems 513–4
 Indian national satellite 513–4
Navigation satellites 614–55
 Applications 650–4
 Compass satellite navigation system 648
 EGNOS satellite navigation system 642–4
 Future trends 654–5
 Galileo satellite navigation system 645–6
 Global positioning system (GPS) 621–37
 GLONASS navigation system 637–41

Navigation satellites (*Continued*)
  GPS-GLONASS integration 641–2
  Hybrid navigation system 648–9
  Indian regional navigational satellite system (IRNSS) 647
  Operational principles 614–21
Network characteristics 433–7
  Availability 434
  Cost 437
  Reliability 434–5
  Scalability 437
  Security 435–6
  Throughput 436
  Topology 437
Networking protocols 450–9
  Asynchronous transfer mode (ATM) 459
  File transfer protocol 457–8
  Hyper text transfer protocol 457
  Internet protocol 456
  Open system interconnect (OSI) model 453–6
  Simple mail transfer protocol 458
  Transmission control protocol 457
  User datagram protocol 458
Network services 438–9
Network technologies 447–50
  Circuit switched networks 447–8
  Packet switched networks 448–9
Network topologies 442–7
  Bus topology 442–3
  Hybrid topology 446–7
  Logical topology 437
  Mesh topology 444–5
  Physical topology 437
  Ring topology 444
  Star topology 443
  Tree topology 445–6
New Horizons 691
Newton's law of gravitation 39–40
Newton's second law of motion 40–1
NOAA satellites 608–11
  Payloads 610–1
  Salient features 609
Noise considerations, satellite link design 324, 342–50
  Noise figure 343–4
  Noise temperature 344–50
  Thermal noise 342–3
Noise figure 343–4
Noise temperature 344–50
  Antennas noise temperature 346–9

  Cascaded stages 345–46
  System noise temperature 350
Non-scanning sensors 540
  ERS-1, 2
  Laser distance meter 540
  Laser water depth meter 540
  Microwave altimeters 540
  Microwave scatterometer 540
Nozomi 689
Nuclear explosion satellites 738
Number of stages 108

OEDIPUS 777
O'Neill cylinder 803
Open system interconnect (OSI) model 453–6
Operational principles (*Navigation satellites*) 614–21
  Doppler effect based systems 615
  Trilateration based systems 615–7
OPSEK 800
OPSNET 202–3
Optical mechanical scanner 536–7
Optical remote sensing 526–8
Orbcomm satellite constellation 464
Orbit 37, 67–75
  Circular 69
  Elliptical 69
  Equatorial 67
  Geostationary earth orbit (GEO) 70
  Low earth orbit (LEO) 70
  Medium earth orbit 70
  Molniya 69–70
  Polar 67
  Prograde 68
  Retrograde 68–69
  Sun synchronous 73–4
Orbital effects 149–50
  Doppler shift 149
  Orbital distance variation 150
  Solar eclipse 150
  Sun transit outrage 150
Orbital parameters 44–51
  Angles defining direction 50
  Apogee 47
  Argument of perigee 50
  Ascending nodes 45
  Descending nodes 45
  Eccentricity 48
  Equinoxes 45–6
  Inclination 49, 80–1, 84
  Perigee 47–8
  Period 83

Index 817

Right ascension of ascending node 48–9, 80, 83
Semi-major axis 48
Semi-minor axis 48
Solstices 47–8
True anomaly of satellite 50
Orbital perturbations 144–6
Orbit control 200–1
Orbits (Weather forecasting satellites) 586–7

Packet switched networks 448–9
Palapa-A 17
Palapa-B 17
Palapa-C 17
Palapa-D 17
Passive remote sensing 526
Passive sensors (*Remote sensing satellites*) 535–9
   Central perspective scanner 538
   Optical mechanical scanner 536–7
   Push broom scanner 539–40
Payload 203–5
   Multispectral scanner 204
   Radiometer 204, 588
   Thematic mapper 204
   Transponder 203–4
Payloads (*Scientific satellites*) 659–65
   Accelerometers 659
   Charged particle detectors 660
   Corner reflectors 659
   Gamma ray telescopes 663
   Imagers 660
   Infrared telescopes 662
   Ionosphere sounders 660
   Laser altimeters 659
   LEGRI (Low energy gamma ray scanner) 664
   LIMS instrument 660
   Optical telescopes 662
   Photometers 660
   Radar altimeters 659
   Spectrographs 660
   Spectrometers 660
   Synthetic aperture radar 659
   TOMS instrument 660
   X-ray telescopes 663
Payloads (*Weather forecasting satellites*) 587–92
   Active payloads 589–90
   Altimeter 590
   Lidar 591–2
   Radiometer 588

Scatterometer 598–9
Sounder 589
Synthetic aperture radar (SAR) 591
Pegasus series 29
Perigee 47–8
   Argument of 50, 81
Perigee distance 82, 87–90
Perigee motor 99
Personal communication services 516
Phase shift keying 269–77
   8PSK 274
   16PSK 274
   Amplitude phase shift keying (APSK) 276–7
   Differential phase shift keying (DPSK) 270–1
   Offset QPSK 273–4
   Quadrature amplitude modulation (QAM) 274–6
   Quadrature phase shift keying (QPSK) 271–3
Phobos 1 and 2 689
Phoenix lander 690
Physical topology (*Networks*) 437
Pico satellites 26
Pioneer Venus orbiter 687
Plasma motor generator (PMG) 777
Point-to-point trunk telephone network 482
Polarization loss 209
Polar satellite launch vehicle (PSLV) 123–4
Power supply subsystem 189–98
   Batteries 195–8
   Solar panels 191–5
   Types 189–90
Pre-assigned multiple access 287, 290–1
Precipitation radar 586
Pre-emphasis 249–50
Principal parameters (*Launch vehicles*) 106–8
   Initial mass 106
   Final mass 106
   Hybrid propellants 107
   Liquid propellants 107
   Mass fraction 106
   Number of stages 108
   Propellant mass fraction 106
   Solid propellants 107
   Specific impulse 107
   Thrust 107
   Thrust-to-weight ratio 107
   Type of propellant 107–8
Progress cargo craft 796

Propagation considerations, satellite link design 324, 330–340
  Cloud attenuation 334
  Free space loss 330–1
  Gaseous absorption 331–2
  Ionospheric effects 335–8
  Rain attenuation 333–4
  Signal fading due to multipath signals 338–9
  Signal fading due to refraction 334–5
Propellant mass fraction 106
Propulsion 177–84
  Basic principle 178
  Electric propulsion 181
  Ion propulsion 182
  Liquid fuel propulsion 180–1
  Solid propulsion 179–80
Protocol translation 466
Proton series rockets 28, 109–10
  Analogue 259–61
  Digital 261–5
Pulsed plasma thruster 182
Push broom scanner 539–40

Quadrature amplitude modulation (QAM) 274–6
Quadrature phase shift keying (QPSK) 271–3
Quality of service (QoS) 466–7

R3E system 237
Raduga series satellites 16
Random multiple access 288
Radarsat 564–65
  Synthetic aperture radar (SAR) 564–5
Raisting earth station 428
Receive-only earth station terminals 379
Receiver figure-of-merit-Earth Station 390
Reconnaissance satellites 728–32
  Electro-optical imaging satellites 729–30
  IMINT satellites 728
  Optical imaging satellites 728–9
  Radar imaging satellites 730
Regional satellite systems 512–3
  EUTELSAT 513
Relay-1 12
Reliability (*Networks*) 434–5
Remote sensing satellites 524–4
  Active remote sensing 526
  Active sensors 540–2
  Aerial remote sensing 525
  Applications 548–58
  Classification 526–31

  Future trends 573–4
  Image classification 545–6
  Image interpretation 546–8
  Microwave remote sensing 529–31
  Missions 558–68
  Optical remote sensing 526–8
  Orbits 531
  Passive remote sensing 526
  Passive sensors 535–9
  Payloads 531–42
  Satellite remote sensing 525
  Thermal (or infrared) remote sensing 528–9
  Types of images 542–5
Repeaters 465
Resourcesat-1 22
Right ascension of ascending node 80, 83
Ring topology 444
RISAT 22
RITA thruster 183
Rotating wheel 803
Routers 466

Salyut 794–5
Sampling theorem 265
Satellite constellations 459–65
  Geometry 459–60
  Major constellations 460–5
Satellite data communication 496–501
  Satellite data broadcasting 496–7
  Very small aperture terminal (VSAT) 497–501
Satellite link design 367–70
  Link budget 368–70
  Procedure 368
Satellite link parameters 324–6
  Frequency considerations 324, 326–9
  Interference related problems 325–6
  Noise considerations 325
  Propagation considerations 324–5
Satellite networks advantages 479–80
Satellite networks disadvantages 480
Satellite phone 383
Satellite radio 496
Satellite remote sensing 525
  Advantages 525–6
Satellite services 438
Satellite stabilization 146–9
  Spin stabilization 146–7, 149
  Station keeping 149
  Three-axis stabilization 147–9
Satellite subsystems 174–220
  Antennas 205–20

Index

Attitude control 199–200
Mechanical structure 175–7
Orbit control 200–1
Payload 203–5
Power supply 189–98
Propulsion 177–84
Reliability 225
Space qualification 224–5
Telemetry, tracking and command 201–3
Thermal control 185–9
Satellite telephony 481–3
  Global mobile personal communication services (GMPS) 483
  Mobile satellite telephony 482–3
  Point-to-point trunk telephone network 482
Satellite television 484–95
  Digital video broadcasting 490–5
  Direct-to-home (DTH) satellite television 487–9
  Satellite cable television 485–6
  Satellite TV network 484–5
Satellite tracking 412–9
  Block diagram 412
  Techniques 412–9
Satellite tracking techniques 412–9
  Conical scan 415
  Intelligent tracking 419
  Lobe switching 413–5
  Monopulse track 416–9
  Sequential lobing 415
  Step track 419
Satish Dhawan space centre 143–4
Scalability (*Networks*) 437
Scanning sensors 540–2
  Synthetic aperture radar (SAR) 540–2
Scientific satellites 658–714
  Applications 665–70
  Astronomical observations 680–6
  Comparison with ground based techniques 658–9
  Earth's observation 670–80
  Future trends 714
  Missions 686–714
  Payloads 659–65
Search and rescue processor (SARP) 611
Search and rescue repeater (SARR) 611
Search and rescue transponder (SARSAT) 603
Sea surface temperature measurement 552
Second cosmic (or orbital) velocity 62
Security (*Network Reliability*) 435–6
Sequential lobing 415
SEVIRI sensor 606

Shanon-Hartley theorem 266
Shared media topology (*Networks*) 437
SIGINT satellites 732–5
  Cerise satellite 735
  Chalet series 734
  Essaim series 735
  GRAB (Galactic radiation and background) 735
  JSSW series (China) 735
  Magnum/Orion satellites 734
  Russian satellites 735
  Tselina series 735
  USA satellites 733
  Vortex 734
Simple mail transfer protocol 458
Single channel per carrier system 293–5
  SCPC/FM/FDMA system 293–4
  SCPC/PSK/FDMA system 294–5
Single function earth stations 384
Sirius XM radio 462–3
Sky Bridge satellite constellation 464
Skylab 795
Slant range 156–8
Slotted ALOHA 501
Small expendable deployer system (SEDS) 776–7
Small satellites 22–7
  Femto satellites 26–7
  Medium satellites 23
  Micro satellites 24–5
  Mini satellites 23–4
  Nano satellites 25–6
  Pico satellites 26
Smokeless propellants 108
Solar backscatter ultraviolet radiometer (SBUV) 611
Solar observatory (SOHO) 707
Solar X-ray imager (SXI) 603
Solid fuel propulsion 179–80
Solid propellants 107
Solid state power amplifiers (SSPA) 477
Solstices 47–8
Sounding rocket 7
Soyuz series 110–3
Space division multiple access 316–9
  Frequency re-use 316–7
  SDMA/CDMA system 319
  SDMA/FDMA system 317–8
  SDMA/TDMA system 318–9
Space environment monitor (SEM) 602
Space docking 802
Space elevator 779–81

Space launch centres 122–44
  Baikonur cosmodrome 133–4
  Constituent parts 128–9
  German aerospace centre 141–3
  Guiana space centre 133–5
  Jiuquan satellite launch centre 138–9
  John F. Kennedy space centre 129–32
  Location considerations 127–8
  Major centres 128–44
  Satish Dhawan space centre 143–4
  Tanegashima space centre 141
  Uchinoura space centre 139–41
  Xichang satellite launch centre 136–8
  Yuri Gagarin cosmonaut training centre 135–6
SPACENET 203
Space shuttle 29, 124–7
Space stations 793–801
  Almaz commercial programme 800
  Bernal sphere 803
  Bigelow commercial space station 799–800
  Commercial space station 801
  Dry dock 802
  Inflatable space modules 802
  International space station (ISS) 797
  Mir 795–6
  Modular 802
  O'Neill cylinder 803
  OPSEK 800
  Progress cargo craft 796
  Rotating wheel 803
  Salyut 794–95
  Skylab 795
  Space docking 802
  Stanford torus 803
  Tiangong-1 798–9
  Tiangong-2 799
  Tiangong-3 799
Space tether experiment (STEX) 778–9
Space tether missions 775–9
  Atmospheric tether mission (ATM) 778
  OEDIPUS 777
  Plasma motor generator (PMG) 777
  Small expendable deployer system (SEDS) 776–7
  Space tether experiment (STEX) 778–9
  STEP-AIRSEDS 778
  Tethered satellite system (TSS) 775–6
  Tether physics and survivability (TiPS) 777–8
Space tethers 769–70
  Applications 774

Electrodynamic tether 772–3
Formation flying 773–4
Hybrid tethers 775
Missions 775–9
Momentum exchange tether 770–2
Space weapons 739–45
  Almaz programme 740
  Classification 740–5
  Earth-to-space weapons 741–4
  Integrated flight experiment (IFX) 745
  Manned orbital laboratory (MOL) 741
  OPS-1, 2 and 3 740
  Space based laser (SBL) 745
  Space-to-earth weapons 744–5
  Space-to-space weapons 740–1
Specific impulse 107, 178–9
Spin stabilization 146–7, 149
Spitzer space telescope 707
SPOT series 21, 561–3
  HRV instrument 562
  HR VIR instrument 562–3
  High resolution stereoscopic instrument 563
  NAOMI 564
  Vegetation instrument 564
Sputnik-1 8
Sputnik-2 8
Sputnik-3 8
Sputnik-5 10
Sputnik-6 10
Stacking 466
Stanford torus 803
Star topology 443
  Unidirectional star networks 499
Station keeping 149
STEP-AIRSEDS 778
Step track 419
Strategic Defence Initiative (SDI) 745–52
  Extended range interception system (ERINT) 749
  Flexible lightweight agile guided experiment (FLAGE) experiment 746–7
  Ground based programmes 746
  Homing overlay experiment (HOE) 748–9
  Kinetic kill vehicle (KKV) 748
SUPARCO satellite ground station 428
Superbird 484
Switches 466
SYNCOM 12–3
System level testing-Earth station 392–8
  Additional tests 397–8
  Mandatory tests 393–7

Tanegashima space centre 141
TDMA burst structure 299–301
  Carrier and clock recovery sequence 299
  Signaling channel 300
  Traffic information 301
  Unique word 299–300
TDMA frame structure 297–9
  Guard time 299
  Reference burst 298
  Traffic burst 298
Teledesic satellite constellation 463
Telemetry, tracking and command 201–3
Teleports 386
Television receive-only satellite systems (TVRO) 474, 488
Telstar 11–2
Terrestrial interface-Earth station 409–11
  Interface 410–1
  Tail options 409
Tether physics and survivability (TiPS) 777–8
Tethered satellite system (TSS) 775–6
Thermal control subsystem 185–9
  Heat transfer mechanism 186
  Thermal inequilibrium 186
  Types of heat transfer 187–9
Thermal (or infrared) remote sensing 528–9
Third cosmic (or orbital) velocity 62
Three-axis stabilization 147–9
Throughput (*Network security*) 436
Thrust 107
Thrust-to-weight ratio 107
Thuraya 512
Tiangong-1 798–9
  Tiangong-2 799
  Tiangong-3 799
Time division multiple access 297–306
  Comparison with FDMA 307–08
  Frame acquisition and synchronization 305–306
  TDMA burst structure 299–301
  TDMA frame efficiency 302–3
  TDMA frame structure 297–9
  Traffic control and coordination 303–4
TIROS-1 to TIROS-10 10, 578
Titan series rockets 29, 117–9
Topology (*Networks*) 437
Total energy detector (TED) 611
Trajectory 37
Transfer rate of information bits (TRIB) (*Network throughput*) 436
Transit-1B 10
Transmission control protocol 457

Transmission delay (*Satellite networks*) 480
Transmission equation 322–3
Transmit-only earth station terminals 379
Transponder 475–9
  Bent pipe transponders 477
  Performance parameters 478–9
  Regenerative transponders 478
  Transparent transponders 477
  Types 477–9
Transponder assignment modes 287–8
  C/N ratio 290–2
  Demand assigned multiple access 287, 290
  Pre-assigned multiple access 287, 290–1
  Random multiple access 288
Traveling wave tube amplifiers (TWTA) 477
Tree topology 445–6
Trojan horse (*Network security*) 435
Tropical rainfall measuring mission (TRMM) 586
True anomaly of satellite 50
Tunneling 466
Type of propellant (*Launch vehicles*) 107–8

Uchinoura space centre 139–41
Ulysses 691
Unidirectional star networks 499
Urban monitoring and development 552
User datagram protocol (UDP) 458

Vanguard-1 9
Vanguard-2 577
Venera atmospheric probe 687
Very small aperture terminal (VSAT) 497–501
Video signals 230–1
Viking 1 and 2 689
Viruses (*Network security*) 435
(VISSR) Visible and infrared spin scan radiometer 602
VISSR atmospheric sounder (VAS) 602
Voice signals 230
Vostok series rockets 28
Voyager-1 and 2 690–1

Water quality management 550
Water vapour images 583
Weather facsimile transponder (WEFAX) 602
Weather forecasting 577–80
Weather satellites 577–612
  Applications 593–9
  Future trends 612
  Images 580–6

Weather satellites (*Continued*)
   Missions   599–611
   Orbits   586–7
   Payloads-587–92
   Weather forecasting   577–80

Xichang satellite launch centre   136–8
X-ray sensor (XRS)   603

Yinghuo-I   690
Yuri Gagarin cosmonaut training centre   135–6

Zenit series rockets   28
Zinc Sulpher propellants   108
Zone antenna   206
Zone lens see *Fresnel lens*

CPSIA information can be obtained
at www.ICGtesting.com
Printed in the USA
BVOW04*2259271016
466119BV00009B/25/P